Brock/Springer Series in Contemporary Bioscience

Autotrophic Bacteria

The reviews presented in this book are
based on the symposium
"Lithoautotrophy, a centenary meeting
in memory of S.N. Winogradsky"
held in Göttingen, August 23-28, 1987
Published as FEMS Symposium No. 42

The symposium was sponsored by the following institutions:

Deutsche Forschungsgemeinschaft (DFG)
Federation of European Microbiological Societies (FEMS)
Ministerium für Wissenschaft und Kunst
des Landes Niedersachsen
Deutscher Akademischer Austauschdienst (DAAD)

Autotrophic Bacteria

Edited by

Hans G. Schlegel and Botho Bowien

Science Tech Publishers
Madison, WI

Springer-Verlag
Berlin Heidelberg New York
London Paris Tokyo

Library of Congress Cataloging-in-Publication Data

Autotrophic bacteria / edited by Hans G. Schlegel and Botho Bowien.
p. cm.—(Brock/Springer series in contemporary bioscience)
Includes bibliographies and index.
ISBN 0-910239-22-3
1. Bacteria, Autotrophic. I. Schlegel, Hans Günter, 1924– .
II. Bowien, Botho. III. Series.
QR88.2.A97 1989
589.9—dc19 88-37175
CIP

Science Tech Publishers, 701 Ridge Street
Madison, Wisconsin 53705 U.S.A.

Sole distribution rights outside of the USA, Canada, and Mexico granted to Springer-Verlag Berlin Heidelberg New York London Paris Tokyo

ISBN 0-910239-22-3 Science Tech Publishers Madison, WI
ISBN 3-540-50665-9 Springer-Verlag Berlin Heidelberg New York

Production supervision: Katherine M. Brock
Editorial supervision: Thomas D. Brock, Ruth Siegel, and Katherine Noonan

Printed in the United States of America

10 9 8 7 6 5 4 3 2 1

Contents

Contributors

E. Bock Institut für Allgemeine Botanik, Abt. Mikrobiologie, Universität Hamburg, Ohnhorststrasse 18, 2000 Hamburg 52, FRG

P. Bos Laboratory of Microbiology, Delft University of Technology, Julianalaan 67a, 2628 BC Delft, The Netherlands

B. Bowien Institut für Mikrobiologie, Georg-August-Universität Göttingen, Grisebachstrasse 8, 3400 Göttingen, FRG

T.D. Brock Department of Bacteriology, University of Wisconsin-Madison, 1550 Linden Drive, Madison, WI 53706, USA

G. Drews Lehrstuhl für Mikrobiologie, Albert-Ludwigs Universität, Schänzlestrasse 1, 7800 Freiburg, FRG

B. Friedrich Institut für Pflanzenphysiologie, Zellbiologie und Mikrobiologie, Freie Universität Berlin, Königin-Luise-Strasse 15-16, 1000 Berlin 33, FRG

G. Fuchs Abt. Angewandte Mikrobiologie, Universität Ulm, Oberer Eselsberg M27, 7900 Ulm, FRG

G. Gottschalk Institut für Mikrobiologie, Georg-August-Universität Göttingen, Grisebachstrasse 8, 3400 Göttingen, FRG

H. Harms Institut für Allgemeine Botanik, Abt. Mikrobiologie, Universität Hamburg, Ohnhorststrasse 18, 2000 Hamburg 52, FRG

A.B. Hooper Department of Genetics and Cell Biology, University of Minnesota, St. Paul, MN 55101, USA

H.W. Jannasch Woods Hole Oceanographic Institution, Woods Hole, MA 02543, USA

B.B. Jørgensen Institute of Ecology and Genetics, University of Aarhus, 8000 Aarhus C, Denmark

D.P. Kelly Department of Biological Sciences, University of Warwick, Coventry CVA 7AL, England

E. Kondratieva Department of Microbiology, Moscow State University, Moscow B-234, USSR

H.-P. Koops Institut für Allgemeine Botanik, Abt. Mikrobiologie, Universität Hamburg, Ohnhorststrasse 18, 2000 Hamburg 52, FRG

J.G. Kuenen Laboratory of Microbiology, Delft University of Technology, Julianalaan 67a, 2628 BC Delft, The Netherlands

B.A. McFadden Biochemistry/Biophysics Program, Washington State University, Pullman, WA 99164, USA

O. Meyer Lehrstuhl für Mikrobiologie, Universität Bayreuth, Postfach 101251, 8580 Bayreuth, FRG

D.C. Nelson Department of Bacteriology, University of California, Davis, CA 95616, USA

N. Pfennig Fakultät für Biologie, Universität Konstanz, Universitätsstrasse 10, 7750 Konstanz, FRG

H.G. Schlegel Institut für Mikrobiologie, Georg-August-Universität Göttingen, Grisebachstrasse 8, 3400 Göttingen, FRG

K.O. Stetter Lehrstuhl für Mikrobiologie, Universität Regensburg, Universitätsstrasse 31, 8400 Regensburg, FRG

R. Steudel Institut für Anorganische und Analytische Chemie, Technische Universität Berlin, Strasse des 17. Juni 135, 1000 Berlin 12, FRG

C.K. Stumm Department of Microbiology, University of Nijmegen, Toernooiveld, 6525 ED Nijmegen, The Netherlands

F.R. Tabita Department of Microbiology, University of Texas at Austin, Austin, TX 78712, USA

R.K. Thauer Fachbereich Biologie-Mikrobiologie, Philipps-Universität Marburg, Auf den Lahnbergen, 3550 Marburg, FRG

H.G. Trüper Institut für Mikrobiologie, Rheinische Friedrich-Wilhelms-Universität Bonn, Meckenheimer Allee 168, 5300 Bonn 1, FRG

G.D. Vogels Department of Microbiology, University of Nijmegen, Toernooiveld, 6525 ED Nijmegen, The Netherlands

H.G. Wood Department of Biochemistry, School of Medicine, Case Western Reserve University, 2109 Adelbert Road, Cleveland, OH 44106, USA

G.A. Zavarzin Institute of Microbiology, USSR Academy of Sciences, Moscow, USSR

Preface

In the minds of most people, the bacteria (procaryotes) are linked either with infectious diseases or with the broad field of biotechnology. However, another significant role of bacteria is in the formation of the earth's crust and in the cycles of matter in nature. Most natural mineral cycles are driven by organic matter, and, for at least the last 3.5 billion years, bacteria have played major roles in the production of this organic carbon. In the biosphere, bacteria have played major roles in the production of molecular oxygen from water, the formation of iron deposits, the deposition of sulfur and limestone, and the formation of oil, coal, and natural gas. Today, bacteria still play major roles in the cycles of matter, especially in the transformations of carbon, nitrogen, sulfur, and phosphorus. In fact, there are several key steps in these nutrient cycles that higher organisms are completely unable to perform; only the procaryotes can do them. Indeed, it is often unrecognized that procaryotes alone could maintain all the nutrient cycles of the earth but that eucaryotes could not. Thus, if all procaryotes were eliminated from the earth, the eucaryotes would soon perish.

A substantial fraction of these indispensable procaryotes are the autotrophic bacteria. An "autotroph" is defined as an organism able to obtain all its carbon for biosynthesis from inorganic carbon in the form of CO_2. The autotrophic bacteria were the first organisms on this planet producing large amounts of organic carbon and hence they provided the starting point for the evolution of other bacteria as well as higher life forms. But even today, some autotrophic bacteria play unique roles in nutrient cycling in the biosphere. Thus, key transformations such as ammonia and sulfur oxidation are carried out in nature, so far as we know, only by autotrophic bacteria. In anaerobic environments, the only photosynthetic process that takes place is carried out by bacterial autotrophs. And another major process, the anaerobic decomposition of organic matter to methane, is also carried out only by methanogenic bacteria, many of which are autotrophs. Without the activity of methanogenic bacteria, the accumulation of high levels of organic matter in lakes and oceans would soon prevent the ecosystem from functioning.

However, the global ecologist is not the only individual interested in autotrophic bacteria. The biochemist has found the unique processes in autotrophic bacteria of great interest. Indeed, by examining the comparative biochemistry of autotrophic bacteria, the first clear idea was obtained of how the photosynthetic process operates in green plants. Comparative biochemistry also led to comparative enzymology and then to molecular biology, and our understanding of evolutionary processes in the living world has been greatly enhanced by these sophisticated studies of a number of autotrophic bacteria.

The roles of autotrophic bacteria in the biosphere, and their unique routes of metabolism, were first recognized by the distinguished bacteriologist S.N. Winogradsky about 100 years ago. Until Winogradsky's work, the only organisms known to grow using CO_2 as their sole carbon source were the green plants. Winogradsky showed that in addition to photosynthesis carried out by green plants, there was another process, chemosynthesis, carried out by certain bacteria. Whereas the green plants used light as energy source, the chemosynthetic bacteria used inorganic substances such as ammonia and sulfur as energy sources. It is difficult today to appreciate the "revolutionary" nature of Winogradsky's concept of chemosynthesis. It showed that the dichotomy between green autotrophs and nongreen heterotrophs, considered to indicate a major evolutionary difference, was untenable. Within a few more years it was demonstrated that animals themselves are able to "fix" CO_2. Soon, rigid ideas about evolutionary distinctions between plants and animals had to be abandoned. Finally, nucleic acid sequence studies have permitted the recognition of a third major group or kingdom of life, the archaebacteria. Many of the autotrophic bacteria are archaebacteria, and the study of these bacteria has greatly extended our ideas about the nature of evolutionary processes in the living world.

In 1987, a distinguished group of scientists met in Göttingen, Federal Republic of Germany, to review the current status of the autotrophic bacteria. This meeting, held on the centenary of Winogradsky's first paper, covered all aspects of the autotrophic bacteria: diversity, ecology, biochemistry, molecular biology, evolution, and genetics. This meeting was sponsored by the German Research Council (DFG), the Ministry of Science and Art of Lower Saxony, the German Academic Exchange Program (DAAD), and the Federation of European Microbiological Societies (FEMS). As a result of the exchange of ideas that occurred at that meeting, it seemed evident that a major book which reviewed the current status of the autotrophic bacteria would be of great value. Consequently, a number of participants at this meeting were invited to prepare thorough reviews of the status of their respective specialties. The response to these invitations was prompt and enthusiastic, and the present volume is the result. As will be evident, this volume, which is much more than a symposium volume, is a treatise which covers all aspects of modern research on autotrophic bacteria. As a major area of modern biological research, it is fitting that such a book should be published as part of the Brock/Springer series on contemporary bioscience. The editors would like to acknowledge the kind cooperation and timely response of the contributors, as well as the extensive collaboration of Thomas D. Brock, whose help in planning and editing this book is gratefully acknowledged.

Hans G. Schlegel
Botho Bowien

Brock/Springer Series in Contemporary Bioscience

Autotrophic Bacteria

1

Introduction

Thomas D. Brock

Department of Bacteriology
University of Wisconsin-Madison
Madison, Wisconsin, U.S.A.

Hans G. Schlegel

Institute for Microbiology
University of Göttingen
Göttingen, Federal Republic of Germany

Autotrophy or *self-feeding* is a concept that has had a long history in biology. The first studies leading to an understanding of plant nutrition were carried out by the Belgian physician Jean Baptiste van Helmont (1577–1644). While attempting to prove that water was the material element underlying the whole of biological existence, he carried out careful quantitative studies on the growth of willow trees in soil in confined pots. In his own words: "I took an earthenware pot, placed in it 200 pounds of earth dried in an oven, soaked this with water, and planted in it a willow shoot weighing 5 pounds. After five years had passed, the tree grown therefrom weighed 169 pounds and about 3 ounces. But the earthenware pot was constantly wet only with rain. . .water. . . .Finally, I again dried the earth of the pot, and it was found to be the same 200 pounds minus about 2 ounces. Therefore, 164 pounds of wood, bark, and root had arisen from the water alone."

This experiment was the first biological experiment based on quantitative calculations. Although perfectly correct in plan and execution, it was only wrong in its conclusion, since water is not the sole source of plant substance. As Joseph Priestley (1733–1804) showed in 1772, plants also exchange gasses with the atmosphere and are able to "purify" the air so that animals could live in it. Priestley, a believer in the phlogiston theory, did not draw the proper interpretation from his experiments, but the demonstration soon after by Antoine Lavoisier (1743–1794) of the existence of oxygen led to a clarification of the situation. A year after Priestley's experiments, John Ingen-Housz (1730–1799), demonstrated that sunlight was necessary for the air purification process, and following Lavoisier concluded that plants produced oxygen. He also showed that only the green parts of the plant could utilize sunlight. In 1787, a Swiss minister Jean Senebier (1742–1809) concluded that carbon dioxide (called at that time "fixed air") was the carbon source for plants. This idea was placed on a quantitative basis by Theodore de Saussure (1767–1845) in experiments published in Paris in 1804. Finally, the relationship of respiration and photosynthesis to energetics was analyzed in 1842 by the German physician and physicist Julius Robert Mayer (1814–1878), one of the founders of the concept of "conservation of energy." Thus, by the middle of the 19th century, the concept that respiration meant combustion and photosynthesis meant biosynthesis was well established, as were the carbon and oxygen cycles. Also, the processes of photosynthesis and respiration were linked to phylogenetic considerations through descriptions of the places of plants and animals in the living world.

1.1 TERMINOLOGY

***The terms* autotrophy *and* chemosynthesis** Although green plant photosynthesis was well understood by the middle of the 19th century—at least as far as the sources of material and energy were concerned—the term autotrophy was not in use before the work of the great Russian bacteriologist Sergei Winogradsky (1856–1953), discussed below, which placed the whole problem of energetics and biosynthesis in a new light. Wilhelm Pfeffer (1845–1920), a German plant physiologist who was well aware of Winogradsky's work, concerned himself with the problems of carbon dioxide assimilation using light energy or chemical energy. He first used the terms *photosynthesis* and *chemosynthesis* as well as *autotrophy* and *heterotrophy* in his influential textbook of plant physiology (Pfeffer, 1897). Pfeffer's concept of autotrophy has continued to be used since this time in essentially the way Pfeffer defined it. According to Pfeffer, autotrophs are *organisms able to synthesize their cell substance from inorganic carbonate as their main source of carbon* (see also Woods and Lascelles, 1954 and Chapman and Gest, 1983). It should be noted that this definition refers to carbon dioxide as the *main* source of carbon; it is of minor importance that many autotrophs require organic growth factors, or that some autotrophs are facultative and can grow heterotrophically when appropriate organic energy sources are present.

***The term* lithotrophy** When Winogradsky (1887) studied the colorless filamentous bacterium *Beggiatoa*, he observed the oxidation of hydrogen sulfide to sulfuric acid, with the transient intracellular accumulation of globules of elemental sulfur. He recognized clearly that this ability to oxidize an inorganic substance was a unique characteristic: "The sulfur bacteria are therefore a clearly defined physiological group, a physiological type, which differs significantly from the general. Their life processes follow a very simple scheme, since they use a purely inorganic chemical process, sulfur oxidation, to maintain their vitality. Because of this, I have named these organisms *Sulfur organisms* or *Sulfur bacteria*."

A year after his first paper on *Beggiatoa*, Winogradsky (1888) reported his observations on the growth of *Gallionella ferruginea* and *Leptothrix ochracea* on ferrous iron salts in the absence of organic carbon. He discerned the low energy yield from the oxidation of ferrous iron and related this to the large amounts of iron oxide produced by the relatively low numbers of iron bacteria found in natural habitats. He concluded that these iron bacteria were able to use carbon dioxide as their source of carbon. This conclusion led Winogradsky to his definitive studies on chemosynthesis with the nitrifying bacteria (discussed below). Considering Winogradsky's work, Pfeffer (1897) recognized the oxidation of sulfur, iron, and other reduced inorganic compounds as a new type of metabolism which he called "Anorgoxydation" (inorganic oxidation): "These studies have shown that the organisms called inorgoxidants can be distinguished by a collection of special physiological properties that separate them sharply from the ordinary bacterial world, namely: 1) They thrive in nature only in selective environments which consist of almost pure mineral media, containing specifically oxidizable inorganic substances. 2) Their existence is linked to the presence of these substances, which by their vital activity are oxidized. 3) This oxidation process is their sole energy source. 4) They require no organic nutrients, neither as carbon sources nor as energy sources. 5) They possess practically no ability to degrade organic substances. On the contrary, their growth is actually inhibited by organic materials. 6) As carbon source, these organisms use exclusively carbon dioxide, which they assimilate by chemosynthesis."

We now know that Pfeffer's categorizations of chemosynthesis and inorgoxidation were much too restrictive, but Pfeffer's concepts have had marked influence on subsequent thinking about bacterial autotrophy.

Lithotrophy, chemosynthesis, and photosynthesis The term *lithotrophy* was not actually used until the reexamination of the terminology at the Cold Spring Harbor Symposium in 1946 (Lwoff, van Niel, Ryan, and Tatum, 1946). Before this, the Dutch school under Kluyver had objected to the use of the term *chemosynthesis* to describe a special type of metabolism, because every organism displays the faculty for chemical synthesis (Kluyver, 1924). However, van Niel, in his influential paper on the purple and green sulfur bacteria (van Niel, 1931) retained the "term *chemosynthesis* to designate that type of metabolism in which the organic matter of the organism is being built up from carbon dioxide as the sole source of carbon without the supply of radiant energy, as contrasted with *photosynthetic* processes." Later, van Niel apparently thought better about this term, and participated in the Cold Spring Harbor reassessment. According to the Cold Spring Harbor committee (Lwoff et al., 1946) it is essential to distinguish between two aspects of cellular nutrition, the source of energy and the source of carbon. The term "chemosynthesis" is actually a composite, since it refers to both the energy source (an inorganic molecule) and carbon source (carbon dioxide). The terminology proposed by Lwoff et al. (1946) emphasized energy source and electron donor. The combining forms "photo" and "chemo" were used to refer to the energy source (light or chemical) and the terms "litho" and "organo" referred to the hydrogen (electron) donor (inorganic or organic material). Thus: *Photolithotroph*—requiring radiant energy, using inorganic oxidizable substrates (water, hydrogen sulfide) as electron donor. Generally autotrophic. Examples: green plants, green and purple sulfur bacteria, cyanobacteria. *Photoorganotroph*—requiring radiant energy, and using organic compounds as electron donors. Not autotrophic. Example: purple nonsulfur bacteria. *Chemolithotroph*—deriving energy from the oxidation of inorganic electron donors. Generally autotrophic. Example: nitrifying bacteria. *Chemoorganotrophic*—deriving energy from the oxidation of organic compounds. Since the organic energy source almost always serves as principal carbon source, these organisms are *heterotrophic*. Examples: most bacteria, animals.

1.2 WINOGRADSKY'S LEGACY: BACTERIAL AUTOTROPHY

"One of the first great discoveries of modern biological science was the fact that green plants are able to transform radiant energy into chemical energy. A sharp antithesis became, therefore, apparent between green organisms and the rest of the living world,—an antithesis that remained in its original form until Sergei Winogradsky's genius enabled him to grasp the wider significance of the autotrophic state. In a remarkable series of classical papers (1887–1891) he established a third group of organisms, a second group of autotrophonts." (Baas-Becking and Parks, 1927.)

Winogradsky and lithotrophy As emphasized in the above quotation, Winogradsky's work was so revolutionary that one can consider it to be a major biological milestone. However, Winogradsky's concept of chemosynthesis (lithotrophic autotrophy) was not developed immediately. In his first paper, published in 1887, Winogradsky described clearly the oxidation of an inorganic energy source, but had not yet discerned that sulfur bacteria were capable of growth using carbon dioxide as sole source of carbon.* Here are the key lines

---

*Winogradsky did not describe autotrophy in his first paper. This idea is a fallacy that was apparently propounded by van Niel (1931) in his influential paper on purple and green sulfur bacteria. This paper, dedicated to Winogradsky, has the following lines in its introduction: "In his first paper on this subject, published in 1887, [Winogradsky] formulated his brilliant concept of what nowadays is known as chemosynthetic metabolism. . . .In the case of the *sulphur bacteria* the oxidation of the hydrogen sulphide to sulphuric acid was considered as the energy furnishing reaction and with the aid of this energy supply these organisms are capable of utilising carbon dioxide as the only source of carbon." This statement is wrong, since Winogradsky did not demonstrate chemosynthesis in his 1887 paper, or, really, at any time with sulfur bacteria. van Niel, writing many years after the event, either had not read the 1887 paper, or did not check to be certain that his statement was correct. Most subsequent workers have read van Niel rather than Winogradsky and have propounded this error.

from the 1887 paper: "1) [The sulfur bacteria] oxidize H_2S and store sulfur in the form of tiny spheres which consist of amorphous sulfur. 2) They oxidize this sulfur to sulfuric acid, which is immediately neutralized by the carbonate which they take up. The oxidized sulfur is therefore excreted in the form of sulfate. 3) Without sulfur, nutrition and movement cannot take place and the cells sooner or later die. . . .4) They can live and reproduce very well in a medium which contains only traces of organic substances, a medium in which other nonchlorophyllous organisms cannot live.—How are we to interpret these remarkable properties, which set these organisms apart from all other creatures of the living world? First, we should think how this sulfur-oxidation process should be linked to life. It seems most reasonable to conclude that this process corresponds to a respiratory one, although because of its chemical character it is quite different from ordinary respiration."

Winogradsky then went on to discuss the fact that in normal respiration CO_2 is produced, and there was no evidence that *Beggiatoa* produced CO_2. He then showed how difficult it would be to demonstrate CO_2 release. After a lengthy discussion of respiration and CO_2 release, Winogradsky returned to the essential point: "If we assume that *Beggiatoa* releases no carbon as CO_2, it is therefore clear why this organism can grow so well in the presence of only traces of organic food. Organisms of ordinary physiological types incorporate into their bodies only a part of the organic carbon which they use; the rest is respired, destroyed, converted into usable energy. . . .The largest part of the organic matter used by an organism serves as its energy source. It is just this large amount of organic carbon which the sulfur bacteria do not use. The energy of these organisms is derived exclusively from the sulfur-oxidation process. It is therefore easy to understand why the sulfur bacteria require such a tiny quantity of organic matter. The organic matter which these organisms take up is used exclusively for the formation of cell material. Once assimilated, the carbon atoms remain in their cells, in the same manner in which nitrogen remains in the cells of higher plants. . . .It must be of no significance for the sulfur bacteria which kind of organic compounds are provided, since they contain carbon only in assimilated form. . . .Our bacteria find no competition from more rapidly growing organisms. This is because they are able to use those substances which for other organisms are unsuitable. Substances which yield very little energy when broken down. . . .Fungi and [conventional] bacteria do not grow on formic, propionic, or butyric acid. . . .Therefore, we see no putrefying bacteria develop in the waters of sulfur springs, the reason being that these waters contain no other organic substances than these aforementioned organic acids (which Fresenius has shown are present in the sulfur water of Weilbacher Spring). These substances are enough for the sulfur bacteria to grow, because these compounds are used only as carbon sources and [not as energy sources]."

These statements seem clear enough to conclude that in 1887 Winogradsky had not yet developed the concept of bacterial autotrophy. This concept arose a few years later, as will be explained below. In this first paper, Winogradsky described the process we now call lithotrophy. It is not the intention to minimize

Winogradsky's 1887 paper, despite its frequent misinterpretation in subsequent years. It was a remarkable paper—especially amazing when one considers the state of technical knowledge at that time and the fact that Winogradsky was only 31 years old (see Chapter 2)!

Winogradsky and chemosynthesis It was only when Winogradsky turned to the nitrifying bacteria that he was led to the concept of bacterial autotrophy (chemosynthesis). The events that led Winogradsky to work on the nitrifying bacteria are discussed in Chapter 2. The process of nitrification was already well known at the time Winogradsky started his work. Studies in soil and sewage had shown that when ammonia was added to one of these materials it was readily oxidized to nitrate. The likely link of nitrification to microorganisms had been shown by Schloesing and Muntz (1877, 1878, 1879), who showed that the process takes place only under conditions appropiate for the activity of living organisms. Suitable temperature and moisture was required, as well as the presence of O_2 and other nutrients. If a soil was treated with chloroform or dried, the process of nitrification ceased. But nothing was known about the causal organism. Winogradsky, aware of the requirements of Koch's postulates, stated: "In order to prove that this process is due to a specific organism, one must isolate the organism and obtain with this pure culture the same process that is seen in soil. Until this is done, it is impossible to conclude that a specific organism is responsible for the specific process of nitrification. . . ." (Winogradsky, 1891.)

The only researcher before Winogradsky who had attempted to isolate the nitrifying organism was Heraeus (1886), who used Koch's gelatin plate technique but never succeeded in isolating the appropriate organism. Reviewing Heraeus' work, Winogradsky concluded that the gelatin plate technique was unsuitable for isolating an organism that used an inorganic energy source. Recalling his work on the sulfur and iron bacteria, which, as inorgoxidants, had only very small requirements for organic carbon, Winogradsky deduced that nitrifying organisms, using ammonia, might also be inorgoxidants. Winogradsky therefore set up a simple enrichment medium, using Lake Zurich water to which he added small amounts of inorganic salts together with ammonium as energy source, but no organic substances. After appropriate incubation, such enrichment cultures exhibited vigorous nitrification, but the cultures were visibly impure. Substituting distilled water for Lake Zurich water, Winogradsky found that the contaminants were greatly reduced whereas the nitrifying organisms continued to grow vigorously. By now Winogradsky was convinced that organic matter should be avoided for the culture of the nitrifying bacteria, which meant that gelatin plates for final purification would be unsuitable. Looking for an inorganic solidifying agent, Winogradsky hit upon silica gel. On a silica gel-containing medium in which only inorganic salts were added, isolated colonies of nitrifying bacteria could be obtained which could be picked and purified.

Once pure cultures were available, Winogradsky could study the physiology of the nitrification process in liquid culture. And he quickly discovered that when ammonium was used as the energy source, the oxidation process did not proceed all the way to nitrate, but stopped at nitrite. "Why is it," Winogradsky asked, "that in soil the oxidation of ammonium is complete, whereas in pure culture, in an aqueous medium, the process is not complete? Is the nitrifying organism perhaps weakened when it is brought into artificial culture?"

Winogradsky readily showed that the nitrification process also stopped at nitrite when the pure culture was placed back into sterile soil. Because in nonsterile soil ammonium oxidation proceeded completely to nitrate, he concluded that there must be a second organism in soil that converted nitrite into nitrate. Unable initially to isolate this second organism, Winogradsky turned to the more interesting question of the organic carbon requirement of the ammonium oxidizer. "How can the nitrobacter live and grow in a completely inorganic medium? One must assume either that nitrification occurs in the absence of growth or that the cells are able to use CO_2 to make cell material. The first hypothesis is unreasonable, since we can see the organisms growing. But to absolutely prove the second hypothesis, we must be certain that all traces of organic carbon are gone from the medium. This could be shown and there was no doubt that the medium was devoid of organic carbon. The nitrobacter are completely devoid of chlorophyll, and the whole process of nitrification takes place in the dark. But the idea of a colorless organism using CO_2 as sole carbon source stands in complete opposition to all we have known about the role of photosynthesis in the carbon cycle of our planet."

Winogradsky, originally trained in Russia as a plant physiologist (see Chapter 2), knew that the only way of proving that the nitrifying bacteria were capable of total synthesis of organic carbon from CO_2 was to carry out quantitative experiments. Developing appropriate analytical techniques, he measured the increase in organic (combustible) carbon in the cultures after appropriate incubation, and found a significant increase. The final step then was to determine the relationship between the amount of organic carbon synthesized and the amount of ammonium oxidized. "My expectations were fully confirmed. In a series of four experiments, in which I determined all of the products of the activity of the nitrobacteria—nitrite, nitrate, and assimilated carbon—the relationship between nitrified nitrogen and amount of assimilated carbon was always the same, as the following table shows:

	I	II	III	IV
N, mg	506.1	722.0	815.4	928.3
C, mg	15.2	19.7	22.4	26.4
Ratio, N/C	33.3	36.6	36.4	35.2

"You see, my friends, that the larger the oxidation of ammonium, the larger the assimilation of carbon. . . .This relationship shows clearly that the assimilation process is linked with the nitrification process. The assimilation process requires energy and the nitrification process is the only energy-generating process available to the organism." (Winogradsky, 1891.)

We can be greatly impressed by Winogradsky's work! In a little over three years, he had moved from his first tentative studies on the oxidation of inorganic compounds to a clear conception of bacterial autotrophy. What were the key steps that Winogradsky took? 1) The discovery that *Beggiatoa* must oxidize an inorganic compound as an energy source—the concept of an inorgoxidant. 2) The distinction between a compound as an energy source and a compound as a carbon source. Winogradsky concluded that *Beggiatoa* could grow on a medium very low in organic carbon and hence was not using organic carbon for energy. 3) The study of nitrifying organisms, which proved to be much more favorable than sulfur bacteria for quantitative studies. Vigorous enrichment cultures could be obtained in purely mineral medium, and pure cultures were eventually isolated using inorganic (silica gel) plates. 4) The demonstration that in such inorganic media, the nitrifying bacteria synthesize organic carbon, and that the only source of this carbon was CO_2. 5) The determination by analytical methods of a direct relationship between the amount of ammonium oxidized and the amount of carbon fixed, placing the whole phenomenon on a convincing quantitative basis.

Winogradsky's initial work on nitrification was so extensive and so complete that all subsequent researches on nitrification have only served to confirm Winogradsky's classical researches, both on the physiology of the bacteria active in the process, and on the chemical progress of the nitrification.

1.3 ENERGETICS OF AUTOTROPHIC BACTERIA

Subsequent to Winogradsky's work, thermodynamic concepts were applied to bacterial metabolism, and especially to the processes found in inorganic oxidations. The first work in this area was by Meyerhof (1916) on the nitrifying bacteria, but this whole field of endeavor was systematized in a superb manner by Baas-Becking and Parks (1927), whose work provides the starting point for all subsequent work on autotrophic bacteria. Baas-Becking and Parks (1927) divided autotrophic bacteria into five distinct groups, depending upon the energy source: 1) Bacteria which oxidize hydrogen; 2) Bacteria which oxidize carbon (carbon monoxide); 3) Bacteria which oxidize nitrogen compounds; 4) Bacteria which oxidize sulfur compounds; 5) Bacteria which oxidize ferrous (or manganous) salts.

In each type of reaction, Baas-Becking and Parks (1927) constructed likely balanced reactions, with appropriate energy sources and electron acceptors, and then carried out free energy calculations. They emphasized the importance of free energy rather than enthalpy, and discussed the significance of entropy. Using reactants that had already been shown to be used by bacteria, they showed that all five categories of energy sources could be oxidized in a thermodynamically favorable manner. Using the free energy calculations, and bacterial yields (calculated as organic carbon), they were able to calculate the efficiencies of the autotrophic bacteria.

Baas-Becking and Parks (1927) also considered inorganic reactions that might occur for which no organisms had yet been discovered. They concluded that a variety of inorganic reactions are thermodynamically possible, and that only a small percentage of these reactions had been studied from a microbiological viewpoint. They suggested that a systematic study would uncover organisms able to carry out many of these reactions (which subsequent research proved to be true).

Baas-Becking and Parks (1927) also considered the purple sulfur bacteria from a thermodynamic standpoint. They clearly distinguished the colorless from the colored sulfur bacteria. Although there was no experimental evidence to support it, Kluyver and Donker (1926) had already suggested that the colored sulfur bacteria might be photosynthetic. Baas-Becking and Parks (1927) therefore constructed a balanced reaction in which CO_2 was reduced to the oxidation state of carbohydrate while H_2S was oxidized to elemental sulfur. This reaction had a positive free energy change, but less than one-third of the radiant energy was required by a purple sulfur bacterium than by a green plant to synthesize an equivalent amount of organic carbon. In the balanced reaction, no O_2 was produced. Based on these calculations, and on the knowledge that the purple bacteria required light for growth, they concluded that the purple sulfur bacteria were photosynthetic autotrophs.

This conclusion was amply confirmed in the careful quantitative studies of van Niel (1931). Following Winogradsky's approach, van Niel determined the amount of H_2S oxidized and the amount of organic matter formed in cultures of purple sulfur bacteria growing in the light. In contrast to the colorless sulfur bacteria, where the quantity of H_2S oxidized was greatly in excess of the amount of CO_2 assimilated, with the purple sulfur bacteria the proportions were about equal. "It needs not be pointed out that this result is wholly incompatible with the assumption of chemosynthetic activity. . . .Thus the observed proportionality. . .of necessity leads to the conclusion that the H_2S is directly involved in the photosynthetic process."

This conclusion led van Niel to formulate his famous equations for photosynthesis:

$$CO_2 + 2H_2A \rightarrow CH_2O + H_2O + 2A$$

$$CO_2 + 2H_2O \rightarrow CH_2O + H_2O + O_2$$

$$CO_2 + 2H_2S \rightarrow CH_2O + H_2O + 2\ S$$

It is not surprising that van Niel dedicated this groundbreaking paper to Winogradsky!

Once the natures of the chemosynthetic and photosynthetic bacteria were clarified, the way was open for extensive systematic studies on the diversity and characteristics of these organisms. Over the more than 50 years since van Niel's paper, a vast new knowledge has developed of the wide array of autotrophic bacteria. The chapters in this book amply attest to the success of the various lines of investigation that have been opened by the work of Winogradsky, van Niel, and others.

1.4 DISCOVERY OF THE MAIN GROUPS OF AUTOTROPHIC BACTERIA

Most of the autotrophic bacteria were described in the years around the end of the 19th and the beginning of the 20th century. In some cases, the descriptions were made earlier but the naming and taxonomic recognition came later. There has been no general historical treatment of the autotrophic bacteria, although the textbook by Thimann (1963) covers most of the pertinent older literature in a succinct and accessible fashion.

Hydrogen bacteria The first enrichment cultures for hydrogen-oxidizing bacteria were developed by Kaserer (1906) and Niklewski (1910). Cultures developed on hydrogen and oxygen mixtures could be readily isolated on heterotrophic medium, showing that the isolates were facultative autotrophs. These organisms were subsequently given the genus name *Hydrogenomonas* by Orla-Jensen, but because these organisms are not restricted to the utilization of hydrogen they have been subsequently distributed into other genera such as *Pseudomonas* and *Alcaligenes* (see Chapter 17).

Carbon monoxide-oxidizing bacteria The history of this group is peculiar. Beijerinck and van Delden (1903), in attempting to isolate algae on a purely mineral medium, obtained the development of thin films of colorless or pink bacteria which grew slowly over a period of months. Because no energy or carbon source had been added, Beijerinck concluded that the organism was utilizing a carbon compound derived from the atmosphere of the laboratory. Beijerinck called this organism *Bacillus oligocarbophilus*. Kaserer (1906) found a similar organism and hypothesized that the atmospheric gas was carbon monoxide, a common gas present in the air (those were times when coal gas, a mixture of H_2 and CO, was used commercially instead of natural gas). Subsequently, Lantzsch (1922) isolated bacteria that could definitely grow on CO. These bacteria, now classified as pseudomonads or related organisms, are facultative heterotrophs and also generally oxidize hydrogen (see Chapter 18).

Iron-oxidizing bacteria Although Winogradsky and others had believed that they had observed iron-oxidizing lithotrophs in the 19th century, the first organism that was definitively shown to oxidize ferrous iron (and grow autotrophically) was *Thiobacillus ferrooxidans*, isolated from acid mine drainage by Colmer et al., (1950) and Temple and Colmer (1951). It is still uncertain whether the "classical" iron-oxidizing bacterium *Leptothrix*, actually oxidizes ferrous iron. The early history of these latter iron bacteria has been reviewed by Cholodny (1926), who also did considerable work on these organisms himself. *Gallionella* was first described by Ehrenberg in 1836 and *Leptothrix* was first published by Kützing in 1843. As we have seen, Winogradsky (1888) in one of his earliest papers studied the process of iron precipitation from ferrous carbonate solutions by these organisms, and categorized the group as "iron bacteria." Although this study of Winogradsky's was important as a stage in the development of his thinking about lithotrophy and bacterial autotrophy, there is no reason to believe that Winogradsky actually demonstrated autotrophic iron oxidation. As is now well known, ferrous iron autooxidizes so rapidly at neutral pH that it is only stable in air when it is chelated with an organic compound. Most of these so-called "iron bacteria" are effective oxidizers of the organic chelate, thus indirectly precipitating ferric iron. Only under acidophilic conditions, where *T. ferrooxidans* lives (Temple and Colmer, 1951), is ferrous iron stable in the presence of oxygen (see Chapter 4).

Colorless sulfur bacteria The organism *Beggiatoa* has been known by phycologists for many years. The organism is named for the Italian physician F. S. Beggiato, who had done pioneering work on the microorganisms of sulfur springs (Breed et al., 1939). Because of the association of *Beggiatoa* with hydrogen sulfide-containing waters, it had been concluded that this organism reduced sulfate to hydrogen sulfide (de Bary, 1885). It was Winogradsky's first major contribution to show that just the opposite was true; *Beggiatoa* oxidizes hydrogen sulfide to sulfate (see Section 1.2).

However, cultivation of *Beggiatoa* is difficult (see Chapter 12). The first pure cultures of sulfur autotrophs (see Chapter 4) were from the genus *Thiobacillus*, isolated on thiosulfate or calcium sulfide by Nathansohn (1902) and Beijerinck (1904). It was Beijerinck who coined the name *Thiobacillus*, and he also described the most well-known species, *T. thioparus* and *T. denitrificans* (Beijerinck, 1904).

Nitrifying bacteria Winogradsky's work on this group has already been discussed briefly earlier in this chapter and a detailed discussion can be found in Chapter 5. Winogradsky recognized that nitrification was a concerted process of two distinct types of organism, one which oxidizes ammonium to nitrite and the other which oxidizes nitrite to nitrate. For the first group, he used the combining form *Nitroso-*, and for the second group the combining form *Nitro-*. Almost all of the genera of nitroso- and nitrobacteria were first described by Winogradsky in his classical work (Winogradsky, 1890, 1892).

Purple and green sulfur bacteria The early history of work on the purple sulfur bacteria has been reviewed by van Niel (1931), and that on the green sulfur bacteria by Larsen (1953). Definitive evidence that these organisms were photosynthetic only arose from van Niel's work (already discussed earlier in this chapter), but because these organisms are of distinctive morphology and frequently occur in mass accumulations in nature, these organisms had been described taxonomically long before pure cultures became available (see Chapter 6). Many of the genus names of purple sulfur bacteria were actually developed by Winogradsky as part of his work on the bacteria of sulfur springs (Winogradsky, 1888), although the name for one of the largest and most distinctive purple sulfur bacterium, *Chromatium*, was coined much earlier (Perty, 1852).

As discussed by Larsen (1953), green sulfur bacteria had also been frequently observed in nature in mass accumulations, but they were misinterpreted (even by Winogradsky) to be oxygen-evolving organisms capable of a conventional photosynthesis. It was Nadson (1912) in Russia who first presented evidence that green bacteria do not produce oxygen, although Nadson concluded that these organisms contained an inactive chlorophyll. Nadson also established the genus name *Chlorobium*. van Niel (1931) obtained the first pure culture of a green bacterium, and demonstrated that *Chlorobium* grows anaerobically with light as the sole energy source when hydrogen sulfide is present.

Purple nonsulfur bacteria Molisch (1907) first recognized a group of purple bacteria that grew anaerobically in the light in the presence of large amounts of organic matter. He concluded that these organisms were distinct from the purple sulfur bacteria and classified them in a new family, the Athiorhodaceae. Molisch established a number of genus names, of which the genus *Rhodospirillum* is still used in current classifications. Because his organisms required organic carbon for growth, Molisch did not believe that they were autotrophic, and because their pigments were similar to those of the purple sulfur bacteria, he also erroneously concluded that these latter organisms were also not autotrophic. He concluded that the purple bacteria assimilate organic compounds in the light. This conclusion, of course, *is* correct for the purple nonsulfur bacteria, although several species are able to grow photolithotrophically with H_2, sulfide, or thiosulfate as electron donor (see Chapter 6).

1.5 TAXONOMIC NOTES

There have been major revisions in the taxonomy of the autotrophic bacteria over the past several decades, as our understanding of these organisms continues to increase. It is likely that many more changes will take place as work on molecular evolution is extended (see Chapter 27). A current genus list of autotrophic, phototrophic, and lithotrophic organisms, can be found in the latest edition of Bergey's Manual of Systematic Bacteriology (summarized in Brock and Madigan, 1988). Over the past several editions of the Bergey's Manual, there have been major realignments of genera, so that the number of genera is now much fewer than it had been in earlier times.

Of considerable interest is the new classification of bacteria based on base sequence comparisons of 16S ribosomal RNA molecules (Woese, 1987). According to this scheme, all of the purple bacteria (sulfur and nonsulfur) are classified together with various lithotrophic bacteria (sulfur, nitrifying, hydrogen, carbon monoxide, etc.) in a group which has been called the "purple phototrophic bacteria group." The green sulfur bacteria classify in a separate group, related to the purple bacteria only at the deepest level of the phylogenetic tree. These distinctions are consistent with the distinctly different structure of the photosynthetic apparatus and the biochemical pathway of CO_2

fixation in these two groups of organisms. Although the purple sulfur, purple nonsulfur, and green sulfur bacteria have been known for many years, two further groups of phototrophic bacteria have been recently described. The green nonsulfur bacteria (*Chloroflexus* and relatives) is in a separate group from the green sulfur bacteria, despite similarities in structure and biochemistry. Finally, the phototrophic genera *Heliobacterium, Heliospirillum,* and *Heliobacillus* fit together with the Gram-positive eubacteria and hence are quite separate from all other phototrophic bacteria.

All of the above organisms are members of the Kingdom Eubacteria. The other Kingdom of bacteria, the Archaebacteria, does not contain any conventional phototrophic organisms, although the halobacteria are able to use light energy via the bacteriorhodopsin system. There are numerous autotrophic Archaebacteria, including the hydrogen-oxidizing autotrophs (methanogens) as well as autotrophs capable of oxidizing ferrous iron and reduced sulfur compounds. Details of the phylogeny of the bacterial autotrophs can be found in Woese (1987).

1.6 PRACTICAL ASPECTS OF THE AUTOTROPHIC BACTERIA

Autotrophic bacteria play major roles in the chemical processes of the global ecosystems and hence are critical in our understanding of biogeochemical processes. The redox cycles of the various elements are intimately related to the transfer of elements from place to place on earth. Autotrophic microorganisms are key actors in these biogeochemical mobilization processes. In the **carbon cycle**, methane is a key gas and the methanogens, many of which are autotrophs, play major roles in carbon cycle transformations. In the **nitrogen cycle**, the nitrification process results in the conversion of the less mobile ammonia into the more mobile nitrate. Further, two important gaseous nitrogen species, nitrous oxide and nitrogen dioxide, are produced, at least transiently, by autotrophic nitrifying bacteria (see Chapter 5).

Autotrophic bacteria play even greater roles in the **sulfur cycle** (see Chapter 4). In the sulfate reduction process, hydrogen-oxidizing sulfur and sulfate-reducing bacteria participate significantly. Lithotrophic sulfur oxidation results in the reverse process, in which hydrogen sulfide and elemental sulfur are oxidized to sulfate.

Iron is one of the most abundant elements in the earth's crust, but is a relatively minor component of ecosystems because of its relative insolubility in water. At low pH values, ferrous iron is stable to autooxidation, but is readily oxidized by the autotrophic iron bacterium *Thiobacillus ferrooxidans*. Because this organism oxidizes both iron and sulfur, it plays a major role in the mobilization of sulfide minerals. Its contribution to the process of acid mine drainage is well known. On the positive side, *T. ferrooxidans* also is involved in the microbial leaching process in which valuable minerals such as copper are recovered from low-grade ores. The sulfur-oxidizing lithotrophs also play major roles in the mobilization of metals such as iron and copper, liberating these elements from sulfide-containing minerals.

Other elements which are mobilized or fixed through the action of autotrophic microorganisms include mercury, arsenic, selenium, antimony, and lead. Another important process in which autotrophic sulfate-reducing bacteria participate is the corrosion of iron and steel. It can truly be said that wherever there are active biospheres on earth the autotrophic bacteria are present and functional.

REFERENCES

Baas-Becking, L.G.M. and **Parks, G.S.**. 1927. Energy relations in the metabolism of autotrophic bacteria. *Physiological Reviews* 7:85–106.

Beijerinck, M. and **van Delden, A.** 1903. On a colourless bacterium, whose carbon-food comes from the atmosphere. *Proceedings of the Section of Sciences Kon. Akademie van Wetenschappen, Amsterdam* 5:389–413.

Beijerinck, M. 1904. Ueber die Bakterien, welche sich im Dunkeln mit Kohlensäre als Kohlenstoffquelle ernähren können. *Zentralblatt für Bakteriologie und Parasitenkinde, II Abteilung*. 11:592–599.

Breed, R.S., Murray, E.G.D., and **Hitchens, A.P.** 1939. *Bergey's Manual of Determinative Bacteriology, 6th edition*. Williams and Wilkins, Baltimore. 1529 pp.

Brock, T.D. and **Madigan, M.T.** 1988. *Biology of Microorganisms, 5th edition*. Prentice-Hall, Englewood Cliffs, NJ. 835 pp. The current Bergey's classification is given on pages 796–799.

Chapman, D.J. and **Gest, H.** 1983. Terms used to describe biological energy conversions, electron transport processes, interactions of cellular systems with molecular oxygen, and carbon nutrition. Schopf, J.W. (editor), *Earth's Earliest Biosphere*, Princeton University Press, pp. 459–463.

Cholodny, N. 1926. *Die Eisenbakterien*. Gustav Fischer, Jena. 162 pp.

Colmer, A.R., Temple, K.L., and **Hinkle,M.E.** 1950. An iron-oxidizing bacterium from the drainage of some bituminous coal mines. *Journal of Bacteriology* 59:317–328.

de Bary, A. 1885. *Vorlesungen über Bacterien*. Wilhelm Engelmann Verlag, Leipzig. 146 pp.

Ehrenberg, C.G. 1836. Vorläufige Mitteilungen über das wirkliche Vorkommen fossiler Infusorien und ihre grosse Verbreitung. *Poggendorfs Annalen Physik und Chemie* 38:213–227.

Heraeus, W. 1886. Ueber das Verhalten der Bacterien im Brunnenwasser, sowie über reducirende und oxydirende Eigenshaften. *Zeitschrift für Hygiene* 1:193–234.

Kaserer, H. 1906. Die Oxydation des Wasserstoffes durch Mikroorganismen. *Zentralblatt für Bakteriologie, II Abteilung* 16:681–696.

Kluyver, A.J. 1924. Eenheid en verscheidenheid in de stofwisseling der microben. *Chemisch Weekblad*, 21:266. English translation in *Albert Jan Kluyver, His Life and Work*, 1959, North-Holland Publishing Company, Amsterdam.

Kluyver, A.J. and **Donker, H.J.L.** 1926. Die Einheit in der Biochemie. *Chem. Zelle Gewebe* 13:134.

Kützing, F.T. 1843. *Phycologia Generalis*. Leipzig.

Lantzsch, K. 1922. *Actinomyces oligocarbophyilus* (*Bacillus oligocarbophilus* Beij.), sein Formwechsel und seine Physiologie. *Zentralblatt für Bakteriologie, II Abteilung* 57:309–319.

Larsen, Helge. 1953. On the microbiology and biochemistry of the photosynthetic green sulfur bacteria. *Det Kgl. Norske Videnskabers Selskabs Skrifter 1953 Nr 1.* F. Bruns Bokhandel, Trondheim. 199 pp.

Lwoff, A., van Niel, C.B., Ryan, F.T., and **Tatum,E.L.** 1946. Nomenclature of nutritional types of microorganisms. *Cold Spring Harbor Symposia on Quantitative Biology* 11:302–303.

Meyerhof, Otto. 1916. Untersuchungen über Atmungsprozesse von Nitritbakterien. *Pflüger's Archiv* 165:353.

Molisch, Hans. 1907. *Die Purpurbakterien nach neuen Untersuchungen.* Gustav Fischer Verlag, Jena. 95 pp.

Nadson, G.A. 1912. Mikrobiologische Studien. I. *Chlorobium limicola,* ein grüner Mikroorganismus mit inaktiven Chlorophyll. *Bull. Imp. Bot. Garden St. Petersburg* 12:55–72.

Nathansohn, A. 1902. Über eine neue Gruppe von Schwefelbakterien und ihren Stoffwechsel. *Mitt. Zool. Stat. Neapel* 15:665.

Niklewski, B. 1910. Über die Wasserstoffoxydation durch Mikroorganismen. *Jahrbücher für Wissenschaftliche Botanik* 48:113–142.

Perty, M. 1852. *Zur Kenntnis kleinster Lebensformen.* Jent and Reinert, Bern I–VIII, pp. 1–228.

Pfeffer, W. 1897. *Pflanzenphysiologie.* 2nd edition. Wilhelm Engelmann Verlag, Leipzig.

Schloesing, T. and **Muntz, A.** 1877. Sur la nitrification par les ferments organisés. *Compt. rend. acad. sci.* 84:301.

Schloesing, T. and **Muntz, A.** 1878. Recherches sur la nitrification par les ferments organiseés. *Compt. rend. acad. sci.,* 86:892.

Schloesing, T. and **Muntz, A.** 1879. Recherches sur la nitrification. *Compt. rend. acad. sci.,* 89:891.

Temple, K.L. and **Colmer, A.R.** 1951. The autotrophic oxidation of iron by a new bacterium *Thiobacillus ferrooxidans. Journal of Bacteriology* 62:605–611.

Thimann, K.V. 1963. *The Life of Bacteria.* 2nd edition. Macmillan, New York. 909 pp.

van Helmont, J.B. 1748. By experiment, that all vegetable matter is totally and materially of water alone. *Ortus Medicinae,* Amsterdam, pp. 108–109. Translation in Gabriel, M.L. and S. Fogel (editors). 1955. *Great Experiments in Biology,* page 155. Prentice-Hall, Englewood Cliffs, NJ.

van Niel, C.B. 1931. On the morphology and physiology of the purple and green sulphur bacteria. *Archiv für Mikrobiologie* 3:1–112.

Waksman, S.A. 1953. *Sergei N. Winogradsky. His Life and Work.* Rutgers University Press, New Brunswick. 150 pp.

Winogradsky, S. 1887. Ueber Schwefelbacterien. *Botanische Zeitung* 45:489–523; 529–539; 545–559; 569–576; 585–594; 606–610.

Winogradsky, S. 1888. Ueber Eisenbacterien. *Botanische Zeitung* 46:261–270.

Winogradsky, S. 1890. Recherches sur les organismes de la nitrification. *Annales de l'Institut Pasteur* 4:213–231; 257–275; 760–771.

Winogradsky, S. 1891. Ueber die Organismen der Nitrification. *Vierteljahrsschrift der Naturforschenden Gesellschaft in Zürich* 16:176–208.

Winogradsky, S. 1892. Contributions á la morphologie des organismes de la nitrification. *Archives des sciences biologiques St. Petersburg* 1:88–137.

Woese, C. 1987. Bacterial evolution. *Microbiological Reviews* 51:221–271.

Woods, D.D. and **Lascelles, J.** 1954. The no-man's land between the autotrophic and heterotrophic ways of life. *Symposium of the Society for General Microbiology* 4:1–27.

2

Sergei N. Winogradsky and the Discovery of Chemosynthesis

Georgi A. Zavarzin

Institute of Microbiology
USSR Academy of Sciences
Moscow, U.S.S.R.

Western scientists can obtain a brief account of Sergei Nikolaevich Winogradsky's life and work from the biography by Waksman (1953) and from Winogradsky's own explanatory material in his collected works *Microbiologie du sol* (Winogradsky, 1949). In the Russian translation of this latter work, Imshenetsky (1952) has written a valuable biographical article, and two short biographies have also been written by Winogradsky's successors in St. Petersburg, V. L. Omeliansky and B. L. Issatchenko. However, to the general public the name of this outstanding Russian naturalist is largely unknown, unlike the name of Elie Metchnikoff; and even those scientists who are engaged in neighboring branches of science and are working within the conceptual framework laid down by Winogradsky know very little about its author and mostly ignore his place in the entire system of scientific knowledge. Although the discovery of chemosynthesis often appears to be an isolated event, there was in fact an integral system standing behind it.

Several well-defined periods in the scientific activity of S. N. Winogradsky can be identified. The years before 1885 were his school and university years. In 1885 began an exceptionally fruitful five-year period of work with chemosynthetic organisms, all conducted in laboratories outside Russia. In 1890 this work was crowned with the discovery of chemosynthesis that brought him international recognition. The next 15-year period of work at the Institute of Experimental Medicine (IEM) in St. Petersburg, conducted together with his students, began with the formulation of the principle of elective media, the discovery of free-living nitrogen fixers, and with investigations on microorganisms involved in the carbon cycle. The number of papers Winogradsky published dropped sharply, and most microbiologists were not aware of what he was doing at the time, although by the end of this period he was the Director of IEM. In 1905 Winogradsky gave up all his research work and retired to his

Sergei Winogradsky, photographed in 1927.

family estates in the Ukraine. He only returned to research 15 years later after he had left Russia for Yugoslavia. After publishing several theoretical papers, he was invited to join the Pasteur Institute where he worked practically alone for the rest of his life at the estate Brie-Comte-Robert, 30 km from Paris. This estate, belonging to the Pasteur Institute, served as its Laboratory of Agricultural Microbiology. There Winogradsky pursued his experimental studies until 1940, the year of the German occupation of France. During this French period he emphasized broad studies in microbial ecology, studying specific groups of microorganisms as well as the application of direct methods to microbial ecology. Dating from 1940, a new period began, which Winogradsky himself, at that time 84 years old, termed "dozhivanie" (winding up). It is during this period that he prepared the edition of his selected works under the title *Microbiologie du sol* and wrote the introductory and often controversial forewords which laid out his conceptions of soil microbiology. The book appeared in 1949 in French (Winogradsky, 1949) and in Russian in 1952 (Winogradsky, 1952). The last scientific paper by Winogradsky was devoted to bacterial taxonomy and appeared in 1952, the year of his death. He spent an unusually long life in science, and the persistence of ideas that he formulated was striking. Along with M. W. Beijerinck and after L. Pasteur, Winogradsky is rightfully acknowledged as the founder of general microbiology, and his name is usually mentioned at the beginning of any course in general microbiology.

To grasp the origins of Winogradsky's views and interests, one has to be, at least to some extent, aware of the trends in Russian social thinking at the end of the last century. The influence of Russian art and literature of this period on the world culture is widely acknowledged and can be associated with the names of the classics of literature (Leo Tolstoy, Feodor Dostoevsky, etc.) and music (Peter Tchaikovsky, Modest Moussorgsky, Alexander Borodin, etc.). Much less realized is the contribution that Russian naturalists made in the last quarter of the 19th century. Until the advent of the era of physics, natural sciences might appear to have lingered in the shadow of great figures such as Charles Darwin and Louis Pasteur, however indifferent they may have been to each other. Nevertheless, a thoughtful analysis of the history of ideas in natural sciences would show that it was in Russia at that time that a particular system of ideas took shape which had a far-reaching impact on the development of sciences and is of great importance now when scientific thinking has turned to global problems. It would be only fair to note that the starting point of this line of Russian scientific thinking is somewhere very close to the universal "cosmos" of Alexander von Humboldt.

Characteristic of Russia in the second half of the 19th century was the existence of many sharply contrasting trends, an evident process of change in the ideology of generations, and the divergence of interests, as opposed to a more traditional development in western Europe and the United States of America. V. I. Vernadsky (1981) put it this way: "Russian intellectual culture of the 19th and beginning of the 20th century can be regarded as a creation of independent thinkers. State institutions were largely alien to them, and there have been periods of time when a merely lukewarm performance by state officers of what was logically their duties could be honored as a historic service." As a consequence, a professorship of natural sciences presented a sort of a challenge and an alternative to a military or bureaucratic career, options mainly open for well-placed people but more difficult to pursue for one of independent thought. This explains why "raznochintsy" (intellectuals not belonging to the gentry in the 19th-century Russian "middle class") were drawn to science, and also why some well-off people, even some very rich ones, pursued academic careers, as was the case for the renowned mycologist M. S. Voronin, who served as a role model for Winogradsky. This independence of mind was probably the initial reason that Winogradsky began work in the natural sciences, an activity absent from his family environment except for a passion for gardening which he had acquired from his father.

S. N. Winogradsky was born in Kiev, September 1, 1856. His father, Nikolai Konstantinovich Winogradsky, was Moldavian by birth. He had a university degree in law and was what we would call now a civil servant. Fast economic growth in the Ukraine at that time had brought Winogradsky's father to prosperity. When Sergei was still a boy, his father became the Director of a private commercial bank in Kiev and made his way even higher later on, becoming Director of the Kiev Zemelny Bank (involved in financing agriculture). Nikolai Winogradsky died in 1876 following a prolonged illness, leaving considerable capital invested in the sugar business. The illness of their father had brought early independence to his sons, making them think for themselves, which as argued by sociologists, is beneficial for the development of one's abilities. Their mother, Nataly Victorovna, belonging to a junior branch of the Skoropadskys, got married when she was 16, and was a "professional beauty," as described by her son. She nurtured her children well, teaching Sergei and his older brother Alexander music, and the music culture played an important role in the childhood and adult lives of both Winogradsky brothers. Alexander had an expansive and volatile character, unlike the reserved disposition of Sergei, and he became eventually a well-known conductor, the Director of the Kiev Conservatory, and in the tradition of his father, the director of a bank. The two brothers studied in the same class of the 2nd Kiev Classic School, which they entered in 1866; they remained among the best pupils throughout their school careers.

Recollecting this time, Elie Metchnikoff, who went to school in 1856, wrote in his memoirs: "During the initial period of the rule of Alexander II one could feel considerable enthusiasm in the society and the animation of intellectual activity. There was a favorable attitude in secondary schools to scientific aspirations of pupils, not suppressed by stringent requirements relating to the classic subjects. Greek was abandoned altogether, while Latin was taken to be no more than an unimportant formality. Instead, natural sciences were introduced into the curriculum, attracting particular attention of the young" (Metchnikoff, 1946). However, the 10 years separating the period when Metchnikoff and Winogradsky went to school had brought considerable changes: when Winogradsky was in school a conservative reform had set in and classics were again emphasized.

Later, Winogradsky was quite critical of his secondary education, saying that all that tedious drilling could do was to improve his memory, but failed to teach him to think and/or to combine pieces of data, for these skills can only be gained through mathematics, which was taught terribly. "Meanwhile it is clear that mathematical logic is utterly important for any educated person, whatever his occupation. It is also clear that if one fails to acquire mathematical knowledge during secondary education, one will be given no second chance as it will be impossible to put aside current work to study secondary and higher mathematics and it will prove too difficult to master for a mind that had not been treated with mathematics in early days. I have been aware of this deficiency through all my life and I am aware of it now. I am perfectly sure that if I had acquired in my school days a sound body of mathematical knowledge, it would have made all the difference for my science work throughout" (unpublished notes of Winogradsky written near the end of his life).

Having finished secondary school, Winogradsky entered the University of Kiev and studied law for two years, attending at the same time lectures delivered in the department of natural sciences and studying music for his own pleasure. As noted by V. I. Vernadsky, who himself took the same route, a desire for a natural science education was common to all graduates of classic schools, despite the conflict with the official line. Piano performances at amateur concerts given by Winogradsky resulted in recognition, and believing he was destined for musical success, he decided to give up law and enter the St. Petersburg Conservatory of Music. He did well at the Conservatory and received high marks from his professors, but saw clearly the difference between his own abilities and those who were musicians "by God's grace." Although music always played an important role in his life, Winogradsky regarded himself as nothing more than an accomplished amateur. As he admitted in his notes, his passion for music was strongest during periods when his scientific creativity was lowest.

Eschewing formal music training, Winogradsky entered the natural sciences division of the University of St. Petersburg. V. I. Vernadsky, who was also studying there at the time, later wrote: "The departments of physics and mathematics of the St. Petersburg University were at the time superb. Mendeleev, Menshutkin, Beketov, Dokuchaev, Famintsyn, M. Bogdanov, Vagner, Sechenov, Ovs'anikov, Kostychev, Inostrantsev, Voejkov, Petrushevsky, Butlerov, and Konovalov are responsible for much of the development of natural sciences in Russia" (Vernadsky, 1981).

At the University, Winogradsky studied under A. S. Famintsyn, now remembered mainly for his symbiogenesis hypothesis, but who was the professor of plant physiology and at that time was studying plant energetics. Elie Metchnikoff had this to say about the state of physiology at that period: "The basic trend in physiology was to reduce different life phenomena to a combination of more elemental physical and chemical processes. . . .This approach was passed to Russia from outside largely at the end of the fifties and in the sixties of the 19th century, when a whole constellation of young and gifted scientists travelled to Europe to learn the new physiological methods" (Metchnikoff, 1946). Famintsyn went to Strasbourg and became a student of the noted botanist Anton deBary. Upon his return to St. Petersburg, Famintsyn started experiments on the effects of light on the alga *Spirogyra*, and because in the winter it was always twilight in the city he was the first to employ artificial light for research. Famintsyn studied the energetics of the alga under a microscope by observing the accumulation and disappearance of starch, a method invented concurrently and independently of the technique that later came to be known as Sach's iodine test.

As for Winogradsky, he felt very lonely studying with Famintsyn but this had more to do with Winogradsky's own character than with Famintsyn's teaching. In fact, many people had studied under Famintsyn, including I. P. Borodin, an outstanding botanist, Kh.Ya. Gobi, a well-known taxonomist of lower organisms, and D. I. Ivanovsky, a renowned founder of virology. Traditional student/professor relations also developed since Famintsyn himself had been a student and a follower of L. S. Tsenkovsky, one of the principle adherents of the theory of bacterial pleomorphism which was also supported by Metchnikoff.

The brief outline given above affords a glimpse of the university life that was shaping Winogradsky as a scientist. Much of what he had learned and perceived at the time remained with him throughout his life. Among other things, one can list his sound knowledge of taxonomy and the feeling that taxonomy is incompatible with the idea of pleomorphism, his interest in the energetics of living organisms, his command of microscopy, and his total abstention from petty professor "politics" of all sorts as well as from involvement in group interests.

At that time, Winogradsky was already married to Zinaida Alexandrovna Tikhotskaya and was a father of three daughters. Zinaida Winogradsky was a loving and devoted wife who was able to appreciate his value and importance as a personality. One might add that he did not lack that feeling himself and knew how to convey it to others. Nevertheless, when Winogradsky graduated from the University, passed his examinations for the M.Sc. degree, and was invited to stay with the University to prepare for a professorship, he was uncertain as to what research to pursue. Like so many outstanding scientists, he experienced a peculiar mixture of tradition and negativity. Botanist by education, Winogradsky had strong objections to the theory of pleomorphism in bacteriology and the realization of the rigidity of bacterial form and function came to be one of the leading ideas spanning all his research work.

To develop the concept of monomorphism, Winogradsky, like his teacher Famintsyn, went to Strasbourg to the laboratory of deBary. The concept of monomorphism is intimately related to taxonomy: thus an organism exhibiting variability with respect to a given framework of characteristics cannot be possibly classified within this system. A deep interest in bacterial diversity, only initially connected with formal taxonomy, is evident in all Winogradsky's works. Even in his first publication, concerned with the taxonomy of sulfur bacteria, Winogradsky demonstrated his skill as a taxonomist, producing a classification system that persisted for quite some time. Later he ignored conventional binomial nomenclature by assigning no species names to the nitrifying bacteria he had discovered. But it was only in the year of his death, in his last work, that he openly rejected conventional bacterial taxonomy, making an ironical remark about an attempt by C. B. van Niel and R. Y. Stanier to build up an evolutionary taxonomy using morphological and physiological characteristics available at the time (unpublished notes).

In Strasbourg, Winogradsky managed to put aside all his worries: the detachment from his family and relatives with their complicated relations, an unfulfilled duty toward his children, vague prospects for an academic career in Russia, all this receded, allowing him to concentrate on his work. In Strasbourg, at this time German, Winogradsky enjoyed the measured work pace of his German colleagues and the benevolence of professors who regarded it desirable to attract foreigners to German universities. Russians constituted a substantial part of probationers working in German laboratories: with few exceptions, almost all Russian scientists of the time attended German universities, studies in France being much less frequent. As a subject of his research, Winogradsky chose the aquatic bacteria *Beggiatoa* and *Thiothrix*, which were favorite examples for pleomorphists. The morphology of these organisms was the main theme of his work, an exploration of their physiology being but the first step, a precondition for solving the main problem: the demonstration of the stability of the form on which their taxonomy was based. This was a time of a rapid development of evolutionary principles in taxonomy, and lower organisms were widely believed to be able to supply the most spectacular confirmation to evolutionary ideas. "The leading idea of most biological explorations carried out in Russia over the last half-century was the teaching of Darwin about the succession of species." (Metchnikoff, 1946). However, no trace of Darwin can be found in Winogradsky's work, either at the beginning of his career or at the end of his life.

The contemporary framework of general microbiology is based on the ideas on bacterial diversity expressed by Winogradsky and by representatives of the Dutch school, and certainly these ideas warrant a centenary celebration as much as the discovery of chemosynthesis.

DeBary's interest was aroused when Winogradsky found that the filaments of *Beggiatoa* in a solution containing hydrogen sulfide became rapidly filled with droplets of sulfur which when the filaments were removed from the hydrogen sulfide medium quickly disappeared without a trace. "Oh, God, why do they need that much sulfur?" de Bary would exclaim, and having no answer would give a shrug and proceed with his rounds. "Then, one day as I was following the Île canal on my way home after a tiring day of chemical work which also involved hydrogen sulfide and sulfur, it suddenly occurred to me that sulfur might well be oxidized by *Beggiatoa* to sulfuric acid. I could at once appreciate all the significance and implications of my conjecture, having no doubts that it offered the solution to my problem. The next day I said to deBary, 'Ich glaube zu wissen wozu sie so viel Schwefel brauchen' [I think I know why they need so much sulfur]. He listened to me and without saying a word went to talk to Fitting, the professor of chemistry, and only after that did he appreciate what I had to say. The work was humdrum, it dragged on and on, sometimes sluggishly, and all of a sudden it developed into an interesting result and was finished. All the beating around suddenly made sense, and I matured in my own eyes. Even so, I could not foresee that my discovery would become an epoch-making discovery, would determine the course of all my future work, and that it would open a new chapter in microbiology and physiology. Though, even then, after a short note by deBary in *Naturforschende Gesellschaft* many people congratulated me with the words, 'Sie haben einem neuen modus vivendi gefunden' [You have discovered a new mode of life]." That is how Winogradsky described the discovery of chemosynthesis in his unpublished memoirs. It is accepted as a textbook example of "scientific insight." This view, however, fails to take into account the methods and approaches that Winogradsky acquired while studying at St. Petersburg University. Following his teacher Famintsyn, Winogradsky employed microscopy in microculture to monitor the balance of stored substances in the organism. Famintsyn's study of starch accumulation of *Spirogyra* cells is similar to Winogradsky's study of the accumulation of sulfur droplets by sulfur bacteria, the only difference being that Winogradsky was able to view sulfur droplets directly without staining in a light microscope. We see, therefore, that the discovery of chemosynthesis was based on observations quite similar to the starch test and was done within the framework of the conventional theory.

However, at the time of Winogradsky's work the classic starch test, quite common among plant physiologists, was no longer considered to be able to provide an undisputed proof. This is quite evident from the textbook by D. I. Ivanovsky, which, while giving a detailed and highly positive account of Winogradsky's work, is notably reticent about his experiments with sulfur bacteria. These microscopic observations were apparently thought insufficient by Ivanovsky, who wrote, "The first instance of chemosynthesis was established by S. Winogradsky in the so-called saltpeter bacteria oxidizing ammonia to nitric acid," (D. I. Ivanovsky, *Plant Physiology*, Rostov, 1919, p. 116, in Russian).

As for Winogradsky, he used to draw a distinction between his own approach and the standard methods prevailing at the time: "I did my first work in the early days of microbiology when it was totally dominated by the ideas and methods of two schools: that of Pasteur and that of Koch. These methods were rigorously standardized and were obligatory for all investigators. I was only a beginner in microbiology but had enough sense to question the wisdom of such an approach: sulfur bacteria were an exception from the usual bacteria and up to that moment had been totally unknown. I decided to follow my own way, a strictly inductive one, which can provide the best possible guarantee against follies and errors of all types."

Although the first work by Winogradsky on chemosynthesis elicited considerable interest, it was not accepted as an ultimate evidence because it failed to satisfy logical criteria even at that time considered quite obligatory in microbiological studies. We mean, of course, pure culture and the application of "Koch's postulates." Although Winogradsky learned microbiology using German manuals and continued his postgraduate studies in Germany, he definitely took a pro-French stand, as stressed in the foreword in his collected works (Winogradsky, 1949). Even in his early work he was closer to Pasteur than to Koch. Winogradsky's main idea, drawn from fermentation studies by Pasteur, was that particular chemical processes are carried out by particular agents. Regarding soil, however, the position adopted by the French school was notably different: it was the environmental conditions that were thought to make this or that microorganism carry out the given reaction. In the 1880s the denitrification process was described by Guyon and Dupetit. Using conventional broth, they succeeded in isolating microorganisms able to denitrify. The discovery of Guyon and Dupetit made a very strong impression on agricultural chemists. In his book on denitrification, Payne (1981) gives a vivid picture of fears that because of this process the application of fertilizer and the anaerobiosis of soil might destroy soil fertility. It was apparently these polemics that prompted Winogradsky to look for representatives of chemosynthesizing bacteria among organisms responsible for the production of nitrates, the reverse process of denitrification. The biological nature of nitrification had already at that time been well established, and the problem was only to discover the particular microorganism.

Still in Strasbourg, it became evident to Winogradsky that he would fail unless he could substantially increase his knowledge of chemical processes. Microscopic observations alone would not advance him any further in his understanding of processes carried out by bacteria in nature.

With the death of deBary, Winogradsky moved to Zurich, mainly because that city had numerous science institutions and, most importantly, an institute of chemistry. "Three professors, vast halls for students and probationers, every instrument one might need. Enticed by all this, I decided to advance my chemistry. . . .Doing my chemical work in a huge laboratory among so many students and probationers, I felt very small and unimportant. But keeping my nose to the grindstone, I acquired enough skill to proceed to more complex chemical manipulations. . . .To complement my chemical exercises, I joined a physical chemistry laboratory headed by Ernst Schultz, a modest and amiable man who gave me a free hand to do whatever I wanted to in his laboratory. . . .I was engaged in determinations of nitrogen in all forms, and occasionally I had to devise my own instruments. I needed these nitrogen determinations in order to effectively study the problem of nitrification. It was with this work in mind, which had been conceived shortly after the work on sulfur bacteria, that I appproached Dr. Roth, for there was no other bacteriological laboratory in Zurich. The laboratory was as ugly as its boss, and it was a branch of the Hygienic Institute. . . . This Roth was running 6-week courses in bacteriology for doctors in the Zurich University, quite similar in programme to those arranged in Germany by Robert Koch's school. All he was doing was to repeat his course term after term with no change, following a syllabus set once and forever. To tell the truth, I have never seen a professor to be such an ass either before or afterwards." This is how Winogradsky described his working conditions and, as one can see, there is no question that he learned little from foreign microbiologists. At the same time, his working conditions inspired in him a strong negative attitude to routine procedures to which the field of sanitary microbiology had been reduced. One may wonder whether Winogradsky's method of isolating microorganisms from those parts of the dish where nothing appeared to have grown using conventional Koch's methods might be viewed as a gesture of irony? It was simultaneously a demonstration of the special properties of bacteria he isolated and a denial of conventional bacteriological methods. What is important, however, is that in his work with nitrifiers Winogradsky showed no deviation from Koch's rules: first show that the microorganism occurs wherever there is a process, then isolate the organism in a pure culture, finally reproduce the process using the culture. In fact, it was the only series of works in which Winogradsky conformed to this principle. At the same time, in his investigation on nitrifiers Winogradsky followed the approach used by Pasteur in his studies on fermentation: chemical analysis first, microscopy second.

The work done by Winogradsky on nitrifiers is discussed in Chapters 1 and 5. In this work, done in a heat of passion, Winogradsky never doubted his inevitable success. Even his illness, his depression and gloomy moods, his raking shortness of temper, could not affect his zeal and capacity to work. "The intellectual work is all that my life is now. And what is more, far from being a duty, it is full of delight and brings utter satisfaction," wrote Winogradsky in his diary. His preparation of the first two reports on nitrifying bacteria was not even impaired by the birth in the adjacent room of his daughter Elena, who would years later become his assistant in describing new nitrifiers. "The papers were the summary of much work and effort and the results seemed to me quite significant. Especially the second memoir, mailed to Paris close to the day of Elena's delivery, seemed to me 'un grand coup' as regards the novelty of conclusions. I expected a response in a few weeks, but getting nothing I gave up waiting. . . ." The response came in June and was quite varied. First there was a letter from Paris from Duclaux and then wide coverage in the French press. "Just think that all this sensation was caused by two short articles published in a scientific journal and signed by a stranger belonging to no school, a beginner, who found shelter in a bad foreign laboratory." This is how those days were described by Winogradsky in his autobiographical notes. The spread of news of a scientific discovery does not seem to be inferior in the last century to that of today.

Winogradsky's concept of chemosynthesis should be reiterated: "*Organic matter on earth is produced by the activity of vital beings not only through photosynthesis but also in the process of chemosynthesis.* To sum up, the microorganism performing nitrification appears to me to have some remarkable properties, allowing it to be regarded as belonging to a new physiological type. These properties are as follows: 1) development on purely mineral medium in the presence of an oxidizable inorganic compound; 2) its vital activity is crucially dependent on the availability of this inorganic compound, ammonia in the case of nitrification; 3) oxidation of this compound is the sole source of energy; 4) lack of demand for organic nutrition as a source of material and energy; 5) inability to degrade organic compounds, the presence of which only hinders the development of the organism; 6) the only source of carbon is assimilation of carbon dioxide in the process of chemosynthesis."

Actually, 1890 is the year of the recognition of chemosynthesis as a firmly established natural phenomenon and also the end of Winogradsky's creative interest in it. Later on he succeeded in solving a few of the remaining problems, the most important of which was the discovery of the organism involved in nitrite oxidation, the second phase of nitrification.

Winogradsky never returned to experimental work with sulfur and iron bacteria, limiting himself to polemical notes in which he defended their capacity for *anorgoxidation*, his word for the process that today is widely known as *lithotrophy*. It was generally typical for Winogradsky to bring all his scientific investigations to an exceptional level of logical completion and to produce a harmonious concept that can be added to and be refined in detail, but not changed.

Let us now take a general look at the five years starting in 1885, when Winogradsky timidly introduced himself to deBary, through June 1890, when he gained international recognition and when a new era in microbiology began. All this time Winogradsky spent in foreign laboratories; but was he in any way influenced by them? True enough, he learned from deBary an orderly way of working in a laboratory. It was here that he first examined large aquatic bacteria, so readily seen in a light microscope, as an object of a preconceived yet true concept of monomorphism. Winogradsky was right in thinking that standard bacteriological methods would not be applicable to such microorganisms. A modern microbiologist might put it in even stronger form: standard bacteriological methods are inherently inapplicable to organisms which are part of a microflora developed as a stable community with all processes counterbalanced, as opposed to colonization on a new substrate taking place after a transfer to a fresh nutrient medium. Undoubtedly, it was drawing on this experience that Winogradsky argued later for the application of direct methods in soil microbiology, foreseeing the existence of an autochthonous microflora, but relating it, conventionally, to the utilization of a specific substrate—humic substances. In fact, his ideas relate to a much wider class of phenomena, still not well enough studied in general microbiology. It can be seen from this example what a continuous chain of internally and logically linked steps creative scientific work is, and how dangerous it is to damage this chain. That is just what happened in the case of Winogradsky and was at the root of his tragedy as a scientist.

It is often argued that all Winogradsky's successes were achieved outside Russia and that at home he failed to do anything. For this reason he should, admittedly, be regarded as an international phenomenon rather than a national figure. However, this idea is not true. At the end of the nineteenth century, the attitude of the Russian government toward science began to change and even before World War I science institutions in Russia were as good as those in Britain and France. This is supported by such competent witnesses as Elie Metchnikoff and V. I. Vernadsky. At the start of this period, the Institute of Experimental Medicine was set up in St. Petersburg by Prince A. P. Oldenburgsky, who was doing his utmost to get Metchnikoff to join it, saving him for Russia. However, Metchnikoff rejected the offer and went off to Paris. The Institute was left with no bacteriologist and so the Prince approached Winogradsky through A. S. Famintsyn. Concurrent with that, Winogradsky had an invitation from Pasteur to move to Paris, this one relayed by Metchnikoff. After examining the conditions in both places, Winogradsky decided for patriotic reasons to return to Russia. He worked at the Institute until 1905, when he resigned as its Director, having achieved, in terms of his career, as much as was possible for a scientist. Yet, the Institute of Experimental Medicine was a medically oriented institute, and this fact underlies both his failure and the eventual close of the laboratory he had founded.

Upon his arrival in St. Petersburg, Winogradsky undertook an investigation that resulted in the isolation of the first free-living nitrogen fixer, *Clostridium pasteurianum* (in 1893), which in its own right was an outstanding discovery. It was also in this same work that Winogradsky formulated his method of *elective media,* which came to be the principle technique spanning all contemporary microbiology, including the genetics of microorganisms where it is used even today for the selection of mutants. In those days, when universal media designed for medical purposes were predominant, the elective method was far from being self-evident. Developed in the course of investigations on the nitrifying bacteria, it was formulated only with the discovery of free-living nitrogen fixers. It is absolutely clear why Winogradsky selected these bacteria as his research objects: by identifying unknown bacterial agents he wanted to complete the nitrogen cycle in nature. Fixed nitrogen could originate only via bacterial nitrogen fixation under the conditions of unlimited supply of energy material. Thus, restated in a deductive form, the problem became readily amenable to investigation, allowing rapid progress in studying the physiological diversity of bacteria.

The method of elective media is a deductive method; to design a medium one has to have a preliminary understanding of the properties of the anticipated organism. That is what Winogradsky had in the case of nitrifiers and nitrogen fixers. Such an understanding, however, can be provided only by a general theory or a concept. This takes us to the fundamental idea that allowed different branches of natural sciences to be linked in a unified system. The first steps in the development of this concept were made in Russia as early as the 1880s. In 1895 V. V. Dokutchaev presented an extensive memorandum entitled "On the opening in Russian universities of departments for instruction in soil and microbiological sciences" in which he connected the state of the soil with the activity of microscopic organisms. As for Winogradsky, he would never agree to deliver "courses," finding public presentations excruciating. But he did have a general approach to teaching. As a matter of fact, Winogradsky's teachings were presented by his student V. L. Omeliansky in his manual on "General microbiology." Browsing through this textbook, it comes as a surprise that with the obvious exception of genetics and biochemistry, it is identical to modern microbiology manuals in its presentation of microbiology.

In a review of Winogradsky's 1949 book, R. Y. Stanier noted that this collection of papers constituted a strikingly harmonious and integral volume and that something like that could not be compiled from the works of M. W. Beijerinck. Where does this integrity of Winogradsky's outlook come from? A clue can be found in his lecture "On the role of microbes in general life processes" delivered at a general meeting of members of the Imperial Institute of Experimental Medicine on December 8, 1896. Beginning his presentation, Winogradsky admitted that general speculations about microbes could hardly be expected to produce anything new. "But there is always a point when not only the scientist, but also every thinking person, comes to feel an irresistible urge to penetrate the general meaning of new facts, to place them in one's world outlook, and to apply to them the kind of logic one had always looked for and found in natural phenomena." True enough, experimental explorations of Winogradsky always had their origin in his world outlook. At that time no one would have demanded quantitative estimates of microbial processes in nature. The important thing was to qualitatively establish their roles in the world. "If even today, when there is a certain uniformity of methods, we have been able to find so many species within such a short time, what then are we in for? If already now each new investigation technique 'not yet in print' is immediately rewarded with the discovery of important new species, then what is to be expected in the future?" One important principle in the world of microbes is that of a *division of labor,* wrote Winogradsky. "We can hardly imagine microbes that are or produce universal reagents. As we go deeper and deeper into the theory in an attempt to solve the mysteries of molecular mechanics, we start to apprehend that every type of action produced by a vital enzyme on a substrate is due to and limited by a close correspondence between the physiological nature of the enzyme and the molecular structure of the substrate in question. . .the functions performed by microbes in nature are highly specific; for each job there is a special microbe adapted, in terms of its chemistry, to do that job. . . .But if we put together several species of one enzyme group, then the substrate will belong to the species most active *under the given conditions.* Sometimes this will be just one species with the others suppressed, ousted, or destroyed. . . .One can readily imagine, with even no recourse to Darwin's theory, that powerful enzymes can be and in fact are produced in nature. . . ." Winogradsky concluded his talk with the following words: "Given such interdependence and interrelation of phenomena, one cannot but view animate nature as an integral whole, as one huge organism borrowing its elements from reservoirs of the inorganic world, expediently controlling all processes of its progressive and regressive metamorphosis, and, finally, giving back to the dead nature all that has been borrowed." Of special interest to us in this brief summary are the words depicting animate nature as one huge organism. In modern terms, that would mean an integral system with the mentioned "expediency" becoming the criterion of its existence. And it is what stands behind the striking integrity of Winogradsky's world outlook.

Winogradsky's concept included investigations of pure cultures of specific microorganisms responsible for the conversion of chemical elements in nature. So, having settled in St. Petersburg, he began with the exploration of the carbon cycle, the most important of all natural cycles. First, together with Fribes, he made a study of the degradation of pectin, the binding substance of plant cells. Then, his assistant V. L. Omeliansky started work on methanogenesis and the anaerobic degradation of cellulose, studies that were to bring him worldwide recognition. By present-day standards, Winogradsky would have been named as the second author of V. L. Omeliansky's work, but standards were different in those days. Many years later, Winogradsky would once again turn to the carbon cycle, undertaking a study of the aerobic destruction of cellulose, only to realize that he had missed his time.

The very productive work on general microbiology which Winogradsky was doing in St. Petersburg was cut short when he became involved in specialized studies for which he was ill prepared and which did not logically originate from his own ideas. This happened when Prince A. P. Oldenburgsky persuaded Winogradsky to undertake a study of plague, a real threat in Russia at that time. Winogradsky acquired Yersin's culture of the plague bacillus and worked for a long time on the development of disinfection methods. He worked in a "pest tower," a somber-looking fortress in the Gulf of Finland, where he was making experiments on animals. But that was in no way creative research, being merely a development of existing approaches, something he was doing for patriotic reasons, with Pasteur as an idol. Winogradsky constitutes a unique case of the dangers of encumbering a scientist with work that does not devolve directly from the inherent logic of his research.

Even without Winogradsky, general microbiology has followed the road that he first selected. New organisms with the properties he predicted were discovered both in Russia and elsewhere. In Odessa, hydrogen bacteria were discovered by A. V. Lebedev, which in addition to carrying out chemosynthesis had the capacity to change over to a heterotrophic nutrition (see Chapter 17). In Naples, sulfur bacteria were described by Nathanson, fully conforming to the criteria set by Winogradsky (see Chapter 4). Beijerinck established in his extensive research the foundations of our knowledge of specific groups of microorganisms. But how much faster would progress have been had Winogradsky continued with his creative work instead of ending up in a state of deep internal crisis, despite his fame and his high position?

In 1905 Winogradsky gave up all scientific work and his directorship at St. Petersburg. He went into agriculture, retiring to the estate inherited from his father. He showed himself to be a shrewd proprietor, and his income from farming exceeded many times his wage as institute director. For fifteen years he seems to have put microbiology completely out of his mind. His new rise in France in the 1920s is a striking instance of a "revival" of a scientist after a prolonged break. The invitation by E. Roux to take the post of the head of a laboratory of agricultural microbiology presented Winogradsky, at the age of 66, with an opportunity to pursue his own research program. This led to the creation of a new scientific discipline, which long before the present-day enthusiasm for ecology he called *ecological microbiology*. In many ways it was a departure from the general microbiology research which had arisen from his first period of work.

A lesson derived from the days of his directorship at the Institute of Experimental Medicine made Winogradsky very prudent in selecting his new occupation. What he was looking for was a place that could provide him with an opportunity for total concentration, impossible when one has to deal with a large group of co-workers with encumbering complicated relations. He found such an opportunity in an isolated, almost "home," laboratory at Brie-Comte-Robert, far away from the large city of Paris. The cost of such an isolation was the absence of any technical assistance whatever, so that nearly everything had to be done with his own hands. As a result, in France too, Winogradsky found himself in a Russian-speaking environment with only his family and no more social contacts than at the start of his scientific work in Switzerland. It is interesting to note that there have been other outstanding Russian microbiologists with equally individualistic character, among them N. G. Cholodny and B. V. Perfiljev, who were followers of Winogradsky's "direct method" in microbiology. On the whole, the example of Winogradsky reminds us once again that any creative personality needs specific conditions for its potential to be fully realized. Thanks to Emile Roux, Winogradsky could enjoy late in life such conditions at Brie-Comte-Robert.

As a first step in laying down the foundations of soil microbiology, Winogradsky proclaimed the restricted utility of studies with pure cultures. He was looking for a truly inductive approach, and this could be only obtained by direct observation of microorganisms in their natural environments. The direct method was seen as a link between the properties of the pure culture and chemical conversions that were occurring in nature. "In a pure culture there is no place for the most important ecological factor, I mean competition," wrote Winogradsky, "because that is what governs the distribution of natural processes performed by bacteria and which automatically determines their succession." In his denial of the significance of the pure culture method, Winogradsky was even ready to sacrifice the method of elective media, "which is very useful and sometimes quite applicable, but, on the whole, the improvements brought about by this method have not made it any more suitable for ecological studies." However, in his rejection of elective media and the use of what he called spontaneous cultures, Winogradsky, in fact, remained well within the elective medium principle, with the modification that his attention was transferred from individual species to isolated parts of the microbial community closely interrelated in their activities. Unfortunately, the microbial association as a functional unit of a microbial community in a natural habitat has not yet received due attention. Researchers try to get over this stage as fast as they can to concentrate on what they perceive as the more rewarding work with pure cultures.

Even a brief discussion of issues raised by Winogradsky in his unpublished notes brings us immediately to present-day scientific problems of crucial importance. Apart from being a spectacular demonstration of the scope and power of his thinking, this also gives clear evidence that Winogradsky's ideas still have a considerable bearing on problems currently under active investigation. These ideas have their origin in the systems approach to natural phenomena, which was so characteristic of the Russian school in natural science in the last quarter of the 19th century, i.e., in the years when Winogradsky was being shaped as a scientist. Indeed, this contribution of the Russian school to world science is comparable to its contributions to literature and music.

REFERENCES

Issatchenko, B.L. 1951. *Selected Works*, v. 2, (Imshenetsky, A.A., editor), USSR Academy of Sciences Press, Moscow-Leningrad. In Russian.

Metchnikoff, Elie. 1946. *Autobiographical Pages*, USSR Academy of Sciences Press. In Russian.

Payne, W.G. 1981. *Denitrification*. John Wiley and Sons, New York.

Vernadsky, V.I. 1981. *Autobiographical Notes*, "Nauka" Publishers, Moscow. In Russian.

Waksman, S.A. 1953. *Sergei N. Winogradsky. His Life and Work.* Rutgers University Press, New Brunswick. 150 pp.

Winogradsky, S. 1949. *Microbiologie du sol. Problèmes et méthodes. Cinquante ans de recherches,* Masson, Paris. 861 pp.
Winogradsky, S.N. 1952. *The microbiology of soil. Problems and methods. Fifty years of research,* USSR Academy of Sciences Press, Moscow. Russian translation of Winogradsky (1949).

3

Past and Present of CO_2 Utilization

H. G. Wood

Department of Biochemistry
Case Western Reserve University
Cleveland, Ohio, U.S.A.

In textbooks written prior to 1940 there is no discussion of the utilization of CO_2 by any forms of life other than chemoautotrophs and photosynthetic organisms. It was considered that CO_2 was completely inert in the metabolism of heterotrophs. Today, of course, textbooks are replete with examples of the utilization of CO_2 by all forms of life and the essential role of CO_2 in metabolism. Because of these earlier misconceptions about CO_2 utilization, some very peculiar proposals were made, one of which was the cleavage of glucose to 4-carbon and 2-carbon compounds (Virtanen, 1925; Virtanen and Karström, 1931; Kluyver, 1931). This was done to provide a fermentative pathway which did not give rise to CO_2 and it was used as an explanation when lower yields of CO_2 were obtained than expected. At that time, information was very rudimentary concerning intermediary metabolism. Methylglyoxal was considered to be a key intermediate and the roles of triose phosphates, 3-phosphoglycerate, and phosphoenolpyruvate were not known. Succinate was thought to be formed by condensation of two molecules of acetic acid ($HOOCCH_3 + H_3CCOOH \rightarrow HOOCCH_2CH_2COOH + 2H^+ + 2e$). We (Wood and Werkman, 1936a) also used some of these reactions to explain our results prior to realizing that the combination of CO_2 with a C_3 compound might be involved in the formation of succinate.

It is not clear why biochemists believed so uniformly and strongly that CO_2 was not used by heterotrophs. It may have arisen from the fact that Winogradsky (1891), in his brilliant studies which showed that nitrifying and sulfur bacteria are autotrophs, reported that the presence of organic compounds actually inhibited the growth of these bacteria (see Chapter 2). Winogradsky's work perhaps encouraged the idea that autotrophs were unique and that they were the only forms that had the ability to utilize CO_2.

Prior to the availability of carbon isotopes as tracers, it was difficult to demonstrate fixation of CO_2 by heterotrophs. Probably the earliest evidence that CO_2 might play an active role in the metabolism of heterotrophs was by Rockwell and Highberger (1927), Winslow et al. (1932), and Gladstone et al. (1935) who found that if CO_2-free air was passed continuously through liquid cultures, it prevented or greatly retarded growth. In fact, Rockwell and Highberger (1927) proposed that heterotrophic organisms could utilize CO_2 in their metabolism. The only other proposal I am aware of, prior to my direct demonstration that heterotrophs do use CO_2, was that by Lebedev (1921). He published a penetrating discussion from which he concluded that the separation of autotrophs and heterotrophs, on the basis of their ability to assimilate CO_2, was not established by known facts. Although Lebedev obtained no experimental evidence to support his view, there is no doubt that he was the first to visualize a broad function of CO_2 in the synthetic processes of all forms of life.

3.1 ORIGINAL PROOF OF THE ASSIMILATION OF CO_2 BY A HETEROTROPH

In general, heterotrophs produce more CO_2 than they utilize, therefore, the overall balance is a net production of CO_2. The propionic acid bacteria, when fermenting glycerol, are an exception to this general rule. If CO_2 is available, they utilize more CO_2 than they produce, and there is thus a net uptake of CO_2. The initial demonstration of heterotrophic CO_2 fixation arose from quantitative determinations of the products formed in the propionic acid fermentation (Wood and Werkman, 1936b). Calcium carbonate was included in the medium to neutralize the acids as they were produced during the fermentation; the liberated CO_2 provided a source of CO_2 for utilization. The CO_2 was collected during the fermentation and the residual $CaCO_3$ was converted to CO_2 by the addition of acid. Surprisingly, less CO_2 was obtained than had been added as $CaCO_3$. The details of the discovery and subsequent developments have been reviewed several times, most recently by Wood (1985). Here we will briefly summarize the results that led to the proposal of Reaction (1).

$$CO_2 + CH_3COCOOH \rightarrow HOOCCH_2COCOOH \qquad (1)$$

A stoichiometric relationship between the CO_2 fixed and succinate formed from glycerol was observed (Wood and Werkman, 1938). It is seen (Table 3.1) that if the assumption is made that one CO_2 is produced along with each acetate, the net CO_2 fixed plus this amount of CO_2 is very close to the amount of succinate formed. We had previously isolated pyruvate from fermentations by the propionic acid bacteria by including bisulfite in the medium to bind the pyruvate as a complex (Wood and Werkman, 1934). We, therefore, proposed (Wood et al., 1940) that CO_2 combined with pyruvate to form oxalacetate which is then reduced to succinate, thus giving the required stoichiometric relationship between CO_2 fixed and succinate formed. Barron (1943), in a review, called this reaction the Wood and Werkman reaction and it carried that name for many years.

Table 3.1 Relationship of CO_2 fixation and the formation of succinate in fermentations of glycerol by the propionic acid bacteria

	Culture*	
	49W	52W
CO_2 utilized per 100 mmol of fermented glycerol	17.2	21.1
Products per 100 mmol of fermented glycerol		
Succinate, mmol	20.2	25.1
Propionate, mmol	76.6	74.2
Acetate, mmol	1.38	2.08
Propyl alcohol, mmol	1.95	0.19
Carbon recovery, %	101	102
Oxidation/reduction balance, O/R	1.03	1.07

*49W, *P. pentosaceum*; 52W, *P. shermanii*.

By using $^{13}CO_2$ as a tracer, it was shown that the fixed CO_2 was exclusively in the carboxyl groups of succinate formed by the propionic acid bacteria and by *Escherichia coli* (Wood et al., 1941). In addition, $^{13}CO_2$ was shown to occur in the carboxyl groups of succinate, malate, fumarate, and the carboxyl group of ketoglutarate formed from pyruvate by pigeon liver (Wood et al., 1942). Evans and Slotin (1941), likewise, showed that $^{11}CO_2$ was fixed in the α carboxyl of ketoglutarate by pigeon liver. Then, Krampitz et al. (1943) demonstrated that an enzyme of *Micrococcus lysodeikticus* catalyzed an exchange of $^{13}CO_2$ with the β carboxyl of oxalacetate. All of these results were in accord with the Wood and Werkman reaction. However, the net conversion of CO_2 and pyruvate to oxalacetate was not demonstrated until 1960 (Utter and Keech, 1960). This reaction is now known as the pyruvate carboxylase reaction.

$$\text{Pyruvate} + HCO_3^- + \text{ATP} \rightarrow \text{oxalacetate} + \text{ADP} + P_i \qquad (2)$$

By 1960, many other CO_2 fixation reactions had been reported (see reviews by Werkman and Wood, 1942, and Krebs, 1943, on the status of CO_2 fixation in the early 40s, and Wood and Stjernholm, 1962, and Moss and Lane, 1971, for later reviews).

There were two other reports in 1936 that showed that CO_2 is not inert in heterotrophs. D. D. Woods (1936) demonstrated that CO_2 is reduced to formate by *E. coli* using gaseous hydrogen and Barker (1936) presented evidence that CO_2 is reduced to methane by methane bacteria. These results did not encounter the doubts expressed concerning those obtained with the propionic acid bacteria since these reactions did not involve formation of a carbon-to-carbon linkage using CO_2.

3.2 PHOTOSYNTHESIS

In the late 1930s, the mechanism of autotrophic CO_2 fixation was very poorly understood. Based on the proposals of Willstätter and Stoll, it was considered that carbonic acid, in combination with chlorophyll, might be converted to formaldehyde-peroxide, yielding oxygen, H_2O, and formaldehyde and that the formaldehyde, by an aldol condensation, was converted to carbohydrate. Warburg and Negelein (1922; 1923) reported that a molecule of CO_2 was utilized with absorption of only 4 quanta of radiant energy. This efficiency made it almost essential that the reaction occur without intermediate steps. If intermediate steps were involved, the energy supplied by 4 quanta would be too small and, therefore, it appeared that the conversion of CO_2 must occur directly on a single site on the chlorophyll. However, Emerson and Lewis (1939) presented evidence that the quantum measurements were not accurate and after much controversy with Warburg and verification by other investigators, a value of 8 quanta was determined. This then opened the way for consideration of various other pathways.

The investigations of bacterial photosynthesis by Gaffron and by van Niel led to clear evidence that there was a complete separation of the light reaction from utilization of CO_2. Van Niel (1941) pictured the events as follows:

Green plants

$$4(H_2O + h\nu \rightarrow H + OH)$$

$$4H + CO_2 \rightarrow (CH_2O) + H_2O$$

$$2(2OH \rightarrow \text{peroxide} + H_2O + O_2)$$

Purple bacteria

$$3(H_2O + h\nu \rightarrow H + OH)$$

$$4H + CO_2 \rightarrow (CH_2O) + H_2O$$

$$2(2OH + H_2A \rightarrow 2H_2O + A)$$

in which H_2A in the case of the purple sulfur bacteria is H_2S or some other sulfur compound and A is H_2SO_4 (see reviews by van Niel, 1941; and Frank and Gaffron, 1941). Van Niel stated regarding the above equation, "The carbon dioxide is here pictured as being reduced as such, and to carbohydrate. This is not necessarily true. The reaction discovered by Wood and Werkman, in which a four-carbon organic compound is formed from a three-carbon substance and carbon dioxide, may serve as the most interesting example of such reactions. The actual reduction process, therefore, may well involve a substance different from carbon dioxide" (van Niel, 1941).

There was little that could be done to dissect the pathway of carbon in photosynthesis until carbon isotopes became available. Ruben et al. (1939) were the first to study photosynthesis with labeled carbon. They showed that CO_2 was fixed in carboxyl groups in the dark and light but found no evidence of the fixation of CO_2 in the usual intermediates of heterotrophic metabolism. These experiments were done under difficult conditions because only the short-lived radioactive isotope of carbon, ^{11}C (half life 22.5 min), was available at that time. When ^{14}C and ^{13}C became available, more detailed experiments were possible and one of the first observations was that CO_2 was fixed predominantly in the 3 and 4 positions of sugars during short-time photosynthesis and that subsequently the concentration of the tracer increased in the remaining positions (Arnoff et al., 1947; Wood and Burr, 1947). This result was similar to that found when rats were given $^{13}CO_2$ and the glucose of the glycogen was degraded (Wood et al., 1945). Except with rats, the ^{13}C remained exclusively in carbons 3 and 4. These results encouraged consideration that CO_2 might be fixed in photosynthesis by reactions similar to those found in heterotrophs. The results clearly showed that CO_2 was not converted to formaldehyde and then to sugar by an aldol condensation. In this case, all carbons of the sugar would have been totally labeled simultaneously.

These preliminary studies were followed by extensive studies by Calvin's group in California and Gaffron's group in Chicago. They established (Benson and Calvin, 1947; Fager et al., 1950) that a major portion of $^{14}CO_2$ was fixed in the carboxyl group of phosphoglyceric acid during the early moments of photosynthesis. There was disagreement about the light reaction. Calvin and Benson (1947) considered that reductants formed by the light reaction were capable of promoting in the dark all the conversions of CO_2 which occur in photosynthesis. Fager et al. (1950), in contrast, interpreted their results as showing that the substance which survived in the dark was not a reducing substance but a C_2 acceptor for carbon dioxide. Thus, both groups presented different schemes for the carbon pathway of photosynthesis but both involved a C_2 compound serving as the CO_2 acceptor which gave rise to phosphoglycerate. C_4 dicarboxylic acids were found to be labeled and this carboxylation was considered to be part of the photosynthetic pathway. Various versions were presented which involved reactions similar to those known in heterotrophs, one of which is shown in Figure 3.1. It was proposed that a C_4 compound is cleaved to two C_2 compounds which serve as the carboxyl acceptor to form phosphoglyceric acid. A number of C_2 compounds were considered, among them vinyl phosphate and glycolate, and among the C_4 dicarboxylic acids, succinate, oxalacetate, malate, dihydroxymaleate, and tartrate.

This view that CO_2 was assimilated in photosynthesis by fixation with C_2 and C_3 compounds was maintained until 1951 (Calvin, 1951). However, sugar phosphates also were rapidly labeled and it became evident that a scheme such as that of Figure 3.1 was not the pathway of carbon in photosynthesis and that phosphoglycerate did not arise by carboxylation of a C_2 compound. It is beyond the scope of this presentation to describe the events leading to what is now called the reductive pentose cycle or the Calvin cycle, which is described in all modern texts (see also Chapter 19). Suffice it to say that it was observed that ribulose bisphosphate was labeled relatively early after addition of radioactive CO_2 and that illumination caused a decrease in its steady state concentration and a simultaneous rise in 3-phosphoglycerate. Calvin and co-workers, based on these observations and on analysis of the distribution of ^{14}C in the sugars during light and dark transitions, presented evidence for a cyclic scheme involving formation of sedoheptulose-P and carboxylation of ribulose-bisphosphate (Bassham et al., 1953). This evidence, along with isolation and characterization of the enzymes of the pathway by Horecker's group, Ochoa's group, and by Racker, placed this pathway on a firm basis; for references and

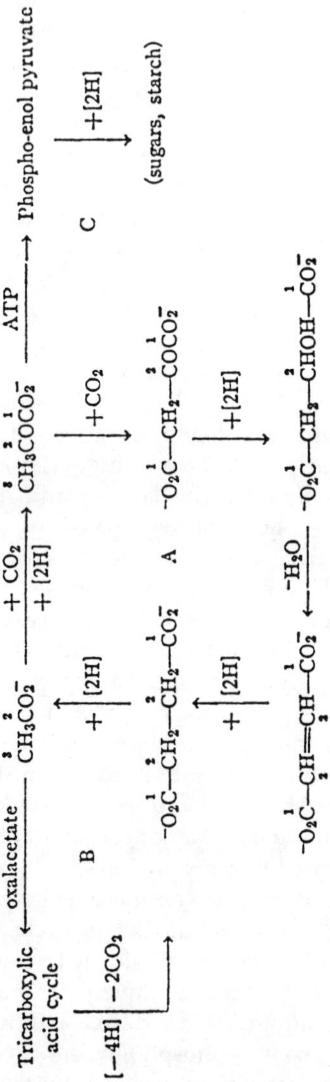

Figure 3.1 Early pathway of photosynthesis proposed by Benson and Calvin (1947).

discussion, see reviews by Vishniac et al. (1957) and Calvin and Bassham (1962). Subsequently, the pathway involving ribulose bisphosphate carboxylase was found in the sulfur bacteria and in many chemoautotrophs in addition to plants and algae. Clearly, it is the major pathway of autotrophic assimilation of CO_2.

3.3 THE ACETYL–CoA PATHWAY OF AUTOTROPHY

While the Calvin cycle was being solved, there were exciting events occurring in other studies of bacterial metabolism. Wieringa (1940) described an organism, *Clostridium aceticum,* which reduced CO_2 to acetate in the presence of hydrogen. At first this organism was considered to be an autotroph but in a subsequent study, Karlsson et al. (1948) demonstrated that it required glucose, glutamic acid, biotin, pyridoxamine, and pantothenic acid for growth. Barker et al. (1940) isolated *Clostridium acidiurici,* which ferments hypoxanthine and produces 1.25 mol of acetate per mol of purine. Since there is only one C-C chain in hypoxanthine, it was clear that the acetate in excess of 1 mol must have been synthesized from C_1 compounds. By the use of radioactive CO_2, they showed that CO_2 was incorporated in both the methyl and carboxyl groups of acetate. Fontaine et al. (1942) isolated *Clostridium thermoaceticum* and demonstrated that it ferments glucose with almost a quantitative conversion to acetate, yielding $\sim$3 mol of acetate per mol of glucose. The question was: does this occur by cleavage of glucose to three two-carbon compounds or is the third acetate synthesized from CO_2? Barker and Kamen (1945) answered this question. In one of the first experiments done with ^{14}C, they showed that $^{14}CO_2$ appeared in both carbons of acetate. Three other organisms were also shown by Barker and co-workers to fix CO_2 in both carbons of acetate, *Butyribacterium rettgeri* (Barker et al., 1945), *Clostridium cylindrosporum* (Barker and Beck, 1942), and *Diplococcus glycinophilus* (Barker et al., 1948).

These reports were very interesting because they indicated there were heterotrophs that differed from others in that they apparently could form carbon-to-carbon bonds from CO_2 *per se,* a property which heretofore had been considered to occur only in autotrophs. In addition, it was clear that the synthesis of acetate from CO_2 had the potential of being an autotrophic pathway since if acetate was synthesized, it could serve as the building block for formation of cellular components.

Before seriously considering that these pathways might function as autotrophic pathways, it was important to be certain that the acetate was, in fact, totally synthesized from CO_2. For these studies *C. thermoaceticum* was used. It was possible that the labeled acetate found by Barker and Kamen (1945) was made up of a mixture of [2-^{14}C] and [1-^{14}C] acetate and that none of the acetate was totally synthesized from CO_2. This seemed possible since Barker and Kamen had found that there was considerably more ^{14}C in the carboxyl group than in the methyl group of the acetate. We, therefore, conducted fermentations of glucose in the presence of $^{13}CO_2$, isolated the acetate, and determined its mass in a mass spectrometer. It was found (Wood, 1952) that acetate was present with a mass two times greater than acetate containing only ^{12}C. This showed that some of the acetate was totally synthesized from CO_2. From the mass ratios, it was calculated that one-third of the acetate had been formed from $^{13}CO_2$. Thus, the results were in accord with Reactions 3 and 4 that had been proposed by Barker and Kamen (1945).

$$C_6H_{12}O_6 + H_2O \rightarrow 2CH_3COOH + 8H + 2CO_2 \quad (3)$$

$$2CO_2 + 8H \rightarrow CH_3COOH + 2H_2O \quad (4)$$

We, therefore, set out to determine the mechanism of this synthesis. Progress was slow and we are still investigating this pathway today. We now know that the synthesis of acetyl–CoA from CO_2 involves a pathway utilized by a number of autotrophic bacteria (see Chapter 20). Since reviews have been written recently concerning the acetyl–CoA pathway (Ljungdahl and Wood, 1982; Zeikus et al., 1985; Fuchs, 1986 and Chapter 19, this volume; Wood et al., 1986a, 1986b, and 1986c; Ljungdahl, 1986), I will only give a general description of highlights leading to an understanding of the mechanism.

There were four major turning points in the development of the pathway for total synthesis of acetate from CO_2. The first was when Poston et al. (1964) found that ^{14}C-methyl vitamin B_{12} served as the source of the methyl group of acetate during fermentations of pyruvate by extracts of *C. thermoaceticum*. We then looked for the formation of methyl corrinoids by pulse labeling with $^{14}CO_2$ for 15 seconds with cells fermenting glucose. The methyl group was very highly labeled (Ljungdahl et al., 1965) and it was clear the methyl of corrinoids might be an intermediate of the pathway. We reasoned by analogy with the known pathway of methionine synthesis, in which the methyl of methyl tetrahydrofolate (CH_3THF) is transferred to homocysteine via a corrinoid enzyme, that something similar might be occurring in acetate synthesis. If this were true the methyl of CH_3THF should become highly labeled. By pulse labeling with $^{14}CO_2$, we found that the methyl became highly labeled in cells of *C. thermoaceticum* fermenting glucose (Parker et al., 1971). The enzymes involved in the synthesis of formate from CO_2 and the formation of formyl tetrahydrofolate and its conversion to CH_3THF have all been purified from *C. thermoaceticum*. Thus, this portion of the pathway has been firmly established.

The second major turning period was when Drake et al. (1981) purified from *C. thermoaceticum* five components that catalyzed the synthesis of 2 mol of acetyl–CoA from CH_3THF, pyruvate, and coenzyme A (CoASH). The five components were phosphotransacetylase, methyl transferase, ferredoxin, pyruvate-ferredoxin oxidoreductase, and a fraction which contained several unidentified proteins. The availability of a soluble system opened the way for the subsequent purification of the corrinoid enzyme (Hu et al., 1984) and carbon monoxide (CO) dehydrogenase disulfide reductase (Pezacka and Wood, 1986).

The third turning point began when Diekert and Thauer (1978) reported that *C. thermoaceticum* contained a high level of CO dehydrogenase (CODH) which catalyzes the following reaction:

$$CO + H_2O \rightleftharpoons CO_2 + 2H^+ + 2e \qquad (5)$$

An important related aspect of these studies was the demonstration by Diekert and Thauer (1980) that CODH activity is increased when nickel is added to the growth medium of *C. thermoaceticum* or *C. formicoaceticum* and when Drake et al. (1980) established that CO dehydrogenase is a nickel enzyme. Hu et al. (1982) then showed, using ferredoxin, methyltransferase and a fraction containing CO dehydrogenase, the corrinoid enzyme, and some other unidentified proteins, that this system catalyzed the following reaction:

$$CH_3THF + CO + CoASH \rightarrow CH_3COSCoA + THF \qquad (6)$$

With CO present, pyruvate was no longer required nor was pyruvate ferredoxin oxidoreductase. Now it was clear that the pathway we were studying could certainly be an autotrophic pathway since the methyl group of CH_3THF could be formed from CO_2 via formate and the CO could be formed from CO_2 by the CODH. By electron spin resonance, Ragsdale and co-workers (1983; 1985) demonstrated that a complex is formed between the CO and nickel and that this is the active site for CO on CODH.

The fourth turning point was the demonstration that CODH *per se* from *C. thermoaceticum* catalyzes an exchange between CO and the carbonyl group of acetyl–CoA (Ragsdale and Wood, 1985). The reaction is illustrated below; X Y Z represents three sites on the CODH:

$$\begin{array}{c} CH_3 \\ {}^{14}| \\ C{-}O \\ | \\ SCoA \end{array} + \left[\begin{array}{l} X \\ Y \\ Z \end{array}\right. \rightleftharpoons \left[\begin{array}{l} X{-}CH_3 \\ Y{-}^{14}CO \\ Z{-}SCoA \end{array}\right. \rightleftharpoons \left[\begin{array}{l} X{-}CH_3 \\ Y \\ Z{-}SCoA \end{array}\right. \begin{array}{c} {}^{14}CO \\ + \upharpoonleft\downharpoonright \\ {}^{12}CO \end{array}$$

The experiment involves the incubation of [1-^{14}C] acetyl–CoA and CODH in a solution in equilibrium with unlabeled CO in the gas phase. It was found that there was no decrease in the concentration of acetyl–CoA but its radioactivity decreased and the CO became radioactive. Clearly, for this to occur the C–C bond and the C–S bond of the [1-^{14}C] acetyl–CoA must be cleaved and the resulting ^{14}CO must equilibrate with the unlabeled CO. Since no CH_3– or –SCoA acceptors were added, there clearly were sites on the CODH which bound them during the exchange. Furthermore, since unlabeled CO is entering the acetyl–CoA, acetyl–CoA obviously is being resynthesized. These observations prompted us to modify our schemes for the pathway. Prior to these findings, it had been considered that the corrinoid enzyme was directly involved in the carboxylation, perhaps by CO_2 reacting with the methyl on cobalt of the corrinoid as in a Grignard reaction. Now it was apparent that the CODH *per se* catalyzes the synthesis and that it in fact is a synthetase as well as a dehydrogenase. It has also been demonstrated that CODH can be methylated with CH_3I or with the CH_3 corrinoid enzyme and with this methylated enzyme,

CO and CoASH [2-^{14}C] acetyl–CoA is synthesized without the presence of any other enzymes (Pezacka and Wood, 1988). These studies also showed that the methyl is linked to a cysteine of CODH. These results confirm the fact that CODH *per se* catalyzes the combination of the methyl, CO, and CoASH groups.

Now we will consider the scheme which has been derived from the observations described above and from others we have not considered but are described in reviews (Ljungdahl and Wood, 1982; Zeikus et al., 1985; Fuchs, 1986; Wood et al., 1986a, 1986b, 1986c; Ljungdahl, 1986).

There are two entries for CO_2 in the pathway (Reactions 2 and 4 of Figure 3.2). H_2 is required for both sequences as a source of electrons via hydrogenase for the reduction of the CO_2 to CO by the CODH and for formation of CH_3THF via formate. These latter reactions have all been demonstrated by Ljungdahl and his collaborators using highly purified enzymes (Ljungdahl and Wood, 1982; Ljungdahl, 1986). When CO is the source of carbon, hydrogen and hydrogenase are not required since the oxidation of CO to CO_2 by CODH supplies the electrons (Reaction 5), thus replacing the requirements for H_2. A portion of the generated CO_2 is converted via formate to CH_3THF (Reaction 2 of Figure 3.2). Since CO can serve as the source of electrons and energy (presumably by electron transfer phosphorylation, see Mackivey and Ljungdahl, 1986), acetogenic bacteria are able to grow with CO as the sole source of carbon and energy. Kerby and Zeikus (1983) have shown that *C. thermoaceticum* can be adapted to use its autotrophic capabilities and to grow with CO or H_2 and CO_2 as the source of carbon and energy. Thus, *C. thermoaceticum* is a facultative autotroph (see Fuchs, 1986, and Chapter 20 for detailed considerations of the energetics).

CO dehydrogenase is the central enzyme of the scheme serving as the acceptor for the CO at the Y site (apparently at the nickel), for the CH_3 group at the X site (apparently a cysteine of the enzyme), and for the CoASH at the Z site. The transfer of the CH_3 group of CH_3THF to the cobalt on the corrinoid enzyme is catalyzed by methyltransferase (Hu et al., 1984) (Reaction 3 of Figure 3.2) and then there is transfer from the corrinoid enzyme to the X site on CODH via Reaction 5 (Pezacka and Wood, 1988). An acetyl group apparently is formed (Reaction 6) since in the absence of CoASH, acetate is slowly released. Addition of CoASH to the Z site is shown as Reaction 7 but little is known about it except that CO dehydrogenase disulfide reductase facilitates the overall reaction (Pezacka and Wood, 1986). The final step is Reaction 8, in which CODH catalyzes formation of acetyl–CoA. The acetyl–CoA is used in part in anabolic reactions but the major portion is converted to acetate with generation of ATP by phosphotransacetylase and acetate kinase.

Our efforts have mainly been concerned with the portion of the scheme involving the conversion of CH_3THF, CO, and CoASH to acetyl–CoA. All the enzymes of this portion of the pathway have been purified and the synthesis of acetyl–CoA has been accomplished from CH_3THF, CO, and CoASH using methyltransferase, corrinoid enzyme, CODH, ferredoxin, and CODH disulfide reductase (Pezacka and Wood, 1986). Thus, after over 30 years of work it is very gratifying that the total synthesis of acetyl–CoA from CO_2 could be ac-

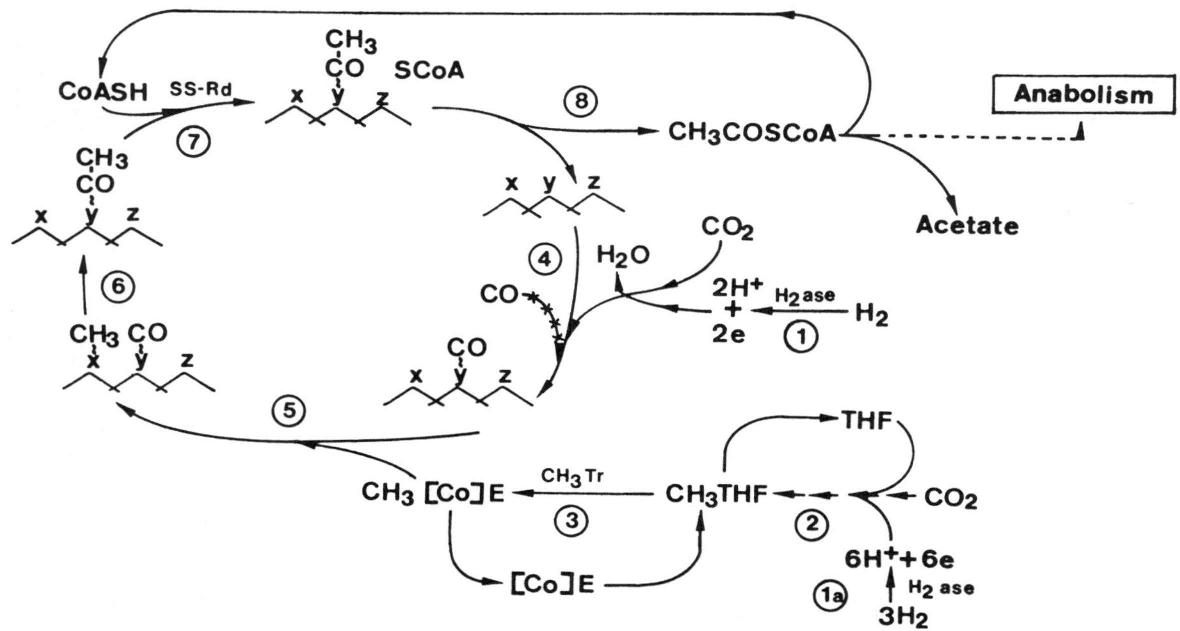

Figure 3.2 The acetyl–CoA pathway of autotrophic growth by acetogenic bacteria with CO or CO_2 and H_2. H_2ase is hydrogenase, /\/\/\ is CO dehydrogenase, CH_3Tr is methyl transferase, [Co] E is corrinoid enzyme, THF is tetrahydrofolate, S–S–Rd is CO dehydrogenase disulfide reductase (from Wood et al., 1986b).

complished with purified enzymes. The development of this pathway has required a long time partly (i) because the intermediates of the reaction are enzyme bound, thus they were not detected in early attempts to use pulse labeling, and (ii) because the enzymes are extremely sensitive to oxygen and therefore early efforts at purification failed. There seems little doubt that this pathway is used by other acetogens and most likely by methanogens although with modifications (see Chapter 19). The evidence from investigations in various laboratories relative to the possible role of this pathway in metabolism has been reviewed recently (Wood et al., 1986c). Almost certainly *Acetobacterium woodii*, which grows with CO_2 and H_2 as the carbon and energy source, utilizes the acetyl–CoA pathway. Recently, Shanmugasandaram et al. (1988) have purified enzymes from this organism and have shown that they catalyze the synthesis of acetyl–CoA and that the CODH has properties identical to those of the enzyme from *C. thermoaceticum*. However, Grahame and Stadtman (1987) have found that the CODH of *Methanosarcina barkeri* has a subunit structure quite different from that of *C. thermoaceticum* and it does not catalyze the exchange of CO with [^{14}C]acetyl–CoA.

3.4 THE GLYCINE SYNTHASE PATHWAY OF SYNTHESIS OF ACETATE FROM CO_2

We have noted that Barker and co-workers found in fermentations of purines by *C. acidiurici* (Barker et al., 1940) and *C. cylindrosporum* (Barker and Beck, 1942) and of glycine by *Diplococcus glycinophilus* (Barker et al., 1948) that $^{14}CO_2$ was fixed in both carbons of acetate. The questions were, do these organisms form acetate from CO_2 by a mechanism differing from that of *C. thermoaceticum*? Is it an autotrophic pathway? As an approach to this problem, the types of acetate synthesized from $^{13}CO_2$ were determined by mass analysis (Schulman et al., 1970). *C. formicoaceticum* was included in the experiments because it ferments fructose with almost a quantitative yield of acetate (Andreesen et al., 1970) and contains enzymes of the acetyl–CoA pathway (O'Brien and Ljungdahl, 1972) and is thought to have a pathway like *C. thermoaceticum*. It was determined from the mass ratios that *C. acidiurici* and *C. cylindrosporum* produced acetate that was totally synthesized from CO_2 but this type of acetate was only 9 percent of the total acetate that was formed in the fermentations of uric acid. With *C. formicoaceticum* about 28 percent of the acetate produced from fructose was totally synthesized from CO_2. *D. glycinophilus* did not incorporate enough $^{13}CO_2$ to permit determination of the types of acetate by mass analysis, although with $^{14}CO_2$ it was shown that both positions were labeled.

The mechanism of the total synthesis of acetate was investigated by Waber and Wood (1979) using *C. acidiurici* grown on uric acid. The results showed that the mechanism in this organism is completely different from that of *C. thermoaceticum*; it was in accord with the scheme shown in Figure 3.3. The key enzyme in this pathway is glycine decarboxylase, also called the glycine

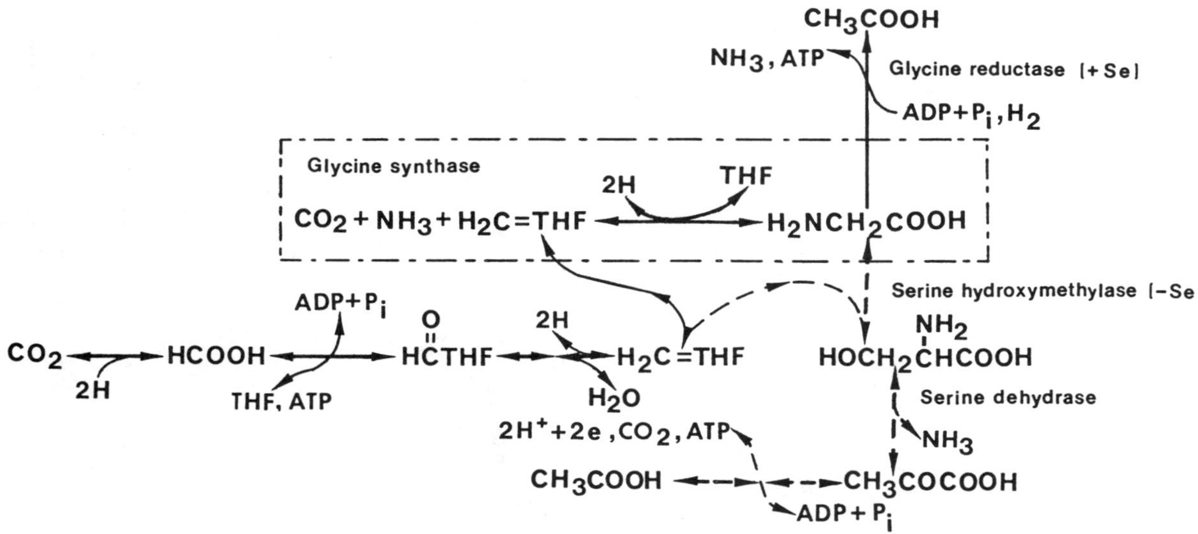

Figure 3.3 The glycine synthase pathway of total synthesis of acetate from CO_2 by *Clostridium acidiurici*. THF is tetrahydrofolate. The box outlines the glycine synthase reaction. The dashed arrows indicate the serine pathway. +Se, pathway when selenium is present; −Se, pathway in the absence of selenium.

synthase system. The reaction catalyzed by this enzyme is enclosed in the box in Figure 3.3. CO_2, NH_3, and methylene tetrahydrofolate (CH_2=THF) combine to form glycine. Since CH_2=THF can be synthesized by reduction of CO_2 to formate, conversion to formyl THF, and reduction to CH_2=THF, the carbon of the resulting glycine thus is formed entirely from CO_2. The glycine can enter one of two pathways: (1) conversion to acetate catalyzed by glycine reductase (Stadtman, 1979), or (2) conversion to serine catalyzed by serine hydroxymethylase with utilization of a second CH_2=THF. The serine is converted to pyruvate by serine dehydrase and then the pyruvate is oxidized to CO_2 and acetic acid. This latter portion of the scheme is indicated by the dashed arrows of Figure 3.3. By either pathway, acetate is totally synthesized from CO_2.

Glycine synthase and glycine reductase are complex systems. The glycine reductase system is made up of three components, one of which is a selenoprotein (Stadtman, 1979). Dürre and Andreesen (1983) have shown that when selenium is present in the medium, the glycine is converted directly to acetate. When selenium is limiting, the conversion to acetate proceeds via the serine route (Dürre and Andreesen, 1983). The latter was the pathway investigated by Waber and Wood (1979). The possibility that glycine decarboxylase (synthase) might be involved in the synthesis of acetate from CO_2 by *Peptococcus glycinophilus (Diplococcus glycinophilus* of Barker), *C. cylindrosporum*, or *C. acidiurici* had been proposed by Vogels and Van Der Drift (1976) in their comprehensive review of purine and pyrimidine degradation by bacteria, but it was the experiments of Waber and Wood (1979) and Motokawa and Kikuchi (1974), showing that glycine can be synthesized by glycine synthase, that placed this pathway for formation of acetate from CO_2 on a firm basis.

The glycine decarboxylase (synthase) system consists of four components, a protein containing pyridoxal phosphate with which glycine forms a Schiff base (called P protein), a heat stable hydrogen carrier protein (H protein), a tetrahydrofolate binding protein (T protein), and a lipoamide dehydrogenase (L protein). The system has been studied extensively by Sagers and co-workers, using *P. glycinophilus* (Robinson et al., 1973), and by Kikuchi (1973) and Hiraga and Kikuchi (1980) using rat liver. Gariboldi and Drake (1984) have purified two components of the glycine synthase system of *C. acidiurici*, one heat labile (designated HLC) and the other heat stable (designated HSC). These proteins have some properties similar to the P protein and H protein from liver. These two components catalyzed exchange of CO_2 with glycine but not net synthesis from CH_2=THF, NH_3 and CO_2. The HLC component has an $\alpha_2\beta_2$ tetramer structure of M_r = 230,000 and does not contain pyridoxal phosphate, but pyridoxal phosphate is required for the exchange reaction. The chicken liver mitochondrial P-protein has an M_r of 200,000 but is an α_2 dimer and contains pyridoxal phosphate (Kikuchi, 1973). Gariboldi and Drake (1984) noted that the reaction in *C. acidiurici* is considered to be in the direction of glycine synthesis rather than degradation and the differences in the systems may reflect the difference in their physiological roles.

There have been extensive investigations of the metabolism of purines by Rabinowitz and co-workers and others (for references, see the review by Vogels and Van Der Drift, 1976). In addition, Dürre and Andreesen (1982) have reported that *Clostridium purinolyticum* degrades uric acid to acetate, formate, glycine, NH_3, and CO_2 by an unknown pathway. We have confined our attention entirely to the mechanism of the total synthesis of acetate; the overall pathway of purine metabolism is beyond the scope of this report. The total synthesis from CO_2 in these fermentations appears to play a minor role in contrast to that of *C. thermoaceticum* in which CO_2, in its conversion to acetate, serves as the major electron acceptor in the fermentation of the glucose.

Thus far, the glycine synthase pathway has not been shown to be utilized by organisms for autotrophic growth. If an organism contained hydrogenase, it appears it could use H_2 and CO_2 to form $CH_2{=}THF$ and by glycine synthetase form glycine which then by the glycine reductase reaction could generate acetate and ATP from which the anabolic pathways could proceed.

3.5 CONCLUDING REMARKS

The discovery that CO_2 is used by heterotrophs widened the concepts not only of the metabolism of heterotrophs but also of autotrophs. It brought to the forefront the idea that CO_2 might be fixed in photosynthesis by addition of CO_2 to a preexisting organic compound in the dark just as occurs in heterotrophs. It turned out that the primary fixation of CO_2 in the reductive pentose cycle is by ribulose bisphosphate carboxylase and not via a C_2 or C_3 intermediate. However, it is by addition of CO_2 to a preexisting organic compound rather than direct binding to chlorophyll (see section 3.2). From the standpoint of the chemistry involved in the carboxylase reaction, there is no reason to think this reaction could not occur in heterotrophs, but no evidence of its occurrence in heterotrophs exists other than in facultative autotrophs or in those that can grow on C_1 compounds such as methanol (Taylor et al., 1980).

The reductive pentose cycle is used by the majority of autotrophs but it is not the only mechanism for total synthesis of compounds from CO_2. The acetyl–CoA pathway occurs not only in heterotrophs, but it also is a pathway in certain autotrophs (see Chapter 20). It is the most straightforward pathway conceivable; it involves the combination of two CO_2 (or CO) to form a 2-carbon compound from which the cellular components are formed by heterotrophic types of CO_2 fixation and other typical reactions of heterotrophs. It is the kind of pathway a chemist would envision for autotrophy. This is a mechanism that uses many metalloenzymes as catalysts and it possesses many challenges for understanding the details of the mechanism of the catalysis by these enzymes. It appears to be a mechanism used by the methane bacteria, though perhaps in modified form. There are exciting investigations ahead to determine what these modifications may be and to learn more about the role this type of autotrophy plays in the overall carbon cycle.

There is a third autotrophic pathway, the reductive tricarboxylic acid cycle (Buchanan, 1979; Fuchs et al., 1980). It occurs by sequential addition of CO_2

beginning with acetyl–CoA ($C_2 \rightarrow C_3 \rightarrow C_4 \rightarrow C_5 \rightarrow C_6$) yielding citrate which is cleaved to acetyl–CoA and oxalacetate and the cycle is repeated. This pathway is also considered in Chapter 20.

The preparation of this paper was supported by research grant GM24913 from the U.S. National Institutes of Health.

REFERENCES

Andreesen, J.R., Gottschalk, G., and **Schlegel, H.G.** 1970. *Clostridium formicoaceticum* nov. spec. Isolation, description and distinction from *C. aceticum* and *C. thermoaceticum. Archiv für Mikrobiologie* 72:154–174.

Aronoff, S., Barker, H.A., and **Calvin, M.** 1947. Distribution of labeled carbon in sugar from barley. *Journal of Biological Chemistry* 169:459–460.

Barker, H.A. 1936. The biochemistry of methane formation. *Archiv für Mikrobiologie* 7:404–419.

Barker, H.A. and **Beck, J.F.** 1942. *Clostridium acidiurici* and *Clostridium cylindrosporum,* organisms fermenting uric acid and some other purines. *Journal of Bacteriology* 43:291–304.

Barker, H.A. and **Kamen, M.D.** 1945. Carbon dioxide utilization in the synthesis of acetic acid by *Clostridium thermoaceticum. Proceedings of the National Academy of Sciences USA* 31:219–225.

Barker, H.A., Ruben, S., and **Beck, F.J.** 1940. Radioactive carbon as an indicator of carbon dioxide reduction. IV. The synthesis of acetic acid from carbon dioxide by *Clostridium acidiurici. Proceedings of the National Academy of Sciences USA* 26:447–482.

Barker, H.A., Kamen, M.D., and **Haas, V.** 1945. Carbon dioxide utilization in the synthesis of acetic acid and butyric acids by *Butyribacterium rettgeri. Proceedings of the National Academy of Sciences USA* 31:355–360.

Barker, H.A., Volcani, B.E., and **Cardon, B.P.** 1948. Tracer experiments on the mechanism of glycine fermentation by *Diplococcus glycinophilus. Journal of Biological Chemistry* 176:803–804.

Barron, E.S.G. 1943. Mechanisms of carbohydrate metabolism. An essay on comparative biochemistry. *Advances in Enzymology* 3:149–189.

Bassham, J.A., Benson, A.A., Kay, L.D., Harris, A.Z., Wilson, A.T., and **Calvin, M.** 1953. The path of carbon in photosynthesis. XXI. The cyclic regeneration of carbon dioxide acceptor. *Journal of the American Chemical Society* 76:1760–1770.

Benson, A.A. and **Calvin, M.** 1947. The dark reductions of photosynthesis. *Science* 105:648–649.

Buchanan, B.B. 1979. Ferredoxin-linked carbon dioxide fixation in photosynthetic bacteria, pp. 416–424, in Gibbs, M. and Lasko, E. (editors), *Photosynthesis II. Photosynthetic carbon metabolism and related processes,* Springer-Verlag, N.Y.

Calvin, M. 1951. The path of carbon in photosynthesis. *Harvey Lectures* 46:218–251.

Calvin, M. and **Bassham, J.A.** 1962. *The photosynthesis of carbon compounds.* W.A. Benjamin, Inc., New York.

Diekert, G.B. and **Thauer, R.K.** 1978. Carbon monoxide oxidation by *Clostridium thermoaceticum* and *Clostridium formicoaceticum. Journal of Bacteriology* 136:597–606.

Diekert, D. and **Thauer, R.K.** 1980. The effect of nickel on carbon monoxide dehydrogenase formation in *Clostridium thermoaceticum* and *Clostridium formicoaceticum. FEMS Microbiology Letters* 7:187–189.

Drake, H.L., Hu, S-I., and **Wood, H.G.** 1980. Purification of carbon monoxide dehydrogenase, a nickel enzyme, from *Clostridium thermoaceticum. Journal of Biological Chemistry* 255:7174–7180.

Drake, H.L., Hu, S-I, and **Wood, H.G.** 1981. Purification of five components from *Clostridium thermoaceticum* which catalyzes synthesis of acetate from pyruvate and methyltetrahydrofolate. Properties of phosphotransacetylase. *Journal of Biological Chemistry* 256:11137–11144.

Dürre, P. and **Andreesen, J.R.** 1982. Anaerobic degradation of uric acid via pyrimidine derivatives by selenium-starved cells of *Clostridium purinolyticum. Archives of Microbiology* 131:255–260.

Dürre, P. and **Andreesen, J.R.** 1983. Purine and glycine metabolism by purinolytic clostridia. *Journal of Bacteriology* 154:192–199.

Emerson, R. and **Lewis, C.M.** 1939. Factors influencing the efficiency of photosynthesis. *American Journal of Botany* 26:808–822.

Evans, E.A., Jr. and **Slotin, L.** 1941. Carbon dioxide utilization by pigeon liver. *Journal of Biological Chemistry* 141:439–450.

Fager, E.W., Rosenberg, J.L., and **Gaffron, H.** 1950. Intermediates in photosynthesis. *Federation Proceedings* 9:535–542.

Fontaine, F.E., Peterson, W.H., McCoy, E., Johnson, M.J., and **Ritter, G.J.** 1942. A new type of glucose fermentation by *Clostridium thermoaceticum* NSP. *Journal of Bacteriology* 43:701–715.

Frank, J. and **Gaffron, H.** 1941. Photosynthesis facts and interpretations. *Advances in Enzymology* 1:199–262.

Fuchs, G. 1986. CO_2 fixation in acetogenic bacteria: variations on a theme. *FEMS Microbiology Review* 39:181–213.

Fuchs, G., Stupperich, E., and **Enden, C.** 1980. Autotrophic CO_2 fixation in *Chlorobium limicola.* Evidence for a reductive tricarboxylic acid cycle in growing cells. *Archives of Microbiology* 128:64–71.

Gariboldi, R.T. and **Drake, H.L.** 1984. Glycine synthase of the purinolytic bacterium, *Clostridium acidiurici.* Purification of the glycine-CO_2 exchange system. *Journal of Biological Chemistry* 259:6085–6089.

Gladstone, G.P., Fildes, P., and **Richardson, G.M.** 1935. Carbon dioxide as an essential factor in the growth of bacteria. *British Journal of Experimental Pathology* 16:335–348.

Grahame, D.A. and **Stadtman, T.C.** 1987. Carbon monoxide dehydrogenase from *Methanosarcina barkeri.* Degradation, purification and physicochemical properties of the enzyme. *Journal of Biological Chemistry* 262:3706–3712.

Hiraga, I. and **Kikuchi, G.** 1980. The mitochondrial glycine cleavage system. Purification and properties of glycine decarboxylase from chicken liver mitochondria. *Journal of Biological Chemistry* 255:11664–11670.

Hu, S-I., Drake, H.L., and **Wood, H.G.** 1982. The synthesis of acetyl coenzyme A from carbon monoxide, methyltetrahydrofolate, and coenzyme A by enzymes from *Clostridium thermoaceticum. Journal of Bacteriology* 149:440–448.

Hu, S-I., Pezacka, E., and **Wood, H.G.** 1984. Acetate synthesis from carbon monoxide by *Clostridium thermoaceticum.* Purification of the corrinoid protein. *Journal of Biological Chemistry* 259:8892–8897.

Karlsson, J.L., Volcani, B.E., and **Barker, H.A.** 1948. The nutritional requirements of *Clostridium aceticum. Journal of Bacteriology* 56:781–782.

Kerby, R. and **Zeikus, J.G.** 1983. Growth of *Clostridium thermoaceticum* on H_2CO_3 or CO as energy source. *Current Microbiology* 8:27–30.

Kikuchi, G. 1973. The glycine cleavage system. Composition, reaction mechanism, and physiological significance. *Molecular and Cellular Biochemistry* 1:169–187.

Kluyver, A.J. 1931. *The chemical activities of microorganisms.* London, University Press.

Krampitz, L.O., Wood, H.G. and **Werkman, C.H.** 1943. Enzymatic fixation of CO_2 in oxalacetate. *Journal of Biological Chemistry* 147:243–253.

Krebs, H.A. 1943. The intermediary stages in the biological oxidation of carbohydrate. *Advances in Enzymology* 3:191–252.

Lebedev, A.F. 1921. *Izvest. Donskovo Gosudarstvenno Universiteta* 3:25. (Original journal not available to author.) English translation: The assimilation of carbon by saprophytes. *American Reviews of Soviet Medicine* 5:15–27 (1947-1948).

Ljungdahl, L.G. 1986. The autotrophic pathway of acetate synthesis in acetogenic bacteria. *Annual Reviews of Microbiology* 40:415–450.

Ljungdahl, L.G. and **Wood, H.G.** 1982. *Acetate biosynthesis in B_{12}*. Dolphin, D., editor, John Wiley and Sons, Inc. pp. 165–202.

Ljungdahl, L.G., Irion, E., and **Wood, H.G.** 1965. Total synthesis of acetate from CO_2. I. Co-methylcobric acid and Co-(methyl)-5-benzimidizoyl-cobamide as intermediates with *Clostridium thermoaceticum. Biochemistry* 4:2771–2779.

Mackivey, D. and **Ljungdahl, L.G.** 1986. Purification and characterization of the F_1-ATPase from *Clostridium thermoaceticum. Journal of Bacteriology* 165:252–257.

Moss, J. and **Lane, M.D.** 1971. The biotin-dependent enzymes. *Advances in Enzymology* 35:321–442.

Motokawa, Y. and **Kikuchi, G.** 1974. Glycine metabolism by rat liver mitochondria. Isolation and some properties of the protein bound intermediate of the reversible glycine cleavage reaction. *Archives of Biochemistry and Biophysics* 164:634–640.

O'Brien, W.E. and **Ljungdahl, L.G.** 1972. Fermentation of fructose and synthesis of acetate from carbon dioxide by *Clostridium formicoaceticum. Journal of Bacteriology* 109:626–632.

Parker, D.J., Wu, T-F., and **Wood, H.G.** 1971. Total synthesis of acetate from CO_2: methyltetrahydrofolate, an intermediate and a procedure for separation of folates. *Journal of Bacteriology* 108:770–776.

Pezacka, E. and **Wood, H.G.** 1986. The autotrophic pathway of acetogenic bacteria. Role of CO dehydrogenase disulfide reductase. *Journal of Biological Chemistry* 261:1609–1615.

Pezacka, E. and **Wood, H.G.** 1988. Acetyl-CoA pathway to autotrophic growth: Identification of the methyl binding site of the CO dehydrogenase. *Journal of Biological Chemistry,* in press.

Poston, J.M., Kuratomi, K., and **Stadtman, E.** 1964. Methyl-vitamin B_{12} as a source of methyl groups for the synthesis of acetate by cell free extracts of *Clostridium thermoaceticum. Annals of the New York Academy of Science* 112:804–806.

Ragsdale, S.W. and **Wood, H.G.** 1985. Acetate biosynthesis of acetogenic bacteria. Evidence that carbon monoxide dehydrogenase is the condensing enzyme that catalyzes the final steps of the synthesis. *Journal of Biological Chemistry* 260:3970–3977.

Ragsdale, S.W., Ljungdahl, L.G., and **DerVartanian, D.V.** 1983. ^{13}C and ^{61}Ni isotope substitutions confirm the presence of a nickel-III-carbon species in acetogenic CO dehydrogenase. *Biochemical Biophysical Research Communication* 115:658–665.

Ragsdale, S.W., Wood, H.G., and **Antholine, W.E.** 1985. Evidence that an iron-nickel-carbon complex is formed by reaction of CO with the CO dehydrogenase from *Clostridium thermoaceticum. Biochemistry* 82:6811–6814.

Robinson, J.R., Klein, S.M., and **Sagers, R.D.** 1973. Glycine metabolism. Lipoic acid as the prosthetic group in the electron transfer protein P_2 from *Peptococcus glycinophilus. Journal of Biological Chemistry* 248:5319–5323.

Rockwell, G.E. and **Highberger, J.H.** 1927. The necessity of carbon dioxide for the growth of bacteria, yeasts and molds. *Journal of Infectious Diseases* 40:438–446.

Ruben, S., Hassid, W.Z., and **Kamen, M.D.** 1939. Radioactive carbon in the study of photosynthesis. *Journal of American Chemical Society* 61:661–663.

Schulman, M., Parker, D., Ljungdahl, L.G., and **Wood, H.G.** 1970. Total synthesis of acetate from CO_2. V. Determination by mass analysis of the different types of acetate formed from $^{13}CO_2$ by heterotrophic bacteria. *Journal of Bacteriology* 109:633–644.

Shanmugasundaram, T., Ragsdale, S.W., and **Wood, H.G.** 1988. Role of carbon monoxide dehydrogenase in acetate synthesis by the acetogenic bacterium, *Acetobacterium woodii. BioFactors* 1: in press.

Stadtman, T.C. 1979. Selenium-dependent enzymes. *Annual Reviews of Biochemistry* 49:93–110.

Taylor, S., Dalton, H., and **Dow, C.** 1980. Purification and initial characterization of ribulose-1,5-bisphosphate carboxylase from *Methylococcus capsulatus* (Bath). *FEMS Microbiology Letters* 8:157–160.

Utter, M.F. and **Keech, D.B.** 1960. Formation of oxalacetate from pyruvate and CO_2. *Journal of Biological Chemistry* 235:PC17–18.

van Niel, C.B. 1941. The bacterial photosyntheses and their importance for the general problem of photosynthesis. *Advances in Enzymology* 1:263–328.

Virtanen, A.I. 1925. Über die Propionsäuregärung. II. *Societas Scientiarum Fennica, Commentationes Physico-Mathematicae* 2:1–13.

Virtanen, A.I. and **Karstrom, H.** 1931. Über die Propionisäuregärung. III. *Acta Chemica Fennica, Series B.* 7:17–19.

Vishniac, W., Horecker, B.L., and **Ochoa, S.** 1957. Enzymatic aspects of photosynthesis. *Advances in Enzymology* 19:1–77.

Vogels, G.D. and **Van Der Drift, C.** 1976. Degradation of purines and pyrimidines by microorganisms. *Bacteriological Reviews* 40:403–468.

Warburg, O. and **Negelein, E.** 1922. Über den Energieumsatz der Kohlensäureassimilation. *Zeitschaft für Physikalische Chemie* 102:235–266.

Warburg, O. and **Negelein, E.** 1923. Über den Einfluss der Wellenlänge auf den Energieumstatz bei der Kohlensäureassimilation. *Zeitschrift für Physikalische Chemie* 106:191–218.

Waber, L. and **Wood, H.G.** 1979. Mechanism of acetate synthesis from CO_2 by *Clostridium acidiurici. Journal of Bacteriology* 140:468–478.

Wieringa, K.T. 1940. The formation of acetic acid from carbon dioxide and hydrogen by an anaerobic spore-forming bacteria. *Antonie van Leeuwenhoek Journal of Microbiology and Serology* 6:251–262.

Werkman, C.H. and **Wood, H.G.** 1942. Heterotrophic assimilation of CO_2. *Advances in Enzymology* 2:135–182.

Winogradsky, S. 1891. Recherches sur les organismes de la nitrification. *Annales de l'Institut Pasteur, Paris* 5:577–646.

Winslow, C.-E.A., Walker, H.H., and **Sutermeister, M.** 1932. The influence of aeration and of sodium chloride upon the growth curve of bacteria in various media. *Journal of Bacteriology* 24:185–208.

Wood, H.G. 1952. A study of carbon dioxide fixation by mass determination of the types of C^{13} acetate. *Journal of Biological Chemistry* 194:905–931.

Wood, H.G. 1985. Then and now. *Annual Review of Biochemistry* 54:1–41.

Wood, H.G. and **Burr, G.O.** 1947. Photosynthesis with $C^{13}O_2$ and the distribution of heavy carbon in the sugars. *Federation Proceedings* 6:303.

Wood, H.G. and **Stjernholm, R.T.** 1962. Assimilation of CO_2 by heterotrophic organisms. *The Bacteria* 3:41–117. Academic Press, New York.

Wood, H.G. and **Werkman, C.H.** 1934. Pyruvate in the dissimilation of glucose by the propionic acid bacteria. *Biochemical Journal* 28:745–747.

Wood, H.G. and **Werkman, C.H.** 1936a. The mechanism of glucose dissimilation by the propionic acid bacteria. *Biochemical Journal* 30:618–623.

Wood, H.G. and **Werkman, C.H.** 1936b. The utilization of CO_2 in the dissimilation of glycerol by the propionic acid bacteria. *Biochemical Journal* 30:48–53.

Wood, H.G. and **Werkman, C.H.** 1938. The utilization of CO_2 by the propionic acid bacteria. *Biochemical Journal* 32 1262–1271.

Wood, H.G., Werkman, C.H., Hemingway, A., and **Nier, A.O.** 1940. Heavy carbon as a tracer in bacterial fixation of CO_2. *Journal of Biological Chemistry* 135:789–790.

Wood, H.G., Werkman, C.H., Hemingway, A., and **Nier, A.O.** 1941. Position of CO_2 in succinic acid synthesized by heterotrophic bacteria. *Journal of Biological Chemistry* 139:377–381.

Wood, H.G., Werkman, C.H., Hemingway, A., and **Nier, A.O.** 1942. Fixation of CO_2 by pigeon liver in the assimilation of CO_2. *Journal of Biological Chemistry* 192:31–45.

Wood, H.G., Lifson, N., and **Lorber, V.** 1945. The position of fixed carbon in glucose from rat liver. *Journal of Biological Chemistry* 159:475–489.

Wood, H.G., Ragsdale, S.W., and **Pezacka, E.** 1986a. A new pathway of autotrophic growth utilizing carbon monoxide or carbon dioxide and hydrogen. *Biochemistry International* 12:421–440.

Wood, H.G., Ragsdale, S.W., and **Pezacka, E.** 1986b. The acetyl–CoA pathway: a newly discovered pathway of autotrophic growth. *Trends in Biochemical Sciences* 11:14–18.

Wood, H.G., Ragsdale, S.W., and **Pezacka, E.** 1986c. The acetyl–CoA pathway of autotrophic growth. *FEMS Microbiology Reviews* 39:345–362.

Woods, D.D. 1936. Hydrogenlyases IV. The synthesis of formic acid by bacteria. *Biochemical Journal* 30:515–527.

Zeikus, J.G., Kerby, R., and **Kryscki, J.A.** 1985. Some carbon chemistry of acetogenic and methanogenic bacteria. *Science* 227:1167–1173.

4

Habitats and Ecological Niches of Chemolitho(auto)trophic Bacteria

J. Gijs Kuenen and Pieter Bos

Laboratory of Microbiology
Delft University of Technology
Delft, The Netherlands

4.1 INTRODUCTION

Nature harbors a large variety of bacteria which can derive energy from the oxidation of inorganic compounds (Table 4.1). Many chemolithotrophs use oxygen as their electron acceptor. A number of them can also use nitrate or nitrite and obligately anaerobic chemolithotrophs can be found among the sulfate reducers and methanogens. Recently, a bacterial species which can derive energy from the disproportionation reaction of thiosulfate has been discovered (Bak and Pfennig, 1987). Table 4.2 gives examples of each of the different types of energy generating reactions.

Many of the chemolithotrophs are able to grow autotrophically, and a number of the well-known representatives are typical "specialists," which were originally designated by Winogradsky as "inorgoxidants," able to derive energy from inorganic compounds only. However, in addition to the obligate chemolithotrophs, many genera also harbor facultative chemolithoautotrophs and chemolithoheterotrophs. The colorless sulfur bacteria may serve as an example. At present, 16 genera of colorless sulfur bacteria are known, as shown in Table 4.3. This list is not exclusive since organisms such as *Paracoccus denitrificans* (Friedrich and Mitrenga, 1981), *Hydrogenobacter thermophilus* (Kawasumi et al., 1984), and many phototrophs (Kondratieva et al., 1976; Kämpf and Pfennig,

Table 4.1 Electron donors for chemolithotrophs

Hydrogen, H_2
Carbon monoxide, CO
S^0 and reduced inorganic sulfur compounds
Reduced inorganic nitrogen compounds
Ferrous, Fe^{2+}
Manganous, Mn^{2+}
Antimony, Sb^{3+}(?)

Table 4.2 Examples of energy generating reactions for chemolithotrophs

O_2 as electron acceptor:
$2NH_4^+ + 3O_2 \rightarrow 2NO_2^- + 2H_2O + 4H^+$
NO_3^- as electron acceptor:
$5S^{2-} + 8NO_3^- + 8H^+ \rightarrow 5SO_4^{2-} + 4N_2 + 4H_2O$
SO_4^{2-} as electron acceptor:
$4H_2 + SO_4^{2-} \rightarrow S^{2-} + 4H_2O$
CO_2 as electron acceptor:
$4H_2 + CO_2 \rightarrow CH_4 + 2H_2O$
Disproportionation:
$S_2O_3^{2-} + H_2O \rightarrow SO_4^{2-} + HS^- + H^+$

Table 4.3 The genera of colorless sulfur-oxidizing bacteria*

Thiobacterium	*Thiobacillus*	*Beggiatoa*	*Thermothrix*
Macromonas	*Thiomicrospira*	*Thioploca*	
Thiospira	*Thiosphaera*	*Thiothrix*	*Sulfolobus*
Thiovulum			*Acidianus*
Bilophococcus	*Thiodendron*	*Achromatium*	

*Adapted from Kuenen (1975).
For further taxonomic information see Kuenen (1988a).

1980) are able to grow chemolithotrophically on reduced sulfur compounds (Kuenen, 1988a). The 16 genera listed in Table 4.3 represent an enormous diversity of physiological and morphological types, and even within one genus one may find a spectrum of metabolic types, including highly specialized, obligate chemolithotrophs, facultative types, and chemolithoheterotrophs. In addition, many heterotrophic bacteria which can oxidize sulfur compounds but do not derive energy from the oxidation, are known (Kelly and Kuenen, 1984; Kuenen et al., 1985). An analogous spectrum exists among the other chemolithotrophs.

This chapter focuses on the habitats of the chemolitho(auto)trophs and on their ecological niches. The ecological niche of an organism can best be described as its "functional status" or function in a particular environment. C. B. van Niel described the ecological niche as the "profession" of an organism.

In the following we will present a generalized picture of the habitats and ecological niches which the (facultatively) aerobic chemolitho(auto)trophs occupy. The specific examples to support this picture will mainly be drawn from work on the sulfur-oxidizing bacteria.

4.2 HABITATS

If the physical and physicochemical requirements for life (e.g., pH, temperature, E_h, ionic strength) are fulfilled, the first obvious need of a chemolithoautotroph is the presence of a suitable energy source and an electron acceptor.

Figure 4.1 summarizes the geological, anthropogenic, and biological sources of reduced inorganic compounds that may support chemolithotrophic life. Important geological sources include volcanic activity, as can be found in hot springs, or at the oceanic spreading centers (see Chapter 8), and geological

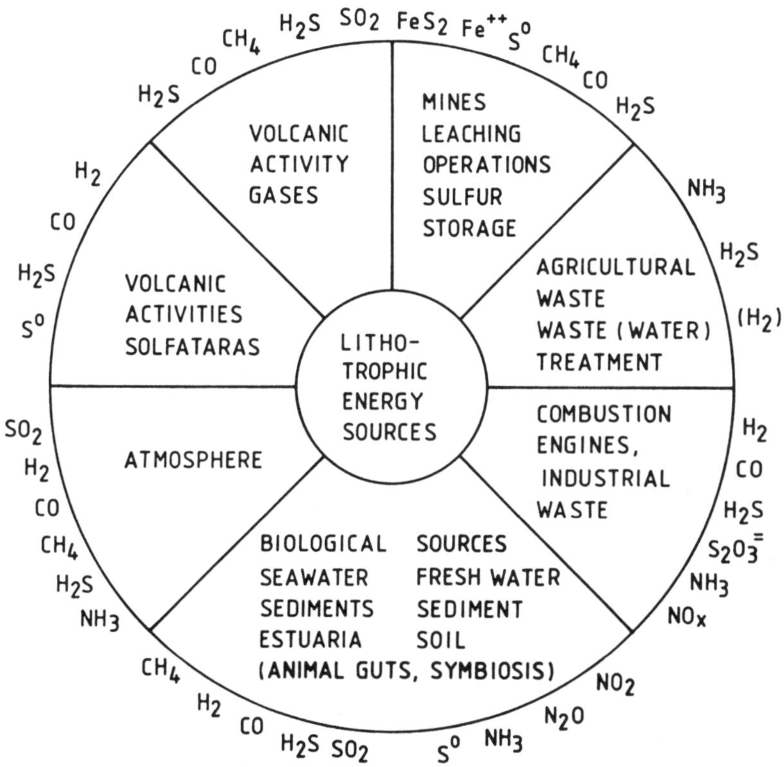

Figure 4.1 Geological, biological and anthropogenic sources of reduced inorganic compounds supporting chemolithotrophic life.

deposits of compounds such as elemental sulfur, and pyrite (FeS_2). Agriculture and industry are also major sources of reduced compounds which can be used as a major source of energy for chemolithoautotrophs. For example, it is estimated that human activity has boosted the oxidative part of the global sulfur cycle at least two fold (Ivanov, 1983). Finally, the chemolitho(auto)trophs have a variety of biological sources of energy available. These are formed in the natural cycling of elements on this globe. Estimates of the quantitative contribution of the different global biological sources and sinks are given in Table 4.4. For further details, the reader is referred to Seiler and Schmidt (1976).

The supply and concentration of a particular inorganic, reduced compound, and the resulting turnover of that compound, can be very high, as observed in certain volcanic springs or during mineralization in sediments of high organic content. On the other hand, the source can be very diffuse and dilute, for example in the production of ammonium from the aerobic mineralization of organic nitrogen compounds in oligotrophic aquatic environments.

Of course, chemolithotrophs also need electron acceptors for growth. This is usually oxygen or nitrate, and therefore many of the chemolithotrophs can be found at the anaerobic/aerobic interface of sediments or stratified water

Table 4.4 Rates of production and consumption of atmospheric gases in 10^{14} g/year*

Compound	Production rate	Consumption rate
CH_4	5.3–8.3	—
H_2	0.01–0.07	1.2
CO	0.2–1.2	5.0
H_2S	0.98	—
SO_2	—	400
NO_x	0.77	6.0
N_2O	1.6–2.6	—
NH_3	11.9	—

*Seiler and Schmidt (1976).

bodies (see Chapter 7). In the anaerobic phases of these environments, reduced inorganic compounds such as hydrogen sulfide, ammonium, and ferrous iron are formed. Due to a lack of suitable electron acceptors, these compounds will often accumulate and diffuse upwards to the interface. Oxygen or nitrate formed by nitrification diffuses downwards, and where the donors and acceptors meet the chemolithotrophs can thrive.

The conditions at these interfaces are not necessarily in steady state with respect to the physical and chemical conditions or other variables. Due to the diurnal, or tidal, changes the interface may shift up and down, causing dramatic changes or fluctuations in the environmental conditions. As a result, strong fluctuations can be observed in the supply of O_2, pH, sulfide, organic nutrients, and many other inorganic nutrients. Solubility, and thus accessibility of nutrients, will change with aerobic and anaerobic condition (e.g., ferrous hydroxide vs. ferric hydroxide). Thus, it may be expected that chemolithotrophic organisms possess mechanisms to cope with these fluctuations.

Growth and survival of any particular organism does not require adaptation only to the abiotic environment, but to the biotic environment as well. Therefore in the study of chemolithotrophs and their habitats we must also pay attention to the many possible microbial interactions such as commensalism, mutualism, predation, and especially *competition* which will be of paramount importance in the survival of the organism.

Hydrogen bacteria, carboxydobacteria, and their habitats Hydrogen (H_2) is produced in different ways and volcanic activity is one of the important sources. Much of this hydrogen ends up, highly diluted, in the troposphere where it will be oxidized in the soil. However, although this oxidation is undoubtedly biological, it is unlikely that, at these low concentrations, it is carried out by known hydrogen bacteria, which have too low an affinity for hydrogen (Conrad, 1984). Sometimes high fluxes of hydrogen can be locally available from volcanic sources. In such areas, hydrogen bacteria such as *Hydrogenobacter* (Bonjour and Aragno, 1986) can be found in large numbers.

Anaerobic decomposition is also an important biological source of hydrogen. However, in most cases this hydrogen is immediately oxidized under anaerobic conditions by methanogens or sulfate reducers. Only when the anaerobic consumption is slower than production or where it is interrupted by natural fluctuations will hydrogen escape to the anaerobic/aerobic interface or to the aerobic environment where it can be used by the aerobic or denitrifying hydrogen bacteria. Finally, during symbiotic nitrogen-fixation in the rhizosphere, large amounts of hydrogen are produced as a by-product of nitrogenase action. Indeed hydrogen bacteria can easily be isolated from the rhizosphere of nitrogen-fixing plants. Also, nitrogen-fixing algal blooms may be rich sources of biological hydrogen. In the few cases that have been investigated, hydrogen bacteria have been found to be associated with algal blooms (Schink and Zeikus, 1984).

At first glance, carbon monoxide might appear a typical nonbiological compound in view of its toxicity, but it is produced during the first stages of the anaerobic breakdown of organic products. In addition, it is a key intermediate of the assimilatory pathway of methane producers, and of the dissimilatory pathway of acetogenic bacteria. An exotic source of carbon monoxide is a jellyfish which contains buoyancy-controlling gas bodies which are filled with 90 percent CO (Pickwell et al., 1964). It may be speculated that this jellyfish should be covered with carboxydobacteria. The diversity of aerobic hydrogen- and carbon monoxide-utilizing bacteria is large. However, no hydrogen- or carbon monoxide-utilizing bacteria are known that can be recognized by their characteristic morphology, as is the case for some of the sulfur bacteria (e.g., *Beggiatoa* or *Thiovulum* with sulfur globules). Many of these organisms are taxonomically totally unrelated (Chapter 13), and have been classified among the heterotrophs, with the exception of the obligately chemolithotrophic *Hydrogenobacter* (Kawasumi et al., 1984).

Nitrifying bacteria and their habitats The most important biological sources of ammonia are mineralization processes, especially ammonification whereby ammonia is liberated from organic nitrogen compounds. The concentration of this ammonia is very low in aerobic environments, since it can directly be reoxidized again by the nitrifiers. Ammonia which is anaerobically produced, either by ammonification or dissimilatory nitrate reduction, diffuses toward the aerobic/anaerobic interface, and this interface is a typical dwelling place for nitrifying bacteria. When ammonia production is too high to be oxidized at the interface, it may diffuse up into the aerobic phase where it can be oxidized under fully aerobic conditions.

Although nitrification is known to peak at the aerobic/anaerobic interface, many well-known nitrifying autotrophs such as *Nitrosomonas* and *Nitrobacter* have relatively high affinity constants (K_s) for oxygen. These are in the order of 10–20 μM, in contrast to the K_s of 1–2 μM possessed by many other aerobic bacteria (Helder and de Vries, 1983). In line with this, nitrite accumulation at low oxygen concentrations has been observed at interfaces (Chapter 7), in tidal mud flats (Helder and de Vries, 1983), and also at low oxygen concentrations in wastewater treatment plants. However, some marine nitrifiers are known to possess higher affinities for oxygen (Gundersen et al., 1966), and thus it may well be that nature harbors other freshwater nitrifiers with higher affinity constants for oxygen.

The diversity among the nitrifying bacteria is discussed in detail in Chapter 5. As with the hydrogen bacteria, no unusual morphological types are known. Among the nitrifiers, a metabolic spectrum analogous to that mentioned for the sulfur oxidizers may exist. Very little is known about the so-called heterotrophic nitrifiers which cooxidize ammonium to nitrite or nitrate, but do not seem to benefit energetically from the oxidation. Further parallels with the sulfur oxidizers will be discussed in section 4.5.

Another important source of ammonia which can lead to atmospheric pollution is, of course, the large amount of manure produced in modern agriculture. The oxidation of ammonia to nitrite and nitrate by nitrifying bacteria not only contributes to the problem of acidification of the environment but also to the deterioration of buildings. As discussed in Chapter 5, an important habitat for nitrifiers is building stone. Large numbers of nitrifying bacteria (50 percent of the total population) have been found a few centimeters deep in porous rock where they thrive on ammonium from the air. The nitric and nitrous acid produced may dissolve the carbonates in the rock and thus speed up the natural deterioration.

Sulfur-oxidizing bacteria and their habitats Hydrogen sulfide, polysulfides, elemental sulfur, thiosulfate, polythionates, and sulfite can all be oxidized by the sulfur-oxidizing bacteria. All of these compounds can be found in nature under various aerobic and anaerobic conditions. As already mentioned, geological sources not only include volcanic gases, but also deposits of elemental sulfur (Sokolova and Karavaiko, 1968) and sulfide ores such as pyrite. Thus the sulfur-oxidizing bacteria can be found wherever these substrates occur, provided that a suitable electron acceptor, in most cases oxygen, is available. Examples of such natural environments are (hot) sulfur springs, both on land and the sea floor, where very high numbers of sulfur oxidizers can be found (Jannasch and Wirsen, 1979 and 1981; Mosser et al., 1974; Sokolova and Karavaiko, 1968). Also mines for sulfur, coal, and sulfidic ores provide a suitable habitat for high numbers of sulfur-oxidizing bacteria (Harrison, 1984; Norris and Kelly, 1982). In this case, the high concentration of substrates leads to the production of significant amounts of sulfuric acid, resulting in pH values even lower than 2. Many sulfur oxidizers require these low pH values for growth, and thus are acidophiles. For example, *Thiobacillus thiooxidans* can grow at a pH as low as 1 and will not grow at pH values higher than 6.

Aerobic biological decomposition of organic sulfur compounds often does not lead to the liberation of reduced inorganic sulfur compounds, because the thiol groups are often oxidized to sulfonic-sulfur in the organic state (Freney and Stevenson, 1966). On the other hand large amounts of hydrogen sulfide are produced under anaerobic conditions from the mineralization of organic matter both directly by desulfuration, and indirectly by dissimilatory sulfate reduction. The latter process is by far more important if sulfate is available in excess, as is the case in marine sediments. Sometimes natural gas deposits contain high concentrations of hydrogen sulfide. Many artificial sources of hydrogen sulfide occur. Industrial activities such as paper making may produce large quantities of hydrogen sulfide and organic sulfides (e.g., dimethyl sulfide) which can serve as energy sources for sulfide-oxidizing bacteria (Sivelä and Sundman, 1975). Finally, hydrogen sulfide production from sulfate-containing wastewater in anaerobic treatment installations can also be significant.

The sulfide-oxidizing bacteria are often predominant at the aerobic/anaerobic interface because of the rapid spontaneous chemical oxidation of reduced sulfur compounds. The half-life of H_2S in the presence of O_2 is in the order of one hour at 20–25°C, and only a few minutes at 60°C (Jørgensen et al., 1978). In order to compete successfully with the chemical oxidation process, the sulfur bacteria must therefore be able to locate themselves at the narrow interface where sulfide and oxygen meet at low concentrations. Indeed, many of the sulfur oxidizers are known to be chemotactic toward such interfaces, and can be seen as massive accumulations (blooms) in thin layers between the aerobic and anaerobic environments. By the rapid bacterial consumption of oxygen and sulfide at the interface, the concentrations of these substances are kept very low so that the contribution of the chemical reaction to the total oxidation decreases dramatically in the presence of sulfur-oxidizing bacteria.

An example may serve to demonstrate this point. Field and laboratory experiments using sulfide and oxygen microelectrodes have shown that *Beggiatoa* species have the ability to position themselves at the sulfide/oxygen interface in sediments. Once established, a layer of *Beggiatoa* keeps the coexisting oxygen and sulfide at concentrations in the order of 10^{-6} mol/l. At these levels, the oxidation of sulfide proceeds almost entirely biologically (see Chapter 12). The data indicate that the K_s for oxygen and sulfide in the strain of *Beggiatoa* studied were 1 μmol/l or less. Similar K_s values have been reported for *T. neapolitanus* and *T. versutus* grown in the chemostat under sulfide limitation. In these fully aerobic cultures, spontaneous sulfide oxidation was negligible (Beudeker et al., 1982). These low K_s values are clear examples of adaptation to life at the aerobic/anaerobic interface and this property seems to be very common among many sulfur bacteria.

It was mentioned in Section 4.1 that the diversity of sulfur-oxidizing bacteria is very large, both taxonomically and physiologically. The typical characteristics of obligate and facultative chemolithotrophs, as well as the chemolithoheterotrophs, are presented in Table 4.5. Apart from the first 3 metabolic types, which are known to derive energy from the oxidation, there are also "incidental" sulfur oxidizers which can oxidize sulfur compounds but do not

Table 4.5 Physiological types among the nonphototrophic sulfur-oxidizing bacteria

	Energy source (electron donor)		Carbon source	
Physiological type	Inorganic sulfur compound	Organic compound	CO_2	Organic compound
Obligate chemolithoautotroph	+	−	+	−
Facultative chemolithoautotroph	+	+	+	+
Chemolithoheterotroph	+	+	−	+
Chemoorganoheterotroph	−	+	−	+

appear to derive energy from it. Examples include numerous *Pseudomonas* species, actinomycetes, and fungi. This group of incidental sulfide oxidizers are, in fact, analogous with the heterotrophic nitrifiers mentioned before.

Iron-oxidizing bacteria and their habitats Iron is one of the major elements in the earth's crust. It plays an essential role in living organisms in the prosthetic groups of a variety of essential cell constituents such as enzymes and components of the respiratory chain. Nevertheless, in contrast to the contribution of living organisms to the global carbon, nitrogen and sulfur cycles, their role in the iron cycle is small (Nealson, 1983a). Apparently the availability and the suitability of iron as an energy source for the autotrophic mode of life is limited. As far as is known, only few kinds of organisms can profit energetically from the oxidation of ferrous to ferric iron.

The marginal role of living organisms in the global iron cycle might be explained in different ways. Firstly, ferrous iron is rarely present at high concentrations in the biosphere. Compounds containing ferric iron, especially insoluble oxides and hydroxides, are more abundant. Although ferrous iron can be generated from ferric iron in anaerobic zones by various microorganisms using the latter as their terminal electron acceptor (Chapter 7), in natural aquatic environments with a pH of around 7 the concentration of free ferric iron is generally extremely low, so that the availability of ferrous iron is also limited. Secondly, the free enthalpy change ($\Delta G'_0$) of ferrous oxidation is low compared to that of other oxidation reactions used by chemolithoautotrophs.

$$2Fe^{2+} + \tfrac{1}{2}O_2 + 2H^+ \rightarrow 2Fe^{3+} + H_2O \qquad \Delta G'_0 = -45.4 \text{ kJ/mol}$$

It is therefore not surprising that the cell yield on ferrous iron is extremely low. Thirdly, at neutral pH iron-oxidizing organisms have to compete with spontaneous ferrous iron oxidation. The spontaneous oxidation rate decreases, however, with decreasing pH, so that at moderate temperatures ferrous iron is chemically stable at pH values below 4. At these pH values, the solubilities of ferric iron species are also higher. Thus one would predict that autotrophic life based on ferrous iron as the energy substrate would occur, primarily, under acidic conditions. Indeed, a number of ferrous iron-oxidizing, acidophilic chemolithoautotrophs are known. They are listed in Table 4.6.

The common habitats of the acidophilic iron oxidizers are geological deposits of sulfide minerals such as pyrite (FeS_2) (one of the most common) and chalcopyrite ($CuFeS_2$). When sulfide ores, or coal deposits containing these iron-containing minerals, are exposed to moisture and oxygen, abundant growth of the iron oxidizers can occur. As has been pointed out in the section on sulfur-oxidizing bacteria and their habitats, the oxidation of the sulfur moiety of the iron-containing minerals leads to a rapid drop in pH, thereby providing optimal conditions for bacterial growth on ferrous iron.

Some of the species mentioned in Table 4.6 have been discussed in the section on sulfur-oxidizing autotrophs and their habitats because they have the ability to oxidize both iron and sulfur (e.g., *T. ferrooxidans* and *Sulfolobus* sp.).

Table 4.6 Acidophilic iron-oxidizing bacteria

Mesophiles:
Thiobacillus ferrooxidans
Leptospirillum ferrooxidans
Moderate thermophiles:
Unnamed *Thiobacillus*-like isolates
(strains m-1, TH1, TH3, ALV, BC, and K, see for details Le Roux et al., 1977; Brierley et al., 1978; Wood and Kelly, 1983; Harrison, 1984)
Sulfobacillus thermosulfidooxidans (Golovacheva and Karavaiko, 1978)
Extreme thermophiles:
Sulfolobus acidocaldarius
Sulfolobus solfataricus
Acidianus brierleyi

Others (e.g., *Leptospirillum* sp.) are only able to oxidize iron, and cannot oxidize reduced sulfur compounds. However, Norris (1983) has shown that *Leptospirillum* sp. can oxidize pyrite to ferric iron and sulfate. It had been assumed previously that a second organism was necessary for the oxidation of the sulfur moiety of the pyrite. However, it is possible that *Leptospirillum ferrooxidans* oxidizes pyrite by the following indirect reaction mechanism: the bacterium would oxidize the free ferrous iron present. The ferric iron thus produced would then act as an oxidizing agent for the pyrite. The chemical reaction between ferric iron and pyrite would thus result in the formation of ferrous iron and sulfate.

$$FeS_2 + 14Fe^{3+} + 8H_2O \rightarrow 15Fe^{2+} + 2SO_4^{2-} + 16H^+$$

It has also been suggested that this reaction mechanism is responsible for the oxidation of pyrite by *Thiobacillus ferrooxidans*. However, Arkesteyn (1980) and Hazeu et al. (1987) have demonstrated that in pyrite oxidation the sulfur-oxidizing capacity of *Thiobacillus ferrooxidans* must play an important role (see also Table 4.9).

Mining activities have uncovered vast amounts of sulfide minerals which thereby were exposed to both water and oxygen. This has led to the pollution of the environment with acid mine water, which is another important habitat of the iron oxidizers (Colmer and Hinkle, 1947). When the concentration of ferrous iron becomes too high or when the acid mine water mixes with natural water of higher pH, precipitates of basic ferric sulfates (jarosites) are formed. These compounds may block drainage systems, and the low pH may cause the death of virtually all higher forms of life.

At the same time, acidophilic iron- and sulfide-oxidizing bacteria have found a "niche" in the large scale leaching operations for copper extraction (Brierley, 1982) where divalent copper is brought into solution and is extracted from the ore both by direct and indirect oxidation of the ores (see above).

In addition to these acidophiles, there are a number of neutrophilic microorganisms which are presumed to profit energetically from ferrous iron oxidation. Most evidence has been found for representatives of the filamentous sheathed bacteria of the genera *Leptothrix, Sphaerotilus,* and *Gallionella.* It is not known how these neutrophilic iron bacteria have "solved" the problem of spontaneous ferrous iron oxidation by oxygen. If one assumes a lithotrophic mode of life among the neutrophiles, one should expect to find them in zones with coexisting low concentrations of ferrous iron and oxygen; a situation which is quite similar to that discussed earlier for the aerobic sulfide-oxidizing bacteria. Figure 4.2 (after Jones, 1986) shows rather sharp ferrous iron and oxygen profiles in a stratified lake. In the zone indicated in this figure, various neutrophilic iron oxidizers can be located. Natural blooms of these neutrophilic bacteria can also be observed in drainage pipes where ferrous-rich anoxic water contacts air. The iron bacteria may coat the pipes with thick biofilms containing large amounts of precipitated ferric hydroxide. A survey of neutrophilic iron-oxidizers is given in Table 4.7. Some of these species have not been isolated in pure culture.

Manganese-oxidizing bacteria and their habitats The ability of bacteria to oxidize divalent manganese has been known for a long time (Beijerinck, 1913). A number of manganese-oxidizing bacteria is known. They are listed in Table 4.8. It has been doubted for a long time that autotrophs existed among the

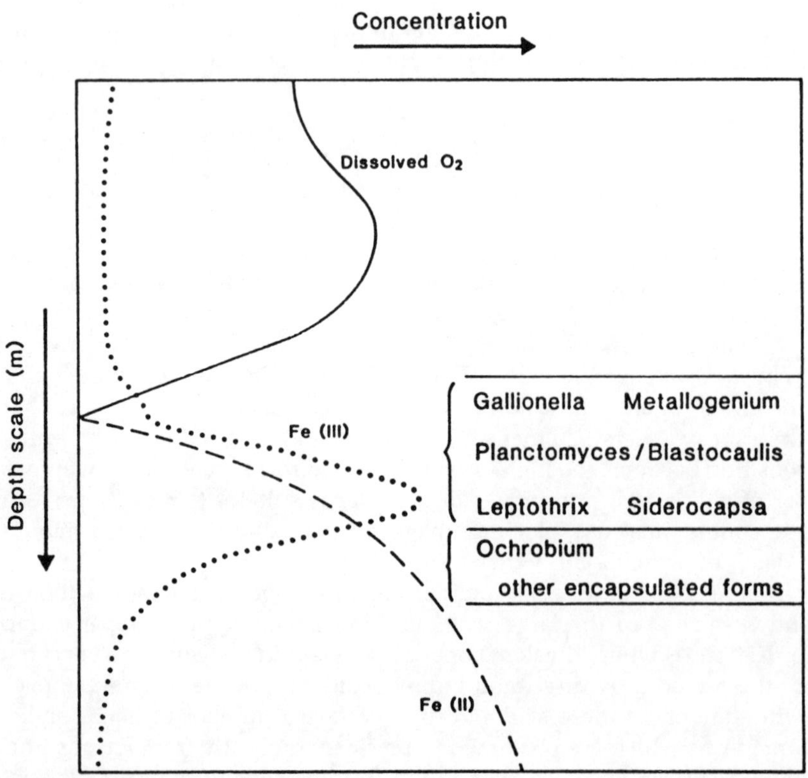

Figure 4.2 Zonation of iron bacteria along gradients of ferrous iron and oxygen in a stratified eutrophic lake (taken from Jones, 1986).

Table 4.7 Neutrophilic iron bacteria associated with the deposition of ferric iron*

	Genus or group	
Morphological type	In axenic culture	Not yet isolated
Filamentous	*Sphaerotilus*	*Clonothrix*
	Leptothrix	*Crenothrix*
		Toxothrix
		Lieskeella
Prosthecate/appendaged	*Planctomyces*	*Gallionella* (?)
	Hyphomicrobium	*Metallogenium*
	Pedomicrobium	*Seliberia?*
Encapsulated/coccoid		*Ochrobium*
		Siderocapsa(/Arthrobacter)
		Naumanniella
		Siderococcus

*From Jones, 1986.

Table 4.8 Bacterial genera with manganese-oxidizing representatives*

Pseudomonas	*Aeromonas*
Arthrobacter	*Flavobacterium*
Oceanospirillum	*Leptothrix*
Clonothrix	*Metallogenium*
Kuznetsovia	*Caulococcus*
Pedomicrobium	*Hyphomicrobium*
Nocardia	*Streptomyces*
Bacillus	*Micrococcus*

*After Nealson, 1983b.

manganese-oxidizing bacteria. The standard free enthalpy change of the oxidation of Mn^{2+} to Mn^{4+} with oxygen is positive. However, due to the extreme insolubility of the final product, MnO_2, the overall reaction has a negative free energy change.

Recently Kepkay and Nealson (1987) presented experimental evidence that a marine *Pseudomonas* species was able to fix carbon dioxide while it was oxidizing Mn^{2+}. Their study demonstrated that the continuous culture technique is appropriate for solving such problems. As the molar cell yields are unexpectedly high (5 g cell dry weight/mol electrons), further studies are essential to substantiate the autotrophic mode of life of this organism.

The presence of notable amounts of Mn^{2+} and of organisms able to oxidize it might be expected in low pH-E_h environments. These could be established by microbial oxygen consumption and acid production. However, in high pH/E_h environments other manganese species such as MnO_2 are dominant. It has been suggested that an important source of manganese in the marine environment might be the deep sea hydrothermal vents (Nealson, 1983b). For a further discussion of the role of manganese as an electron carrier at the aerobic/anaerobic interface, the reader is referred to Chapter 7.

4.3 NICHES OF CHEMOLITHO(AUTO)TROPHS

In the foregoing, various types of chemolithotrophs and their habitats have been discussed and some details on their nutritional requirements and on the environmental conditions of their habitats have been presented. Although these data may indicate the functional role a particular organism has in its environment, additional knowledge is necessary to identify its ecological niche.

One should expect that elucidation of the niche of a microorganism requires, preferably, studies *in situ*. Quantitative data on the occurrence of the organism in its habitat and the population dynamics caused by fluctuations of environmental factors are important in this respect. Once the *in situ* behavior of the organism in its habitat is known, further physiological studies can shed greater light on its ecological niche in that particular habitat. However, in many cases the connection between a given observed microbial activity in the field and the organism responsible cannot be made because the organism cannot be recognized *in situ*. In only a few cases will the characteristic morphology of the organism allow direct observation. A promising alternative method is the use of labeled specific antibodies, by which bacteria can be "tagged" (Ward and Carlucci, 1985; Muyzer et al., 1987). Other methods have been suggested for use in unravelling the composition of a mixed population *in situ*, such as DNA probes (Yates and Holmes, 1986) and 16S RNA sequencing (Pace et al., 1986). However, even if the reliable enumeration of a special type of microorganism *in situ* is possible, it will be very difficult to link the observed phenomena to the metabolism of a particular organism. Only if that particular organism is dominant in a habitat can this connection be made, but the habitats where this happens are comparatively scarce. These habitats often have exceptional conditions such as a high temperature, high salinity, or an extreme pH. For example, the thermophilic acidophilic *Sulfolobus* dominates hot sulfur springs (pH 2.0 and $T = 85\,°C$) and accounts for $2.5–7.4 \times 10^7$ cells/ml (Mosser et al., 1974). Acidophilic *Thiobacillus* species are abundant at pH 1.5–3.0 in acid mine drainage water, or in metal leaching operations.

In many cases, ecological studies *in situ* aimed at the elucidation of the functional role of an organism are hampered by the complexity of the microbial population. For instance, representatives of the different metabolic types of sulfur-oxidizing bacteria (see Table 4.5) are often found to coexist in a habitat. In other cases, organisms with seemingly identical physiology occur in the same habitat. To identify the ecological niche of a particular organism in its habitat, it is sometimes possible to use an ecophysiological approach. In this procedure, the physiology of the organism is studied in pure culture in order to understand the general principles of its metabolism. This is then followed by studies using simple mixed cultures in model systems in order to reveal the effects of microbial interactions such as competition and mutualism on the performance and the composition of the community. Such studies may help to give answers to questions concerning the contribution of different organisms to the observed activities in their natural environment.

In the following sections, both ecological and ecophysiological approaches will be discussed. Firstly, examples of ecological studies on the role *in situ* of the chemolithotrophs on *Beggiatoa* and *Thermothrix* will be presented. Secondly, examples of ecophysiological work, in particular on the colorless sulfur-oxidizing bacteria and related organisms, will be given.

4.4 ECOLOGICAL STUDIES *IN SITU*

Beggiatoa species belong to the colorless sulfur bacteria which, due to their characteristic long filamentous, gliding cells and the typical presence of sulfur globules in the cells, have attracted the attention of many investigators, beginning with Winogradsky. *Beggiatoa* can be found in samples from many habitats such as sediments of lakes and streams, in brackish marshes, and in association with cyanobacteria in microbial mats (Larkin and Strohl, 1983). It can frequently be observed as the predominant organism at the interface of sulfide-rich anoxic and aerobic zones in sediments. These interfaces are characterized by their extremely sharp sulfide and oxygen concentration profiles. The autotrophic growth of *Beggiatoa* was predicted by Winogradsky, and has indeed been established recently (Chapter 12). However, in spite of the obvious dominance of *Beggiatoa,* its quantitative role in the turnover of sulfide at interfaces had not been established. The recent development of microelectrodes for sulfide and oxygen has now allowed direct measurements of the *in situ* activity of *Beggiatoa* mats. It is now well established that *Beggiatoa* mats can oxidize virtually all the sulfide at the aerobic/anaerobic interface of (dark) marine sediments, and that these mats also play a key role in the diurnal changes of the sulfide concentrations in light-exposed sediments or microbial mats. More details on these studies can be found in Chapters 7 and 12.

An example by which it can be demonstrated that the niche of a predominant organism can be derived by collecting experimental data *in situ* is the study on *Thermothrix thiopara* by Caldwell et al. (1984). *Thermothrix thiopara* is a thermophilic facultatively autotrophic sulfur-oxidizing bacterium which may be the dominant organism in hot, neutral sulfur springs (Caldwell et al., 1976). *Thermothrix thiopara* forms large bundles, called "streamers," of filamentous cells attached to solid surfaces of such springs. It was shown that these streamers were located at well-defined zones representing the sulfide/oxygen interface. The sulfide serves as the energy source for the organism. Measurements *in situ* revealed that the dilution rate in the spring exceeded the maximal specific growth rate of *Thermothrix thiopara,* and attachment was thus necessary to prevent wash out. The filamentous cell bundles were positioned precisely at the sulfide/oxygen interface. The elongation of the streamers continued until they extended beyond the interface. Sulfide then became limiting for the protruding portion and the filaments fragmented. The unicells formed localized patches at other locations in the spring where the appropriate conditions existed and new filaments formed. By using a plastic cover to restrict oxygen access, Caldwell and co-workers were able to extend the zone of reducing conditions at the mouth of their spring. This resulted in a downstream displacement of the *Thermothrix* streamers. It was also demonstrated that the flagellated unicells preferentially attached to pyrite crystals present in the reduced part of the spring rather than to calcite crystals which were present in both the anoxic and aerobic zones. Probably the growth of the filamentous form of *T. thiopara* was induced by oxygen limitation.

4.5 ECOPHYSIOLOGICAL STUDIES

In this section ecophysiological studies will be presented to demonstrate their potential for identifying ecological niches of organisms. Attention will be mainly focused on sulfur-oxidizing bacteria and, where appropriate, some parallels will be drawn with the other chemolithotrophs. Firstly, attention will be paid to the physiology of obligately chemolithotrophic, facultatively chemolithoautotrophic and chemolithoheterotrophic bacteria. The emphasis will be on comparative properties. Secondly, details of competition experiments with mixed populations consisting of the different metabolic types as mentioned before (see also Table 4.5) will be given.

Physiology of obligately chemolithoautotrophic sulfur-oxidizing bacteria
The obligately chemolithoautotrophic sulfur oxidizers, such as *Thiobacillus neapolitanus* and *T. thioparus*, are highly specialized organisms with a metabolism that is geared to the specialist way of life. Their rate of oxidation of reduced sulfur compounds is high and their maximum specific growth rates may be as high as 0.6 h^{-1}. Respiration rates can be up to 6000 nmol O_2/min/mg protein, which is an order of magnitude higher than in common heterotrophs. When grown at submaximal growth rates under energy limitation in the chemostat, they may display a high respiratory overcapacity (up to 300–400 percent) for the oxidation of sulfur compounds. This implies that these organisms can respond immediately to a sudden increase of the concentration of the energy source. Some of these specialists are not limited to obtaining energy from sulfur compounds, but can also grow on other inorganic compounds such as iron (*Thiobacillus ferrooxidans*) or hydrogen (*Hydrogenobacter* sp.). Under limitation of growth by the energy source, more than one compound can be used simultaneously. An example of this type of growth, sometimes referred to as mixolithotrophy (Bonjour and Aragno, 1986), is given in Table 4.9 which shows yields of *Thiobacillus ferrooxidans* grown in the chemostat on a mixture of ferrous iron and tetrathionate (Hazen et al., 1987). In order to make the yields comparable, the data have been normalized to the dry weight in gram per mol electrons. It is evident that when normalized to equal numbers of electrons the yield on sulfur compounds is higher than on ferrous iron. For further comparison, the yield of batch growth on a natural substrate, pyrite, is also given. It can be seen that the yield on pyrite is significantly higher than that on ferrous

Table 4.9 Yields of *Thiobacillus ferrooxidans* grown in the chemostat under various conditions*

Limiting substrate(s)	Conc. (mM)	Dilution rate (h^{-1})	pH	Yield (g dry weight/mol electron)
Fe^{2+}	160	0.034	1.6	0.23
$S_4O_6^{2-}$	10	0.030	3.0	0.92
$Fe^{2+}/S_4O_6^{2-}$	15/10	0.030	1.6	0.76
$Fe^{2+}/S_4O_6^{2-}$	70/15	0.032	1.6	0.55
Pyrite	(1/14)	Batch	1.8	0.35–0.5

*Hazen et al. (1987).

iron. This implies that energy must be obtained directly from the oxidation of the sulfur moiety of pyrite. As was mentioned in the section on iron-oxidizing bacteria and their habitats, this is an important observation as it has frequently been claimed that pyrite is primarily oxidized chemically by ferric ion followed by reoxidation by *T. ferrooxidans* of the ferrous iron thus generated. In that case the yield on pyrite should have been the same as that given for iron in Table 4.9. Thus, also in nature the sulfur-metabolizing properties of *T. ferrooxidans* must be expected to be of major importance in the energy metabolism of the organism.

The obligate chemolithotrophs are unable to derive energy from exogenously supplied organic compounds (Kelly, 1971; Matin, 1978) due to their very limited capacity to take up these compounds. An exception is the use of dimethyl sulfide by a *T. thioparus* strain (Kanagawa and Kelly, 1986). The dimethyl sulfide serves as an energy source for autotrophic growth and is oxidized to carbon dioxide and sulfate.

Although the obligate chemolithotrophs use carbon dioxide as their major source of carbon under all growth conditions, they can assimilate up to 20–30 percent of their carbon from exogenous organic substrates. When these organisms are grown under energy limitation in the presence of an organic carbon source such as acetate, the cell yields are higher than in the absence of acetate, demonstrating that mixotrophy of carbon sources causes an energy-sparing effect (Kuenen and Veldkamp, 1973).

True heterotrophic energy metabolism in these organisms occurs primarily under starvation conditions at the expense of internal reserve materials, such as polyglucose. These reserves can be oxidized under aerobic conditions, or even fermented anaerobically (Beudeker et al., 1981). It is a common observation that the obligate chemolithotrophs are very resistant to starvation. *T. neapolitanus* can be starved for 50 days without appreciable loss of activity (Beudeker et al., 1981 and 1982). During aerobic starvation periods, the rate of endogenous respiration is extremely low, supporting this observation. In terms of ecophysiology, the resistance to starvation appears to be a logical property of a specialized organism which is able to use only a very limited number of substrates in its environment.

The physiology of facultative chemolithoautotrophic sulfur-oxidizing bacteria In contrast to the relatively restricted metabolism of the specialist sulfur oxidizers, facultative species appear to have a metabolism which is very versatile and can be accurately adjusted to the required turnover of available (mixtures of) substrates. Their growth rates on single substrates, including sulfur compounds, are relatively low, compared with those of the specialists. For example, the μ_{max} of the versatile *T. versutus* (formerly named *Thiobacillus* A2) during autotrophic growth on thiosulfate is 0.1 h^{-1}, as compared to 0.4–0.6 h^{-1} for many specialists. On acetate its μ_{max} is around 0.2 h^{-1}, while specialist acetate utilizers are known to possess rates between 0.4–0.9 h^{-1}. When provided with mixtures of organic and inorganic compounds, the facultative chemolithotrophs will show diauxic or biphasic growth on the two substrates. However, when these compounds are limiting the growth of the organism, as will often be the case in nature, several substrates are utilized simultaneously. During this mixotrophic growth, an efficient use of the available resources is displayed, and in the cases investigated CO_2 fixation is only induced when the ratio of the inorganic sulfur compound and the organic compound has become so high that the organic compound no longer meets the demand for an organic carbon source. When the ratio is low, the organic compound can serve both as carbon and energy source. Under these conditions, the sulfur compound serves as a supplementary energy source. The versatile, facultative chemolithotrophs can also grow effectively under alternating supplies of inorganic and organic substrates. Under these conditions, the organism must grow alternately heterotrophically and autotrophically. Experiments with *Thiobacillus versutus* have shown that the organism can do this very effectively during alternate growth limitation by acetate and thiosulfate in the chemostat, provided that the cycling time is not more than 4 hours (Gottschal et al., 1981). When the periods of acetate and thiosulfate supply were 8 hours or more, the organism repressed so much of its autotrophic potential during the heterotrophic period that at the onset of the thiosulfate period the capacity of the organism to grow autotrophically was no longer sufficient. As we will see in a later section of this chapter, this property makes the organism less competitive.

The chemolithoheterotrophic sulfur bacteria Similar studies on chemolithoheterotrophs have been limited to a few experiments. The chemolithoheterotrophic *Thiobacillus* Q can grow mixotrophically on mixtures of acetate and thiosulfate, provided these compounds are growth limiting (Gommers, 1987). Figure 4.3 shows the results of a series of these experiments. This organism lacks the ability to grow autotrophically, and consequently thiosulfate can be used only as a supplementary energy source. Accordingly, its maximum capacity to oxidize thiosulfate was limited. The observed oxidation rate was roughly 300 nmol/min/mg protein, which is about 30 percent of that observed in the facultative chemolithotrophs studied thus far, and even less than 10 percent of that in the specialist thiosulfate oxidizers.

A second example of chemolithoheterotrophic growth is that of *Hyphomicrobium* EG growing on dimethyl sulfide (Suylen et al., 1986). It has been shown that this compound is initially metabolized to formaldehyde and hydrogen sulfide. The formaldehyde is then used as a source of carbon and energy, while the sulfide is used as a supplementary energy source. It should be noted that there was no evidence for autotrophic CO_2 fixation under any growth conditions. The organism could also be grown on mixtures of methylamine and sulfide, showing very similar mixotrophic behavior. Thus, chemolithoheterotrophic metabolism of this type, rather than partial metabolism and excretion of one of the intermediates as a waste product, must be expected to confer an advantage whenever the energy is in limiting supply.

Parallels with other chemolithotrophs With the knowledge of the (eco)physiology of the metabolic spectrum of the sulfur-oxidizing bacteria, one

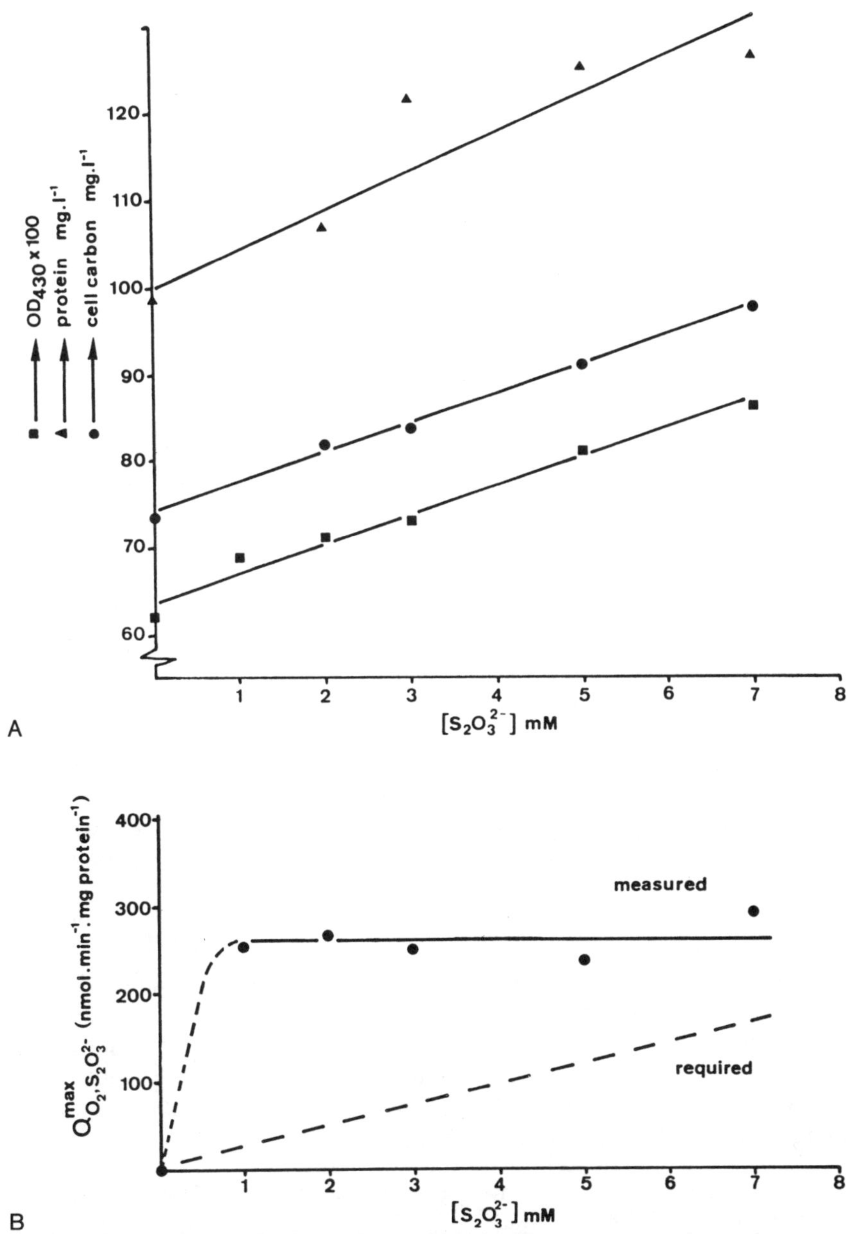

Figure 4.3 (A) Growth of chemostat cultures of the chemolithoheterotrophic colorless sulfur bacterium *Thiobacillus* Q. Measurements of biomass, protein, and total cell carbon under limitation by acetate (10 mmol/l) with increasing concentrations of thiosulfate as indicated on the x-axis. Dilution rate was 0.09 h^{-1}, the temperature 37°C, and the pH 7.5. (B) Maximal thiosulfate oxidation rate as a function of the thiosulfate concentration. The dotted line is the thiosulfate oxidation potential required to oxidize the thiosulfate present in the medium.

may wonder how far parallels with other chemolithotrophs can be drawn. Table 4.10 shows that this is the case to various extents. The majority of the autotrophic hydrogen bacteria are of the facultative type. In Chapter 16, some of their relevant properties are discussed and it emerges that their metabolism and its control has much in common with that of the sulfur bacteria. Many chemolithoheterotrophic hydrogen bacteria are known and, as with the sulfur bacteria, hydrogen can serve as a supplementary energy source via an (inducible) hydrogenase. *Hydrogenobacter*, which can also oxidize sulfur compounds, is, at present, the only known obligate chemolithoautotroph, and CO_2 fixation occurs, like in *Sulfolobus*, via a reversed citric acid cycle (Shiba et al., 1985; Kandler and Stetter, 1981). Very little is known about the general physiology of *Hydrogenobacter* and thus at present it is not possible to make further comparisons with other specialist chemolithotrophs. From Chapter 5 it will appear that among the ammonium and nitrite oxidizers, the most common organisms seem to be the obligate chemolithoautotrophs. Facultatively chemolithoautotrophic ammonium oxidizers are not known. However, the specialist ammonium oxidizers have much in common with the typical obligate sulfur oxidizers. Their metabolism appears to be geared completely to the specialist way of life and their behavior towards organic compounds is very similar indeed.

At present, a few facultative nitrite oxidizers have been described, (e.g., *Nitrobacter hamburgensis*, which occurs commonly in soil). It has been shown that *Nitrobacter agilis* and *Nitrobacter winogradsky* can grow very slowly on organic compounds. Apparently this is only a survival mechanism and these organisms should be considered as typical specialists.

It is particularly interesting that some nitrifiers may be able to denitrify under anaerobic conditions on nitrate or nitrite, whereby endogenous or in some cases exogenous organic compounds serve as electron donors (Chapter 5). This seems to be a survival mechanism that is most appropriate for organisms living at the aerobic/anaerobic interface.

Among the iron oxidizers a spectrum of metabolic types also appears to exist. Typical specialists include the neutrophilic *Gallionella* and the acidophilic *T. ferrooxidans* and *Leptospirillum ferrooxidans*. Only the physiology of *T. ferrooxidans* has been studied in much detail. It is hardly surprising that it is very similar to that of the other specialist sulfur oxidizers. The only facultative iron oxidizer so far described, *Acidianus*, is also a sulfur oxidizer (Segerer et al., 1986; Wood et al., 1987). This organism was originally described as a strain of *Sulfolobus*. However, because of a lack of detailed knowledge of the phys-

Table 4.10 Metabolic spectrum of hydrogen/carboxydo-, sulfur-oxidizing, nitrifying, and iron-oxidizing bacteria

	Obligate or specialist	Facultative	Chemolitho-heterotroph	Incidental
H_2/CO	+	+++	+++	−(?)
S^{2-}/S^0/$S_2O_3^{2-}$	+++	+++	++(+)	+
NH_4^+/NO_2^-	+++	+	−(?)	+++
Fe^{2+}	++	++	?	+++

− = absent, + = low frequency, ++ = frequent, +++ = majority.

iology of most iron oxidizers, it is at present too early to say how far the parallels with other chemolithotrophs can be drawn.

A final point concerning the ecophysiology of the chemolithotrophs which has already been mentioned for *T. neapolitanus* and the nitrifiers is their ability to survive under anaerobic conditions. This would seem a particularly important property for organisms living at the dynamic aerobic/anaerobic interface. In four of the examples given in Table 4.11 *Nitrosomas* (Poth, 1986), *T. ferrooxidans* (Brock and Gustafson, 1976), *Beggiatoa* (Nelson and Castenholz, 1981), and *Acidianus* (Segerer et al., 1986), the organisms use the products of lithotrophic oxidation as their electron acceptors for anaerobic metabolism. *T. neapolitanus* shows heterofermentative lactic acid production (Beudeker et al., 1981) and *Alcaligenes eutrophus* shows a mixed acid fermentation (Vollbrecht et al., 1978).

In the following sections we will discuss how the characteristic physiological properties of the different types of organisms may play a role in the competition with other organisms.

Ecological niches The discussion of one of the typical habitats of the chemolithotrophs, the aerobic/anaerobic interface, has given some clues as to which variables might be important in the selective forces exerted on different organisms. Of paramount importance will be how successful a particular organism will be in the competition with other organisms for growth limiting substrates. If it is realized that the whole spectrum of metabolic types presented in Table 4.5 can often be isolated from one and the same habitat, the question as to how these organisms can compete and coexist in that habitat may be posed.

Table 4.11 Anaerobic (survival) metabolism by aerobic lithotrophs

Nitrosomonas europaea (Poth, 1986)

$$NH_4^+ \xrightarrow{O_2} NO_2^- \xrightarrow{\text{anaerobic}} N_2$$

Thiobacillus ferrooxidans (Brock and Gustafson, 1976)

$$Fe^{2+} \xrightarrow{O_2} Fe^{3+};\quad Fe^{3+} + S^0 \xrightarrow{\text{anaerobic}} Fe^{2+} + SO_4^{2-}$$

Beggiatoa (Nelson and Castenholz, 1981)

$$S^{2-} \xrightarrow{O_2} S^0;\quad S^0 + \text{organic acid} \xrightarrow{\text{anaerobic}} H_2S + CO_2$$

Acidianus (Segerer et al., 1986)

$$S^{2-} \xrightarrow{O_2} S^0;\quad S^0 + H_2 \xrightarrow{\text{anaerobic}} H_2S$$

Thiobacillus neapolitanus (Beudeker et al., 1981)

$$\text{internal polyglucose} \longrightarrow \text{glucose} \xrightarrow{\text{fermentation}} \text{lactate} + \text{ethanol}$$

Alcaligenes eutrophus (Vollbrecht et al., 1978)

$$\text{glucose} \xrightarrow[\text{(or } NO_3^- \rightarrow N_2)]{\text{fermentation}} \text{organic acids}$$

Clearly each of them must, by definition, occupy a different ecological niche. Ecophysiological knowledge should help both in the making of educated predictions and in the design of laboratory experiments to prove that these predictions can hold in simple or complex mixed cultures. For such experiments, the continuous culture techniques appear to be most appropriate (Veldkamp et al., 1984). Such experiments can reveal competitive differences, and in particular the growth kinetics of the respective metabolic types. In the chemostat, the combination of K_s and μ_{max} for a certain (combination of) substrate(s) will determine whether the organism is able to outcompete other organisms. At low substrate concentrations, where the actual substrate concentration is low compared to K_s, the affinity (i.e., μ_{max}/K_s), will determine the outcome of the competition.

Competition for single substrates If both the μ_{max} and the K_s value of an organism are known, the outcome of competition for a growth-limiting compound can be predicted. However, in the case of the sulfur oxidizers, the K_s is so low that it cannot be measured accurately enough. Thus, only an actual competition experiment between pure cultures of the organisms to be compared will provide evidence as to their relative K_s values. Such experiments have been carried out between two specialist marine thiosulfate oxidizers, *T. thioparus* and *Thiomicrospira pelophila,* both of which had been isolated from the same habitat and had many properties in common (Kuenen et al., 1977). It was predicted that *T. thioparus,* which has a higher μ_{max} than *T. pelophila,* might dominate when relatively high substrate concentrations were available, and *T. pelophila* at low substrate concentrations. An obvious growth-limiting compound might be the energy substrate, thiosulfate or sulfide. When competition experiments were carried out at pH 7.5 with thiosulfate as the growth-limiting substrate, the organisms appeared to coexist for prolonged periods in the chemostat, but when the pH was raised to 7.8 or lowered to 6.0 *T. pelophila* and *T. thioparus,* respectively, became dominant. Another variable that might play a role in the competition of these organisms might be iron, which is highly insoluble in the marine environment, and in fact is only present either as the hydroxide or sulfide. Indeed when a mixture of ferrous and ferric iron was the growth-limiting substrate, *T. thioparus* dominated at high dilution rates and *T. pelophila* dominated at low dilution rates, showing that under these conditions the organisms had crossing substrate saturation curves. In line with these observations was a field observation indicating that the ratio of *T. pelophila* to *T. thioparus* was higher in sulfide-rich (relatively high pH and low soluble iron) mussel banks than in neighboring mud samples with lower sulfide concentrations. The higher affinity for sulfide of *T. pelophila* may be related to its greater surface to volume ratio (Kuenen and Veldkamp, 1972; Kuenen et al., 1977).

Competition for mixed substrates According to the mathematical models for competing organisms, the number of coexisting species can never be more than the number of variables that play a role in the competition (Veldkamp et al., 1984). The simplest example is competition for two substrates. In this case the maximum number of organisms that can coexist is two. In view of the metabolic spectrum of organisms (Table 4.5), the role of organic compounds in the competition of thiobacilli and similar organisms for more than one substrate has been studied in detail in the laboratory. This work has been reviewed in several recent articles (Kuenen and Beudeker, 1982; Kuenen and Gottschal, 1982; Kuenen et al., 1985; Gottschal, 1986) and we will summarize the main points. The work was based on the ecophysiological studies described in the preceding paragraph. In order to study the competitiveness of the different metabolic types, competition experiments were performed in nutrient-limited chemostats fed with single substrates or mixtures. The experiments were carried out by pregrowing the specialist and facultative sulfur oxidizers in the chemostat under the required conditions, followed by mixing of the cultures. The progress of the competition was then monitored by a number of counting procedures. The dominance of one or another organism could usually be established after 10 to 12 volume changes when a stable new situation had been reached. It was shown that during single substrate limitation, specialists dominated over the versatile organisms. For example, *T. neapolitanus* accounted for more than 95 percent of the population after competition for thiosulfate with *T. versutus* at various dilution rates. This implies that, as expected, the affinity of *T. neapolitanus* for thiosulfate is higher than that of *T. versutus*. When mixtures of thiosulfate and organic substrates (e.g., acetate) were provided, the percentage of the versatile organism in the chemostat culture increased. Above a certain ratio of acetate to thiosulfate, the versatile *T. versutus* completely outcompeted the specialist. Similar competition experiments were also carried out between a specialist acetate utilizer and *T. versutus* with analogous results. When different ratios of thiosulfate and acetate were presented to mixed cultures containing all three organisms, the versatile species dominated in the range where the turnover of the two substrates was of the same order of magnitude. These experiments show that mixotrophic metabolism of thiosulfate and acetate enables the versatile *T. versutus* to compete successfully with specialist species.

In order to simulate the changes of nurient fluxes as they may be expected to take place at the aerobic/anaerobic interface in nature, competition experiments were also carried out under alternating supplies of thiosulfate and acetate. Under these conditions, the versatile organism was outcompeted by the specialists. It appeared that this was due to the metabolic adaptation of *T. versutus*. As discussed previously, *T. versutus* repressed some of its autotrophic potential during the heterotrophic period, rendering it less reactive to the appearance of thiosulfate. In contrast, the two specialists starved during either the thiosulfate or the acetate period. While starving, the specialists did not reduce their relatively high (over)capacity for their respective substrate. As a result they could respond immediately with full capacity when their substrate supply was resumed. However, when different mixtures of the two substrates were presented alternately, coexistence of the three species was possible. These experiments show that not only is metabolic flexibility an important property for survival in nature, but that the ability to respond rapidly to a changing environment (i.e., the reactivity) may also be a decisive factor for the survival of the various metabolic types (Beudeker et al., 1982).

Prediction and testing of ecological niches of different metabolic types The above results, complemented with other experiments, have allowed the construction of an ecophysiological profile of the specialist and versatile organisms, as summarized in Table 4.12 (from Beudeker et al., 1982).

On the basis of these experiments, a working model to predict the ecological niches of the different metabolic types has been developed (Kuenen and Beudeker, 1982). Specialist sulfur oxidizers would become dominant when the ratio of turnover rates (= fluxes) of reduced inorganic sulfur compounds to fluxes of organic compounds was high. When this ratio was of the same order of magnitude, the versatile species would dominate. If the ratio dropped further, chemolithoheterotrophs would become dominant. The latter prediction is based on the fact that below certain ratios, the possession of an autotrophic potential (i.e., CO_2 fixing potential) would no longer be advantageous. If only traces of sulfur compounds were available, incidental sulfur oxidizers might become responsible for the oxidation of the sulfur compounds. Of course, it should be expected that these ranges would overlap and that in many cases the coexistence of different types should be possible. Indeed the first three types could easily be enriched in a chemostat fed with different ratios of thiosulfate and acetate (Table 4.13). *Thiobacillus* Q, discussed in a previous paragraph, was the dominant organism (85 percent of the community) in a culture

Table 4.12 Differences between a metabolically versatile and a specialized chemolithotrophic *Thiobacillus**

	Specialist *T. neapolitanus*	Versatile *T. versutus*
1	Few energy substrates utilized (S^{2-}, S^0, $S_2O_3^{2-}$, $S_4O_6^{2-}$, etc.)	Many different inorganic and organic energy substrates utilized
2	High specific growth rate on single substrate (μ_{max} on thiosulfate = 0.35 h^{-1})	Low specific growth rate on single substrate (μ_{max} on thiosulfate = 0.10 h^{-1}) Relatively high specific growth rate on mixed substrates (μ_{max} on thiosulfate plus acetate = 0.22 h^{-1})
3	High affinity for reduced sulfur compounds	Relatively low affinity for reduced sulfur compounds
4	High overcapacity of respiratory capacity during substrate-limited growth	Low overcapacity of respiratory capacity during substrate-limited growth
5	Low "flexibility" with respect to energy generation and organic carbon assimilation; "constitutive" enzymes	High "flexibility"; "inducible" enzymes for energy generation and carbon assimilation
6	High "reactivity" toward few substrates	Low "reactivity" toward many substrates
7	Metabolic "lesions"	Many pathways, often overlapping
8	Low endogenous respiration	High endogenous respiration in autotrophically and heterotrophically grown cells
9	Very resistant to starvation	Less resistant to starvation
10	Ecological niche: in environments with continuous or fluctuating supply of reduced sulfur compounds and a low turnover of organic compounds	Ecological niche: in environments with simultaneous presence of both inorganic and organic substrates ("mixotrophic" conditions)

*Adapted from Beudeker et al. (1982).

Table 4.13 Results of chemostat enrichments with different sulfide/acetate ratios

Ratio sulfide/acetate	Dominating species	Respiratory overcapacity	Sulfide oxidation
High	Specialist autotroph	High	Constitutive
~1	Versatile (oxidizes sulfide and/or acetate)	Low	Inducible
Low	Chemolithoheterotroph (oxidizes acetate and sulfide)	Low	Inducible

fed with 15 mmol/l of acetate and 10 mmol/l of thiosulfate (Kuenen et al., 1984). It should be stressed, however, that satellite populations of one or more of the other types were always present.

The above model was put to the test during an investigation into the microbial community within a denitrifying waste water treatment plant fed with sulfide and organic acids such as acetate from a methanogenic reaction (Robertson and Kuenen, 1983b). On the basis of the composition of the influent, it was predicted that a facultative sulfide oxidizer should be the dominant type. Indeed, more than 90 percent of the sulfide oxidizing denitrifiers were of this type. This observation supports the model, but it should be stressed that in a complex waste water treatment system many unknown variables might influence the results.

One of the dominant facultative chemolithotrophs was isolated for further study. This organism, *Thiosphaera pantotropha,* was shown to be able to grow on a large variety of organic compounds, and in addition to sulfide and thiosulfate it could also grow autotrophically on hydrogen. Not surprisingly, the organism was a facultative anaerobe which was able to denitrify (Robertson and Kuenen, 1983a). Quite unexpectedly, however, the organism was also found to be a constitutive denitrifier which continued to denitrify even in the presence of oxygen concentrations near air saturation (aerobic denitrification). When grown at 80 percent air saturation on acetate medium, the organism oxidized about half of the acetate with oxygen and half with nitrate as electron acceptor (Robertson and Kuenen, 1984b). Other isolates from this and other waste water treatment systems also appeared to be aerobic denitrifiers. There is a whole range of denitrifying organisms with different pO_2 cut-off points above which denitrification is no longer possible. In the case of *T. pantotropha,* one might wonder why such an organism would have come to the fore. This species was one of the dominant organisms in a fluidized bed reactor where the biomass grew in a biofilm attached to sand particles which were kept suspended by the flowing influent. The influent contained sulfide, acetate, and nitrate as the major substrates, and, in addition, a few μmol/l of oxygen were introduced at the bottom of the bed. This oxygen was rapidly depleted over the first few centimeters at the bottom of the reactor. The growing particles circulated up and down through the system. This implies that the organisms were exposed to alternating aerobic and anaerobic conditions. Our hypothesis is that the competitive advantage of a constitutive denitrifier is that it will not require any adaptation lags in contrast to an organism such as *Paracoccus denitrificans* which would switch off its denitrifying capacity on exposure to oxygen

Table 4.14 Examples of opportunistic or incidental oxidation of chemolithotrophic energy sources

Opportunistic organism	Growth substrate	Competing with specialist	For (substrate)
Methanotrophs	CH_4/O_2	*Nitrosomonas*	NH_4^+/O_2
Heterotrophic nitrifiers	Organics	*Nitrosomonas*	NH_4^+/O_2
Chromatium	S^{2-}/light anaerobic	*Thiobacillus* *Beggiatoa*	S^{2-}/O_2

and then needs to reinduce it (Robertson and Kuenen, 1984a, b). Indeed, when the enrichments were repeated in a laboratory scale fluidized bed, the enrichment of constitutive denitrifying populations was obtained. The conclusion is that not only fluctuations or mixtures of electron donors, but also of electron acceptors play a very important role in the survival strategy of the chemolithotrophs. This would seem particularly important for the sulfide oxidizers, which have to compete with the spontaneous oxidation of sulfide with oxygen (Kuenen and Robertson, 1987).

The model predicting the dominance of a certain metabolic type as a function of the relative turnover is, of course, only one mechanism that may come into play in the competition of organisms for limited resources. Extrapolating from the model, the appearance of opportunistic or incidental oxidation of the chemolithotrophic energy source by populations of bacteria that have built up on other resources, but possess the ability to oxidize the reduced inorganic substrate, can be predicted. Examples are given in Table 4.14. Incidental oxidation of ammonium occurs both by methane oxidizers and by heterotrophs which do not appear to benefit from the oxidation. Yet, if present in sufficiently high numbers, they may be able to contribute substantially to the total turnover of this compound. Specific activities of the heterotrophic nitrifiers can be of the order of 20–60 nmol NH_4^+/min/mg protein. This is one of two orders of magnitude lower than that of *Nitrosomonas* cultures, but if the heterotrophic nitrifiers outnumbered the autotrophic specialists by two orders of magnitude their contribution to the total oxidation might be substantial, assuming that their K_s for ammonium was the same.

Another interesting example of opportunistic oxidation is that carried out by the phototrophic sulfur bacterium *Chromatium*, and other sulfide-utilizing phototrophs. These organisms can grow chemolithoautotrophically in the dark under microaerophilic conditions. A comparison of affinities (μ_{max}/K_s) of the phototrophs with those of the specialist thiobacilli shows that once the phototrophs have built up a dense population (bloom) in the light, they may successfully compete with the thiobacilli for limiting sulfide in the dark, and, in fact, even outcompete these organisms (Kuenen, 1988b).

For critically reading and correcting the English text the help of Ms. Lesley A. Robertson, M.Sc. M.I.Biol., is gratefully acknowledged.

REFERENCES

Arkesteyn, G.J.M.W. 1980. *Contribution of microorganisms to the oxidation of pyrite.* Thesis Agricultural University Wageningen, The Netherlands. 80 pp.

Bak, F. and **Pfennig, N.** 1987. Chemolithotrophic growth of *Desulfovibrio sulfodismutans* sp. nov. by disproportionation of inorganic sulfur compounds. *Archives of Microbiology* 147:184–189.

Beijerinck, M.W. 1913. Oxidation of manganocarbonate by microbes. *Proceedings of the Koninklijke Nederlandse Akademie van Wetenschappen* 15:397–401.

Beudeker, R.F., de Boer, W., and **Kuenen, J.G.** 1981. Heterolactic fermentation of intracellular polyglucose by the obligate chemolithotroph *Thiobacillus neapolitanus* under anaerobic conditions. *FEMS Microbiology Letters* 12:337–342.

Beudeker, R.F., Gottschal, J.C., and **Kuenen, J.G.** 1982. Reactivity versus flexibility in thiobacilli. *Antonie van Leeuwenhoek Journal of Microbiology and Serology* 48:39–51.

Bonjour, F. and **Aragno, M.** 1986. Growth of thermophilic obligatorily chemolithoautotrophic hydrogen-oxidizing bacteria related to *Hydrogenobacter* with thiosulfate and elemental sulfur as electron and energy source. *FEMS Microbiology Letters* 35:11–15.

Brierley, C.L. 1982. Microbiological mining. *Scientific American* 247:44–51.

Brierley, J.A., Norris, P.R., Kelly, D.P., and **Le Roux, N.W.** 1978. Characteristics of a moderately thermophilic and acidophilic iron-oxidizing *Thiobacillus. European Journal of Applied Microbiology and Biotechnology* 5:291–299.

Brock, T.D. and **Gustafson, J.** 1976. Ferric iron reduction by sulfur and iron oxidizing bacteria. *Applied and Environmental Microbiology* 32:567–571.

Caldwell, D.E., Kieft, T.L., and **Brannan, D.K.** 1984. Colonization of sulfide-oxygen interfaces on hot spring tufa by *Thermothrix thiopara. Geomicrobiological Journal* 3:181–200.

Caldwell, D.E., Caldwell, S.J., and **Laycock, D.P.** 1976. *Thermothrix thioparus* gen. et sp. nov. a facultatively anaerobic facultative chemolithotroph living at neutral pH and high temperature. *Canadian Journal of Microbiology* 22:1509–1517.

Colmer, A.R. and **Hinkle, M.E.** 1947. The role of microorganisms in acid mine drainage. *Science* 106:253–256.

Conrad, R. 1984. Capacity of aerobic organisms to utilize and grow on atmospheric trace gases (H_2, CO, CH_4), pp. 461–467, in Klug, M.J. and Reddy, C.A. (editors), *Current Perspectives in Microbial Ecology,* American Society for Microbiology, Washington.

Freney, J.R. and **Stevenson, F.J.** 1966. Organic sulfur transformations in soils. *Soil Science* 101:307–316.

Friedrich, C.G. and **Mitrenga, G.** 1981. Oxidation of thiosulfate by *Paracoccus denitrificans* and other hydrogen bacteria. *FEMS Microbiology Letters* 10:209–212.

Golovacheva, R.S. and **Karavaiko, G.I.** 1978. A new genus of thermophilic spore-forming bacteria, *Sulfobacillus. Mikrobiologiya* 47:815–822.

Gommers, P.J.F. 1987. *Microbiological oxidation of sulfide and acetate in a denitrifying fluidized bed reactor.* Thesis Delft University of Technology. 207 pp.

Gottschal, J.C. 1986. Mixed substrate utilization by mixed cultures, pp. 261–292, in Poindexter, J.S. and Leadbetter, E.R. (editors), *Bacteria in Nature* Vol. 2: *Methods and Special Applications in Bacterial Ecology,* Plenum Press, New York.

Gottschal, J.C., Nanninga, H.J., and **Kuenen, J.G.** 1981. Growth of *Thiobacillus* A2 under alternating growth conditions in the chemostat. *The Journal of General Microbiology* 126:85–96.

Gundersen, K., Carlucci, A.F., and **Boström, K.** 1966. Growth of some chemoautotrophic bacteria at different oxygen tensions. *Experientia* 22:229–230.

Harrison, A.P. 1984. The acidophilic thiobacilli and other acidophilic bacteria that share their habitat. *Annual Review of Microbiology* 38:265–292.

Hazeu, W., Schmedding, D.J., Goddijn, O., Bos, P., and **Kuenen, J.G.** 1987. The importance of the sulphur-oxidizing capacity of *Thiobacillus ferrooxidans* during leaching of pyrite, pp. 497–499, in Neijssel, O.M., van der Meer, R.R., and Luyben, K.Ch.A.M. (editors), *Proceedings 4th European Congress on Biotechnology,* Vol. 3, Elsevier, Amsterdam.

Helder, W. and **de Vries, R.T.P.** 1983. Estuarine nitrite maxima and nitrifying bacteria (Ems-Dollard estuary). *Netherlands Journal of Sea Research* 17:1–18.

Ivanov, M.V. 1983. Major fluxes of the global biogeochemical cycle of sulphur, pp. 449–463, in Ivanov, M.V. and Freney, J.R. (editors), *The Global Biogeochemical Sulphur Cycle,* J. Wiley & Sons, Chichester.

Jannasch, H.W. and **Wirsen, C.O.** 1979. Chemosynthetic primary production at East Pacific sea floor spreading centers. *BioScience* 29:492–498.

Jannasch, H.W. and **Wirsen, C.O.** 1981. Morphological survey of microbial mats near deep-sea thermal vents. *Applied and Environmental Microbiology* 41:528–538.

Jones, J.G. 1986. Iron transformation by freshwater bacteria. *Advances in Microbial Ecology* 9:149–185.

Jørgensen, B.B., Kuenen, J.G., and **Cohen, Y.** 1978. Microbial transformations of sulfur compounds in a stratified lake (Solar Lake, Sinai). *Limnology and Oceanography* 24:799–822.

Kämpf, C. and **Pfennig, N.** 1980. Capacity of Chromatiaceae for chemotrophic growth. Specific respiration rates of *Thiocystus violacea* and *Chromatium vinosum. Archives of Microbiology* 127:125–137.

Kanagawa, T. and **Kelly, D.P.** 1986. Breakdown of dimethyl sulphide by mixed cultures and by *Thiobacillus thioparus. FEMS Microbiology Letters* 34:13–19.

Kandler, O. and **Stetter, K.O.** 1981. Evidence for autotrophic CO_2 assimilation in *Sulfolobus brierleyi* via a reductive carboxylic acid pathway. *Zentralblatt für Bakteriologie Mikrobiologie und Hygiene. 1 Abt. Originale C* 2:111–121.

Kawasumi, T., Igarashi, Y., Kodama, T., and **Minoda, Y.** 1984. *Hydrogenobacter thermophilus* gen. nov., sp. nov., an extremely thermophilic, aerobic, hydrogen-oxidizing bacterium. *International Journal of Systematic Bacteriology* 34:5–10.

Kelly, D.P. 1971. Autotrophy: concepts of lithotrophic bacteria and their organic metabolism. *Annual Review of Microbiology* 25:177–210.

Kelly, D.P. and **Kuenen, J.G.** 1984. Ecology of the colourless sulphur bacteria, pp. 211–240, in Codd, G.A. (editor), *Aspects of Microbial Metabolism and Ecology,* Academic Press, London.

Kepkay, P.E. and **Nealson, K.H.** 1987. Growth of a manganese oxidizing *Pseudomonas* sp. in continuous culture. *Archives of Microbiology* 148:63–67.

Kondratieva, E.N., Zhukov, V.G., Ivanovsky, R.N., Petushkova, Y.P., and **Monosov, E.Z.** 1976. The capacity of phototrophic sulphur bacterium *Thiocapsa roseopersicina* for chemosynthesis. *Archives of Microbiology* 108:287–292.

Kuenen, J.G. 1975. Colorless sulfur bacteria and their role in the sulfur cycle. *Plant and Soil* 43:49–76.

Kuenen, J.G. 1988a. Section 20. Aerobic chemolithotrophic bacteria and associated organisms. B. Colorless sulfur bacteria, in Holt, J.G. (editor in chief) *Bergey's Manual of Systematic Bacteriology, Vol. 3,* Williams and Wilkins, Baltimore (in press).

Kuenen, J.G. 1988b. Comparative ecophysiology of the non-phototrophic sulfide oxidizing bacteria, in *Microbial Mats: Physiological Ecology of Benthic Microbial Communities,* American Society for Microbiology, Washington (in press).

Kuenen, J.G. and **Beudeker, R.F.** 1982. Microbiology of thiobacilli and other sulphur-oxidizing autotrophs, mixotrophs and heterotrophs. *Philosophical Transactions of the Royal Society of London,* Series B 298:473–497.

Kuenen, J.G., Boonstra, J., Schröder, H.G.J., and **Veldkamp, H.** 1977. Competition for inorganic substrates among chemoorganotrophic and chemolithotrophic bacteria. *Microbial Ecology* 3:119–130.

Kuenen, J.G. and **Gottschal, J.C.** 1982. Competition among chemolithotrophs and methylotrophs and their interactions with heterotrophic bacteria, pp. 153–187, in Bull, A.T. and Slater, J.H. (editors), *Microbial Interactions and Communities, Vol. 1,* Academic Press, London.

Kuenen, J.G. and **Robertson, L.A.** 1987. Ecology of nitrification and denitrification, pp. 161–218, in Cole, J. and Ferguson, S. (editors), *Nitrogen-and sulphur-cycles,* Society for General Microbiology, Cambridge University Press, Cambridge.

Kuenen, J.G., Robertson, L.A., and **van Gemerden, H.** 1985. Microbial interactions among aerobic and anaerobic sulfur-oxidizing bacteria. *Advances in Microbial Ecology* 8:1–59.

Kuenen, J.G., Robertson, L.A., Scheulderman-Suylen, G.M.H., and **Gommers, P.J.F.** 1984. Microbial desulphurization of industrial wastes, pp. 9–15, in *Proceedings Third European Congress on Biotechnology München,* Vol. III, Verlag Chemie, Weinheim.

Kuenen, J.G. and **Veldkamp. H.** 1972. *Thiomicrospira pelophila* nov. gen., nov. sp. a new obligately chemolithotrophic colourless sulfur bacterium. *Antonie van Leeuwenhoek Journal of Microbiology and Serology* 38:241–256.

Kuenen, J.G. and **Veldkamp, H.** 1973. Effects of organic compounds on growth of chemostat cultures of *Thiomicrospira pelophila, Thiobacillus thioparus,* and *Thiobacillus neapolitanus. Archiv für Mikrobiologie* 94:173–190.

Larkin, J.M. and **Strohl, W.R.** 1983. *Beggiatoa, Thiothrix,* and *Thioploca. Annual Review of Microbiology* 37:341–367.

Le Roux, N.W., Wakerley, D.S., and **Hunt, S.D.** 1977. Thermophilic *Thiobacillus*-type bacteria from Icelandic thermal areas. *The Journal of General Microbiology* 100:197–201.

Matin, A. 1978. Organic nutrition of chemolithotrophic bacteria. *Annual Review of Microbiology* 32:433–468.

Mosser, J.L., Bohlool, B.B., and **Brock, T.D.** 1974. Growth rates of *Sulfolobus acidocaldarius* in nature. *Journal of Bacteriology* 118:1075–1081.

Muyzer, G., de Bruyn, J.C., Schmedding, D.J.M., Bos, P., Westbroek, P., and **Kuenen, J.G.** 1987. A combined immunofluorescence-DNA-fluorescence staining technique for enumeration of *Thiobacillus ferrooxidans* in a population of acidophilic bacteria. *Applied and Environmental Microbiology* 53:660–664.

Nealson, K.H. 1983a. The microbial iron cycle, pp. 159–190, in Krumbein, W.E. (editor), *Microbial Geochemistry,* Blackwell, Oxford.

Nealson, K.H. 1983b. The microbial manganese cycle, pp. 191–221, in Krumbein, W.E. (editor), *Microbial Geochemistry,* Blackwell, Oxford.

Nelson, D.C. and **Castenholz, R.W.** 1981. Use of reduced sulfur compounds by *Beggiatoa* sp. *Journal of Bacteriology* 147:140–154.

Norris, P.R. 1983. Iron and mineral oxidation with Leptospirillum-like bacteria, pp. 83–96, in Rossi, G. and Torma, A.E. (editors), *Recent Progress in Biohydrometallurgy,* Associazione Mineraria Sarda, Iglesias.

Norris, P.R. and **Kelly, D.P.** 1982. The use of mixed microbial cultures in metal recovery, pp. 443–474, in Bull, A.T. and Slater, J.H. (editors), *Microbial Interactions and Communities,* Vol. 1, Academic Press, London.

Pace, N.R., Stahl, D.A., Lane, D.J., and **Olsen, G.J.** 1986. The analysis of natural microbial populations by ribosomal RNA sequences. *Advances in Microbial Ecology* 9:1–55.

Pickwell, G.V., Barham, E.G., and **Wilton, J.W.** 1964. Carbon monoxide production by a bathypelagic siphonophore. *Science* 144:860–862.

Poth, M. 1986. Dinitrogen production from nitrite by a *Nitrosomonas* isolate. *Applied and Environmental Microbiology* 52:957–959.

Robertson, L.A. and **Kuenen, J.G.** 1983a. *Thiosphaera pantotropha* gen. nov. sp. nov. a facultatively anaerobic, facultatively autotrophic sulphur bacterium. *The Journal of General Microbiology* 129:2847–2855.

Robertson, L.A. and **Kuenen, J.G.** 1983b. Anaerobic and aerobic denitrification by sulphide-oxidizing bacteria from waste water, pp. 3–12, in van der Brink, W.J. (editor), *Proceedings of the European Symposium Anaerobic Waste Water Treatment Noordwijkerhout,* AWWT Symposium Secretariat, The Hague.

Robertson, L.A. and **Kuenen, J.G.** 1984a. Anaerobic denitrification—old wine in new bottles? *Antonie van Leeuwenhoek Journal of Microbiology and Serology* 50:525–544.

Robertson, L.A. and **Kuenen, J.G.** 1984b. Aerobic denitrification: a controversy revived. *Archives of Microbiology* 139:351–354.

Schink, B. and **Zeikus, J.G.** 1984. Ecology of aerobic hydrogen-oxidizing bacteria in two freshwater lake ecosystems. *Canadian Journal of Microbiology* 30:260–265.

Segerer, A., Neuner, A., Kristjansson, J.K., and **Stetter, K.O.** 1986. *Acidianus infernus* gen. nov., sp. nov., and *Acidianus brierleyi* comb. nov.: Facultatively aerobic, extremely acidophilic thermophilic sulfur-metabolizing archaebacteria. *International Journal of Systematic Bacteriology* 36:559–564.

Seiler, W. and **Schmidt, U.** 1976. The role of microbes in the cycle of atmospheric trace gases, especially of hydrogen and carbon monoxide, pp. 35–45, in Schlegel, H.G., Gottschalk, G., and Pfennig, N. (editors), *Microbial Production and Utilization of Gases,* Goltze KG, Göttingen.

Shiba, H., Kawasumi, T., Igarashi, Y., Kodama, T., and **Minoda, Y.** 1985. The CO_2 assimilation via the reductive tricarboxylic acid cycle in an obligately autotrophic aerobic hydrogen-oxidizing bacterium, *Hydrogenobacter thermophilus. Archives of Microbiology* 141:198–203.

Sivelä, S. and **Sundman, V.** 1975. Demonstration of *Thiobacillus*-type bacteria which utilize methyl sulphides. *Archives of Microbiology* 103:303–304.

Sokolova, G.A. and **Karavaiko, G.I.** 1968. *Physical and geochemical activity of thiobacilli.* Israel Program for Scientific Translations, Jerusalem. 283 pp.

Suylen, G.M.H., Stefess, G.C., and **Kuenen, J.G.** 1986. Chemolithotrophic potential of a *Hyphomicrobium* species, capable of growth on methylated sulphur compounds. *Archives of Microbiology* 146:192–198.

Veldkamp, H., van Gemerden, H., Harder, W., and **Laanbroek, H.J.** 1984. Competition among bacteria: an overview, pp. 279–290, in Klug, M.J. and Reddy, C.A. (editors), *Current Perspectives in Microbial Ecology,* American Society for Microbiology, Washington.

Vollbrecht, D., El Nawawy, M.A., and **Schlegel, H.G.** 1978. Excretion of metabolites by hydrogen bacteria. I. Autotrophic and heterotrophic fermentations. *European Journal of Applied Microbiology and Biotechnology* 6:145–155.

Ward, B.B. and **Carlucci, A.F.** 1985. Marine ammonia- and nitrite-oxidizing bacteria: serological diversity determined by immunofluorescence in culture and the environment. *Applied and Environmental Microbiology* 50:194–201.

Wood, A.P. and **Kelly, D.P.** 1983. Autotrophic and mixotrophic growth of three thermoacidophilic iron-oxidizing bacteria. *FEMS Microbiology Letters* 20:107–112.

Wood, A.P., Kelly, D.P., and **Norris, P.R.** 1987. Autotrophic growth of four *Sulfolobus* strains on tetrathionate and the effect of organic nutrients. *Archives of Microbiology* 146:382–389.

Yates, J.R. and **Holmes, D.S.** 1986. Construction and use of molecular probes to identify and quantitate bioleaching micro-organisms, pp. 409–418, in Lawrence, R.W., Branion, R.M.R., and Ebner, H.G. (editors), *Fundamental and Applied Biohydrometallurgy,* Elsevier, Amsterdam.

5

Nitrifying Bacteria

Eberhard Bock, Hans-Peter Koops, and Heinz Harms

Department of Microbiology
University of Hamburg
Hamburg, Federal Republic of Germany

The biological conversion of reduced forms of nitrogen to nitrate is called *nitrification*. The most important groups of microorganisms involved in this process are the lithoautotrophic ammonia and nitrite oxidizing bacteria placed in the family *Nitrobacteraceae* (Watson, 1974; Watson et al., 1981, 1987). With one exception, these organisms are obligately lithoautotrophic, although they can assimilate organic compounds to a limited extent.

Another process, *heterotrophic nitrification*, involves the oxidation of reduced nitrogen compounds (predominantly organic nitrogen compounds). This process, carried out by various species of bacteria, fungi, and even some algae (Focht and Verstraete, 1977; Killham, 1986), is a cooxidation process which is not coupled to growth. None of the heterotrophic nitrifiers can grow autotrophically. Compared to the lithoautotrophic nitrification process, only small amounts of nitrite or nitrate are formed, but heterotrophic nitrification may be of importance if autotrophic nitrification is inhibited by environmental conditions. Furthermore, some methane-oxidizing bacteria cooxidize ammonia to nitrite, but these organisms cannot use ammonia as an energy source (Dalton, 1977; O'Neill and Wilkinson, 1977; Yoshinari, 1985). A possible significance of methane-oxidizing bacteria in the nitrification process of natural ecosystems is discussed by Hall (1986).

5.1 NITROGEN CYCLE

Table 5.1 describes the geochemical distribution of nitrogen in the lithosphere, atmosphere, hydrosphere, and biosphere. Most of the nitrogen is held in primary rocks of the lithosphere. This value is 50 times that found in the atmosphere and only relatively small amounts of nitrogen are found in the hy-

Table 5.1 Nitrogen content in the four spheres of earth

Region of the earth	Nitrogen content
Lithosphere	
Rocks of the crust	100
Rocks of the mantle	16,200
Fossil N (sediments)	35–55
Coal	0.007
Organic matter, sea bottom	0.054
Organic matter, soil	0.022
Clay, fixed NH_4^+	0.002
Atmosphere	386
Hydrosphere	
Dissolved N_2	2.19
Combined nitrogen	0.11
Biosphere	0.028–0.065

Units, kg, all $\times 10^{16}$.
Modified from Burns and Hardy, 1975; Söderlund and Svensson, 1976; Stevenson, 1986.

drosphere and the biosphere. Combined nitrogen of the hydrosphere and biosphere originates from nitrogen of the atmosphere which is believed to have built up from rocks primarily in the form of NH_3.

The nitrogen of the hydrosphere and the biosphere, present in the forms of NH_3, NO_2^-, NO_3^-, and organic compounds, belongs to the transient nitrogen fraction which is characterized by quick gains and losses. However, the values given in Table 5.1 are only approximate and some data suggest an unbalanced global nitrogen cycle. The transfer of ammonia to microbial habitats via volatilization is the most important process from the view of nitrification, with 91–186 $\times 10^9$ kg NH_3 plus NH_4^+ being deposited per year on the surfaces of aqueous habitats, soils, rocks, and concrete or natural building stones. Furthermore, ammonia forms salts by reaction with acid components of the atmosphere. Sulfates are the main constituents of the dust caused by anthropogenic air pollution (Shaw, 1987).

The nitrogen cycle of the soil, one of the best understood internal cycles, is shown in Figure 5.1. Nitrification is the main process by which nitrogen flux to nitrate occurs. Nitrogen fluxes vary with the type of ecosystem. Generally, the data for inputs via biologically N_2 fixation are more or less accurate, whereas those for nitrogen losses via denitrification and ammonia volatilization are less reliable.

Because of the voluminous literature an exhaustive coverage of relevant data of the global flux of nitrogen is not possible. Table 5.2 presents the best-accepted values for transfers from one compartment to another.

5.2 ECOLOGICAL SIGNIFICANCE OF NITRIFICATION

The activity of life on earth is closely associated with the element nitrogen. Nitrogen exists in a number of oxidation states from −3 to +5 with the most important being $-3(NH_3)$, $0(N_2)$, $+3(NO_2^-)$, and $+5(NO_3^-)$. Because of biological nitrogen transformations, any form of the nitrogen atom has only a

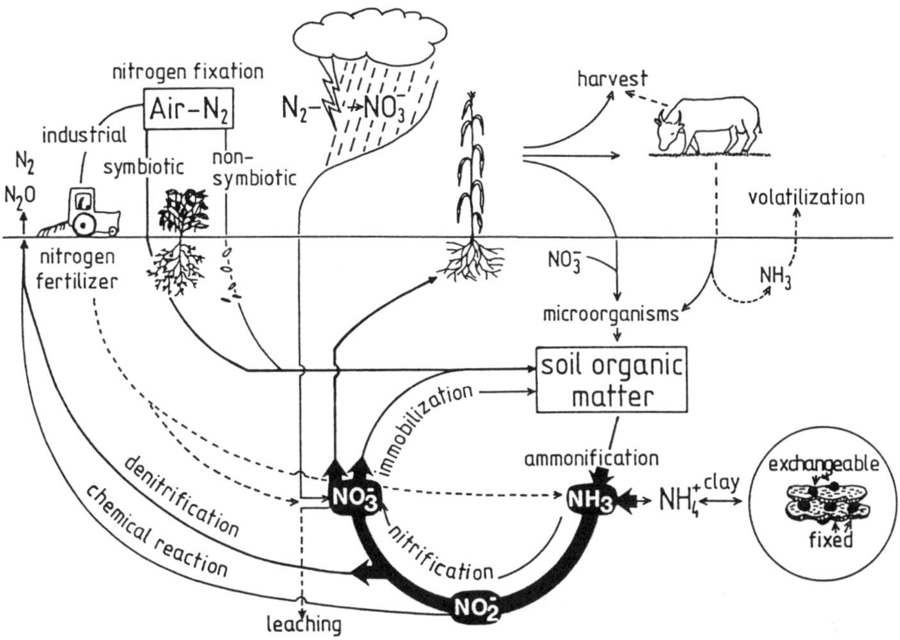

Figure 5.1 The nitrogen cycle in soil.

transient existence in nature. In nature the nitrogen cycle is furthermore characterized by immobilization and mobilization (= mineralization). Thus, immobilized nitrogen in form of organic nitrogen is converted to mobile ammonia through the activity of proteolytic organisms.

Depending on environmental conditions, nitrification can prevent nitrogen loss (if ammonia volatilization is a major process) or will lead to nitrogen loss

Table 5.2 Global fluxes for nitrogen

Process	Nitrogen flux, kg/year × 10^9
Emissions and losses:	
NH_3 volatilization by	
a) Animals and coal burning	26–53
b) Others (assumed)	87–191
Denitrification	107–161
Total	220–385
Additions:	
Biological N_2 fixation	139
Industrial fertilizers	36
NH_3/NH_4^+ deposition	91–186
NO_x deposition	32–83
Organic nitrogen deposition	10–100
Total	328–534
Transfer: land/river/ocean	
NH_4^+, NO_3^-, NO_x, organic nitrogen	18–35

NO_x, various gaseous nitrogen oxides.
Modified according to Söderlund and Svensson (1976) and Stevenson (1986).

in one of the following three ways: 1) *Denitrification*. This process, the microbial reduction of NO_3^- via NO_2^- and N_2O to N_2, leads to the loss of nitrogen from the environment. It is likely that N_2, the principal constituent of the atmosphere, has been mostly produced by the nitrification/denitrification cycle. 2) Nitrite serves as a substrate for *chemodenitrification*. In most soils the nitrite-oxidizing activity seems to proceed with a faster rate than the conversion of ammonia to nitrite. Therefore nitrite concentrations above 1 ppm (El-Demerdash and Ottow, 1983) are rarely found except in the presence of ammonia fertilizers. This is presumably due to the inhibition of *Nitrobacter* by high concentrations of ammonia. Nitrite is undesirable because of its chemical instability and toxicity at low pH. Chemodenitrification is the nonenzymatic loss of nitrogen from the dismutation of nitrous acid to nitric acid and NO. NO reacts with oxygen to form NO_2, which in the presence of water is transformed to nitrous and nitric acid. 3) *Leaching*. The transfer of mobilized nitrogen to rivers, lakes, seawater, and ground water occurs through leaching of the very water-soluble nitrate anion. It is obvious that nitrification is an integral part of the internal cycles mentioned above.

5.3 HABITATS AND ECOLOGICAL NICHES

Nitrifying bacteria are present in most aerobic environments where organic matter is mineralized and on/in rocks and natural building stones. In their natural environment, substrate availability depends most of all on the rate of ammonification and the rate of ammonia deposition. Urea can also serve as a primary substrate (Figure 5.2) for most of the ammonia oxidizers (Watson et al., 1971; Harms and Koops, unpublished data). In soil, NH_4^+ is absorbed to negatively charged clay particles. Whenever nitrifiers are present, the absorbed ammonia is transformed to nitrite and nitrate, which can be leached. Therefore, in agriculture nitrification inhibitors such as nitrapyrin (N-Serve) are used to minimize losses of nitrogen (Huber et al., 1977; Nelson and Huber, 1980; Keeney, 1982, 1986; Slangen and Kerkhoff, 1984; Lipschultz et al., 1981; Poth and Focht, 1985).

In eutrophic environments, nitrification might lead to significant oxygen deletion and might contribute to the onset of conditions favorable for denitrification. At low oxygen concentrations, nitric oxide and nitrous oxide, which play an important role in the atmospheric chemistry, are observed to be products of denitrification as well as nitrification (Yoshida and Alexander, 1970; Goreau et al., 1980).

Nitrification in nature proceeds under a broad range of conditions, even those that are not observed to be tolerable for pure cultures. For example, strains of ammonia oxidizers were isolated from acid tea soils with a pH of 4.6 (Walker and Wickramasinghe, 1978), from low-temperature environments (Jones and Morita, 1985), and from acid surfaces of sandstone (Bock, 1987). Even at temperatures of 50–60°C, enrichment cultures could be developed from hot spring samples (Golovacheva, 1976; Harms and Koops, unpublished data).

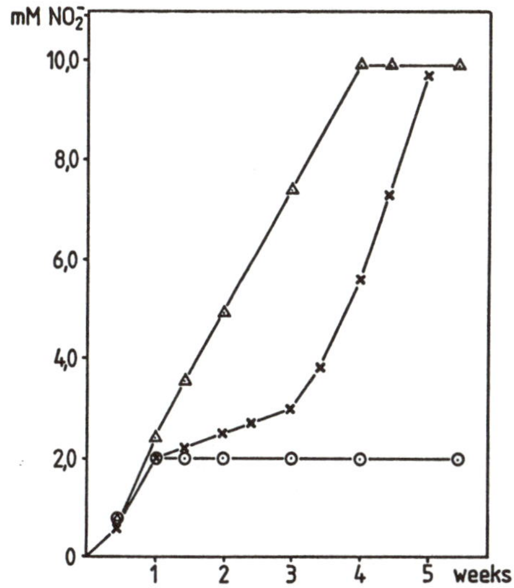

Figure 5.2 Nitrite formation by autotrophically grown cells of *Nitrosomonas* sp. Substrate: 2 mM NH_4Cl (O—O), 10 mM NH_4Cl (Δ—Δ), and 2 mM NH_4Cl plus 4 mM urea (X—X).

Many species of the lithoautotrophic nitrifiers prefer special environments. For example, all species of ammonia oxidizers isolated from brackish or marine environments seem to be restricted to their specific ecosystems, being obligately halophilic (Watson, 1965; Watson and Mandel, 1971; Watson and Waterbury, 1971; Koops et al., 1976). Other species were observed to be adapted to low temperatures (Jones and Morita, 1985; Macdonald, 1986), substrate availability (Ward, 1986), and substrate concentrations (Harms and Koops, unpublished data).

When studying nitrification in nature, one should keep in mind that special ecological niches go together with highly specialized nitrifiers. Differences in sensitivity to environmental factors seem to be attributed to species differences. Light was observed to be especially inhibitory to nitrite oxidizers (Schön and Engel, 1962; Bock, 1965). Thus, visible light is an important determinant of the distribution of nitrifiers in aquatic systems (Hooper and Terry, 1974; Olson, 1981; Ward, 1986).

Sometimes nitrifiers are endolithic organisms. Samples taken from the rock walls of the churches of Cologne, Regensburg, Gellenhausen, the Alte Pinakothek Museum in Munich, and the castle in Arolsen revealed high numbers of nitrifying bacteria. In the outermost layers to a depth of 5 mm the sandstone was heavily contaminated. When penetrating 15 cm into the Schlaitdorfer sandstone of the Cologne cathedral, the number of nitrifiers and the concentration of nitrate were observed to decrease, but the organisms were present at any depth. At the stone surface, the pH had decreased to 7.0 because the alkaline building material—the dolomite—was used up by acid attack; at a depth of 5 cm the pH increased to 8.0 (Figure 5.3).

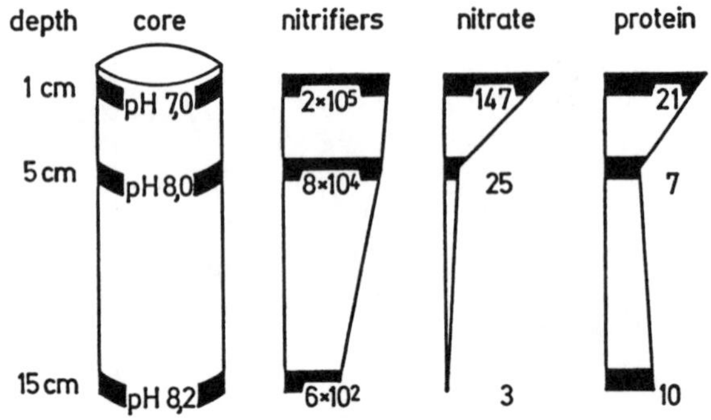

Figure 5.3 Decrease of cell number as cells per mg protein (MPN-technique) of nitrifying organisms with increasing depth of the Schlaitdorfer sandstone. The concentrations of nitrate and protein are given in mg per kg stone.

The most abundant ammonia oxidizing organism in stone was found to be a *Nitrosovibrio* (Meincke, unpublished data). In 160 samples from the Cologne cathedral, this was the only ammonia oxidizer present. The strains are characterized morphologically by carboxysomes and intracytoplasmic membranes and physiologically by the tolerance to high concentration of NH_4Cl (up to 160 mM). The nitrite oxidizers are obligate lithoautotrophic strains of *Nitrobacter).* Simulation experiments (Bock, unpublished data) have shown that nitrifying bacteria bring about a deterioration of stone.

In natural ecosystems a coexistence of different species of nitrifying bacteria often occurs (Fliermans et al., 1974; Belser and Schmidt, 1978a, b). At this time serotyping of species and strains seems to be the only helpful technique in population analysis of natural ecosystems, but the use of fluorescent antisera is restricted to immunologically characterized species (Belser and Schmidt, 1978b; Harms and Koops, unpublished data).

5.4 GROWTH CHARACTERISTICS OF LITHOAUTOTROPHIC NITRIFYING BACTERIA

The first ammonia oxidizer was isolated in pure culture in 1890 by Frankland and Frankland (Frankland and Frankland, 1890), the same year Winogradsky claimed to have isolated an ammonia-oxidizing nitrifying organism (see Chapter 1). Since he regarded nitric acid to be the direct endproduct of a single stage reaction, the culture must have been contaminated (Winogradsky, 1890). In 1891, Winogradsky purified the first isolate of a nitrite oxidizer, thus confirming the early idea of Warrington (1878) that nitrification includes two separate stages. Winogradsky was the first to recognize that nitrifiers could grow in media lacking organic substances, indicating that they grow autotrophically. Nitrifying organisms are a physiologically defined group of bacteria which consists of phylogenetically heterogenous genera. Most of the nitrifying microorganisms are highly specialized obligately lithotrophic bacteria, growing only in the presence of an inorganic energy source.

The most comprehensive investigations on physiology and biochemistry have been conducted with members of the genera *Nitrosomonas* and *Nitrobacter*. The overall reactions for the two stages of nitrification are as follows:

1. **Ammonia oxidation**

a) $NH_3 + \frac{1}{2}O_2 \rightarrow NH_2OH$

b) $NH_2OH + H_2O \rightarrow HNO_2 + 4H^+ + 4e^-$

c) $4H^+ + 4e^- + O_2 \rightarrow 2H_2O$

2. **Nitrite oxidation**

a) $NO_2^- + H_2O \rightarrow NO_3^- + 2H^+ + 2e^-$

b) $2H^+ + 2e^- + \frac{1}{2}O_2 \rightarrow H_2O$

The mechanism by which nitrifiers generate energy from the above reactions is reviewed in detail in Chapter 13.

CO_2 is the major carbon source, being assimilated via the Calvin cycle. The early belief that organic compounds are indiscriminately toxic to these obligate chemolithotrophs is no longer accepted, although growth sometimes is inhibited by organic compounds (Bock, 1976). Many, if not all species, can assimilate organic carbon to a limited extent and organic carbon stimulates growth and increases cell yield (Steinmüller and Bock, 1976; Matin, 1978; Krümmel and Harms, 1982; Watson et al., 1986). Growth with ammonia or nitrite as the only energy sources and CO_2 and organic compounds as carbon sources is termed mixotrophic growth.

The reason for the obligate chemolithotrophy is not understood up to now. The incorporation of organic carbon compounds is restricted with regard to the extent of assimilation as well as to the contribution to cell carbon (Smith and Hoare, 1977; Martiny and Koops, 1982). Deficiencies in the tricarboxylic cycle and the lack of the NADH oxidase system coupled to ATP synthesis are the reasons most commonly discussed (Matin, 1978).

Lithotrophic growth of the nitrifying bacteria is controlled by substrate availability, temperature, oxygen concentration, and pH. Most species grow optimally at substrate concentrations up to 10 mM, a pH of 7.5–8.0, and at a temperature between 25–30°C. Nitrifying organisms are slow-growing bacteria, the generation times of *Nitrosomonas* and *Nitrobacter* varying between 7 to 13 hours when grown in batch cultures. The metabolic end products nitrite and nitrate are inhibitory at concentrations of about 300 mg/l for *Nitrosomonas* (Wullenweber et al., 1978) and 4000 mg/l for *Nitrobacter* (Schön, 1965). The cell yield obtained by lithoautotrophic growth is low with $3\text{–}5 \times 10^7$ cells being formed per ml when 1 mg/ml nitrite is oxidized. The energy charge, (ATP + 0.5ADP) ÷ (ATP + ADP + AMP), of logarithmically growing *Nitrobacter* cells is 0.37 and increases to 0.5 after nitrite has been consumed (Eigener, 1975). The energy efficiency varies from strain to strain (Belser and Schmidt, 1980) and also depends on growth conditions (Krümmel and Harms, 1982). The understanding of the metabolism of the nitrite oxidizer, *Nitrobacter*, has gradually changed during the past years. The progress of knowledge is marked by the investigations of Ida and Alexander (1965), Delwiche and Finstein (1965), and Rittenberg (1972). *Nitrobacter* was shown to be a facultative lithoautotroph (Smith and Hoare, 1968; Steinmüller and Bock, 1976; Bock, 1976) and many strains were able to grow effectively with organic substances such as acetate, pyruvate and glycerol. Freitag et al. (1987) were able to demonstrate growth of *Nitrobacter* in the absence of dissolved oxygen with acetate or pyruvate as electron donor and nitrate as terminal electron acceptor. From these results it must be concluded that representatives of the genus *Nitrobacter* are able to oxidize nitrite to nitrate but are also able to reduce nitrate to nitrite, NO, and N_2O. The technical and ecological consequences of being able to reverse the oxidation pathway are obvious (Painter, 1986). In wastewater treatment plants, the loading of biological reactors with organic matter is kept limited so that efficient nitrification can occur. The extent of anaerobic growth of *Nitrobacter* in wastewater treatment plants and in natural habitats has to be investigated in the future.

5.5 TAXONOMY OF THE LITHOAUTOTROPHIC NITRIFIERS

The lithoautotrophic nitrifying bacteria are Gram-negative and are classified in the family *Nitrobacteraceae* (Watson, 1971a, 1974; Watson et al., 1981, 1987; Bock et al., 1986). This is a physiologically defined group of ammonia and nitrite-oxidizing genera which are differentiated on morphological characteristics. However, investigations on the phylogeny of these bacteria indicate that the lithoautotrophic nitrifiers are taxonomically more diverse than expected. Comparing oligonucleotides of the 16S ribosomal RNA, Woese et al. (1984a, b, 1985) showed that members of the *Nitrobacteraceae* grouped together with other nonphotosynthetic species among three major phylogenetic subdivisions of the purple photosynthetic bacteria. The nitrite-oxidizing *Nitrobacter* species grouped in the alpha subdivision with high similarity to *Rhodopseudomonas palustris* (Seewaldt et al., 1983). The ammonia oxidizers thus far studied, with one exception, group in the beta subdivision (Woese et al., 1984a). *Nitrosococcus oceanus* has a phylogenetically isolated position among the ammonia oxidizers, grouping in the gamma subdivision, the *Chromatiaceae* branch of the purple bacteria (Woese et al., 1985).

5.6 AMMONIA OXIDIZERS

The names of all genera of the ammonia oxidizers have the prefix *Nitroso-* (signifying "nitrite" formation). At this time five genera are differentiated on morphological criteria.

Genus Nitrosomonas Cells are rod-shaped to ellipsoidal, 0.8–0.9 by 1.0–2.0 μm (Winogradsky, 1892). Sometimes coccoidal forms are observed. Intracytoplasmic membranes are arranged as flattened vesicles in the peripheral region of the cytoplasm. Carboxysomes (see Chapter 19) and polyphosphate granules are observed as intracytoplasmic inclusions. Motile cells possess polar flagella. *N. europaea* (Winogradsky, 1892) is the only described species of this genus. It is distinguishable from other species by DNA homology experiments (Watson and Mandel, 1971; Dodson et al., 1983; Koops and Harms, 1985), and base composition of DNA, which is 51.0 ± 0.4 percent G + C in *N. europaea*.

In some but not all cases observed genotypic diversity of the genus *Nitrosomonas* correlated with differences in the shape and size of the cells and in physiological adaptations to special environments (Watson and Mandel, 1971; Harms and Koops, unpublished data). To define different species of the genus by apparent phenotypic characteristics, further investigations are required.

Genus Nitrosococcus Included are at least three species (Winogradsky, 1892). Cells are spherical to ellipsoidal. The species with one exception are differentiable genotypically as well as by size and ultrastructure of their cells. *N. nitrosus* (Buchanan, 1925), the type species, was isolated by Winogradsky (1892) but not preserved in laboratory culture, so that information on the genotypic and ultrastructural characteristics are not available. This species was isolated from a soil sample from Quito (Ecuador). The spherical cells are 1.5–1.7 μm in diameter. Motility was not observed but suggested by Winogradsky (1892).

Strains of *N. oceanus* (Watson, 1971a) have been isolated exclusively from marine environments. Recently, strains similar in shape, size, and ultrastructure, but distinguishable genotypically, have been isolated from salt lakes (Koops and Harms, 1985). Cells of *N. oceanus* are spherical to ellipsoidal, 1.8–2.2 μm in diameter, and motile by means of 1–20 polar flagella (Watson, 1965). In mixed cultures, cysts have been occasionally observed and therefore this species originally was described as *Nitrosocystis oceanus* (Watson, 1965). Outside the typical Gram-negative cell envelope two additional cell wall layers are present, composed of subunits arranged in a macromolecular array (Murray and Watson, 1965). *N. oceanus* is distinguished from all the other ammonia oxidizers by presence of a centrally located stack of parallel, flattened membrane vesicles. The DNA base composition is 50.8 percent G + C.

Cells of *N. mobilis* (Koops et al., 1976), isolated from brackish water, are spheres, 1.2–1.9 μm in diameter when grown in 2:3 diluted seawater. They are motile by means of a tuft of flagella. Like other marine and estuarine strains of ammonia oxidizers, they have an additional cell wall layer outside the Gram-negative envelope (Watson, personal communication). Cell membranes are arranged as in *Nitrosomonas*. The DNA base composition is 49.2 ± 0.3 percent G + C.

Genus Nitrosospira The three genera *Nitrosospira, Nitrosolobus,* and *Nitrosovibrio* are phylogenetically closely related (Woese et al., 1984b), although they are easy to distinguish morphologically. All isolates originate from soils or fresh waters; marine or brackish-water strains have never been observed. In the genus *Nitrosospira*, the species *N. briensis* (Winogradsky and Winogradsky, 1933) is the only accepted species (Watson, 1971a), but at least five species distributed in two groups have been recognized based on the DNA base composition (Koops and Harms, 1985).

Genus Nitrosolobus The genus *Nitrosolobus* is characterized by the pleomorphic, lobate shape of the cells (Watson et al., 1971). One species, *N. multiformis*, has been described. Cells are 1.0–1.5 μm wide and 1.0–2.5 μm long and motile by peritrichous flagella. Cells have 1–4 central compartments surrounded by 5–20 peripheral compartments. Glycogen inclusions, located primarily in the peripheral compartments, are observed. The base composition of the DNA is 54.5 percent G + C. Recently a second species, distinguishable by base composition 56.4 percent G + C was isolated (Koops and Harms, 1985).

Genus Nitrosovibrio Cells of *N. tenuis* are slender, curved rods, 0.3–0.4 μm wide and 1.1–3.0 μm long (Harms et al., 1976). They are motile by 1–4 polar to lateral flagella. An increasing number of spherical forms, 1.0–1.2 μm in diameter, are observed in the early stationary phase of growth. Extensive intracytoplasmic membrane systems are lacking but some invaginations are observed. A second species was detected by DNA:DNA hybridization experiments (Koops and Harms, 1985). Both species have a base composition of DNA of 53.9 ± 0.3 percent G + C.

5.7 NITRITE OXIDIZERS

The names of the genera of nitrite oxidizers have the prefix *Nitro*-, signifying "nitrate" formation. Four morphologically distinct genera have been described but the phylogenetic relationship between these genera has not yet been investigated.

Genus Nitrobacter Two species have been described (Winogradsky, 1892). Cells of both species are rod- to pear-shaped, dividing by either budding or binary fission. Average size is 0.6–0.8 × 1.2–2.0 μm. If motile, they possess a single polar or lateral flagellum. The cell envelope is atypical for Gram-negative bacteria since a separate peptidoglycan layer of the cell wall is not detectable. An extensive intracytoplasmic membrane system is arranged as a polar cap of 4–6 layers of paired membranes. The cytoplasmic membrane as well as the intracytoplasmic membranes are asymmetric in cells grown lithotrophically. The cytoplasmic, more electron-dense side of the membranes is stacked with particles 7–9 nm in diameter (Remsen and Watson, 1972; Sundermeyer and Bock, 1981), which are reduced in resting cells and absent in cells grown heterotrophically (Bock and Heinrich, 1971). Cytoplasmic inclusions are carboxysomes, poly-β-hydroxybutyrate, glycogen, and polyphosphate (Seewaldt et al., 1983). The DNA in both *N. winogradskyi* (Winslow et al., 1917) and *N. hamburgensis* (Bock et al., 1983) is 60–62 percent G + C. Three plasmids were observed in one culture of *N. hamburgensis*.

A third species of the genus *Nitrobacter* was distinguished by its DNA base composition (59.2 ± 0.3 percent G + C) and by DNA:DNA hybridization experiments (Koops et al., unpublished data). Mixotrophic growth of this species, in contrary to the other two *Nitrobacter* species, is biphasic (Bock et al., 1986). Nitrite is oxidized first and after a lag phase, organic compounds are metabolized. Most of the *Nitrobacter* strains thus far studied can grow heterotrophically.

Genus Nitrococcus One species, *N. mobilis*, was isolated from marine environments (Watson and Waterbury, 1971). Cells are spheres, 1.5–1.8 μm wide following division and 1.8 × 3.5 μm just prior to division. Tubular cytomembranes, randomly arranged throughout the cytoplasm, are a characteristic of this species. Cells are motile by 1 to 2 flagella. They are obligate halophilic, growing optimally in 70–100 percent seawater. Carboxysomes and occasionally poly-β-hydroxybutyrate granules and electron dense bodies, believed to be glycogen storage products, were observed as cytoplasmic inclusions. The DNA is 61.2 percent G + C.

Genus Nitrospina *N. gracilis*, the only species, was isolated from marine environments (Watson and Waterbury, 1971). Cells are nonmotile, rod-shaped, and 0.3–0.4 by 1.7–6.6 μm in size. In old cultures, spherical forms 1.35–1.45 μm in diameter were observed. Cells lack an extensive intracytoplasmic membrane system but some bleblike invaginations were observed. The DNA is 58 percent G + C.

Genus Nitrospira One species, *N. marina*, was isolated from marine environments (Watson et al., 1986). Cells are nonmotile and vibrioid or helically formed (up to 20 turns). The spiral amplitude is of 0.8–1.0 μm and the cell width 0.3–0.4 μm. Cells are lacking an intracytoplasmic membrane system. The width of the periplasmic space is approximately twice that found in other Gram-negative bacteria. Tight clumps of cells are observed in enrichment cultures. The DNA is 50.5 ± 0.5 percent G + C. *Nitrospira* strains seem to be not restricted to marine environments since similar forms were observed in enrichment cultures from soil samples from Namibia (Watson et al., 1987).

5.8 CONCLUSIONS

Investigations on phylogeny have demonstrated that the *Nitrobacteraceae* are a physiological but not taxonomically definable group. Furthermore, by means of the DNA/DNA hybridization technique it is possible to distinguish between physiologically and morphologically similar strains on the species level. First results have indicated a very wide diversity of species within different genera. The taxonomical and physiological diversity of the nitrifiers makes ecological studies complicated. Methods useful in population analysis in natural habitats are rare. For instance, the fluorescent antibody technique works well with pure cultures but its practicability in natural environments is still doubtful.

Since Winogradsky's pioneering work in the last century, our knowledge of the physiology of the ammonia and nitrite-oxidizing organisms has increased significantly, but many questions of basic interest remain unanswered. The energy-conserving system is not totally understood (see Chapter 13). The link between ammonia or nitrite oxidation and proton motive force is controversial (Sone et al., 1983; Hooper and DiSpirito, 1985; Wetzstein and Ferguson, 1985; Wood, 1986). In the facultative *Nitrobacter*, oxygen and/or nitrate may function as an electron acceptor in aerobic or anaerobic respiration. As shown by Tanaka et al. (1983) and Sundermeyer-Klinger et al. (1984), the membrane-bound nitrite oxidoreductase also functions as a nitrite oxidase as well as a nitrate reductase. In both cases, ATP is generated by electron transport phosphorylation. Investigations are in progress suggesting *Nitrobacter* to be a denitrifying organism even in the presence of oxygen (Bock, unpublished data). We still do not know why there are no organisms capable of oxidizing ammonia all the way to nitrate.

REFERENCES

Belser, L.W. and **Schmidt, E.L.** 1978a. Diversity in the ammonia-oxidizing population of a soil. *Applied and Environmental Microbiology* 36:584–586.

Belser, L.W. and **Schmidt, E.L.** 1978b. Serological diversity within a terrestrial ammonia-oxidizing population. *Applied and Environmental Microbiology* 36:589–593.

Belser, L.W. and **Schmidt, E.L.** 1980. Growth and oxidation kinetics of three genera of ammonia oxidizing nitrifiers. *Federation of European Microbiological Societies, Microbiology Letters* 7:213–216.

Bock, E. 1965. Vergleichende Untersuchungen über die Wirkung sichtbaren Lichtes auf *Nitrosomonas europaea* und *Nitrobacter winogradskyi. Archiv für Mikrobiologie* 51:18–41.

Bock, E. 1976. Growth of *Nitrobacter* in the presence of organic matter. II. Chemoorganic growth of *Nitrobacter agilis. Archives of Microbiology* 108:305–312.

Bock, E. 1978. Lithoautotrophic and chemoautotrophic growth of nitrifying bacteria, pp. 310–314, in Schlessinger, D. (editor), *Microbiology—1978.* American Society for Microbiology, Washington, D.C.

Bock, E. 1987. Biologisch induzierte Korrosion von Natursteinen—starker Befall mit Nitrifikanten. *Bautenschutz/Bautensanierung* 1:24–27.

Bock, E. and **Heinrich, G.** 1971. Struktur- und Funtionsänderungen in reaktivierenden Zellen von *Nitrobacter winogradskyi* Buch. *Archiv für Mikrobiologie* 77:349–365.

Bock, E., Koops, H.-P., and **Harms, H.** 1986. Cell biology of nitrifying bacteria, pp. 17–38, in Prosser, J.I. (editor), *Nitrification,* Special Publications of the Society for General Microbiology, Volume 20, IRL Press, Oxford, Washington, DC.

Bock, E., Sundermeyer-Klinger, H., and **Stackebrandt, E.** 1983. New facultative lithoautotrophic nitrite-oxidizing bacteria. *Archives of Microbiology* 136:281–284.

Buchanan, R.E. 1925. *General Systematic Bacteriology.* Williams and Wilkins Co., Baltimore, 398 pp.

Burns, R.C. and **Hardy, R.W.F.** 1975. *Nitrogen Fixation in Bacteria and Higher Plants*, Springer-Verlag, New York.

Dalton, H. 1977. Ammonia oxidation by the methane oxidizing bacterium *Methylococcus capsulatus* strain Bath. *Archives of Microbiology* 114:273–279.

Delwiche, C.C. and **Finstein, M.S.** 1965. Carbon and energy sources for the nitrifying autotroph *Nitrobacter. Journal of Bacteriology* 90:102–107.

Dodson, M.S., Mangan, J., and **Watson, S.W.** 1983. Comparison of deoxyribonucleic acid homologies of six strains of ammonia-oxidizing bacteria. *International Journal of Systematic Bacteriology* 33:521–524.

Eigener, U. 1975. Adenine nucleotide pool variations in intact *Nitrobacter winogradskyi* cells. *Archives of Microbiology* 102:233–240.

El-Demerdash, M.E. and **Ottow, J.C.G.** 1983. Einfluß einer hohen Nitratdüngung auf Kinetik und Gaszusammensetzung der Denitrifikation in unterschiedlichen Böden, pp. 138–150, in *Zeitschrift für Pflanzenernährung und Bodenkunde* 146, Verlag Chemie GmbH, Weinheim.

Fliermans, C.B., Bohlool, B.B., and **Schmidt, E.L.** 1974. Autecological study of the chemotroph *Nitrobacter* by immunofluorescence. *Applied Microbiology* 27:199–208.

Focht, D.D. and **Verstraete, W.** 1977. Biochemical ecology of nitrification and denitrification. *Advances in Microbial Ecology* 1:135–214.

Frankland, P.F. and **Frankland, G.C.** 1890. The nitrifying process and its specific ferment. Part I. *Philosophical Transactions of the Royal Society of London. Series B* 181:107–128.

Freitag, A., Rudert, M., and **Bock, E.** 1987. Growth of *Nitrobacter* by dissimilatory nitrate reduction. *Federation of European Microbiological Societies, Microbiology Letters* 14:105–109.

Golovacheva, R.S. 1976. Thermophilic nitrifying bacteria from hot springs. *Microbiologiya* 45:329–331.

Goreau, T.J., Kaplan, W.A., Wofsy, S.C., McElroy, M.B., Valois, F.W., and **Watson, S.W.** 1980. Production of NO_2^- and N_2O by nitrifying bacteria at reduced concentrations of oxygen. *Applied and Environmental Microbiology* 40:526–532.

Hall, G.H. 1986. Nitrification in lakes, pp. 127–156, in Prosser, J.I. (editor), *Nitrification*, Special Publications of the Society for General Microbiology, Volume 20, IRL Press, Oxford, Washington, DC.

Harms, H., Koops, H.-P., and **Wehrmann, H.** 1976. An ammonia-oxidizing bacterium, *Nitrosovibrio tenuis* nov. gen. nov. sp. *Archives of Microbiology* 108:105–111.

Hooper, A.B. and **DiSpirito, A.A.** 1985. In bacteria which grow on simple reductants, generation of a proton gradient involves extracytoplasmic oxidation of substrate. *Microbiological Reviews* 49:140–157.

Hooper, A.B. and **Terry, K.R.** 1974. Photoinactivation of ammonia oxidation in *Nitrosomonas. Journal of Bacteriology* 119:899–906.

Huber, D.M., Warren, H.L., Nelson, D.W., and **Tsai, C.Y.** 1977. Nitrification inhibitors—new tools for food production. *BioScience* 27:523–529.

Ida, S. and **Alexander, M.** 1965. Permeability of *Nitrobacter agilis* to organic compounds. *Journal of Bacteriology* 90:151–156.

Jones, RT.D. and **Morita, R.Y.** 1985. Low-temperature growth and whole-cell kinetics of a marine ammonium oxidizer. *Marine Ecology–Progress Series* 21:239–243.

Keeney, D.R. 1982. Nitrogen management for maximum efficiency and minimum pollution, pp. 605–649, in Stevenson, F.J. (editor), *Nitrogen in Agricultural Soils*, American Society of Agronomy, Crop Science Society of America, Soil Science Society of America, Madison, Wisconsin.

Keeney, D.R. 1986. Inhibition of nitrification in soils, pp. 99–115, in Prosser, J.I. (editor), *Nitrification, Special Publications of the Society for General Microbiology*, Volume 20, IRL Press, Oxford, Washington, DC.

Killham, K. 1986. Heterotrophic nitrification, pp. 117–126, in Prosser, J.I. (editor), *Nitrification*, Special Publications of the Society for General Microbiology, Volume 20, IRL Press, Oxford, Washington, DC.

Koops, H.-P. and **Harms, H.** 1985. Deoxyribonucleic acid homologies among 96 strains of ammonia-oxidizing bacteria. *Archives of Microbiology* 141:214–218.

Koops, H.-P., Harms, H., and **Wehrmann, H.** 1976. Isolation of a moderate halophilic ammonia-oxidizing bacterium, *Nitrosococcus mobilis* nov. sp. *Archives of Microbiology* 107:277–282.

Krümmel, A. and **Harms, H.** 1982. Effect of organic matter on growth and cell yield of ammonia-oxidizing bacteria. *Archives of Microbiology* 135:50–54.

Lipschultz, F., Zafiriou, O.C., Wofsy, F.C., McElroy, M.B., Valois, F.W., and **Watson, S.W.** 1981. Production of NO and N_2O by soil nitrifying bacteria. *Nature* 294:641–643.

Macdonald, R.M. 1986. Nitrification in soil: an introductory history, pp. 1–16, in Prosser, J.I. (editor), *Nitrification*, Special Publications of the Society for General Microbiology, Volume 20, IRL Press, Oxford, Washington, DC.

Martiny, H. and **Koops, H.-P.** 1982. Incorporation of organic compounds into cell protein by lithotrophic, ammonia-oxidizing bacteria. *Antonie van Leeuwenhoek Journal of Microbiology and Serology* 48:327–336.

Matin, A. 1978. Organic nutrition of chemolithotrophic bacteria. *Annual Review of Microbiology* 32:433–468.

Murray, R.G.E. and **Watson, S.W.** 1965. Structure of *Nitrosocystis oceanus* and comparison with *Nitrosomonas* and *Nitrobacter. Journal of Bacteriology* 89:1594–1609.

Nelson, D.W. and **Huber, D.M.** 1980. Performance of nitrification inhibitors in the midwest (east), pp. 75–88, in Meisinger, J.J., Randall, G.W., and Vitosh, M.L. (editors), *Nitrification Inhibitors—Potential and Limitations*, American Society of Agronomy, Soil Science Society of America, Madison, Wisconsin.

Olson, R.J. 1981. Differential photoinhibition of marine nitrifying bacteria: a possible mechanism for the formation of the primary nitrite maximum. *Journal of Marine Research* 39:227–238.

O'Neill, J.G. and **Wilkinson, J.F.** 1977. Oxidation of ammonia by methane oxidizing bacteria and the effects of ammonia on methane oxidation. *Journal of General Microbiology* 100:407–412.

Painter, H.A. 1986. Nitrification in the treatment of sewage and wastewaters, pp. 185–211, in Prosser, J.I. (editor), *Nitrification*, Special Publications of the Society for General Microbiology, Volume 20, IRL Press, Oxford, Washington, DC.

Poth, M. and **Focht, D.D.** 1985. ^{15}N kinetic analysis of N_2O production by *Nitrosomonas europaea*: an examination of nitrifier denitrification. *Applied and Environmental Microbiology* 49:1134–2241.

Remsen, C.C. and **Watson, S.W.** 1972. Freeze-etching of bacteria. *International Review of Cytology* 33:253–296.

Rittenberg, S.C. 1972. The obligate autotroph—the demise of a concept. *Antonie van Leeuwenhoek Journal of Microbiology and Serology.*

Schön, G. 1965. Untersuchungen über den Nutzeffekt von *Nitrobacter winogradskyi* Buch. *Archiv für Mikrobiologie* 50:111–132.

Schön, G. and **Engel, H.** 1962. Der Einfluß des Lichtes auf *Nitrosomonas europaea* Win. *Archiv für Mikrobiologie* 42:415–428.

Seewaldt, E., Schleifer, K.-H., Bock, E., and **Stackebrandt, E.** 1983. The close phylogenetic relationship of *Nitrobacter* and *Rhodopseudomonas palustris. Archives of Microbiology* 131:287–290.

Shaw, R.W. 1987. Luftverschmuzung durch Stäube, pp. 100–107, in *Spektrum der Wissenschaft* 10/87, Internationale Ausgabe in deutscher Sprache von Scientific American.

Slangen, J.H.G. and **Kerkhoff, P.** 1984. Nitrification inhibitors in agriculture and horticulture: a literature review. *Fertilizer Research* 5:1–76.

Smith, A.J. and **Hoare, D.S.** 1968. Acetate assimilation by *Nitrobacter agilis* in relation to its "obligate autotrophy." *Journal of Bacteriology* 95:844–855.

Smith, A.J. and **Hoare, D.S.** 1977. Specialist phototrophs, lithotrophs, and methylotrophs: a unity among a diversity of procaryotes? *Bacteriological Reviews* 41:429–448.

Söderlund, R. and **Svensson, B.H.** 1976. The global nitrogen cycle, pp. 23–73, in Svensson, B.H. and Söderlund, R. (editors), *Nitrogen, Phosphorus, and Sulphur-Global Cycles.* SCOPE Report 7, Ecological Bulletin 22, Stockholm.

Sone, N., Yanagita, Y., Hon-Nami, K., Fukomori, Y., and **Yamanaka, T.** 1983. Proton-pump activity of *Nitrobacter agilis* and *Thermus thermophilus* cytochrome *c* oxidase. *Federation of European Biochemical Societies Letters* 155:150–154.

Steinmüller, W. and **Bock, E.** 1976. Growth of *Nitrobacter* in the presence of organic matter. I. Mixotrophic growth. *Archives of Microbiology* 108:299–304.

Stevenson, F.J. 1986. *Cycles of Soil—Carbon, Nitrogen, Phosphorus, Sulfur, Micronutrients.* John Wiley and Sons, New York.

Sundermeyer, H. and **Bock, E.** 1981. Characterization of the nitrite-oxidizing system in *Nitrobacter*, pp. 317–324, in Bothe, H. and Trebst, A. (editors), *Biology of Inorganic Nitrogen and Sulfur,* Springer-Verlag, Berlin, Heidelberg, New York.

Sundermeyer-Klinger, H., Meyer, W., Warninghoff, B., and **Bock, E.** 1984. Membrane-bound nitrite oxidoreductase of *Nitrobacter*: evidence for a nitrate reductase system. *Archives of Microbiology* 140:153–158.

Tanaka, Y., Fukumori, Y., and **Yamanaka, T.** 1983. Purification of cytochrome a_1c_1 from *Nitrobacter agilis* and characterization of nitrite oxidation system of the bacterium. *Archives of Microbiology* 135:265–271.

Walker, N. and **Wickramasinghe, K.N.** 1978. Nitrification and autotrophic nitrifying bacteria in acid tea soils. *Soil Biology and Biochemistry* 11:231–236.

Ward, B.B. 1986. Nitrification in marine environments, pp. 157–184, in Prosser, J.I. (editor), *Nitrification,* Special Publications of the Society for General Microbiology, Volume 20, IRL Press, Oxford, Washington, DC.

Warrington, R. 1878. On nitrification. *Journal of the Chemical Society* 33:44–51.

Watson, S.W. 1965. Characteristics of a marine nitrifying bacterium, *Nitrosocystis oceanus* sp. n. *Limnology and Oceanography* (Supplement) 10:R274–R289.

Watson, S.W. 1971a. Taxonomic considerations of the family *Nitrobacteraceae* Buchanan. Requests for opinions. *International Journal of Systematic Bacteriology* 21:254–270.

Watson, S.W. 1971b. Reisolation of *Nitrosospira briensis* S. Winogradsky and H. Winogradsky 1933. *Archiv für Mikrobiologie* 75:179–188.

Watson, S.W. 1974. Family I. *Nitrobacteraceae* Buchanan, pp. 450–456, in Buchanan, R.E. and Gibbons, N.E. (editors), *Bergey's Manual of Determinative Bacteriology,* 8th Edition, The Williams and Wilkens Company, Baltimore.

Watson, S.W. and **Mandel, M.** 1971. Comparison of the morphology and deoxyribonucleic acid composition of 27 strains of nitrifying bacteria. *Journal of Bacteriology* 107:563–569.

Watson, S.W., Bock, E., Harms, H., Koops, H.-P., and **Hooper, A.B.** 1987. Family I. *Nitrobacteraceae* Buchanan 1917, in *Bergey's Manual of Determinative Bacteriology, 9th Edition,* Volume 3, The Williams and Wilkins Company, Baltimore (in press).

Watson, S.W., Bock, E., Valois, F.W., Waterbury, J.B., and **Schlosser, U.** 1986. *Nitrospira marina* gen. nov., sp. nov.: a chemolithotrophic nitrite-oxidizing bacterium. *Archives of Microbiology* 144:1–7.

Watson, S.W., Graham, L.B., Remsen, C.C., and **Valois, F.W.** 1971. A lobular, ammonia-oxidizing bacterium, *Nitrosolobus multiformis* nov. gen. nov. sp. *Archiv für Mikrobiologie* 76:183–203.

Watson, S.W. and **Mandel, M.** 1971. Comparison of the morphology and deoxyribonucleic acid composition of 27 strains of nitrifying bacteria. *Journal of Bacteriology* 107:563–569.

Watson, S.W., Valois, F.W., and **Waterbury, J.B.** 1981. The family *Nitrobacteraceae,* pp. 1005–1022, in Starr, M.P., Stolp, H., Trüper, H.G., Balows, A., and Schlegel, H.G. (editors), *The Prokaryotes. Volume 1. A Handbook on Habitats, Isolation, and Identification of Bacteria.* Springer-Verlag, Berlin, Heidelberg, New York.

Watson, S.W. and **Waterbury, J.B.** 1971. Characteristics of two marine nitrite oxidizing bacteria, *Nitrospina gracilis* nov. gen. nov. sp. and *Nitrococcus mobilis* nov. gen. nov. sp. *Archiv für Mikrobiologie* 77:203–230.

Wetzstein, H.G. and **Ferguson, F.J.** 1985. Respiration-dependent proton translocation and the mechanism of proton motive force generation in *Nitrobacter winogradskyi*. *Federation of European Microbiological Societies, Microbiology Letters* 30:87–92.

Winogradsky, S. 1890. Recherches sur les organismes de la nitrification. *Annales de l'Institut Pasteur* 4:213–231.

Winogradsky, S. 1891. Recherches sur les organismes de la nitrification. *Annales de l'Institut Pasteur* 5:577–616.

Winogradsky, S. 1892. Contributions à la morphologie des organismes de la nitrification. *Archives des Sciences Biologiques (St. Petersbourg)* 1:86–137.

Winogradsky, S. and **Winogradsky, H.** 1933. Etudes sur la microbiologie du sol. VII. La nitrification. Recherches nouvelles. *Annales de l'Institut Pasteur* 50:291–343.

Winslow, C.E.A., Broadhurst, J., Buchanan, R.E., Krumwiede, Jr., C., Rogers, L.A., and **Smith, G.H.** 1917. The families and genera of the bacteria. Preliminary report of the Committee of the Society of American bacteriologists on characterization and classification of bacterial types. *Journal of Bacteriology* 2:505–566.

Woese, C.R., Stackebrandt, E., Weisburg, W.G., Paster, B.J., Madigan, M.T., Fowler, V.J., Hahn, C.M., Blanz, P., Gupta, R., Nealson, K.H., and **Fox, G.E.** 1984a. The phylogeny of purple bacteria: the alpha subdivision. *Systematic and Applied Microbiology* 5:315–326.

Woese, C.R., Weisburg, W.G., Paster, B.J., Hahn, C.M., Tanner, R.S., Krieg, N.R., Koops, H.-P., Harms, H., and **Stackebrandt, E.** 1984b. The phylogeny of purple bacteria: the beta subdivision. *Systematic and Applied Microbiology* 5:327–336.

Woese, C.R., Weisburg, W.G., Hahn, C.M., Paster, B.J., Zablen, L.B., Lewis, B.J., Macke, T.J., Ludwig, W., and **Stackebrandt, E.** 1985. The phylogeny of purple bacteria: the gamma subdivision. *Systematic and Applied Microbiology* 6:25–33.

Wood, P.M. 1986. Nitrification as a bacterial energy source, pp. 39–62, in Prosser, J.I. (editor), *Nitrification*, Special Publications of the Society for General Microbiology, Volume 20, IRL Press, Oxford, Washington, DC.

Wullenweber, M., Koops, H.-P., and **Martiny, H.** 1978. Der Einfluß von Nitrit auf den Verlauf des Wachstums von *Nitrosomonas* Stamm Nm_1. *Mitteilungen des Instituts für Allgemeine Botanik Hamburg* 16:159–164.

Yoshida, T. and **Alexander, W.** 1970. Nitrous oxide formation by *Nitrosomonas europaea* and heterotrophic microorganisms. *Soil Science Society of America, Proceedings.* 34:880–882.

Yoshinari, T. 1985. Nitrite and nitrous oxide production by *Methylosinus trichosporium*. *Canadian Journal of Microbiology* 31:139–144.

6

Ecology of Phototrophic Purple and Green Sulfur Bacteria

Norbert Pfennig

Biology, University of Konstanz
Konstanz, Federal Republic of Germany

The purple and green sulfur bacteria are anaerobic phototrophic bacteria that carry out anoxygenic photosynthesis. In contrast to the oxygenic photosynthesis of algae and green plants, anoxygenic photosynthesis is dependent on exogenous substrates which provide the reducing equivalents for CO_2 assimilation. Hydrogen sulfide, elemental sulfur, or thiosulfate are the most common electron donors; in the light they become oxidized anaerobically to sulfate. The capacity to oxidize reduced sulfur compounds without oxygen consumption is ecologically and geobiochemically the most important feature of the purple and green sulfur bacteria. As a result, both groups find highly selective environmental conditions in anoxic hydrogen sulfide-containing fresh- or saltwaters where light is present.

The purple sulfur bacteria comprise two phylogenetically related families, the Chromatiaceae and the Ectothiorhodospiraceae, each of which includes closely related species and genera (Fowler et al., 1984; Stackebrandt et al., 1984). In contrast, the green sulfur bacteria, Chlorobiaceae, are not phylogenetically related to the other phototrophic sulfur bacteria nor to any other group of the Gram-negative eubacteria (Stackebrandt and Woese, 1981; Trüper, 1987). The species and genera of the Chlorobiaceae are physiologically and cytologically fairly similar but it is not yet known whether they are related among each other.

Phototrophic sulfur bacteria form conspicuous pink to purple red or green mass accumulations in stagnant water and on the mud surfaces of aquatic habitats. This is certainly one of the reasons for the attention paid to them during the last century. Winogradsky (1888) was the first investigator to present comprehensive taxonomic descriptions of the various morphological types that he encountered in samples from nature and raw cultures. Based on his own observations and the older literature, Bavendamm (1924) published the first detailed account on the ecology of the colorless and purple sulfur bacteria. The first defined enrichment and isolation techniques for certain species of purple and green sulfur bacteria were developed by van Niel (1931). He also demonstrated unequivocally the anaerobic, photosynthetic nature of these bacteria. An improved synthetic culture medium containing vitamin B_{12} was developed much later (Pfennig, 1967; Pfennig and Trüper, 1981); it provided rather nonselective growth conditions for most phototrophic sulfur bacteria of freshwater, estuarine, and marine habitats.

Using such culture media, many long-known and new species of purple and green sulfur bacteria were isolated in pure culture and studied in detail (Trüper and Pfennig, 1981). Only with this knowledge at hand were reliable and meaningful ecological studies feasible. Consequently, in the first part of this chapter, the species and genera of the purple and green sulfur bacteria known from pure culture studies will be presented, and their major ecologically significant characteristics will be emphasized. In the following parts, various natural habitats with their special physical and chemical properties will be described. It will be considered as much as possible how these properties constitute the selective conditions for certain species of the sulfur bacteria. A number of review articles on the ecology of the phototrophic bacteria have appeared during the past ten years (Pfennig, 1978; Pfennig and Trüper, 1981; van Gemerden and Beeftink, 1983; Gorlenko et al., 1983; Tindall and Grant, 1986; Lindholm, 1987) in which various aspects have been treated in more detail than in the present contribution.

6.1 SPECIES OF PURPLE AND GREEN SULFUR BACTERIA STUDIED IN PURE CULTURE

For various kinds of ecological studies it is necessary to designate the particular organisms that are encountered in a habitat. Due to the lack of pure cultures in former times, phototrophic bacteria were named and identified exclusively on the basis of morphological features. However, without knowledge of the major physiological characteristics of a particular microbe, little can be concluded from the presence of this microbe in a given habitat. Also, different metabolic types may be indistinguishable morphologically. Today, many species of purple and green sulfur bacteria from various different habitats can be grown in pure culture and well studied with respect to physiological, biochemical, and cytological features (Pfennig, 1977). We are in the good position, therefore, to present their species names assembled in groups based on certain features that are of major ecological significance (see Tables 6.1, 6.2, and 6.3). For full species descriptions the reader may consult Pfennig and Trüper (1974), Clayton and Sistrom (1978) or Trüper and Pfennig (1981).

For the species of all three families the following characteristics will be discussed briefly: a) salinity requirements, b) type of movement, c) capacity for chemotrophic growth, d) light absorption characteristics, and e) utilization of hydrogen sulfide.

Table 6.1 Purple sulfur bacteria of the family Chromatiaceae

Obligately anaerobic, phototrophic	Phototrophic and facultatively chemotrophic
1. Cells contain no gas vesicles	
Chromatium okenii	Chromatium violascens
Chromatium weissei	*Chromatium vinosum*
Chromatium warmingii	*Chromatium minus*
Chromatium buderi	*Chromatium gracile*
Chromatium purpuratum	*Chromatium salexigens* (halophilic)
Chromatium tepidum	*Thiocystis violacea*
Thiospirillum jenense	*Thiocapsa roseopersicina* (non)
Thiocapsa pfennigii (b) (non)	
2. Cells contain gas vesicles, planktonic development likely	
Lamprocystis roseopersicina	*Amoebobacter roseus* (non)
Thiodictyon elegans (non)	*Amoebobacter pendens* (non)
Thiodictyon bacillosum (non)	*Amoebobacter pedioformis* (non)
Thiopedia rosea (non)	*Amoebobacter purpureus* (non)
	Lamprobacter modestohalophilus

Only species studied in pure culture are indicated. Long wavelength absorption maxima of bacteriochlorophylls: bchl a (800–900 nm), bchl b (1020 nm). Most species contain bchl a; presence of bchl b is indicated (b). Flagellar motility is present in many species, nonmotile species are indicated (non). Species are found in either freshwater or marine habitats.

Table 6.2 Purple sulfur bacteria of the family Ectothiorhodospiraceae

Obligately anaerobic, phototrophic	Phototrophic and facultatively chemotrophic
Ectothiorhodospira halophila (11–32%)	*Ectothiorhodospira mobilis* (2–10%)
Ectothiorhodospira halochloris (b) (14–27%)	*Ectothiorhodospira shaposhnikovii* (1–7%)
Ectothiorhodospira abdelmalekii (b) (12–18%)	*Ectothiorhodospira vacuolata* (1–6%) (gv)

Only species studied in pure culture are indicated. All species contain bchl a; presence of bchl b is indicated (b). Flagellar motility is present in all species. Optimal salt concentration ranges are given in parentheses. Species are found in either marine or hypersaline and alkaline habitats. Species with gas vesicles are indicated by (gv).

Table 6.3 Green sulfur bacteria of the family Chlorobiaceae

Cells without gas vesicles	Cells contain gas vesicles, planktonic development likely
1. Species of freshwater habitats	
Chlorobium limicola	*Ancalochloris perfilievii*
Chlorobium phaeobacteroides (3)	*Pelodictyon luteolum*
Symbiotic consortia not in culture:	*Pelodictyon clathratiforme*
Chlorochromatium	
Pelochromatium (e)	
2. Species of marine and moderately saline habitats	
Chlorobium vibrioforme	*Pelodictyon phaeum* (e)
Chlorobium chlorovibriodes	*Chloroherpeton thalassium* (gm)
Chlorobium phaeovibrioides (e)	
Prosthecochloris aestuarii	
Prosthecochloris phaeoasteroidea (e)	

Only species studied in pure culture are given. Long wavelength absorption maxima of bacteriochlorophylls: bchl c (~750 nm), bchl d or e (710–730 nm). Most species contain bchl c or d, only bchl e is idicated (e). No flagellar motility, only one species is motile by gliding (gm). All species are strictly anaerobic and obligately phototrophic.

Occurrence in freshwater and saltwater The species of the Chromatiaceae and Chlorobiaceae characteristically thrive in freshwater, brackish water or seawater habitats (Table 6.1 and 6.3). Many species occur under all these conditions, while some species are known from freshwater only (e.g., *Chromatium okenii, Chlorobium limicola*) and others from seawater only (e.g., *Chromatium buderi, Prosthecochloris aestuarii*). In both families only one species, *Chromatium salexigens,* has so far been encountered that grows at salt concentrations much above seawater (10 percent NaCl; Caumette et al., 1988). In contrast, such saltwater environments are the common domain of the Ectothiorhodospiraceae (Table 6.2), several species of which typically thrive in hypersaline and alkaline salt or soda lakes with reddish water (Trüper and Imhoff, 1981).

Motility and presence of buoyancy-conferring gas vesicles Many species of the Chromatiaceae and all species of Ectothiorhodospira are motile by means of polar flagella. These species are capable of photo- and chemotactic responses to environmental conditions. As a result, vertical migrations occur that may result in cell accumulations at optimal growth conditions in sulfide and light gradients. Several species of the Chromatiaceae (e.g., *Thiocapsa* species) and all Chlorobiaceae (except the gliding *Chloroherpeton;* Gibson et al., 1984) are nonmotile. Some of these nonmotile species possess gas vesicles that make the cells buoyant in the water column of stagnant water bodies. In fact, most of the planktonic purple and green sulfur bacteria possess gas vesicles (e.g., *Thiopedia rosea, Amoebobacter* species, *Pelodictyon* species).

Capacity for chemotrophic growth in the dark and utilization of organic compounds While all species of the green sulfur bacteria (Table 6.3) are strictly anaerobic and obligately phototrophic, many species of the purple sulfur bacteria tolerate low oxygen concentrations and are facultatively chemoautotrophic. Sulfide, thiosulfate or sulfur are preferentially oxidized in energy conserving respiration (Kämpf and Pfennig, 1986; De Wit and van Gemerden, 1987), while organic substrates are only poorly oxidized if at all. This is a major difference from the purple nonsulfur bacteria, of which nearly all species are facultatively chemoorganotrophic with organic compounds serving as respiratory substrates.

The capacity for chemoautotrophic growth is, of course, ecologically significant since such species are metabolically more flexible than obligate phototrophic ones. They will occupy habitats with changing conditions (e.g., anoxic, oxic; Nicholson et al., 1987) and may continue to grow with oxygen in the absence of light.

Many species of the Chromatiaceae are potentially photoorganotrophic, while all green sulfur bacteria can assimilate organic substrates only if bicarbonate and sulfide are metabolized simultaneously.

In vivo ***light absorption characteristics*** In addition to species with bacteriochlorophyll a (bchl a) both families of purple sulfur bacteria include some species with bchl b. Whereas bchl a has *in vivo* long wavelength absorption maxima between 800 and 900 nm, bchl b exhibits a maximum at 1020 nm (Figure 6.1). This difference is ecologically as relevant as the differences among the long wavelength absorption maxima between 710 and 755 nm of the bchl's

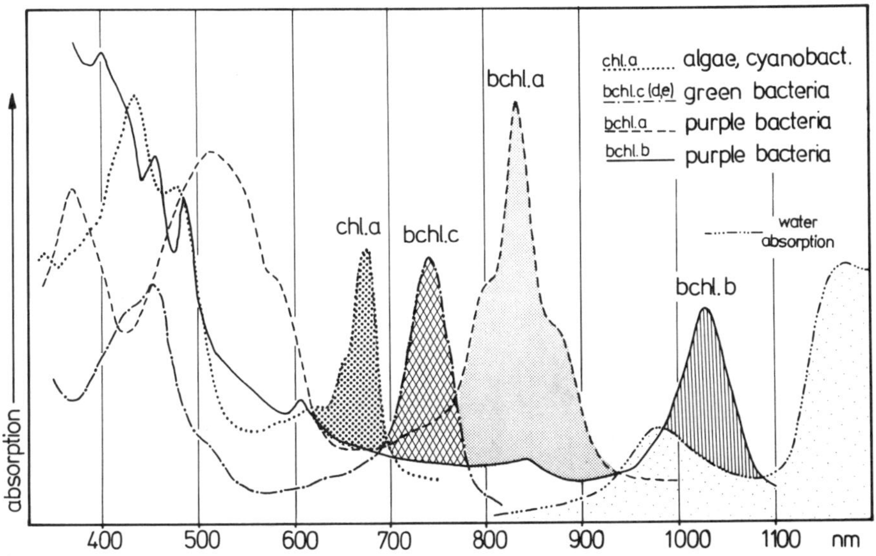

Figure 6.1 Absorption spectra of intact cells of the green alga *Chlorella* (chl a) and the phototrophic sulfur bacteria *Chlorobium limicola* (bchl c), *Chromatium okenii* (bchl a), and *Thiocapsa pfennigii* (bchl b).

c, d, and e of the green sulfur bacteria (Pierson et al., 1987). Algae and cyanobacteria with chlorophyll a absorb at 680 nm so that light absorption by these organisms in shallow waters or algal mats does not interfere with the radiation absorption of the purple and green bacteria. In deeper layers of lakes where only bluegreen radiation is available, light absorption by the carotenoid pigments is of primary importance, since infrared radiation is strongly absorbed in the uppermost water layers (Figure 6.1). The *in vivo* absorption spectrum of a green *Chlorobium* species with bchl c and the carotenoid chlorobactene is given in Figure 6.2. For comparison, the additional absorption contributed by isorenieratene carotenoids in a brown *Chlorobium* species with bchl e from a deep lake is indicated.

The green sulfur bacteria with their chlorosomes as light harvesting units are unique among all phototrophic organisms in that they require only about one-fourth of the light intensity of the purple bacteria in order to grow at comparable growth rates (Biebl and Pfennig, 1978). In natural habitats with various phototrophic organisms, the green sulfur bacteria regularly occur below the layers of all other phototrophs and closest to the sulfide zone.

Utilization of sulfide In the presence of sufficient sulfide and light, the phototrophic sulfur bacteria form globules of sulfur as an intermediate oxidation product of sulfide. The three families differ with respect to the location at which the globules occur: in the case of the Chlorobiaceae and Ectothiorhodospiraceae the sulfur globules arise outside the cells, while characteristically for the Chro-

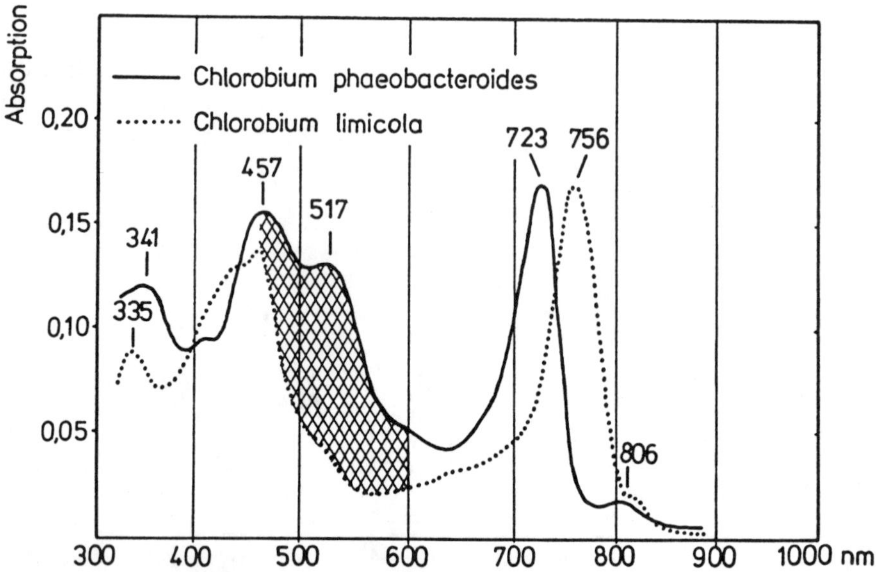

Figure 6.2 Absorption spectra of intact cells of the green bacterium *Chlorobium limicola* with bchl c and the carotenoid chlorobactene: ······, and of brown *Chlorobium phaeobacteroides* with bchl e and isorenieratene carotenoids: ———. The additional absorption region of the brown species is indicated.

matiaceae the globules form inside the cells. In all cases the sulfur globules are oxidized further to sulfate, particularly if sulfide is exhausted. Using chemostat cultures, van Gemerden (1984) determined both the saturation coefficient for sulfide at half maximum growth rate (K_s) as well as the maximum specific growth rate (μ_{max}) for various phototrophic sulfur bacteria. From these two data of each species, van Gemerden calculated the sulfide affinity as the quotient of $\mu_{max} \cdot K_s^{-1}$. Interestingly, the green sulfur bacteria and the Ectothiorhodospiraceae showed the highest affinities for sulfide (between 65 and $120 \cdot h^{-1} \cdot mM^{-1}$) while the Chromatiaceae exhibited lower affinities (between 2 and $18 \cdot h^{-1} \cdot mM^{-1}$). Apparently there is a correlation between the affinity for sulfide and the location of the sulfide oxidation sites on or inside the cells. The lower sulfide affinity of the Chromatiaceae may be compensated for by the fact that the cells carry the electron donor substrate inside their cells, thus being temporarily independent of exogenous sulfide.

6.2 FRESHWATER HABITATS

In nature, growth conditions for purple and green sulfur bacteria are established by the simultaneous presence of hydrogen sulfide and sunlight. Hydrogen sulfide has essentially two sources: 1) sulfide of geochemical origin present in warm and cold sulfur springs, and 2) sulfide formed as an endproduct of the dissimilatory sulfate reduction of sulfate-reducing bacteria. These bacteria are active in anoxic habitats as terminal decomposers of organic substances with sulfate as electron acceptor (Pfennig et al., 1984; Postgate, 1984; Widdel, 1988). Sulfate reduction may be limited by the availability of sulfate in freshwater habitats, while it is generally the dominant terminal oxidation process in anoxic water and in sediments rich in organic matter in brackish and seawater habitats with their high sulfate concentrations (up to 28 mM sulfate).

As a rule, both sulfide and sunlight occur in gradients of opposite directions: sulfide production is strongest in the sediments and may extend upward into the overlying anoxic water, while light intensities decrease strongly with depth in the water body or sediment. Wherever the two countercurrent gradients are overlapping, they provide suitable growth conditions for the phototrophic sulfur bacteria. All of the various habitats can be considered as variations on this general theme.

In lakes with medium and high productivity, the phototrophic sulfur bacteria may form blooms just below the oxic/anoxic transition zone of the stratified water mass. The productivity of such blooms has been determined by the $^{14}C\text{-}CO_2$ fixation technique. During the time of maximum development in summer, the daily photosynthetic production in 18 lakes studied ranged from 20 to 5000 mg $C \cdot m^{-3} \cdot day^{-1}$ (Wetzel, 1975; Biebl and Pfennig, 1979). These values accounted for 3 to 90 percent of the total production in these lakes. On the average the carbon assimilation was in the range of 100–200 mg $C \cdot m^{-3} \cdot day^{-1}$, accounting for about 20 to 40 percent of the total photosynthetic production of a lake.

Although the phototrophic sulfur bacteria use light as energy source and assimilate CO_2 into cell material, their biomass production must be considered secondary rather than primary. True primary production occurs only if the reducing equivalents for CO_2 assimilation are derived from water. The sulfide used by the phototrophic sulfur bacteria is formed, however, by dissimilatory sulfate reduction which depends on the degradation of organic substances that ultimately stem from primary production. In spite of its nature as secondary production, the phototrophic bacterial biomass may contribute to the whole productivity of a habitat. This is so because dissolved reduced substances that would otherwise be lost from the food web are converted into particulate organic matter that can reenter the food chain. Many authors have observed predation on phototrophic sulfur bacteria in stratified lakes by rotifers, copepods, and cladocerans (Caumette et al., 1983). The intestines of the zooplankters can be purple or red from ingested cells. Gophen et al. (1974) demonstrated that *Cereodaphnia* from Lake Kinneret preferentially ingested *Chlorobium phaeobacteroides* instead of algae as food source. Quantitative field studies of the grazing of cladocera on a bloom of *Chromatium okenii* in Lake Belovod were performed by Sorokin (1970). His data show that phototrophic sulfur bacteria may contribute significantly to zooplankton nutrition in a lake.

Sulfur springs In various hot spring areas all over the world a few springs exist that contain about 0.03–0.06 mM sulfide in pH 6–10 spring source water.

Due to the sulfide, thermophilic cyanobacteria that would otherwise thrive in the temperature range from 66°C to about 55°C are lacking here. Instead, mats of a dull deep green color on the surface develop that consist of a green colored *Chloroflexus* which differs from typical *Chloroflexus aurantiacus* in being unable to grow aerobically (Castenholz, 1984). Below the mat of several mm thickness, orange layers of *C. aurantiacus* develop. The genus *Chloroflexus* belongs to the multicellular filamentous gliding green bacteria with bacteriochlorophylls, the phototrophic flexibacteria, which are only very distantly related to the green sulfur bacteria (Stackebrandt and Woese, 1981).

In hot springs of Mammoth Terrace, Yellowstone National Park, the top green mat of *Chloroflexus* sp. was found to have a sulfide-dependent photoautotrophic metabolism (Castenholz, 1973; Madigan and Brock, 1977) while the orange *C. aurantiacus* mat appeared to be photoorganotrophic as in other hot springs.

In the same hot spring area of Yellowstone Park, Madigan (1986) discovered and isolated the first moderately thermophilic purple sulfur bacterium, *Chromatium tepidum*. This bacterium developed in thin reddish mats that were firmly embedded in carbonaceous sinter in the Stygian Springs at a temperature range between 40 and 50°C. Pure cultures of *C. tepidum* had a temperature optimum at 48–50°C; no growth was obtained below 34°C or above 57°C. A similar *Chromatium* was isolated by Madigan (1986) from a hot spring in New Mexico. At present no other species of the purple or green sulfur bacteria are known to thrive at temperatures above 40°C.

The outflow streams of sulfur springs below 30°C are the preferred habitats of chemotrophic colorless sulfur bacteria since both sulfide and oxygen are present in the well aerated stream water. The white trichome bundles of *Thiothrix* and *Beggiatoa* often densely cover the pebbles of the stream bed. In these aerobic habitats, purple and green sulfur bacteria are rarely found in visible accumulations although reddish layers may sometimes be detected well below the whitish trichome bundles.

Waste stabilization lagoons In rural areas of many countries various kinds of waste water are treated in so-called stabilization ponds or lagoons before the water is discharged into streams or rivers. Such shallow lagoons (0.5–2 m deep) may receive waste water from cattle feedlot (Wenke and Vogt, 1981), poultry manure, piggery waste, slaughterhouse waste water (Sletten and Singer, 1971), or fellmongery waste water (Cooper et al., 1975). Due to the high load of organic material, the lagoon water is aerobic only in the uppermost water layers. Most investigators have reported temporal changes in the pigmentation of the lagoons. During winter the water is brownish and turns more greenish in spring. Due to the subsequent mass development of purple sulfur bacteria, the water becomes pink to rose-red during summer and fall. The special conditions of the waste water lagoons, the high content of organic matter, pH values between 7.5 and 9, and high light intensities and temperatures in the upper water layers, are highly selective for certain species of the Chromatiaceae, e.g., *Thiocapsa roseopersicina, Amoebobacter roseus, A. pendens,* and *A. pediofor-*

mis (Eichler and Pfennig, 1986). Due to its growth in rectangular platelets and the possession of gas vacuoles, the latter species has often been mistaken for *Thiopedia rosea* (Wenke and Vogt, 1981), a very fastidious purple bacterium that is unable to multiply under the conditions prevailing in waste lagoons. The mass development of the purple sulfur bacteria in these sewage lagoons is highly desirable since the bacteria oxidize the hydrogen sulfide that has been formed by protein decomposition and dissimilatory sulfate reduction.

Stratified lakes Most freshwater lakes of the temperate climatic zones of the earth which are deeper than about 10 m show temperature- or density-stratification. During summer and fall only the warmer upper part of the water is well mixed by wind while the lower and colder part, the hypolimnion, remains almost stagnant. In the case of the so-called holomictic lakes, the whole water body becomes completely mixed in late fall or winter, while in meromictic lakes complete mixing is permanently prevented by the high specific density of the lower water body, usually caused by high solute concentration. In many cases the anoxic salt water may be relict seawater or recent seawater in lakes that are connected to the sea by a small inlet (Guerrero et al., 1987). Particularly in mesotrophic and eutrophic lakes, the oxygen-consuming degradation of organic substances as well as the oxidation of methane from the sediment cause oxygen depletion in the stagnant hypolimnion water (Anagnostidis and Overbeck, 1966). The subsequent anaerobic degradation processes result in the formation of sulfide by dissimilatory sulfate reduction. The sulfide concentration is highest in the sediment surface and gradually decreases toward the upper oxygen-containing water body. The oxic/anoxic transition zone, called the chemocline, is an interface of particular high microbial activity. In or just below the chemocline populations of Chromatiaceae, Chlorobiaceae or mixed populations of both families may bloom very profusely. This becomes apparent if water samples from that depth are brought to the lake surface. They may be pink to rose-red, brownish-red to magenta pink or more or less intensely green, depending which species is predominant (Trüper and Genovese, 1968; Caldwell and Tiedje, 1975; Takahashi and Ishimura, 1968; Parkin and Brock, 1981; Gorlenko et al., 1983).

Stratified lakes are of interest in two ways: 1) The anoxic hypolimnic water is nonturbulent and stagnant in holomictic lakes during summer stratification, in meromictic lakes permanently. 2) The spectral composition of the sunlight that reaches down into the chemocline is more or less restricted by scattering and absorption depending on the depth. As a result, these habitats are particularly selective for species of phototrophic bacteria with gas vesicles that confer buoyancy in nonturbulent water (Walsby, 1978). In addition, these habitats are also selective for species that can thrive in extremely dim light of the wavelength range between 450 and 550 nm and have relatively low temperature optima for growth (psychrotrophic species). The following species were often described to occur under these conditions (Figure 6.3): *Thiopedia rosea, Lamprocystis roseopersicina, Thiodictyon elegans, Amoebobacter purpureus* as well as *Chloronema giganteum, Chloronema spiroideum, Ancalochloris perfilievii* and *Pe-*

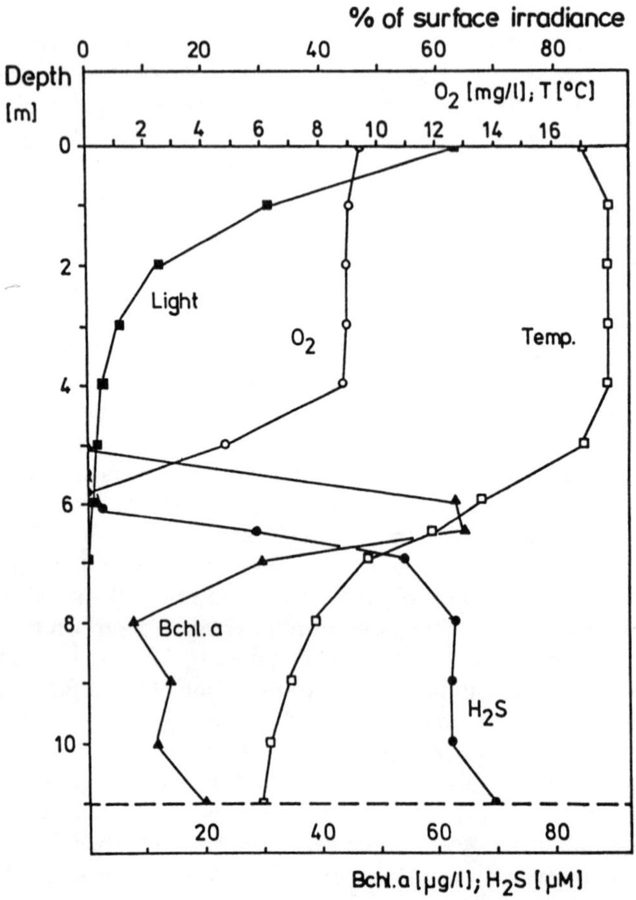

Figure 6.3 Physicochemical conditions and distribution of bacteriochlorophyll a in Lake Schleinsee (Lake Constance area) on September 11, 1985, during summer stagnation. The bchl a maximum at the chemocline between 5.5 and 6.5 m depth is caused by *Thiopedia rosea, Amoebobacter purpureus* and *Lamprocystis roseopersicina* (results from B. Eichler, unpublished).

lodictyon luteolum (Bavendamm, 1924; Utermöhl, 1925; Vetter, 1937; Anagnostidis and Overbeck, 1966; Czezuga, 1968; Dubinina and Gorlenko, 1975; Eichler and Pfennig, 1988).

Since oxygen, light, sulfide, and other factors occur in gradients within the colored water layer, it is not surprising that the phototrophic species are often unevenly distributed. The ecologically more flexible Chromatiaceae are usually found above the green sulfur bacteria (Caldwell and Tiedje, 1975; Caldwell, 1977; Gorlenko et al., 1983). Guerrero et al. (1987) pointed out that many species of purple sulfur bacteria that thrive in stratified holomictic and meromictic lakes at low light intensities have okenone as the main carotenoid. Such species are *Thiopedia rosea, Amoebobacter purpureus, Chromatium okenii,*

C. weissei, and *C. minus*. From pure culture studies we know that these species are in fact sensitive to light intensities which are optimal for growth of species with spirilloxanthin as the major carotenoid (Eichler and Pfennig, 1986, 1988). It remains to be determined whether this has something to do with a more efficient energy transfer from the carotenoids to bacteriochlorophyll by carotenoids like okenone with an aromatic ring at one end of the carbon chain. Among phototrophic bacteria, the only other carotenoids with aryl end groups are chlorobactene and the isorenieratenes (Liaaen-Jensen, 1978). These carotenoids occur in the green and brown Chlorobiaceae which are known to be best adapted to develop in extremely dim light (Biebl and Pfennig, 1978; Montesinos et al., 1983).

A number of relatively deep meromictic lakes have been found to be highly selective for the development of the brown-colored green sulfur bacterium *Chlorobium phaeobacteroides*. In these lakes, the chemocline occurs at 10 to 25 m depth and, therefore, the bacteria must be able to grow at exceptionally low light intensities (about 0.3 $\mu Ein \cdot m^{-2} \cdot s^{-1}$) that are too low for any other phototrophic organism (Pfennig, 1968; Trüper and Genovese, 1968; Culver and Brunskill, 1969; Dubinia et al., 1973; Matheron and Baulaigue, 1977; Bergstein et al., 1979; Veldhuis and van Gemerden, 1986). Due to the presence of the light-harvesting carotenoids isorenieratene and β-isorenieratene (Liaaen-Jensen, 1978) the bchl e-containing brown *C. phaeobacteroides* has a broad absorption band between 450 and 550 nm (Figure 6.2). This is exactly the wavelength range of radiation that penetrates deepest in the water column (Pfennig, 1967; Culver and Brunskill, 1969) and can effectively be absorbed and utilized by the brown Chlorobiaceae.

Montesinos et al. (1983) studied the ratio between the green *Chlorobium limicola* and the brown *Chlorobium phaeobacteroides* in eight Spanish lakes. They included their results in a list of altogether 30 freshwater lakes from around the world in which the relation between the dominant species of the Chlorobiaceae and the depth at which they developed is presented. From their compilation it is quite obvious that in lakes with Chlorobiaceae blooms near the surface (2–4 m), the green Chlorobiaceae are dominant. If the blooms occurred in 5–9 m depth, green and/or brown species developed, while in lakes with blooms between 9 and 25 m depth the brown Chlorobiaceae occurred almost exclusively. For Lake Cisó, Montesinos et al. (1983) recognized that a developing bloom of the purple sulfur bacterium *Chromatium* near the lake surface caused a decrease of the *Chlorobium* population in the lower part of the lake due to the strong light absorption in the carotenoid region. As expected, the green *C. limicola* population was less strongly affected by the shading effect than the brown *C. phaeobacteroides* population. All these field data together with laboratory experiments show convincingly that light quality selectively controls the species composition of green and brown Chlorobiaceae in stratified lakes. It is also apparent that in lakes of some depth the remarkable differences in the long wavelength absorption maxima above 700 nm of the groups of phototrophic sulfur bacteria (see Figure 6.1) are virtually insignificant as compared to the *in vivo* absorption characteristics in the wavelength range between 400 and 600 nm. It will be shown later that this is quite different in shallow multilayered microbial mats.

6.3 SALTWATER ENVIRONMENTS

The sulfate concentration of seawater is close to 28 mM, about 200 times higher than usually present in freshwater habitats (~150 μM). This is certainly one of the reasons why dissimilatory sulfate-reducing bacteria are so prominently represented as terminal oxidizers of organic substances in anoxic marine habitats. As soon as plant material starts to decay, hydrogen sulfide production becomes noticeable even under aerobic conditions. While in organic rich freshwater habitats sulfide production is confined to anoxic stagnant water and sediments and is generally limited by the availability of sulfate, the opposite is true for all coastal marine environments. Here, sulfate reduction will almost exclusively be limited by the presence of degradable organic matter. And if photosynthetic production is high, as in shallow brackish lagoons in summer (e.g., at the French Mediterranean coast; Caumette, 1986), the decay of algal blooms (Ulva, Enteromorpha spp.) can cause a dystrophic crisis or "malaigues" with completely anoxic, sulfide-containing red water loaded with purple sulfur bacteria. Other more or less productive marine habitats are isolated sheltered beaches, coastal lagoons, estuaries of the salt marsh type, regular inter- and supra-tidal salt marshes and intertidal sand flats. In all these marine environments, pink to purple-red mass developments of phototrophic sulfur bacteria have been observed on decaying algae or on and above sulfide-rich sediments. Such habitats, which are dominanted by the microbes of the sulfur cycle, have been called sulfureta (Jørgensen, 1982).

Microbial mats of salt marshes Protected from the open sea by beach dunes but connected to the sea by tidal channels and streams, salt marshes may extend over quite large and flat areas of land. In drier parts, marsh eelgrass (*Spartina*) covers the sandy or silty flats, while the lower sand flats may get submerged in several inches of seawater during occasional high tides. Such sandy or silty flats are the preferred places for the development of microbial mats. The mats consist of vertically laminated benthic communities of photosynthetic microorganisms. Generally the mat thickness does not exceed 5–10 mm. The most important and stabilizing component for the development of a mat are filamentous gliding cyanobacteria. If the cyanobacterial layer is well developed it has a high sediment-binding capacity by the abundant filaments and their extracellular polymers (slime) and eventually forms a leathery consistency. Thus, the surface of the sand flat becomes protected against erosion by wind and high tidal seawater. During summer and early fall, the original cyanobacterial mat may have developed into a multilayered mat system between the anoxic, sulfide-producing sediment and the fully aerobic and sunlight-exposed surface layer. Steep physiochemical gradients occur in these mats which profoundly influence the spatial distribution of the individual physiological groups of microbes that compose the mat system. Since the various groups of microbes are characteristic in color, the laminated ecosystem was called (in German) "Farbstreifen-Sandwatt" (Schulz, 1936; Stal et al., 1984, 1985). In the multi-layered mat, dense populations of diatoms, cyanobacteria, purple and green sulfur bacteria coexist and compete for light, CO_2, and sulfide (Jørgensen, 1982). Depending on the habitat, filamentous sulfide-oxidizing bacteria like *Beggiatoa* and/or mesophilic *Chloroflexus*-type green bacteria may be present in varying numbers between the cells of the other groups. The following example of a fully developed multilayered microbial mat was given by Nicholson et al. (1987) and Pierson et al. (1987) and is based on detailed studies of a mat at Great Sippewissett Salt Marsh, Cape Cod, Massachusetts, U.S.A. (see Figure 6.4).

The surface layer (1) was yellowish-brown in color and contained mostly diatoms of the genus *Navicula* and cyanobacteria of the genera *Lyngbya, Nostoc* and *Phormidium*. The subsequent blue-green layer (2) was dominated by large filaments of *Oscillatoria;* in addition *Phormidium* and *Microcoleus* (infrequently) were observed. Below these two layers of chlorophyll a-containing organisms with oxygenic photosynthesis, was a central pink layer (3) of bchl a-containing purple sulfur bacteria with sulfur globules inside the cells. Two nonmotile species dominated: *Thiocapsa roseopersicina* and *Amoebobacter roseus*. Below the pink layer followed a distinctive peach-colored or orange-brown layer (4) that consisted of the bchl b-containing *Thiocapsa pfennigii*. The lowermost yellowish green to olive-green layer (5) of phototrophic bacteria was very thin and not always discernible; it was formed by the green sulfur bacterium *Prosthecochloris aestuarii*. The sandy sediment below the green layer was dark gray to blackish from precipitated iron sulfide.

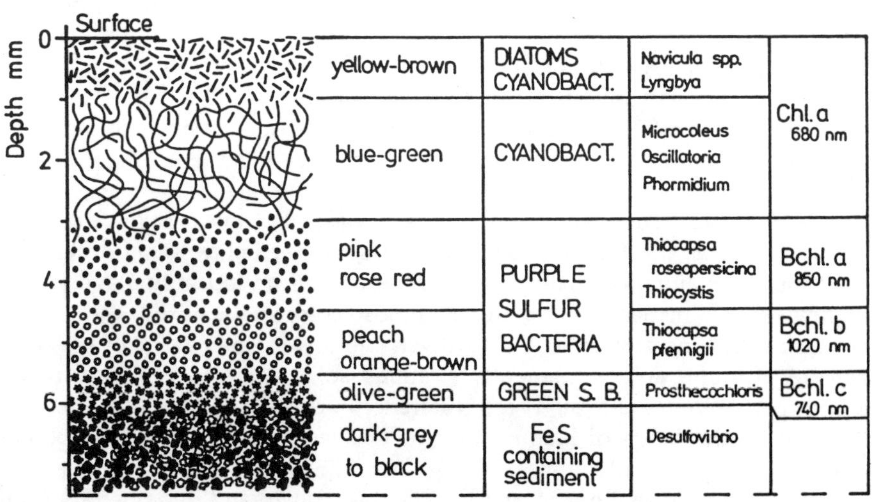

Figure 6.4 Schematic cross-section through the laminated microbial mat at Great Sippewissett Salt Marsh; modified after Nicholson et al. (1987). Explanation in the text.

The order of the described 5 layers of vertically stratified groups of microbes can be understood if the relationships of the organisms toward oxygen, sulfide, and light are considered (Figure 6.4). Layers 1 and 2 contain aerobic photosynthetic diatoms and cyanobacteria with oxygen-evolving photosynthesis. The diatoms are known to be least sensitive to direct sunlight; they thrive topmost and shield the more sensitive cyanobacteria from photooxidative damage. Layers 3–5 consist of purple and green sulfur bacteria that use sulfide for anoxygenic photosynthesis, indicated by the presence of sulfur globules inside the cells (Nicholson et al., 1987). In layer 3 both *Thiocapsa roseopersicina* and *Amoebobacter roseus* are facultatively chemotrophic species that can tolerate and even consume oxygen from layer (see Table 6.1). In contrast, the sulfur bacteria of layers 4 and 5 and obligately anaerobic, phototrophic and sulfide-dependent. They thrive next to the sulfide-producing sediment and are protected from oxygen by the facultatively aerobic purple bacteria of layer 3. If we now consider the light intensities and the spectral light penetration in the mat, the measurements of Jørgensen and Des Marais (1986) are of particular interest. These authors determined spectral light gradients in similar microbial mats at Guerrero Negro, Baja California, using their fiber optic microprobe. From their data, it is apparent that the thickness of layers 1 plus 2 is determined by the self-shading effect of the chlorophyll a organisms in the mat. Under layer 2 the intensity of the radiation below 700 nm is too low to support oxygenic photosynthesis. However, radiation above 700 nm and particularly above 1000 nm is still available, because the depth of light penetration into the sandy sediment increases with the wavelength of the radiation. Layer 3 with bchl a absorbs the radiation between 800 and 900 nm but weakens only little the radiation beyond 950 nm. Consequently, the bchl b-containing *Thiocapsa pfennigii* is able to thrive in layer 4. The lowermost *Prosthecochloris aestuarii* in layer 5, known to belong to the most efficient phototrophs, must be satisfied with the still available radiation between 700 and 800 nm (Figure 4 of Jørgensen and Des Marais, 1986). The multilayered microbial mat of Sippewissett salt marsh is a surprisingly good example of the fact that the long wavelength absorption maxima of the phototrophic organisms do in fact have a selective advantage in certain types of natural habitats.

Coastal lagoons There are various kinds of coastal lagoons or lakes. Common to all of them is the fact that they contain seawater below some depth and brackish water of various salinities up to freshwater on top. If the lagoons are shallow (0.5–2 m deep) the water is almost completely mixed and mostly brackish. With increasing depth, however, the lower water body of seawater remains stagnant and permanently stratified so that the lake is only partly mixed, perhaps meromictic. These differences are so profound that they are reflected by the kinds of blooms and the species composition of the phototrophic sulfur bacteria.

A very typical shallow coastal lagoon was studied and described by Caumette (1986). Like most water bodies of this type, Prevost Lagoon, France, turns anoxic during warm summers after the decay of blooming algae. At this time, the water is pink to reddish-brown due to purple sulfur bacteria. During the rest of the year the various kinds of purple bacteria thrive and survive in the upper sediment layers. Since these coastal lagoons are never much deeper than about 2 m, sufficient radiation penetrates to the bottom. The phototrophic sulfur bacteria are represented by a number of species which are quite characteristic for this type of environment: *Thiocapsa roseopersicina, Amoebobacter pendens, Chromatium vinosum, C. gracile, Thiocystis violacea,* and as minor components *Chromatium buderi* and the green sulfur bacteria *Chlorobium vibrioforme* and *Prosthecochloris aestuarii* (Trüper, 1970; Matheron and Baulaigue, 1972; Caumette, 1986; Eichler and Pfennig, 1986). In contrast to purple sulfur bacteria of stratified freshwater lakes, most of these species are facultatively chemotrophic, tolerate high light intensities and temperatures up to 40°C. Also, most species readily tolerate a wide and rapid variation of the salt concentrations from near freshwater to up to 6 percent NaCl.

Quite different from these shallow lagoons are the deeper meromictic coastal lakes where blooms of phototrophic sulfur bacteria occur. Typical examples are Lake Faro, Messina, Sicily (Trüper and Genovese, 1968), Lake Mogil'noe, Kil'din island, USSR (Gorlenko et al., 1978), the Spanish Mediterranean coastal lakes Massona, Cullera, and Remolar (Guerrero et al., 1978), and Bietri Bay of Ebrie Lagoon, Ivory Coast (Caumette, 1984). In these lakes, the blooms of phototrophic sulfur bacteria occurred almost permanently throughout the year at the interface between the stagnant anoxic and sulfide-containing seawater and the less salty overlying water. The dominant species of the reddish-brown to magenta-pink bloom water were *Chlorobium phaeobacteroides* (Lake Faro, Bietri Bay) and *C. phaeovibrioides* (Spanish coastal lakes, Lake Mogil'noe). Together with the brown *Chlorobium* species small numbers of *Chromatium violascens* and in Lake Massona *Thiocapsa roseopersicina* were found.

Hypersaline lakes As seawater has a rather uniform composition worldwide, coastal lagoons or lakes and brines in marine salterns also show a fairly uniform chemical composition (thalassohaline). In contrast, landlocked lakes in the arid climatic zones, such as desert lakes, may have a very different ion composition. The neutral salt lakes (e.g., Great Salt Lake, U.S.A.; Dead Sea, Israel) have a low bicarbonate content (<1 g/l), while the alkaline or soda lakes (e.g., Wadi Natrun, Egypt) contain high concentrations of carbonate which lead to magnesium- and calcium-free brines of pH values up to 11.

The purple and green sulfur bacteria that thrive in brackish or seawater habitats with pH values between 7 and 7.8 were mentioned above. If the salt concentrations and pH values are somewhat higher (up to 10 percent NaCl and pH 9), three species of the Ectothiorhodospiraceae (Table 6.2) may form the main population of purple sulfur bacteria in anoxic, sulfide-containing salt water habitats: *Ectothiorhodospira shaposhnikovii, E. vacuolata* and *E. mobilis* (Trüper and Imhoff, 1981). Although the optimal salt concentrations for pure cultures of the latter species are between 2 and 10 percent NaCl, this species was shown to occur in salt lakes with up to 25 percent salt (Imhoff et al., 1978). The two *Ectothiorhodospira* species with bchl b, *E. halochloris* and *E. abdelmalekii,* are moderately to extremely halophilic bacteria. Imhoff et al. (1979) reported that these green species occurred separately from the red species in greenish-colored brine pools of the Wadi Natrun. The most widely distributed red *Ectothiorhodospira* species in red water or red sediment layers is *E. halophila*. This species was found at salt concentrations between 9–32 percent and pH values between 7.5 and 11 in salt lakes in Oregon, U.S.A., and soda lakes in Egypt and the East African Rift Valley. Like in most halophilic or halotolerant Gram-negative eubacteria, the enzymes of *Ectothiorhodospira* species exhibit maximal activities at low ionic strength. In order to counterbalance the outer osmotic pressure, these species form organic osmoregulants that are compatible with the cellular enzymatic activities. Trüper and Galinski (1986) identified as the main compatible solute in the extreme halophilic species glycine betaine; in addition small amounts of trehalose and ectoin were also present.

Although other morphological types of purple sulfur bacteria (e.g., *Chromatium*-type cells) were encountered microscopically in hypersaline salt and soda lakes, only the *Ectothiorhodospira* species could so far be isolated and studied in pure culture (Imhoff et al., 1979).

Dedicated to Prof. Dr. H.-J. Elster on the occasion of his 80th birthday.

REFERENCES

Anagnostidis, K. and **Overbeck, J.** 1966. Methanoxidierer und hypolimnische Schwefelbakterien. *Berichte der Deutschen Botanischen Gesellschaft* 79:163–174.

Bavendamm, W. 1924. *Die farblosen und roten Schwefelbakterien des Süß- und Salzwassers.* Fischer Verlag, Jena, pp. 156.

Bergstein, T., Henis, Y., and **Cavari, B.Z.** 1979. Investigations on the photosynthetic sulfur bacterium *Chlorobium phaeobacteroides* causing seasonal blooms in Lake Kinneret. *Canadian Journal of Microbiology* 25:999–1007.

Biebl, H. and **Pfennig, N.** 1978. Growth yields of green sulfur bacteria in mixed culture with sulfur and sulfate reducing bacteria. *Archives of Microbiology* 117:9–16.

Biebl, H. and **Pfennig, N.** 1979. Anaerobic CO_2 uptake by phototrophic bacteria. A review. *Archiv für Hydrobiologie, Beihelft* (Ergebnisse der Limnologie) 12:48–58.

Caldwell, D.E. 1977. The planktonic microflora of lakes. *CRC Critical Reviews of Microbiology* 5:305–370.

Caldwell, D.E. and **Tiedje, J.M.** 1975. The structure of anaerobic bacterial communities in the hypolimnion of several Michigan lakes. *Canadian Journal of Microbiology* 21:377–385.

Castenholz, R.W. 1973. The possible photosynthetic use of sulfide by the filamentous phototrophic bacteria of hot springs. *Limnology and Oceanography* 18:863–876.

Castenholz, R.W. 1984. Composition of hot spring microbial mats: a summary, pp. 101–119, in Cohen, Y., Castenholz, R.W., and Halvorson, H.O. (editors) *Microbial mats: stromatolites*, Alan R. Liss, Inc., New York.

Caumette, P. 1984. Distribution and characterization of phototrophic bacteria isolated from the water of Bietri Bay (Ebrie Lagoon, Ivory Coast). *Canadian Journal of Microbiology* 30:273–284.

Caumette, P. 1986. Phototrophic sulfur bacteria and sulfate-reducing bacteria causing red waters in a shallow brackish coastal lagoon (Prévost Lagoon, France). *FEMS Microbiology Ecology* 38:113–124.

Caumette, P., Baulaigue, R., and **Matheron, R.** 1988. Characterization of *Chromatium salexigens* sp. nov., a halophilic Chromatiaceae isolated from Mediterranean salinas (Salins de Giraud, France). *Systematic and Applied Microbiology* 10:284–292.

Caumette, P., Pagano, M. and **Saint-Jean, L.** 1983. Répartition verticale du phytoplankton, des bacteries et du zooplancton dans un milieu stratifié en Baie de Bietri (Lagune Ebrié, Cote d'Ivoire). Relations trophiques. *Hydrobiologia* 106:135–148.

Clayton, R.K. and **Sistrom, W.R.** 1978. *The photosynthetic bacteria.* Plenum, New York, pp. 946.

Cooper, D.E., Rands, M.B., and **Woo, Ch.-P.** 1975. Sulfide reduction in fellmongery effluent by red sulfur bacteria. *Journal of the Water Pollution Control Federation* 47:2088–2100.

Culver, D.A. and **Brunskill, G.J.** 1969. Fayetteville Green Lake, New York. V. Studies of primary production and zooplankton in a meromictic marl lake. *Limnology and Oceanography* 14:862–873.

Czezuga, B. 1968. Primary production of the purple sulphuric bacteria, *Thiopedia rosea* Winogr. (Thiorhodaceae). *Photosynthetica* 2:161–166.

De Wit, R. and **van Gemerden, H.** 1987. Chemolithotrophic growth of the phototrophic sulfur bacterium *Thiocapsa roseopersicina. FEMS Microbiology Ecology* 45:117–126.

Dubinina, G.A. and **Gorlenko, V.M.** 1975. New filamentous photosynthetic green bacteria with gas vacuoles. *Microbiologiya* 44:511–517.

Dubinina, G.A., Gorlenko, V.M., and **Suleimanov, Ya.I.** 1973. A study of microorganisms of the manganese, iron and sulphur cycle in the meromictic lake Gek-Gel'. *Microbiologiya* 42:918–924.

Eichler, B. and **Pfennig, N.** 1986. Characterization of a new platelet-forming purple sulfur bacterium, *Amoebobacter pedioformis* sp. nov. *Archives of Microbiology* 146:295–300.

Eichler, B. and **Pfennig, N.** 1988. A new purple sulfur bacterium from stratified freshwater lakes, *Amoebobacter purpureus* sp. nov. *Archives of Microbiology* 149:395–400.

Fowler, V.J., Pfennig, N., Schubert, W., and **Stackebrandt, E.** 1984. Toward a phylogeny of phototrophic purple sulfur bacteria—16S rRNA oligonucleotide cataloging of 11 species of Chromatiaceae. *Archives of Microbiology* 139:382–387.

Gibson, J., Pfennig, N., and **Waterbury, J.B.** 1984. *Chloroherpeton thalassium* gen. nov. et spec. nov., a nonfilamentous, flexing and gliding green sulfur bacterium. *Archives of Microbiology* 138:96–101.

Gophen, M., Cavari, B.Z., and **Berman, T.** 1974. Zooplankton feeding on differentially labelled algae and bacteria. *Nature, London,* 247:393–394.

Gorlenko, V.M., Dubinina, G.A., and **Kuznetsov, S.I.** 1983. *The ecology of aquatic microorganisms.* Schweizerbart'sche Verlagsbuchhandlung, Stuttgart. pp. 252.

Gorlenko, V.M., Vainshtein, M.B., and **Kachalkin, V.I.** 1978. Microbiological characteristic of Lake Mogyl'noe. *Archiv für Hydrobiologie* 81:475–492.

Guerrero, R., Pedrós-Alió, C., Esteve, I., and **Mas, J.** 1987. Communities of phototrophic sulfur bacteria in lakes of the Spanish Mediterranean region. *Acta Academiae Aboensis* 47:125–151.

Imhoff, J.F., Hashwa, F., and **Trüper, H.G.** 1978. Isolation of extremely halophilic phototrophic bacteria from the alkaline Wadi Natrun, Egypt. *Archiv für Hydrobiologie* 84:381–388.

Imhoff, J.F., Sahl, H.G., Soliman, G.S.H., and **Trüper, H.G.** 1979. The Wadi Natrun: Chemical composition and microbial mass developments in alkaline brines of eutrophic desert lakes. *Geomicrobiology Journal* 1:219–234.

Jørgensen, B.B. 1982. Ecology of the bacteria of the sulphur cycle with special reference to anoxic-oxic interface environments. *Philosophical Transactions of the Royal Society London,* B. 298:543–561.

Jørgensen, B.B. and **Des Marais, D.J.** 1986. Competition for sulfide among colorless and purple sulfur bacteria in cyanobacterial mats. *FEMS Microbiology Ecology* 38:179–186.

Kämpf, Ch. and **Pfennig, N.** 1986. Chemoautotrophic growth of *Thiocystis violacea, Chromatium gracile,* and *C. vinosum* in the dark at various O_2-concentrations. *Journal of Basic Microbiology* 26:517–531.

Liaaen Jensen, S. 1978. Chemistry of carotenoid pigments, pp. 233–248, in Clayton, R.K. and Sistrom, W.R. (editors), *The photosynthetic bacteria.* Plenum Press, New York.

Lindholm, T. 1987. *Ecology of photosynthetic prokaryotes with special reference to meromictic lakes and coastal lagoons.* ÅBO Academy Press, ÅBO, Finland. pp. 151.

Madigan, M.T. 1986. *Chromatium tepidum* sp. nov., a thermophilic photosynthetic bacterium of the family Chromatiaceae. *International Journal of Systematic Bacteriology* 36:222–227.

Madigan, M.T. and **Brock, T.D.** 1977. CO_2 fixation in photosynthetically grown *Chloroflexus aurantiacus. FEMS Microbiology Letters* 1:301–304.

Matheron, R. and **Baulaigue, R.** 1972. Bactéries photosynthétiques sulfureuses marines. *Archiv für Mikrobiologie* 86:291–304.

Matheron, R. and **Baulaigue, R.** 1977. Influence de la pénétration de la luminère solaire sur le développement des bactéries phototrophes sulfureuses dans les environments marins. *Canadian Journal of Microbiology* 23:267–270.

Montesinos, E., Guerrero, R., Abella, C., and **Esteve, I.** 1983. Ecology and physiology of the competition for light between *Chlorobium limicola* and *Chlorobium phaeobacteroides* in natural habitats. *Applied and Environmental Microbiology 46: 1007–1016.*

Nicholson, J.A.M., Stolz, J.F., and **Pierson, B.K.** 1987. Structure of a microbial mat at Great Sippewissett Marsh, Cape Cod, Massachusetts. *FEMS Microbiology Ecology* 45:343–364.

Parkin, T.B. and **Brock, T.D. 1981.** The role of phototrophic bacteria in the sulfur cycle of a meromictic lake. *Limnology and Oceanography* 26:880–890.

Pfennig, N. 1967. Photosynthetic bacteria. *Annual Reviews of Microbiology* 21:285–324.

Pfennig, N. 1968. *Chlorobium phaeobacteroides* nov. spec. und *C. phaeovibrioides* nov. spec., zwei neue Arten der grünen Schwefelbakterien. *Archiv für Mikrobiologie* 63:224–226.

Pfennig, N. 1977. Phototrophic green and purple bacteria: A comparative, systematic survey. *Annual Reviews of Microbiology* 31:275–290.

Pfennig, N. 1978. General physiology and ecology of photosynthetic bacteria, pp. 3–18, in Clayton, R.K. and Sistrom, W.R. (editors), *The photosynthetic bacteria.* Plenum, New York, pp. 946.

Pfennig, N. and **Trüper, H.G.** 1974. The phototrophic bacteria, pp. 24–64, in Buchanan, R.E. and Gibbons, N.E. (editors), *Bergey's Manual of Determinative Bacteriology,* 8th ed., Williams and Wilkins Co., Baltimore, pp. 1246.

Pfennig, N. and **Trüper, H.G.** 1981. Isolation of members of the families Chromatiaceae and Chlorobiaceae, pp. 279–289, in Starr, M.P., Stolp, H., Trüper, H.G., Balows, A., and Schlegel, H.G. (editors), *The Prokaryotes.* Springer-Verlag, Berlin.

Pfennig, N., Widdel, F. and **Trüper, H.G.** 1984. The dissimilatory sulfate-reducing bacteria, pp. 926–940 in Starr, M.P., Stolp, H., Trüper, H.G., Balows, A. and Schlegel, H.G. (editors), *The Prokaryotes.* Springer-Verlag, Berlin.

Pierson, B., Oesterle, A., and **Murphy, G.L.** 1987. Pigments, light penetration, and photosynthetic activity in the multilayered microbial mats of Great Sippewissett Salt Marsh, Massachusetts. *FEMS Microbiology Ecology* 45:365–376.

Postgate, J.R. 1984. *The sulphate-reducing bacteria.* Second ed., University Press, Cambridge, pp. 208.

Schulz, E. 1936. Das Farbstreifen-Sandwatt und seine Fauna. *Kieler Meeresforschung* 1:359–378.

Sletten, O. and **Singer, R.H.** 1971. Sulfur bacteria in red lagoons. *Journal of the Water Pollution Control Federation* 43:2118–2122.

Sorokin, Yu.I. 1970. Interrelations between sulphur and carbon turnover in meromictic lakes. *Archiv für Hydrobiologie* 66:391–446.

Stackebrandt, E., Fowler, V.J., Schubert, W. and **Imhoff, J.F.** 1984. Towards a phylogeny of phototrophic purple sulfur bacteria—the genus *Ectothiorhodospira. Archives of Microbiology* 137:366–370.

Stackebrandt, E. and **Woese, C.R.** 1981. The evolution of prokaryotes, pp. 1–32, in Carlile, M.I., Collins, I.F., Moseley, B.E.B. (editors), *Molecular aspects of microbial evolution*, University Press, Cambridge.

Stal, L.J., Krumbein, W.E., und **van Gemerden, H.** 1984. Das Farbstreifen-Sandwatt—Ein laminiertes mikrobielles Ökosystem im Wattenmeer. *Veröffentlichungen der Naturforschenden Gesellschaft zu Emden* 1984, 7:1–60.

Stal, L.J., van Gemerden, H., and **Krumbein, W.** 1985. Structure and development of a benthic marine microbial mat. *FEMS Microbiology Ecology* 31:111–125.

Takahashi, M. and **Ishimura, S.** 1968. Vertical distribution and organic matter production of photosynthetic sulfur bacteria in Japanese lakes. *Limnology and Oceanography* 31:644–655.

Tindall, B.J. and **Grant, W.D.** 1986. The anoxygenic phototrophic bacteria, pp. 115–155, in Barnes, E.M. and Mead, G.C. (editors), *Anaerobic bacteria in habitats other than man.* Blackwell Science Publishing, Oxford, pp. 444.

Trüper, H.G. 1970. Culture and isolation of phototrophic sulfur bacteria from the marine environment. *Helgoländer wissenschaftliche Meeresuntersuchungen* 20:6–16.

Trüper, H.G. 1987. Phototrophic bacteria (an incoherent group of prokaryotes). A taxonomic versus phylogenetic survey. *Microbiologia SEM* 3:71–89.

Trüper, H.G. and **Galinski, E.A.** 1986. Concentrated brines as habitats for microorganisms. *Experientia* 42:1182–1187.

Trüper, H.G. and **Genovese, S.** 1968. Characterization of photosynthetic sulfur bacteria causing red water in Lake Faro (Messina, Sicily). *Limnology and Oceanography* 13:225–232.

Trüper, H.G. and **Imhoff, J.F.** 1981. The genus *Ectothiorhodospira*, pp. 274–278, in Starr, M.P., Stolp, H., Trüper, H.G., Balows, A., and Schlegel, H.G. (editors), *The Prokaryotes.* Springer-Verlag, Berlin.

Trüper, H.G. and **Pfennig, N.** 1981. Characterization and identification of the anoxygenic phototrophic bacteria, pp. 299–312, in Starr, M.P., Stolp, H., Trüper, H.G., Balows, A., and Schlegel, H.G. (editors), The Prokaryotes. Springer-Verlag, Berlin.

Utermöhl, H. 1925. Limnologische Phytoplanktonstudien. Die Besiedlung ostholsteinischer Seen mit Schwebpflanzen. *Archiv für Hydrobiologie* 5 (Suppl.):1–527.

van Gemerden, H. 1984. The sulfide affinity of phototrophic bacteria in relation to the location of elemental sulfur. *Archives of Microbiology* 139:289–294.

van Gemerden, H. and **Beeftink, H.H.** 1983. Ecology of phototrophic bacteria, pp. 146–185, in Ormerod, J.G. (editor), *The phototrophic bacteria: Anaerobic life in the light.* Blackwell Science Publishing, Oxford.

van Niel, C.B. 1931. On the morphology and physiology of the purple and green sulphur bacteria. *Archiv für Mikrobiologie* 3:1–112.

Veldhuis, M.J.W.. and **van Gemerden, H.** 1986. Competition between purple and brown phototrophic bacteria in stratified lakes: sulfide, acetate and light as limiting factors. *FEMS Microbiology Ecology* 38:31–38.

Vetter, H. 1937. Limnologische Untersuchungen über das Phytoplankton und seine Beziehungen zur Ernährung des Zooplanktons im Schleinsee bei Langenargen am Bodensee. *Internationale Revue der Gesamten Hydrobiologie und Hydrographie* 34:499–561.

Walsby, A.E. 1978. The gas vesicles of aquatic prokaryotes, pp. 327–357. *Symposium of the Society for General Microbiology* 28, Cambridge University Press.

Wenke, Th.L. and **Vogt, J.C.** 1981. Temporal changes in a pink feedlot lagoon. *Applied and Environmental Microbiology* 41:381–385.

Wetzel, R.G. 1975. *Limnology.* W.B. Saunders Co., Philadelphia. pp. 743.

Widdel, F. 1988. Microbiology and ecology of sulfate- and sulfur-reducing bacteria, in Zehnder, A.J.B. (editor), *Environmental microbiology of anaerobic bacteria*, Chapter 10. J. Wiley & Sons, New York.

Winogradsky, S. 1888. *Beiträge zur Morphologie und Physiologie der Bakterien*. Vol. 1. Zur Morphologie und Physiologie der Schwefelbakterien. Verlag A. Felix, Leipzig, pp. 120.

7

Biogeochemistry of Chemoautotrophic Bacteria

Bo Barker Jørgensen

Institute of Ecology of Genetics
University of Aarhus
Aarhus, Denmark

Chemoautotrophic bacteria play important roles in the biogeochemical cycles on Earth. From a geochemical point of view, they function as catalysts of reactions which would otherwise proceed slowly on an ecological or geological time scale. This kinetic effect has important geochemical consequences, e.g., by accelerating the formation of mineral deposits or the mobilization of elements from rocks. By catalyzing various specific reactions, the chemoautotrophs also affect the pathways and the coupling of the different element cycles, especially in relation to the carbon cycle. Chemoautotrophs are finally an additional source of "primary" production in the ecosystems, although this production is energetically derived from organic carbon which was fixed photosynthetically at an earlier stage (see Chapter 8).

Microbial environments are generally heterogeneous and unstable. Many or perhaps even most of the chemolithotrophic bacteria which oxidize inorganic compounds in nature are therefore not obligate autotrophs. Rather than depending on a narrow range of potential substrates, they have a more flexible metabolism which allows them to survive and grow in the presence of alternating substrate. These bacteria are facultative autotrophs or mixotrophs, alternating between organic and inorganic energy and carbon sources (see Chapter 4). Some are even obligate heterotrophs, using inorganic compounds only as supplementary energy sources. These chemolithoheterotrophs play similar ecological and geochemical roles as the true autotrophs by utilizing inorganic reduced compounds for their energy metabolism. Although they do not synthesize new organic carbon from CO_2, they incorporate the assimilated organic matter into their cells much more efficiently than do conventional heterotrophs. Lithoheterotrophy thus conserves biomass and makes it available for the bacterial grazers and the microbial food chains.

Chemoautotrophy, which leads to the synthesis of new cells from CO_2, would seem to be an effective mechanism of converting inorganic reducing power into biomass production for the food chains. With the exception of hydrogen-oxidizing bacteria, however, chemoautotrophic CO_2 assimilation has a low efficiency of energy conservation. Some 80–90 percent of the reducing power is spent on ATP production in order to generate reduced NADPH by reverse electron flow (see Chapter 11). Chemoautotrophic production is therefore of limited significance for the ecosystem as a whole. Only in special environments, such as the oxic/anoxic interface of stratified water bodies, may the chemoautotrophs be a local, major source of organic matter. A notable exception to this is the luxurious bacterial and animal life surrounding hot vents in the deep sea, where sulfide-rich hydrothermal fluid emanates from the ocean floor into the ambient, oxic seawater. These fascinating ecosystems, which are fueled by true chemoautotrophic primary production, are described in Chapter 8.

Reduced inorganic compounds, the potential substrates for chemoautotrophic bacteria, are mostly produced within anoxic environments in sediments, soil, or stagnant water bodies. Through molecular diffusion or convective transport these compounds come in contact with more oxidized species, which can function as electron acceptors in a chemical or biological oxidation reaction.

In some reactions, e.g., between ferrous ions and molecular oxygen, the chemical process at neutral pH is very rapid. In this case it is doubtful whether bacteria can effectively compete with the autocatalytic reaction and thus conserve sufficient energy for chemoautotrophic growth (see Chapter 4). Other compounds, such as ammonia, are more stable under oxic conditions and bacterial catalysis helps to speed up the oxidation reaction by many orders of magnitude. The chemoautotrophs thereby alter not only the rates but also the pathways of the nitrogen cycle. In the (hypothetical) absence of such nitrifying bacteria, nitrogen would mostly remain in a reduced state and cycle between free ammonia and organically bound nitrogen. This would have dramatic effects on the global nitrogen cycle. If biological nitrate formation stopped, N_2 formation from denitrification and thus also nitrogen fixation would be affected accordingly. The atmospheric nitrogen level would depend more on photochemical reactions in the stratosphere between ozone, hydroxyl radicals, and ammonium, which had been emitted into the air from terrestrial and marine ecosystems. Ultraviolet (UV) radiation and atmospheric chemistry could presumably substitute for the chemolithotrophs in many reactions, both for nitrogen and several other elements with gaseous products, but the rates of cycling would be slowed significantly due to slow vertical transport in the troposphere.

A fraction of the potential substrates for chemoautotrophs never reach into more oxidized environments but are buried within soils and sediments. These inorganic species are sources of mineral deposits in sedimentary rocks. Over geological periods, deposits such as pyrite in coal or metal-rich sulfide ores may form. Such deposits may subsequently become exposed and available again to the chemoautotrophs through uplifting, folding, and weathering of rock material or through draining and mining activities. Some of the mineral formations which are exploited today were laid down during the early Precambrian period, even before photosynthetic oxygen evolution had established an oxic atmosphere on Earth.

The following discussion will focus on the role of chemoautotrophic bacteria in mineral cycling of modern ecosystems. The quantitatively most important pathways will be emphasized. Examples will be taken mostly from aquatic environments which are less heterogeneous than terrestrial systems and where the geochemical zonation of lithotrophic reactions is consequently better described.

7.1 ANAEROBIC MINERALIZATION

Reducing power which provides the energy source for chemoautotrophy is originally derived from photosynthetically reduced carbon. When the dead organic matter from the photoautotrophic ecosystems is deposited and becomes buried in anoxic soils and sediments, it is degraded through anaerobic microbial food chains. The small organic molecules which are released from complex detrital structures are gradually oxidized to CO_2 through pathways of fermentation and anaerobic respiration. It is characteristic for these anaerobic food chains that each physiological type of bacterium and each step during the degradation process utilizes only a fraction of the energy of the substrate. The remaining potential chemical energy is left in the excreted metabolic products which are assimilated and further metabolized by other bacteria. It is also characteristic of anaerobic food chains that the pathways of carbon and energy become partly separated during the degradation. This is due to electron transfer into inorganic ions such as H^+ or SO_4^{2-} which are reduced to H_2 and H_2S, respectively. These reduced inorganic compounds become the energy substrates of chemolithotrophic bacteria.

Metabolic zonations The chemical and microbial zonations in marine sediments provide a good example of the sequence of metabolic reactions through which the chemolithotrophic substrates are formed (Figure 7.1). Below the oxic surface layer, dissolved nitrate functions as a terminal electron acceptor for denitrifying bacteria. The partially reduced product, N_2, is not used as an energy source by chemolithotrophs, however, because of the high activation energy required to break the triple bond between the nitrogen atoms. Dissimilatory reduction of nitrate may also lead to formation of ammonia by two different types of bacterial metabolism. Nitrate is used by some fermenters, e.g., clostridia, as an electron sink to reoxidize NADH (Hasan and Hall, 1975). Reduction of nitrite to ammonia is catalyzed by many bacteria. A few of the known species of sulfate-reducing bacteria can use nitrate as an alternative electron acceptor for respiration (Keith and Herbert, 1983; Mitchel et al., 1986; Seitz and Cypionka, 1986). The ecological role of these types of dissimilatory ammonia production is not clear, but several studies of nitrate reduction in natural environments have demonstrated the simultaneous formation of N_2 and NH_4^+,

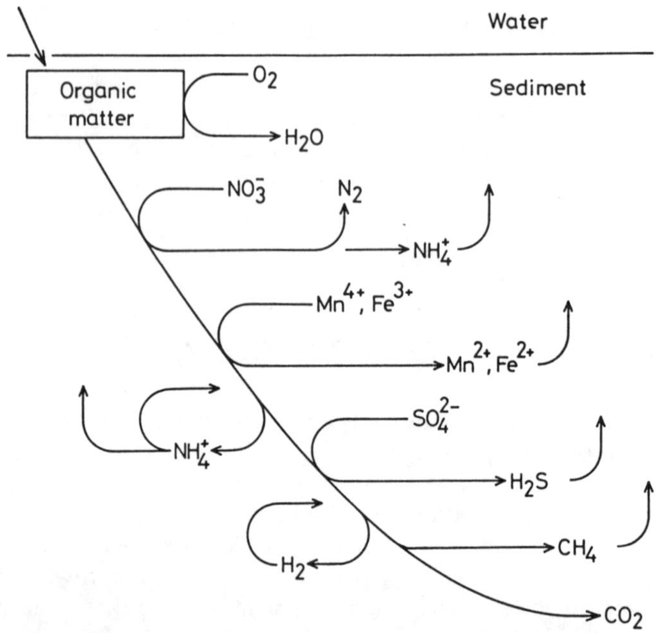

Figure 7.1 Vertical zonation of the dominant mineralization processes in a sediment. With depth and time, as the organic matter is buried deeper in the anoxic environment, a sequence of electron acceptors are used. As these ultimately become depleted, the anaerobic pathways lead to methanogenesis. Ascending arrows indicate reduced inorganic compounds which accumulate and migrate upward along a diffusion gradient. As they meet suitable electron acceptors, these compounds become potential substrates for chemolithotrophs. Hydrogen is generally recycled where it is produced. Ammonia production on the left side of the figure is due to degradation of organic nitrogen compounds.

with the latter accounting for 4–60 percent of the total nitrate reduction (Koike and Hattori, 1978; Sørensen, 1978; MacFarlane and Herbert, 1984; King and Nedwell, 1985; Jørgensen and Sørensen, 1985). Ammonification from nitrate seems to take place under more reducing conditions than denitrification.

Most of the ammonia production in anaerobic systems is due to the degradation of organic nitrogen compounds such as proteins and nucleic acids (Figure 7.1). The net amount of ammonia released during the bacterial mineralization process depends on the relative composition of carbon and nitrogen in the degraded material. This is due to the biomass production of bacteria which grow on organic matter. If the C:N ratio of the substrate is high, greater than 20, there may be no net release of ammonia but rather a net uptake due to assimilation by the bacteria. Tracer studies using ^{15}N-labeled ammonia have shown that the balance of this recycling in sediments may vary seasonally between net release and net uptake (Blackburn, 1980). Although such ammonia becomes sorbed to clay particles, it also accumulates in the pore water and gradually diffuses up to oxic surface layers where it is utilized by nitrifying bacteria.

Manganese and iron minerals also become reduced by heterotrophic bacteria and the reduced metals, being more soluble, become mobilized for further cycling in the environment. Manganese reductase activity is present in many different bacteria which use manganese oxides as electron acceptors in a respiratory metabolism or just as an electron sink (Trimble and Ehrlich, 1968; Burdige and Nealson, 1985). In sulfide-rich environments, manganates and ferric compounds are reduced chemically by reaction with H_2S produced by sulfate-reducing bacteria (Berner, 1970; Godhaber and Kaplan, 1974; Burdige and Nealson, 1986). Many heterotrophic bacteria and fungi can reduce iron (Ottow, 1969; Ottow and Klopotek, 1969), which in some species is an alternative electron acceptor for nitrate reductase (Ottow, 1968). The formation of ferrous iron thus becomes inhibited by addition of nitrate in marine sediments. This is consistent with the observation that the maximal Fe^{2+} concentration occurs below the NO_3^- maximum (Sørensen, 1982; Lyle, 1983; Jørgensen and Sørensen, 1985). In the presence of abundant ferric iron, iron reduction may even be a dominant pathway of organic matter oxidation in sediments (Lovley and Phillips, 1986, 1987; Aller et al., 1986). Iron can also be reduced by sulfur- or hydrogen-oxidizing bacteria (Brock and Gustafson, 1976; Balashova and Zavarzin, 1980). The metal transformations are thus closely coupled to the major ecological element cycles.

Sulfate reduction to H_2S is the terminal process in anaerobic pathways of degradation in marine sediments as well as in surface layers of freshwater sediments and soils. Due to the high sulfate concentration in seawater, 2.7 g SO_4^{2-} per liter, sulfate generally penetrates a meter or more down into marine sediments (Goldhaber and Kaplan, 1974). The sulfate-reducing bacteria are responsible for up to 50 percent of the total mineralization of organic matter in coastal marine sediments (Jørgensen, 1982b). The sulfate zone is usually only a few cm deep in lake sediments because the sulfate concentration is 2–3 orders of magnitude lower in freshwater than in seawater. Sulfate reduction is consequently of less significance than methanogenesis in most lakes (Ingvorsen and Brock, 1982; Lovley and Klug, 1982).

Role of sulfide In general, only 5–20 percent of the H_2S produced in marine sediments becomes trapped and permanently buried due to reaction with iron and other metals (Jørgensen, 1977). Pyrite, FeS_2, is the dominant sulfur mineral. Due to sulfate limitation in freshwater, lake sediments have a lower content of reduced sulfur than marine sediments in which the sulfur content is rather constant relative to the organic carbon (mean C:S ratio = 2.8 by weight; Burner and Raiswell, 1983). The burial rate of reduced carbon and sulfur in the sea bed is important for the global biogeochemical cycles because it balances the buildup of O_2 in the atmosphere (Garrels and Lerman, 1984; Berner, 1987):

Carbon: $CO_2 + H_2O \rightarrow CH_2O + O_2$

Sulfur: $2Fe_2O_3 + 16Ca^{2+} + 16HCO_3^- + 8SO_4^{2-} \rightarrow$
$FeS_2 + 16CaCO_3 + 8H_2O + 15O_2$

When pyrite-rich sedimentary rocks become weathered and exposed to oxygen later during the geological cycle, the reverse processes take place. Pyrite is then reoxidized by lithotrophic bacteria, often resulting in acidification of the environment. The iron is here of little importance relative to sulfur as an electron reservoir and energy source. Since iron oxidation involves only a 1 electron shift (oxidation step +2 to +3) while the sulfur oxidation involves a 7 electron shift (−1 to +6), there is 1 oxidation equivalent of Fe and 14 oxidation equivalents of S in FeS_2.

The large amounts of sulfide, which are not permanently buried, ultimately become exposed to more oxidized environments and may thus become available to sulfide-oxidizing bacteria. Howarth (1984) has made an estimate of how much energy from organic matter, produced in coastal ecosystems, flows through the sulfur cycle and potentially becomes available to sulfur bacteria: 1) The fraction of organic carbon which is oxidized by sulfate-reducing bacteria is 25–50 percent of the total organic input to the sediment; 2) The net burial of reduced sulfur is 7.5–20 percent of the total produced H_2S (i.e., 80–92.5 percent of the H_2S is reoxidized); 3) The free energy of the organic substrate, which becomes available by sulfate reduction, is 25 percent relative to the free energy by complete oxidation with O_2 (i.e., 75 percent of the energy is stored in H_2S); 4) Thus, from the total degradation of organic matter in the sediment, 15 to 35 percent ($25 \times 0.80 \times 0.75$ to $50 \times 0.925 \times 0.75$) of the energy is released upon reoxidation of sulfide at the sediment surface.

In the absence of respiratory electron acceptors, methanogenesis is the terminal process in the anaerobic degradation of organic matter. In most ecosystems, the reoxidation of methane is therefore a very important process. Similar to ammonia and sulfide, methane diffuses up through the sediment and carries energy from the reduced environment. Although of much less quantitative significance, several other organic and inorganic C_1 compounds are produced and degraded in the environment; e.g., carbon monoxide, methylated mercaptans, and methylamines.

7.2 RAPID HYDROGEN RECYCLING

Hydrogen (H_2) has a unique position among the chemolithoautotrophic substrates. Together with sulfide, it is the most important inorganic energy carrier in anaerobic pathways in nature. In contrast to sulfide and the other lithotrophic substrates, however, hydrogen is an excellent electron donor in all environments, even under the most reducing conditions. H_2 therefore does not accumulate but is rapidly recycled and remains at very low concentrations in soils, sediments, and aquatic environments. The typical interface environments, where an inorganic electron donor meets an appropriate electron acceptor in opposed diffusion gradients, are therefore not important for H_2 cycling.

Distribution and turnover H_2 can be detected in all environments including oxic waters and the atmosphere, given sufficiently sensitive analytical techniques. H_2 concentrations in oxic ocean water, 0.05–0.5 nM (Schropp et al., 1987), are even below the atmospheric equilibrium value. In lake water, the H_2 level is 10–100-fold higher (Conrad et al., 1983; Figure 7.2), while in marine and freshwater sediments concentrations of 5–30 nM and 1–>150 nM, respectively, have been measured (Strayer and Tiedje, 1978; Phelps and Zeikus, 1984; Conrad et al., 1985; Novelli et al., 1987). H_2 levels are generally lower in marine sediments than in freshwater sediments and lowest, <1 nM, in deep-sea sediments (Novelli et al., 1987).

Due to the low concentrations of H_2, it has been difficult to measure its turnover in natural environments. Several of the experimental setups in which vials incubated contained a headspace have probably underestimated the turnover rate because of delayed H_2 equilibration over the gas/water interface. In a system without a headspace, Conrad et al. (1985) estimated a turnover time in anaerobic lake sediment of about 2 minutes.

Hydrogen production Hydrogen is first of all a product of the main anaerobic fermentation pathways. It may also be produced anaerobically by cyanobacteria or by heterotrophic bacteria as a side reaction of the nitrogenase enzyme system. Root nodules are thus a source of hydrogen in soil. It is not clear how H_2 is produced in oxic waters, although it may be produced within anaerobic microenvironments such as the gut of zooplankton and fish or in fecal pellets and marine snow particles. Geothermally produced H_2 is only of biological

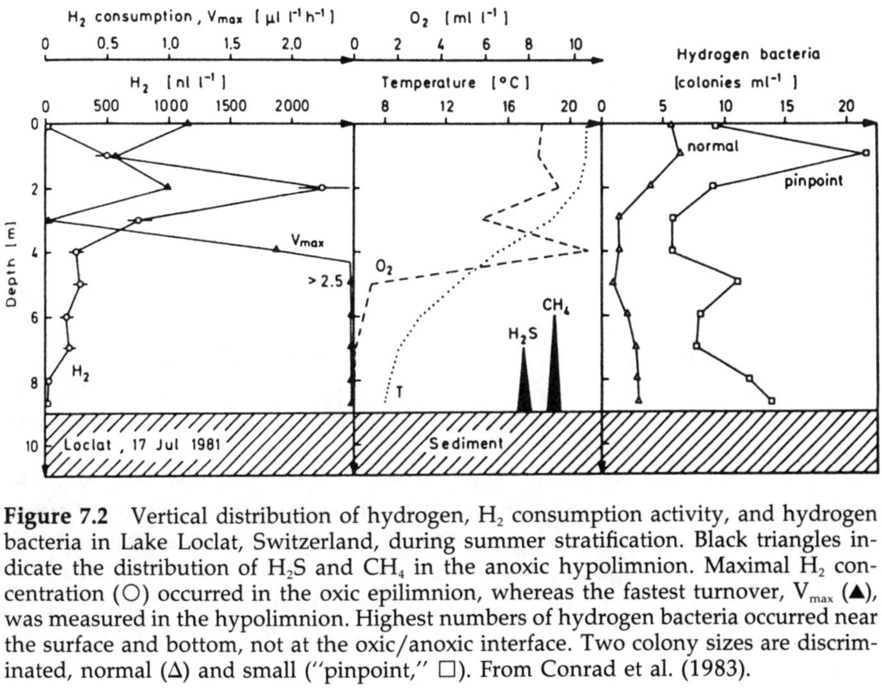

Figure 7.2 Vertical distribution of hydrogen, H_2 consumption activity, and hydrogen bacteria in Lake Loclat, Switzerland, during summer stratification. Black triangles indicate the distribution of H_2S and CH_4 in the anoxic hypolimnion. Maximal H_2 concentration (○) occurred in the oxic epilimnion, whereas the fastest turnover, V_{max} (▲), was measured in the hypolimnion. Highest numbers of hydrogen bacteria occurred near the surface and bottom, not at the oxic/anoxic interface. Two colony sizes are discriminated, normal (Δ) and small ("pinpoint," □). From Conrad et al. (1983).

significance in very restricted environments (Zeikus et al., 1980; Baross et al., 1982).

During the anaerobic degradation of organic matter, hydrogen is an obligatory product of those bacteria which degrade intermediate fermentation products such as alcohols or fatty acids to acetate. The enzymatic evolution of H_2 involves a reduction of protons ($2H^+ \rightarrow H_2 + 2e^-$), which is thermodynamically favorable only under very low H_2 partial pressures of 10^{-4} to 10^{-3} atm (about 70–700 nM). Among the important H_2-producing fermentations are:

Lactate degradation:

$$CH_3CHOHCOO^- + 2H_2O \rightarrow CH_3COO^- + HCO_3^- + H^+ + 2H_2$$

Ethanol degradation:

$$CH_3CH_2OH + H_2O \rightarrow CH_3COO^- + H^+ + 2H_2$$

Butyrate degradation:

$$CH_3CH_2CH_2COO^- + 2H_2O \rightarrow 2CH_3COO^- + H^+ + 2H_2$$

Propionate degradation:

$$CH_3CH_2COO^- + 3H_2O \rightarrow CH_3COO^- + HCO_3^- + H^+ + 3H_2$$

The Gibbs free energies of reaction of these fermentations can be calculated given the ambient organic substrate and hydrogen concentrations, pH, and temperature. In the sediment of the eutrophic Lake Mendota, Wisconsin, the fermentation of lactate and ethanol were exergonic reactions even at high H_2 partial pressures, whereas butyrate and especially propionate fermentation required extremely low H_2 levels of <100 and <20 nM, respectively (Conrad et al., 1986). The latter was below the measured hydrogen concentration of 31 nM, so that theoretically acetogenesis from propionate should not occur. At elevated hydrogen concentrations the fermentative pathways will consequently yield relatively more propionate and butyrate in systems where these can be utilized, e.g., by sulfate-reducing bacteria.

Chemolithotrophic hydrogen consumption H_2 is utilized as an energy source by many aerobic and anaerobic bacteria. The list of autotrophic H_2 oxidizers (Table 1) comprises first of all aerobes (the so-called "knallgas" bacteria which use $H_2 + O_2$), the anaerobic methanogens, homoacetogens, and sulfate reducers. Nitrate, elemental sulfur, Fe^{3+}, and possibly also Mn^{4+} may also play roles as electron acceptors for H_2 oxidation in environments where these electron acceptors are available.

Most of the biogenic H_2 in aquatic environments is rapidly recycled, and the H_2 oxidation rates are therefore strictly limited by the rates of H_2 production. There does not seem to be any particular H_2 oxidation activity associated with oxic/anoxic interfaces (Figure 7.2). Thus, Novelli et al. (1987) found that all H_2 within the marine sediments studied was reoxidized below the oxic surface, and Conrad et al. (1983) found in a eutrophic lake that the reducing sediment was a sink rather than a source of H_2. The mechanism of aerobic H_2 oxidation

Table 7.1 Examples of aerobic and anaerobic H_2 oxidation by chemolithotrophic bacteria. Both obligate and facultative autotrophs are included

Reaction	Organisms	References
$2H_2 + O_2 \rightarrow 2H_2O$	Hydrogen bacteria, e.g., *Hydrogenomonas* sp.	Bowien and Schlegel (1981)
$4H_2 + CO_2 \rightarrow CH_4 + 2H_2O$	Most methanogens, e.g., *Methanobacterium* sp.	Zeikus (1977)
$4H_2 + 2CO_2 \rightarrow CH_3COOH + 2H_2O$	Homoacetogenic bacteria, e.g., *Acetobacterium woodii*	Balch et al. (1977), Braun et al. (1981)
$5H_2 + 2NO_3^- + 2H^+ \rightarrow 6H_2O + N_2$	*Paracoccus denitrificans*	Verhoeven et al. (1954)
$4H_2 + SO_4^{2-} + 2H^+ \rightarrow 4H_2O + H_2S$	Some sulfate reducers, e.g., *Desulfobacterium autotrophicum*	Brysch et al. (1987)
$H_2 + S^0 \rightarrow H_2S$	*Campylobacter* sp.	Laanbroek et al. (1978)
$H_2 + 2Fe(OH)_3 + 4H^+ \rightarrow 6H_2O + 2Fe^{2+}$	Unnamed iron reducers	Balashova and Zavarzin (1980)

in soil or water is not clear. Although autotrophic hydrogen bacteria are found in these environments, kinetic and inhibition experiments point toward H_2 oxidation by cell-free hydrogenases or perhaps by mixotrophic or chemolithoheterotrophic bacteria (Conrad and Seiler, 1981; Conrad et al., 1983).

In sulfate-rich marine sediments, the sulfate-reducing bacteria are the main H_2 oxidizers. Some sulfate reducers among the genera *Desulfonema, Desulfosarcina, Desulfococcus,* and *Desulfobacterium* are known to grow autotrophically on H_2, while the classical *Desulfovibrio* strains grow lithoheterotrophically on H_2 with acetate as the main carbon source (Brysch et al., 1987; see also Chapter 22). The role of H_2 as an energy source for sulfate reduction in marine sediments has been analyzed by measuring the initial accumulation rate of H_2 after blocking the sulfate reduction with molybdate (Sørensen et al., 1981). The H_2 contributed 5–10 percent of the electrons while acetate, propionate, and butyrate accounted for 40–50 percent, 10–20 percent, and 10–20 percent, respectively. Although even a slight accumulation of H_2 may affect the fermentative pathways, these seem less sensitive to H_2 concentration in marine sediments than in lake sediments. It is mostly the propionate and butyrate fermentations which are dependent on low H_2 concentrations. These two substrates need not be fermented in marine sediments, in contrast to the methanogenic lake sediments, but can be directly respired by sulfate reducers.

In the absence of sulfate, methanogenesis is the dominant H_2-oxidizing reaction. Next to acetate, which contributes about 70 percent of the methane, H_2 is the most important substrate for methane formation. Of the other potential substrates, methanol and methylamines were the precursors for less than 5 and 1 percent, respectively, of the total methane produced in the sediments of eutrophic Wintergreen Lake, Michigan (Lovley and Klug, 1983b). Hydrogen may also be consumed by homoacetogenic bacteria (Table 7.1). These oxidized less than 5 percent of the H_2 in the same lake sediment and thus less than 2 percent of the acetate was derived from CO_2 reduction. The insignificant role of the homoacetogens relative to the methanogens is in accordance with the lower energy yield of the acetogenic reaction (Conrad et al., 1986). The homoacetogens also seem to occur in very low numbers relative to the methanogens, e.g., in the ratio of 1:100 in lake sediment (Braun et al., 1979). It is not clear where the H_2-oxidizing homoacetogens may play their main role, possibly in sediments and soils under slightly more acidic conditions where H_2-oxidizing methanogens are inhibited. In a lake sediment at pH 6.2, Phelps and Zeikus (1984) found that only 4 percent of the methane was derived from CO_2 while acetate played a completely dominant role.

Competition for H_2 A competition between sulfate reducers and methanogens for the common substrates, acetate and H_2, was demonstrated by Winfrey and Zeikus (1977) and has since been confirmed for a range of marine and freshwater sediments (Abram and Nedwell, 1978; Sørensen et al., 1981; Zeiss, 1981; Oremland and Polcin, 1982; Lovley and Klug, 1983a; King, 1984). Kinetic studies with pure cultures have shown that sulfate reducers have higher affinities (lower K_m) for both H_2 and acetate than do the methanogens (Kristjansson et al., 1982; Schönheit et al., 1982; Robinson and Tiedje, 1984; Lupton and Zeikus, 1984). Also, thermodynamically the sulfate reducers are at an advantage as they have a higher growth yield.

Competitive exclusion thus probably explains the depression of methanogenesis in anaerobic sulfate-containing environments. In lake sediments, a sulfate gradient generally penetrates only one or a few cm below the surface, but often a low, residual sulfate concentration of 10–40 μM remains in the pore water of deeper layers. Lovley and Klug (1986) developed a model for the competitive exclusion under *in situ* physiological conditions in freshwater sediments. Their calculations indicated that sulfate limitation at concentrations below 30 μM prevented the sulfate reducers from growing due to substrate competition from the methanogens.

Interspecies H_2 transfer The fermentation pathways depend strongly on the ambient H_2 concentration, especially in methanogenic environments where acetate and H_2 are obligatory intermediates. The H_2-*producing* acetogens, which ferment propionate and butyrate, require a low H_2 partial pressure which can only be maintained by methanogenic or sulfate-reducing bacteria. Syntrophic partnerships established between H_2 producers and H_2 consumers enable rapid recycling of H_2. This interspecies H_2 transfer alters the fermentation toward more oxidized products and thus a more complete usage of energy in the organic substrates. The obligate requirement of syntrophy among the acetogens is reflected in their names, e.g., *Syntrophobacter* and *Syntrophomonas* (Boone and Bryant, 1980; McInerney et al., 1981).

An additional requirement of physical proximity of the syntrophic partners was suggested by Conrad et al. (1985) based on studies in lake sediments. The measured turnover rate of the main H_2 pool in the sediment accounted for only 5–6 percent of the expected H_2 utilization for methanogenesis. This could be explained if the interspecies H_2 transfer took place between juxtapositioned bacteria within particles or other microenvironments without equilibration with the ambient H_2 pool (Figure 7.3). If a lower H_2 partial pressure could be established within such microenvironments, this mechanism would also help to

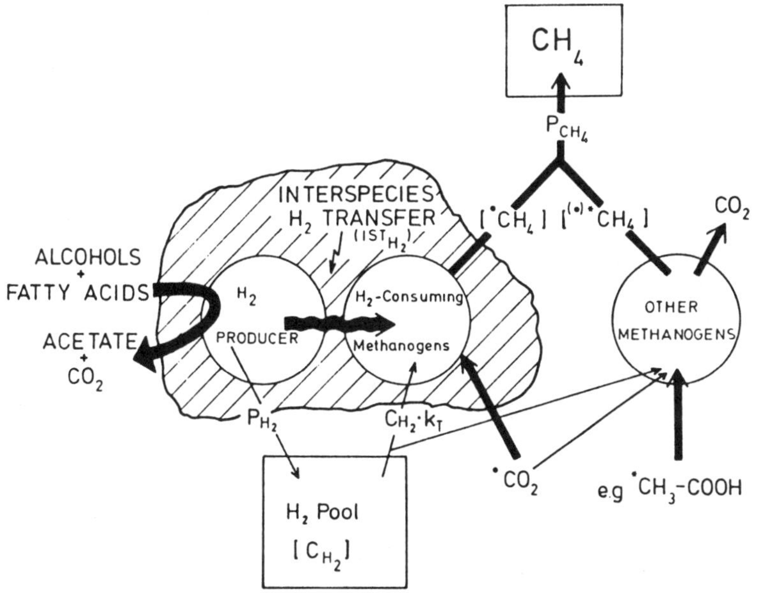

Figure 7.3 Diagram showing the principles of interspecies hydrogen transfer in anaerobic environments and the importance of juxtapositioning of H_2 producers and methanogens. The syntrophic partners are closely located within particles or flocs in which H_2 is rapidly cycled at low partial pressure and with limited equilibration with the ambient H_2 pool. This mechanism may explain how the total rate of methanogenesis can highly exceed the accumulated rates measured by radiolabeling the external pools of CO_2 and acetate. From Conrad et al. (1985).

explain the occurrence of acetogenesis from propionate in spite of a too high ambient H_2 level for this reaction to be exergonic (Conrad et al., 1986). It is an interesting possibility that a significant H_2 oxidation may take place in some environments within symbiotic associations of methanogenic bacteria and protozoa (see Chapter 10).

7.3 MINERAL CYCLING AND REDOX ZONATIONS

Similar to the zonation of different electron acceptors in sediments, there is a zonation in the reoxidation of different end products from anaerobic degradation. Most of these oxidizable products, NH_4^+, Mn^{2+}, Fe^{2+}, H_2S, CH_4 etc., accumulate in the pore water and migrate up toward the surface along diffusion gradients (Figure 7.1). They may also be transiently trapped in the particulate phase within reducing sediments and only become available for chemical or bacterial oxidation when the sediment is exposed to oxygen by animals (infaunal irrigation and bioturbation). Thus, much of the ammonia in sediment and soil is sorbed to clay particles and the manganese and iron are partly bound as metal sulfides or other minerals.

The chemical zonations of these free or bound elements is important for both geochemists and microbiologists because they give clues as to which interactions and processes are important in the element cycling. Although oxygen is the electron acceptor of paramount importance, many of the oxidation reactions take place within the anoxic environment. The electron flow is thus a cascade of reactions from the most reduced compounds toward the more oxidized ones. Some of the reactions are well-known lithotrophic processes, while others could possibly be either chemically or enzymatically catalyzed, although their microbiology is not yet understood.

In order to discriminate between biological and chemical reaction, several studies of oxygen uptake in sediments have compared the oxygen consumption with and without poisoning. This, however, is generally not a realistic approach because: 1) it interferes with the competition between lithotrophic and spontaneous chemical reactions; 2) it inhibits also the production of inorganic, reduced compounds; and 3) several of the poisons used react chemically with the potential substrates of chemolithotrophs (e.g., formaldehyde with sulfides). The shape of oxygen gradients in sediments, however, often indicates that a

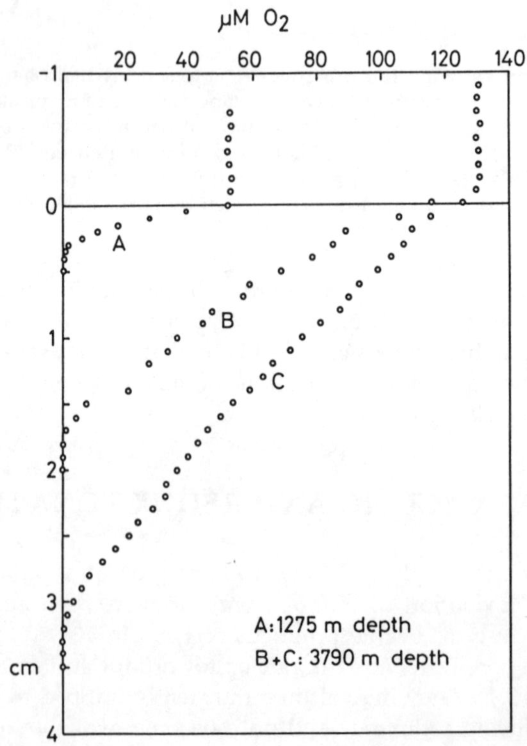

Figure 7.4 Oxygen distribution in the sea bed off the coast of California at 1275 and 3790 m water depth. Measurements were made with O_2-microelectrodes, either *in situ* with a free-vehicle instrument (B+C) or on board a ship (A). Redrawn from Reimers et al. (1986).

significant part of the oxygen consumption takes place right at the bottom of the oxygen zone. The gradients here have the steepest change in slope (curvature) with depth (Figure 7.4). According to simple diffusion laws, the curvature is a measure of reaction rate in one-dimensional, state-state diffusion gradients:

$$D \cdot d^2C/dz^2 = dC/dt$$

where C is the O_2 concentration, t is time, z is depth and D is the diffusion coefficient of O_2 (assumed to be constant). The second derivative of C versus depth, i.e., the curvature, times D thus equals the rate of oxygen consumption, dC/dt, at a given depth. Thus, much of the oxygen uptake must be due to reduced compounds diffusing up from below rather than to heterotrophic degradation and respiration of organic matter within the oxic zone. Although it has not yet been directly demonstrated that the lithoautotrophic activity is associated with this oxic/anoxic interface in sediments or soils, it would constitute a parallel to the similar interfaces which have been well studied in stratified water bodies.

7.4 NITROGEN

Ammonia, which is released from organic nitrogen during anaerobic degradation, is mostly reoxidized via nitrite to nitrate by the aerobic and autotrophic, nitrifying bacteria:

$$NH_4^+ + 1.5O_2 \rightarrow NO_2^- + H_2O + 2H^+$$

$$NO_2^- + 0.5O_2 \rightarrow NO_3^{2-}$$

Some heterotrophic nitrification may also take place, especially in soils and under acidic conditions, but the bacteria and fungi which carry out heterotrophic nitrification probably do not conserve energy from the reaction (Painter, 1970; Witzel and Overbeck, 1979). Under low oxygen conditions, N_2O is also a product of ammonia oxidation, up to 25 percent in marine sediments (Jørgensen et al., 1984). Nitrous oxide is a side product of the biochemical nitrification pathway rather than an intermediate as in denitrification.

Nitrite is a chemically reactive compound. It can be chemically reduced (chemodenitrification) or oxidized by surface reactions on ferric oxyhydroxides. The product depends mostly on the pH and the Fe^{2+}/Fe^{3+} ratio (Chao and Kroontje, 1966; Moraghan and Buresh, 1977). The quantitative significance of these chemical reactions is not known, but they may interfere with the biological transformations of the NO_3^-/NO_2^- and Fe^{3+}/Fe^{2+} systems, even in pure cultures (Obuekwe et al., 1981).

Due to the lack of a suitable radioisotope of nitrogen, it is difficult to make reliable measurements of nitrification rates in natural systems. The diversity of approaches which have been applied to measure this well-known process gives an impression of the problems generally associated with measuring lithoautotrophic activities: 1) Measurement of $^{14}CO_2$ assimilation with and without N-Serve (2-chloro-6-(trichloromethyl)pyridine), which blocks the first step in nitrification, $NH_4^+ + 0.5O_2 \rightarrow NH_2OH + H^+$; the difference in fixed ^{14}C is taken as a measure of chemoautotrophy by nitrifiers (Billen, 1976); 2) Measurement of changes in ammonia pools over time in intact samples with and without N-Serve; the difference indicates accretion of ammonia, which would have been nitrified in the absence of N-serve (Henriksen, 1980); 3) Measurement of potential nitrification rates in the presence of excess oxygen and ammonia and recalculation to *in situ* conditions (Hansen et al., 1981); the maximum activity per cell is relatively constant (Belser, 1979) and thus the potential nitrification is a measure of the number of nitrifiers; 4) Measurement of changes in nitrite pools over time with and without chlorate, which inhibits the second nitrification step as well as the reduction of nitrite; the difference indicates accretion of nitrite, which would have been nitrified in the absence of chlorate (Belser and Mays, 1980); 5) Mathematical modeling based on chemical gradients and mass balance calculations (e.g., Vanderborght and Billen, 1975; Christensen and Rowe, 1984).

Although nitrification in sediments is associated with the oxic surface layers, nitrifying bacteria are distributed throughout the anoxic zone down to reducing layers. The bacteria may be inactive in the absence of oxygen but can survive for long periods if they are not exposed to sulfide (Painter, 1970; Hansen et al., 1981). They readily oxidize ammonia when oxygen is introduced.

Figure 7.5 shows this situation for a coastal marine sediment where nitrifying bacteria were abundant down to 5 cm depth, although oxygen apparently penetrated only down to 1.5 mm. The deeper nitrifiers may receive oxygen only occasionally when the sediment is disturbed by burrowing animals. Due

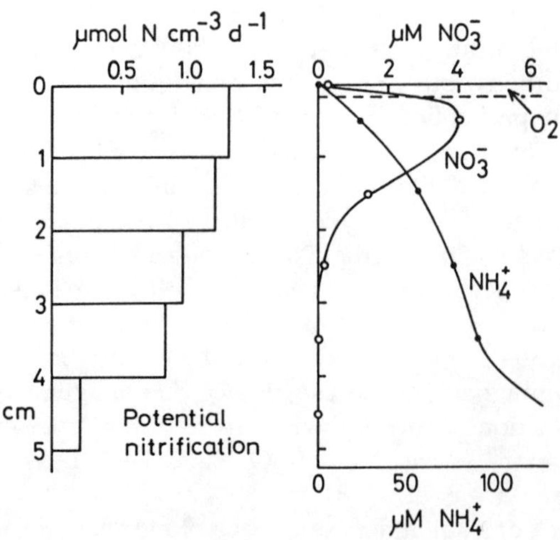

Figure 7.5 Distribution of potential nitrification, as a relative measure of the population density of nitrifying bacteria, in relation to the chemical zonation in a coastal marine sediment. Redrawn from Henriksen et al. (1981), Station VIII, July 1979.

to the large inactive population, the potential nitrification rate of the whole sediment column was 20–30-fold higher than the actual rate (Henriksen et al., 1981). The population size of nitrifiers must give an indication of their possibilities for growth, although their number is not a reliable indication of their activity. Direct counting techniques have been developed for nitrifying bacteria based on specific staining with fluorescent antibodies. Such counts yield numbers 2–3 orders of magnitude higher than the traditional viable counting techniques (Belser and Schmidt, 1978; Belser and Mays, 1982).

Because of the combined need for oxygen and ammonia, the nitrifying bacteria in sediments require a certain overlap and residence time of the two energy substrates in order to compete with the diffusive loss of ammonia to the overlying water. This explains why nitrification, and thus denitrification, is not always highest during summer, but rather during spring and fall or even winter, when the oxic zone is deeper (Hansen et al., 1981; Jenkins and Kemp, 1984). During summer, ammonia may be released directly into the overlying water where it can be oxidized by planktonic nitrifying bacteria.

The nitrification process is consequently a relatively more efficient and dominant process in low-organic systems with a deep oxidized zone. It has been calculated for deep-sea sediments that ammonia oxidation alone consumed 1/4–1/3 of the total oxygen uptake (Christensen and Rowe, 1984). The chemoautotrophic biomass production by the denitrifying bacteria is not significant in spite of such a high contribution because of their low growth yield (*Nitrosomonas*: 0.56–1.82 g cells/mol NH_4^+; *Nitrobacter*: 0.28–0.98 g cells/mol NO_2^-; Painter, 1970). In the deep-sea sediment their CO_2 assimilation corresponded to only 1.6 percent of the total carbon turnover by heterotrophic bacteria (Christensen and Rowe, 1984).

An important role of nitrification is the production of nitrite and nitrate for denitrification. The two processes are often tightly coupled in sediments so that denitrification is limited by internally produced $NO_2^- + NO_3^-$, with less exchange of nitrate across the sediment-water interface (Seitzinger et al., 1984). The two processes often seem to take place within the same environment in oxidized sediments although they are generally considered to be mutually exclusive due to different oxygen requirements. This may be due to the occurrence of anoxic microenvironments within the generally oxic zone. It could perhaps also be due to aerobic denitrification. Bacteria which can denitrify in the presence of oxygen have only recently been isolated and their role in natural environments is still uncertain (Robertson and Kuenen, 1984).

7.5 MANGANESE AND IRON

Manganese and iron are among the most abundant elements in the Earth's crust, ranking number 12 and 4 in total abundance, respectively. The two metals are generally found together, occurring mainly in rocks and minerals where they may accumulate in large enough concentrations to become valuable ore deposits. Massive deposits include the Precambrian banded iron formations and red beds as well as ferromanganese nodules on the sea floor, bog ores, and hydrothermal deposits.

The role of chemoautotrophic bacteria in the modern biogeochemical cycling of these metals is still not well understood in most environments. Thermodynamically, reduced manganese and iron should be good energy substrates, but their chemical characteristics are unfavorable at neutral pH: They form precipitates of oxides and hydroxides under oxic conditions and they are readily oxidized in the presence of inorganic catalysis. Because of precipitation, most aquatic manganese and iron cycling takes place within sediments although some processes do occur in the chemoclines of stratified waters.

Manganese (II) can be chemically oxidized by O_2 to manganese (IV), although the average oxidation state of the precipitated product is often lower (MnO_x, x = 1.3–2.0):

$$Mn^{2+} + \frac{1}{2}O_2 + 2OH^- \rightarrow MnO_2 + H_2O$$

The manganese oxidation reaction has high activation energy, which is lowered by binding to surfaces. MnO_2 or FeOOH act as inorganic surface catalysts, speeding up the rate of reaction by several orders of magnitude (Morgan, 1964; Sung and Morgan, 1981). Inorganic surface catalysis is probably an important mechanism for the formation and growth of ferromanganese nodules at sediment-water interfaces. Also, the bacterial catalysis involves a surface binding before oxidation. The known manganese-oxidizing bacteria are heterotrophs and it is uncertain whether they conserve energy from the oxidation (Nealson, 1983). A report of Mn-dependent CO_2 fixation by a marine *Pseudomonad* sp. grown in chemostat culture has not been confirmed later (Kepkay et al., 1984). The ability to oxidize manganese occurs among many different genera of bacteria and may in many cases be an unspecific reaction, which inhibits further bacterial growth because of the mineral coating by MnO_2.

Kinetic and inhibitor studies of manganese oxidation in freshwater and seawater have shown that the oxidation is primarily catalyzed by bacteria and that the rate is limited by the number of bacterial binding sites (Emerson et al., 1979, 1982; Chapnick et al., 1982; Tebo and Emerson, 1985, 1986). Manganese oxidation rates in sediments have mostly been calculated indirectly from steady-state diffusion-reaction models based on the vertical gradients of manganese chemistry (e.g., Robbins and Callender, 1975; Elderfield et al., 1981; Burdige and Gieskes, 1983). Although the process is much faster in sediments than in the water column, it is not known to what extent it is microbially catalyzed.

In contrast to manganese, reduced iron is highly unstable under oxic conditions at neutral pH. The half-life of Fe^{2+} in oxygenated water is in the range of one to seven minutes, and the oxidation seems to be a homogeneous reaction with little catalysis from metal oxide surfaces (Murray and Gill, 1978; Davison and Seed, 1983; Millero et al., 1987).

Many bacteria are known to be associated with iron oxidation at neutral pH. These iron bacteria include the filamentous, sheathed forms such as *Sphaerotilus* and *Leptothrix,* as well as *Gallionella, Metallogenium, Hyphomicrobium,* etc. There is no unequivocal proof, however, that any of these bacteria can grow autotrophically. There is also no direct evidence that bacteria play a quantitative role for the iron oxidation kinetics in oxic environments (Davison and Seed, 1983). Several explanations of this are possible. Perhaps the chemical iron oxidation is too rapid for the bacteria to compete effectively. With a chemical half-life of a few minutes, this may be true for oxidation in the water column.

A more favorable situation would occur in environments where iron and oxygen met in opposed diffusion gradients. Such a situation may exist in sediments, marshes, mineral springs, bogs, and where ferromanganese concretions are formed (Mustoe, 1981; Lyle, 1983; Giblin and Howarth, 1984). By analogy with *Beggiatoa,* which grows autotrophically in gradients of coexisting oxygen and sulfide at a 100-fold shorter half-life (see section 7.6), the rapid spontaneous iron oxidation kinetics should not in itself exclude autotrophic utilization. Perhaps it is the heavy precipitation of iron oxyhydroxides which exerts a physical constraint on the continuous bacterial growth in O_2/Fe^{2+} diffusion gradients. The chemical microenvironment of iron bacteria should be examined more closely with microelectrodes to analyze this question.

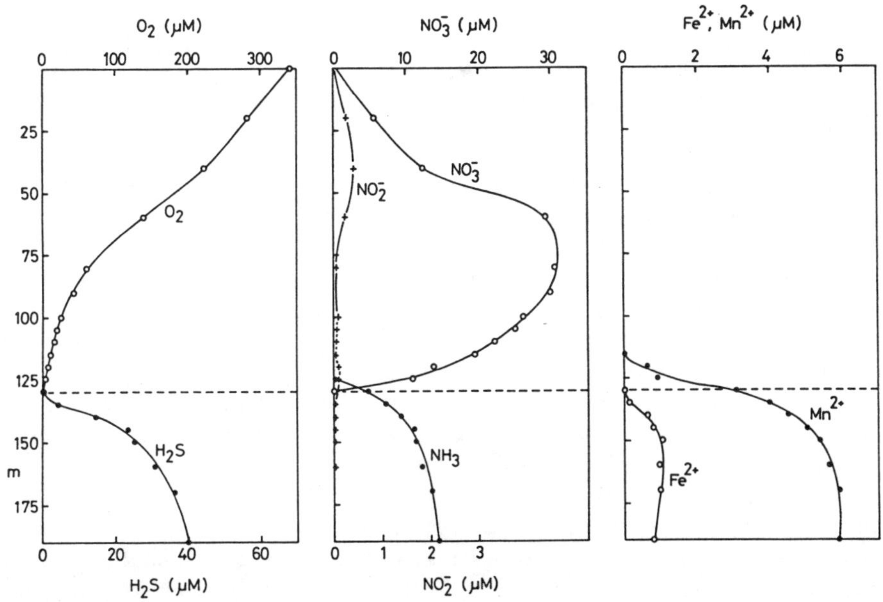

Figure 7.6 Chemical zonations through the water column of a 200-m deep, stratified fjord, Saanich Inlet. Oxidation of sulfide, ammonia, iron, and manganese takes place at the oxic/anoxic interface. The less reactive species reach further up into the oxic zone. Redrawn from Emerson et al. (1979).

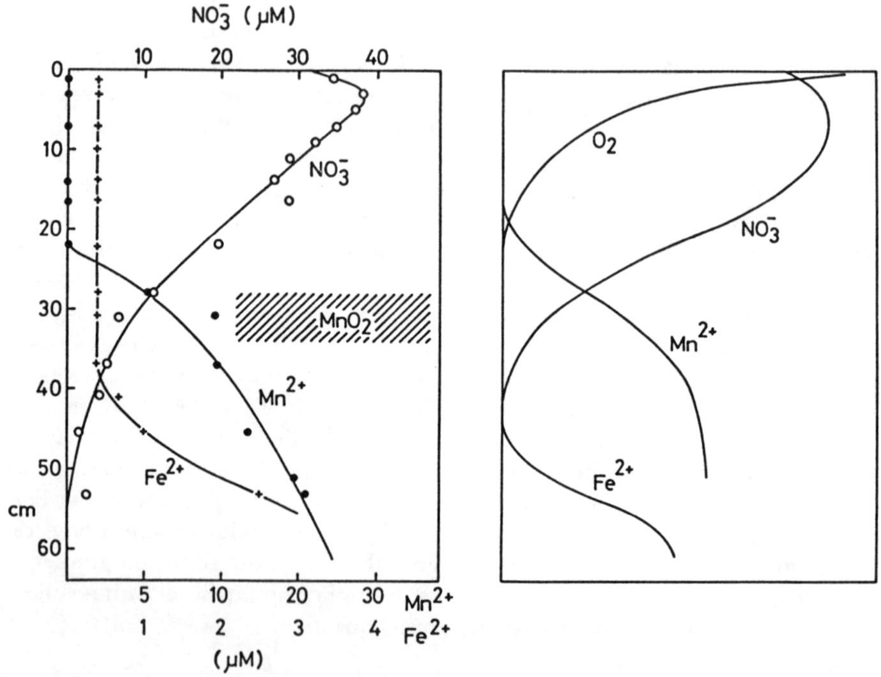

Figure 7.7 Chemical zonations within the upper 60 cm of a deep-sea sediment from the eastern equatorial Atlantic at 5000 m water depth. Dissolved Fe^{2+} and Mn^{2+} diffuse upward in the pore water to the interfaces where nitrate and oxygen, respectively, penetrate from above. A band of precipitated manganese oxide was observed at 30 cm depth. Redrawn from Froelich et al. (1979).

The difference in reactivity of dissolved Fe^{2+} and Mn^{2+} is evident from their gradients in the chemoclines of stratified water bodies. An example from an intermittently anoxic fjord in British Columbia is shown in Figure 7.6. Manganese reached further up than iron into the oxygenated zone where both metals existed in dynamic equilibrium with their oxidation and the precipitation and settling of the oxides (Emerson et al., 1979). As the metal oxides sink into the anoxic water they can again be reduced and recycle.

The oxic/anoxic interface was also the lower boundary of nitrate in the water, which thus coincided with the upper boundary of iron. There is now accumulating evidence from marine sediments that Fe^{2+} oxidation takes place at the interface toward nitrate rather than oxygen (Froelich et al., 1979; Klinkhammer, 1980; Lyle, 1983). An example of the chemical zonations in a deep-sea sediment is shown in Figure 7.7 together with the generalized pattern of distribution. Nitrate penetrates down to the light brown/green color transition in these sediments, which marks the Fe^{3+}/Fe^{2+} redox boundary (Lyle, 1983). The color change is due to iron-rich clay minerals, smectites, which are green when the iron is reduced and yellow-brown when it is oxidized. Closer to the surface the sediment is dark brown due to precipitated manganese oxide. Similar zonations, including oxygen gradients measured with microelectrodes, were found also in marine shelf sediments (Sørensen and Jørgensen, 1987).

These studies open new perspectives on the microbiology of iron oxidation. Specialized anaerobic iron oxidizers are not known although a pure culture study of the iron bacterium, *Gallionella ferruginea*, concluded that Fe^{2+} was oxidized with nitrate under anaerobic conditions (Gouy, 1984). However, this oxidation may not have been enzymatic but rather a chemical reaction with nitrite excreted by the organism:

$$2HNO_2 + 6Fe^{2+} + 6H^+ \rightarrow N_2 + 6Fe^{3+} + 4H_2O$$

Chemical reactions between inorganic nitrogen compounds and ferrous iron are known from laboratory studies (Buresh and Moraghan, 1976), but their role in nature is uncertain. A reaction between ferrous iron and manganese oxide is also thermodynamically favorable and is potentially significant if such a process was biologically catalyzed.

Bacterial iron oxidation at low pH is well known from acid mine waters or acid hot springs where ferrous iron is relatively stable in the presence of oxygen and where iron hydroxides do not precipitate. The chemolithoautotrophic *Thiobacillus ferrooxidans* grows in these environments together with other acidophilic genera such as *Sulfolobus* and *Metallogenium* with Fe^{2+} as the sole source of energy (e.g., Brock et al., 1972; Lundgren et al., 1974; Zavarzin, 1982).

7.6 SULFUR

Anaerobic sulfide oxidation As mentioned in section 7.1, up to 35 percent of the energy in deposited organic matter may be transferred to H_2S in coastal marine sediments. The reoxidation of this H_2S back to sulfate requires the equivalent of half the total oxygen uptake of the sediment (Jørgensen, 1977). Yet, the oxidation seems in most environments not to be a direct reaction between H_2S and O_2, because the two chemical species do not meet. Instead, they are separated by an oxidized but anoxic zone (Figure 7.8). This is the brown layer of sediments in which ions such as NO_3^-, Fe^{3+}, and Mn^{4+} cause a positive redox potential, but which only contains O_2 in a much thinner layer below the sediment-water interface (Revsbech et al., 1980). The terms "suboxic zone" or "Fe-Mn zone" have been used for the oxidized, anoxic sediment to distinguish it from the underlying reduced or sulfidic sediment (Froelich et al., 1979; Sørensen and Jørgensen, 1987).

Since nitrate penetrates further down than oxygen, it is a potential electron acceptor for sulfide oxidation. Denitrifying sulfide oxidizers such as *Thiobacillus denitrificans* and *Thiomicrospira denitrificans* occur widespread in sediments. A comparison between actual denitrification rates and sulfide oxidation rates in marine sediments have shown, however, that nitrate could only account for a smaller fraction of the sulfide oxidation during most seasons (Sørensen et al.,

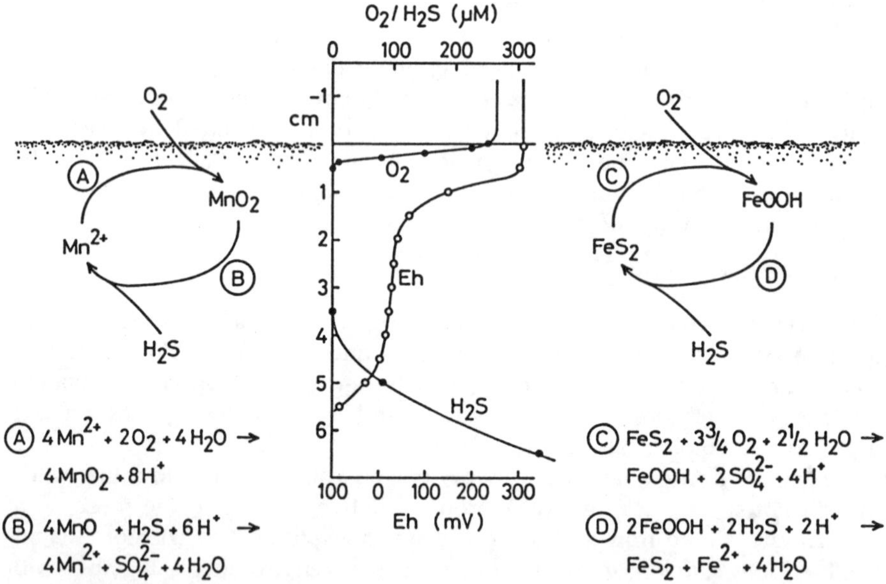

Figure 7.8 Chemical zonations and metal cycling in marine sediments. A slightly oxidized but anoxic zone separates the oxic surface layer from the deeper sulfidic sediment. Iron and manganese in the sediment may function as intermediate electron carriers between H_2S and O_2. Chemical data from Jørgensen (1983). Partly redrawn from Howarth (1984).

1979; Jørgensen and Sørensen, 1985; Sørensen and Jørgensen, 1987). Thus, although up to 10^4–10^6/g of facultatively anaerobic thiobacilli can be found from viable counts of such sediments, their quantitative role in sulfide oxidation is still not clear.

Iron and manganese are alternative electron acceptors for sulfide oxidation. Sulfide reacts spontaneously with oxidized iron minerals such as haematite and is oxidized to elemental sulfur (Goldhaber and Kaplan, 1974):

$$2FeOOH + H_2S + 4H^+ \rightarrow 2Fe^{2+} + S^0 + 4H_2O$$

A similar chemical reaction takes place with oxidized manganese at an even faster rate (Pyzik and Sommer, 1981; Burdige and Nealson, 1986):

$$MnO_2 + H_2S + 2H^+ \rightarrow Mn^{2+} + S^0 + 2H_2O$$

In sulfidic environments, most of the ferrous iron reacts with H_2S, S^0, and polysulfides to form FeS and pyrite, FeS_2. The corresponding manganese sulfides are more soluble and Mn^{2+} therefore tends to stay in solution.

The elemental sulfur can be either further oxidized or reduced. Seasonal studies of the S^0 distribution in sediments have indicated that it is indeed a dynamic pool in the sulfur cycle (Troelsen and Jørgensen, 1982). A complete oxidation in sediments to sulfate may take place with ferric iron (Aller et al., 1986):

$$6FeOOH + S^0 + 10H^+ \rightarrow 6Fe^{2+} + SO_4^{2-} + 8H_2O$$

Although this process could proceed chemically at a slow rate (Berner, 1970), it may also be catalyzed by bacteria. *Thiobacillus ferrooxidans, T. thiooxidans,* and *Sulfolobus acidocaldarius* are known from acidic environments to carry out the process (Brock and Gustafson, 1976; Kino and Usami, 1982). It is not yet known to what extent bacteria are involved at neutral pH. A similar oxidation with manganese is also thermodynamically feasible (Aller and Rude, 1988):

$$3MnO_2 + S^0 + 4H^+ \rightarrow 3Mn^{2+} + SO_4^{2-} + 2H_2O$$

Reducing power from H_2S is thus gradually transferred to iron and manganese. When the reduced metals diffuse up into the oxic zone, they are reoxidized as the electrons are transferred to O_2. Solid-phase metal sulfides such as pyrite accumulate to high concentrations and are only slowly reoxidized as the sediment is mixed up toward the oxic surface layers by burrowing animals or as oxygen is carried down, e.g., by respiratory ventilation of worm tubes or by plant roots in marsh sediments. In this way the cycling of manganese and iron in the suboxic zone may be a vehicle for the electron transfer between sulfide and oxygen (Figure 7.8)

While these pathways are potentially important for lithotrophic bacteria, their microbiology in nature is largely unknown. Aller and Rude (1988) have thus demonstrated that manganese oxide caused a complete oxidation to sulfate of solid phase sulfide in a marine sediment and that the process was biologically catalyzed:

$$4MnO_2 + FeS + 8H^+ \rightarrow 4Mn^{2+} + SO_4^{2-} + Fe^{2+} + 4H_2O$$

This reaction was inhibited by 2,4-dinitrophenol, an uncoupler of oxidative phosphorylation, and by sodium azide, a cytochrome oxidase inhibitor. Further studies of these important and potentially lithoautotrophic reactions are strongly needed.

Aerobic sulfide oxidation The role of microorganisms in aerobic sulfide oxidation is much more conspicuous although not necessarily more simple to study. Free H_2S reacts spontaneously with O_2 to produce mostly $S_2O_3^{2-}$, SO_4^{2-}, and S^0. In oxic waters, H_2S has a half-life in the order of an hour to a day depending on the temperature, the O_2/H_2S ratio, and the presence of catalysts (Millero, 1986; Millero et al., 1987). The aerobic sulfide oxidizers therefore have to compete with this chemical reaction.

A comparison of different O_2/H_2S interface environments has shown that the contribution of bacteria to the aerobic sulfide oxidation is largely determined by the scale and by the physical mixing of the two substrates (Jørgensen, 1982). Where oxygen and sulfide meet in stratified waters such as the Black Sea or meromictic lakes, eddy diffusion is the dominant transport mechanism. The two compounds here coexist in dynamic equilibrium between supply and consumption within a zone of a few centimeters to many meters in thickness.

The largest body of sulfidic water is found in the Black Sea, which is anoxic from the chemocline at 150–200 m down to a depth of 2000 m. Sorokin (1972) found a peak of dark CO_2 fixation associated with the O_2/H_2S interface in the Black Sea which he attributed to the activity of lithoautotrophic thiobacilli. The initial reaction of sulfide was, however, not catalyzed by bacteria but was rather a chemical oxidation to thiosulfate and sulfate. The bacteria then oxidized the thiosulfate completely to sulfate:

$$3H_2S + 4O_2 \rightarrow S_2O_3^{2-} + SO_4^{2-} + 4H^+ + H_2O \qquad \text{chemical}$$

$$S_2O_3^{2-} + 2O_2 + H_2O \rightarrow 2SO_4^{2-} + 2H^+ \qquad \text{bacterial}$$

Later studies in the Black Sea chemocline have indicated that the sulfide oxidizers may not be strict autotrophs but rather lithoheterotrophs or mixotrophs with a more flexible metabolism (Tuttle and Jannasch, 1973; Kuenen, 1975). Such organisms could better take advantage of the diverse energy and carbon sources which become available during fluctuating conditions in their environment (see Chapter 4).

Similar studies in meromictic lakes have shown that high dark CO_2 fixation is generally associated with the O_2/H_2S interface (Sorokin, 1970). In the hypersaline Solar Lake, the interface was only 10 cm thick and coexisting oxygen and sulfide had half-lives of 5–10 min (Jørgensen et al., 1979). In this very dynamic system, sulfide oxidation was significantly slowed when bacteria were removed by filtration, or was speeded up when bacteria were added in larger numbers in chemocline samples. The kinetic studies indicated that 30–50 percent of H_2S oxidation was here carried out by bacteria.

Optimal conditions for bacteria to compete with chemical H_2S oxidation are provided in microenvironments where chemical transport is dominated by molecular diffusion. The motile bacteria are here positioned within a very narrow zone of coexisting oxygen and sulfide. The filamentous *Beggiatoa* are typical of such gradient organisms (see Chapter 12). They often form dense mats on sediments which are rich in decomposing organic matter or they may occur together with cyanobacteria and phototrophic bacteria in sulfureta.

An example of the chemical and bacterial microzonation within a cyanobacterial mat community is shown in Figure 7.9. *Beggiatoa* occurred in a well-defined band at 1.5 mm depth below the mat surface. Due to intensive photosynthesis in the upper 0.5 mm, a peak of 1 atm pO_2 developed from which oxygen diffused up into the water and down toward the sulfide zone. Oxygen and sulfide coexisted in a 50–100 μm thick layer with an estimated half-life of a few seconds. Only bacterial catalysis can oxidize H_2S so rapidly and the *Beggiatoa* thus have all the reducing power of the sulfide available for their energy metabolism. Pure culture studies of marine *Beggiatoa* have shown that the organisms can grow chemoautotrophically with sulfide as the sole energy source. Good autotrophic growth, however, requires O_2/H_2S microgradients in which the organisms are exposed to a high supply rate but low concentration

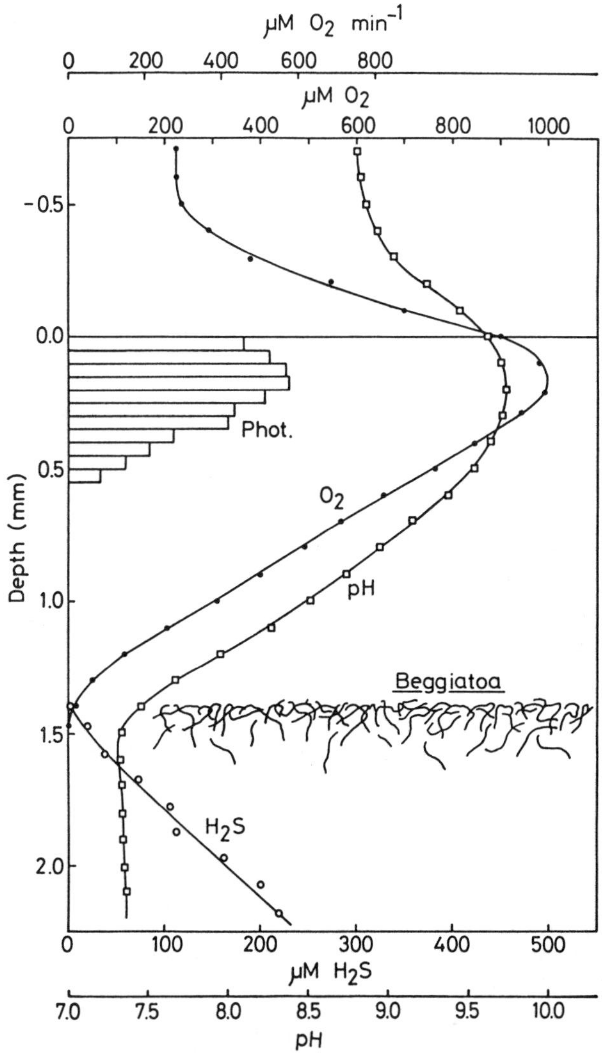

Figure 7.9 Distribution of oxygenic photosynthesis and of oxygen, sulfide, and pH in a cyanobacterial mat. The mat grew in a shallow hypersaline pond in Baja California, Mexico. A sharp band of *Beggiatoa* occurred at the O_2/H_2S interface. From Jørgensen and Des Marais (1986).

of both substrates (Jørgensen and Revsbech, 1983; Nelson et al., 1986; see Chapter 12).

Due to efficient sulfide oxidation, a *Beggiatoa* film of only 0.5 mm thickness on a sediment surface can be sufficient to prevent release of H_2S to the water above. This is important ecologically because H_2S is toxic to most aerobic organisms. The huge mats of another filamentous bacterium, *Thioploca*, may have a similar function. These benthic mats grow in upwelling areas where the oxygen minimum zone of the water column reaches the shelf sediments at 50–300 m depth, notably along the coast of Chile and Peru (Gallardo, 1977). Although the organisms transiently store sulfur globules just like *Beggiatoa*, their energy metabolism and association with sulfide is not yet clear. They may be lithoheterotrophs which conserve energy from sulfide oxidation but assimilate acetate and other organic substrates (Maier and Gallardo, 1984). Due to the large-scale accumulation of organic matter as well as nitrogen and phosphorus associated with such mats, they may be of geological significance, e.g., for the formation of phosphorite deposits (Williams and Reimers, 1983).

REFERENCES

Abram, J.W. and **Nedwell, D.B.** 1978. Inhibition of methanogenesis by sulfate-reducing bacteria competing for transferred hydrogen. *Archives of Microbiology* 117:89–92.

Aller, R.C. and **Rude, P.D.** 1988. Complete oxidation of solid phase sulfides by manganese and bacteria in anoxic marine sediments. *Geochimica et Cosmochimica Acta* 52:751–765.

Aller, R.C., **Mackin, J.E.** and **Cox Jr., R.T.** 1986. Diagenesis of Fe and S in Amazon inner shelf muds: apparent dominance of Fe reduction and implications for the genesis of ironstones. *Continental Shelf Research* 6:263–289.

Balashova, V.V. and **Zavarzin, G.A.** 1980. Anaerobic reduction of ferric iron by hydrogen bacteria. *Microbiology* 48:635–639.

Balch, W.E., **Schoberth, S.**, **Tanner, R.S.**, and **Wolfe, R.S.** 1977. *Acetobacterium*, a new genus of hydrogen-oxidizing, carbon dioxide-reducing anaerobic bacteria. *International Journal of Systematic Bacteriology* 27:355–361.

Baross, J.A., **Lilley, M.D.**, and **Gordon, L.I.** 1982. Is the CH_4, H_2 and CO venting from submarine hydrothermal systems produced by thermophilic bacteria? *Nature* 298:366–368.

Belser, L.W. 1979. Population ecology of nitrifying bacteria. *Annual Review of Microbiology* 33:309–333.

Belser, L.W. and **Mays, E.L.** 1980. Specific inhibition of nitrite oxidation by chlorate and its use in assessing nitrification in soils and sediments. *Applied and Environmental Microbiology* 39:505–510.

Belser, L.W. and **Mays, E.L.** 1982. Use of nitrifier activity measurements to estimate the efficiency of viable nitrifier counts in soil and sediments. *Applied and Environmental Microbiology* 43:945–948.

Belser, L.W. and **Schmidt, E.L.** 1978. Serological diversity within a terrestrial ammonia-oxidizing population. *Applied and Environmental Microbiology* 36:589–593.

Berner, R.A. 1970. Sedimentary pyrite formation. *American Journal of Science* 268:1–23.

Berner, R.A. 1987. Models for carbon and sulfur cycles and atmospheric oxygen: application to paleozoic geologic history. *American Journal of Science* 287:177–196.

Berner, R.A. and **Raiswell, R.** 1983. Burial of organic carbon and pyrite sulfur in the modern ocean: its geochemical and environmental significance. *American Journal of Science* 282:451–473.

Billen, G. 1976. A method for evaluating nitrifying activity in sediments by dark ^{14}C-bicarbonate incorporation. *Water Research* 10:51–57.

Blackburn, T.H. 1980. Seasonal variations in the rate of organic-N mineralization in anoxic marine sediments. *Colloque International C.N.R.S. (Marseille)* 293:173–183.

Boone, D.R. and **Bryant, M.P.** 1980. Propionate-degrading bacterium, *Syntrophobacter wolinii* sp. nov., gen. nov., from methanogenic ecosystems. *Applied and Environmental Microbiology* 40:626–632.

Bowien, B. and **Schlegel, H.G.** 1981. Physiology and biochemistry of anaerobic hydrogen-oxidizing bacteria. *Annual Review of Microbiology* 35:405–452.

Braun, M., Mayer, F., and **Gottschalk, G.** 1981. *Clostridium aceticum* (Wieringa), a microorganism producing acetic acid from molecular hydrogen and carbon dioxide. *Archives of Microbiology* 128:288–293.

Braun, M., Schoberth, S., and **Gottschalk, G.** 1979. Enumeration of bacteria forming acetate from H_2 and CO_2 in anaerobic habitats. *Archives of Microbiology* 120:201–204.

Brock, T.D. and **Gustafson, J.** 1976. Ferric iron reduction by sulfur- and iron-oxidizing bacteria. *Applied and Environmental Microbiology* 32:567–571.

Brock, T.D., Brock, K.M., Belly, R.T., and **Weiss, R.L.** 1972. *Sulfolobus*: a new genus of sulfur-oxidizing bacteria living at low pH and high temperature. *Archiv für Mikrobiologie* 84:54–68.

Brysch, K., Schneider, C., Fuchs, G., and **Widdel, F.** 1987. Lithoautotrophic growth of sulfate-reducing bacteria, and description of *Desulfobacterium autotrophicum* gen. nov., sp. nov. *Archives of Microbiology* 148:264–274.

Burdige, D.J. and **Gieskes, J.M.** 1983. A pore water/solid phase diagenetic model for manganese in marine sediments. *American Journal of Science* 283:29–47.

Burdige, D.J. and **Nealson, K.H.** 1985. Microbial manganese reduction by enrichment cultures from coastal marine sediments. *Applied and Environmental Microbiology* 50:491–497.

Burdige, D.J. and **Nealson, K.H.** 1986. Chemical and microbiological studies of sulfide-mediated manganese reduction. *Geomicrobiology Journal* 4:361–387.

Buresh, R.J. and **Moraghan, J.T.** 1976. Chemical reduction of nitrate by ferrous iron. *Journal of Environmental Quality* 5:320–325.

Chao, T.-T. and **Kroontje, W.** 1966. Inorganic nitrogen transformations through the oxidation and reduction of iron. *Soil Science Society of America Proceedings* 30:193–196.

Chapnick, S.C., Moore, W.S., and **Nealson, K.H.** 1982. Microbially mediated manganese oxidation in a freshwater lake. *Limnology and Oceanography* 27:1004–1014.

Christensen, J.P. and **Rowe, G.T.** 1984. Nitrification and oxygen consumption in northwest Atlantic deep-sea sediments. *Journal of Marine Research* 42:1099–1116.

Conrad, R. and **Seiler, W.** 1981. Decomposition of atmospheric hydrogen by soil microorganisms and soil enzymes. *Soil Biology and Biochemistry* 13:43–49.

Conrad, R., Aragno, M., and **Seiler, W.** 1983. Production and consumption of hydrogen in a eutrophic lake. *Applied and Environmental Microbiology* 45:502–510.

Conrad, R., Phelps, T.J., and **Zeikus, J.G.** 1985. Gas metabolism evidence in support of the juxtaposition of hydrogen-producing and methanogenic bacteria in sewage sludge and lake sediments. *Applied and Environmental Microbiology* 50:595–601.

Conrad, R., Schink, B., and **Phelps, T.J.** 1986. Thermodynamics of H_2-consuming and H_2-producing metabolic reactions in diverse methanogenic environments under in situ conditions. *FEMS Microbiology Ecology* 38:353–360.

Davison, W. and **Seed, G.** 1983. The kinetics of the oxidation of ferrous iron in synthetic and natural waters. *Geochimica et Cosmochimica Acta* 47:67–79.

Elderfield, H., Luedtke, N., McCaffrey, R.J., and **Bender, M.** 1981. Benthic flux studies in Narragansett Bay. *American Journal of Science* 281:768–787.

Emerson, S., Cranston, R.E., and **Liss, P.S.** 1979. Redox species in a reducing fjord: equilibrium and kinetic considerations. *Deep-Sea Research* 26A:859–878.

Emerson, S., Kalhorn, S., Jacobs, L., and **Tebo, B.M.** 1982. Environmental oxidation rate of manganese(II): bacterial catalysis. *Geochimica et Cosmochimica Acta* 46:1073–1079.

Froelich, P.N., Klinkhammer, G.P., Bender, M.L., Leudtke, N.A., Heath, G.R., Cullen, D., Dauphin, P., Hammond, D., Hartman, B., and **Maynard, V.** 1979. Early oxidation of organic matter in pelagic sediments of the eastern equatorial Atlantic: suboxic diagenesis. *Geochimica et Cosmochimica Acta* 43:1075–1090.

Gallardo, V.A. 1977. Large benthic microbial communities in sulphide biota under Peru-Chile subsurface countercurrent. *Nature* 268:331–332.

Garrels, R.M. and **Lerman, A.** 1984. Coupling of the sedimentary sulfur and carbon cycles—an improved model. *American Journal of Science* 284:989–1007.

Giblin, A.E. and **Howarth, R.W.** 1984. Pore water evidence for a dynamic sedimentary iron cycle in salt marshes. *Limnology and Oceanography* 29:47–63.

Goldhaber, M.B. and **Kaplan, I.R.** 1974. The sulfur cycle, pp. 569–655, in Goldberg, E.D. (editor), *The Sea, Vol. 5,* John Wiley and Sons, New York.

Gouy, J.-L., Bergé, P., and **Labroue, L.** 1984. *Gallionella ferruginea,* facteur de dénitrification dans les eaux pauvres en matière organique. *Comptes Rendus Academie des Sciences, Paris* 298 (III):153–156.

Hansen, J.I., Henriksen, K., and **Blackburn, T.H.** 1981. Seasonal distribution of nitrifying bacteria and rates of nitrification in coastal marine sediments. *Microbial Ecology* 7:297–304.

Hasan, S.M. and **Hall, J.B.** 1975. The physiological function of nitrate reduction in *Clostridium perfringens. Journal of General Microbiology* 87:120–128.

Henriksen, K. 1980. Measurement of in situ rates of nitrification in sediment. *Microbial Ecology* 6:329–337.

Henriksen, K., Hansen, J.I., and **Blackburn, T.H.** 1981. Rates of nitrification, distribution of nitrifying bacteria, and nitrogen fluxes in different types of sediment from Danish waters. *Marine Biology* 61:299–304.

Howarth, R.W. 1984. The ecological significance of sulfur in the energy dynamics of salt marsh and coastal marine sediments. *Biogeochemistry* 1:5–27.

Ingvorsen, K. and **Brock, T.D.** 1982. Electron flow via sulfate reduction and methanogenesis in the anaerobic hypolimnion of Lake Mendota. *Limnology and Oceanography* 27:559–564.

Jenkins, M.C. and **Kemp. W.M.** 1984. The coupling of nitrification and denitrification in two estuarine sediments. *Limnology and Oceanography* 29:609–619.

Jørgensen, B.B. 1977. The sulfur cycle of a coastal marine sediment (Limfjorden, Denmark). *Limnology and Oceanography* 22:814–832.

Jørgensen, B.B. 1982a. Ecology of the bacteria of the sulphur cycle with special reference to anoxic-oxic interface environments. *Philosophical Transactions of the Royal Society of London, Series B* 298:543–561.

Jørgensen, B.B. 1982b. Mineralization of organic matter in the sea bed—the role of sulphate reduction. *Nature* 296:643–645.

Jørgensen, B.B. 1983. The microbial sulphur cycle, pp. 91–124, in Krumbein, W.E. (editor), *Microbial Geochemistry,* Blackwell, Oxford.

Jørgensen, B.B. and **Des Marais, D.J.** 1986. Competition for sulfide among colorless and purple sulfur bacteria in cyanobacterial mats. *FEMS Microbiology Ecology* 38:179–186.

Jørgensen, B.B. and **Revsbech, N.P.** 1983. Colorless sulfur bacteria, *Beggiatoa* spp. and *Thiovulum* spp., in O_2 and H_2S microgradients. *Applied and Environmental Microbiology* 45:1261–1270.

Jørgensen, B.B. and **Sørensen, J.** 1985. Seasonal cycles of O_2, NO_3^-, and SO_4^{2-} reduction in estuarine sediments: the significance of an NO_3^- reduction maximum in spring. *Marine Ecology—Progress Series* 24:65–74.

Jørgensen, B.B., Kuenen, J.G., and **Cohen, Y.** 1979. Microbial transformations of sulfur compounds in a stratified lake (Solar Lake, Sinai). *Limnology and Oceanography* 24:799–822.

Jørgensen, K.S., Jensen, H.B., and **Sørensen, J.** 1984. Nitrous oxide production from nitrification and denitrification in marine sediment at low oxygen concentrations. *Canadian Journal of Microbiology* 30:1073–1078.

Keith, S.M. and **Herbert, R.A.** 1983. Dissimilatory nitrate reduction by a strain of *Desulfovibrio desulfuricans. FEMS Microbiology Letters* 18:55–59.

Kepkay, P.E., Burdige, D.J., and **Nealson, K.H.** 1984. Kinetics of bacterial manganese binding and oxidation in the chemostat. *Geomicrobiology Journal* 3:245–262.

King, D. and **Nedwell, D.B.** 1985. The influence of nitrate concentration upon the end products of nitrate dissimilation by bacteria in anaerobic salt marsh sediment. *FEMS Microbiology Ecology* 31:23–28.

King, G.M. 1984. Utilization of hydrogen, acetate and "noncompetitive" substrates by methanogenic bacteria in marine sediments. *Geomicrobiology Journal* 3:275–306.

Kino, K. and **Usami, S.** 1982. Biological reduction of ferric iron by iron- and sulfur-oxidizing bacteria. *Agricultural Biology and Chemistry* 46:803–805.

Klinkhammer, G.P. 1980. Early diagenesis in sediments from the eastern equatorial Pacific, II. Pore water metal results. *Earth and Planetary Science Letters* 49:81–101.

Koike, I. and **Hattori, A.** 1978. Denitrification and ammonia formation in anaerobic coastal sediments. *Applied and Environmental Microbiology* 35:278–282.

Kristjansson, J.K., Schönheit, P., and **Thauer, R.K.** 1982. Different K_s values for hydrogen of methanogenic bacteria and sulfate reducing bacteria: An explanation for the apparent inhibition of methanogenesis by sulfate. *Archives of Microbiology* 131:278–282.

Kuenen, J.G. 1975. Colourless sulphur bacteria and their role in the sulphur cycle. *Plant and Soil* 43:49–76.

Laanbroek, H.J., Stal, L.J., and **Veldkamp, H.** 1978. Utilization of hydrogen and formate by *Campylobacter* spec. under aerobic and anaerobic conditions. *Archives of Microbiology* 119:99–102.

Lovley, D.R. and **Klug, M.J.** 1982. Intermediary metabolism of organic matter in the sediments of a eutrophic lake. *Applied and Environmental Microbiology* 43:552–560.

Lovley, D.R. and **Klug, M.J.** 1983a. Sulfate reducers can outcompete methanogens at freshwater sulfate concentrations. *Applied and Environmental Microbiology* 45:187–192.

Lovley, D.R. and **Klug, M.J.** 1983b. Methanogenesis from methanol and methylamines and acetogenesis from hydrogen and carbon dioxide in the sediments of a eutrophic lake. *Applied and Environmental Microbiology* 45:1310–1315.

Lovley, D.R. and **Klug, M.J.** 1986. Model for the distribution of sulfate reduction and methanogenesis in freshwater sediment. *Geochimica et Cosmochimica Acta* 50:11–18.

Lovley, D.R. and **Phillips, E.J.P.** 1986. Organic matter mineralization with reduction of ferric iron in anaerobic sediments. *Applied and Environmental Microbiology* 51:683–689.

Lovley, D.R. and **Phillips, E.J.P.** 1987. Competitive mechanisms for inhibition of sulfate reduction and methane production in the zone of ferric reduction in sediments. *Applied and Environmental Microbiology* 53:2636–2641.

Lundgren, D.G., Vestal, J.R., and **Tabita, F.R.** 1974. The iron oxidizing bacteria, pp. 457–472, in Neilands, J.B. (editor), *Microbial Iron Metabolism*, Academic Press, New York.

Lupton, F.S. and **Zeikus, J.G.** 1984. Physiological basis for sulfate-dependent hydrogen competition between sulfidogens and methanogens. *Current Microbiology* 11:7–12.

Lyle, M. 1983. The brown-green color transition in marine sediments: A marker of the Fe(III)-Fe(II) redox boundary. *Limnology and Oceanography* 28:1026–1033.

Maier, S. and **Gallardo, V.A.** 1984. Nutritional characteristics of two marine thioplocas determined by autoradiography. *Archives of Microbiology* 139:218–220.

Macfarlane, G.T. and **Herbert, R.A.** 1984. Dissimilatory nitrate reduction and nitrification in estuarine sediments. *Journal of General Microbiology* 130:2301–2308.

McInerney, M.J., Bryant, M.P., and **Hespell, R.B.** 1981. *Syntrophomonas wolfei* gen. nov., sp. nov., an anaerobic, syntrophic, fatty acid-oxidizing bacterium. *Applied and Environmental Microbiology* 41:1029–1039.

Millero, F.J. 1986. The thermodynamics and kinetics of the hydrogen sulfide system in natural waters. *Marine Chemistry* 18:121–147.

Millero, F.J., Hubinger, S., Fernandez, M., and **Garnett, S.** 1987. Oxidation of H_2S in seawater as a function of temperature, pH, and ionic strength. *Environmental Science and Technology* 21:439–443.

Millero, F.J., **Sotolongo, S.**, and **Izaguirre, M.** 1987. The oxidation kinetics of Fe(II) in seawater. *Geochimica et Cosmochimica Acta* 51:793–801.

Mitchell, G.J., **Jones, J.G.**, and **Cole, J.A.** 1986. Distribution and regulation of nitrate and nitrite reduction by *Desulfovibrio* and *Desulfotomaculum* species. *Archives of Microbiology* 144:35–40.

Moraghan, J.T. and **Buresh, R.J.** 1977. Chemical reduction of nitrate and nitrous oxide by ferrous iron. *Soil Science Society of American Journal* 41:47–50.

Morgan, J.J. 1964. *Chemistry of Aqueous Manganese II and IV*. Ph.D. Thesis, Harvard University, Cambridge, Massachusetts, 244 pp.

Murray, J.W. and **Gill, G.** 1978. The geochemistry of iron in Puget Sound. *Geochimica et Cosmochimica Acta* 42:9–19.

Mustoe, G.E. 1981. Bacterial oxidation of manganese and iron in a modern cold spring. *Geological Society of America Bulletin, I* 92:147–153.

Nealson, K.H. 1983. The microbial manganese cycle, pp. 191–221, in Krumbein, W.E. (editor), *Microbial Geochemistry*, Blackwell Scientific Publications, Oxford.

Nelson, D.C., **Jørgensen, B.B.**, and **Revsbech, N.P.** 1986. Growth pattern and yield of a chemoautotrophic *Beggiatoa* sp. in oxygen-sulfide microgradients. *Applied and Environmental Microbiology* 52:225–233.

Novelli, P.C., **Scranton, M.I.**, and **Michener, R.H.** 1987. Hydrogen distributions in marine sediments. *Limnology and Oceanography* 32:565–576.

Obuekwe, C.O., **Westlake, D.W.S.**, and **Cook F.D.** 1981. Effect of nitrate on reduction of ferric iron by a bacterium isolated from crude oil. *Canadian Journal of Microbiology* 27:692–697.

Oremland, R.S. and **Polcin, S.** 1982. Methanogenesis and sulfate reduction: competitive and noncompetitive substrates in estuarine sediments. *Applied and Environmental Microbiology* 44:1270–1276.

Ottow, J.C.G. 1968. Evaluation of iron-reducing bacteria in soil and the physiological mechanisms of iron reduction in *Aerobacter aerogenes. Zeitschrift für Allgemeine Mikrobiologie* 8:441–443.

Ottow, J.C.G. 1969. The distribution and differentiation of iron-reducing bacteria in gley soils. *Zentralblatt für Bakteriologie, II* 123:600–615.

Ottow, J.C.G. and **von Klopotek, A.** 1969. Enzymatic reduction of iron oxide by fungi. *Applied Microbiology* 18:41–43.

Painter, H.W. 1970. A review of literature on inorganic nitrogen metabolism in microorganisms. *Water Research* 4:393–450.

Phelps, T.J. and **Zeikus, J.G.** 1984. Influence of pH on terminal carbon metabolism in anoxic sediments from a mildly acidic lake. *Applied and Environmental Microbiology* 48:1088–1095.

Pyzik, A.J. and **Sommer, S.E.** 1981. Sedimentary iron monosulfides: kinetics and mechanisms of formation. *Geochimica et Cosmochimica Acta* 45:687–698.

Reimers, C.E., **Fischer, K.M.**, **Merewether, R.**, **Smith Jr., K.L.**, and **Jahnke, R.A.** 1986. Oxygen microprofiles measured in situ in deep ocean sediments. *Nature* 320:741–744.

Revsbech, N.P., **Jørgensen, B.B.**, and **Blackburn, T.H.** 1980. Oxygen in the sea bottom measured with a microelectrode. *Science* 207:1355–1356.

Robbins, J.A. and **Callender, E.** 1975. Diagenesis of manganese in Lake Michigan sediments. American Journal of Science 275:512–533.

Robertson, L.A. and **Kuenen, J.G.** 1984. Aerobic denitrification—old wine in new bottles? *Antonie van Leeuwenhoek Journal of Microbiology and Serology* 50:525–544.

Robinson, J.A. and **Tiedje, J.M.** 1984. Competition between sulfate-reducing and methanogenic bacteria for H_2 under resting and growing conditions. *Archives of Microbiology* 137:26–32.

Schönheit, P., **Kristjansson, J.K.**, and **Thauer, R.K.** 1982. Kinetic mechanism for the ability of sulfate reducers to out-compete methanogens for acetate. *Archives of Microbiology* 132:285–288.

Schropp, S.J., Scranton, M.I., and **Schwarz, J.R.** 1987. Dissolved hydrogen, facultatively anaerobic, hydrogen-producing bacteria, and potential production rates in the western North Atlantic Ocean and Gulf of Mexico. *Limnology and Oceanography* 32:396–402.

Seitz, H.-J. and **Cypionka, H.** 1986. Chemolithotrophic growth of *Desulfovibrio desulfuricans* with hydrogen coupled to ammonification of nitrate or nitrite. *Archives of Microbiology* 146:63–67.

Seitzinger, S.P., Nixon, S.W., and **Pilson, M.E.Q.** 1984. Denitrification and nitrous oxide production in a coastal marine ecosystem. *Limnology and Oceanography* 29:73–83.

Sørensen, J. 1978. Capacity for denitrification and reduction of nitrate to ammonia in a coastal marine sediment. *Applied and Environmental Microbiology* 35:301–305.

Sørensen, J. 1982. Reduction of ferric iron in anaerobic, marine sediment and interaction with reduction of nitrate and sulfate. *Applied and Environmental Microbiology* 43:319–324.

Sørensen, J., Christensen, D., and **Jørgensen, B.B.** 1981. Volatile fatty acids and hydrogen as substrates for sulfate-reducing bacteria in anaerobic marine sediments. *Applied and Environmental Microbiology* 42:5–11.

Sørensen, J. and **Jørgensen, B.B.** 1987. Early diagenesis in sediments from Danish coastal waters: microbial activity and Mn-Fe-S geochemistry. *Geochimica et Cosmochimica Acta* 51:1583–1590.

Sørensen, J., Jørgensen, B.B., and **Revsbech, N.P.** 1979. A comparison of oxygen, nitrate and sulfate respiration in a coastal marine sediment. *Microbial Ecology* 5:105–115.

Sorokin, Y.I. 1970. Interrelations between sulphur and carbon turnover in meromictic lakes. *Archiv für Hydrobiologie* 66:391–446.

Sorokin, Y.I. 1972. The bacterial population and the process of hydrogen sulfide oxidation in the Black Sea. *Journal du Conseil International pour l'Exploration de la Mer* 34:423–455.

Strayer, R.F. and **Tiedje, J.M.** 1978. Kinetic parameters of the conversion of methane precursors to methane in a hypertrophic lake sediment. *Applied and Environmental Microbiology* 36:330–340.

Sung, W. and **Morgan, J.J.** 1981. Oxidative removal of Mn(II) from solution catalyzed by the γ-FeOOH (lepidocrocite) surface. *Geochimica et Cosmochimica Acta* 45:2377–2383.

Tebo, B.M. and **Emerson, S.** 1985. Effect of oxygen tension, Mn(II) concentration, and temperature on the microbially catalyzed Mn(II) oxidation rate in a marine fjord. *Applied and Environmental Microbiology* 50:1268–1273.

Tebo, B.M. and **Emerson, S.** 1986. Microbial manganese(II) oxidation in the marine environment: a quantitative study. *Biogeochemistry* 2:149–161.

Trimble, R.B., and **Ehrlich, H.L.** 1968. Bacteriology of manganese nodules. III. Reduction of MnO_2 by strains of nodule bacteria. *Applied Microbiology* 16:695–702.

Troelsen, H. and **Jørgensen, B.B.** 1982. Seasonal dynamics of elemental sulfur in two coastal sediments. *Estuarine and Coastal Shelf Science* 15:255–260.

Tuttle, J.H. and **Jannasch, H.W.** 1973. Sulfide and thiosulfate-oxidizing bacteria in anoxic marine basins. *Marine Biology* 20:64–70.

Vanderborght, J.P. and **Billen, G.** 1975. Vertical distribution of nitrate concentration in interstitial water of marine sediments with nitrification and denitrification. *Limnology and Oceanography* 20:953–961.

Verhoeven, W., Koster, A.L., and **van Nievelt, M.G.A.** 1954. Studies on true dissimilatory nitrate reduction. III. *Micrococcus denitrificans* Beijerinck, a bacterium capable of using molecular hydrogen in denitrification. *Antonie van Leeuwenhoek Journal of Microbiology and Serology* 20:273–284.

Williams, L.A. and **Reimers, C.** 1983. Role of bacterial mats in oxygen-deficient marine basins and coastal upwelling regimes: Preliminary report. *Geology* 11:267–269.

Winfrey, M.R. and **Zeikus, J.G.** 1977. Effect of sulfate on carbon and electron flow during microbial methanogenesis in freshwater sediments. *Applied and Environmental Microbiology* 33:275–281.

Witzel, K.P. and **Overbeck, J.G.** 1979. Heterotrophic nitrification by *Arthrobacter* sp. (strain 9006) as influenced by different cultural conditions, growth state and acetate metabolism. *Archives of Microbiology* 122:137–143.

Zaiss, U. 1981. Seasonal studies of methanogenesis and desulfurication in sediments of the river Saar. *Zentralblatt für Bakteriologie und Hygiene I. Abteiling Originale C* 2:76–89.

Zavarzin, G.A. 1982. The genus *Metallogenium*, pp. 524–528, in Starr, M. et al. (editors), *The Prokaryotes*, Springer-Verlag, Berlin.

Zeikus, J.G. 1977. The biology of methanogenic bacteria. *Bacteriological Reviews* 41:514–541.

Zeikus, J.G., Ben-Bassat, A., and **Hegge, P.W.** 1980. Microbiology of methanogenesis in thermal, volcanic environments. *Journal of Bacteriology* 143:432–440.

8

Chemosynthetically Sustained Ecosystems in the Deep Sea

Holger W. Jannasch

Woods Hole Oceanographic Institution
Woods Hole, Massachusetts, U.S.A.

8.1 INTRODUCTION

Chemolithotrophy appears to be the only type of metabolism that may result in primary production of organic carbon in a permanently dark environment. This fact was implicit in Winogradsky's (1887) postulation of chemoautotrophy (see Chapter 1), i.e., the reduction of carbon dioxide to organic carbon by energy derived from the oxidation of an inorganic compound. In place of "chemolithoautotrophy" the following discussion will often apply the term "chemosynthesis," making use of its contrast to "photosynthesis." The potential importance of chemosynthesis for organisms existing in the absence of light was never seriously considered because life on our planet is overwhelmingly supported by photosynthesis.

This was generally held to be true also for the deep ocean where living organisms, although in decreasing numbers, have been observed down to the greatest depths of 11 km. About 65 percent of this globe is covered by water deep enough to eliminate all light from reaching the sea floor. This immense volume of permanently lightless seawater below a depth of about 300 m represents approximately 80 percent of the earth's biosphere. The scarcity and the distribution of organisms in this environment clearly reflect the photosynthetic food input from the uppermost layers of water and the high rate of dilution and consumption on its way down to the bottom. The strategies of survival of the deep-sea fauna have much in common with those of desert communities.

It was quite a sensation when another characteristic feature of deserts was found in the deep oceans: oases. During geological studies of zones of active submarine volcanism known as "tectonic ocean spreading centers," these oasis sites were actually visited by manned research submersibles to depths between 1800 and 3700 m. On the first dives to particularly active centers, hot springs were discovered discharging "hydrothermal fluid" into the surrounding seawater (Corliss et al., 1979; Jannasch and Wirsen, 1979). The immediate vicinity of these "hydrothermal vents" was inhabited by dense populations of marine invertebrates: mussels, clams, and tube worms of unusual size and representing new species, genera, and families (reviewed by Grassle, 1986). It was quite clear from the first observations that these immense amounts of biomass could not be explained by a supply of photosynthetically produced food particles. Instead, the energy resource was theorized, and later found, to be terrestrial (chemical) rather than solar (reviewed by Jannasch, 1985).

It seems evident that physical energy (high temperature and pressure) is converted into potential chemical energy by geothermal processes within the earth's crust. The resulting highly reduced hot hydrothermal fluid is carried upward to the sea floor where the release of energy on contact with cold oxygenated seawater takes place chemically or, if mediated microbiologically, is accompanied by the fixation of CO_2 into organic carbon. Thus, in an unexpected manner the geophysical/geological exploration of plate tectonics resulted in the discovery of a major microbiological phenomenon: the chemolithotrophic sustenance of thriving communities of organisms in the deep sea—a timely contribution to the 100th anniversary of Winogradsky's observation on microbial autotrophy in the absence of light.

8.2 DEEP SEA SOURCES OF ENERGY

The tectonic plate movements at spreading centers involve an intermittent penetration of seawater several km into the earth's crust. Through heating by underlying magma chambers it may reach temperatures of 1200°C before ascending again and reaching the seafloor as hydrothermal fluid with temperatures of 350°–360°C. These "hot" hydrothermal vents have flow rates of 1–2 m/s. The instant precipitation of black polymetal sulfides gives them a characteristic appearance that has been termed "black smokers." The precipitation process also produces anhydrite ($CaSO_4$) which forms part of these chimney-like structures (Edmond and Von Damm, 1983). More important for the occurrence of rich animal populations are the "warm" vents with temperatures around 25°C and flow rates of 1–5 cm/s, where a premixing of hydrothermal fluid with ambient seawater takes place in the porous sub-seafloor lava beds prior to emission (Figure 8.1). An overview of the hydrothermal processes involved is given in a symposium volume edited by Rona et al. (1983).

"Hydrothermal fluid" is thermally altered seawater which is anoxic, acid, and highly reduced. Its chemical composition is the key for chemolithotrophic processes. Certain ions are lost during subsurface precipitations or exchange reactions; others are enriched through leaching processes. The chemical composition of most offshore vent emissions resembles that presented for the 21°N site (about 100 miles off the Pacific coast of northern Mexico) in Table 8.1. At the Guaymas Basin vent site (in the Gulf of California), on the other hand, the actual basalt lava emissions are overlayed by several hundred meters of sed-

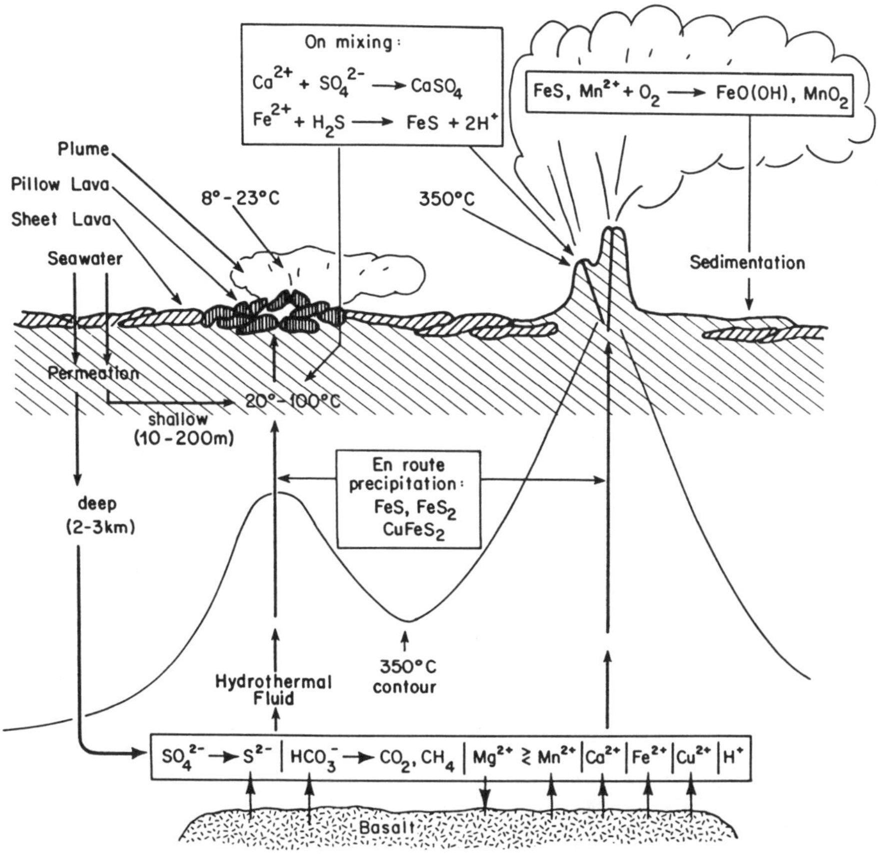

Figure 8.1 The major geochemical processes occurring within the oceanic crust and on the deep sea floor. As seawater penetrates several km into the crust, it is heated to 350–400°C, reacts with basaltic rocks and leaches various chemical species into solution. The highly reduced "hydrothermal fluid" rises and reaches the seafloor either directly (hot vents) or after mixing with cold, oxygenated seawater before emission (warm vents). On mixing, polymetal sulfides and calcium sulfate (anhydrite) precipitate either within subsurface lava conduits or as "chimneys" and suspended particulate matter in the "black smokers" (modified from Jannasch and Mottl, 1985).

iment which is penetrated by the hydrothermal fluid (data in Table 8.1). Geochemical processes result in the difference of ion composition in the sea floor emissions at these two sites (Lonsdale et al., 1980; Edmond and Von Damm, 1985).

Most of the heavy metal enrichment found in the offshore "bare lava" vent emissions is stripped during the passage of hydrothermal fluid through sediments at the Guaymas Basin site (Table 8.1). As a result, the typical 350°C black smokers are here transparent. Intrusion of magnesium indicates the degree of seawater mixing with the originally magnesium-free hydrothermal fluid. The calcium enrichment of hydrothermic fluid produces, at the point of mixing

Table 8.1 Chemical composition of hydrothermal emissions in the deep sea are compared with ambient sea water

Chemical	Unit of concentration	Guayamas Basin	21°N	Sea water
Fe	(μM)	56	1664	0.001
Mn	(μM)	139	960	0.001
Co	(nM)	5	213	0.03
Cu	(μM)	1	35	0.007
Zn	(μM)	4.2	106	0.01
Ag	(nM)	230	38	0.02
Pb	(nM)	265	308	0.01
Na	(mM)	489	432	464
K	(mM)	48.5	28	9.79
Li	(μM)	1054	891	26
Mg	(mM)	29	0	52.7
Ca	(μM)	29	15.6	10.2
Sr	(μM)	202	81	87
Ba	(μM)	12	7	0.14
Cl	(mM)	601	489	541
SO_4	(mM)	0	0	27.9
Si	(mM)	12.9	17.6	0.16
Al	(μM)	0.9	5.2	0.01
NH_3	(mM)	15.6	0	0
H_2S	(mM)	5.82	7.3	0
H_2	(mM)	8	50	0.001
CH_4	(mM)	1	1.6	0.001
pH		5.9	3.4	8.0
alk meq		10.6	−0.4	0.16

The samples were collected at vents in the Guaymas Basin (depth 2003 m) and the site at 21°N (depth 2610 m). (Data from Welhan and Craig, 1983; Edmond and Von Damm, 1985; Jannasch and Mottl, 1985).

with the sulfate-containing cold seawater, fast-growing cones of anhydrite (calcium sulfate) blackened by traces of metal sulfides. On cooling, anhydrite redissolves in seawater, a process that also applies to the above mentioned anhydrite constituent of hot vent chimneys (Edmond et al., 1982). The resulting porosity of the chimney walls becomes an important feature for the slow passage of sulfide-containing hydrothermal fluid and growth of microbial mats.

The amount of sulfide ions, as the most important electron donor for chemolithotrophy at deep sea vents, derives only in part from chemical reduction of the original sulfate content of the entrained seawater. Some of the geothermal sulfate reduction and metal sulfide deposition takes place during the downward movement of seawater into the earth's crust, and part of the sulfide content of the recirculating hydrothermal fluid stems from sulfur leaching of basalt at high temperature (Figure 8.1). The additional metal sulfide depositions from the ascending hydrothermal fluid in the Guaymas Basin sediments lowers the final sulfide concentration at this vent site.

At the same time, the organic constituents of these sediments, derived from the deposition of diatoms in the photosynthetically highly productive waters of the Gulf of California, provides the vent emissions of the Guaymas Basin site with a high ammonia content. This is, again, very different from the "bare lava" vents of the 21°N site where the photosynthetic input is negligible and no ammonia is found in the hydrothermal fluid (Table 8.1).

The amounts of dissolved hydrogen and methane and other potential electron donors for chemosynthesis, are highly variable at every vent site studied so far and, therefore, undistinguishable for the two vent sites recorded in Table 8.1. Their background quantities in ambient seawater are equally nonconservative, i.e., subject to biological conversion. Methane, being organic, is not commonly considered a substrate for chemolithotrophy. Dealing with deep sea vent emissions, however, where the bulk of methane is of geothermic and not of biological origin (Welhan and Craig, 1983), one is tempted to treat it like an inorganic compound and a source of energy and carbon at the same time in a mixed chemolithomethylotrophy.

8.3 *IN SITU* CHEMOSYNTHETIC ACTIVITY

The first observations of dense bacterial suspensions in the plumes of "warm" vents (around 23°C) at the Galapagos Rift ocean spreading center (Corliss et al., 1979; Jannasch and Wirsen, 1979) provided evidence to support the assumed microbial base of the food chain leading to the rich animal communities. During the years following these initial observations, studies on chemolithotrophic *in situ* activities, physiology of microbial isolates, and measurement of the transfer of "chemosynthate" to the vent animals have been limited by the rare occasions to visit the deep sea sites by submersible vessels and by the relatively restricted operations that are possible on the deep sea floor.

Attempts to measure the total biomass of natural microbial populations have advanced during the last decade from plate counts and microscopic cell counts, using epifluorescence optics on acridine orange-stained cells, to ATP and total adenylate determinations. In addition, the ratio of GTP:ATP (GTP, guanosine 5'triphosphate) has been used as a determination of microbial growth activity at certain warm vent sites (Karl et al., 1980).

Such general growth or biomass assessments (Karl, 1980) do not estimate, however, the part played by chemolithotrophic organisms. Chemosynthesis was measured, therefore, in much the same manner that field studies of photosynthetic CO_2-fixation are conducted. Sets of six 120 ml syringes, precharged with ^{14}C-labeled bicarbonate, were filled (using the mechanical arm of the submersible) with the bacterial suspension from a warm vent (around 23°C). One set was incubated *in situ* in ambient water of temperature around 3°C; two other sets were incubated in the ship's laboratory at 3° and 23°C. Rates of CO_2 fixation of a typical experiment are presented in Figure 8.2. The hatched columns show data from syringes that were also supplemented with 1 mM thiosulfate. Placing a syringe holder within a vent plume for incubation at *in situ* temperature was not possible and may also be irrelevant in the absence of accurate temperature recording.

Three major characteristics of chemolithoautotrophic activity in warm vent emissions emerge from the data presented in Figure 8.2: 1) the effect of *in situ*

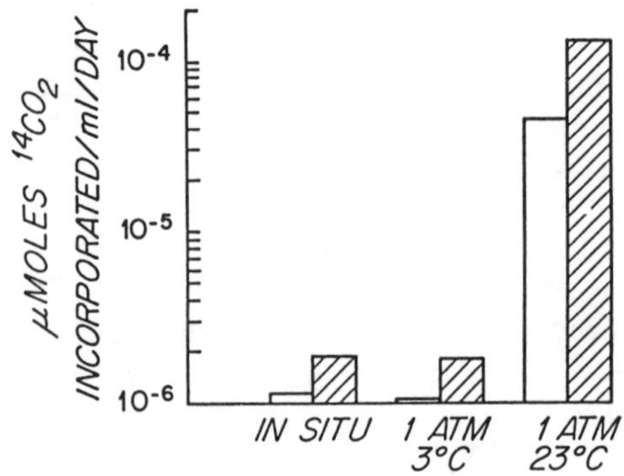

Figure 8.2 Rate of carbon dioxide incorporation by the natural population of bacteria suspended in "warm vent" water incubated *in situ* (260 atm, 3°C) and in the ship's laboratory at two temperatures. Hatched columns indicate a 1 mM thiosulfate supplement (from Jannasch, 1984).

pressure (around 250 atm) is negligible; 2) the response to temperature indicates that the organisms are mesophiles; 3) the addition of thiosulfate resulted in at least a doubling of the CO_2-fixation rate. A more detailed study (Tuttle et al., 1983; Wirsen et al., 1986) confirmed this general scheme. Experiments conducted at 1 atm under otherwise equal conditions resulted in similar data, i.e., within the range of error, as in the *in situ* experiments. The CO_2-fixation rate slowed abruptly when the samples were cooled from 23°C to ambient seawater temperature (2–3°C). In addition to thiosulfate, tetrathionate was also oxidized by the natural microbial warm vent populations. Experiments supplementing ammonia, nitrite, hydrogen, or methane have not been done. Only the latter two are present in the water issuing from bare lava vents. In separate experiments, uptake of the ^{14}C-label from acetate and glucose was found to be quite rapid (Tuttle et al., 1983; Wirsen et al., 1986).

One disturbing result of these studies was the early notion that the entire production of bacteria within a warm bare lava vent field would not be sufficient to support the growth of those massive invertebrate populations that are found outside of the areas where the bacterial cell suspensions occurred. The explanation emerged later when the assumption of a new type of endosymbiotic chemolithotrophy was substantiated (see below).

An unusual feature of the sediment-covered Guaymas Basin vent site is the massive occurrence of mats of essentially monocultures of *Beggiatoa* (Nelson et al., 1988). Chemoautotrophy was demonstrated in these mats by measuring ribulose bisphosphate carboxylase (RuBPC) and $^{14}CO_2$-uptake activities on freshly collected material; the results were similar to those obtained earlier with pure cultures of marine *Beggiatoa* (Nelson and Jannasch, 1983).

8.4 CHEMOLITHOTROPHIC ISOLATES

Isolates of thiobacilli-like organisms were readily obtained in large numbers. A representative list of strains in Table 8.2 is subdivided according to metabolic characteristics such as acid formation as well as growth and CO_2 fixation in the presence and absence of an organic substrate (Ruby et al., 1981). The wide range of organisms from obligate chemolithotrophs, through the mixotrophs and facultative chemolithotrophs to the chemolithoheterotrophs—and certain true heterotrophs that are able to oxidize sulfur compounds with no measurable gain of energy—is not different from other marine habitats and from what is known of aerobic microbial sulfur oxidation in general (see Chapter 4, and Kuenen and Beudeker, 1982).

The acid-producing and slightly acidophilic obligate chemolithotrophs may only be able to grow competitively in dense mats on rock surfaces of subseafloor lava chambers. Many of the mixotrophs and facultative chemolithotrophs grow well at the pH of seawater (Figure 8.1) and may even raise the pH slightly by producing polythionates instead of sulfate. Their growth is favored by the occurrence of variable concentrations of dissolved organic compounds in the sulfide- and thiosulfate-containing vent water surrounding the rich animal communities. Gottschal et al. (1979) have emphasized the ecological significance of the metabolic flexibility of these facultative sulfide-oxidizing organisms from results of continuous culture studies on competitive population dynamics (discussed in detail in Chapter 4). This principle appears to be directly applicable to the microbial vent populations considering, of course, that no homogeneous mixing exists and that part of the various metabolic types are spatially separated. However, the result appears to be the same: the larger portion of the total population of aerobic sulfur-oxidizing bacteria consists of nonobligate chemolithotrophs.

Tests for low-temperature adaptation (the psychrophile character, shown by growth optima between 8° and 16°C) and high-temperature adaptation

Table 8.2 Growth of bacterial isolates (thiobacilli-like) from vent water samples on thiosulfate and yeast extract media[a]

		Growth[b]		$\mu g\ HCO_3^-$ fixed per 10^7 cells		
Strain	Final pH in T-ASW	Y-ASW	T-ASW	Y-ASW	T-ASW	TY-ASW
L-12	4.7	.7	67	IG[c]	3.1	3.2
RTPMB	4.7	.6	56	IG	2.7	3.1
TB-49	4.7	1.4	86	IG	6.1	3.7
SS-T	3.9	2.0	199	IG	3.1	3.6
NF-18	6.3	203	12	.1	2.5	.1
AG-33	6.5	218	7	.2	.6	.2
TB-66	5.7	106	4	.3	.6	.2
RTRG	6.2	ND[d]	6	.1	.7	.2
AG-25	8.4	351	0.4	0.1	IG	0.1
NF-13	8.8	115	.7	.2	IG	.2
TB5.5-9	8.7	245	.6	.1	.1	.1
TB-A	7.9	ND	ND	.1	.1	.1
K-12	7.8	ND	ND	.0	IG	.1

[a]T, thiosulfate (10 mM); ASW, artificial seawater; Y, yeast extract [0.01% wt (vol)]; initial pH was 7.0.
[b]Ratio of maximum cell density (as determined by epifluorescence direct counts) over that in the unsupplemented control
[c]IG, insufficient growth for determination
[d]ND, no data
Data from Ruby et al., 1981.

(thermophily, growth optima above 40°C) were negative: all isolates were mesophilic with optima between 28° and 35°C. Growth studies at various pressures relevant to deep-sea origin of the isolates revealed them to be barotolerant rather than barophilic, i.e., highest growth rates occurred at 1 atm while pressures of 200–300 atm lowered the rates no more than 20 percent. No growth occurred at pressures above 500–600 atm.

Two physiologically typical isolates, listed in Table 8.2, were studied in more detail: strains L-12 and NF-18. The obligately chemolithotrophic strain L-12 was found to belong to the genus *Thiomicrospira* (Ruby and Jannasch, 1982) and appeared not to be different enough from *T. pelophila* (Kuenen and Veldkamp, 1972) to create a new species. A typical microaerophile, it could be grown to unusually large colonies on agar plates when care was taken to reduce the oxygen concentration of the atmosphere. Also unusual was the high tolerance of this strain (*Thiomicrospira* L-12) for hydrogen sulfide (300 μM). Its growth response toward temperature and pressure fell within the ranges given above for natural populations of sulfur-oxidizing bacteria. During later studies, a *Thiomicrospira* strain was isolated from another vent site (21°N) that was indeed different from *T. pelophila,* mainly by its G+C mol percent content and its higher growth rate and optimal pH in the same thiosulfate medium. It was described as a new species, *T. crunogena* (Jannasch et al., 1985).

The second strain (NF-18 in Table 8.2) studied specifically represents a facultative or mixotrophic thiosulfate oxidizer. The stimulation of CO_2-fixation by thiosulfate is almost 10-fold (Figure 8.3) The addition of an organic substrate, such as glucose or yeast extract, eliminated or obscured any growth advantage due to thiosulfate oxidation (Ruby et al., 1981). By virtue of their catabolic flexibility and their ability to grow optimally at a large range of relatively high pH values (6.5–8.5), many organisms in this category are likely to constitute the major population of sulfur-oxidizing bacteria at the hydrothermal vents.

Attempts to isolate extremely thermoacidophilic, aerobic sulfur oxidizers of the *Sulfolobus*-type from the deep sea hydrothermal vents have so far been unsuccessful. It is quite possible that the gradient between the two zones, the hot anoxic hydrothermal fluid on the one side and the cold oxygenated mixture of hydrothermal fluid and ambient seawater on the other, is simply too steep to permit the establishment of a *Sulfolobus* population. Fluctuations at this interface may also be highly destructive to such populations. If such organisms exist, they do not seem to be extensive and might simply have escaped detection.

The first isolation of an anaerobic chemolithotroph (Jones et al., 1983) originated from the outside of a 21°N black smoker and turned out to be an extremely thermophilic methanogen. It grew optimally at a temperature of 86°C with a doubling time of 28 min, and was identified as a new species of the genus *Methanococcus*. Very similar organisms were isolated later from the hot Guaymas Basin sediments independently by W. J. Jones, Georgia Institute of Technology at Atlanta, and by Zhao et al. (1988). This extremely thermophilic methanogen appears to be a common constituent of the microbial hydrothermal vent population.

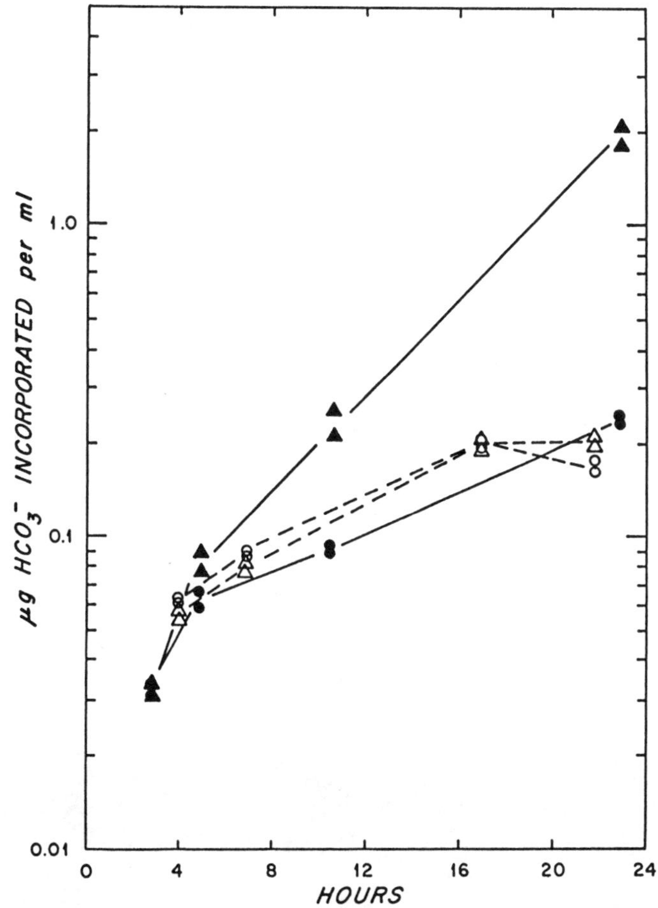

Figure 8.3 Incorporation of ^{14}C-labeled bicarbonate by the mixotrophic thiosulfate-oxidizer, strain N-18, in artificial seawater medium (ASW, dots) and in ASW plus 10 mM thiosulfate (filled triangles), as compared to the base-producing strain TB5.5-9 (see Table 8.2) in ASW (circles) and in ASW plus 10 mM thiosulfate (triangles) (from Ruby et al., 1981).

These archaebacteria were recently joined by another one: extremely thermophilic sulfate-reducing bacteria (Stetter et al., 1987). At least in its capability to use molecular hydrogen as a source of electrons, this organism might also be found as part of the lithotrophic microbial deep-sea vent population. So far the predominant group of extremely thermophilic organisms isolated from various vent sites is represented by fermentative and sulfur-reducing archaebacteria (Jannasch, 1988).

8.5 SYMBIOTIC CHEMOLITHOAUTOTROPHY

Three major species of invertebrates constitute the bulk of the rich animal populations observed in the immediate vicinity of the "bare lava" hydrothermal vents: mussels (*Bathymodiolus thermophilus*, Kenk and Wilson, 1985), the "giant" white clams (*Calyptogena magnifica*, Boss and Turner, 1980, Figure 8.4), and the vestimentiferan tube worms (*Riftia pachyptila*, Jones, 1981, Figure 8.5). This is in contrast to the sediment-covered vents at certain ocean spreading or subduction zones (Lonsdale et al., 1980; Kulm et al., 1986) where the presence of photosynthetically provided organic substrates limit the significance of a primary chemosynthetic production of organic carbon and where a much larger variety of partly well-known deep sea invertebrates are observed.

Of the above three groups of animals, only some mussels occur within the dense microbial suspensions of the immediate warm vent plumes. The rest, as do most of the white clams and the vestimentiferan tube worms, occur outside the direct vent emissions in clear water where the concentration of particulate organic matter is much too low to serve as a food source. Subsequently it was found by electron microscopy and by enzymatic studies that the gill tissue of the bivalves contain procaryotic endosymbionts that are involved in the production of ATP through the oxidation of inorganic sulfur compounds and in the reduction of CO_2 to organic carbon (Cavanaugh et al., 1981; Felbeck, 1981). Specifically indicative of microbial metabolism is the presence of ribulose bisphosphate carboxylase (RuBPC; see Chapter 19) and adenosine 5′phosphosulfate reductase (APSR; see Chapter 14). While microbial symbionts may be absent from mussels or occur in variable quantity, the white clams appear to depend on their presence.

Figure 8.4 Populations of white clams (*Calyptogena magnifica*, Boss and Turner, 1980), 20–30 cm in length, located within cracks between boulders of "pillow lava" at the 21°N East Pacific Rise vent site (depth 2610 m, from H. W. Jannasch, 1984).

Figure 8.5 Populations of vestimentiferan tube worms (*Riftia pachyptila*, Jones, 1981), about 40 cm to 1 m in length, brachyuran crabs (*Bythograea thermydron*, Williams, 1980), limpets (Neomphalidae), and the small white housings of certain polychaetes (Serpulidae) attached to the rock surface. Also shown is one specimen of the so far undescribed "vent fish." Galapagos Rift vent site (depth 2550 m, photo J. M. Edmond).

Dependence on microbial symbionts is also true for the vestimentiferan tube worms (Figure 8.5) which reach lengths of more than 2 m and can occur at densities that appear to exceed the highest concentration of biomass per area known anywhere in the biosphere. The high biomass alone is astounding, especially considering that hydrogen sulfide, a potential metabolic poison, is serving as the major source of potential energy (electron donor). Detailed anatomical studies (Jones, 1981) revealed the absence of a mouth, stomach, and intestinal tract, in short: the absence of any ingestive and digestive system. Instead, the animals contain in their body cavity the so-called "trophosome" (Figure 8.6), a tissue consisting of coccoid procaryotic cells interspersed with the animals' blood vessels (Cavanaugh et al., 1981; Cavanaugh, 1983). A "chemoautotrophic potential" in these worms was described by Felback (1981) based on assays of enzymes catalyzing the synthesis of ATP via sulfur oxidation (rhodanese, APSR and ATP sulfurylase) as well as the Calvin cycle enzymes RuBPC and ribulose 5'phosphate kinase (APK) (Figure 8.7). ADP sulfurylase (ADPS) and phosphenolpyruvate carboxylase (PEPC) were also found (Felbeck,

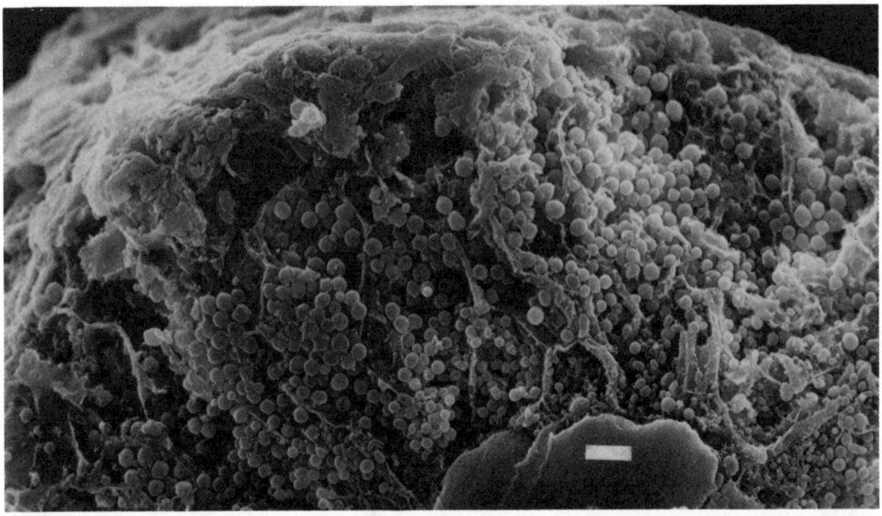

Figure 8.6 Trophosome "tissue" of *Riftia pachyptila* consisting of coccoid procaryotic cells, 3–5 μm in diameter, and blood vessels as eucaryotic animal tissue; pieces of elemental sulfur (bottom left center) form during contact with free oxygen (scale bar 10 μm, from Jannasch, 1984).

et al., 1981). None of these enzymes were detected within the worm tissue proper.

The trophosome may represent more than half of the wet weight of these worms. The necessary simultaneous transport of oxygen and hydrogen sulfide from the retractable plume of gill tissue (Figure 8.5) to the trophosome is carried out by an extracellular hemoglobin of the annelid-type blood system (Arp and Childress, 1981; Wittenberg et al., 1981; Terwilliger et al., 1983). Respiratory poisoning appears to be prevented by the presence of a sulfide-binding protein (Arp and Childress, 1983).

No chemolithotrophic symbiont has so far been cultured from any of the above-mentioned vent invertebrates. A study on concentrated and purified cell suspensions obtained from mussel gill tissue and from tube worm trophosomes by density centrifugation (Belkin et al., 1986) revealed that the symbionts of these two types of animals are distinctly different. While the symbionts of the mussel *Bathymodiolus thermophilus* use thiosulfate as the only electron donor and are clearly psychrophilic, i.e., they show an optimal CO_2-uptake activity at 16°C with a maximum just beyond 22°C (Figure 8.8), the trophosome symbionts of *Riftia pachyptila* oxidize hydrogen sulfide and are active at temperatures up to 35°C. In addition to differences in cell size, the mussel symbionts use free oxygen while the worm symbionts depend on its uptake from oxidized hemoglobin. From the analyses of 5S rRNA nucleotide sequences (Pace et al., 1985), and more recently from 16S rRNA sequences (D. L. Distel, personal communication), it became apparent that the procaryotic symbionts of all chem-

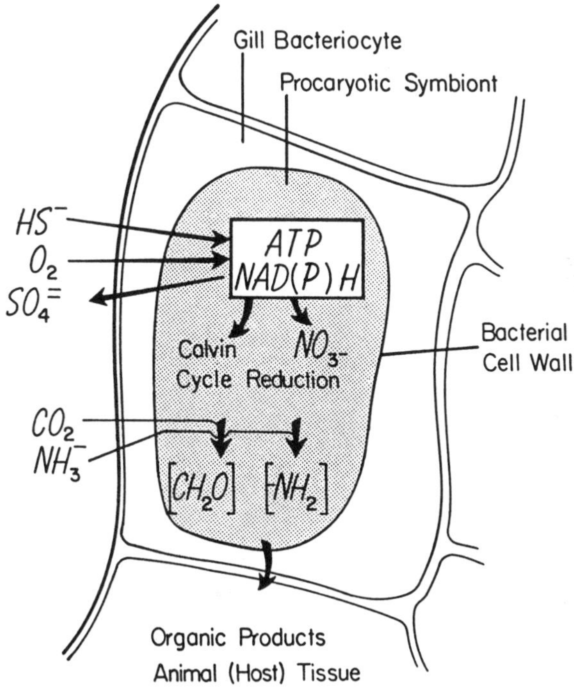

Figure 8.7 Scheme of metabolic processes in procaryotic symbionts within bacteriocytes in the gills of vent bivalves; the presence of the enzymes catalyzing ATP-production, the Calvin cycle, and nitrate reduction have been demonstrated (adapted from Felbeck et al., 1983).

olithoautotrophically existing marine invertebrates so far investigated belong in different groups.

8.6 TOTAL CHEMOSYNTHETIC PRODUCTION IN THE DEEP SEA

It has been a classical enigma for geochemists that the globe's heat flow balance could not be explained from conductivity calculations nor does the actual mineral composition of seawater come anywhere near the sum of weathering processes, river input, and dissolution/precipitation reactions (Edmond and Van Damm, 1983). These apparent anomalies can be resolved by the assumption that the total volume of the oceans' seawater (1.37×10^{21} liters) percolates through the earth's crust about once every 8 million years. While this hydrothermal cycling only amounts to about 1/200 of the oceans' river input, the mineral enrichment of the vent flow is 1000 times that of land runoff. These geochemical calculations lead to a total annual entrainment of 120 million tons of SO_4^{2-}-sulfur. Three-quarters of this amount is deposited as polymetal sul-

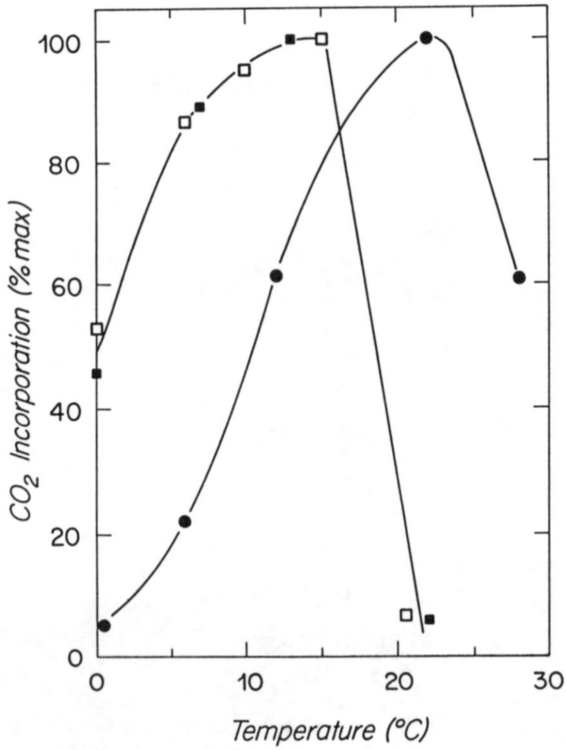

Figure 8.8 Incorporation of ^{14}C-labeled bicarbonate by homogenates of mussel (*Bathymodiolus thermophilus*) gill tissue and *Riftia pachyptila* trophosome at various temperatures. Open squares: gill homogenate incubated with 200 μM $Na_2S_2O_3$; filled squares: purified fraction of mussel gill homogenate incubated with 400 μM $Na_2S_2O_3$; dots: trophosome homogenate incubated with 600 μM Na_2S. Maximal incorporation rates in these three experiments were 6.0, 1.0, and 3.3 nmol CO_2 mg^{-1} protein h^{-1}, respectively (from Belkin et al., 1986).

fides on both sides of the spreading axis while the rest, 30 million tons of S^{2-}-sulfur, is emitted on the seafloor by hydrothermal venting (Edmond et al., 1982).

It is the general consensus of geochemists that most of the H_2S emission into oxygenated deep seawater takes place by diffusion through hydrothermal conduits or by slow mixing in warm vents. Only the smaller portion of hydrothermal fluid is emitted by the hot vents or black smokers, the chemosynthetic potential of the reduced inorganic compounds being lost by a quick dilution and subsequent slow chemical oxidation in the water column. On these grounds, it may be a generous but reasonable assumption that half of the available S^{2-}-sulfur at hydrothermal vents, i.e., 15 million tons, may be used for chemosynthesis. If the ensuing amount of organic carbon (in a 1:1 stoichiometric proportion of C:S) is compared to the annual photosynthetic production (global: 77.6×10^9 tons, oceanic: 18.7×10^9 tons, Woodwell et al.,

1978), it follows that the global chemosynthetic production, based on geothermally reduced sulfur, does not amount to more than 0.02 percent, and the oceanic production alone to about 0.1 percent of photosynthesis. However, this negligible proportion increases considerably, namely to 10 percent, if we consider the deep sea environment only where not more than about 1 percent of the surface-produced photosynthetic food materials arrive in the form of sedimenting particles.

Other reduced inorganic compounds in hydrothermal fluid are NH_3, H_2, CH_4, Fe^{2+}, and Mn^{2+} (Table 8.1). Their microbial oxidation has been demonstrated (Jannasch, 1985), but the largely qualitative data do not permit an estimation of their significance with respect to *in situ* chemosynthesis. The considerable concentrations of ammonia at the sediment-covered vents of the Guaymas Basin suggest a substantial rate of nitrification-based chemosynthesis which, however, has not yet been measured.

The concentrations of hydrogen and methane are highly variable in vent emissions, and their contributions to the *in situ* production of microbial cell material is equally difficult to determine as that of nitrification. Methylotrophy was first indicated by the characteristic morphology of methane-oxidizing bacteria occurring in microbial mats that cover surfaces within a vent plume (Jannasch and Wirsen, 1981). More recently, symbiotic methylotrophy has been suspected on the basis of ^{13}C-isotope depletion in the tissue of certain invertebrates (Kulm et al., 1986) and has been later demonstrated enzymatically (Childress et al., 1986; Schmaljohann and Flügel, 1987; Cavanaugh et al., 1987). In the latter case, however, the methane is not a product of geothermal activity but occurs in the neighborhood of bituminous deposits or results from the decomposition of organic matter.

Although the occurrence of extremely thermophilic methanogenic bacteria at hot vent environments (see above) suggests biogenic production of methane, studies on stable carbon isotope ratios indicate that most of the methane contained in the hydrothermal fluid is of geothermal origin (Welhan and Craig, 1983). This abiogenic origin may define it as an inorganic compound and, in turn, its microbial oxidation a chemolithotrophic gain of energy. The coupling to autotrophy is insignificant, however (Whittenbury and Dalton, 1981).

Microbial iron oxidation can neither occur at the high pH of oxygenated seawater nor in the acidic but oxygen-free hydrothermal fluid. It may occur at certain limited localities in acidified microbial mats (Jannasch and Wirsen, 1981). The successful isolation of manganese-oxidizing bacteria on the other hand, and observations of their *in vitro* activity render their *in situ* growth more likely (Ehrlich, 1983).

8.7 DISCUSSION

The term "primary production" can be used for the aerobic chemolithotrophic fixation of CO_2 in the deep sea, keeping in mind that free oxygen is a product of photosynthesis (Jannasch and Wirsen, 1979). This qualification is not necessary in the case of anaerobic chemosynthesis. The same applies when contrasting the "terrestrial" (chemical) source of energy for the support of life at the deep sea vents with the "solar" energy for the support of life at the continental and sea surface (Figure 8.9). In this context the term "source of energy" is applied to an electron donor or reductant, taking into account that a suitable electron acceptor or oxidant is present for the liberation of the potential energy. The suitability of the electron donor/acceptor couple determines the amount of energy released.

Anaerobic chemosynthesis at the vents, e.g., methanogenesis, can truly be termed a primary production of organic carbon based on terrestrial energy because both the electron donor, H_2, and the electron acceptor, CO_2, are of geothermal origin. However, its actual contribution to organic carbon production might be small in comparison to the aerobic chemolithotrophic production in accordance with the yield of free energy of the oxidations involved:

$$H_2 + \tfrac{1}{4}CO_2 \rightarrow \tfrac{1}{4}CH_4 + \tfrac{1}{2}H_2O \qquad -8.3 \text{ kcal}$$

$$HS^- + 2O_2 \rightarrow SO_4^{2-} + H^+ \qquad -190.4 \text{ kcal}$$

Thermodynamically free oxygen plays the key role in the efficiency of chemosynthesis at the deep sea vents. It is derived from photosynthesis and is commonly present in deep seawater at half-air-saturation values. Its store in the world's ocean is enormous when compared to the amounts consumed by deep sea chemosynthesis. It has been theorized, therefore, that a catastrophic darkening of the earth's surface and the resulting temporary obstruction of

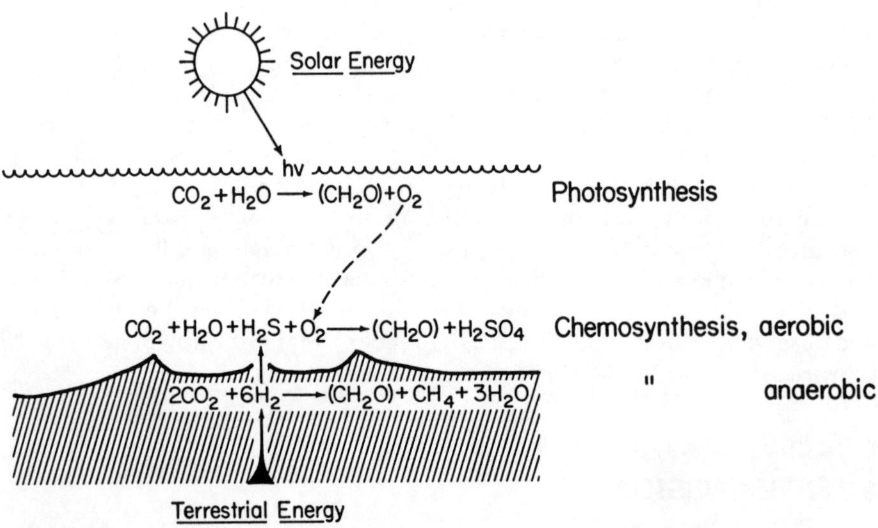

Figure 8.9 Scheme of energy supply for photo- and chemosynthesis and the role of free oxygen at deep sea hydrothermal vents.

photosynthesis would not necessarily affect the light-independent vent ecosystems and would hence give them highest survival chances (Jannasch and Mottl, 1985). Anaerobic chemosynthesis is, of course, entirely independent of the photosynthetically produced oxygen, but is limited to procaryotic ecosystems. It should be pointed out that the reaction between the two most prevalent constituents of hydrothermal fluid, HS^- and CO_2, is endergonic:

$$HS^- + 2HCO_3^- + H^+ \rightarrow 2\text{"}CH_2O\text{"} + SO_4^{2-} \qquad +4.6 \text{ kcal}$$

Any speculation or calculation of a production of free energy by this reaction at high temperatures and pressures rests on the rather independent problem of the existence and functioning of biological systems under these conditions (Jaenicke, 1987).

The predominantly symbiotic chemosynthesis, based upon the oxidations of H_2S as well as CH_4, has now also been found in anoxic marine sediments of nontectonic origin in shallow waters. The detour via the discovery of deep-sea vents was apparently necessary to detect this metabolic way of life in certain animals that have been known for a long time (Cavanaugh, 1983; Schmaljohann and Flügel, 1987; Southward, 1982; Giere et al., 1987). No symbiotic nitrification has yet been demonstrated.

The most surprising characteristic of the chemolithotrophic sustenance of the copious deep-sea communities is its efficiency. Considering the point source of the geothermally provided energy for bacterial growth in the normal food chain, filter feeding on bacterial cells from the quickly dispersing vent plumes appears highly wasteful. Evolution overcame this problem by transferring the chemosynthetic production of organic carbon to a site within the animal where the electron donor as well as the acceptor are made available with the aid of the respiratory system. This symbiotic association combines the metabolic versatility of the procaryotes with the genetic and differentiative capabilities of the eucaryotes and takes advantage of a most direct and efficient transfer of the chemosynthate to the animal tissue. These transfer processes and the biochemical interactions between the microbial and animal metabolisms are presently the focus of research in this area.

In 1965 the eminent ecologist G. E. Hutchinson, as cited by Grassle (1986), stated—with an admiringly intuitive foresight over a decade before the deep sea hydrothermal vent ecosystems were discovered: "The internal heat of a planet, mostly of radioactive origin, in theory would provide an alternative to incoming radiation though we have little precedent as to how an organism could use it."

My thanks are due to S. J. Molyneaux and J. M. Peterson for help in the preparation of this manuscript. The work was supported by the National Science Foundation Grants OCE-838631 and OCE-87581. Contribution No. 6642 of the Woods Hole Oceanographic Institution.

REFERENCES

Arp, A.G. and **Childress, J.J.** 1981. Blood function in the hydrothermal vent vestimentiferan tubeworm. *Science* 213:342–344.

Arp, A.G. and **Childress, J.J.** 1983. Sulfide binding by the blood of the hydrothermal vent tube worm *Riftia pachyptila. Science* 219:295–297.

Belkin, S., Nelson, D.C., and **Jannasch, H.W.** 1986. Symbiotic assimilation of CO_2 in two hydrothermal vent animals, the mussel *Bathymodiolus thermophilus* and the tube worm *Riftia pachyptila. Biological Bulletin* 170:110–121.

Boss, K.J. and **Turner, R.D.** 1980. The giant white clam from the Galapagos Rift, *Calyptogena magnifica* species novum. *Malacologia* 20:161–194.

Cavanaugh, C.M. 1983. Symbiotic chemotrophic bacteria in marine invertebrates from sulfide-rich habitats. *Nature* 302:58–61.

Cavanaugh, C.M., Gardiner, S.L., Jones, M.L., Jannasch, H.W., and **Waterbury, J.B.** 1981. Procaryotic cells in the hydrothermal vent tube worm *Riftia pachyptila* Jones: possible chemoautotrophic symbionts. *Science* 213:340–341.

Cavanaugh, C.M., Levering, P.R., Maki, J.S., Mitchell, R., and **Lidstrom, M.E.** 1987. Symbiosis of methylotrophic bacteria and deep sea mussels. *Nature* 325:346–348.

Childress, J.J., Fisher, C.R., Brooks, J.M., Konnikut II, M.C., Bidigare, A., and **Anderson, A.E.** 1986. A methanotrophic marine molluscan (Bivalvia, Mytilidae) symbiosis: mussels fueled by gas. *Science* 233:1306–1308.

Corliss, J.B., Dymond, J., Gordon, L.I., Edmond, J.M., von Herzen, R.P., Ballard, R.D., Green, K., Williams, D., Bainbridge, A., Crane, K., and **van Andel, T.H.** 1979. Submarine thermal springs on the Galapagos Rift. *Science* 203:1073–1083.

Edmond, J.M. and **Von Damm, K.L.** 1983. Hot springs on the ocean floor. *Scientific American* 248:78–93.

Edmond, J.M. and **Von Damm, K.L.** 1985. Chemistry of ridge crest hot springs. *Biological Society Washington Bulletin* 6:43–47.

Edmond, J.M., Von Damm, K.L., McDuff, R.E., and **Measures, C.I.** 1982. Chemistry of hot springs on the East Pacific Rise and their effluent dispersal. *Nature* 297:187–191.

Ehrlich, H. 1983. Manganese oxidizing bacteria from a hydrothermally active area on the Galapagos Rift. *Ecological Bulletin* 35:357–366.

Felbeck, H. 1981. Chemoautotrophic potentials of the hydrothermal vent tube worm, *Riftia pachyptila* (Ventimentifera). *Science* 213:336–338.

Felbeck, H., Childress, J.J., and **Somero, G.N.** 1981. Calvin-Benson cycle and sulphide oxidation enzymes in animals from sulphide-rich habitats. *Nature* 293:291–293.

Felbeck, H., Childress, J.J., and **Somero, G.N.** 1983. Biochemical interactions between molluscs and their algal and bacterial symbionts, pp. 331–358 in *The Mollusca, Vol. 2, Environmental Biochemistry and Physiology,* Academic Press, New York.

Giere, O., Wirsen, C.O., Schmidt, C., and **Jannasch, H.W.** 1987. The contrasting effect of sulfide and thiosulfate on the symbiotic CO_2-assimilation in *Phallodrilus leukodermatus* (annelida). *Marine Biology*: 97:413–419.

Gottschal, J.C., de Vries, S., and **Kuenen, J.G.** 1979. Competition between the facultatively chemotrophic *Thiobacillus* A2, an obligately chemolithotrophic *Thiobacillus* and a heterotrophic spirillum for inorganic and organic substrates. *Archives of Microbiology* 121: 241–249.

Grassle, J.F. 1986. The ecology of deep sea hydrothermal vent communities. *Advances in Marine Biology and Ecology* 23:301–362.

Hutchinson, G.E. 1965. *The Ecological Theater and the Evolutionary Play.* Yale University Press, New Haven.

Jaenicke, R. 1987. Cellular components under extremes of pressure and temperature: structure-function relationship of enzymes under pressure, pp. 257–272, in Jannasch, H.W., Marquis, R.E., and Zimmerman, A.M. (editors), *Current Perspectives in High Pressure Biology,* Academic Press, London.

Jannasch, H.W. 1984. Microbes in the oceanic environment, pp. 97–122, in Kelly, D.P. and Carr, N.G. (editors), *The Microbe 1984, Part II, Prokaryotes and Eukaryotes,* Cambridge University Press, Cambridge.

Jannasch, H.W. 1985. The chemosynthetic support of life and the microbial diversity at deep sea hydrothermal vents. *Proceedings of the Royal Society (London)* B225:277–297.

Jannasch, H.W. 1988. Isolation of extremely thermophilic, fermentative archaebacteria from deep sea geothermal sediments, in Wise, D.L. (editor), *Biotechnology Applied to Fossil Fuels,* CRC Press, Boca Raton (in press).

Jannasch, H.W. and **Mottl, M.J.** 1985. Geomicrobiology of deep-sea hydrothermal vents. *Science* 229:717–725.

Jannasch, H.W. and **Wirsen, C.O.** 1979. Chemosynthetic primary production at East Pacific sea floor spreading centers. *BioScience* 29:592–598.

Jannasch, H.W. and **Wirsen, C.O.** 1981. Morphological survey of microbial mats near deep-sea thermal vents. *Applied and Environmental Microbiology 41:528–538.*

Jannasch, H.W., Wirsen, C.O., Nelson, D.C. and **Robertson, L.A.** 1985. *Thiomicrospira crunogena* sp. nov., a colorless sulfur-oxidizing bacterium from a deep-sea hydrothermal vent. *International Journal of Systematic Bacteriology 35:422–424.*

Jones, M.L. 1981. *Riftia pachyptila* Jones: observations on the vestimentiferan worm from the Galapagos Rift. *Science* 213:333–336.

Jones, W.J., Leigh, J.A., Mayer, F., Woese, C.R., and **Wolfe, R.S.** 1983. *Methanococcus jannaschii* sp. nov., an extremely thermophilic methanogen from a submarine hydrothermal vent. *Archives of Microbiology* 136:254–261.

Karl, D.M. 1980. Cellular nucleotide measurements and applications in microbial ecology. *Microbiological Reviews* 44:739–796.

Karl, D.M., Wirsen, C.O., and **Jannasch, H.W.** 1980. Deep sea primary production at the Galapagos hydrothermal vents. *Science* 207:1345–1347.

Kenk, V.C. and **Wilson, B.R.** 1985. A new mussel (Bivalvia, Mytilidae) from hydrothermal vents in the Galapagos Rift zone. *Malacologia* 26:253–271.

Kuenen, J.G. and **Buedeker, R.F.**. 1982. Microbiology of thiobacilli and other sulphur oxidizing autotrophs, mixotrophs and heterotrophs. *Philosophical Transactions Royal Society of London* B298:473–497.

Kuenen, J.G. and **Veldkamp. H.** 1972. *Thiomicrospira pelophila,* gen. n., sp. n., a new obligately chemolithotrophic colourless sulfur bacterium. *Antonie van Leeuwenhoek* 38:241–256.

Kulm, L.D., Suess, E., Moore, J.C., Carson, B., Lewis, B.T., Ridger, S.D., Katko, D.C., Thornburg, T.M., Embley, R.W., Rough, W.D., Massoth, G.J., Langseth, M.G., Cochrane, G.R., and **Scamman, R.L.** 1986. Oregon subduction zone: venting, fauna and carbonates. *Science* 231:561–566.

Lonsdale, P.F., Bischoff, J.L., Burns, V.M., Kastner, M., and **Sweeney, R.E.** 1980. A high-temperature hydrothermal deposit on the seabed at the Gulf of California spreading center. *Earth and Planetary Science Letters* 49:8–20.

Nelson, D.C. and **Jannasch, H.W.** 1983. Chemoautotrophic growth of a marine *Beggiatoa* in sulfide gradient cultures. *Archives of Microbiology* 136:262–269.

Nelson, D.C., Wirsen, C.O., and **Jannasch, H.W.** 1988. Massive occurrence of autotrophic *Beggiatoas* at hydrothermal vents of the Guaymas Basin. *Applied and Environmental Microbiology* (in press).

Pace, N.R., Stahl, D.A., Lane, D.J., and **Olsen, G.J.** 1985. Analyzing natural microbial populations by rRNA sequences. *ASM News* 51:4–12.

Rona, P.A., Boström, K., Laubier, L., and **Smith, K.L.** (editors). 1983. *Hydrothermal Processes at Seafloor Spreading Centers.* Plenum Press, New York. 796 pp.

Ruby, E.G. and **Jannasch, H.W.** 1982. Physiological characteristics of *Thiomicrospira* sp. L-12 isolated from deep sea hydrothermal vents. *Journal of Bacteriology* 149:161–165.

Ruby, E.G., Wirsen, C.O., and **Jannasch, H.W.** 1981. Chemolithotrophic sulfur-oxidizing bacteria from the Galapagos Rift hydrothermal vents. *Applied and Environmental Microbiology 42:317–324.*

Schmaljohann, R. and **Flügel, H.** 1987. Methane-oxidizing bacteria in Pogonophora. *Sarsia* 72:91–98.

Southward, E.C. 1982. Bacterial symbionts in Pogonophora. *Journal of Marine Biology of the Association of the United Kingdom* 62:889–906.

Stetter, K.O., Lauerer, G., Thomm, M., and **Neuner, A.** 1987. Isolation of extremely thermophilic sulfate reducers: evidence for a novel branch of archaebacteria. *Science* 236:822–824.

Terwilliger, R.C., Terwilliger, N.B., and **Arp, A.** 1983. Thermal vent clam (*Calyptogena magnifica*) hemoglobin. *Science* 219:981–983.

Tuttle, J.H., Wirsen, C.O., and **Jannasch, H.W.** 1983. Microbial activities in the emitted hydrothermal waters of the Galapagos Rift vents. *Marine Biology* 73:293–299.

Welhan, J.A. and **Craig, H.** 1983. Methane, hydrogen, and helium in hydrothermal fluids at 21°N on the East Pacific Rise, pp. 391–409, in Rona, P.A., Boström, K., Laubier, L., and Smith, K.L. (editors), *Hydrothermal Processes at Seafloor Spreading Centers*, Plenum Press, New York.

Whittenbury, R. and **Dalton, H.** 1981. The methylotrophic bacteria, pp. 894–902, in Starr, M.P., Stolp, H., Trüper, H.G., Balows, A., and Schlegel, H.G. (editors), *The Prokaryotes*, Springer-Verlag, Berlin.

Williams, A.B. 1980. A new crab family from the vicinity of submarine thermal vents on the Galapagos Rift (Crustacea, Decapoda, Brachyura). *Biological Society of Washington Bulletin* 93:443–472.

Winogradsky, S. 1887. Über Schwefelbakterien. *Botanische Zeitung* 45:489–507, 513–523, 529–539, 545–559, 569–576, 585–594, 606–610.

Wirsen, C.O., Tuttle, J.H., and **Jannasch, H.W.** 1986. Activities of sulfur-oxidizing bacteria at the 21°N East Pacific Rise vent site. *Marine Biology* 92:449–456.

Wittenberg, J.B., Morris, R.J., Gibson, Q.H., and **Jones, M.L.** 1981. Hemoglobin kinetics of the Galapagos Rift vent tube worm *Riftia pachyptila* Jones (Pogonophora, Vestimentifera). *Science* 213:244–246.

Woodwell, G.M., Whittaker, R.H., Reiners, W.A., Likens, G.E., Delwiche, C.C., and **Botkin, D.B.** 1978. The biota and the world carbon budget. *Science* 119:141–146.

Zhao, H., Wood, A.G., Widdel, F., and **Bryant, M.P.** 1988. An extremely thermophilic *Methanococcus* from a deep sea hydrothermal vent and its plasmid. *Archives of Microbiology* 150:178–183.

9

Extremely Thermophilic Chemolithoautotrophic Archaebacteria

Karl O. Stetter

Department of Microbiology
University of Regensburg
Regensburg, Federal Republic of Germany

9.1 GENERAL REMARKS

During the last 20 years several extremely thermophilic bacteria have been isolated, growing best at temperatures between 80 and 105°C (Brock et al., 1972; Brock, 1978; Zillig et al., 1981; Stetter, 1982; Stetter and Zillig, 1985; Stetter, 1986a). As a rule, these organisms do not grow at temperatures below 60°C, and some do not even grow below 80°C (Stetter et al., 1983). Such extremely thermophilic bacteria are designated as "hyperthermophiles" to distinguish them from thermophiles with lower optimal temperatures. The mode of nutrition is obligately or facultatively chemolithoautotrophic or obligately heterotrophic. Hyperthermophiles are important organisms for studies on the evolution of life (Woese, 1987; Lake, 1988) and for biotechnology (Stetter, 1986b). Except *Thermotoga*, which is a very unusual heterotrophic eubacterium (Huber et al., 1986; Achenbach-Richter et al., 1987a), all hyperthermophiles isolated up to now belong to the archaebacteria, the third kingdom of life (Woese et al., 1978). Within the archaebacteria, hyperthermophiles are found in both of the two main evolutionary lineages as determined by 16S rRNA sequence comparisons: most of them belong to the branch of the sulfur metabolizers, while a few are within the branch of methanogens and their relatives (*Thermococcales; "Archaeoglobales"*; Achenbach-Richter et al., 1987b; Woese, 1987). Hyperthermophilic archaebacteria are very diverse, representing several different orders (Woese, 1987). Within the branch of sulfur metabolizers, some recent isolates have not yet been classified phylogenetically due to strong post-transcriptional base modification of the 16S rRNA, which complicates base sequencing (C. R. Woese, personal communication). This chapter shall give an overview of the habitats and the physiological and biochemical properties of the autotrophic hyperthermophilic archaebacteria.

9.2 HABITATS

All hyperthermophilic bacteria known at present have been isolated from solfataras and from submarine hydrothermal areas rich in CO_2, H_2 and sulfur compounds originating from the magma chambers situated below (Table 9.1). The pH of the surface of solfataras is usually strongly (pH 0.5 to 3) or mildly (pH 4 to 6) acidic and rich in sulfate (Brock, 1978; Stetter and Zillig, 1985). In the depth, solfataras are less acidic (pH 5 to 7; Stetter et al., 1986). Some solfataras may also contain neutral to weakly alkaline (pH 7–9) springs rich in NaCl. Also artificial hot environments such as the boiling outflows of geothermal power plants are suitable habitats for hyperthermophiles (Huber et al., 1987). Submarine hydrothermal systems are slightly acidic to neutral (pH 5 to 8) and contain normally high amounts of NaCl and SO_4^{2-} due to the presence of seawater (Table 9.1). Sometimes, there are also hot springs with low salinity occurring at the sea floor, as is the case close to Cape Reykjanes in the southwest of Iceland and at Sangeang Island, Indonesia (Stetter, unpublished). Due to the low solubility of oxygen at high temperatures and to the presence of reducing gases, most environments bearing hyperthermophiles are anaerobic. Within solfataras, soil profiles show a distinct ochre-colored surface layer containing some oxygen, as indicated by the presence of ferric iron above the blackish reduced zone in the depth, which contains ferrous iron (Stetter et al., 1986).

Hyperthermophilic autotrophic bacteria are the primary producers of organic matter at high temperatures. Most likely, they are providing organic material for heterotrophic hyperthermophiles sharing the habitat (Stetter, 1986b). Although proliferating only well above 60°C, hyperthermophiles can survive for years at low temperatures. For anaerobic hyperthermophiles, oxygen is usually extremely toxic at temperatures allowing growth (Stetter et al., 1983; Lauerer et al., 1986; Huber et al., 1987). However, some anaerobic hyperthermophiles are able to survive long times in the presence of oxygen at low temperatures. This property may be important in dispersal of these organisms through the air or in cold oxygen-rich water.

Table 9.1 Habitats of chemolithoautotrophic hyperthermophiles

	Type of geothermal area	
Characteristics	Solfataras	Submarine hydrothermal systems
Sites	Steam-heated surface waters, mud holes and soils; deep hot springs; hot waste water of geothermal power plants	Hot sediments and vents ("black smokers")
Temperatures	Up to 100°C*	Up to about 374°C**
pH	0.9 to 9	6 to 8.5
Presence of O_2	Surface, oxic; depth, anoxic	Anoxic
Major gases and sulfur compounds	CO_2, CO, CH_4, H_2, NH_4^+, S^{2-}, S^0, SO_3^{2-}, SO_4^{2-}***	

*At the surface, depending on the altitude above sea level.
**Critical point of water (no living organisms found at these temperatures).
***Seawater contains about 30 mmol/l of sulfate.

9.3 EXTREMELY ACIDOPHILIC HYPERTHERMOPHILIC AUTOTROPHS

The aerobic and facultatively aerobic hyperthermophiles are lobed coccoid organisms belonging to the genera *Sulfolobus* and *Acidianus* (Table 9.2; Brock et al., 1972; Segerer et al., 1985, 1986). They are extreme acidophiles and due to their requirement of acidic pH, they are mainly found within the upper layer of solfataras. A less extremely thermophilic archaebacterium occurring in acidic habitats is *Thermoplasma*. Members of this genus are facultative anaerobic heterotrophic sulfur respirers (Ohba and Oshima, 1982; Segerer et al., 1988).

Members of the genus *Sulfolobus* are strict aerobes which grow autotrophically by oxidation of S^{2-} and S^0, forming sulfuric acid (Table 9.3). Many *Sulfolobus* isolates are also able to grow heterotrophically, and a few are even strictly heterotrophic (Stetter, unpublished). The type strain of *Sulfolobus acidocaldarius* (DSM 639) shows excellent growth on organic matter (Brock et al., 1972). Autotrophically on S^0 and S^{2-}, however, this strain shows only extremely weak growth after a long period of adaptation (Marsh et al., 1983; Huber et al., 1986; Huber, 1988). Some *Sulfolobus*-like strains are able to oxidize sulfidic ores like pyrite, chalcopyrite, and sphalerite, forming sulfuric acid and solubilizing the heavy metal ions ("ore leaching"; Table 9.3). They are obligate autotrophs also able to grow by S^0 oxidation (strains VE2; HV5; Table 9.3). Many *Sulfolobus* strains are able to oxidize ferrous iron (Brock, 1978; Brierley and Brierley, 1973). However, no significant growth could be observed, possibly due to extensive precipitates of jarosite even at pH 2 and in uninoculated controls (Brock, 1978; Huber and Stetter, unpublished). Under microaerophilic conditions, *Sulfolobus* is able to reduce ferric iron and molybdate, which therefore act anaerobically as electron acceptors (Brock and Gustafson, 1976; Brierley and Brierley, 1982). Strains of the genus *Sulfolobus* grow only at very low ionic strength and may therefore not be able to grow within marine hydrothermal systems (Brock, 1972; Segerer and Stetter, unpublished).

Members of the genus *Acidianus*, similar to *Sulfolobus*, are able to grow by oxidation of S^0, forming sulfuric acid (Segerer et al., 1985, 1986). Some strains are also able to grow by oxidation of sulfidic ores, but grow much slower and more inefficiently than do the ore-leaching *Sulfolobus* strains (Table 9.3; Huber et al., 1986; Huber, 1988). Under anaerobic conditions, members of *Acidianus* are able to grow by anaerobic oxidation of H_2, using S^0 as electron donor (Table 9.3; Segerer et al., 1985). Therefore, S^0 can be used as both an electron donor and electron acceptor by the same organism, depending on the redox potential of the environment. Brierley's isolate, which was described as *Sulfolobus brierleyi* (Zillig et al., 1980) shows this metabolic versatility, too, and shares further features with *Acidianus*. Therefore, it has been reclassified and described as *Acidianus brierleyi* (Segerer et al., 1986). Another isolate which was tentatively named "*Sulfolobus ambivalens*" (Zillig et al., 1985) was later described as *Desulfurolobus ambivalens*, representing a new genus (Zillig et al., 1987). By DNA:DNA hybridization, *Desulfurolobus ambivalens* shows 60 percent DNA

Table 9.2 Morphological, physiological, and biochemical features of chemolithoautotrophic hyperthermophiles

Species	Growth conditions					Habitat marine (m)/ solfataric (s)	DNA mol%G+C	Morphology
	Temperature (°C)			pH	Aerobic (ae)/ anaerobic (an)			
	Minimum	Optimum	Maximum					
Sulfolobus acidocaldarius	60	80	85	1–5	ae	s	37	Lobed cocci
"*Sulfolobus*" isolate VE2	60	82	85	1.5–5	ae	s	37	
"*Sulfolobus*" isolate HV5	60	90	95	1.5–5	ae	s	30	
Acidianus infernus	60	88	95	1–6	ae/an	s/m	31	
Thermoproteus tenax	70	88	97	3–6	an	s	56	Rods with terminally protruding spheres
Pyrobaculum islandicum	74	100	103	5–7	an	s	46	
Pyrodictium occultum	82	105	110	5–7	an	m	62	Discs with fibers
Archaeoglobus fulgidus	60	83	95	5.5–7.5	an	m	46	Irregular cocci
Methanococcus jannaschii	50	85	86	5.5–6.5	an	m	31	Irregular cocci
Methanothermus sociabilis	65	88	97	5.5–7.5	an	s	33	Rods in clusters

Table 9.3 Metabolism of chemolithoautotrophic hyperthermophiles

Species	Energy-yielding reaction	Obligately (o)/facultatively (f) lithoautotrophic	Heterotrophic growth substrates
Sulfolobus acidocaldarius	Oxidation of S^{2-} and S^0; $\rightarrow H_2SO_4$	f	Oxidation of cell extracts; sugars
"*Sulfolobus*" isolate VE2	Oxidation of sulfidic ores (leaching) and $S^0 \rightarrow H_2SO_4$	o	no
"*Sulfolobus*" isolate HV5			
Acidianus infernus	Oxidation of sulfidic ores (leaching) and $S^0 \rightarrow H_2SO_4$; reduction of S^0 by $H_2 \rightarrow H_2S$	o**	no**
*Thermoproteus tenax**	Reduction of S^0 by $H_2 \rightarrow H_2S$	f	S^0, $S_2O_3^{2-}$, SO_3^{2-}; respiration of cell extracts, peptone
Pyrobaculum islandicum			$S_2O_3^{2-}$; respiration of (H_2 + cell extracts)
Pyrodictium occultum			
Archaeoglobus fulgidus	Reduction of $S_2O_3^{2-}$ by $H_2 \rightarrow H_2S$	f	SO_4^{2-}, $S_2O_3^{2-}$, SO_3^{2-}; respiration of formate, lactate, sugars, proteins, cell extracts
Methanococcus jannaschii	Reduction of CO_2 by $H_2 \rightarrow CH_4$	o	no
Methanothermus sociabilis			

Thermoproteus neutrophilus is an obligate chemolithoautotroph.
***Acidianus brierleyi* is a facultative heterotroph (growth by oxidation of yeast extract).

homology with the type species of *Acidianus, A. infernus* (Huber et al., 1987), indicating that it is another species of *Acidianus* (Steigerwaldt et al., 1976; Schleifer and Stackebrandt, 1983). Members of the genus *Acidianus* are able to grow in the presence of up to 4 percent salt. In agreement with this result, they also occur in marine hydrothermal systems (Segerer et al., 1986).

9.4 NEUTROPHILIC AND SLIGHTLY ACIDOPHILIC HYPERTHERMOPHILIC AUTOTROPHS

Within solfataras, the slightly acidophilic (opt pH $\approx$ 6) members of the genera *Thermoproteus* and *Pyrobaculum* are found (Table 9.2; Zillig et al., 1981; Huber et al., 1987). Cells are characteristically stiff rods with protruding spheres ("golf clubs") mainly at their ends, a morphology arising most likely as a specific mode of budding. Isolates of *Thermoproteus* and *Pyrobaculum* are strict anaerobes (Table 9.2). Most of them are able to grow autotrophically by anaerobic reduction of S^0 by H_2 (Fischer et al., 1983; Huber et al., 1987). *Pyrobaculum organotrophum* is an obligate heterotroph (Huber et al., 1987). While S^0 is obligately required as electron acceptor for autotrophic growth by members of both genera, it can be replaced by sulfite, thiosulfate, and cystine for heterotrophic growth (Table 9.3; Zillig et al., 1981; Huber et al., 1987). Members of the genus *Pyrobaculum* exhibit a G + C content of their DNA which is 10 percent lower than that of *Thermoproteus* and show a much higher optimal growth temperature, at around 100°C (Table 9.2). Similar to *Sulfolobus*, strains of *Thermoproteus* and *Pyrobaculum* are only able to grow at low ionic strength and may therefore not be found in marine hot environments (Huber et al., 1987).

The organisms with the highest growth temperatures known are the two species of *Pyrodictium, Pyrodictium occultum* and *Pyrodictium brockii*, both growing up to 110°C (Stetter, 1982; Stetter et al., 1983). Cells of *Pyrodictium* are so well adapted to high temperatures that they do not even grow below 82°C. Cultures of *Pyrodictium* grow in flocs, often forming a cobweb-like covering of sulfur particles. Cells are disc-shaped and are connected by a network of very thin hollow fibers which, similar to the cells, show a positive DNA-specific staining (DAPI, modified; Huber et al., 1985) and may therefore contain DNA. The mode of cell propagation of *Pyrodictium* is not as yet understood. *Pyrodictium* is a strict anaerobic autotrophic organism, gaining its energy by reduction of S^0 by H_2. Alternatively, *Pyrodictium* can grow mixotrophically on H_2 and thiosulfate in the presence of bacterial cells or yeast extract (Table 9.3). Although growing within a wide salt range from 0.1 to 12 percent NaCl, *Pyrodictium* has up to now only been found within submarine hydrothermal vents off Vulcano, Italy (Table 9.2; Stetter et al., 1983). Within the sulfur-metabolizing archaebacteria, *Pyrodictium* represents a separate branch, indicating a different order (Woese, 1987).

A group of coccoid hyperthermophilic marine sulfate-reducing archaebacteria is represented by *Archaeoglobus fulgidus* (Stetter et al., 1987; Stetter,

1988). It is facultatively autotrophic. During autotrophic growth, *Archaeoglobus fulgidus* gains its energy by reduction of thiosulfate by H_2 (Table 9.3; Stetter et al., 1987). Only very little growth is obtained when thiosulfate is replaced by sulfate under autotrophic conditions. During heterotrophic growth, various substrates like formate, lactate, glucose, starch, and proteins can be used as electron donors. As electron acceptors, sulfate, thiosulfate, and sulfite can be used heterotrophically. S^0 is inhibitory during autotrophic and heterotrophic growth. In addition to H_2S, low amounts of methane (up to 0.1 μmol/ml) are formed via an unknown pathway (Stetter et al., 1987). Similar to the methanogens, cells of *Archaeoglobus* show a blue-green fluorescence at 420 nm under the fluorescence microscope which is due to the possession of factor 420 (Stetter et al., 1987). Furthermore, *Archaeoglobus* cells contain methanopterin and tetrahydromethanopterin, which have been found up to now only in methanogens. In contrast to methanogens, coenzyme M and methyl CoM reductase are lacking in *Archaeoglobus,* while essential enzymes of the dissimilatory sulfate reduction pathway like adenosine triphosphate sulfurylase, adenylyl-sulfate reductase, and bisulfite reductase are present (Stetter, 1988; Trüper, personal communication). Due to its unique type of RNA polymerase and its 16S rRNA sequence, *Archaeoglobus* forms a separate phylogenetic branch (the order: *"Archaeoglobales"*) situated between the thermophilic sulfur metabolizers and the methanogens (Achenbach-Richter et al., 1987; Stetter et al., 1987; Stetter, 1988).

A few solfataric and hydrothermal hyperthermophilic autotrophs belong to the methanogenic bacteria, forming methane by reduction of CO_2 by H_2 (Table 9.3). They are obligate autotrophs belonging to the coccoid-shaped *Methanococcales* and the rod-shaped *Methanobacteriales* (Balch et al., 1979). Within the *Methanococcales,* there are the extremely thermophilic species *Methanococcus thermolithotrophicus* (Huber et al., 1982) and the hyperthermophilic species *Methanococcus jannaschii* (Table 9.2), both living within marine hydrothermal systems. Other relatives within the same genus are mesophiles (Balch et al., 1979). *Methanococcus thermolithotrophicus* was isolated from different shallow hydrothermal systems in Naples and Vulcano (both: Italy) and Obock (Dschibouti, Africa), while *Methanococcus jannaschii* has up to now been isolated only from deep sea vents (Huber et al., 1982; Stetter, unpublished; Jones et al., 1983). From solfataras in the southwest of Iceland, rod-shaped methanogens growing at temperatures up to 97°C (Table 9.2) were isolated (Stetter et al., 1981; Lauerer et al., 1986). The isolates are extremely sensitive to oxygen, even in comparison with other methanogens. They occur in the lower reduced neutral zone of solfatara fields. The isolates represent the new family *Methanothermaceae.* Two species are known: *Methanothermus fervidus* (Stetter et al., 1981) and *Methanothermus sociabilis,* the latter growing in clusters up to 3 mm in diameter (Lauerer et al., 1986). Members of the *Methanothermaceae* are Gram-positive and possess a rigid cell wall composed of pseudomurein which is covered with an outer S-layer of glycoprotein subunits (Lauerer et al., 1986; Nusser et al., 1988). Cells of *Methanothermus* are motile due to bipolar polytrichous flagellation. Little channels are visible at the tips of the cells where the flagella pass through the pseudomurein layer (Stetter, 1986).

9.5 CONCLUSIONS

The isolation of hyperthermophilic chemolithoautotrophic bacteria from geothermally heated environments demonstrates primary production of organic matter even at temperatures around 100°C. In spite of the unexpected variety and phylogenetic diversity of these organisms, the energy-yielding reactions are only based on reduction or oxidation of sulfur compounds by H_2 and O_2, respectively, or in the case of methanogens on reduction of CO_2 by H_2. Since methanogens are also powerful sulfur reducers (Stetter and Gaag, 1983), the ability to metabolize sulfur compounds appears to be a general property of hyperthermophilic archaebacteria. This may be due to an early adaptation to habitats abundant in sulfur compounds since the Archaean age. The anaerobic members of chemolithoautotrophic hyperthermophiles are even independent of a sun, since they do not require oxygen. They could therefore grow on any planet where an aqueous hot volcanic environment exists.

The upper temperature border of microbial life is still unknown. The existence of *Pyrodictium* and other hyperthermophiles demonstrates that still unrecognized thermostabilizing principles must exist. Since stability of biomolecules at temperatures above 100°C decreases rapidly (White, 1984; Bernhardt et al., 1984; Trent et al., 1984), the maximal growth temperature at which life can exist may possibly be found between 110 and 150°C, temperatures at which heat-sensitive biomolecules could still be resynthesized at biologically feasible rates. Whether or not organisms capable of growth at temperatures of 150°C exist still has to be demonstrated by culture experiments.

REFERENCES

Achenbach-Richter, L., Gupta, R., Stetter, K.O., and **Woese, C.R.** 1987a. Were the original eubacteria thermophiles? *Systematic and Applied Microbiology* 9:34–39.

Achenbach-Richter, L., Stetter, K.O., and **Woese, C.R.** 1987b. A possible biochemical missing link among archaebacteria. *Nature* 327:348–349.

Balch, W.E., Fox, G.E., Magrum, L.J., Woese, C.R., and **Wolfe, R.S.** 1979. Methanogens: Reevaluation of a unique biological group. *Microbiological Reviews* 43:260–296.

Bernhardt, G., Lüdemann, H.D., Jaenicke, R., König, H., and **Stetter, K.O.** 1984. Biomolecules are unstable under "black smoker" conditions. *Naturwissenschaften* 71:583–585.

Brierley, C.L. and **Brierley, J.A.** 1973. A chemolithoautotrophic and thermophilic microorganism isolated from an acidic hot spring. *Canadian Journal of Microbiology* 19:183–188.

Brierley, C.L. and **Brierley, J.A.** 1982. Anaerobic reduction of molybdenum by *Sulfolobus* species. *Zentralblatt für Bakteriologie und Hygiene,* I. Abteilung Originale C 3:289–294.

Brock, T.D., Brock, K.M., Belly, R.T., and **Weiss, R.L.** 1972. *Sulfolobus*: a new genus of sulfur-oxidizing bacteria living at low pH and high temperature. *Archiv für Mikrobiologie* 84:54–68.

Brock, T.D. and **Gustafson, J.** 1976. Ferric iron reduction by sulfur- and iron-oxidizing bacteria. *Applied and Environmental Microbiology* 32:567–571.

Brock, T.D. 1978. *Thermophilic microorganisms and life at high temperatures.* Springer-Verlag, Berlin, Heidelberg, New York.

Fischer, F., Zillig, W., Stetter, K.O., and **Schreiber, G.** 1983. Chemolithoautotrophic metabolism of anaerobic extremely thermophilic archaebacteria. *Nature* 301:511–513.

Huber, H., Thomm, M., König, H., Thies, G., and **Stetter, K.O.** 1982. *Methanococcus thermolithotrophicus,* a novel thermophilic lithotrophic methanogen. *Archives of Microbiology* 132:47–50.

Huber, H., Huber, G., and **Stetter, K.O.** 1985. A modified DAPI fluorescence staining procedure suitable for the visualization of lithotrophic bacteria. *Systematic and Applied Microbiology* 6:105–106.

Huber, G., Huber, H., and **Stetter, K.O.** 1986a. Isolation and characterization of new metal-mobilizing bacteria. *Biotechnology and Bioengineering Symposia* 16:239–251.

Huber, R., Langworthy, T.A., König, H., Thomm, M., Woese, C.R., Sleytr, U.B., and **Stetter, K.O.** 1986b. *Thermotoga maritima* sp. nov. represents a new genus of unique extremely thermophilic eubacteria growing up to 90°C. *Archives of Microbiology* 144:324–333.

Huber, R., Kristjansson, J.K., and **Stetter, K.O.** 1987. *Pyrobaculum* gen. nov., a new genus of neutrophilic, rodshaped archaebacteria from continental solfataras growing optimally at 100°C. *Archives of Microbiology* 149:95–101.

Huber, G. 1988. Isolierung, Charakterisierung und taxonomische Einordnung neuer thermophiler, metallmobilisierender Archaebakterien. Thesis, University of Regensburg, W. Germany.

Jones, W.L., Leigh, J.A., Mayer, P., Woese, C.R., and **Wolfe, R.S.** 1983. *Methanococcus jannaschii* sp. nov., an extremely thermophilic methanogen from a submarine hydrothermal vent. *Archives of Microbiology* 136:254–261.

Lake, J. 1988. Origin of the eukaryotic nucleus determined by rate-invariant analysis of r-RNA sequences. *Nature* 331:184–186.

Lauerer, G., Kristjansson, J.K., Langworthy, T.A., König, H., and **Stetter, K.O.** 1986. *Methanothermus sociabilis* sp. nov., a second species within the Methanothermaceae growing at 97°C. *Systematic and Applied Microbiology* 8:100-n105.

Marsh, R.M., Norris, P.R., and **Le Roux, N.W.** 1983. Growth and mineral oxidation studies with *Sulfolobus*, pp. 71–81, in Rossi, G. and Torma, A. (editors), *Recent Progress in Biohydrometallurgy,* Associazione Mineraria Sarda, Iglesias, Italy.

Nußer, E., Hartmann, E., Allmeier, H., König, H., Paul, G., and **Stetter, K.O.** 1988. A glycoprotein surface layer covers the pseudomurein sacculus of the extreme thermophile *Methanothermus fervidus.* **Proceedings of the Second International Workshop on S-Layers in Prokaryotes, Vienna.** Springer-Verlag, Berlin (in press).

Ohba, M. and **Oshima, T.** 1982. Some biochemical properties of the protein synthesizing machinery of acidothermophilic archaebacteria isolated from Japanese hot springs, p. 353, in Kandler, O. (editor), *Archaebacteria,* G. Fischer Verlag, Stuttgart.

Schleifer, K. and **Stackebrandt, E.** 1983. Molecular systematics of prokaryotes. *Annual Reviews of Microbiology* 37:143–187.

Segerer, A., Stetter, K.O., and **Klink, F.** 1985. Two contrary modes of chemolithotrophy in the same archaebacterium. *Nature* 313:787–789.

Segerer, A., Neuner, A., Kristjansson, J.K., and **Stetter, K.O.** 1986. *Acidianus infernus* gen. nov., sp. nov., and *Acidianus brierleyi* comb. nov.: Facultatively aerobic, extremely acidophilic thermophilic sulfur-metabolizing archaebacteria. *International Journal of Systematic Bacteriology* 36:559–564.

Segerer, A., Langworthy, T.A., and **Stetter, K.O.** 1988. *Thermoplasma acidophilum* and *Thermoplasma volcanium* sp. nov. from solfatara fields. *Systematic and Applied Microbiology* 10:161–171.

Steigerwaldt, A.G., Fanning, G.R., Fife-Asbury, M.A., and **Brenner, D.J.** 1976. DNA relatedness among species of Enterobacter and Serratia. *Canadian Journal of Microbiology* 22:121–137.

Stetter, K.O., Thomm, M., Winter, J., Wildgruber, G., Huber, H., Zillig, W., Janecovic, D., König, H., Palm, P., and **Wunderl, S.** 1981. *Methanothermus fervidus,* sp. nov., a novel extremely thermophilic methanogen isolated from an Icelandic hot spring. *Zentralblatt für Bakteriologie und Hygiene,* I. Abteilung Originale C 2:166–178.

Stetter, K.O. 1982. Ultrathin mycelia-forming organisms from submarine volcanic areas having an optimum growth temperature of 105°C. *Nature* 300:258–260.

Stetter, K.O., König, H., and **Stackebrandt, E.** 1983. *Pyrodictium* gen. nov., a new genus of submarine disc-shaped sulphur reducing archaebacteria growing optimally at 105°C. *Systematic and Applied Microbiology* 4:535–551.

Stetter, K.O. and **Gaag, G.** 1983. Reduction of molecular sulphur by methanogenic bacteria. *Nature* 305:309–311.

Stetter, K.O. and **Zillig, W.** 1985. *Thermoplasma* and the thermophilic sulfur-dependent archaebacteria, pp. 85–170, in Wolfe, R.S. and Woese, C.R. (editors), *The Bacteria,* Vol. VIII, Academic Press, New York.

Stetter, K.O. 1986a. Diversity of extremely thermophilic archaebacteria, pp. 39–74, in Brock, T.D. (editor), *Thermophiles: General, Molecular and Applied Microbiology,* J. Wiley and Sons, Inc., New York, London, Sydney, Toronto.

Stetter, K.O. 1986b. Extreme thermophile Mikroorganismen, pp. 59–75, in Crueger, W., Esser, K., Präve, P., Schlingmann, M., Thauer, R., and Wagner, F. (editors), *Jahrbuch der Biotechnologie,* Carl Hanser Verlag, München, Wien.

Stetter, K.O., Segerer, A., Zillig, W., Huber, G., Fiala, G., Huber, R., and **König, H.** 1986. Extremely thermophilic sulfur-metabolizing archaebacteria. *Systematic and Applied Microbiology* 7:393–397.

Stetter, K.O., Lauerer, G., Thomm, M., and **Neuner, A.** 1987. Isolation of extremely thermophilic sulfate reducers: Evidence for a novel branch of archaebacteria. *Science* 236:822–824.

Stetter, K.O. 1988. *Archaeoglobus fulgidus* gen. nov., sp. nov.: A new taxon of extremely thermophilic archaebacteria. *Systematic and Applied Microbiology* 10:172–173.

Trent, J.D., Chastain, R.A., and **Yayanos, A.A.** 1984. Possible artefactual basis for apparent bacterial growth at 250°C. *Nature* 207:737–740.

White, R.H. 1984. Hydrolytic stability of biomolecules at high temperatures and its implication for life at 250°C. *Nature* 310:430–432.

Woese, C.R., Magrum, L.J., and **Fox, G.E.** 1978. Archaebacteria. *Journal of Molecular Evolution* 11:245–252.

Woese, C.R. 1987. Bacterial evolution. *Microbiological Reviews* 51:221–271.

Zillig, W., Stetter, K.O., Wunderl, S., Schulz, W., Priess, H., and **Scholz, I.** 1980. The *Sulfolobus-"Caldariella"*-group: Taxonomy on the basis of the structure of DNA-dependent RNA polymerases. *Archives of Microbiology* 125:259–269.

Zillig, W., Stetter, K.O., Schäfer, W., Janekovic, D., Wunderl, S., Holz, I., and **Palm, P.** 1981. *Thermoproteales*: a novel type of extremely thermoacidophilic anaerobic archaebacteria isolated from Icelandic solfataras. *Zentralblatt für Bakteriologie und Hygiene,* I. Abteilung Originale C 2:205–227.

Zillig, W., Yeats, S., Holz, I., Böck, A., Gropp, F., Rettenberger, M., and **Lutz, S.** 1985. Plasmid-related anaerobic autotrophy of the novel archaebacterium *Sulfolobus ambivalens. Nature* 313:789–791.

Zillig, W., Yeats, S., Holz, I., Böck, A., Gropp, F., and Simon, G. 1987. *Desulfurolobus ambivalens* gen. nov. sp. nov., an autotrophic archaebacterium facultatively oxidizing or reducing sulfur. *Systematic and Applied Microbiology* 8:197–203.

10

Autotrophic Bacteria in Protozoa

Claudius K. Stumm and Godfried D. Vogels

Department of Microbiology
University of Nijmegen
Nijmegen, The Netherlands

Every scientist who is studying the intricate details of protozoan cells will eventually detect structures which do not fit in the long list of "normal" organelles. Attached to the protozoan pellicle or dispersed through its cytoplasm (or even through its nucleus), spirals, rods or globules can be found; besides their shape these structures possess all the morphological qualities which are characteristic for bacteria. Thus, Kirby (1941) described episymbiotic spirochetes adherent to certain flagellates living in the gut of termites, Greeff (1874) mentioned rod-like endosymbiotic bacteria in the giant amoeba *Pelomyxa* which was earlier discovered by him, and Liebmann (1937) pointed to the observation that many ciliates living in polluted water harbor large numbers of bacterial cells. Liebmann already speculated about the function of the bacteria: possibly the symbionts would be digested by the ciliates when external food was scarce.

After these early observations, a stream of publications (many of them reviewed by Lee and Corliss, 1985) demonstrated continued interest in this topic. The problem, however, concerning the part which the symbiont plays in the protozoan cell (e.g., whether the relation is parasitic, mutualistic, or merely commensalistic) could not be solved in most cases because of the difficulties of cultivating protozoa under defined conditions both with and without the symbiont. Also, very little is known about the metabolism of the bacteria concerned, because most of them have not yet been isolated and grown in pure culture. Thus it is not known whether a bacterium symbiotic in a protozoan cell gains its energy from organic or inorganic sources or light, and whether it is a hetero- or autotroph. Exceptions, however, exist when the symbiont possesses morphological or chemical markers which are also found in free-living bacteria whose autotrophic metabolism is proven by cultures. Markers of this type are 1) photosynthetic membranes and pigments present in the blue-green cell constituents (cyanelles) found in certain protozoa and well-known from studies in cyanobacteria (Reisser, 1984); 2) the sulfur globules in the unnamed episymbionts of the ciliate *Kentrophoros*, also found in hydrogen sulfide-oxidizing bacteria (Fauré-Fremiet, 1950); and 3) fluorescent compounds in the symbionts of many anaerobic protozoa, indicating (besides other criteria) that they are methanogenic bacteria (Van Bruggen et al., 1983).

10.1 PHOTOAUTOTROPHIC BACTERIA AS SYMBIONTS IN PROTOZOA

Cyanobacteria are photoautotrophic procaryotes containing chlorophyll *a* and characteristic blue-green pigments called phycobiliproteins. Most cyanobacteria are free living, but a number of species are involved in a variety of symbiotic inter- or intracellular associations: with plants (e.g., *Gunnera, Azolla*), fungi (as lichens), and animals (e.g., sponges, bryozoa, shrimps) (reviewed by Geitler, 1959; Whitton, 1973; Whitton and Carr, 1982; Reisser, 1984). Also in eucaryotic microorganisms blue-green cell constituents strongly resembling cyanobacteria have been occasionally described; they are called cyanelles. Eucaryotic microorganisms which harbor cyanelles are, among others, *Paulinella chromatophora*, a thecamoeba devoid of food vacuoles, and the alga-like organisms *Cyanophora paradoxa, Glaucocystis nostochinearum, Glaucosphaera vacuolata*, and *Gloeochaete wittrockiana* (Reisser, 1984). The taxonomic position of these hosts is uncertain because no sexual processes or morphological traits were found so far by which they could be linked with other microbial groups.

With respect to physiology and morphology, the cyanelle system of *Cyanophora paradoxa* is well studied. The host is a small ovoid freshwater flagellate (Figure 10.1) containing two or more globular cyanelles to which the name *Cyanocyta korschikoffiana* was conferred by Hall and Claus (1963). In naming these globular elements, the authors wanted to stress that the cyanelle of *Cyanophora* should be considered as an endosymbiotic organism and not as a chloroplast—the discussion about this problem has continued until today (Kremer and Kies, 1987). On the one hand, these cyanelles are indistinguishable from the cyanobacteria in a number of characteristics. Thus, like cyanobacteria, the organelles have constituents of the photosynthetic apparatus (thylakoids, phycobiliproteins) and carry out a conventional oxygenic photosynthesis. Also a cyanelle has its own rudimentary cell wall containing peptidoglycan, is localized within a host cell vacuole, and undergoes division out of synchrony with the host cell. On the other hand, the cyanelle genome contains no more than about 10 percent of the DNA content of a comparable free-living cyanobacterium, and cycloheximide (an inhibitory of eucaryotic but not procaryotic protein synthesis) inhibits chlorophyll synthesis of the cyanelle, demonstrating that—at least on the level of chlorophyll biosynthesis—a dependency of *Cyanocyta* on the protein synthesis of the host does exist.

In contrast to the situations in protozoa or animals in which the endosymbiont is a eucaryotic alga, in a cyanelle system neither of the partners can be grown without the other, and generally one can say now (as Trench and

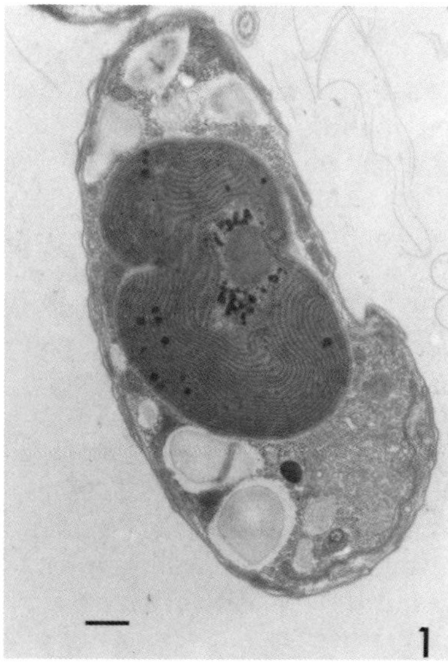

Figure 10.1 Electron photomicrograph of a thin section of *Cyanophora paradoxa*. A cyanelle with conspicuous thylakoids is visible in the center of the flagellate cell. Bar is 0.5 μm. Photo: L. Kies, from Kremer and Kies (1987).

his co-workers showed in *Cyanophora*; Trench et al., 1978) that there is a close mutual interdependence of host and cyanelle, and that the symbiotic association represents a well-integrated, obligate relation. The question, however, do cyanelles really represent the "missing link" between cyanobacteria and chloroplasts, has still to be answered (Whatley and Whatley, 1984).

10.2 PURPLE SULFUR BACTERIA AS EPISYMBIONTS OF *KENTROPHOROS*

The conspicuous sulfur grains described by Fauré-Fremiet in 1950 as cell constituents of episymbiotic bacteria living on the ciliate *Kentrophoros*, together with their dark pigmentation, are strong indications that these bacteria are purple sulfur bacteria (Chromatiaceae). This is a group of morphologically diverse, anaerobic bacteria with a predominantly photoautotrophic mode of metabolism (see Chapter 6). Chromatiaceae are able to oxidize hydrogen sulfide to sulfur, which in most species is deposited in the cytoplasm and which can be further oxidized to sulfate. These bacteria flourish in anoxic marine or freshwater habitats, such as sediments or sulfur springs.

This association of purple sulfur bacteria with protozoa is to our knowledge the only well-documented example; it was found by Kahl in 1935 and investigated by Fauré-Fremiet (1950) and Raikov (1974). In the initial work, Kahl described a number of marine ciliates which are "psammophilic," living in the interstices of sandy sediments where they are highly adapted. The body shape of these ciliates is thread- or ribbon-like, they are extremely flexible, and crawl like worms between the sand grains. The organisms described as *Centrophorella* (later *Kentrophoros*) were found to be ciliated only on the side of the cell body which is turned to the substrate (ventral side). The dorsal side, instead, was covered with "obviously symbiotic sulfur bacteria." Within the sand habitat, *Kentrophoros* species are confined to a narrow layer between the oxygenated and anoxic regions of the sediment. In this "redox potential discontinuity" layer (Fenchel and Riedl, 1970), oxidizing processes are followed by reducing processes. The *Kentrophoros* cells have no mouth (cytostome)—a peculiar feature, since the other ciliate species which share with them the same habitat are hunters or grazers with well-developed cytostomata. Fauré-Fremiet was the first who studied this symbiosis. He found the bacteria in parallel rows firmly attached to the naked dorsal side of *Kentrophoros fasciolatum* and *K. fistulosum*. The edges of the ribbon-like cells are curled upward, so that in fact the episymbionts are fixed to the inner wall of a long tube (Figure 10.2). The bacteria are nonmotile and show a virtually unique mode of division by longitudinal fission. Besides rather large quantities of polysaccharides, indicated by iodine treatment, they contain characteristic refractive inclusion granules in various quantities. These inclusions are soluble in chloroform, toluene, etc., characteristics which can be attributed to sulfur grains. The length of the bacterium is host-specific: the rods from *K. fistulosum* measured 4 μm, those of *K.*

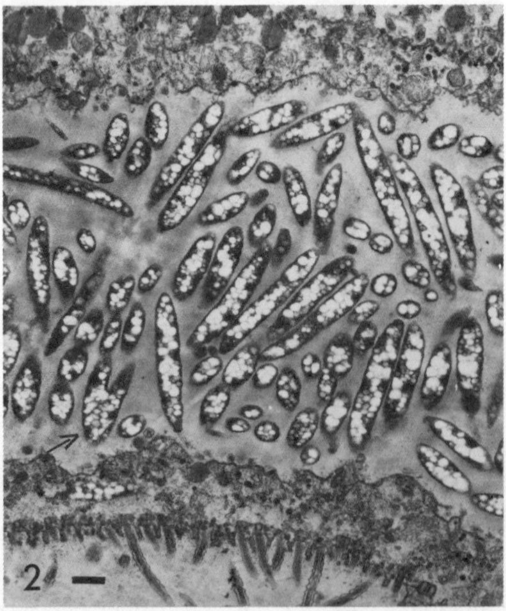

Figure 10.2 *Kentrophoros fistulosum*, the lumen of the tube-like body being filled with episymbiotic sulfur bacteria. The arrow points to a bacterium dividing longitudinally. Bar is 1 μm. From I. B. Raikov, *Tsitologiya* (Leningrad) (1971) 13:1514–1517.

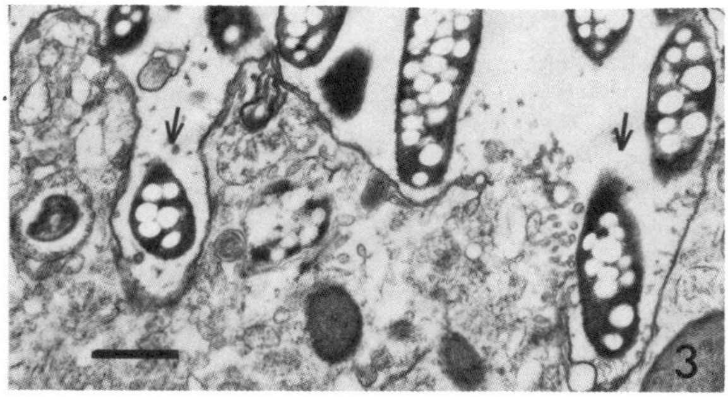

Figure 10.3 Section through *Kentrophoros latum* showing two sulfur bacteria being ingested by the ciliate (arrows). Bar is 1 μm. From Raikov (1974).

fasciolatum 3 μm. The fact that both types are exclusively found on ciliates and neither on the accompanying sand grains nor on added pieces of glass, points to an adaptation for the attachment on the *Kentrophoros* pellicle.

An electron microscopical study by Raikov (1974) revealed more details on the relation of the purple sulfur bacteria with the ciliate hosts, *K. fistulosum* and *K. latum*. The cells of the latter species are not tubular but flat; they also were found to bear symbionts on the cilia-free dorsal side, attached tightly as bristles of a brush. The food vacuoles of the two *Kentrophoros* species contain exclusively bacteria which show the morphological markers of the episymbionts (sulfur grains, longitudinal division clefts). Thus, it is obvious that the symbionts form an important, if not the only, food source for the ciliates. Raikov was able to determine how the mouthless *Kentrophoros* cells ingest their food. Everywhere on the pellicle surface, small invaginations or pockets are formed randomly. Episymbionts present at the place of such an invagination are "trapped" (Figure 10.3) and enclosed by a membrane, which eventually detaches from the inner side of the pellicle and as a food vacuole migrates to the endoplasm, where digestion proceeds (Figure 10.4). This unique manner of food uptake by a ciliate was called "random phagocytosis" ("phagocytose non localisée"). The *Kentrophoros*/sulfur bacteria symbiosis in the only example hitherto described of an apparent complete dependency of a ciliate on its bacterial

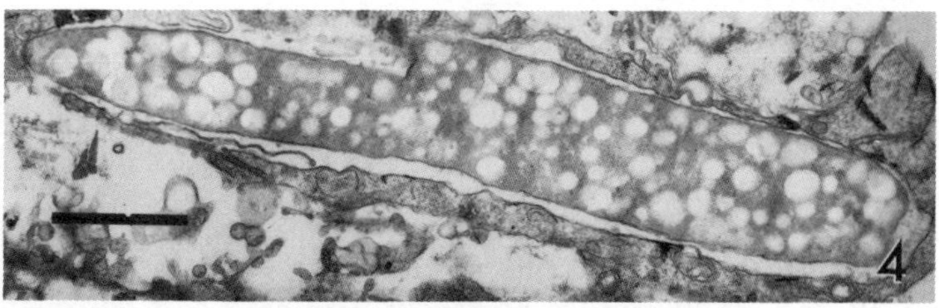

Figure 10.4 Section through *Kentrophoros latum* with an ingested, dividing sulfur bacterium. Bar is 1 μm. From Raikov (1974).

symbiont which, on its part, is strongly adapted to the specific attachment on the ciliate surface.

10.3 SYMBIOSIS OF METHANOGENIC BACTERIA WITH PROTOZOA

Although animal life requires oxygen, anaerobic habitats exist within animals, such as the intestinal tract, rumen, and the guts of many insects (for example, termites). Also, sewage plants and the sediments of many freshwater and marine sediments are anaerobic (see Figure 10.5). By living in these surroundings, the microorganisms contribute to the gradual degradation and mineralization of the organic material.

Prerequisites for growth of a living organism are—also in anaerobic habitats—oxidative reactions. If oxygen is absent, the electrons which are released during energy-generating processes are transported to either nitrate, sulfate (anaerobic respiration), organic compounds (fermentation), or to protons (proton reduction with release of hydrogen gas). Production of hydrogen (H_2) by proton reduction involves the enzyme hydrogenase. However, proton reduction functions well only if the concentration of hydrogen, generated by the metabolic activities of many organisms—procaryotes as well as protozoa—does not exceed a pressure of about 10^{-5} atmosphere. The most important organisms in keeping hydrogen concentrations low are the methanogenic bacteria, which perform the reaction

$$4H_2 + CO_2 \rightarrow CH_4 + 2H_2O$$

As a result, by the activity of methanogens, oxidative reactions of hydrogen-producing anaerobic protozoa become possible, since the release of gaseous methane carries electrons out of the anoxic habitat, thus making life feasible for other organisms.

The process of methanogenesis is discussed in Chapter 21. Of significance in methanogenesis is the number of unique enzymes and coenzymes. Two of them, the deazaflavin coenzyme F_{420} and compound F_{342} (now known as 7-methylpterin; Keltjens et al., 1983), show fluorescence when irradiated with short-wave-length blue light. Thus fluorescence microscopy makes it relatively easy to recognize methanogens even if they are mixed with other bacteria or are attached to nonfluorescent substrates. If a sample from an anaerobic habitat is examined under the fluorescence microscope, one perceives against a black background bluish light-emitting rods, platelets, or globules, indicative of various types of methanogens.

The rumen of herbivorous animals contains a real microcosm of organisms consisting of bacteria, fungi, and ciliates (Hungate, 1975): in 1 ml rumen fluid live about 10^{11} bacteria, 10^5 ciliates, and about 10^3 fungi. These microorganisms have in common an adaptation to life without oxygen, and furthermore to high concentrations of dissolved and solid nutrients, as well as to a constant tem-

Figure 10.5 Two typical habitats in which symbiosis of methanogenic bacteria with protozoa occurs: ruminants (water buffaloes) in a muddy pool.

perature of about 40°C. By the various metabolic activities of these microbes, ingested plant material is partially degraded to volatile fatty acids which form an important energy source for the ruminant.

A small fraction of the rumen bacteria are methanogens, able to utilize carbon dioxide and hydrogen produced by the nonmethanogenic organisms. Among the nonmethanogens, the ciliates are most striking, by their size (some types are large enough to be seen with the naked eye) and by their curious shape, since many of them bear long spines. By means of fluorescence microscopy, it can be shown that numerous methanogenic bacteria are attached to the cell surfaces of many ciliates (Figures 10.6–10.8). In rumen samples from slaughtered cows, all ciliate species with a partly naked cell surface (the so-called entodiniomorphids) showed this episymbiotic interactions with methanogens (probably *Methanobrevibacter ruminantium*) in various frequencies (Vogels et al., 1980). Experiments with a fistulated sheep revealed a decrease of the association frequency after food uptake, if the animal had fasted before, and also when the rumen of a fasted animal was flushed via the fistula with hydrogen gas; however, the association frequency increased during fasting or by flushing of the rumen with molecular nitrogen (Stumm et al., 1982). Flushing experiments performed with an artificial rumen system built in the laboratory showed a similar negative correlation of association frequency with hydrogen concentration.

The phenomenon of the attachment of the methanogens to the ciliate surface can be explained by the release of hydrogen by the ciliates. Yarlett et al. (1983) showed that a certain cell fraction from a rumen ciliate homogenate contained small organelles called hydrogenosomes, showing strong hydrogenase activity. By these experiments it could be demonstrated that the rumen ciliates, like some rumen bacterial species, generate hydrogen, which in fact is unfavorable for the cell if its concentration is not sufficiently lowered. Attachment of the hydrogen-consuming methanogens evidently facilitates hydrogen removal, bringing advantage to both ciliates and episymbionts. In the case of the fasting animal, hydrogen concentration in the rumen fluid decreases due to the decreasing metabolic activity of hydrogen-producing bacteria which are

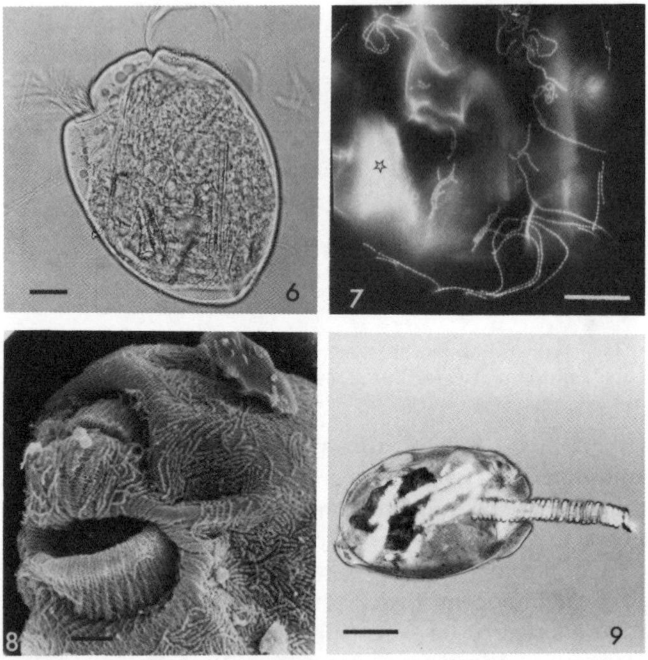

Figures 10.6–10.9 Methanogens associated with rumen ciliates.
Figure 10.6. The ciliate *Eudiplodinium maggii* is a common host of attached methanogens. Bright field, bar is 25 μm.
Figure 10.7. Fluorescence microscopy reveals the methanogens (probably *Methanobrevibacter ruminantium*) as autofluorescent chains of bacteria on the distal part of the same cell. The light, out-of-focus areas (✫) are remnants of cellulosic particles ingested by the ciliate. Bar is 25 μm.
Figure 10.8. *Eudiplodinium maggii* from an artificial rumen system, covered with methanogens. Scanning electron micrograph; bar is 10 μm.
Figure 10.9. A rumen ciliate engulfing plant fiber material. Bright field, polarized light; bar is 50 μm.

dependent on easily degradable substrates. Consequently, methanogens may attach to ciliates because the latter probably still produce hydrogen: also a fasting rumen still contains a mass of cellulose fibers on which the ciliates thrive (Figure 10.9). In case of a well-fed rumen, the situation is reversed: hydrogen concentration in the rumen fluid goes up and thus free-living (non-attached) methanogens can thrive. Flushing experiments with nitrogen or hydrogen give support to this hypothesis and simulate the fasting and well-fed animal, respectively.

The sediments at the bottom of stagnant waters rich in organic material are called sapropel (Lauterborn, 1901) or sulfuretum (Baas Becking, 1925). The sapropel represents, next to the rumen, one of the best-studied anaerobic habitats. It is inhabited by large numbers of methanogenic bacteria and anaerobic protozoa. The numbers, however, are much lower than in the rumen, evidently as a consequence of the lower nutrient concentrations. A survey of sapropelic ciliates by means of fluorescent microscopy revealed the presence of methanogenic bacteria *inside* the cells, spread throughout the cytoplasm in considerable quantities (Figures 10.10 and 10.11). During the course of the survey it was found that not only sapropelic ciliates invariably harbored fluorescent bacteria, but also sapropelic amoebae and flagellates (Table 10.1).

Conclusive evidence that the autofluorescent particles were indeed methanogenic bacteria could be gained from various experiments: 1) Analysis of extracted protozoan cells by means of such sensitive biochemical techniques as high performance liquid chromatography and thin layer chromatography revealed the presence of coenzyme F_{420} and 7-methylpterin. 2) A single cell of the giant sapropelic amoeba *Pelomyxa* excreted measurable amounts of methane. 3) Methane excretion from a washed suspension of the ciliate *Metopus striatus* was inhibited by bromoethanesulfonic acid, a specific inhibitor of methanogenesis. 4) Endosymbionts from several sapropelic protozoa were isolated and characterized as methanogens. These were *Methanobacterium formicicum* from several freshwater ciliates as well from the amoeba *Pelomyxa*, and a new methanogen (*Methanoplanus endosymbiosus*) from the marine ciliate *Metopus contortus* (Figure 10.12).

The number of methanogenic bacteria was found to be high in most protozoan species, e.g., a single *Mastigella* amoeba, which is rather small compared to the large *Pelomyxa*, contained about 2×10^4 methanogens; taking into account the number and distribution of the various sapropel-inhabiting protozoa, an estimate of about 1.5×10^6 methanogens per ml sapropel was made (Van Bruggen et al., 1985). From these results one may conclude that a prominent fraction—if not all—of the methane originating from the upper layer of sapropel could be produced by methanogen-inhabiting ("methanophorous") protozoa. This hypothesis is supported by the work of several authors (Cappenberg and Verdouw, 1982; Zaiss, 1981; Zeikus and Winfrey, 1976) who had earlier determined the number of methanogenic bacteria in the mud by different microbiological methods. They found about 10^4–10^5 methanogens per ml. Taking into account the technical difficulties in handling anaerobic bacteria and determining their numbers by these methods, the results are in fairly good agreement with those reported by Van Bruggen et al. (1985).

Electron microscopical studies of the sapropelic protozoa revealed the absence of mitochondria in all species studied so far. The lack of this cell organelle, which is otherwise found in all free-living protozoa, and which is involved in electron transfer to oxygen, might be the result of an adaptation to anaerobic conditions and to the transfer of electrons to endosymbiotic methanogens. Another result of fine structure analysis of the sapropelic protozoa was the detection of microbodies which morphologically strongly resembled the hydrogenosomes known from anaerobic rumen ciliates and certain parasitic flagellates. In sapropelic ciliates such as *Metopus contortus* and *Plagiopyla nasuta*, the microbodies were found to lie side by side with the methanogens (Figure 10.13) or even to surround them like a jacket (Figure 10.14). In the ciliate *Metopus striatus* the association of microbodies with methanogens was less intimate, and in the ciliates *Discomorphella pectinata* and *Saprodinium dentatum*,

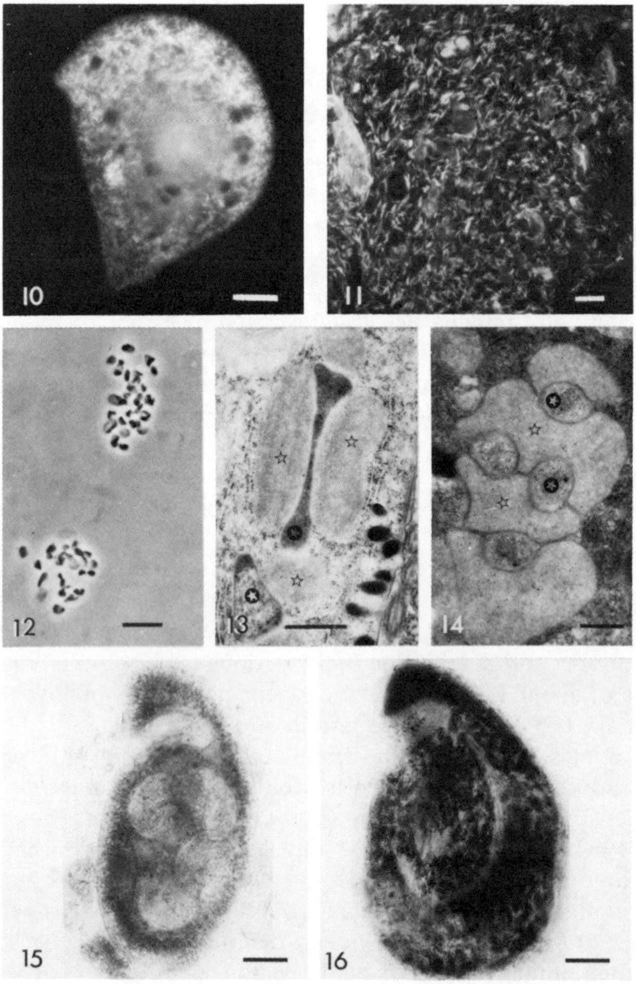

Figures 10.10–10.16 Endosymbiotic methanogens with sediment-dwelling (sapropelic) protozoa.
Figure 10.10. In the ciliate *Metopus striatus*, innumerable methanogens are present as fluorescent, needle-like structures. Bar is 15 μm.
Figure 10.11. Fluorescent methanogens in a small part of the amoeba *Mastigella*. Bar is 15 μm. Stumm and Zwart (1986).
Figure 10.12. *Methanoplanus endosymbiosus* from the marine ciliate *Metopus contortus* in pure culture. Phase contrast; bar is 5 μm.
Figure 10.13. Section of *Metopus contortus; Methanoplanus* cells (✪) and adjacent hydrogenosomes (✫). Transmission electron micrograph (TEM), bar is 0.5 μm.
Figure 10.14. Section of the freshwater ciliate *Plagiopyla nasuta* with methanogens (✪) and surrounding hydrogenosomes (✫). TEM; bar is 0.5 μm. Stumm and Zwart (1986).
Figure 10.15; 10.16. *Plagiopyla nasuta*, anaerobically incubated with 2-(2′-benzothiazolyl)-5-styryl-3-(4′-phthalhydrazidyl)- tetrazolium chloride (BSPT) + N_2 (control; Fig. 10.15) and BSPT + H_2 (Fig. 10.16), respectively. Black precipitate in Fig. 10.16 reveals presence of hydrogenase. Bright field; bar is 15 μm.

Table 10.1 Protozoa Containing Endosymbiotic Methanogenic Bacteria

Ciliates:		**Amoebae:**
Bothrostoma undulans	*Lacrymaria cucumis*	*Pelomyxa palustris*
B. mirabilis	*L. elegans*	*Mastigella* spec.
Brachonella caduca	*Metopus es*	
B. campanula	*M. contortus*	**Flagellates:**
B. darwini	*M. fuscus*	*Cercoboda* spec.
B. galeata	*M. laminarius*	2 more unidentified species
B. spiralis	*M. minimus*	
Caenomorpha lata	*M. pulcher*	
C. levanderi	*M. striatus*	
C. medusula	*Plagiopyla frontata*	
C. sapropelica	*P. nasuta*	
C. uniserialis	*Saprodinium dentatum*	
Discomorphella pectinata	*Trimyema compressum*	
Epalxella antiquorum		

no microbodies were found at all. Because the methanogens isolated from protozoa thus far were found to consume hydrogen, it can be hypothesized that the methanogen-associated microbodies are hydrogenosomes. By the close association of hydrogenosome and methanogen an efficient hydrogen transfer would be achieved, and via the production of methane, oxidative reactions by the ciliate would be favored.

A direct proof for the hydrogenosome character of the microbodies was obtained by the *in situ* localization of hydrogenase by means of the tetrazolium compound 2-(2′-benzothiazolyl)-5-styryl-3-(4′-phthalhydrazidyl)—tetrazolium chloride (BSPT) as electron acceptor. With this indicator, which had earlier been used successfully in the flagellate *Trichomonas,* hydrogenase activity could be shown to be present in microbodies of the sapropelic ciliates *Plagiopyla nasuta* and *Trimyema compressum* (Figures 10.15, 10.16) (Zwart et al., 1988).

In the context of symbiosis of methanogens with sapropelic protozoa, two amoebae should be introduced, both mentioned shortly before: *Pelomyxa palustris* (Figure 10.17) and *Mastigella spec.* (Figure 10.18). *Pelomyxa* is an amoeba of the freshwater sapropel with a diameter of up to 5 mm. It was first described by R. Greeff in 1874 and thoroughly studied by C. Chapman-Andresen about 100 years later (Chapman-Andresen, 1971). This remarkable organism is not only devoid of mitochondria but also lacks a Golgi apparatus, contractile vacuoles, and a normal endoplasmic reticulum. The cells are multinucleated; no mitosis-like division stage could be detected, though several scientists put special attention to that point. Curiously enough, three different types of bacterial endosymbionts live inside this amoeba, a thick rod with a characteristic longitudinal cleft (cell division, however, is transversal) and two thin rods which can be discriminated by cell-wall structure and by Gram stain. Both thin rods are methanogens, and one was recently isolated and identified as *Methanobacterium formicicum* (Figures 10.19, 10.20) (Van Bruggen et al., 1988). The thick rod has not so far been cultivated or identified. No microbodies could be found in *Pelomyxa;* the methanogens are scattered throughout the cytoplasm, whereas the thick rods are preferentially concentrated around the nuclei (Figure 10.21).

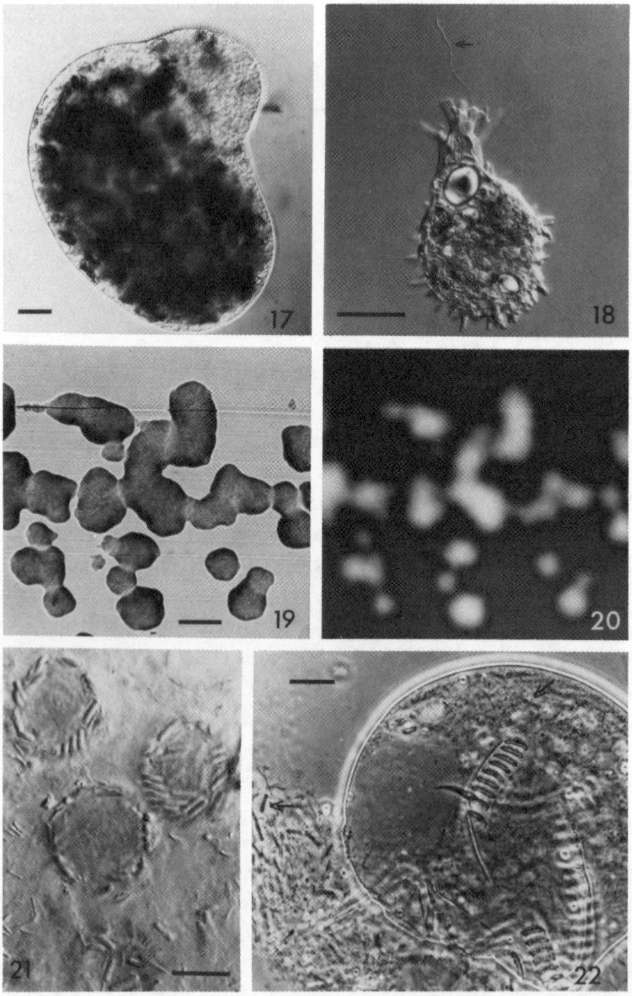

Figures 10.17–10.21 Endosymbiotic methanogens in anaerobic amoebae. **Figure 10.17**. Three types of endosymbiotic bacteria inhabit the sluggish giant amoeba *Pelomyxa*. The dark particles inside the cell consist of ingested food material. Bright field; bar is 50 μm.
Figure 10.18. *Mastigella* is an amoeba with a flagellum (arrow) and, like *Pelomyxa*, with large numbers of methanogenic and nonmethanogenic bacteria (see also Figure 10.11). Differential interference contrast (DIC); bar is 50 μm.
Figures 10.19; 10.20. *Methanobacterium formicicum* (isolated from *Pelomyxa*) on agar, illuminated with visible light (Figure 10.19) and ultraviolet (Figure 10.20), showing the fluorescence of the colonies. Bar is 250 μm.
Figure 10.21. Nuclei of *Pelomyxa* showing the associated thick, nonmethanogenic bacteria. Methanogens (thin rods) are scattered throughout the cytoplasm. DIC; bar is 10 μm.
Figure 10.22. A flattened cell of the ciliate *Discomorphella* showing numerous nonfluorescent, thick bacteria (arrows) inside the cell and in the extruded cytoplasm. Phase contrast; bar is 10 μm.

Mastigella is clearly distinguished from *Pelomyxa* not only by its smaller size (about 120 μm) and different cell shape, but also by the presence of a short flagellum and a single, large nucleus. However, striking similarities exist between the two amoebae: the equipment of *Mastigella* with organelles is as scarce as that of *Pelomyxa*, no description of a mitosis stage has been found, methanogenic bacteria are present (only Gram-positive thin rods), and around the nucleus a layer of thick, clefted rods can be seen. Neither *Pelomyxa* nor *Mastigella* could be cultivated under defined conditions. This is curious since they are frequently encountered in pools and ponds rich in plant-derived sediments, and even in muddy laboratory aquaria where they engulf decaying algae and plant particles.

Recently two highly differentiated anaerobic ciliates (*Discomorphella pectinata* and *Saprodinium dentatum*) were found to lack mitochondria and to harbor methanogens, but at the same time to harbor thick, clefted bacteria (Figure 10.22) and obviously lack hydrogenosomes (Broers, unpublished results). Thus, in anaerobic protozoa one finds either hydrogenosomes or thick, clefted bacteria, one appearing to exclude the other, while methanogens are present in all cases studied so far. The presence of methanogens can be explained by the fact that in the sapropel the concentrations of dissolved substrates outside the protozoa are much lower than in the rumen. Under such circumstances, endosymbiosis offers the possibility to profit optimally from metabolic interactions, and the hydrogenosomes play the role of a bridge in hydrogen transfer between protist and methanogen. One may speculate that the thick rods might be hydrogen-generating bacteria which function as hydrogenosomes if the latter are absent.

The presence of bacterial autotrophs symbiotic in animals of the deep-sea vents has been discussed in Chapter 8. We see here, with the anaerobic protozoa, another example of how eucaryotes have entered into symbiosis with autotrophic bacteria. Since studies on eucaryotes/autotrophic bacteria symbioses are only relatively new, it is possible that many more examples will be discovered as research continues.

REFERENCES

Baas Becking, L.G.M. 1925. Studies on the sulphur bacteria. *Annals of Botany* 39:613–650.

Cappenberg, T.E. and **Verdouw, H.** 1982. Sedimentation and breakdown kinetics of organic matter in the anaerobic zone of Lake Vechten. *Hydrobiologia* 95:165–179.

Chapman-Andresen, C. 1971. Biology of the large amoebae. *Annual Review of Microbiology* 25:27–48.

Fauré-Fremiet, E. 1950. Caulobactéries epizoïques associées aux *Centrophorella* (Ciliés Holotriches). *Bulletin Societé Zoologique de France* 75:134–137.

Fenchel, T.M. and **Riedl, R.J.** 1970. The sulfide system: a new biotic community underneath the oxidized layer of marine sand bottoms. *Marine Biology* 7:255–268.

Geitler, L. 1959. Syncyanosen, pp. 530–545, in Ruhland, E. (editor), *Handbuch der Pflanzenphysiologie*, Springer, Berlin, Heidelberg, New York.

Greeff, R. 1874. *Pelomyxa palustris* (Pelobius), ein amöbenartiger Organismus des süssen Wassers. *Archiv der Mikrokopischen Anatomie* 10:53–73.
Hall, W.T. and **Claus, G.** 1963. Ultrastructural studies on the blue-green algal symbiont in *Cyanophora paradoxa* Korschikoff. *Journal of Cell Biology* 19:551–564.
Hungate, R.E. 1975. The rumen microbial ecosystem. *Annual Review of Ecology and Systematics* 6:39–66.
Kahl, A. 1930–1935. Wimpertiere oder Ciliata (Infusoria), in Dahl, F. (editor), *Die Tierwelt Deutschlands,* Fischer, Jena, part 30.
Keltjens, J.T., van Beelen, P., Stassen, A.M., and **Vogels, G.D.** 1983. 7-Methylpterin in methanogenic bacteria. *FEMS Microbiology Letters* 20:259–262.
Kirby, H. 1941. Devescovinid flagellates of termites. *University of California Publications in Zoology* 45:1–92.
Kremer, B.P. and **Kies, L.** 1987. Cyanellen: Grenzgänger in der Zelle. *Mikrokosmos* 76:357–363.
Lauterborn, R. 1901. Die "sapropelische" Lebewelt. *Zoologischer Anzeiger* 24:50–55.
Lee, J.J. and **Corliss, J.O.** (editors) 1985. Symposium on "Symbiosis in Protozoa." *Journal of Protozoology* 32:371–403.
Liebmann, H. 1937. Bakteriensymbiose bei Faulschlammziliaten. *Biologisches Zentralblatt* 57:442–445.
Raikov, J.B. 1974. Étude ultrastructurale des bactéries épizoïques et endozoïques de *Kentrophoros latum* Raikov, cilié holotriche mésopsammique. *Cahiers de Biologie Marine* 15:379–393.
Reisser, W. 1984. Endosymbiotic cyanobacteria and cyanellae, pp. 91–112, in Pirson, A. and Zimmermann, M.H. (editors), *Encyclopedia of Plant Physiology Vol. 17: Cellular Interactions,* Springer, Berlin, Heidelberg, New York.
Stumm, C.K., and **Zwart, K.B.** 1986. Symbiosis of protozoa with hydrogen-utilizing methanogens. *Microbiological Sciences* 3:100–105.
Stumm, C.K., Gijzen, H.J., and **Vogels, G.D.** 1982. Association of methanogenic bacteria with ovine rumen ciliates. *British Journal of Nutrition* 47:95–99.
Trench, R.K., Pool, R.R., Logan, M., and **Engelland, A.** 1978. Aspects of the relation between *Cyanophora paradoxa* (Korschikoff) and its endosymbiotic cyanelles *Cyanocyta korschikoffiana* (Hall and Claus). I. Growth, ultrastructure, photosynthesis and the obligate nature of the association. *Proceedings of the Royal Society* London B 202:423–443.
Van Bruggen, J.J.A., Stumm, C.K., and **Vogels, G.D.** 1983. Symbiosis of methanogenic bacteria and sapropelic protozoa. *Archives of Microbiology* 136:89–95.
Van Bruggen, J.J.A., Stumm, C.K., Zwart, K.B., and **Vogels, G.D.** 1985. Endosymbiotic methanogenic bacteria of the sapropelic amoeba *Mastigella. FEMS Microbiology Ecology* 31:187–192.
Van Bruggen, J.J.A., van Rens, G.L.M., Geertman, E.J.M., Zwart, K.B., Stumm, C.K., and **Vogels, G.D.** 1988. Isolation of a methanogenic endosymbiont of the sapropelic amoeba *Pelomyxa palustris* Greeff. *The Journal of Protozoology* 35:20–23.
Vogels, G.D., Hoppe, W.F., and **Stumm, C.K.** 1980. Association of methanogenic bacteria with rumen ciliates. *Applied and Environmental Microbiology* 40:608–612.
Whatley, J.M. and **Whatley, F.R.** 1984. Evolutionary aspects of the eukaryotic cell and its organelles, pp. 18–58, in Pirson, A. and Zimmermann, M.H. (editors), *Encyclopedia of Plant Physiology Vol. 17: Cellular Interactions,* Springer-Verlag, Berlin, Heidelberg, New York.
Whitton, B.A. 1973. Interactions with other organisms, pp. 415–433, in Carr, N.G. and Whitton, B.A. (editors), *The Biology of Blue-green Algae,* Blackwell, Oxford.
Whitton, B.A. and **Carr, N.G.** 1982. Cyanobacteria: current perspectives, pp. 1–8, in Carr, N.G. and Whitton, B.A. (editors), *The Biology of Cyanobacteria,* Blackwell, Oxford.
Yarlett, N., Hann, A.C., Lloyd, D. and **Williams, A.G.** 1983. Hydrogenosomes in a mixed isolate of *Isotricha prostoma* and *Isotricha intestinalis* from ovine rumen contents. *Comparative Biochemical Physiology* 74b:357–364.

Zaiss, U. 1981. Seasonal studies of methanogenesis and desulfurication in sediments of the river Saar. *Zentralblatt für Bakteriologie und Hygiene*. I. Abt. Orig. C 2:76–89.

Zeikus, J.G. and **Winfrey, M.R.** 1976. Temperature limitation of methanogenesis in aquatic sediments. *Applied and Environmental Microbiology* 31:99–107.

Zwart, K.B., Goosen, N.K., van Schijndel, M.W., Broers, C.A.M., Stumm, C.K., and **Vogels, G.D.** 1988. Cytochemical localization of hydrogenase activity in the anaerobic protozoa *Trichomonas vaginalis, Plagiopyla nasuta* and *Trimyema compressum*. *Journal of General Microbiology* 134: in press.

11

Physiology and Biochemistry of Unicellular Sulfur Bacteria

Don P. Kelly

Department of Biological Sciences
University of Warwick
Coventry, England

This chapter is concerned with those nonphotosynthetic bacteria that are able to obtain all the energy they need for growth from the oxidation of inorganic sulfur compounds or elemental sulfur, and can obtain carbon for the synthesis of all their cell constituents from carbon dioxide. Such bacteria are thus defined as chemolithotrophic autotrophs, obtaining energy and carbon in a manner distinct from that seen in the heterotrophic (or chemoorganotrophic) bacteria. The sulfur-oxidizing chemolithoautotrophs share some metabolic features with the photosynthetic sulfur-oxidizing bacteria (see Chapters 6 and 14) in that both groups obtain reducing equivalents (NADH and NADPH) from oxidizing inorganic sulfur compounds and both can obtain carbon from carbon dioxide by means of fixation pathways unique to autotrophic organisms (Postgate and Kelly, 1982; Kelly, 1988).

11.1 DEFINING THE "UNICELLULAR SULFUR BACTERIA"

Unicellular colorless sulfur bacteria were probably unknown to Winogradsky when he established the principles of sulfur-dependent chemolithotrophy from his studies on filamentous sulfur bacteria (see Chapters 1 and 12). The first small unicellular typical Gram-negative eubacteria obtaining energy for autotrophic growth from the oxidation of thiosulfate ($S_2O_3^{2-}$) were described by Nathansohn (1902) and the first to be named was *Thiobacillus thioparus* (Beijerinck, 1904). Over the following decades, numerous metabolically similar bacteria were isolated and described, and a number of different generic names appeared, including *Thiobacterium, Sulfomonas* and other genera, about which little is currently known, and whose names have in some cases subsequently disappeared from use. The genus *Thiobacillus* became for many years the taxonomic designation of a diverse range of Gram-negative bacteria all characterized by their ability to grow using sulfur as an energy source.

Until recent years microbiologists were all quite clear in their understanding of what was meant when "thiobacilli" or the genus *Thiobacillus* was mentioned: they knew that they were talking about Gram-negative, rod-shaped small bacteria that grew in the dark, fixing carbon dioxide by means of the Calvin cycle, and obtaining all their energy by oxidizing sulfur and its compounds and producing sulfuric acid (Vishniac and Santer, 1957). This definition is still valid, but its taxonomic justification, and exclusivity of these properties, are much less certain.

The problems in thinking that "thiobacillus" is a taxonomically exact term are twofold. First is the realization that the species of the genus *Thiobacillus* are physiologically and biochemically extraordinarily disparate. Secondly, it has been found that the ability to grow as a sulfur chemolithoautotroph is not the unique preserve of bacteria that can unequivocally be called *Thiobacillus.* Historically, the formulation of this genus resulted as a natural consequence of the technique of elective enrichment culture. Thiobacilli have virtually always been obtained from enrichment cultures in which samples from the natural environment (or in some cases artificial habitats rich in sulfur compounds, such as sewage treatment plants, gas works effluents, mineral leach dumps, or wood pulping plants) have been incubated in mineral media with sulfur or thiosulfate as the principal or only energy-yielding substrate. This technique led to the isolation of "thiobacilli" from habitats of diverse temperature, pH, and other physico-chemical characteristics. The feature unifying all such isolates was the "unique" ability to grow autotrophically and to obtain energy from the oxidative dissimilation of reduced sulfur. This principle perhaps first became questionable with the isolation in the 1930s of *Thiobacillus novellus,* which was not obligately chemolithoautotrophic and could grow both on complex and defined heterotrophic media. This facultative character, subsequently found in other thiobacilli such as *T. versutus, T. acidophilus,* and *T. intermedius,* immediately altered the technical and conceptual boundaries for classifying a "thiobacillus." When an isolate is an obligate chemolithotroph, dependent absolutely on sulfur oxidation for energy generation, virtually all the tests and diagnostic characters employed in standard bacterial taxonomy are useless or inapplicable. An obligate chemolithotroph gives negative reactions in assays for gas production, acid production, or growth on sugars, organic acids and so forth. The conventional bacterial taxonomist is left with a very small number of diagnostic physiological characters, all of which revolve round autotrophy and sulfur metabolism, and features such as pH or temperature optima, which are not traditional taxonomic characters. In contrast, a facultative chemolithoautotroph, also able to grow heterotrophically, is amenable to the whole range of standard "biochemical tests" used in routine diagnostic bacteriology. Exposing an organism such as *T. versutus* to identification as an "unknown soil organism" will result in its description as a "*Pseudomonas* of uncertain species." Having established the principle that the sulfur-oxidizing chemolithotrophs can include facultatively heterotrophic types makes it conceptually reasonable to expect that heterotrophic bacteria that have been isolated by means *other than* elective enrichment culture designed to select for "thiobacilli," could perhaps also behave like members of the genus *Thiobacillus* if given the appropriate cultural conditions. This argument applies also to the *photosynthetic* anaerobic sulfur-oxidizing bacteria, which might also be able to grow aerobically in the dark as chemolithoautotrophs. Both these possibilities are now known to occur in several organisms.

The ability to grow with inorganic sulfur oxidations is, however, not a common attribute of normally heterotrophic bacteria. The heterotrophs for which sulfur chemolithotrophy is an option are predominantly also facultatively hydrogen-oxidizing autotrophs. The important principle brought out by the diversity of bacteria that can obtain energy from the nonphotosynthetic oxidation of sulfur compounds is that there is currently no *a priori* reason to presume that the mechanisms of sulfur compound oxidation, or indeed other fundamentals of their growth physiology, are necessarily the same in a range of bacteria able to grow as sulfur chemolithoautotrophs, even if all are currently classified as species of *Thiobacillus*. This realization is reinforced when the growing number of archaebacteria, such as *Sulfolobus*, *Acidianus*, and *Sulfurococcus*, capable of autotrophic growth on sulfur or tetrathionate, are also added to the ranks of the sulfur chemolithotrophs (see Chapter 9).

Table 11.1 lists a selection of bacteria that are capable of sulfur-dependent lithoautotrophic growth ("thiotrophy"). This list includes all the *Thiobacillus* spp. currently recognized in *Bergey's Manual* (Kelly and Harrison, 1988) and a number of other genera and species whose thiotrophic properties are essentially the same as those of the thiobacilli. These include the obligately chemolithoautotrophic *Thiomicrospira* spp., the versatile facultative *Thiosphaera*, and the extremely thermophilic eubacterium, *Thermothrix*, all of which behave thiotrophically like thiobacilli. No phototrophs are included in Table 11.1 as these are dealt with in Chapter 15.

The bulk of this chapter is devoted to those bacteria currently classified as *Thiobacillus*, as much more is known of the "thiotrophy" of these bacteria than of the other bacteria listed in Table 11.1.

11.2 PHYSIOLOGICAL DIVERSITY AMONG THIOBACILLI

The "physiological lumping" of the thiobacilli on the grounds that they are all autotrophic sulfur-oxidizing chemolithotrophs has been immensely useful to comparative microbial biochemistry, because it has meant that the research had a seemingly clear problem to address when investigating their metabolism. The concept of a genus "*Thiobacillus*" with species whose metabolic properties and problems would be common to all the constituent members of the genus is one that encouraged concentrated work on individual species. The justifi-

Table 11.1 Unicellular bacteria capable of growth as chemolithotrophic autotrophs on reduced sulfur compounds

Organism*	Sulfur substrates supporting chemolithoautotrophic growth**	Optimal physicochemical conditions for growth: pH	Temperature	%G + C of DNA	Maximum observed % DNA homology with several *Thiobacillus* spp.	Principal respiratory ubiquinone	Contain carboxysomes	Ability to grow on: Organic nutrients	One-carbon compounds	Autotrophically with hydrogen
Thiobacillus thioparus	S, S^{2-}, $S_2O_3^{2-}$, SCN–, $S_nO_6^{2-}$, CS_2, $(CH_3)_2S$	6.6–7.2	28°	62–63	22	Q-8	+	–	–	–
T. neapolitanus	S, S^{2-}, $S_2O_3^{2-}$, $S_nO_6^{2-}$	6.5–6.9	28–32°	56	0	Q-8	+	–	–	–
T. tepidarius	As above	6.8–7.5	43–45°	66–67	ND	Q-8	ND	–	–	–
*T. denitrificans**	As above	6.8–7.4	28–32°	63–64	29	Q-8	–	–	–	–
T. novellus	$S_2O_3^{2-}$, $S_4O_6^{2-}$	7.0	25–30°	67–68	14	Q-10	–	+	+	–
T. versutus	$S_2O_3^{2-}$, S^{2-}	7.5–7.9	30–35°	67–68	14	Q-10	–	+	+	–
T. intermedius	S, S^{2-}, $S_2O_3^{2-}$, $S_4O_6^{2-}$	5.5–6.0	30–35°	65–67	31	Q-8	+	(+)	–	–
T. perometabolis	S, $S_2O_3^{2-}$, $S_4O_6^{2-}$	5.5–6.0	35–37°	65–66	35	Q-10	+	+	–	–
*T. delicatus**	As above	5.5–6.0	30–35°	66–67	25	Q-8	+	(+)	–	–
T. ferrooxidans	As above (Also oxidizes Fe^{2+}, FeS_2 and metal sulfides)	2.5	30–35°	58–59	10	Q-8	+	–	–	–
T. thiooxidans	S, $S_2O_3^{2-}$, $S_4O_6^{2-}$	2.0–3.0	28–30°	52	0	Q-8	+	–	–	–
T. albertis	As above	3.5–4.0	28–30°	61–62	ND	ND		–	–	–
T. acidophilus	S^0, $S_2O_3^{2-}$, $S_4O_6^{2-}$, $S_3O_6^{2-}$	2.5–3.0	27–30°	63–64	8	Q-10	+	+	–	–
T. aquaesulis	As above	7.6–8.2	40–50°	65–66	ND	Q-8	ND	+	–	–
Thiomicrospira pelophila	S, S^{2-}, $S_2O_3^{2-}$, $S_4O_6^{2-}$	6.5–7.5	28–30°	43–44	ND	ND	ND	–	–	–
*Thiomicrospira denitrificans**	$S_2O_3^{2-}$	7.0	25°	32	ND	ND	ND	–	–	–
Thiomicrospira crunogena	S, S^{2-}, $S_2O_3^{2-}$	7.5–8.0	28–32°	41–43	ND	ND	ND	–	–	–
*Thiosphaera pantotropha**	S^{2-}, $S_2O_3^{2-}$	8.0	37°	66	ND	ND	–	+	–	+
Thermothrix thiopara	S, S^{2-}, $S_2O_3^{2-}$, $S_3O_6^{2-}$, $S_4O_6^{2-}$	6.7–6.9	72°	ND	ND	ND	ND	(+)	–	–

Acidianus and *Sulfolobus*	S, S^{2-}, $S_2O_3^{2-}$, $S_4O_6^{2-}$	1.5–1.8	66–74°	29–31	ND	(caldariellaquinone)	–	+	–	–
Paracoccus denitrificans	$S_2O_3^{2-}$	8.0	30°	66.5	ND	ND	ND	+	+	+
Aquaspirillum autotrophicum	$S_2O_3^{2-}$	8.0	30°	57–59	ND	ND	ND	+	ND	+
Pseudomonas palleronii (362)	$S_2O_3^{2-}$	8.0	30°	67	ND	ND	ND	+	ND	+
P. pseudoflava (GA3)	$S_2O_3^{2-}$	8.0	30°	66.5–68	ND	ND	ND	+	ND	+
Xanthobacter autotrophicus	$S_2O_3^{2-}$	8.0	30–37°	68.6–69.7	ND	ND	ND	+	ND	+
Sulfobacillus	S (Fe^{2+} and mineral sulfides)	1.9–2.4	50°	53–54	ND	ND	ND	+	–	–
Achromatium (Thiophysa) Macromonas Thiobacterium Thiospira Thiovulum }	At least some of these are probably capable of chemolithoautotrophic growth with sulfide. None exists in pure culture and knowledge of their physiology is sparse									
Unnamed new isolates:										
BC13	S, $S_2O_3^{2-}$, $S_4O_6^{2-}$	ca 2.5	45–50°	61–62	ND	ND	ND	–	ND	ND
LM2	S only (Also Fe^{2+})		45–50°	59–61	ND	ND	ND	+	ND	ND
ALV			45–50°	55–58	ND	ND	ND	+	ND	ND
Autotrophic symbiont from gills of *Thyasira flexuosa* ("*Thiobacillus* TG2")	$S_2O_3^{2-}$, $S_4O_6^{2-}$ (Best growth mixotrophically with yeast extract or acetate)	ca 7.8	25°	51–52	ND	Q-10(Q-8)	ND	+	–	ND

*The species marked with an asterisk are also able to grow anaerobically on thiosulfate, coupling its oxidation to denitrification. *Thiomicrospira denitrificans* is reportedly *obligately* anaerobic.

**In addition to the organisms listed here, some heterotrophs can oxidize thiosulfate only to tetrathionate and do not grow on the energy released. Some methylotrophic hyphomicrobia can obtain energy from the oxidation of thiosulfate to sulfate while growing on methylamine, which provides cell carbon by means of the serine pathway, not autotrophically by the Calvin or any other cycle (Suylen et al., 1986).

ND: not determined.

This table is based on a very large amount of source literature. Further details and literature can be found in Aragno and Schlegel, 1981; Brannan and Caldwell, 1983; Codd and Marsden, 1984; Friedrich and Mitrenga, 1981; Golovacheva and Karavaiko, 1978; Harrison, 1983; Jackson et al., 1968; Jannasch et al., 1985; Katayama-Fujimura et al., 1982, 1983, 1984; Kelly, 1982, 1988; Kelly and Harrison, 1988; Kuenen and Veldkamp, 1972; Kuenen and Tuovinen, 1981; Marsh and Norris, 1983; Norris et al., 1986; la Rivière and Schmidt, 1981; Robertson and Kuenen, 1983; Segerer et al., 1986; Stetter and Zillig, 1985; Timmer-ten Hoor, 1975.

cation of this approach lay in the belief that there would be vindication among the thiobacilli of the principle of Unity in Biochemistry ("Die Einheit der Biochemie") that has been something of a "Holy Grail" in biochemistry since its consecration by Kluyver and Donker in the 1920s (see Chapter 1). As discussed above, it now seems that this conception may have been a delusion—but an extremely useful and profitable delusion—and one can now begin to put together a broader view of just what are the common and disparate features of the thiobacilli. It is, however, essential to bear in mind that the seemingly simple and convenient genus of *Thiobacillus* is itself a heterogeneous collection of bacteria, individual species of which may have as much in common with each other at the molecular genetic and biochemical level as with species of *Thiomicrospira, Rhodobacter, Aquaspirillum* or even sulfide-oxidizing methylotrophic *Hyphomicrobium* species, with which they share superficially similar physiological abilities with regard to sulfur lithotrophy.

The reader is referred to other reviews for further general consideration of inorganic sulfur biochemistry, chemolithotrophy, and the diversity of sulfur-oxidizing processes (Kelly, 1978, 1982, 1988; Kelly and Kuenen, 1984; Kelly and Harrison, 1988).

As summarized in Table 11.1, *Thiobacillus* species range in optimum temperature and pH for growth between about 30–50°C and pH 2–8, different species producing or tolerating pH values of 0.5 to 10. Most species can oxidize polythionates (such as tetrathionate) and may produce polythionates during growth on sulfide or thiosulfate. *T. versutus* cannot use polythionates. Some use thiocyanate or methylated sulfides or carbon disulfide, some do not. Distinct fatty acid compositions and presence in the cells of either ubiquinone Q-8 or Q-10 are found, and these properties are taxonomically useful. Only *T. denitrificans* is capable of excellent autotrophic growth as an anaerobic nitrate-reducing sulfur chemolithotroph, but some others such as *T. versutus* can denitrify if growing autotrophically on formate or heterotrophically on substrates such as glucose or acetate. Alone among the *Thiobacillus* species, *T. ferrooxidans* can obtain energy from the oxidation of ferrous iron, as well as from inorganic sulfur compounds.

$$4FeSO_4 + O_2 + 2H_2SO_4 \rightarrow 2Fe_2(SO_4)_3 + 2H_2O$$

This property, along with the ability to obtain energy from the oxidation of sulfide minerals is shared by *T. ferrooxidans* with *Sulfolobus, Acidianus,* and diverse moderately thermophilic nonsulfur chemolithotrophs, as well as the iron-oxidizing *Leptospirillum*. Even the individual strains of the single species *T. ferrooxidans* are proving so diverse in DNA homology studies that this species may well need subdivision of its strains into species as well as new *genera*.

Among the other *Thiobacillus* species there is a range of different DNA base compositions from about 50 to 70 mol percent GC, and DNA–DNA homology among the species is almost universally so low that very little genetic similarity can be inferred. Studies on ribosomal RNA sequences in thiobacilli also emphasize their phylogenetic heterogeneity.

Physiologically, the thiobacilli exhibit strikingly species-dependent differences in autotrophic growth yields when grown in chemostat culture on thiosulfate or other inorganic sulfur compounds (Kelly, 1982, 1988; Mason et al., 1987; Wood and Kelly, 1987). Thus, maximum growth yields (Y_{max} expressed as g dry wt of bacteria per mol thiosulfate oxidized to sulfate) range from 6 to 8 for *T. neapolitanus, T. thiooxidans, T. ferrooxidans* and *T. versutus,* 10 to 17 for *T. tepidarius* and *T. aquaesulis,* 14 to 15 for *T. denitrificans* and can be compared to values of 4.5 to 7 for *Paracoccus denitrificans, Thiosphaera pantotropha,* or *Thiomicrospira* and of about 23 for *Thermothrix thiopara* (see also Chapter 4 for a discussion of growth yields in these organisms). The significance of these observations is that external *physical* conditions (at each organism's optimum pH and temperature) do not influence Y_{max}, thus virtually identical values are found for *T. versutus* at pH 7.8 and *T. ferrooxidans* at pH 2.5 and 30°C, while *T. denitrificans* at pH 7 and 30° gives a yield of 14 or more. The Y_{max} values thus reflect the inequality of efficiency of energy conservation from the same substrate by different organisms, and thereby indicate the probability of different substrate oxidation pathways, electron transport mechanisms, and modes of energy conservation.

11.3 CARBON ASSIMILATION BY THE UNICELLULAR SULFUR BACTERIA

It has been firmly established that many pure cultures of thiobacilli, as well as *Thiosphaera* and *Thermothrix,* use the Calvin cycle to fix carbon dioxide to supply all their biosynthetic carbon requirements. The operation of the whole Calvin cycle has been demonstrated in some species, while the presence of the key enzyme, ribulose 1,5-bisphosphate carboxylase (see Chapter 19), indicates its role in others. In some species very large amounts of this enzyme are present in polyhedral inclusion bodies up to 100 nm in diameter. The bodies appear to be composed predominantly of the carboxylase enzyme and are called carboxysomes (Figure 11.1). Their presence is a species-specific characteristic that can be used as a taxonomic character in the thiobacilli (see Table 11.1).

In facultatively autotrophic thiobacilli, this enzyme is repressed during heterotrophic growth. A few thiobacilli (*T. novellus, T. versutus, T. acidophilus*) have been studied in the chemostat under mixotrophic growth conditions, in which an organic substrate such as glucose or acetate has been supplied as well as thiosulfate or tetrathionate and carbon dioxide (see Chapter 4). Under such conditions, the bacteria exhibit activities of ribulose bisphosphate carboxylase intermediate between those seen in autotrophic growth and the low (negligible or absent) levels seen during heterotrophic growth. Mixotrophically grown organisms obtain carbon simultaneously from the fixation of carbon dioxide and from the organic substrate. They also obtain energy from the oxidation of both the inorganic and organic nutrients. They thus represent an interesting system for the study of regulation of the autotrophic, mixotrophic,

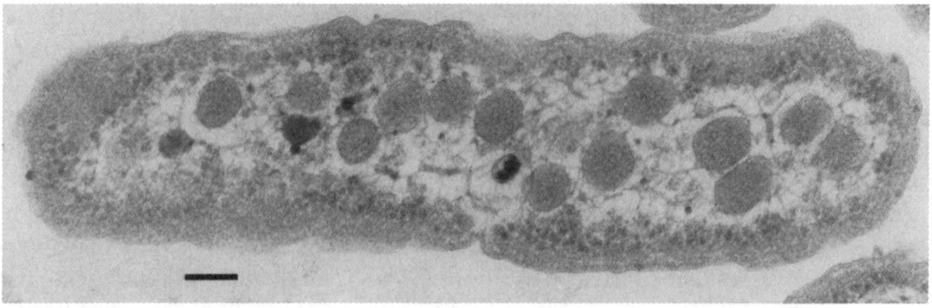

Figure 11.1 Thin section of *Thiobacillus neapolitanus* showing carboxysomes (polyhedral bodies), which are composed primarily of ribulose bisphosphate carboxylase. Bar represents 0.1 μm. Photograph kindly provided by Dr. Jessup P. Shively.

chemolithotrophic, and heterotrophic modes of metabolism. The outcome of such studies to date is the demonstration that 1) the steady state level of ribulose bisphosphate carboxylase in a mixotrophic culture is determined by the relative concentrations of the chemolithotrophic and chemoorganotrophic substrates when both are limiting to the steady state biomass production; 2) the proportions of carbon assimilated from carbon dioxide and the organic substrate are also substrate-concentration dependent and suggest that the regulatory mechanisms of the organisms tend to make most efficient use of metabolic energy available for maximizing total biomass production; 3) mixotrophy lends selective and survival advantage to a potentially mixotrophic organism when it is competing in mixed culture (and presumably also in the natural environment) with obligate heterotrophs (for the organic substrate) or with obligate chemolithotrophs (for the inorganic sulfur compound) (Gottschal et al., 1979; Kuenen et al., 1977, Chapter 4).

It is also important to remember that even the obligately chemolithotrophic bacteria like *Thiobacillus neapolitanus, T. thioparus,* and *T. denitrificans* can assimilate organic nutrients, such as acetate, amino acids and some aliphatic acids, but that these substances seldom become a major source of cell carbon and that their assimilation is apparently absolutely dependent on a supply of energy from the oxidation of inorganic sulfur compounds.

Carbon dioxide is presumed to be fixed by means of the Calvin cycle in *Thiomicrospira,* but no detailed analysis has yet been published for these bacteria. The facultative hydrogen-oxidizing heterotrophs such as *Xanthobacter* and *Paracoccus* which can also grow as sulfur chemolithoautotrophs are, however, well established as Calvin cycle autotrophs.

In contrast, *Sulfolobus* and *Acidianus* are probably not Calvin cycle autotrophs and seem to fix carbon dioxide by an as yet unestablished pathway that may be a reductive carboxylic acid cycle, possibly similar to that shown in the green photosynthetic sulfur bacteria or other unusual nonCalvin-cycle autotrophs such as the more recently studied sulfate-reducing bacteria (see Chapter 20).

Many of the bacteria that can grow as sulfur chemolithotrophs are also capable of heterotrophic growth on a very wide range of organic nutrients. These include the hydrogen-oxidizing *Aquaspirillum, Paracoccus, Xanthobacter, Pseudomonas,* and *Thiosphaera,* as well as several thiobacilli, *Thermothrix, Sulfobacillus, Sulfolobus, Acidianus* and the less well characterized moderately thermophilic acidophiles obtained from mineral leaching environments. Discussion of the heterotrophic metabolism of these organisms is not needed here, as the same fundamental pathways and processes occur in these as in other bacteria. Perhaps worthy of comment, however, is the demonstration that two of the thiobacilli exhibit unusual features of glucose metabolism. *T. versutus,* when grown on glucose, oxidizes it by the simultaneous use of the Embden-Meyerhof and Entner-Doudoroff pathways and the oxidative pentose phosphate cycle (Wood and Kelly, 1980; Wood et al., 1977). *T. novellus* uses a phosphoketolase pathway, involving glucose oxidation and decarboxylation to a pentose sugar that is subsequently cleaved before further oxidation (Greenley and Smith, 1979). The latter is probably rare in aerobic bacteria.

11.4 METABOLIC PROBLEMS OF THE SULFUR CHEMOLITHOTROPHS

For an organism to grow with the oxidation of sulfur and its reduced compounds as sole source of energy, a number of metabolic specializations have to be present. The first of these is obviously the ability to oxidize inorganic sulfur compounds at a rate sufficient to support autotrophic growth. Numerous heterotrophs have the ability to oxidize thiosulfate to tetrathionate, and in some cases sulfite to sulfate, or the ability to oxidize sulfide to elemental sulfur, but in general these oxidations do not lead to energy being conserved and used by the organisms. The sulfur chemolithotrophs, in contrast, can convert sulfur and compounds like thiosulfate completely to sulfate and conserve the energy released. The second problem is dictated by the chemical thermodynamics of these oxidations. Like all organisms, the sulfur chemolithotrophs depend on ATP and reduced NAD(P) as the "energy currency" for their biosynthetic processes. The biosynthetic process of CO_2 fixation by the Calvin cycle is "energy expensive," requiring 3ATP and 2NADH$_2$ for each CO_2 molecule converted to hexose sugar. One consideration is that relatively small amounts of energy are released during sulfur oxidations and consequently there is a need to oxidize large amounts of substrate to support reasonable quantity of growth. This is reflected in the growth yields mentioned in section 11.2: a yield of about 7 g dry wt of bacteria per g mol of thiosulfate oxidized for example, means that a kilogram of anhydrous sodium thiosulfate has to be oxidized to bring about conversion of 21 g of carbon from CO_2 into cell material. In contrast, oxidation of a kilogram of glucose (as an energy *and* carbon source) can produce 300 g of new cell-carbon. The more serious metabolic problem is that reducing equivalents (electrons and protons) released during the oxidative steps leading to sulfate are mostly of relatively high redox potential. The potential of the $NAD^+/NADH$ couple is −340 mV and none of the steps of sulfur oxidation appears to be able to be linked directly to the reduction of NAD^+ *in vivo*. It is the cytochromes (cytochrome b and c) that are the principal acceptors of electrons from the intermediate oxidation steps in sulfur metabolism. Consequently NAD^+ reduction requires the energy-consuming transport of electrons from the reduced cytochromes (or possibly quinones or flavoprotein carriers) in the electron transport chain to NAD^+, by what is essentially a reversal of the "downhill" flow of electrons seen in the normal respiratory chain in the oxidation of NADH. The final question to be resolved in explaining sulfur chemolithotrophy is whether the means by which energy is conserved as ATP is mechanistically similar to that in the oxidation of organic substrates by heterotrophs, or whether there are any unique processes.

11.5 THE OXIDATION MECHANISMS FOR SULFUR AND ITS COMPOUNDS

Based mainly on work with members of the genus *Thiobacillus*, there was for a long period a controversy concerning the role of polythionates ($S_nO_6^{2-}$, where n = 3, 4, 5 or more) as intermediates in the conversion of sulfur, sulfide, and thiosulfate to sulfate (see Kelly, 1982, 1985, 1988, for a detailed discussion). The contrary view was that the oxidation of thiosulfate, for example, involved its initial scission to [S–] and [$-SO_3$] units, which underwent oxidation without polythionate involvement (Peck, 1960). Polythionate accumulation in this scheme was thought to result from artifactual side reactions or processes of little physiological significance. Regarding the "polythionate" and "scission" mechanisms as mutually exclusive alternatives of which only one represented the "common pathway" was of course a result of the view that the basic mechanism would be common to all the sulfur chemolithotrophs (Kelly, 1968, 1982). With the acquisition of more precise data and the successful preparation of cell-free extracts and enzyme systems capable of sulfur compound oxidation in the last twenty years, it has become clear that Schlegel's (1975) comment ["Presumably different pathways operate in different organisms"] may indeed be true for many different sulfur chemolithotrophs. While this is a disappointment from the "unity in biochemistry" viewpoint, "it is a bad plan that admits of no modification" (Publius Syrus, 42B.C.), and the most productive hypothesis for current and future work is that each organism must be analyzed on the assumption that it will exhibit at least some unique oxidative features. In this way the stages of the oxidative processes that are common, if such exist, can eventually be identified. This will then enable analysis of whether there is any phylogenetic conservation of primordial characters across diverse genera and species or whether the existing sulfur oxidizing processes have little or no common ancestry and have arisen by convergent evolution to superficially similar physiological types. Whatever the details of the processes, the same problems are posed by all the organisms: how are sulfur, sulfide, and thiosulfate initially attacked? What intermediates occur in the pathway to sulfate formation? How is the very "reduced" (sulfane) atom of thiosulfate ($^{-}S{-}SO_3^{-}$), and of sulfide, converted to sulfite or sulfate? Is sulfite inevitably the intermediate immediately preceding sulfate? Are polythionates formed or oxidized by all sulfur bacteria? What similarities and differences may exist between the aerobic and anaerobic (denitrifying) sulfur oxidizers?

The overall stoichiometries for the oxidation of sulfur compounds by thiobacilli, *Thermothrix, Sulfolobus,* etc. are as follows:

$$Na_2S_2O_3 + 2O_2 + H_2O \rightarrow Na_2SO_4 + H_2SO_4$$

$$5Na_2S_2O_3 + 8NaNO_3 + H_2O \rightarrow 9Na_2SO_4 + H_2SO_4 + 4N_2$$

$$Na_2S_2O_3 + 4N_2O + H_2O \rightarrow Na_2SO_4 + H_2SO_4 + 4N_2$$

$$Na_2S + 2O_2 \rightarrow Na_2SO_4$$

$$Na_2S_3O_6 + 2O_2 + 2H_2O \rightarrow Na_2SO_4 + 2H_2SO_4$$

$$Na_2S_4O_6 + 3.5O_2 + 3H_2O \rightarrow Na_2SO_4 + 3H_2SO_4$$

$$S_8 + 8H_2O + 12O_2 \rightarrow 8H_2SO_4$$

$$NaSCN + 2H_2O + 2O_2 \rightarrow NaHSO_4 + CO_2 + NH_3$$

$$Na_2SO_3 + \tfrac{1}{2}O_2 \rightarrow Na_2SO_4$$

***Oxidation of thiosulfate by* Thiobacillus versutus** This organism does not produce or metabolize polythionates and oxidizes thiosulfate to sulfate without the formation of sulfur or other intermediates. Cell-free extracts of the organism carry out the cytochrome-dependent oxidation of thiosulfate completely to sulfate and the enzyme system involved has been identified in some detail (Lu, 1986; Lu et al., 1985). It catalyzes a sequence of oxidations in which oxygen from water is introduced into combination with the sulfur atoms of thiosulfate and protons and electrons are released. No actual intermediates have been identified or demonstrated but the process believed to be catalyzed by the complex is illustrated in Figure 11.2. The whole sequence of oxidations takes place in the periplasmic space of the organism and the components of the system are shown in Figure 11.3. Thiosulfate metabolism appears to be initiated by its binding to enzyme A, which may also initiate the oxidation process. Enzyme B and the multiheme, multiple redox center cytochromes catalyze the production of sulfate, without releasing free intermediates, and transfer electrons to membrane-bound c-type cytochromes. Terminal oxygen reduction is effected by an aa_3-type cytochrome oxidase.

***Oxidation of polythionates and thiosulfate by* Thiobacillus tepidarius** Unlike *T. versutus,* in which polythionates are not metabolized, tetrathionate ap-

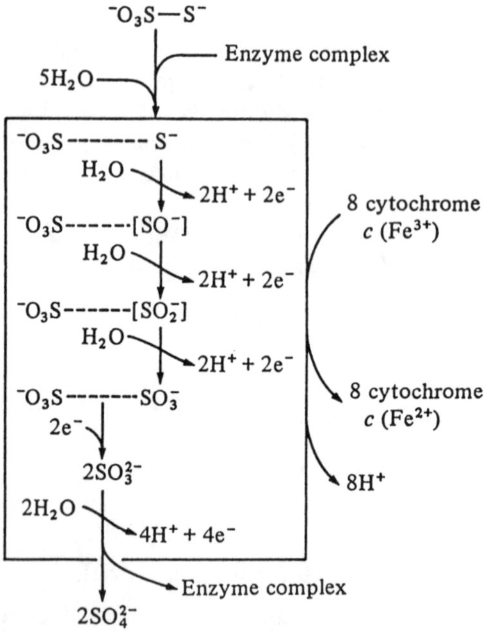

Figure 11.2 Hypothetical sequence of reactions proposed to operate during the oxidation of thiosulfate to sulfate by *Thiobacillus versutus* (Kelly, 1988).

pears to be an obligatory intermediate in the oxidation of both thiosulfate and trithionate by *T. tepidarius*. Trithionate metabolism appears to begin with its enzymatic hydrolysis to thiosulfate and sulfate:

$$Na_2S_3O_6 + H_2O \rightarrow Na_2S_2O_3 + H_2SO_4$$

Thiosulfate is oxidized to tetrathionate by a cytochrome-c-dependent enzyme:

$$2S_2O_3^{2-} + \tfrac{1}{2}O_2 + 2H^+ \rightarrow S_4O_6^{2-} + H_2O$$

Both these reactions probably take place in the periplasm of the cell, and further oxidation of tetrathionate to sulfate seems to require its prior transport across (or at least intimate association with) the cytoplasmic membrane. The evidence for this is that energy-uncoupling inhibitors such as carbonylcyanide *p*-trifluoromethoxyphenylhydrazone (FCCP) inhibit tetrathionate oxidation and result in tetrathionate accumulation when thiosulfate is being oxidized. This suggests that tetrathionate oxidation depends on the presence of an energetically functional cytoplasmic membrane. The oxidation of tetrathionate to sulfate involves electron transport to cytochrome b. This is demonstrable by spectrophotometric analysis and by the fact that 2-heptyl-4-quinolinol-1-oxide (HQNO), which inhibits the transfer of electrons at the ubiquinone/cytochrome b level of the respiratory chain, inhibits tetrathionate oxidation. The involvement of cytochrome b in substrate oxidation is in contrast with *T. versutus* and *T. nea-*

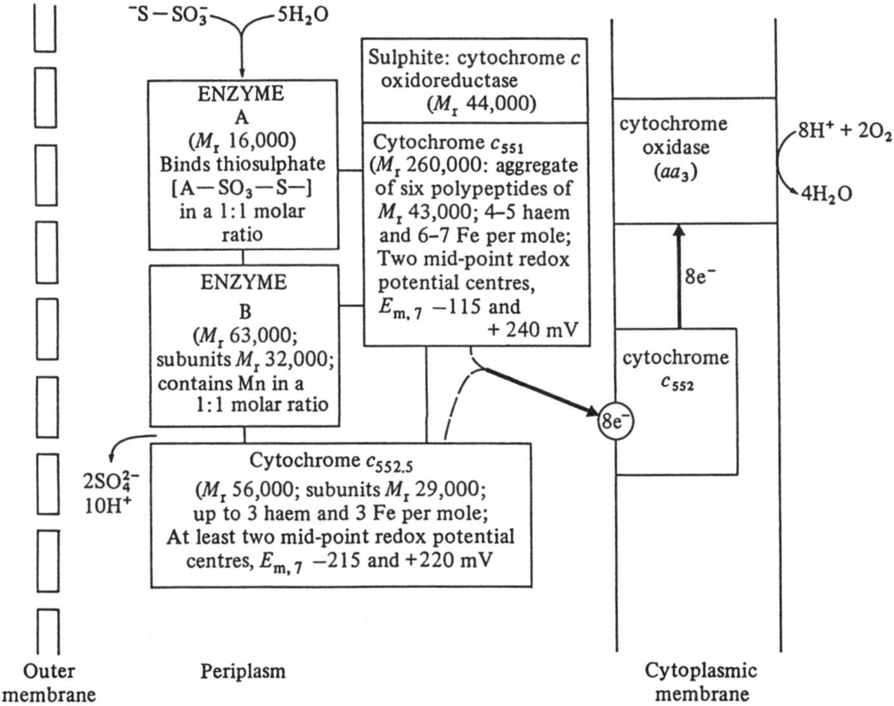

Figure 11.3 The periplasmic thiosulfate-oxidizing multienzyme system of *Thiobacillus versutus*. (Reproduced from Kelly, 1988, with permission of the Society for General Microbiology.)

politanus in which only c-type cytochromes have been shown to be involved in thiosulfate oxidation. It also further supports the view that tetrathionate oxidation by *T. tepidarius* is likely to occur on the inner surface of the cytoplasmic membrane, where cytochrome b is predominantly located.

The mechanism of oxidation of sulfur compounds by *T. tepidarius*, as currently understood, is shown in Figure 11.4. It is probable that an HQNO-sensitive step in tetrathionate oxidation is the conversion of sulfite to sulfate by an HQNO-sensitive oxidoreductase, and that the metabolism of tetrathionate requires its cleavage to sulfane-sulfur and sulfite moieties by an as yet uncharacterized enzyme system.

Oxidation of thiosulfate by **Thiobacillus novellus** A cell-free system from this thiobacillus was obtained that would convert thiosulfate stoichiometrically to sulfate (Oh and Suzuki, 1977), as in the example of *T. versutus*. The *T. novellus* system differed markedly from that of *T. versutus* in being associated with the cell membrane, rather than the periplasmic space. As outlined in Figure 11.5, it consists of the enzymes (a) *rhodanese*:

$$^{-}S\text{–}SO_3^{-} + [R\text{–}] \rightarrow SO_3^{2-} + RS^{-}$$

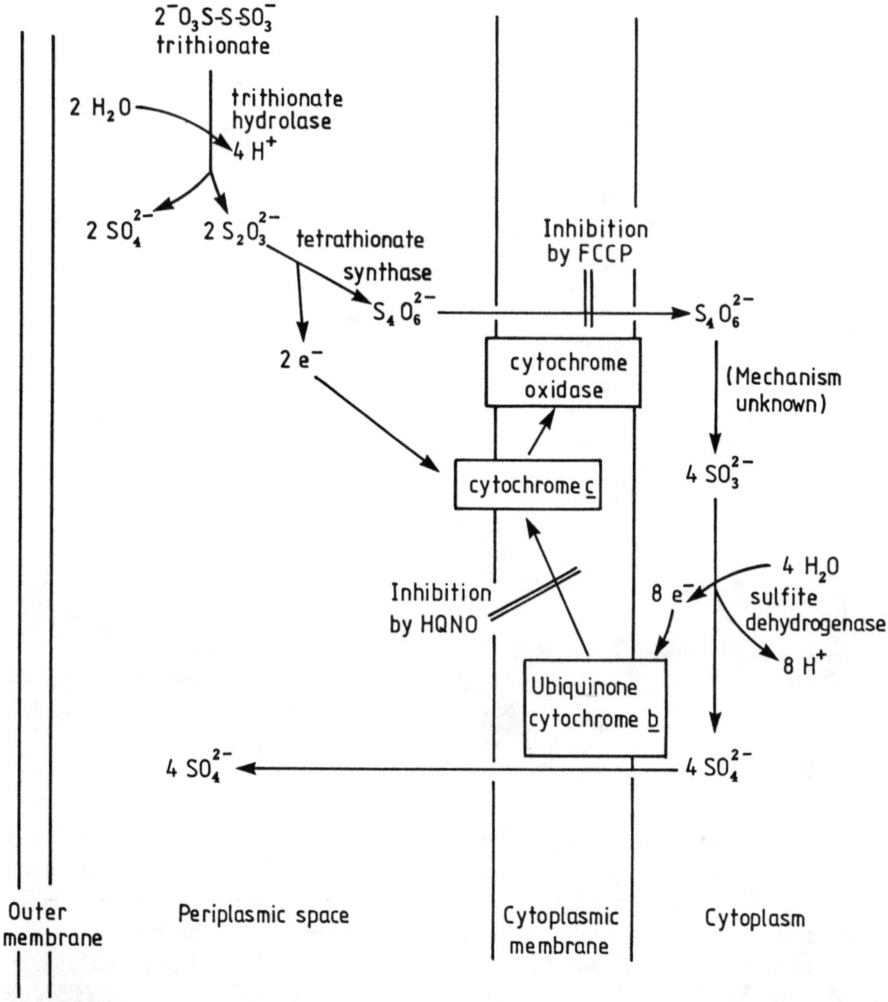

Figure 11.4 The metabolism of trithionate, thiosulfate, and tetrathionate by *Thiobacillus tepidarius.* (Based on studies by W-P. Lu, D. P. Kelly, and R. K. Poole.)

where (R can be cyanide, lipoic acid, lipoamide, or an alternative physiological acceptor); (b) *sulfur oxygenase*:

$$S + H_2O + O_2 \rightarrow H_2SO_3$$

and (c) *sulphite-cytochrome c oxidoreductase*:

$$SO_3^{2-} + H_2O + 2 \text{ (oxidized cytochrome c)} \rightarrow$$
$$SO_4^{2-} + 2H^+ + 2 \text{ (reduced cytochrome c)}$$

The system also contained (d) *cytochrome c oxidase*, coupling to oxygen reduction. This could explain thiosulfate oxidation by the scheme shown in Fig. 11.5.

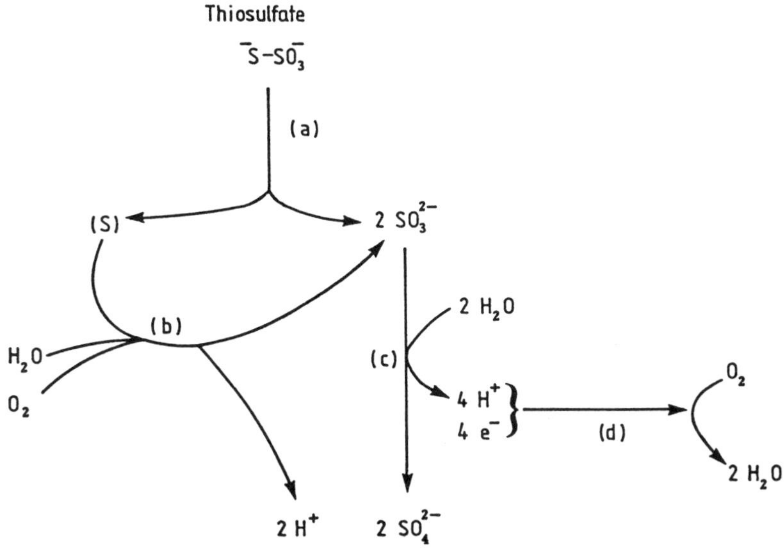

Figure 11.5 A mechanism for the oxidation of thiosulfate as proposed for *Thiobacillus novellus*. See text for names of enzymes indicated by a–d.

This scheme contains sulfur and the sulfur oxygenase enzyme as essential components that have not been mentioned previously. The significant feature of sulfur oxygenase is that it can explain the conversion of elemental sulfur (or of sulfur derived from sulfide or the sulfur groups of thiosulfate and tetrathionate) to sulfite by a process that does not involve electron transport and does require the direct incorporation of molecular oxygen (O_2) (rather than oxygen solely from water) into combination with sulfur. This means that there would seem to be no possibility of energy conservation in the transformation of sulfur to sulfite, meaning for example that a substrate such as thiosulfate should provide twice as much metabolically useful energy as sulfide, even though each compound requires two molecules of oxygen for conversion to sulfate, because the only sites for energy conservation would be during the oxidation of sulfite to sulfate, and sulfite is presumably formed in a 2:1 ratio from thiosulfate and sulfide. The sulfur oxygenase does not function as part of the multienzyme system that oxidizes thiosulfate in *T. versutus* since eight molecules of cytochrome c are reduced for each thiosulfate oxidized: this means that all the oxygen incorporated to make sulfate is derived from water and eight electrons are consequently released for transport to elemental oxygen by the respiratory chain (Figure 11.2).

***Sulfur and tetrathionate oxidation by* Thiobacillus ferrooxidans *and* Acidianus brierleyi** Both these acidophiles (Table 11.1) can grow autotrophically on sulfur and tetrathionate. A sulfur oxygenase has been shown to be present in both (Silver and Lundgren, 1968; Emmel et al., 1986), and sulfur has been postulated as a true intermediate, showing transitory accumulation during tetrathionate (or sulfide) oxidation by *T. ferrooxidans* (Hazeu et al., 1988). Tetrathionate oxidation by *Acidianus* is inhibited by HQNO and by uncoupling agents such as carbonyl-m-chlorophenylhydrazone (CCCP) and dinitrophenol (B. Pike and D. P. Kelly, unpublished data) thus suggesting the involvement of quinone/cytochrome b and the cytoplasmic membrane in the oxidation process, but the mechanism is not yet known. In *T. ferrooxidans* an enzyme is known that dismutes tetrathionate, resulting in sulfur formation, resulting in a process similar to that postulated above for *T. tepidarius.*

***Thiosulfate oxidation by* Thiobacillus denitrificans** The process of sulfur oxidation in this anaerobically growing organism is of interest. It is clear that no sulfur oxygenase can operate in an organism that grows in the absence of molecular oxygen (O_2). The enzyme believed to be involved in the oxidation of sulfide (and possibly elemental sulfur), as well as the S-group of thiosulfate and presumably of tetrathionate, to sulfite is a "siroheme sulfite reductase" operating in the oxidative direction to produce sulfite:

$$S^{2-} + 3H_2O \rightarrow SO_3^{2-} + 6H^+ + 6e^-$$

This reaction would be expected to proceed in three successive hydration and dehydrogenation steps, releasing a pair of electrons and protons each time to as yet unknown electron acceptors. Inhibitor studies suggest that cytochrome b is an electron carrier in the sulfur-dependent respiration of *T. denitrificans.* There could thus be similarities between its thiosulfate oxidation pathway and that of *T. tepidarius.* As *T. denitrificans* can also grow aerobically, it is clear that at least three variants of sulfur-oxidizing aerobic pathways are likely, involving or not involving polythionates, sulfur oxygenase, and "reverse sulfite reductase."

The conversion of sulfite to sulfate All the sulfur-oxidizing bacteria studied in detail show activity of sulfite: cytochrome oxidoreductases, catalyzing the reaction:

$$SO_3^{2-} + H_2O \rightarrow SO_4^{2-} + 2H^+ + 2e^-$$

Cytochrome b and cytochrome c have variously been reported as the electron acceptor for the enzymes from different species. The redox couple of SO_3^{2-}/SO_4^{2-} is certainly sufficiently electronegative to couple to the respiratory chains at the flavin-quinone-cytochrome b, but in some species it seems clear that cytochrome c is the physiological electron acceptor and point of entry into the chain (Yamanaka et al., 1981; Tano et al., 1982; Kelly, 1988).

Some, but not all, thiobacilli also contain adenylsulfate reductase (also called adenosine phosphosulfate (APS) reductase):

$$SO_3^{2-} + AMP \rightarrow APS + 2e^-$$

The adenylyl-sulfate formed can give rise to sulfate by desulfurylation catalyzed either by ADP or ATP sulfurylase:

$$APS^{2-} + PO_4^{3-} \rightarrow ADP^{3-} + SO_4^{2-}$$

$$APS^{2-} + P_2O_7^{4-} \rightarrow ATP^{4-} + SO_4^{2-}$$

These reactions exchange the sulfate group for an orthophosphate or pyrophosphate one, and produce ADP or ATP, thus constituting an energetic gain to species using these reactions to convert sulfite to sulfate.

Generalizations about sulfur oxidations It would seem that all sulfur-oxidizing bacteria are likely to produce sulfite as a final intermediate, whatever the substrate used, even though free sulfite is not always detectable. The initial reactions of thiosulfate are species-dependent, thus in *T. versutus* no polythionate intermediates are produced while tetrathionate is an obligatory intermediate for *T. tepidarius*. The ability to produce sulfur is also variable and its role as a free intermediate is still under debate. The biggest gap in understanding the sulfur oxidation process is still the exact biochemical nature of the reactions converting the sulfane groups of the thionates (and sulfide) to sulfite. In some cases this seems to be by means of the oxygen-dependent sulfur oxygenase but in other cases (for instance, *T. denitrificans*) this enzyme certainly does not play a part. What is certain is that the basic scheme to describe sulfur oxidation is now simpler and less confused than in the period up to the 1980s. Part of the conceptual simplification comes from the recognition that a simple unifying hypothesis is untenable and several pathways exist, just as several distinct pathways exist, for example, to convert glucose to pyruvate. As with glucose catabolism, there may be steps or enzymes common to two or more pathways, but there can equally be sufficient fundamental differences to allow description of several unique oxidative systems.

11.6 ELECTRON TRANSPORT, PROTON TRANSLOCATION, AND ENERGY CONSERVATION DURING SULFUR OXIDATIONS

As discussed above, some sulfur bacteria couple their substrate oxidations to the respiratory chain only at the level of cytochrome c, while others can couple at cytochrome b or even higher in the chain. This is of great energetic significance as the sulfur chemolithotroph able to feed electrons to the chain at cytochrome b gains in two direct ways: it is able to conserve energy at the coupling site for ATP synthesis that is located in the cytochrome b/cytochrome c sector of the chain (coupling site II) as well as in the cytochrome c/oxidase sector, which is the only site available to organisms using cytochrome c as the initial carrier. It is also able to reduce NAD^+ more readily and less expensively in energy terms, as it has only to overcome the thermodynamic "gap" from quinone/cytochrome b to NAD^+ rather than also having to expend energy in transferring electrons to cytochrome b from cytochrome c.

These differences in electron transport pathways, and consequently the amount of energy conserved from substrate oxidation, are probably the principal explanation for the greater growth yields seen with "cytochrome b organisms" like *Thiobacillus denitrificans* and *T. tepidarius* than with "cytochrome c organisms" like *T. versutus* and *T. neapolitanus*.

In cytochrome c organisms, several c-type cytochromes are involved, presumably as sequential electron carriers, but their kinetic interrelationships and spatial organization in the respiring cell are as yet virtually unstudied. In anaerobic *T. denitrificans*, electron transport from sulfite oxidation passes through flavoprotein, quinone, cytochrome b and cytochrome c_{554} to reduce nitrate to nitrite, while electrons from sulfide oxidation (to sulfite) are probably carried by flavin to cytochrome c_{551} to a cytochrome cd nitrite reductase (see Kelly, 1982 for details). In *T. tepidarius* there are several cytochrome b components in the electron transport chain. These are distinguished by having different midpoint electrode potentials, ranging from 8 to 322 mV. The hypothetical path of electron transport through the cytochromes of this organism is shown in Figure 11.6. The nature of the terminal oxidase is as yet unproved but is known to be inhibited by cyanide and carbon monoxide, and may involve the CO-binding cytochrome c observed in this bacterium.

Energy coupling in bacterial respiration is a chemiosmotic process dependent on proton translocation and electron transport. This principle applies also to the sulfur chemolithotrophs, although differences in the means by which transmembrane proton gradients are established have been demonstrated. In *Thiobacillus versutus*, where complete oxidation of thiosulfate to sulfate takes place in the periplasmic space, there is no proton extrusion across the cytoplasmic membrane and protons are generated in the periplasm. ATP synthesis is effected by a proton-translocating ATPase situated on the inner surface of the cytoplasmic membrane (Figure 11.7). This results in the production of one ATP per pair of electrons transported ($P/O = 1.0$). In contrast, *T. tepidarius*, which oxidizes tetrathionate and sulfite on the inner surface of the cytoplasmic membrane, carries out proton translocation and extrudes four protons for each oxygen atom reduced. This contrasts with *T. versutus* whose periplasmic substrate oxidation produces only 2.5 protons for each oxygen atom consumed. The proton translocation carried out by *T. tepidarius* is believed to be driven at the ubiquinone/cytochrome b level in the electron transport system and not to be associated with the terminal oxidase. The system in *T. tepidarius* is believed to produce a P/O ratio of 2.0 and is illustrated in Figure 11.8.

It can thus be seen that while fundamentally similar mechanisms exist to generate ATP, there are again differences among different sulfur chemolithotrophs that reflect the different mechanism they use for sulfur oxidation. The P/O ratio can be one or two ATP formed per oxygen atom reduced, meaning that thiosulfate oxidation could produce four or eight ATP per thiosulfate used. This, coupled with the amount of energy needed to reduce NAD^+ in different species, easily explains the differences seen in growth yields of "cytochrome

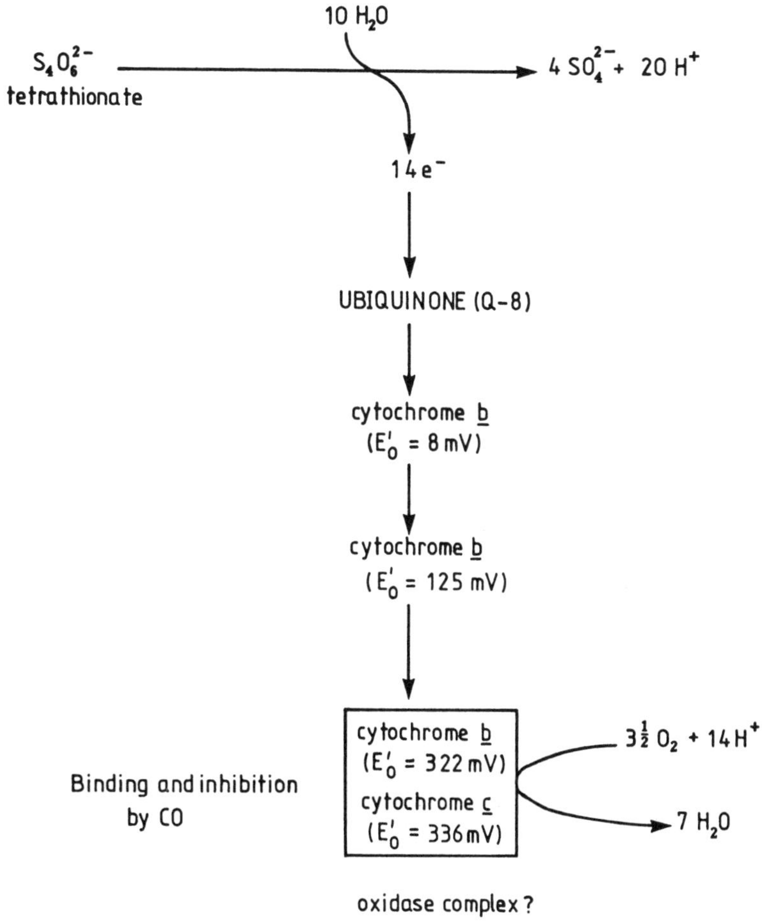

Figure 11.6 A possible sequence of electron transport reactions during tetrathionate oxidation by *Thiobacillus tepidarius*. The nature of the terminal oxidase and the relative roles of the cytochrome c and the 322 mV cytochrome b are uncertain (see also Figure 11.4). (Based on work by W-P. Lu, R. K. Poole, and D. P. Kelly.)

b" and "cytochrome c" organisms. In addition, if an organism produces ADP or ATP by the substrate-level phosphorylations brought about by the APS reductase and sulfurylase enzymes described above, the total energy conserved will be further increased, resulting in overall P/O ratios that differ fractionally from 1.0 or 2.0, and thereby complicating energetic analysis.

Consideration of energetic mechanisms and growth yields also provides further insight into the question of whether or not a sulfur oxygenase is important in a particular strain. Thus if [S] is converted to sulfite by an oxygenase, thiosulfate would be expected to give almost twice the growth yield of sulfide, but *T. versutus* and *T. tepidarius* give virtually identical growth yields on each substrate in chemostat culture. Similarly trithionate should give a greater yield

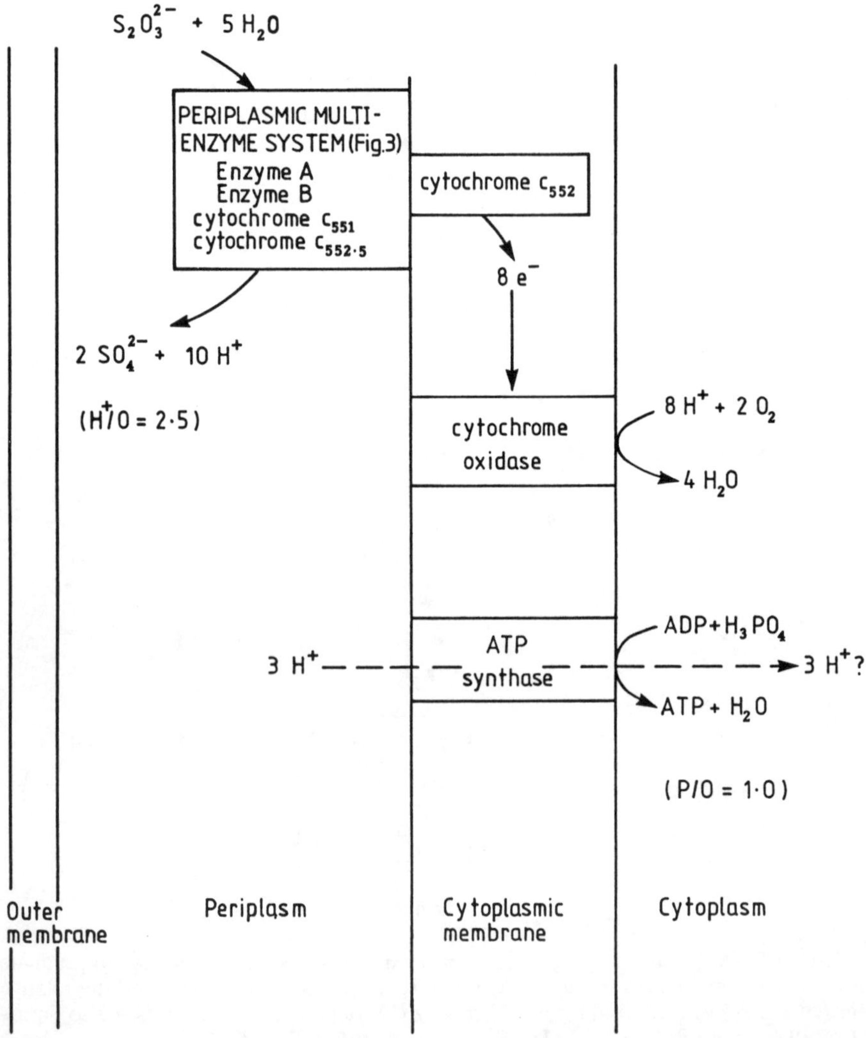

Figure 11.7 Proton translocation, generation of a pH gradient (ΔpH), and ATP synthesis during thiosulfate oxidation by *Thiobacillus versutus*.

than thiosulfate. In fact, yields of *T. tepidarius* on sulfide, thiosulfate, and trithionate are similar, indicating that the prime determinant of the amount of energy conserved is the use of all the oxygen consumed (two molecules per molecule of each substrate) for the terminal reduction of the energy-yielding electron transport chain, and not for any oxygenase reaction. The similarity of yield of different organisms on the same substrate (e.g., *T. ferrooxidans* and *T. versutus* on thiosulfate) could indicate that they use the same pathways. The reported activity of the sulfur oxygenase in *T. ferrooxidans* and not in *T. versutus* could mean that the similarity in yields was fortuitous and could result if one organism was a "cytochrome b type" (*T. ferrooxidans*) and gained more energy

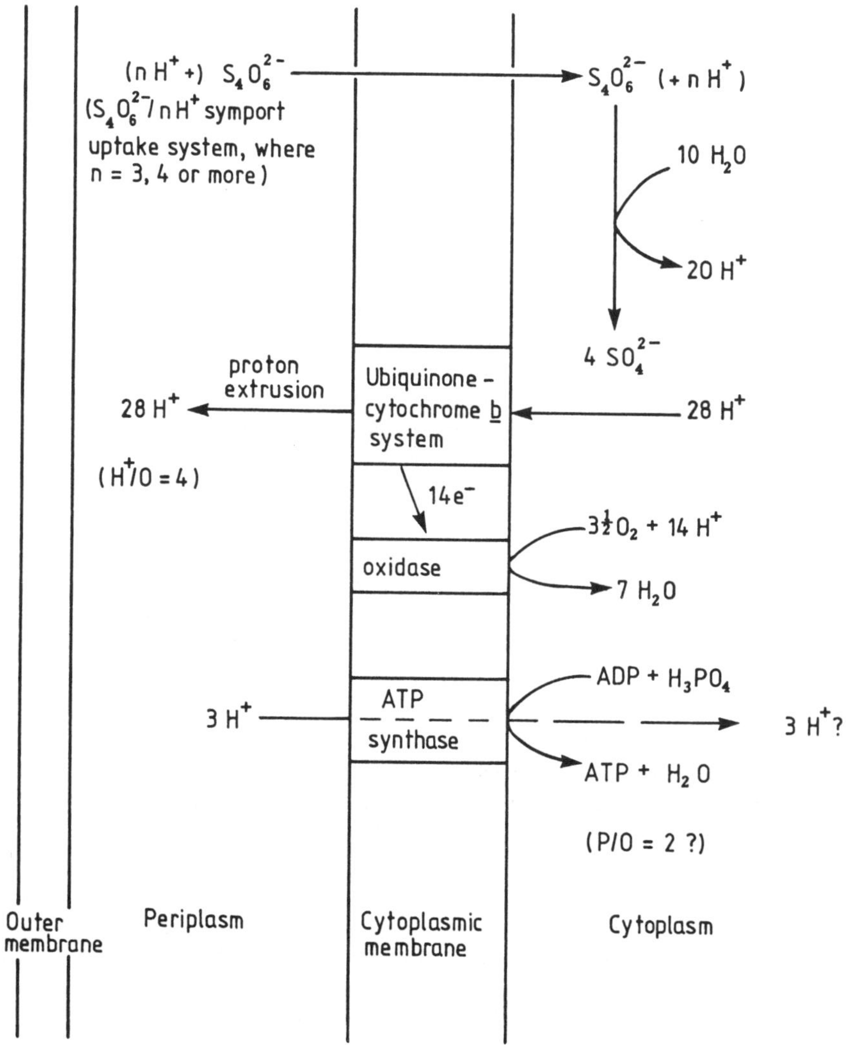

Figure 11.8 Generation of ΔpH, proton translocation, and ATP synthesis during tetrathionate oxidation by *Thiobacillus tepidarius.*

than the other (*T. versutus*), which was a "cytochrome c type," from sulfite oxidation. The alternative explanation could of course be that the sulfur oxygenase functioned *in vivo* in neither. This is an area in which there is still uncertainty in defining pathways and energy conserving processes.

11.7 NAD$^+$ REDUCTION AND ITS RELATION TO ATP PRODUCTION DURING SULFUR OXIDATION

It was elegantly demonstrated by Roth et al. (1973) that ATP was synthesized and NAD$^+$ and NADP$^+$ simultaneously reduced during sulfide and thiosulfate oxidation by *Thiobacillus neapolitanus*, and that these processes were all inhibited by the uncoupler CCCP. Their experiments demonstrated that $NAD(P)^+$ reduction was an energy-dependent process. The nature of the coupling between energy generation from sulfur oxidation and the energy-requiring movement of electrons from cytochromes to NAD^+ in the respiratory chain was, however, not established. Early work on the effect of the uncoupler dinitrophenol on carbon dioxide fixation by *T. neapolitanus* showed that fixation was inhibited even when the cells were still able to generate ATP by substrate-level phosphorylation and intracellular ATP concentration was high (Kelly and Syrett, 1966). This was suggested to indicate that the factor limiting CO_2-fixation was the availability of NAD(P)H and that the effect of dinitrophenol was to prevent the energy-dependent reduction of NAD^+ by preventing energy coupling at a stage closer to the primary energy-conserving process than ATP, and that ATP as such was insufficient to drive NAD^+ reduction. This hypothesis was recently confirmed by the demonstration that low concentrations of FCCP (which as a proton-conducting uncoupler allows protons to equilibrate across this membrane, collapsing the ΔpH and $\Delta\psi$ components of the proton potential, Δp, and thus causing phosphorylation to cease) caused immediate and severe inhibition of thiosulfate-oxidation-dependent NAD^+ and $NADP^+$ reduction in intact *T. tepidarius* and *T. versutus*, this effect being established over a much shorter time scale than any effect on ATP synthesis. This indicated that ATP synthesis *per se* does not drive $NAD(P)^+$ reduction and that inhibition of reduction of $NAD(P)^+$ by a proton-conducting uncoupler is a consequence of dissipating the membrane proton potential. This was confirmed in *T. tepidarius* by showing that dicyclohexylcarbodiimide (DCCD; an inhibitor of ATPase) abolished thiosulfate-dependent ATP synthesis without affecting NAD^+ or $NADP^+$ reduction.

The current view of chemiosmotic energy coupling in the sulfur chemolithotrophs is thus that oxidation of sulfur compounds establishes an energized membrane state that is susceptible to the same inhibitors of ΔpH and $\Delta\psi$ that affect mitochrondrial and heterotrophic bacterial membrane systems, and which drives ATP synthesis by a proton-transporting ATP synthase and the flow of electrons from reduced ubiquinone/cytochromes to NAD^+ and $NADP^+$ at the expense of the membrane proton potential.

11.8 CONCLUDING REMARKS

It must be self-evident from this description of the diversity of sulfur chemolithotrophs that there is still some degree of uncertainty surrounding some aspects of their most important biochemical characteristic: the ability to oxidize sulfur and its compounds. We can expect in the near future to see further resolution of mechanisms and pathways in those bacteria on which much work has already been done, such as some of the thiobacilli, *Acidianus*, and *Sulfolobus*, but much more comparative work is needed on the more recently isolated genera like *Thiomicrospira, Thiosphaera,* and *Thermothrix.* Perhaps of even more interest are those unicellular sulfur bacteria about which little at all is known, and in which even chemolithoautotrophic growth may require proof or at least optimization. These include particularly the "morphologically conspicuous" sulfur bacteria (La Rivière and Schmidt, 1981), such as *Achromatium, Thiobacterium, Thiospira,* and *Thiovulum,* some of which were described morphologically before the first *Thiobacillus* was named. Some of these organisms are morphologically extraordinary—the spirals of *Thiospira* may be 14 μm long and *Thiovulum* may be up to 25 μm long and ovoid in shape, while *Achromatium* cells can be 100 μm in length! The chemolithoautotrophy of *Thiovulum* has been indicated by $^{14}CO_2$ studies, but the biochemistry and ecological importance of these unusual bacteria is in great need of thorough study.

REFERENCES

Aragno, M. and **Schlegel, H.G.** 1981. The hydrogen-oxidizing bacteria, pp. 875–912, in Starr, M.P., Stolp, H., Trüper, H.G., Balows, A., and Schlegel, H.G. (editors). *The Prokaryotes,* Springer-Verlag, Berlin.

Beijerinck, M.W. 1904. Über Bakterien welche sich im Dunkeln mit Kohlensäure ernähren können. *Zentralblatt für Bakteriologie und Parasitenkunde* Abt.II, 11:593–599.

Brannan, D.K. and **Caldwell, D.E.** 1983. Growth kinetics and yield coefficients of the extreme thermophile *Thermothrix thiopara* in continuous culture. *Applied and Environmental Microbiology* 45:169–173.

Codd, G.A. and **Marsden, W.J.N.** 1984. The carboxysomes (polyhedral bodies) of autotrophic prokaryotes. *Biological Reviews* 59:389–422.

Emmel, T., Sand, W., König, W.A., and **Bock, E.** 1986. Evidence for the existence of a sulfur oxygenase in *Sulfolobus brierleyi. Journal of General Microbiology* 132:3415–3420.

Friedrich, C. and **Mitrenga, G.** 1981. Oxidation of thiosulfate by *Paracoccus denitrificans* and other hydrogen bacteria. *FEMS Microbiology Letters* 10:209–212.

Golovacheva, R.S. and **Karavaiko, G.I.** 1978. *Sulfobacillus,* a new genus of thermophilic spore-forming bacteria. *Mikrobiologiya* (Moscow) 47:815–822.

Gottschal, J.C., deVries, S., and **Kuenen, J.G.** 1979. Competition between the facultatively chemolithotrophic *Thiobacillus* A2, an obligately chemolithotrophic *Thiobacillus* and a heterotrophic spirillum for inorganic and organic substrates. *Archives of Microbiology* 121:241–249.

Greenley, D.E.N. and **Smith, D.W.** 1979. A novel pathway of glucose catabolism in *Thiobacillus novellus. Archives of Microbiology* 122:257–261.

Harrison, A.P. 1983. Genomic and physiological comparisons between heterotrophic thiobacilli and *Acidiphilium cryptum, Thiobacillus versutus* sp. nov., and *Thiobacillus acidophilus* nom. rev. *International Journal of Systematic Bacteriology* 33:211–217.

Harrison, A.P. 1986. Characteristics of *Thiobacillus ferrooxidans* and other iron-oxidizing bacteria, with emphasis on nucleic acid analyses. *Biotechnology and Applied Biochemistry* 8:249–257.

Hazeu, W., Steudel, R., Batenburg-van de Vegte, W.H., Bos, P., and **Kuenen, J.G.** 1988. pp. 111–118, in Norris, P.R. and Kelly, D.P. (editors), *Biohydrometallurgy,* Science and Technology Letters, Kew, England, (in press).

Jackson, J.F., Moriarty, D.J.W., and **Nicholas, D.J.D.** 1968. Deoxyribonucleic acid base composition and taxonomy of thiobacilli and some nitrifying bacteria. *Journal of General Microbiology* 53:53–60.

Jannasch, J.W., Wirsen, C.O., Nelson, D.C., and **Robertson, L.A.** 1985. *Thiomicrospira crunogena* sp. nov., a colorless, sulfur-oxidizing bacterium from a deep-sea hydrothermal vent. *International Journal of Systematic Bacteriology* 35:422–424.

Katayama-Fujimura, Y., Tsuzaki, N., and **Kuraishi, H.** 1982. Ubiquinone, fatty acid and DNA base composition determination as a guide to the taxonomy of the genus *Thiobacillus. Journal of General Microbiology* 128:1599–1611.

Katayama-Fujimura, Y., Enokizono, Y., Kaneko, T., and **Kuraishi, H.** 1983. Deoxyribonucleic acid homologies among species of the genus *Thiobacillus. Journal of General and Applied Microbiology* 29:287–295.

Katayama-Fujimura, Y., Tsuzaki, N., Hirata, A., and **Kuraishi, H.** 1984. Polyhedral inclusion bodies (carboxysomes) in *Thiobacillus* species with reference to the taxonomy of the genus *Thiobacillus. Journal of General and Applied Microbiology* 30:211–222.

Kelly, D.P. 1968. Biochemistry of oxidation of inorganic sulphur compounds. *Australian Journal of Science* 31:165–173.

Kelly, D.P. 1978. Bioenergetics of chemolithotrophic bacteria, pp. 363–386, in Bull, A.T. and Meadow, P.M. (editors), *Companion to Microbiology,* Longman, London.

Kelly, D.P. 1982. Biochemistry of the chemolithotrophic oxidation of inorganic sulphur. *Philosophical Transactions of the Royal Society of London* B298:499–528.

Kelly, D.P. 1985. Physiology of the thiobacilli: elucidating the sulphur oxidation pathway. *Microbiological Sciences* 2:105–109.

Kelly, D.P. 1988. Oxidation of sulphur compounds. *Society for General Microbiology Symposium* 42:65–98.

Kelly, D.P. and **Syrett, P.J.** 1966. Energy coupling during sulphur compound oxidation by *Thiobacillus* sp. strain C. *Journal of General Microbiology* 43:109–118.

Kelly, D.P. and **Kuenen, J.G.** 1984. Ecology of the colourless sulphur bacteria, pp. 211–240, in Codd, G.A. (editor). *Aspects of microbial metabolism and ecology,* Academic Press, London.

Kelly, D.P. and **Harrison, A.P.** 1988. Genus *Thiobacillus* Beijerinck, Staley, J.T. (editor), *Bergey's Manual of Systematic Bacteriology,* Vol. 3, Williams and Wilkins, Baltimore (in press).

Kuenen, J.G. and **Tuovinen, O.H.** 1981. The genera *Thiobacillus* and *Thiomicrospira,* pp. 1023–1036, in Starr, M.P., Stolp, H., Trüper, H.G., Balows, A., and Schlegel, H.G. (editors), *The Prokaryotes,* Springer-Verlag, Berlin.

Kuenen, J.G. and **Veldkamp, H.** 1972. *Thiomicrospira pelophila* gen. n., sp. n., a new obligately chemolithotrophic colourless sulfur bacterium. *Antonie van Leeuwenhoek Journal of Microbiology and Serology* 38:241–256.

Kuenen, J.G., Boonstra, J., Schröder, H.G.Y., and **Veldkamp, H.** 1977. Competition for inorganic substrates among chemoorganotrophic and chemolithotrophic bacteria. *Microbial Ecology* 3:119–130.

Lu, W-P. 1986. A periplasmic location for the thiosulphate-oxidizing multienzyme system from *Thiobacillus versutus. FEMS Microbiology Letters* 34:313–317.

Lu, W-P., Swoboda, B.E.P., and **Kelly, D.P.** 1985. Properties of the thiosulphate-oxidizing multienzyme system from *Thiobacillus versutus. Biochimica et Biophysica Acta* 828:116–122.

Marsh, R.M. and **Norris, P.R.** 1983. The isolation of some thermophilic, autotrophic, iron- and sulphur-oxidizing bacteria. *FEMS Microbiology Letters* 17:311–315.

Mason, J., Kelly, D.P., and **Wood, A.P.** 1987. Chemolithotrophic and autotrophic growth of *Thermothrix thiopara* and some thiobacilli on thiosulphate and polythionates, and a reassessment of the growth yields of *T. thiopara* in chemostat culture. *Journal of General Microbiology* 133:1249–1256.

Nathansohn, A. 1902. Über eine neue Gruppe von Schwefelbakterien und ihren Stoffwechsel. *Mittheilungen aus der zoologischen Station zu Neapel* 15:655–680.

Norris, P.R., Marsh, R.M., and **Lindström, E.G.** 1986. Growth of mesophilic and thermophilic acidophilic bacteria on sulfur and tetrathionate. *Biotechnology and Applied Biochemistry* 8:318–329.

Oh, J.K. and **Suzuki, I.** 1977. Isolation and characterization of a membrane-associated thiosulphate-oxidizing system of *Thiobacillus novellus. Journal of General Microbiology* 99:397–412.

Peck, H.D. 1960. Adenosine 5′-phosphosulfate as an intermediate in the oxidation of thiosulfate by *Thiobacillus thioparus. Proceedings of the National Academy of Sciences (USA)* 46:1053–1057.

Postgate, J.R. and **Kelly, D.P.** 1982. *The sulphur bacteria.* The Royal Society, London.

Publius Syrus, 44B.C. Maxim 469 in Barlett, J., *Familiar Quotations,* 10th edition (1917), Macmillan, London.

la Rivière, J.W.M. and **Schmidt, K.** 1981. Morphologically conspicuous sulfur-oxidizing eubacteria, pp. 1037–1048, in Starr, M.P., Stolp, H., Trüper, H.G., Balows, A., and Schlegel, H.G. (editors), *The Prokaryotes,* Springer-Verlag, Berlin.

Robertson, L.A. and **Kuenen, J.G.** 1983. *Thiosphaera pantotropha* gen. nov. sp. nov., a facultatively anaerobic facultatively autotrophic sulphur bacterium. *Journal of General Microbiology* 129:2847–2855.

Roth, C.W., Hempfling, W.P., Conners, J.N., and **Vishniac, W.** 1973. Thiosulfate- and sulfide-dependent pyridine nucleotide reduction and gluconeogenesis in intact *Thiobacillus neapolitanus. Journal of Bacteriology* 114:592–599.

Schlegel, H.G. 1975. Mechanisms of chemoautotrophy, pp. 9–60, in Kinne, O. (editor), *Marine Ecology,* vol. 2, part 1, Wiley, London.

Segerer, A., Neuner, A., Kristjansson, J.K., and **Stetter, K.O.** 1986. *Acidianus infernus* gen. nov., sp. nov., and *Acidianus brierleyi* comb. nov.: facultatively aerobic, extremely acidophilic thermophilic sulfur-metabolizing archaebacteria. *International Journal of Systematic Bacteriology* 36:559–564.

Silver, M. and **Lundgren, D.G.** 1968. Sulfur-oxidizing enzyme of *Ferrobacillus ferrooxidans (Thiobacillus ferrooxidans). Canadian Journal of Biochemistry* 46:457–461.

Stetter, K.O. and **Zillig, W.** 1985. *Thermoplasma* and the thermophilic sulfur-dependent archaebacteria, pp. 85–170, in Woese, C.R. and Wolfe, R.S. (editors), *The Bacteria,* vol. 8, Academic Press, Orlando.

Suylen, G.M.H., Stefass, G.C., and **Kuenen, J.G.** 1966. Chemolithotrophic potential of a *Hyphomicrobium* species, capable of growth on methylated sulphur compounds. *Archives of Microbiology* 146:192–198.

Tano, T., Ito, T., Takesue, H., Sugio, T., and **Imai, K.** 1982. B-type cytochrome, an electron carrier in the sulfite-oxidation system of *Thiobacillus thiooxidans. Journal of Fermentation Technology* 60:181–187.

Timmer-ten Hoor, A. 1975. A new type of thiosulphate oxidizing, nitrate reducing microorganism: *Thiomicrospira denitrificans* sp. nov. *Netherlands Journal of Sea Research* 9:344–350.

Vishniac, W. and **Santer, M.** 1957. The thiobacilli. *Bacteriological Reviews* 21:195–213.

Wood, A.P. and **Kelly, D.P.** 1980. Carbohydrate degradation pathways in *Thiobacillus* A2 grown on various sugars. *Journal of General Microbiology* 120:333–345.

Wood, A.P. and **Kelly, D.P.** 1987. Chemolithotrophic metabolism of the newly-isolated moderately thermophilic, obligately autotrophic *Thiobacillus tepidarius. Archives of Microbiology* 144:71–77.

Wood, A.P., Kelly, D.P., and **Thurston, C.F.** 1977. Simultaneous operation of three catabolic pathways in the metabolism of glucose by *Thiobacillus* A2. *Archives of Microbiology* 113:265–274.

Yamanaka, T., Yoshioka, T., and **Kimura, K.** 1981. Purification of sulphite: cytochrome c reductase of *Thiobacillus novellus* and reconstitution of its sulphite oxidase system with the purified constituents. *Plant and Cell Physiology* 22:613–622.

12

Physiology and Biochemistry of Filamentous Sulfur Bacteria

Douglas C. Nelson

Department of Bacteriology
University of California
Davis, California, U.S.A.

12.1 INTRODUCTION

The colorless sulfur bacteria can be divided into two broad groups based on morphology: unicellular and filamentous. An overview of the habitats and niches of all the colorless surfur bacteria has been given in Chapter 4. The physiology and biochemistry of the unicellular sulfur bacteria is covered in Chapter 11. In the present chapter, the filamentous sulfur bacteria are discussed. Historically, the filamentous sulfur bacteria are of interest because they were the first organisms studied by Winogradsky in his pioneering work on chemolithotrophy (see Chapters 1 and 2).

Niche differences Filamentous sulfur bacteria form some of the most conspicuous microbial aggregations seen in nature. The various genera, most prominent in marine environments but also well known from freshwater environments, occupy specific niches which can apparently be differentiated based on the availability of O_2 and H_2S. Mass occurrences of *Beggiatoa* spp. reveal them to be gradient organisms. They can occur at a sediment surface or within a sediment as a dense bacterial "plate" with sharply defined upper and lower boundaries. Such plates rarely exceed one millimeter in vertical dimension and occur precisely at the H_2S/O_2 interface where the concentration of each of these nutrients is low and relatively constant (Jørgensen and Revsbech, 1983 and Chapter 7). By contrast, *Thiothrix* spp. occur as attached organisms in flowing waters that contain both H_2S and O_2. Waving in flowing water, these filaments probably experience H_2S and O_2 at concentrations which will vary significantly on a short time scale. The maximum concentrations encountered appear to be higher than found in the niche of *Beggiatoa* (Larkin and Strohl, 1983).

The niche of *Thioploca* is quite different. In marine sediments off the western coast of Peru and Chile at water depths of 100 to 700 meters, this organism can occur at biomass densities ranging from 0.1 to 1.0 kg wet wt m^{-2} (Gallardo, 1977; Rosenberg et al., 1983). Vertical distribution of *Thioploca* is relatively uniform over the top 10 to 20 cm of sediment. For much of the year, the concentration of H_2S in this portion of the sediment is below the limit of detection (Henrichs and Farrington, 1984), but filaments possess significant stores of S^0 nonetheless. Based on the extremely low oxygen concentrations present in overlying waters (Henrichs and Farrington, 1984), it is not possible during much of the year for oxygen to be present in these sediments at a depth of even a few centimeters. Curiously, when individual bundles of *Thioploca* filaments are removed from similar sediments for study, they appear to require the presence of small amounts of oxygen before inorganic or organic carbon compounds are incorporated (Maier and Gallardo, 1984b).

Early physiological studies Winogradsky's pioneering studies of the physiology of lithoautotrophic bacteria of the genus *Beggiatoa* are amply discussed elsewhere in this volume (Chapters 1 and 2). The broad impact of his early discoveries with *Beggiatoa* are all the more remarkable when one considers that he worked solely with enrichment cultures.

Twenty-five years after Winogradsky's studies, Kiel (1912) isolated from a freshwater sulfur spring a strain of *Beggiatoa* which he reported grew as a lithoautotroph. It grew in medium based upon spring water, but only under microaerophilic conditions within a narrowly defined range of H_2S partial pressures. This nonquantitative proof of lithoautotrophy gradually lost credibility when later researchers could only isolate *Beggiatoa* strains possessing an apparently nonautotrophic metabolism. Cataldi (1940) discovered that isolates could be obtained which readily grew as aerobic organoheterotrophs in medium lacking reduced sulfur compounds. This observation was repeated by others (Faust and Wolfe, 1961; Scotten and Stokes, 1962; Pringsheim, 1964), and there was no evidence that any of these isolates could grow as lithoautotrophs. Eventually, it was demonstrated that lithoautotrophic *Beggiatoa* strains do, indeed, exist (Nelson and Jannasch, 1983); however, to date only marine strains have been demonstrated to possess this capacity.

Although concluding with brief sections on the other filamentous sulfur bacteria, this chapter will deal primarily with *Beggiatoa* and will consider marine and freshwater strains separately. This dichotomy, when viewed in the future from some improved phylogenetic perspective, will almost certainly prove to be an artificial one. For now it serves to highlight the two distinctly different physiological types of *Beggiatoa* currently known as well as two distinctly different approaches to the study of their physiology.

12.2 MARINE *BEGGIATOA*—LITHOAUTOTROPHIC GRADIENT ORGANISMS

Features defining the genus In spite of numerous pure culture studies of various *Beggiatoa* strains, we still determine whether a filamentous bacterium belongs to this genus based on criteria virtually unchanged since the time of Winogradsky. Although other unifying features will be pointed out in this chapter, formally recognized common features are as follows: 1) *Beggiatoa* spp. are "colorless"; that is they contain no carotenoids or photosynthetic pigments. They do, however, appear pink when concentrated in dense pellets due to the presence of cytochrome pigments (Strohl et al., 1986); 2) *Beggiatoa* spp. grow as filaments. These are up to one centimeter or more in length and composed of long chains of disc or cylindrical shaped cells. A variety of width classes are recognized in natural samples ranging from approximately 1 to greater than 50 μm in width (Larkin and Strohl, 1983). Previously, different width classes were given separate species status. Only the narrower filaments (1 to 4 μm) have been isolated in pure culture, and their width is largely invariant over time. Currently only a single freshwater species is formally recognized (Skerman et al., 1980); 3) *Beggiatoa* spp. deposit periplasmic inclusions (globules) of elemental sulfur (S^0) in the presence of a source of hydrogen sulfide. Those strains tested also deposit S^0 when presented with thiosulfate; 4) Filaments of *Beggiatoa* spp. are capable of gliding motility. They glide at approximately 2–5 μm/s on a solid or semisolid surface and appear to do so constantly, with periodic reversals of direction, throughout their entire growth cycle.

Conditions of isolation and cultivation Six strains, the only available marine representatives of the genus *Beggiatoa*, were isolated from a New England salt marsh (Nelson et al., 1982). They are routinely cultivated in 60–75 percent strength natural seawater (NSW) and will not grow at salinities below approximately 10 percent NSW (Nelson, unpublished). These strains were isolated under microaerophilic conditions (oxygen concentration, $pO_2 \leq 1$ kPa) and all appear to be obligate microaerophiles. In semisolid medium, filaments only grow in regions where the partial pressure of O_2 is below 1.2 kPa (15 μM O_2). This occurrence in a very circumscribed microniche reflects both the negative chemotactic responses of motile filaments to excess O_2 and probably to excess H_2S (Møller et al., 1985). It also reflects the limited range of conditions which permit growth.

Demonstration of autotrophy in gradient cultures All marine *Beggiatoa* strains grow as a thin "plate" in agar-stabilized mineral medium containing oppositely sloped gradients of oxygen and hydrogen sulfide (Figure 12.1A). This bacterial plate, with very sharply defined upper and lower boundaries, never exceeded approximately 1 mm in vertical thickness. Because growth was prolific within this limited zone and because the medium did not contain added organic materials, autotrophic growth—fueled by sulfide oxidation—was strongly suggested. Three experimental findings support this view (Nelson and Jannasch, 1983; Nelson, unpublished). The first employed long-term growth studies (>6 doublings) in a closed system consisting of gradient medium plus a head space of air (Figure 12.2). There, carbon balance studies using ^{14}C-bicarbonate as a tracer confirmed that at least 90 percent of the total carbon in *Beggiatoa* was derived from CO_2. Secondly, activities of D-ribulose-1,5-bisphosphate carbox-

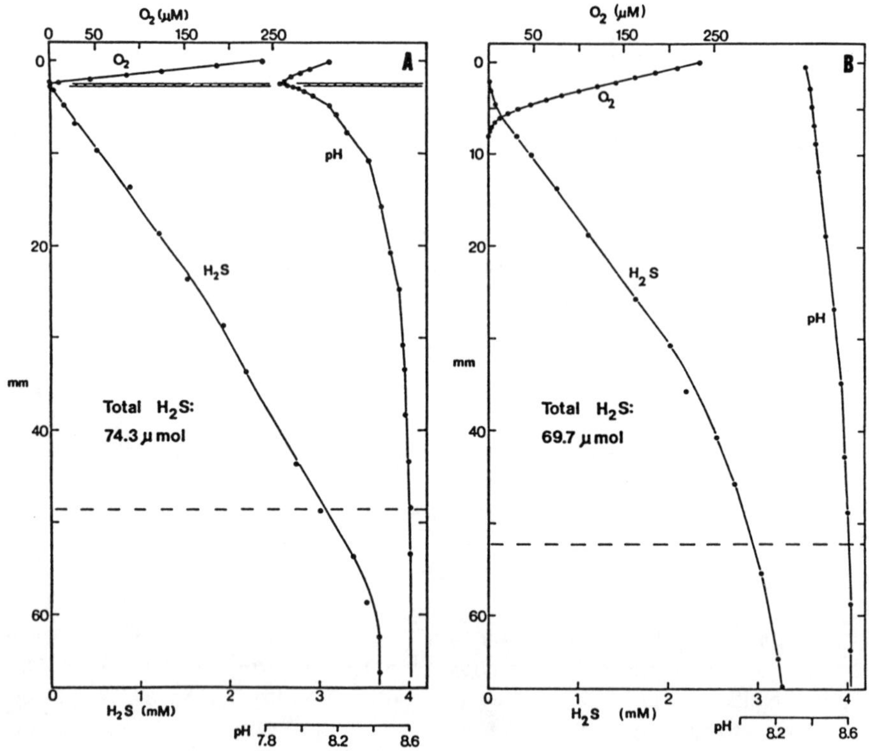

Figure 12.1 O_2, H_2S, and pH profiles in inoculated cultures incubated for 72 hours (A) or in control medium incubated for 94 hours (B). The *Beggiatoa* plate (marine strain 81-6) occurred at 2.5 mm depth (panel A). Zero depth indicates the air-agar interface. Dashed lines indicate the initial interface between H_2S agar and top agar (see Figure 12.2). All tubes initially contained 10.0 ml of H_2S agar (initial concentration 8.5 mM) and 30.0 ml of top agar. Total H_2S values shown were obtained by integrating the H_2S profile over the entire depth of the tube. Slight differences in dimensions of tubes resulted in differences in vertical extent of gradients. Inoculation was at time zero after initial step gradients had been stored for 41 hours (see text for procedure for formation of step gradients). (Redrawn from Nelson et al., 1986b.)

ylase/oxygenase (RuBisCO) and ATP:D-ribulose-5-phosphate-1-phosphotransferase (phosphoribulokinase) as high as 60 and 1,500 nmol substrate converted·mg $prot^{-1}$·min^{-1} (at 30°C), respectively, were measured in crude cell-free extracts. These activities are comparable to those of other known chemoautotrophs (Beudeker and Kuenen, 1981). Thirdly, the yield in gradient medium was proportional to the initial reservoir concentration of Na_2S.

Growth dynamics in sulfide-oxygen gradient medium Because gradient culture methods are quite new, a more detailed description of this approach will be presented. The actual details of measuring gradients using microelectrodes and of other chemical methods can be found elsewhere (Nelson et al., 1986a,b).

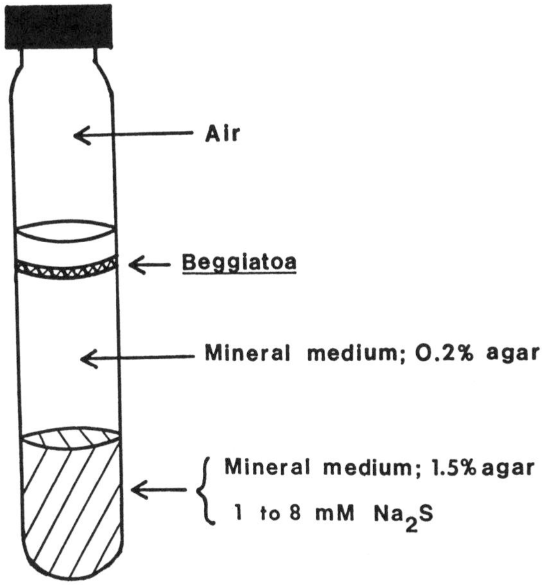

Figure 12.2 Initial geometry of marine gradient medium immediately after construction. In addition to macroelements, trace elements and vitamins, mineral medium contained 2.5 mM sodium bicarbonate as a carbon source. Crosshatched region shows where *Beggiatoa* proliferated as a "plate" following inoculation. An air-filled headspace volume twice the volume of bottom agar (8 mM Na_2S) provided sufficient O_2 for complete oxidation of H_2S in a closed system. See Nelson et al., 1986a,b for details.

Construction of gradient medium begins with the formation of a step gradient of hydrogen sulfide.* Agar-gelled mineral medium containing Na_2S (typically 8 mM) is allowed to solidify in the bottom of a culture tube and is then overlaid with a volume of agar-containing mineral medium lacking H_2S. This initial geometry (Figure 12.2) changes over time due to molecular diffusion of H_2S, diffusion of O_2 from the headspace of air above the medium, and by nonbiological reaction of H_2S and O_2 in the overlap zone. The net result of these processes after 6 days in uninoculated medium is shown in Figures 12.1B and 12.3B. Gradient medium requires several days of "curing" prior to inoculation to ensure that a region of H_2S/O_2 overlap is present. Molecular diffusion is slow enough so that the H_2S in the reservoir is not exhausted; medium cured for 30 days will still support lithoautotrophic growth of strain 81-6 for approximately one week.

Upon inoculation of properly cured gradient medium, filaments immediately migrate to the region of hydrogen sulfide-oxygen overlap. Within 8 hours of inoculation, even when the vertical dimension of overlap is roughly 6 mm, the filaments form a well-defined plate no more than 400 μm in thickness.

*The sum of H_2S + HS^- + $S^=$ will be designated collectively here as hydrogen sulfide or H_2S, even though HS^- is the dominant form at the pH of these experiments.

Whenever measured, the top of a *Beggiatoa* plate is characterized by an O_2 concentration below 15 μM. Over the next few days, growth of the *Beggiatoa* filaments had a profound impact on the shape of the gradients. In a three-day-old culture (Figure 12.3A), the sulfide-oxygen overlap region was 270 μm and both gradients terminated within the plate. By contrast, in uninoculated controls the overlap region ranged from 6–7 mm in medium assayed 2, 6, and 13 days after formation of the initial step gradient (e.g., Figure 12.3B).

The linearity of a hydrogen sulfide or an oxygen gradient just below or above a *Beggiatoa* plate makes it possible to calculate the flux of each compound into the plate. These calculations use Fick's diffusional laws and require knowl-

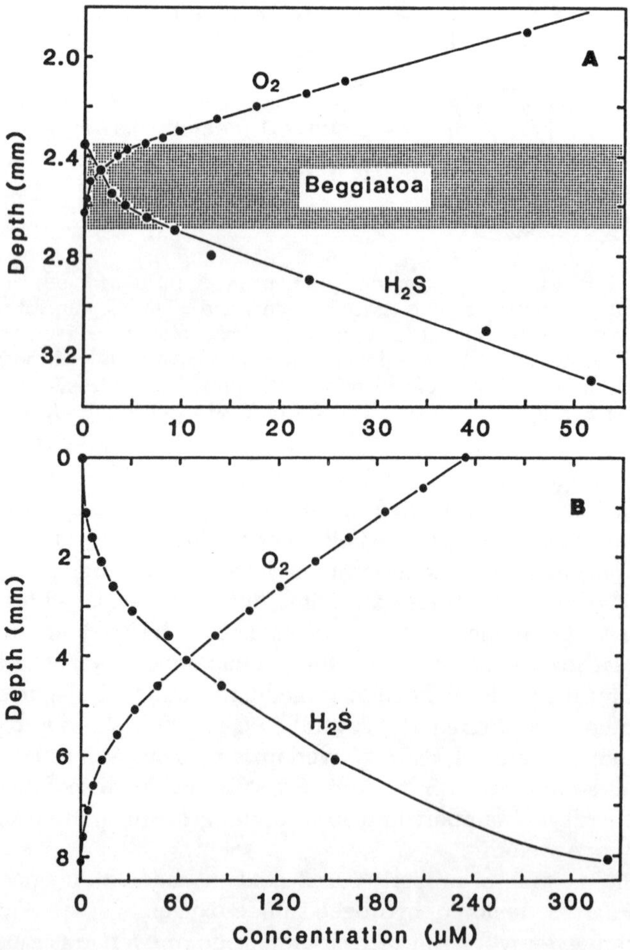

Figure 12.3 Expanded view of regions of overlap between O_2 and H_2S for corresponding panels of Figure 12.1. Shaded area (panel A) shows the *Beggiatoa* plate. Note higher resolution of the depth and concentration scales for panel A than B. Redrawn from Nelson et al., 1986b.

edge of appropriate diffusion coefficients (Nelson et al., 1986b). Analyses of the data from Figure 12.1A indicate a downward O_2 flux of 19.0 pmol·cm^{-2}·s^{-1} and an upward H_2S flux of 11.1 pmol·cm^{-2}·s^{-1}. Integrating the total O_2 or H_2S present in the zone of overlap (e.g., 270 μm vertical dimension, Figure 12.3A) and dividing this quantity by the appropriate flux value yields an average "turnover" time for O_2 (2.7 s) or H_2S (5.2 s). Turnover due to nonbiological oxidation (controls) was approximately three orders of magnitude slower in all cases. Thus, in a well-established *Beggiatoa* culture the observed H_2S and O_2 fluxes are direct measures of the rates of bacterial consumption; nonbiological oxidation is negligible (Nelson et al., 1986b).

As evidence of how sensitive this type of analysis can be, the calculated ratio of H_2S/O_2 fluxes during a time of sustained growth (e.g., Figure 12.1A) is 0.58/1.00, very close to the theoretical ratio of 0.61/1.00. This theoretical ratio was derived by assuming complete oxidation of H_2S to sulfate except for a fraction of H_2S-derived electrons (predicted from growth yield data) which must provide reductant for CO_2 fixation. Details of the calculations are presented elsewhere (Nelson et al., 1986b). Throughout the range of conditions tested, the H_2S/O_2 consumption ratio observed for strain 81-6 could be as low as 0.48/1.00, which was observed in stationary phase ten days after growth was initiated. This ratio is consistent with complete oxidation of H_2S to sulfate (theoretical ratio of 0.50/1.00) as is expected to occur when all available energy is utilized for maintenance (Nelson et al., 1986b). On the other hand, during early exponential growth (growth constant, μ = .065 h^{-1}, 23°C), when the H_2S flux was nonlimiting, an H_2S/O_2 consumption ratio of approximately 1.0/1.0 was transiently observed. This is explained by assuming that approximately 50 percent of the H_2S flux is oxidized to the level of S^0 and stored as internal globules; the remainder of the H_2S is assumed to be completely oxidized. This ratio of H_2S/O_2 consumption cannot prevail during steady state lithoautotrophic growth because it would require that over 75 percent of the cell biomass be S^0 (Nelson et al., 1986b). However, for natural diurnal cycles, transiently high H_2S fluxes are anticipated (Jørgensen, 1982). Flexibility in its ratio of H_2S/O_2 consumption allows *Beggiatoa* to more completely "harvest" these temporally variable fluxes while storing a considerable fraction of the available energy as S^0.

For marine *Beggiatoa* strains, two or more days of growth in gradient medium always produced biomass of sufficient abundance and metabolic activity that all O_2 was consumed before it reached the bottom of the plate (Nelson et al., 1986b). Figure 12.3A demonstrates this phenomenon early in the growth cycle of strain 81-6. As growth continued, approximately the top 200 μm of the plate remained microoxic, but the anoxic portion increased steadily in thickness until, after ten days of growth, it included more than 80 percent of the plate. H_2S was consumed in a narrow region of roughly 100 μm vertical dimension, which was the only place where O_2 coexisted with it. Typically, H_2S does not reach the upper surface of a plate. *Beggiatoa* filaments never show any obvious vertical zonation within such a plate, and they are invariably and immediately observed to be gliding at normal speed whenever sampled (Nelson et al., 1986a,b). It is reasonable to postulate that because of chemotactic movements, individual cells of gliding filaments are exposed to three types of microniches: low O_2/no H_2S; low H_2S/low O_2; low H_2S/no O_2. It also seems likely that individual cells are shifted from one to another of these environments on a time scale measured in minutes or seconds. This must have important implications for their metabolism. For example, cells in the middle microniche can oxidize H_2S or S^0 while those in the upper microniche are restricted to S^0 oxidation.

Molar growth yield Because it competes so successfully against nonbiological oxidation of H_2S, strain 81-6 grown in gradient cultures provided a unique opportunity to determine the growth yield of a sulfide-oxidizing lithoautotroph. A number of replicate gradient cultures, all inoculated equally and simultaneously, were analyzed 1.5 to 10 days after inoculation to yield a growth curve (Nelson et al., 1986b). By integrating H_2S profiles (e.g., Figure 12.1A), it was also possible to compare incremental consumption of H_2S with incremental production of biomass. These calculations indicated 8.4 g (dry wt) of *Beggiatoa* formed per mole of H_2S oxidized. This value is somewhat higher than the value of approximately 5 g dry wt/mole computed for liquid-grown cultures of other aerobic H_2S oxidizers (Kelly, 1982 and Chapter 11). A significant fraction of the available H_2S in those other studies would have been chemically oxidized and therefore unavailable to support growth. This suggests that *Thiobacillus* growth yields should be reevaluated using gradient culture techniques.

Enzymology of hydrogen sulfide utilization The gradient culture method does not provide the quantity of cells required for most enzymological assays. Therefore, methods for bulk cultivation of the lithoautotrophic marine *Beggiatoa* strains must be developed. Growth-yield studies assure us that lithotrophic energy conservation linked to the oxidation of hydrogen sulfide or thiosulfate must be occurring; however, we know nothing of the enzymes specifically responsible for this process in any marine strain. Nonetheless, in a study of native *Beggiatoa* there is some evidence regarding its hydrogen sulfide-oxidizing enzymes. Extremely dense populations of *Beggiatoa* sp., discovered in the Gulf of California at deep sea hydrothermal vents (Chapter 8), have been investigated for their cytochrome content (Prince et al., 1988). They showed that this virtual monoculture of 30 μm wide filaments contained an extremely high concentration of a high molecular weight c-type cytochrome (M_r 210,000). This cytochrome contains at least four thermodynamically distinct c-type hemes, some of which have oxidation-reduction midpoint potentials appropriate for involvement in the oxidation of H_2S. No a-, b-, or d-type cytochromes were detected. Likewise there was no evidence that this *Beggiatoa* sp. possessed a flavocytochrome or a CO-binding heme.

Organic and mixed nutrition Strain 81-6 grew microaerophilically in liquid medium with acetate plus vitamins as the sole source of organic carbon (Nelson and Jannasch, 1983). Since sulfate was the only source of sulfur in the medium, this is strong evidence of facultative organoheterotrophy. Whether other organic compounds will support similar growth is not known. For exploring mixotrophy, on the other hand, the gradient medium technique described above proved very useful. This is especially true when one considers problems inherent in attempting by other methods to continually maintain very low concentrations of O_2 and H_2S in the presence of an organic compound.

The method used to investigate mixotrophy is as follows: In the absence of added organic compounds, strain 81-6 adjusts its vertical position in gradient medium to balance H_2S and O_2 fluxes. Furthermore, because of the linearity of the O_2 gradient above the plate and the very low pO_2 within the plate, an O_2 flux (equivalent to the O_2 consumption rate) can be determined with only a distance measuring device (Nelson et al., 1986a). By studying biomass increase and plate depth as functions of time in H_2S/O_2 gradient medium (controls), compared to identical medium supplemented with various organic compounds, it is possible to group organic compounds into three classes. Compounds in the first class have no influence on plate position or on final biomass achieved, i.e., they are not metabolized or incorporated significantly. Organic compounds of the second class produce a significant biomass increase (1.3–1.6 times the control) and a lesser increase (1.2–1.3 times) in the rate of oxygen consumption. Organics of the third class produce a major increase in both biomass (2.0–4.0 times) and O_2 consumption (1.8–3.0 times). Sugars, alcohols, and amino acids are in the first class. Most tricarboxylic acid (TCA) cycle intermediates are in the second class of compounds. Acetate, lactate, pyruvate, propionate, and oxalacetate are the only identified compounds in this third class (Nelson, unpublished). Simultaneous quantification of gradients of organic compounds, O_2, and H_2S will be necessary to prove which types of organic compounds are only used as biosynthetic precursors and which are also useful in energy generation. Without this analysis one cannot exclude, for example, the possibility that organic compounds of the third class simply fuel organoheterotrophic growth and inhibit H_2S oxidation entirely. Using analyses of all three gradients, it should be possible to identify examples of lithoheterotrophy or mixotrophy (with respect to either carbon or energy) in the above-mentioned classes.

12.3 FRESHWATER *BEGGIATOA* STRAINS

Strains and methods of cultivation A number of strains of freshwater *Beggiatoa* have been available for study in recent years. All strains tested to date are catalase negative and show a strong preference for a microoxic environment (Burton and Morita, 1964; Strohl and Larkin, 1978; Nelson et al., 1986a). Nonetheless, most studies were with fully aerated homogenous liquid cultures. When cultures were started from large inocula, growth in such medium typically resulted in formation of "tufts" or clumps of intertwined filaments one or more millimeters in diameter. It was assumed that respiration within a tuft produced a low-oxygen microenvironment which minimized the formation of toxic peroxides (Nelson and Castenholz, 1981a). The existence of microaerophilic conditions was recently confirmed for strain B18LD using microelectrodes (Schmidt et al., 1987), and it was further discovered that the inner-most portion of a clump became anoxic due to vigorous O_2 consumption. Presumably, the gliding motility invariably observed in healthy *Beggiatoa* filaments exposes individual cells to an alternation of anoxic and microoxic conditions during "aerobic" liquid cultivation. (This is not unlike what must occur in the plate formed by a marine or freshwater strain in gradient culture.) When grown under conditions of extremely vigorous aeration which minimize tuft formation, strain B18LD deposited poly-β-hydroxybutyrate (PHB) granules as approximately 50 percent of total dry weight. This indicates unbalanced growth under more uniformly aerobic conditions.

Strain B18LD, which is the type strain of *Beggiatoa alba*, and *Beggiatoa* sp. strain OH-75-2a (Nelson and Castenholz, 1981b), are those which will be most thoroughly discussed in the following sections. They were isolated from different localities, and are certainly the two best characterized freshwater strains. In general they have very similar phenotypes and appear to be closely related (Table 12.1).

***Heterotrophic carbon metabolism of freshwater* Beggiatoa** There is no evidence that organic polymers can support the growth of *Beggiatoa* (Mezzino et al., 1984). The range of organic compounds capable of supplying all carbon and energy requirements is extremely limited. None of the 18 strains for which data were collected possessed the ability to utilize sugars or sugar derivatives (Table 12.2). In general, single amino acids did not support growth of freshwater strains (Nelson and Castenholz, 1981b; Mezzino et al., 1984; Williams and Unz, 1988). A few of the 13 strains tested by Pringsheim (1964) were reported

Table 12.1 Comparison of two freshwater *Beggiatoa* strains

	B18LD[a]	OH-75-2a[b]
Phylogenetic affinity	Purple bacteria, gamma subgroup[c]	Purple bacteria, gamma subgroup[d]
Mol % G+C	37.1[e]	38.5[e]
Cell width	3 μm	2 μm[h]
Organic compounds supporting growth	See Table 2	See Table 2
Krebs cycle	yes[f]	probably[h]
S^0 aerobically deposited:		
from sulfide	yes	yes
from thiosulfate	yes	yes
S^0 reduced to H_2S (anaerobic)	yes[g]	yes
Autotrophic growth	no	no

[a]Data from Mezzino et al., 1984, except as noted.
[b]Data from Nelson and Castenholz, 1981a, except as noted.
[c]5S rRNA sequencing, Stahl et al., 1987.
[d]16S rRNA oligonucleotide analysis, Woese et al., 1985.
[e]Thermal denaturation, Nelson, unpublished.
[f]Larkin and Strohl, 1983.
[g]Schmidt et al., 1987.
[h]Nelson and Castenholz, 1981b.

Table 12.2 Organic compounds used as sole source of carbon and energy

Organic compound	B18LD[b]	OH-75-2a[c]	U1[d]	B15LD and B25RD[b]	Pringsheim strains
	Beggiatoa strain				
Alcohols					
Ethanol	+	+	+	+	ND
n-Propanol	ND	ND	+	ND	ND
n-Butanol	ND	ND	+	ND	ND
TCA cycle intermediates					
Fumarate	+	−[a]	+	+	(4/13)[e]
Malate	+	−	−	+	(1/13)
Succinate	+	−	−	+	(10/13)
Oxalacetate	ND	−	+	ND	ND
Organic acids					
Acetate	+	+	+	+	(11/13)
Lactate	+	+	+	+	(11/13)
Pyruvate	+	+	+	+	(7/13)
Propionate	ND	ND	+	ND	ND
Hydroxybutyrate	ND	ND	+	ND	ND
Butyrate	ND	ND	+	ND	ND

[a]ND, not determined; +, growth; −, no growth.
[b]Positive if turbidity or tufts present after 2nd transfer in test medium; data of Mezzino et al., 1984.
[c]Positive if yield >20 g/mol; negative if yield <2 g/mol; data of Nelson and Castenholz, 1981b.
[d]Positive if absorbance 4× that of minus-organic control; data of Williams and Unz, 1989.
[e]Numerator is number out of 13 freshwater strains which Pringsheim, 1964 reported as +.

to grow with asparagine, aspartate, or glutamate as the sole source of carbon and energy. These observations were, however, nonquantitative, and in the absence of any published methods we cannot assess the possible influence of acetate carryover or utilization of PHB stores on this reported growth.

Strain B18LD has been shown to possess all of the enzymes necessary for a functional TCA cycle with a glyoxylate bypass (Larkin and Strohl, 1983). This strain and strain OH-75-2a showed molar growth yields of 16 to 21 (g dry wt/mol acetate), comparable to yields of other bacteria growing at the expense of complete oxidation of acetate via the TCA cycle (Nelson and Castenholz, 1981b; Güde et al., 1981). The almost universal ability to utilize lactate, acetate, pyruvate, and ethanol (Table 12.2) is consistent with the assumption that these compounds are metabolized via the TCA cycle. Examination of the rest of Table 12.2 indicates strain-specific differences in the ability of various organic acids to serve as sole source of carbon and energy. Although all of the published evidence supporting use of malate, succinate, fumarate, and oxalacetate is at best semiquantitative, additional studies have confirmed some of these strain-specific differences (Nelson, unpublished).

***Assimilation of nitrogen and sulfur by freshwater* Beggiatoa** Of the twenty-two amino acids tested, only cysteine, aspartate, and glutamine could serve as a source of combined nitrogen for acetate-supported heterotrophic growth of a single strain (Williams and Unz, 1988). Vargas and Strohl (1985a) found that aspartate, asparagine, and alanine but not gutamate or methionine could serve as the combined nitrogen source in acetate-grown cultures of the type strain. In general, urea, salts of ammonia, and nitrate salts could serve as assimilatory nitrogen sources (Vargas and Strohl, 1985a; Nelson et al., 1982). Strain U1 could not utilize urea and also lacked the ability to reductively assimilate NO_2^-. The type strain and three others tested by Mezzino et al. (1984) possessed the capacity to assimilate NO_2^-.

Studies by Vargas and Strohl (1985a) indicated that strain B18LD assimilates ammonia via the glutamine synthetase/glutamate synthase (GS/GOGAT) pathway in batch cultures, regardless of the initial ammonia concentration (0.1 mM and 2.0 mM). There was no evidence that GS and GOGAT activity were regulated by repression/derepression in response to changing ammonia levels. There was evidence that GS may be partially regulated by adenylylation/deadenylylation mechanisms known to operate in other Gram-negative organisms, but the findings require further confirmation. Interestingly, GOGAT activity was only detectable in the presence of NADPH not NADH. Glutamate dehydrogenase, an alternative enzyme involved in ammonia assimilation in some bacteria under conditions of high nitrogen, could not be detected.

All 12 strains of *Beggiatoa* tested (6 freshwater, 6 marine) showed the capacity for nitrogen fixation as measured by the acetylene reduction assay (Nelson et al., 1982). Specific rates of acetylene reduction were typical of other diazotrophs and were only observed under microaerophilic conditions. Growth in medium containing combined nitrogen (NO_3^- or NH_4^+) completely abolished this activity. For strain OH-75-2a, the growth yield in the absence of combined nitrogen was approximately 60 percent of the ammonia-supplemented yield on the same limiting quantity of acetate.

Beggiatoa strains have no demonstrated biosynthetic requirement for reduced sulfur compounds. All strains listed in Table 12.2 could grow in medium containing sulfate as the sole source of sulfur. Thiosulfate also served this function in strain 75-2a, although it may have been assimilated after disproportionation resulted in sulfite production.

***Mixotrophic utilization of reduced sulfur compounds by freshwater* Beggiatoa** After failing to detect lithoautotrophic growth, Pringsheim (1967) instead suggested his strains were capable of "mixotrophic" growth. Although no quantitative data were presented, the metabolism he postulated was lithoheterotrophy with acetate serving as the carbon source and hydrogen sulfide oxidation providing energy.

Whenever tested under S^0-depositing growth conditions, strain OH-75-2a contained this element as 20–30 percent of its total dry weight (Nelson and Castenholz, 1981a). Experimental manipulations were simplified when it was discovered that S^0 globules were deposited from thiosulfate, and this strain was then quantitatively tested for lithoheterotrophic growth. In thiosulfate-supplemented organic medium, the growth yield (g dry wt/mol of acetate consumed) was typically 20 to 30 percent above that of the control in which thiosulfate was absent. However, upon correction for the S^0 content of cells, the apparent stimulatory effect of thiosulfate disappeared (Nelson and Castenholz, 1981a). The consumption of thiosulfate in these experiments took place only after the onset of stationary phase had occurred in response to acetate depletion. High activity of the enzyme rhodanese has been detected in this strain (Yoon and Nelson, unpublished). Disproportionation of thiosulfate to sulfur and sulfite via this enzyme could occur with virtually no energy gain or loss to the cells. These cells maintained a high S^0 content for the duration of stationary phase and no activity of APS reductase or sulfite-cytochrome c oxidoreductase was detected (Yoon and Nelson, unpublished). These data are consistent with the lack of a detectable yield increment from thiosulfate consumption.

These studies leave open the possibility that oxidation of thiosulfate or sulfide in the presence of a limiting quantity of organic compounds could support lithoheterotrophic growth of *Beggiatoa* sp. Such studies were undertaken in a chemostat with the type strain of *B. alba,* but yielded equivocal results. Comparison was made in parallel experiments in organic medium (6.1 mM acetate) and in the same medium supplemented with 1 mM thiosulfate or hydrogen sulfide. After correcting for the contribution of internal S^0 and expressing the data per unit of acetate consumed, "mixotrophic" yields which were 1.6 to 2.8 times the corresponding heterotrophic yields were reported (Güde et al., 1981). Even assuming that all of the reduced sulfur compound was oxidized to sulfate, the yield increments reported in this study were too high (Kuenen and Beudeker, 1982). We now know that this strain does not oxidize its internal S^0 at any appreciable rate (Schmidt et al., 1987). It can be further computed that only a small fraction of the sulfur compounds available in the experiments of Güde et al. were oxidized even to the level of S^0. Rather than invoke impossibly high efficiencies of energy conservation linked to sulfide or thiosulfate oxidation one is forced to consider that the data of Güde et al. resulted from artifacts. They apparently used a single numerical factor to convert absorbance data to dry weight data, regardless of whether or not cells contained highly refractile sulfur globules, and this offers at least a partial explanation for these findings.

It cannot be concluded that freshwater strains lack mixotrophic potential but only that further experimentation is necessary. The experiments of Schmidt et al. (1987) demonstrate rates of sulfide oxidation by this strain which are comparable to those of the lithoautotrophic marine strain, MS-81-6. An enzyme postulated to be responsible for H_2S oxidation in strain B18LD has been identified as a flavocytochrome-c, but it is not known whether this cytochrome is present in cells grown in the absence of H_2S (T. Schmidt, personal communication). The suggestion that this enzyme is involved in H_2S oxidation by B18LD is based on the finding that a flavocytochrome may function as a sulfide oxidase in *Chromatium vinosum.* There it is postulated to act as a sulfide:cytochrome c oxidoreductase (Gray and Knaff, 1982). Alternatively, Schmidt and co-workers postulate that the flavocytochrome might be involved in the reduction of S^0 to H_2S (see below). (Consistent with its postulated metabolic similarity to the purple sulfur bacteria, strain B18LD as well as strain OH-75-2a cluster phylogenetically with the gamma subdivision of the purple bacteria; Table 12.1.) Other electron transport components including a-, b-, and d-cytochromes have also been identified in strain B18LD (Strohl et al., 1986).

However, the evidence for those components is regarded as tenuous (Prince et al., 1988).

Ultimately, the question of whether energy gain can be coupled to sulfide oxidation may not be readily answered by chemostat studies because it may not be possible to supply a limiting concentration of organic compounds and H_2S in this way. In the studies of Güde et al. (1981) it is not obvious which factor limited growth. The concentration of acetate in the culture vessel never fell below 0.85 mM, significantly above the initial concentration in some of the batch culture experiments of Nelson and Castenholz (1981a). Ultimately the issue of mixotrophic growth might be resolved by gradient culture studies in which a reservoir could simultaneously supply acetate and sulfide in very low concentrations but with continual renewal of both by fluxes from the reservoir.

***Autotrophy of freshwater* Beggiatoa** There is no evidence in the literature that freshwater *Beggiatoa* strains grown as aerobic organoheterotrophs or "mixotrophs" possess autotrophic potential. The maximum rate of CO_2 fixation for strains OH-75-2a and B18LD was very low. In rapidly growing cultures of strain OH-75-2a (μ = 0.10), the per hour rate of CO_2 fixation never equaled 2 percent of the biomass. In this strain as well as in B18LD, maximal CO_2 fixation was invariably observed in medium containing organic compounds, and all attempts at culturing these strains in medium devoid of organic compounds failed (Nelson and Castenholz, 1981a; Strohl et al., 1981a). The recent application of gradient culture techniques to the cultivation of these two strains has not changed this conclusion. As long as acetate was present in sulfide-oxygen gradient medium, growth was good. When cells were transferred to acetate-free medium, growth ceased. Gradual transfer to progressively lower acetate concentration did not change this finding nor did the addition of vitamins (Nelson, unpublished).

Heterotrophically and mixotrophically grown cultures of the type strain were assayed for the presence of diagnostic CO_2-fixing enzymes. Ribulose-1,5-bisphosphate carboxylase (RuBisCO) activity was not detected, but some CO_2 was fixed via the "reverse" operation of isocitrate dehydrogenase, one step of the reductive citric acid cycle (Strohl et al., 1981a). Although other enzymes of this cycle were also detected (Larkin and Strohl, 1983), two essential CO_2-fixing enzymes, α-ketoglutarate synthase and phosphoenolpyruvate carboxylase, have never been observed in strain B18LD (W. R. Strohl, personal communication).

Using a more sensitive assay than previously employed, it was recently determined that strains OH-75-2a and B18LD do possess low but highly regulated activities of RuBisCO and phosphoribulokinase (Nelson, unpublished). For chemoheterotrophically grown exponential phase cultures, the RuBisCO activity of strain OH-75-2a ranged from 0.002 to 0.02 nmol CO_2 fixed·mg prot^{-1}·min^{-1}. In acetate depleted stationary phase cultures, the activity was 0.3 to 0.7 units. Within a single growth cycle, this regulatory mechanism could increase the specific activity 100-fold. For strain B18LD, the pattern was similar except that RuBisCO activity changed by at most 10-fold during a growth cycle. Throughout the growth cycle of both strains, phosphoribulokinase activity was approximately 10-fold higher than the corresponding RuBisCO activity. For comparison, the autotrophic marine strain, 81-6, had a RuBisCO specific activity of 30–60 units and a phosphoribulokinase activity of 500–1500 units (Nelson, unpublished). A probe derived from the gene encoding the RuBisCO large unit of *Anacystis nidulans* hybridized strongly with fragments from several restriction enzyme digests of the DNA of strain OH-75-2a (J. M. Shively, personal communication).

***Anaerobic metabolism of freshwater* Beggiatoa** Strain OH-75-2a exhibited the ability to convert much of its aerobically deposited store of S^0 to hydrogen sulfide when shifted to strictly anoxic conditions in acetate-containing medium. Cells remained viable under these anaerobic conditions for five days, and there was apparently a modest increase in biomass. An identical culture tested for anaerobic viability in the absence of acetate showed significant lysis over the same test period (Nelson and Castenholz, 1981a).

Strain B18LD also exhibited the ability to anaerobically reduce S^0. In short-term experiments, its rate of H_2S production was comparable to that of a *Chromatium vinosum* control incubated anaerobically in the dark (Schmidt et al., 1987). *Chromatium* is known to possess a mechanism for anaerobic ATP production at the expense of S^0 reduction. In a two-hour experiment, this *Beggiatoa* strain showed no need for an exogenous organic electron donor in this process; apparently stored PHB served this function. In separate short-term anoxic incubations, molecular hydrogen was observed to stimulate the rate of H_2S production. This suggests the presence of a hydrogenase capable of participating in the reduction of S^0. This hydrogenase is apparently separate from the enzyme associated with nitrogenase (Schmidt et al., 1987).

Further evidence of the anaerobic capacity of strain B18LD came from anoxic incubations in the presence of organic compounds plus nitrate. Significant quantities of nitrite and ammonia were excreted into the medium, while no such activity was observed under aerobic conditions. Over a two-hour experiment, the rate of nitrate-dependent acetate oxidation was comparable to the oxygen-dependent rate (Vargas and Strohl, 1985b). There was no evidence of N_2 or N_2O accumulation, so nitrate respiration rather than denitrification is suggested (Jeter and Ingraham, 1981). Attempts at sustained anaerobic growth with nitrate or nitrite as an electron acceptor were not successful (Vargas and Strohl, 1985b). Evidence of nitrate respiration, taken together with evidence of S^0 respiration, suggests that multiple physiological mechanisms have evolved for dealing with the anoxic niche component of *Beggiatoa*.

Similarity to marine strains The phylogenetic affinity of marine and freshwater strains should be established by ribosomal RNA sequencing. However, in addition to morphology there are numerous metabolic similarities between these two groups which suggest a close relationship. In gradient medium both occupy a niche which is microoxic ($pO_2 < 15\ \mu M$) in the upper portion of the plate and anoxic in the lower portion of the plate. There is an almost universal capacity for diazotrophy among both freshwater and marine strains, and both consume a highly restricted and similar range of organic acids.

All marine strains tested are apparently facultatively autotrophic at the expense of the Calvin cycle (Nelson, unpublished). Both of the freshwater strains tested showed activities of two key Calvin cycle enzymes. These activities were regulated by the presence/absence of organic compounds in a fashion appropriate to facultative autotrophs, but the maximum specific activities were only a small fraction of the corresponding activities of marine strains. By contrast, enzymatic studies of a relatively pure natural population of freshwater *Beggiatoa* demonstrated RuBisCO activity comparable to the autotrophic marine strains (Nelson, unpublished). This native material was collected from the same warm spring which several years previously had been the source of strain OH-75-2a. Taken together these results suggest strongly that OH-75-2a was derived from an autotrophic strain and is still perhaps capable of autotrophic growth under an undiscovered set of conditions.

12.4 OTHER FILAMENTOUS SULFUR BACTERIA

The genera *Thiothrix* and *Thioploca,* like *Beggiatoa,* contain colorless, S^0-depositing filaments which are motile during at least a portion of the growth cycle. For *Thiothrix* only the hormogonial stage is motile. Growing filaments of this genus also attach to surfaces via terminal "holdfasts" and their clustered growth, apparently facilitated by fimbriae (Williams et al., 1987), leads to the formation of rosettes. Details of the life cycle of *Thiothrix* spp. are described elsewhere for both slide cultures of natural material and for pure cultures (Bland and Staley, 1978; Larkin and Shinabarger, 1983).

Recent pure culture studies showed that in all *Thiothrix* strains tested, S^0 globules are deposited in the presence of thiosulfate and sulfide (Larkin and Shinabarger, 1983; Williams and Unz, 1985). When examined by electron microscopy, these globules were found to be deposited within the periplasm surrounded by invaginated pockets of the cytoplasmic membrane. A "sulfur inclusion envelope" consisting of a single densely stained layer approximately 4 nm thick was present between the globule and the cytoplasmic membrane (Williams et al., 1987). This arrangement is also the most common one documented for *Beggiatoa* strains (Strohl et al., 1981b).

The metabolism of the neotype strain of *Thiothrix nivea* has been examined most thoroughly. It has never been successfully cultivated as a lithoautotroph or an organoheterotroph, but it is suggested to be an obligate mixotroph because it requires both an organic compound (acetate, pyruvate, malate, or oxalacetate) and a reduced sulfur compound for growth (Larkin and Shinabarger, 1983). The high rate of sulfide oxidation noted in this strain (Strohl and Schmidt, 1984) and the observation that the resultant S^0 globules were oxidized readily to sulfate (Schmidt et al., 1987) suggest that sulfide and sulfur oxidation can provide the organism with useful energy. In a separate study, another apparently obligately mixotrophic *Thiothrix* strain increased its growth yield in proportion to the amount of thiosulfate present (Williams and Unz, 1988). These data do not preclude the possibility that the "obligate" need for a reduced sulfur compound actually represents a biosynthetic requirement. Since literature evidence of obligate mixotrophy is rare, other possibilities should be thoroughly explored.

Other members of the genus *Thiothrix* can grow as strict organoheterotrophs using a variety of single organic compounds including sugars, polysaccharides, and sugar alcohols (Williams and Unz, 1988). In addition to obtaining biosynthetic carbon from organic compounds, *T. nivea* is believed to incorporate a significant amount of inorganic carbon. The neotype strain assimilated CO_2 at a rate approximately five-fold higher than the freshwater *Beggiatoa* strain B18LD. The enzymes apparently responsible for CO_2 fixation are phosphoenolpyruvate carboxylase and, perhaps, the malic enzyme (Strohl and Schmidt, 1984). Interestingly, two separate genes coding for the large subunit of RuBisCO have been detected in *T. nivea* using heterologous DNA probes (Shively et al., 1986). To date, no evidence concerning the activity of this enzyme has been published for any *Thiothrix* strain.

In the absence of pure cultures, the genus *Thioploca* is known only from native material. One species is formally recognized (Skerman et al., 1980), and three other species have been recently described (Maier, 1984; Maier and Gallardo, 1984a). The two obligately marine *Thioploca* species described appear to be very similar morphologically to a large *Beggiatoa* sp. seen but not cultivated at the deep-sea hydrothermal vents (Nelson, unpublished). Only the presence of a common sheath surrounding bundles of parallel or braided *Thioploca* filaments clearly distinguishes the two. Numerous sulfur globules occur within the peripheral region of the *Thioploca* cytoplasm and along cross-walls between cells (Maier and Gallardo, 1984a). It has not yet been established whether the globules are, in parallel with findings for *Beggiatoa* and *Thiothrix*, actually within the periplasm.

Based on studies with native marine *Thioploca*, Maier and Gallardo (1984a) concluded that the organism sustained motility longest under microaerophilic conditions (5 percent air) and required a low concentration of H_2S (approximately 100 μM). Autoradiographic methods were also employed to determine which carbon compounds among bicarbonate and a variety of organic compounds were preferentially incorporated into filaments incubated in the presence of appropriate levels of H_2S and oxygen. The approximately equivalent silver grain density which resulted from inclusion of radiolabeled bicarbonate, acetate, glycine, glucose, or an amino acid mixture was interpreted as evidence that mixotrophy was a preferred mode of growth (Maier and Gallardo, 1984b). If, however, one reexamines these published data taking account of the very low specific activity of bicarbonate due to dilution from the bicarbonate already present in natural seawater (roughly 2.5 mM), a different picture emerges. The per filament rate of total bicarbonate incorporation (radioactive plus cold) appears to be at least 20 times higher than the total incorporation rate for any organic compound tested. This strongly suggests a lithoautotrophic mode of growth.

Also deserving further investigation is the possibility that *Thioploca* can utilize electron acceptors other than oxygen. Henrichs and Farrington (1984) have shown that this organism is abundant throughout the top 20 cm or more of organic-rich sediments off the coast of Peru. Since the water overlying these sediments is already low in dissolved oxygen ($pO_2 < 5\ \mu M$), it is extremely unlikely that oxygen penetrates even one centimeter into the sediment. The emergence of oxidized manganese and iron as possibly important electron acceptors in marine sediments (see Chapter 7) should be given serious consideration in this context.

REFERENCES

Beudeker, R.F. and **Kuenen, J.G.** 1981. Carboxysomes: "Calvinosomes"? *FEBS Letters* 131:269–274.

Bland, J.A. and **Staley, J.T.** 1978. Observations on the biology of *Thiothrix*. *Archives of Microbiology* 117:79–87.

Burton, S.D. and **Morita, R.Y.** 1964. Effect of catalase and cultural conditions on growth of *Beggiatoa*. *Journal of Bacteriology* 88:1755–1761.

Cataldi, M.S. 1940. Aislamiento de *Beggiatoa alba* en cultivo puro. *Revista del Instituto Bacteriologica—Departamento Nacional de Higiene (Buenos Aires)* 9:393–423.

Faust, L. and **Wolfe, R.S.** 1961. Enrichment and cultivation of *Beggiatoa alba*. *Journal of Bacteriology* 81:99–106.

Gallardo, V.A. 1977. Large benthic microbial communities in sulphide biota under Peru-Chile subsurface countercurrent. *Nature* 268:331–332.

Gray, G.O. and **Knaff, D.B.** 1982. The role of a cytochrome c-552-cytochrome c complex in the oxidation of sulfide in *Chromatium vinosum*. *Biochimica et Biophysica Acta* 680:290–296.

Güde, H., Strohl, W.R., and **Larkin, J.M.** 1981. Mixotrophic and heterotrophic growth of *Beggiatoa alba* in continuous culture. *Archives of Microbiology* 129:357–360.

Henrichs, S.M. and **Farrington, J.W.** 1984. Peru upwelling region sediments near 15 degrees/S. 1. Remineralization and accumulation of organic matter. *Limnology and Oceanography* 29:1–19.

Jeter, R.M. and **Ingraham, J.L.** 1981. The denitrifying prokaryotes. pp. 913–925, in Starr, M.P., Stolp, H., Trüper, H.G., Balows, A., and Schlegel, H.G. (editors), *The Prokaryotes*, Vol. I, Springer-Verlag, Berlin, Heidelberg, New York.

Jørgensen, B.B. 1982. Ecology of the bacteria of the sulphur cycle with special reference to anoxic-oxic interface environments. *Philosophical Transactions of the Royal Society of London* B. 298:543–561.

Jørgensen, B.B. and **Revsbech, N.P.** 1983. Colorless sulfur bacteria, *Beggiatoa* spp. and *Thiovulum* spp. in O_2 and H_2S microgradients. *Applied and Environmental Microbiology* 45:1261–1270.

Keil, F. 1912. Beiträge zur Physiologie der farblosen Schwefelbakterien. *Beiträge zur Biologie der Pflanzen (Breslau)* 11:335–372.

Kelly, D.P. 1982. Biochemistry of chemolithotrophic oxidation of inorganic sulphur. *Philosophical Transactions of the Royal Society of London* B 298:499–528.

Kuenen, J.G. and **Beudeker, R.F.** 1982. Microbiology of thiobacilli and other sulphur-oxidizing autotrophs, mixotrophs, and heterotrophs. *Philosophical Transactions of the Royal Society of London* B 298:473–497.

Larkin, J.M. and **Shinabarger, D.L.** 1983. Characterization of *Thiothrix nivea. International Journal of Systematic Bacteriology* 33:841–846.
Larkin, J.M. and **Strohl, W.R.** 1983. *Beggiatoa, Thiothrix,* and *Thioploca. Annual Review of Microbiology* 37:341–367.
Maier, S. 1984. Description of *Thioploca ingrica* sp. nov., nom. rev. *International Journal of Systematic Bacteriology* 34:344–345.
Maier, S. and **Gallardo, V.A.** 1984a. *Thioploca aieraucae* sp. nov. and *Thioploca chileae* sp. nov. *International Journal of Systematic Bacteriology* 34:414–418.
Maier, S. and **Gallardo, V.A.** 1984b. Nutritional characteristics of two marine thioplocas determined by autoradiography. *Archives of Microbiology* 139:218–220.
Mezzino, M.J., Strohl, W.R., and **Larkin, J.M.** 1984. Characterization of *Beggiatoa alba. Archives of Microbiology* 137:139–144.
Møller, M.M., Nielsen, L.P., and **Jørgensen, B.B.** 1955. Oxygen responses and mat formation by *Beggiatoa. Applied and Environmental Microbiology* 50:373–382.
Nelson, D.C. and **Castenholz, R.W.** 1981a. Use of reduced sulfur compounds by *Beggiatoa* sp. *Journal of Bacteriology* 147:140–154.
Nelson, D.C. and **Castenholz, R.W.** 1981b. Organic nutrition of *Beggiatoa* sp. *Journal of Bacteriology* 147:236–247.
Nelson, D.C., Waterbury, J.B., and **Jannasch, H.W.** 1982. Nitrogen fixation and nitrate utilization by marine and freshwater *Beggiatoa. Archives of Microbiology* 133:172–177.
Nelson, D.C. and **Jannasch, H.W.** 1983. Chemoautotrophic growth of a marine *Beggiatoa* in sulfide-gradient cultures. *Archives of Microbiology* 136:262–269.
Nelson, D.C., Revsbech, N.P., and **Jørgensen, B.B.** 1986a. The microoxic/anoxic niche of *Beggiatoa* spp.: microelectrode survey of marine and freshwater strains. *Applied and Environmental Microbiology* 52:161–168.
Nelson, D.C., Jørgensen, B.B., and **Revsbech, N.P.** 1986b. Growth pattern and yield of a chemoautotrophic *Beggiatoa* sp. in oxygen-sulfide microgradients. *Applied and Environmental Microbiology* 52:225–233.
Prince, R.C., Stokley, K.E., Haith, C.E., and **Jannasch, H.W.** 1988. The cytochromes of a marine *Beggiatoa. Archives of Microbiology* 150:193–196.
Pringsheim, E.G. 1964. Heterotrophism and species concepts in *Beggiatoa. American Journal of Botany* 51:898–913.
Pringsheim, E.G. 1967. Die Mixotrophie von *Beggiatoa. Archiv für Mikrobiologie* 59:247–254.
Rosenberg, R., Arntz, W.E., de Flores, E.C., Flores, L.A., Carbajal, G., Finger, I., and **Tarazona, J.** 1983. Benthos biomass and oxygen deficiency in the upwelling system off Peru. *Journal of Marine Research* 41:263–279.
Schmidt, T.M., Arieli, B., Cohen, Y., Padan, E., and **Strohl, W.R.** 1987. Sulfur metabolism in *Beggiatoa alba. Journal of Bacteriology* 169:5466–5472.
Scotten, H.L. and **Stokes, J.L.** 1962. Isolation and properties of *Beggiatoa. Archiv für Mikrobiologie* 42:353–368.
Shively, J.M., Devore, W., Stratford, L., Porter, L., Medlin, L., and **Stevens, S.E., Jr.** 1986. Molecular evolution of the large subunit of ribulose-1,5-bisphosphate carboxylase/oxygenase (RuBisCO). *FEMS Microbiology Letters* 37:251–257.
Skerman, V.B.D., McGowan, V., and **Sneath, P.H.A.** 1980. Approved lists of bacterial names. *International Journal of Systematic Bacteriology* 30:225–420.
Stahl, D.A., Lane, D.J., Olsen, G.J., Heller, D.J., Schmidt, T.M., and **Pace, N.R.** 1987. Phylogenetic analysis of certain sulfide-oxidizing and related morphologically conspicuous bacteria by 5S ribosomal ribonucleic acid sequences. *International Journal of Systematic Bacteriology* 37:116–122.
Strohl, W.R. and **Larkin, J.M.** 1978. Enumeration, isolation, and characterization of *Beggiatoa* from freshwater sediments. *Applied and Environmental Microbiology* 36:755–770.
Strohl, W.R., Cannon, G.C., Shively, J.M., Güde, H., Hook, L.A., Lane, C.M., and **Larkin, J.M.** 1981a. Heterotrophic carbon metabolism by *Beggiatoa alba. Journal of Bacteriology* 148:572–583.

Strohl, W.R., Geffers, I., and **Larkin, J.M.** 1981b. Structure of the sulfur inclusion envelopes from four beggiatoas. *Current Microbiology* 6:75–79.

Strohl, W.R. and **Schmidt, T.M.** 1984. Mixotrophy of colorless, sulfide-oxidizing gliding bacteria *Beggiatoa* and *Thiothrix.* pp. 79–95 in Strohl, W.R. and Tuovinen, O.H. (editors), *Microbial Chemoautotrophy,* Ohio State University Press, Columbus.

Strohl, W.R., Schmidt, T.M., Vinci, V.A., and **Larkin, J.M.** 1986. Electron transport and respiration in *Beggiatoa* and *Vitreoscilla. Archives of Microbiology* 145:71–75.

Vargas, A. and **Strohl, W.R.** 1985a. Ammonia assimilation and metabolism by *Beggiatoa alba. Archives of Microbiology* 142:275–278.

Vargas, A. and **Strohl, W.R.** 1985b. Utilization of nitrate by *Beggiatoa alba. Archives of Microbiology* 142:279–284.

Williams, T.M. and **Unz, R.F.** 1985. Filamentous sulfur bacteria of activated sludge: Characterization of *Thiothrix, Beggiatoa* and Eikelboom type 021N strains. *Applied and Environmental Microbiology* 49:887–898.

Williams, T.M., Unz, R.F., and **Doman, J.T.** 1987. Ultrastructure of *Thiothrix* spp. and "Type 021N" bacteria. *Applied and Environmental Microbiology* 53:1560–1570.

Williams, T.M. and **Unz, R.F.** 1989. The nutrition of *Thiothrix,* type 021N, *Beggiatoa,* and *Leucothrix* strains. *Water Research,* (in press).

Winogradsky, S. 1887. Über Schwefelbacterien. *Botanische Zeitung* 45:489–507, 513–523, 529–539, 545–559, 569–576, 585–594, 606–610.

Woese, C.R., Weisburg, W., Hahn, C.M., Paster, B., Zablen, L.B., Lewis, B.J., Macke, T.J., Ludwig, W., and **Stackebrandt, E.** 1985. The phylogeny of purple bacteria: the gamma subdivision. *Systematic and Applied Microbiology* 6:25–33.

13

Biochemistry of the Nitrifying Lithoautotrophic Bacteria

Alan B. Hooper

Department of Genetics and Cell Biology
University of Minnesota
St. Paul, Minnesota, U.S.A.

The taxonomy, ecology and biogeochemistry of the nitrifying chemolithotrophic bacteria (Nitrobacteraceae) are discussed in Chapter 5. The focus of the present chapter is the biochemistry of the energy-conserving pathways of these bacteria. The Nitrobacteraceae include the genera *Nitrosomonas*, which is able to grow autotrophically on ammonium as sole energy source, and *Nitrobacter*, which is able to grow on nitrite. The use of these energy sources is the subject of an excellent review by Wood (1986). The mechanism of oxidation of nitrite by *Nitrobacter* has been most recently reviewed by Aleem and Sewell (1984).

13.1 NITROGEN TRANSFORMATIONS OF *NITROSOMONAS*

Overall reactions of nitrogen oxidation The reactions of nitrogen oxidation thought to occur in *Nitrosomonas* are shown below:

$$2H^+ + NH_3 + 2e^- + O_2 \rightarrow NH_2OH + H_2O \qquad (1)$$
Ammonia monooxygenase (AMO)

$$NH_2OH + H_2O \rightarrow HONO + 4e^- + 4H^+ \qquad (2)$$
Hydroxylamine oxidoreductase (HAO)

$$2H^+ + \tfrac{1}{2}O_2 + 2e^- \rightarrow H_2O \qquad \text{Terminal oxidase} \qquad (3)$$

$$NH_3 + 1\tfrac{1}{2}O_2 \rightarrow HONO + H_2O \qquad \text{Summation} \qquad (4)$$

Ammonia is oxidized by a mixed-function *ammonia monooxygenase* (AMO) to hydroxylamine (Equation 1). Hydroxylamine is oxidized to nitrite (Equation 2) by *hydroxylamine oxidoreductase* (HAO), which is essentially a dehydrogenase. This enzyme is also able to oxidize hydrazine to dinitrogen (a reaction which is presumed to be of no significance in nature). HAO removes four electrons from substrate. Subsequently two electrons are returned to the AMO reaction and two electrons pass to the terminal oxidase (Equation 3) thereby completing the overall nitrogen-oxidizing sequence (Equation 4). We will now consider the evidence for these reactions.

Ammonia to hydroxylamine Hydroxylamine is as good a substrate for aerobic nitrite production by *Nitrosomonas* as ammonia (Lees, 1952) and cell extracts catalyze the active oxidation of hydroxylamine to nitrite in the presence of substrate amounts of an artificial electron acceptor (Nicholas and Jones, 1960). In the presence of hydrazine, which inhibits the oxidation of hydroxylamine to nitrite, cells of *Nitrosomonas* produce small amounts of hydroxylamine (Hofman and Lees, 1953). These observations suggest that hydroxylamine is an intermediate in the oxidation of hydroxylamine to nitrite. Dioxygen (O_2), rather than water, is the source of the oxygen of hydroxylamine, i.e., the reaction is catalyzed by an oxygenase. Hydroxylamine can be trapped as an oxime (Dua et al., 1979). To show that oxygen was the substrate for AMO, Hollocher et al. (1981) trapped, as the $^{15}N,^{18}O$-cyclohexanone oxime ($>C{=}N{-}OH$), the hydroxylamine produced from $^{15}NH_3$ in the presence of hydrazine, $^{18}O_2$, and $H_2{}^{16}O$. The oxime was more than 90 percent enriched with ^{18}O, indicating convincingly that the oxygen in the hydroxylamine produced under these conditions had come from dioxygen. Therefore, unless the production of hydroxylamine under these conditions is by a dead-end side reaction, the "first" oxygen incorporated into nitrite is from dioxygen.

Because the ammonia oxygenase enzyme has not been isolated, there has not yet been a direct demonstration that it is a water-producing monooxygenase (Equation 1) and not a dioxygenase, where both atoms of oxygen are incorporated into a nitrogen compound (two atoms in two molecules of ammonia or one atom in ammonia and one atom in HNO, for example; see Hooper, 1984). Nevertheless, the best evidence suggests that the reaction is a monooxygenase: Cell-free systems or whole cells utilize substrate amounts of a reductant such as ferrous cytochrome, hydrazine, or tetramethylhydroquinone for the oxidation of ammonia or other AMO substrates (Suzuki et al., 1981; Tsang and Suzuki, 1982; Shears and Wood, 1986).

Hydroxylamine to nitrite Demonstration that water rather than dioxygen was the source of the "second-added" oxygen in the oxidation of ammonia to nitrite was difficult. Andersson et al. (1982) took advantage of the ^{15}N isotope-shift induced by ^{18}O in the NMR spectrum of nitrite to analyze the oxygen isotope composition of $^{15}NO_2^-$ produced by cells from $^{15}NH_3$ in the presence of $H_2{}^{18}O$ and $^{16}O_2$. The species $^{15}N^{16,16}O_2^-$, $^{15}N^{16,18}O_2^-$, and $^{15}N^{18,18}O_2^-$ each give a separate NMR peak. The height of each peak is proportional to the concentration of that species. ^{14}N species are undetectable by NMR. Unfortunately, during the oxidation of ammonia (or hydroxylamine), *Nitrosomonas* catalyzes the rapid exchange of oxygen atoms between nitrite and water:

$$HN^{18}O_2^- + H_2O \rightarrow HNO_2 + H_2{}^{18}O \tag{5}$$

Thus, regardless of the O-isotopic composition of the nitrite initially produced from ammonia or hydroxylamine, the exchange reaction quickly imposes on nitrite the isotopic composition of the solvent water. The exchange between water and product ^{15}N-nitrite was inhibited by the inclusion of an excess of competing ^{14}N-nitrite or, for unknown reasons, carboxyl cyanide *m*-chlorophenylhydrazone (CCCP). The final analysis of the data involved the prediction, based on the size of the $^{15}N^{16\cdot 16}O_2^-$ and $^{15}N^{16\cdot 18}O_2^-$ peaks and the kinetics of the oxygen exchange, that a significant $^{15}N^{18\cdot 18}O_2^-$ peak would have been observed had the latter species been the initial product of nitrite synthesis from $^{15}NH_3$. A $^{15}N^{18\cdot 18}O_2^-$ peak was not observed. Thus, assuming that the "first-added" oxygen came from dioxygen, the "second-added" oxygen must have come from water and HAO catalyzes the reaction of Equation (2) (Andersson and Hooper, 1983).

Intermediates of the oxidation states of HNO and NO N_2O, which is assumed to have arisen by the chemical decomposition of HNO:

$$2HNO \rightarrow N_2O + H_2O \tag{6}$$

is produced during oxidation of hydroxylamine by extracts (Falcone et al., 1963). Nitric oxide is produced anaerobically from hydroxylamine by cells or extracts with methylene blue as electron acceptor (Anderson, 1965). With phenazine methosulfate as electron acceptor, purified HAO catalyzes the oxidation of NH_2OH to a mixture of nitrite and nitrate (Hooper et al., 1977). This reaction, which does not occur *in vivo,* is best explained by the enzymatic oxidation of hydroxylamine to NO followed by the chemical oxidation of NO to nitrite and nitrate:

$$2NO + O_2 \rightarrow 2NO_2 \rightarrow N_2O_4 \tag{7}$$

$$N_2O_4 + H_2O \rightarrow HNO_2 + HNO_3 \tag{8}$$

Importantly, oxidation of NH_2OH catalyzed by purified HAO in the presence of $^{15}NO_2^-$ does not yield ^{15}N-nitrous or nitric oxide, indicating that these oxides are not produced by reduction of nitrite (Hooper and Terry, 1979).

***Production of nitrous and nitric oxide by cells of* Nitrosomonas** The existence of an enzyme activity from *Nitrosomonas* which catalyzes the reduction of nitrite to N_2O and NO was first reported by Hooper (1968). Leucopyocyanine or hydroxylamine could serve as electron donors. Ritchie and Nicholas (1972) showed that N_2O was produced by suspensions of intact cells and, using $^{15}NO_2^-$ and $^{14}NH_2OH$, that much of the N_2O was produced by reduction of nitrite. This conclusion is supported by a kinetic analysis of N_2O production by Poth and Focht (1985). Miller and Wood (1983) purified a soluble copper protein (molecular weight, 120,000) with cytochrome oxidase and nitrite reductase activity. This protein has both Type I and II copper (DiSpirito, et al., 1985b) and, importantly, is elevated in cells grown at low oxygen concentrations and has a high affinity for O_2 (Miller and Nicholas, 1985). It is not known whether the reduction of nitrite during the oxidation of ammonia is coupled to generation of an electrochemical gradient and ATP synthesis so as to result in energy gain for the organism. The possibility is discussed below.

At low levels of oxygen (0.5 percent), cultures of *Nitrosomonas* convert almost 10 percent of the ammonia oxidized to N_2O (Goreau et al., 1980). Under some conditions, the amount of NO produced exceeds the N_2O produced (Lipschultz et al., 1981). The concentration of N_2O in the atmosphere is increasing due to human activities: increased combustion of fossil fuels, increased cultivation of plants, and thus increased application of nitrogen fertilizers. This is of global significance because N_2O in the stratosphere undergoes photooxidation to NO which, in turn, catalyzes the decomposition of ozone. The relative contribution to global N_2O production of the heterotrophic denitrifiers and the ammonia-oxidizing nitrifying bacteria remains unknown. Anderson and Levine (1986) point out that, on a per cell basis, N_2O (or NO) production by the nitrifying bacteria can exceed production by species of denitrifying bacteria. The latter species, however, occur at much greater concentrations in soils or waters.

To summarize, N_2O and NO are produced by reduction of nitrite by the denitrifying bacteria. *In vivo* in the nitrifying bacteria they are produced by reduction of nitrite and possibly also by the direct oxidation of NH_2OH. Production may be catalyzed by HAO and/or the blue copper oxidase/nitrite reductase. Because the latter is inducible at low oxygen concentrations, it is the prime candidate for catalysis of the reduction of nitrite.

13.2 HYDROXYLAMINE OXIDOREDUCTASE

Structure of the enzyme The enzyme hydroxylamine oxidoreductase or "HAO," which catalyzes the oxidation of hydroxylamine to nitrite is a large (M_r approximately 180,000), soluble protein which is one of the most complex hemoproteins known (Hooper et al., 1978). Because of its great size and complexity, many aspects of the structure are not known with certainty. It consists of 3 α subunits (M_r 63,000) and 3 β subunits (M_r 11,000) (Terry and Hooper, 1981). Enzyme lacking the β cytochrome is fully active but less stable. HAO contains approximately 24 *c*-type hemes (protoheme bound by a thioether linkage to the protein) and approximately 3 residues of a heme-like iron macrocycle named P460 (Erickson and Hooper, 1972). P460 has been observed in nature only in the ammonia-oxidizing chemolithotrophic bacteria (Hooper and Nason, 1965); its structure has not been determined. The *c*-hemes and P460 occur in the approximate ratio of 8–11 to 1 in a twice or thrice repeated structure in HAO (see Figure 13.1). The *c*-hemes can be differentiated from each other on the basis of α-absorbancy maxima at 553 or 559 nm and by their oxidation-reduction midpoint potentials which, remarkably, range from +295 to −390 mV (Prince et al., 1983). To add to the complexity, several pairs of the metal centers interact electronically with each other as determined by their electron paramagnetic resonance (EPR) spectra (Vickery and Hooper, 1981; Lipscomb and Hooper, 1982; Prince and Hooper, 1987). That is, to varying extents they behave as if electrons are shared between them. One of these electronically "coupled" pairs is a combination of the lowest potential *c*-heme and an unknown group; a second pair involves P460 and an unknown group (possibly a *c* heme), and a third pair consists of two hemes *c*-553 (+295 mV) (Figure 13.1). A yet higher level of complexity involves allosteric-like interactions between the hemes. This is revealed by observing the EPR spectrum of the enzyme as the ferric irons are progressively reduced. To illustrate, as

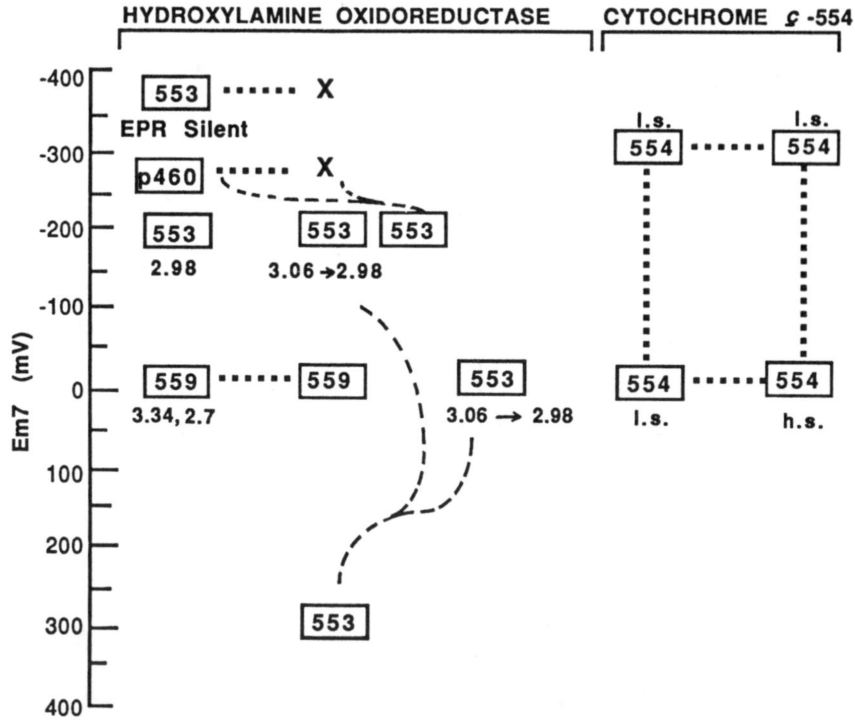

Figure 13.1 *c*-Hemes and P460 of hydroxylamine oxidoreductase and cytochrome *c*-554 of *Nitrosomonas*. Boxes indicate metal centers. EPR, electron paramagnetic resonance. EPR properties (g values) are under boxes; an arrow between g-values indicates a transition from one form to another; l.s. or h.s. indicates low- or high-spin, respectively. (▪▪▪▪▪) designates coupling between metal centers. (— — — —) indicates that reduction of the higher potential center causes change in the EPR properties of the lower potential center(s) connected by the line.

the highest potential *c*-hemes change from ferric to ferrous they cause the EPR peaks of one *c*-553 (−200 mV) and one *c*-553 (0 mV) to shift from g = 3.06 to g = 2.98; as the hemes *c* (−200 mV) are reduced, the P460 group becomes detectable by EPR at g = 6. There is no doubt that future data will call for modification of the model shown in Figure 13.1.

Mechanism of action of HAO At equilibrium or in the steady state of turnover in the presence of hydroxylamine, most of the four hemes with midpoint potentials of 0 mV or greater are reduced. If the P460 chromophore is destroyed by reaction of the oxidized enzyme with hydrogen peroxide, the *c*-hemes (which are not affected by peroxide) are no longer reducible by substrate (Hooper and Terry, 1977). P460, therefore, is considered to be a part of the active site. The function of the low potential hemes is not clear; they are presumably necessary to the activity but probably do not undergo oxidation and reduction. Electrons are thought to flow from the substrate to *c*-hemes of HAO and then, probably, through an "electron exit" site to the soluble tetraheme "cytochrome *c*-554" (Yamanaka and Shinra, 1974). The electronic interactions between the four hemes of this cytochrome are also complex (Andersson et al., 1986).

Hydroxylamine in the active site is thought to be successively dehydrogenated to NO^+:

$$NH_2OH \rightarrow NHO + 2e^- + 2H^+ \qquad (9)$$

$$NHO \rightarrow NO + e^- + H^+ \qquad (10)$$

$$NO \rightarrow NO^+ + e^- \qquad (11)$$

A ferrous nitrosyl derivative of the enzyme might then reversibly hydrolyze to form nitrous acid:

$$Fe^{II}\text{–}NO^+ + H_2O \Leftrightarrow Fe^{II} + HONO + H^+ \qquad (12)$$

The latter reaction, which is based on analogy with a model for nitrite reductase (Averill and Tiedje, 1982), could explain the rapid exchange of oxygen between nitrite and water catalyzed by cells of *Nitrosomonas*. It is, of course, also possible that the nitrite/water exchange is catalyzed by a separate nitrite reductase.

Reaction of ferrous HAO with O_2 Ferrous P460 of HAO is rapidly reoxidized by O_2 or H_2O_2 and binds CO (see Hooper and Balny, 1982). These are properties of O_2-reactive hemoproteins and supported the (incorrect) hypothesis (see Hooper, 1978, 1984) that hydroxylamine is oxidized by an oxygenase. It appears more likely that the reactivity of these compounds (and NH_2NH_2) is simply related to their similarity in structure to NH_2OH.

13.3 ELECTRON TRANSFER AND ENERGY TRANSDUCTION

Electron transfer chain Evidence in support of energy transduction coupled directly to the ammonia monooxygenase does not exist. Thus, all energy generation in *Nitrosomonas* is probably derived from the oxidation of hydroxylamine. The cell membrane of *Nitrosomonas* is extensively invaginated (Murray and Watson, 1965; Hooper et al., 1984), creating a large periplasmic compartment. Cells are a rich red color due to the high content of cytochrome. For every molecule of HAO, the periplasm contains 1.6 molecules of cytochrome *c*-554 and 10 molecules of the monoheme cytochrome *c*-552 whereas the membrane contains 50 molecules of ubiquinone-8, 14 heme equivalents of cytochrome *c*-553 and 9 molecules of cytochrome aa_3 (DiSpirito et al., 1985a; see Figure 13.2). The terminal oxidase, once thought to be a cytochrome a_1, is a cytochrome aa_3 (Yamazaki et al., 1985; DiSpirito et al., 1986). For each hydroxylamine oxidized, two electrons pass from HAO to dioxygen by a path which has not been experimentally established. As pointed out by Wood (1986), the values of oxidation-reduction potential for the ubiquinol/ubiquinone and hydroxylamine/nitrite couples (+90 and +60 mV, respectively) are close enough to allow participation of ubiquinone. The carriers may be in the fol-

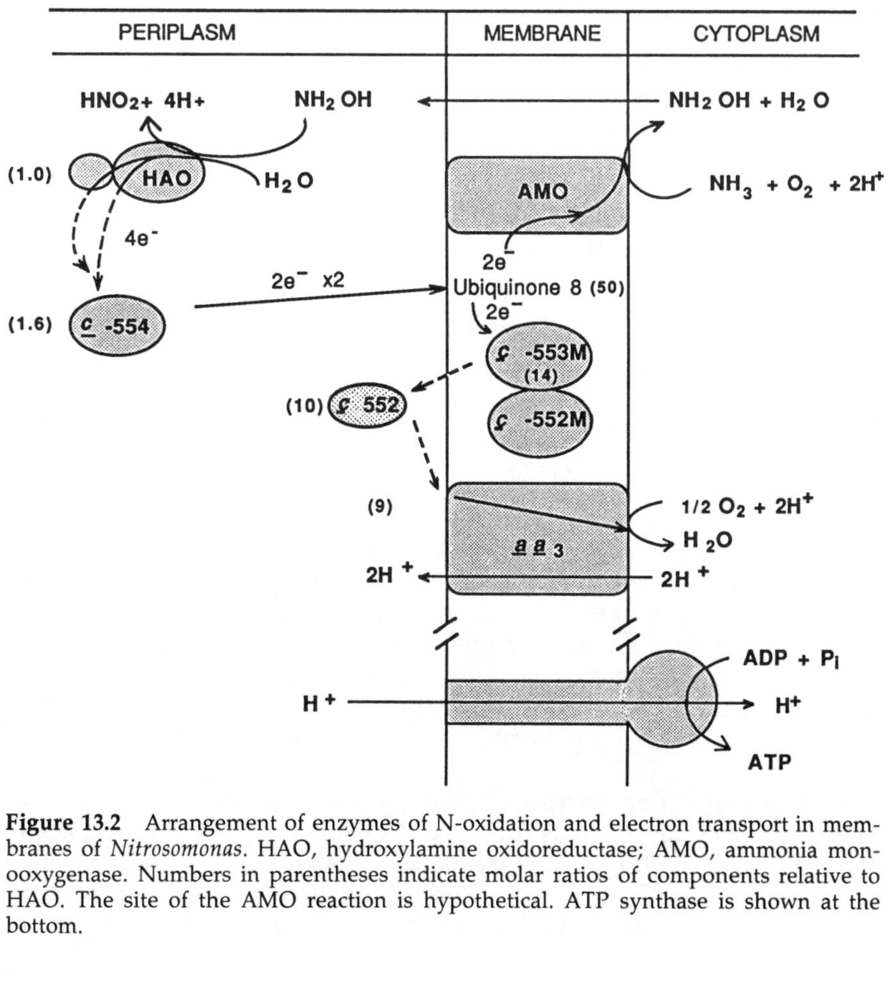

Figure 13.2 Arrangement of enzymes of N-oxidation and electron transport in membranes of *Nitrosomonas*. HAO, hydroxylamine oxidoreductase; AMO, ammonia monooxygenase. Numbers in parentheses indicate molar ratios of components relative to HAO. The site of the AMO reaction is hypothetical. ATP synthase is shown at the bottom.

lowing sequence: cytochrome *c*-554, ubiquinone, membrane cytochrome *c*-553, periplasmic cytochrome *c*-552 and the membrane cytochrome aa_3 terminal oxidase.

Generation of a proton gradient and synthesis of ATP The enzyme ATP synthase is able to phosphorylate ADP under the impetus of a greater concentration of protons in the periplasm (Kumar and Nicholas, 1982a) and protons are translocated into the periplasm coupled to oxidation of ammonia or hydroxylamine (Drozd, 1976; Hollocher et al., 1982). Although it has not been demonstrated with the enzyme from *Nitrosomonas*, by analogy with other cytochrome aa_3 terminal oxidases (Solioz et al., 1982), it is likely that two protons are pumped into the periplasm by cytochrome aa_3 for each two electrons passing through the enzyme to dioxygen (Figure 13.2). Additional periplasmic protons arise directly from the reaction of Equation (2) (Olson and Hooper, 1983). It may be that most bacteria which derive energy for growth by the oxidation of a simple reductant (such as hydroxylamine, hydrogen, water, formate, methanol, reduced sulfur compounds, trimethylamine, glucose, or iron) produce protons from oxidation of the substrate in the periplasm (Hooper and DiSpirito, 1985). In these organisms, utilization of protons in the reduction of the terminal electron acceptor (dioxygen, in the case of *Nitrosomonas*) contributes to the proton gradient.

We can only speculate on the possible contribution to the energy economy of the cell of the copper-containing nitrite reductase activity under low concentrations of dioxygen. The successive enzymatic and nonenzymatic reactions are probably:

$$HNO_2 + 2e^- + 2H^+ \rightarrow HNO + H_2O \qquad (13)$$

$$HNO \rightarrow 0.5\ H_2N_2O_2 \qquad (14)$$

$$0.5H_2N_2O_2 \rightarrow 0.5N_2O + 1.5H_2O \qquad (15)$$

$$HNO_2 + 2e^- + 2H^+ \rightarrow 0.5N_2O + 1.5H_2O \qquad (16)$$

The reaction for oxygen would be as in Equation (3). The reaction with either nitrite or dioxygen uses two protons per two electrons. If the reaction occurs in the cytoplasm, it still allows for the "substrate" utilization of protons but does not provide the transmembrane pumping of protons by the terminal oxidase. If the reaction occurs in the periplasm, the proton pumping and cytoplasmic proton utilization of the dioxygen-reducing terminal oxidase is lost. Yet, if ammonia oxidation occurs on the cytoplasmic surface of the membrane (Figure 13.2), a net yield of protons could occur in the periplasm.

Reduction of pyridine nucleotide Reduced NADP is required for reductive biosynthetic reactions such as fixation of CO_2 in *Nitrosomonas*. Aleem (1966) showed that pyridine nucleotide is reduced by "reverse electron transfer"; NAD reduction by extracts in the presence of hydroxylamine was dependent on the addition of ATP. It was assumed that electrons pass "uphill" against a gradient of oxidation-reduction potential toward pyridine nucleotide under the impetus of a proton/ionic gradient (positive inside the cell) created by hydrolysis of ATP. This electron transfer reaction must occur infrequently relative to the turnover of HAO.

13.4 AMMONIA MONOOXYGENASE

Nature of the enzyme system Based on the pH optimum for oxidation of ammonia by extracts, the preferred substrate of ammonia monooxygenase (AMO) seems to be unprotonated ammonia, NH_3 (Suzuki et al., 1974). Experiments which have been carried out sometimes with cells and sometimes with extracts show that *Nitrosomonas* is able to oxidize many substrates other than ammonia (see Wood, 1986). These reactions are listed in Table 13.1. Many substrates, including methane, carbon monoxide, and methanol, are competitive inhibitors of the oxidation of ammonia (Suzuki et al., 1976), suggesting that their oxidation is catalyzed by AMO. In no case is there evidence that oxidation of compounds other than ammonia supports growth.

Ammonia oxidation in cells is sensitive to inhibition by copper-binding agents (Lees, 1952), and the activity is destroyed by light in the wavelength range 400–430 nm (Hooper and Terry, 1974). Shears and Wood (1985) found that cells exhibit a loss of absorbancy at approximately 380 nm when photoinactivated. They noted that oxygenated tyrosinase exhibits a somewhat similar change in absorbancy when exposed to UV light and proposed that AMO might have a similar structure. The tyrosinase-like oxidation of ammonia which they hypothesized would involve the two-electron reduction of dioxygen by a dicuprous ion center, loss of one oxygen as water, and simultaneous reaction of the other oxygen with ammonia. A cycle would be completed by the reduction of the cupric centers of the enzyme by two electrons which originate from hydroxylamine.

Acetylene apparently inhibits the oxidation of ammonia (Hynes and Knowles, 1978) by acting as a suicide substrate (Hyman and Wood, 1985). The latter authors labeled a 28 kilodalton membrane protein by incubation with ^{14}C-acetylene and suggested that this protein is a component of AMO. Suzuki

Table 13.1 Alternate reactions catalyzed by ammonia monooxygenase of *Nitrosomonas*

Substrate	Product	Reference
CO	CO_2	Tsang and Suzuki, 1982
CH_3	CH_2OH	Jones and Morita, 1983 Hyman and Wood, 1983
CH_2OH	CH_2O	Wood, 1986
Propene	Epoxypropane	Drozd, 1980
Cyclohexane	Cyclohexanol	Drozd, 1980
Benzene	Phenol	Drozd, 1980
Phenol	Hydroquinone	Hyman et al., 1985
Ethylene	Ethylene oxide	Hyman and Wood, 1984a
Bromoethane	Formaldehyde	Hyman and Wood, 1984b

and Kwok (1981) had fractionated the cell-free system which catalyzes the oxidation of ammonia to nitrite into a membrane component, HAO, and a soluble *c*-cytochrome thought to be cytochrome *c*-554. The path of electrons from HAO (or cytochrome *c*-554) to membrane AMO is not known. In Figure 13.2, ubiquinone is shown as intermediate. It is not known whether the proton-utilizing oxidation of ammonia occurs on the cytoplasmic or periplasmic side of the membrane. If it is cytoplasmic as depicted in Figure 13.2, the cell has the advantage of utilizing cytoplasmic protons.

Similarity to methane-oxidizing bacteria Ammonia and methane oxidizers have extensive membrane invaginations (Davies and Whittenbury, 1970) and similar fatty acid composition (Blumer et al., 1969; Smith et al., 1970; Makula, 1978). They are both able to oxidize a broad spectrum of compounds including ammonia and methane (O'Neil and Wilkinson, 1977; Burrows et al., 1984). Both systems are sensitive to a broad range of inhibitors (Hooper and Terry, 1973; Hubley et al., 1975). Methanotrophs produce either a membrane or soluble form of methane monooxygenase when grown on high or low copper concentrations, respectively (Stanley et al., 1983; Burrows et al., 1984). The AMO of *Nitrosomonas* shares with the membrane form of methane monooxygenase of methylotrophs the cellular location, sensitivity to inhibitors, and the use of an electron donor other than NADH.

13.5 ASSIMILATION OF CARBON AND NITROGEN

Although acetate and amino acids are used by *Nitrosomonas*, most carbon is assimilated as carbon dioxide by the ribulose bisphosphate (RuBP) carboxylase pathway (Nicholas, 1978; Chapter 19). A lesser amount is assimilated by the phosphoenolpyruvate carboxylase pathway (Rao and Nicholas, 1966; Chapter 3). Carbon dioxide (rather than carbonate or bicarbonate) can be a limiting nutrient as suggested by the inducibility of carbonic anhydrase at low CO_2 concentrations (Jahnke et al., 1984). The presence of carbonic anhydrase and CO_2-concentrating mechanisms are, in fact, indicators of the importance of this nutrient in autotrophic bacteria (Aizawa and Miyachi, 1986). It is possible that autotrophy in *Nitrosomonas* is obligatory because cells lack α-ketoglutarate dehydrogenase (Hooper, 1969). This conclusion is debated by Williams and Watson (1968) who found enzyme activity in *Nitrosocystis*. Smith et al. (1967) first suggested that the absence of α-ketoglutarate dehydrogenase might be a general phenomenon with obligate autotrophs; the catabolism of carbon compounds by the citric acid cycle is prevented and yet enough of the cycle is present to allow the necessary shuffling of carbon skeletons for biosynthesis. Polyphosphate is accumulated as a phosphate storage compound by *Nitrosomonas* (Terry and Hoooper, 1970).

The nitrogen atom of ammonia, hydroxylamine, or nitrite, but not nitrate, is incorporated into cell material (Wallace and Nicholas, 1968). Assimilation of ammonia may be through a very active glutamate dehydrogenase (Hooper et al., 1967).

13.6 *NITROBACTER*

The substrate-oxidizing and energy-transducing system of the ammonia oxidizers is apparently incompatible with having, simultaneously, an NADH oxidase system. In contrast, the nitrite oxidizers are able to grow using energy from either the oxidation of nitrite or organic substrates such as pyruvate, formate, glycerol, or acetate (Bock, 1976; Chapter 5). With organic substrates, the growth is slow and follows a lengthy period of adaptation. Interestingly, the nitrite-oxidizing capability is retained even after adaptation to heterotrophic growth.

13.7 THE NITRITE-OXIDIZING SYSTEM

Nature of the nitrite-oxidizing reaction Overall, *Nitrobacter* catalyzes the reaction:

$$HNO_2 + O_2 \rightarrow HNO_3 + H_2O \qquad (17)$$

The oxidation of nitrite involves the net extraction of two electrons and two protons from nitrite and water:

$$HNO_2 + H_2O \rightarrow HNO_3 + 2e^- + 2H^+ \qquad (18)$$

In keeping with Equation (18), Aleem et al., 1965 (and later Kumar et al., 1983) demonstrated that ^{18}O was incorporated into nitrate from H_2O rather than O_2, and Kiesow (1967) showed that extracts of *Nitrobacter* catalyzed the stoichiometric reduction of ferricyanide by nitrite. The enzyme catalyzing this reaction is called "nitrite oxidase" or, more appropriately, "nitrite dehydrogenase." Electrons pass from nitrite by an electron transport chain to a terminal oxidase (Equation 3).

The components of the nitrite-oxidizing system Nitrite dehydrogenase and the terminal oxidase are in membranes of *Nitrobacter* (Figure 13.3). Membranes contain a cytochrome aa_3-type cytochrome oxidase (Yamanaka et al., 1981). The redox potential of the hemes of this terminal oxidase are +250 and +400 mV (Sewell et al., 1972). The membranes also contain two heme *a*-containing cytochromes a_1 with midpoint potentials of +352 and +100 mV and a cytochrome *o*. Membranes contain at least two *c*-cytochromes distinguishable by their different redox potentials (Ingledew et al., 1974). A soluble cytochrome *c*-550 has been purified and sequenced (Yamanaka et al., 1982). The electron paramagnetic resonance spectrum of membranes of *Nitrobacter* (Ingledew and Halling, 1976) exhibits a molybdenum-like signal (midpoint potential +340

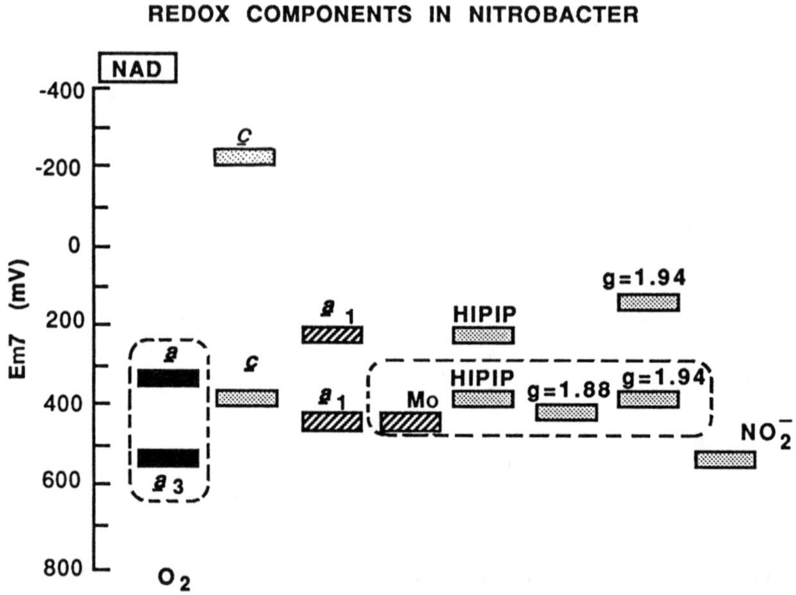

Figure 13.3 Redox components of the nitrite-oxidizing system of *Nitrobacter*. Adapted from Ingledew and Halling (1976) and Sewell et al. (1972). HIPIP, high potential iron protein. g values are from EPR measurements and indicate iron centers. *a* and *c* are hemes.

mV) and signals resembling the iron sulfur centers of ferredoxin (midpoint potentials +60 and +320 mV) and HiPIP (midpoint potentials +280 and +130 mV).

The nature of the nitrite dehydrogenase Study of the biochemistry of nitrite dehydrogenase continues to be limited by the fact that only a small amount (approximately 4 percent) of the initial activity is recovered as purified enzyme and the specific activity of the purified fraction is only 3–10 times greater than the intact cells. Thus, we cannot ompletely rule out the possibility that apparent heme or metal components or apparent component activities such as nitrate reductase are actually contaminants of the true nitrite dehydrogenase.

Nitrite dehydrogenase activity is measured as the oxidation of nitrite in the presence of substrate quantities of an electron acceptor such as mammalian cytochrome *c*. The enzyme samples which have been isolated have nitrate reductase activity as well as nitrite dehydrogenase (Tanaka et al., 1983; Sundermeyer-Klinger et al., 1984) suggesting that the reaction is reversible. The preparation from the Yamanaka laboratory was stimulated by a low-molecular-weight factor. An interesting preparation isolated by the group (Fukuoka et al., 1987) contains the following components in the molar ratios indicated: 1 heme *a*; 4–5 nonheme iron; 2–5 acid labile sulfur; 1–2 Mo; 1 heme *c*. A molybdenum-containing component was separated from the enzyme and shown to activate apo-nitrate reductase of *N. crassa*. This indicates that the preparation contains a Mo-pterin-like component which would likely resemble "bactopterin" found in CO dehydrogenase (Krüger and Meyer, 1986; see Chapter 18).

In nitrate reductases which have been studied in other species, the molybdenum-pterin cofactor is the nitrate-binding component. By analogy, it is logical to imagine that the putative bactopterin is the nitrite-binding moiety of nitrite dehydrogenase of *Nitrobacter*. The reverse reaction could then involve a mechanism similar to the nitrate reductase enzyme. A model chemical mechanism for the reduction of nitrate by a Mo^{V} center involves the binding of nitrate to Mo, reduction of the nitrate and formation of a Mo-O intermediate (Wieghardt et al., 1985). It is perhaps by a similar mechanism that cells of *Nitrobacter* catalyze the transfer of oxygen from one nitrate to another:

$$^{14}N^{18}O_3^- + {}^{15}N^{16}O_3^- \rightarrow {}^{14}N^{16}O_3^- + {}^{15}N^{18}O_3^- \quad (19)$$

without exchange with water at a corresponding rate:

$$^{15}N^{18}O_3^- + H_2O \rightarrow {}^{15}N^{16}O^- + H_2{}^{18}O \quad (20)$$

(DiSpirito and Hooper, 1986; Friedman et al., 1986). Thus an oxygenated intermediate appears to be involved in the nitrite dehydrogenase reaction:

$$HNO_3 + E \rightleftharpoons HNO_2 + E{-}O \quad (21)$$

It should be noted that, although it is likely that nitrite dehydrogenase catalyzes reduction of nitrate and the exchange of oxygen between molecules of nitrate, workers in the field have not completely ruled out the possibility that one or both of these activities are, in fact, catalyzed by a separate nitrate reductase enzyme in *Nitrobacter* and thus have no bearing on the mechanism of nitrite oxidation.

Because it is well studied, we look to nitrate reductase for precedent regarding the mechanism of nitrite oxidation by nitrite dehydrogenase of *Nitrobacter*. Yet, interestingly, if there is a direct evolutionary relationship between nitrite dehydrogenase of *Nitrobacter* and the ubiquitous nitrate reductase found in many species, the latter is likely to have evolved from the former. It is unlikely that the evolutionary predecessor of *Nitrobacter* evolved to occupy the aerobic nitrite-containing niche by the conversion of a nitrate reductase into a nitrate dehydrogenase. The primitive earth atmosphere was anoxic but contained ammonia and this ammonia could have supported the demand of early life forms for nitrogen. Only after the advent of oxygenic photosynthesis and the resulting aerobic atmosphere could the nitrifiers have evolved the oxygen-utilizing ammonia- and nitrite-oxidizing systems. Moreover, only after the advent of the nitrifiers would there have been enough of a decrease in ammonia and increase in nitrate in the environment to promote the evolution of nitrate reductase enzymes. One might look to other Mo-containing dehydrogenases such as CO or xanthine oxidase as progenitors of nitrite dehydrogenase.

13.8 ELECTRON TRANSPORT AND ATP SYNTHESIS COUPLED TO NITRITE DEHYDROGENASE

The addition of nitrite to membrane fragments of *Nitrobacter* in the presence of oxygen results in reduction of cytochrome a_1 and c within a few seconds (Kiesow, 1967). At 5 °C, reduction of cytochrome a_1 is faster than c-cytochrome, consistent with the electron transfer sequence: $a_1 \rightarrow c$. However, it should be noted that the data are also consistent with the initial passage of electrons to a c-heme which is so rapidly reoxidized that it does not come into a steady-state of reduction until its electron acceptor, heme a_1, is nearly completely reduced. A confusing aspect of the nitrite oxidizing and electron transfer system of *Nitrobacter* is that the steady-state level of reduction of heme is *increased* in the presence of oxygen or, under anaerobic conditions, in the presence of ATP. This has led to the conclusion that reduction of this cytochrome is "energy dependent."

The study of the mechanism of energy transduction in *Nitrobacter* presents us with several unusual, conflicting, and paradoxical effects and unanswered questions. We can at least say with certainty that the system is unique. The mechanism must account for the apparent energy-dependent electron transfer mentioned above as well as a series of unusual responses (discussed below) to compounds such as carbonyl cyanide phenylhydrazone (CCCP) which are uncouplers of heterotrophic or mitochondrial electron transport and oxidative phosphorylation. In Figure 13.4 and Table 13.2 are shown some of the hypothesized mechanisms for generation of a proton gradient during oxidation of nitrite by *Nitrobacter*.

***Does the oxidation of nitrite result in a proton gradient in* Nitrobacter?** Although it has not yet been demonstrated directly, it has been assumed that

Table 13.2 Models for proton translocation coupled to nitrite oxidation by *Nitrobacter*[1]

Scheme	H^+ pump in terminal oxidase[10]	Location of nitrite binding	Form of H transported	Net H^+ translocation
a[2]	–	C	2H	2
b[3]	+	C	2H, $2H^+$	4
c[4]	–	C	0	0
d[5]	+	C	$2H^+$	2
e[6]	–	C	H^-	1
f[7]	+	C	H^-, $2H^+$	3
g[8]	–	P	0	2
h[9]	+	P	$2H^+$	4

[1]The net transfer of protons per nitrite oxidized is given. All reactions involve the reduction of an atom of oxygen on the cytoplasmic surface. When nitrite oxidation occurs in the cytoplasm, nitrite is assumed to enter in exchange for an exiting nitrate.
[2,3]Suggested by Aleem, M.I.H. or Mitchell, P. (see Ferguson, 1982).
[4,5]Ferguson, 1982; Aleem, 1984.
[6]Cobley, 1976b.
[7]Ferguson, 1982, modified from Cobley, 1976.
[8,9]Modified from Hooper and DiSpirito, 1985.
[10]–, absent, +, present; C, cytoplasm; P, periplasm; ND, no data.

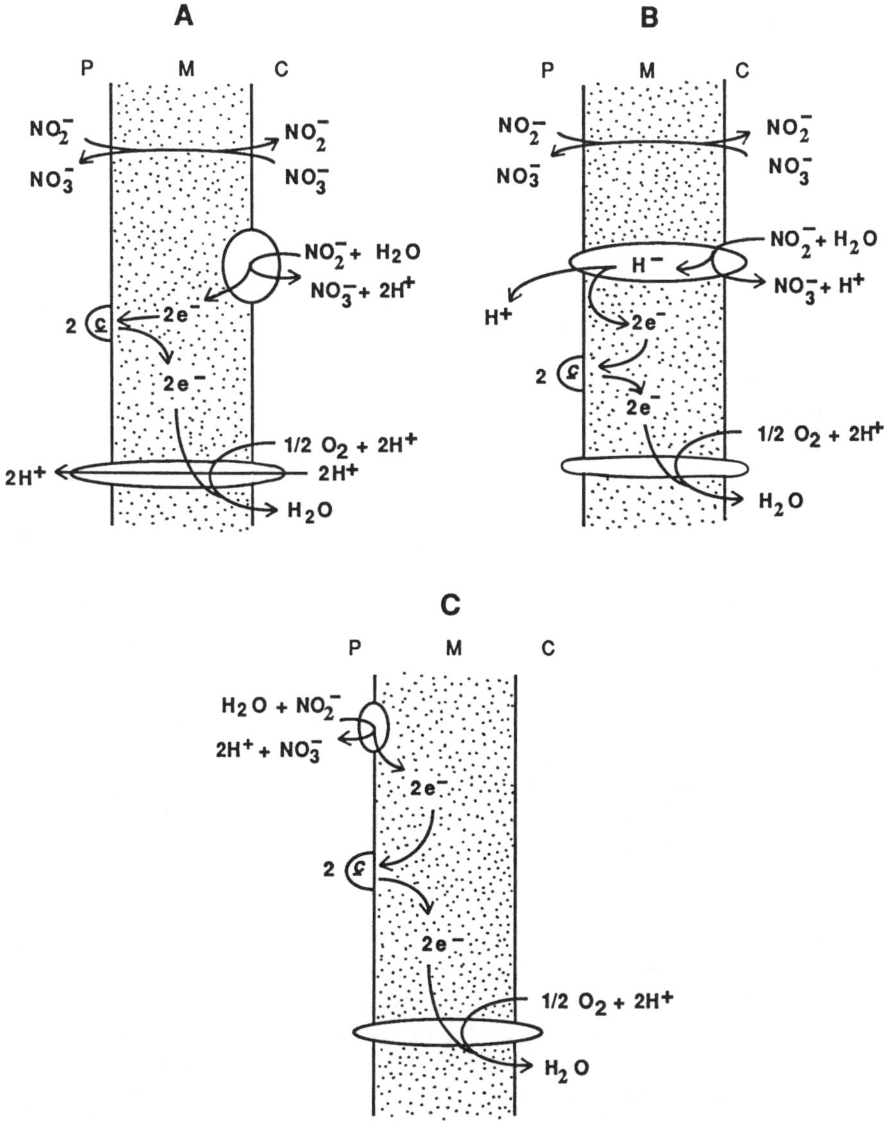

Figure 13.4 Models for generation of a proton gradient coupled to oxidation of nitrite by *Nitrobacter*. P, periplasm; M, membrane; C, cytoplasm. A) Ferguson (1982); Aleem (1984); B) Cobley, 1976; C) Modified from Hooper and DiSpirito, 1985.

phosphorylation of ADP is carried out by a membrane ATP synthase which is driven by a membrane proton- or ion-gradient. Yet some uncertainty regarding even this assumption is suggested by Hollocher and co-workers (1982), who were unable to show by the oxygen-pulse technique that proton translocation accompanied nitrite oxidation by *Nitrobacter* cells.

Does the membrane cytochrome oxidase of* Nitrobacter *pump protons? In reconstituted systems, several cytochrome aa_3 terminal oxidases of bacteria have been shown to transport two protons from cytoplasm to periplasm for each two electrons reacting with dioxygen (Solioz et al., 1982). In the case of *Nitrobacter*, this would represent the net transfer of two protons per nitrite oxidized. However, Sone and co-workers (1983) reported that reconstituted cytochrome oxidase from *Nitrobacter* did not catalyze proton pump activity. To control for possible artifacts of preparation of enzyme, they demonstrated proton pump activity with enzyme isolated by the same procedure from *Thermus thermophilus*. As these authors noted, the results suggested that 1) cytochrome oxidase from *Nitrobacter* does not pump protons or 2) that the pump activity is especially susceptible to loss during purification. Wetzstein and Ferguson (1985), on the other hand, find evidence for proton pumping during dye oxidation by intact membranes.

Is the site of nitrite oxidation (and thus proton release) on the periplasmic or cytoplasmic side of the membrane? The identification of the site of binding of substrate with a membrane enzyme is extremely difficult when the substrate is small and possibly permeable and a natural or artificial product of the reaction is not insoluble. The site of binding of nitrite has not yet been identified in *Nitrobacter*. Note that the location of the site on the cytoplasmic surface of the membrane may not require expenditure of energy for transport of nitrite if cells contain a NO_3^-(in)/NO_2^-(out) "antiport" transferase system analogous to the system thought to occur in denitrifying bacteria (Boogerd et al., 1983). If the nitrite-oxidizing site is periplasmic and is also the site of dissimilatory nitrate reductase, this molybdenum nitrate reductase would be notably different in vectoral orientation (and thus mode of synthesis) from the nitrate reductase enzymes of other bacteria, which are on the cytoplasmic surface (see Boogerd et al., 1983).

Unexpected effects of uncouplers and ADP on oxidation of nitrite in cells and vesicles of* Nitrobacter*: The hydride-transfer hypothesis As first noted by Kiesow (1967) and expanded upon by Cobley (1976a and b), nitrite oxidation assayed as the rate of utilization of dioxygen is *inhibited* by compounds which *stimulate* the corresponding oxidation of NADH. The compounds include the uncouplers 2,4-dinitrophenol (2,4 DNP), CCCP and carbonyl cyanide p-trifluoromethoxyphenylhydrazone (FCCP), and cyclohexylamine as well as oligomycin, ADP, and inorganic phosphate. These effects on activity are not pronounced when an artificial electron acceptor such as ferricyanide is used, i.e., when presumably, all or part of the electron transfer chain is bypassed. One may conclude that the effects do not result from direct binding of the agent to nitrite dehydrogenase. The early observations of this phenomenon led workers to propose that proton-motive force or hydrolysis of ATP was required to "prime" oxidation of nitrite. Somewhat elaborate schemes were then required in order to account for a net gain of energy or ATP associated with the oxidation of nitrite. It should be noted that an inevitable difficulty with these studies was that interpretation depended on knowing that the mechanism of an inhibitor or fate of a permeant compound in *Nitrobacter* was the same as in other organisms.

By analogy with other bacterial vesicles, the vesicles extracted from *Nitrobacter* are presumed to be inside out. This is an important assumption with regard to interpretation of the observations. Cobley (1976a) measured a brief decrease in proton concentration in the suspending medium during dioxygen-stimulated oxidation of nitrite by vesicles. Thus, assuming that vesicles are inside out, oxidation of nitrite results in the apparent disappearance of protons from the cytoplasm.

Both Kiesow and Cobley noted that the "paradoxically acting" compounds also affected the steady-state level of reduction of cytochromes in a manner opposite to that shown for the oxidation of NADH. Cobley (1976b) noted, in particular, that conditions which decreased the transmembrane potential led to inhibition of the rate of oxidation of nitrite or to a decrease in the steady-state level of reduction of cytochromes. Taken with the observed pH changes and the assumption that vesicles are inside out, it appears that causing the periplasm to be less protonic or more negative in charge inhibits nitrite dehydrogenase. Cobley rationalized this observation by proposing that nitrite oxidation (hypothesized to occur on the cytoplasmic surface of the membrane) is accompanied by the translocation across the membrane of the negatively charged hydride ion (Figure 13.4B)

The hydride transfer hypothesis for energy transduction in *Nitrobacter* is the most novel and interesting of the models under consideration and therefore deserves scrutiny. The hypothesis depends on the assumptions: 1) that the inhibitory effect on nitrite dehydrogenase of treatment with compounds such as ADP or CCCP is mediated by a change in the transmembrane gradient; and 2) that the nitrite dehydrogenase reaction itself responds directly to the state of the membrane gradient. Is it possible that these assumptions are not valid? It should be noted that compounds such as CCCP which, in some systems, do act as uncouplers by dissipating the membrane gradient, can also directly inhibit redox reactions. For example, CCCP acts in this way on photosystem II by catalyzing the decay of one of the redox states of the reaction center (Hanssum et al., 1985). An effect on the transmembrane gradient does not appear to be involved in the mechanism of action of these "ADRY" (*A*ccelleration of *D*eactivation *R*eactions of the water splitting enzyme systems *Y*) reagents. The compounds 2,4 DNP, CCCP, and "1799" (2,2'-bis(hexafluoroacetonyl)acetone) inhibit the oxidation of ammonia (though not the oxidation of hydroxylamine) by cells of *Nitrosomonas* (Hooper and Terry, 1973) or inhibit the reduction of nitrous oxide, nitrite, or nitrate by *Pseudomonas* species (Walter et al., 1978). The mechanism of inhibition in either case is not known. In the former case, the reaction is also inhibited by autooxidizable redox dyes such as phenazine methosulfate, suggesting that the reaction is susceptible to promotion of the decay of an important redox poise or simply to loss of electrons which come from hydroxylamine and are needed for the ammonia monooxygenase reaction (Equation 1). Evidence in the case of nitrite reductase indicated that the inhibition did not depend on the collapse of the proton gradient. With the alternate possible modes of actions of compounds commonly known to be uncouplers, we cannot rule out the possibility that some compounds such as CCCP act on nitrite dehydrogenase of *Nitrobacter* by causing the decay of a redox steady state needed for active catalysis. Although it is difficult to imagine that ADP will directly promote decay of a redox state of nitrite dehydrogenase, Ferguson (1982) has noted that a critical redox poise of the nitrite-oxidizing system might itself depend on the membrane potential which, in turn, could be influenced by the concentration of ADP. Herweijer et al. (1986) note that inhibitors of electron transfer by NADH-ubiquinone oxidoreductase decrease the effectiveness of uncouplers such as 2,4 DNP on reverse electron transport in submitochondrial particles. They interpret this observation in terms of a "collision hypothesis" involving a direct physical interaction between the ATP synthase and the NADH oxidoreductase. If an interaction of that sort occurs in *Nitrobacter*, the effect of membrane potential on the rate of nitrite oxidation could possibly have been mediated by the direct interaction of ATP synthase and nitrite dehydrogenase.

***Models for generation of proton gradient in* Nitrobacter** In Figure 13.4, three of the possible models for energy transduction in *Nitrobacter* are shown. In Table 13.2 are summarized the expected net number of protons translocated for eight possible models. The scheme of Figure 13.4A, favored by Ferguson (1982) and Aleem and Sewell (1984), nitrite dehydrogenase is a cytoplasmic surface, proton-releasing enzyme which donates electrons to a cytoplasmic terminal oxidase. This scheme results in a net proton gradient only if the terminal oxidase is proton-pumping (Table 13.2d) (Ferguson, 1982). An earlier version of this hypothesis (Table 13.2a, b) involved translocation of 2H from the cytoplasm to the periplasm by ubiquinone. Although translocating a net two or four protons per nitrite oxidized (presence or absence, respectively, of proton-pumping activity in the terminal oxidase), this model is considered unlikely to apply because of the unfavorable redox potentials of the nitrite/nitrate couple (+430 mV) and the ubiquinone couple (+90 mV) (see Wood, 1984). Figure 13.4B (see also Table 13.2e) depicts the hydride-transfer model of Cobley which results in the net transport of one proton. Ferguson has suggested that the collected observations also might support the hydride transfer mechanism together with a proton-pumping terminal oxidase (Table 13.2f), so that a net three protons are pumped. Hooper and DiSpirito (1985), noting that in nature oxidation of small molecules in proton-producing reactions frequently occurs in the periplasm, suggest that the nitrite dehydrogenase might be on the periplasmic surface of the membrane (Figure 13.4C; Table 13.2g, h). This scheme would involve net translocation of two or four protons, depending on the proton-pumping ability of the terminal oxidase. Note that, in inside-out vesicles, nitrite would have to pass across the membrane prior to reaction with an enzyme site on the inside (periplasmic) surface. Perhaps, in the studies of Cobley (1976b), it is the passage of an ionic nitrite rather than hydride anion that is inhibited by reagents which promote breakdown of the potential gradient.

***Comparison of the efficiency of* Nitrosomonas *and* Nitrobacter** As noted by Wood (1986), the oxidation-reduction potential of the two substrates ammonia and nitrite are similar (approximately +400 mV), yet approximately 2.8 times more CO_2 is fixed per ammonia oxidized to nitrite in *Nitrosomonas* than per nitrite oxidized to nitrate in *Nitrobacter* (Baas Becking and Parks, 1927). Note that two electrons per substrate enter the electron transport chain in both cases (two of the four electrons removed from hydroxylamine are used in the AMO reaction; see Equation 2). Since the ATP- and NADH-utilizing reactions of the ammonia- and nitrite-oxidizing bacteria are probably similar, the difference in CO_2-fixing ratios might reflect a difference in the extent of proton gradient generated per substrate oxidized. Depending on whether the site of utilization of protons during oxidation of ammonia is cytoplasmic or periplasmic, four or six protons might be translocated per ammonia oxidized. If that is the case, the differences in CO_2-fixing ratios would suggest it is unlikely that a net three or four protons are translocated per nitrite oxidized by *Nitrobacter* (Table 13.2a, f or h).

To summarize, the mechanism of energy transduction in *Nitrobacter* remains a fascinating mystery. For reasons noted above, one of the three schemes shown in Figure 13.4 seems to be the most likely possibility.

***NADH-related electron transfer in* Nitrobacter** In contrast to *Nitrosomonas*, and in keeping with its ability to grow heterotrophically with acetate, formate, glycerol, or pyruvate as substrate, *Nitrobacter* has an active NADH oxidase system:

$$NADH + H^+ + \tfrac{1}{2}O_2 \rightarrow NAD^+ + H_2O \qquad (22)$$

As noted above, this system is apparently coupled to phosphorylation of ADP and is *stimulated* by uncouplers such as 2,4 DNP or CCCP (see Kiesow, 1967; Aleem and Sewell, 1981; Aleem, 1984). A CO-sensitive cytochrome aa_3 is the terminal oxidase for this system.

NADH oxidation also occurs with nitrate as terminal electron acceptor:

$$NADH + H^+ + NO_3^- \rightarrow NAD^+ + NO_2^- + H_2O \qquad (23)$$

(see Aleem and Sewell, 1981, 1984).

The last reaction in this electron-transfer chain is a nitrate reductase of the "dissimilatory" type. As noted above, it is not known whether this reaction is catalyzed by the enzyme nitrite dehydrogenase. This nitrate reductase activity can apparently support growth: Freitag and co-workers (1987) have observed growth of *Nitrobacter* in biofilms under nearly anaerobic conditions with either glycerol or a mixture of nitrite and pyruvate as reductant and nitrate as electron acceptor. Nitrite, ammonia, nitrous oxide, and other volatile nitrogen compounds were produced. The authors proposed that the culture contained two microenvironments and populations of cells: aerobic zones containing chemolithotrophic cells (which oxidized nitrite) and anaerobic zones containing heterotrophic cells (which reduced nitrate).

Formate oxidase The oxidation of formate:

$$HCOOH + \frac{1}{2}O_2 \rightarrow H_2O + CO_2 \quad (24)$$

also supports growth. Workers have wondered whether the nitrite and formate dehydrogenase enzymes are identical. Van Gool and Laudelot (1967) found that the electron transport components of the two partially purified oxidase systems seemed to be the same. Yet O'Kelley and Nason (1970) reported that, although the ratio of the two systems was similar in several strains of *Nitrobacter*, the formate oxidase system was inhibited more easily by metal binding agents than the nitrite oxidase. Cobley (1984), noting the chemical similarity between formate and nitrite, reported that properties of the vesicle-catalyzed oxidation of nitrite or formate with O_2 as electron acceptor were similar and differed from corresponding properties of NADH oxidation. The studies reported thus far involve formate oxidation coupled to an electron transport chain and terminal oxidase. The similarities in the two systems could therefore indicate either identity of substrate-binding site or a shared portion of the electron transport chain. The question of whether formate and nitrite dehydrogenase are the same enzyme might be resolved by studies with the purified enzymes.

Reduction of pyridine nucleotide during chemolithotrophic growth The work of Kiesow (1963) and Aleem and co-workers (see Aleem, 1984) showed that NAD or NADP was reduced by electrons coming from nitrite as long as hydrolysis of ATP occurs, that is, by reverse electron transport.

13.9 ASSIMILATION OF CARBON AND NITROGEN

This topic was reviewed by Nicholas (1978). Carbon is assimilated by the RuBP carboxylase pathway (Malavolta et al., 1960; see Chapter 19). The enzyme RuBP carboxylase is found in polyhedral particles called "carboxysomes" (Bock, 1978). *Nitrobacter* readily incorporates compounds such as acetate into cell carbon. The labeling pattern of citric acid cycle intermediates following assimilation of ^{14}C-acetate indicates that α-ketoglutarate dehydrogenase is active in heterotrophically grown cells but not active in autotrophically grown cells (Weining and Nicholas, 1986). This is in keeping with the hypothesis (Smith et al., 1967) that the presence of α-ketoglutarate dehydrogenase is incompatible with a carbon economy based on fixation of CO_2 by the energy-requiring RuBP carboxylase pathway. *Nitrobacter* accumulates the reserve polymers poly-β-hydroxybutyrate, glycogen, and polyphosphate (Bock and Heinrich, 1969; Van Gool et al., 1971).

Nitrogen as nitrate, nitrite, hydroxylamine, or ammonia can be assimilated by *Nitrobacter* (Wallace and Nicholas, 1968). The assimilatory nitrate reductase appears to be soluble and separable from the nitrite dehydrogenase enzyme (Faull et al., 1969). Cells contain glutamate dehydrogenase, glutamine synthetase, and low levels of glutamate synthase. Glutamate dehydrogenase appears to participate in the primary path of assimilation of nitrogen (Kumar and Nicholas, 1982b).

How many nitrite or formate oxidoreductases are there? There may be as many as three separate enzymes and yet it appears possible that a single enzyme catalyzes nitrite dehydrogenase, formate dehydrogenase, and "dissimilatory" nitrate reduction. A separate, soluble, enzyme appears to catalyze assimilatory nitrate reductase activity. Chlorate is apparently reduced by one of the nitrate reductase enzymes. The product, chlorite, is an inhibitor of the oxidation of ammonia by *Nitrosomonas* or nitrite by *Nitrobacter* (Hynes and Knowles, 1983).

Work on this manuscript was supported by research grants USDA-86-CRCR-1-2044 from the U.S. Department of Agriculture and NSF/DMA-8308766 from the U.S. National Science Foundation.

REFERENCES

Aizawa, K. and **Miyachi, S.** 1986. Carbonic anhydrase and CO_2 concentrating mechanisms in microalgae and cyanobacteria. *FEMS Microbiology Reviews* 39:215–233.

Aleem, M.I.H. 1966. Generation of reducing power in chemosynthesis. II. Energy linked reduction of pyridine nucleotides in the chemoautotroph *Nitrosomonas europaea. Biochimica Biophysica Acta* 113:216–224.

Aleem, M.I.H., Hoch, G.E. and **Varner, J.E.** 1965. Water as the source of oxidizing and reducing power in bacterial chemosynthesis. *Proceedings of the National Academy of Sciences USA* 54:869–873.

Aleem, M.I.H. and **Sewell, D.L.** 1981. Mechanism of nitrite oxidation and oxidoreductase systems in *Nitrobacter agilis. Current Microbiology* 5:267–272.

Aleem, M.I.H. and **Sewell, D.L.** 1984. Oxidoreductase systems in *Nitrobacter agilis*, pp. 185–210, in Strohl, W.R. and Tuovinen, O.H. (editors), *Microbial Chemoautotrophy*, Ohio State University Press, Columbus.

Anderson, I.C. and **Levine, J.S.** 1986. Relative rates of nitric oxide and nitrous oxide production by nitrifiers, denitrifiers, and nitrate respirers. *Applied and Environmental Microbiology* 51:938–945.

Anderson, J.H. 1965. Estimation of the nitric oxide formed from hydroxylamine by *Nitrosomonas. Biochemical Journal* 94:236–239.

Andersson, K.K. and **Hooper, A.B.** 1983. O_2 and H_2O are each the source of one O in NO_2^-:^{15}N-NMR evidence. *FEBS Letters* 164:236–240.

Andersson, K.K., Philson, S.B., and **Hooper, A.B.** 1982. ^{18}O isotope shift in ^{15}N NMR analysis of biological N-oxidations: H_2O-NO_2-exchange in the ammonia-oxidizing bacterium *Nitrosomonas. Proceedings of the National Academy of Science USA* 79:5871–5875.

Andersson, K.K., Lipscomb, J.D., Valentine, M., Munck, E., and **Hooper, A.B.** 1986. Tetraheme cytochrome *c*-554 from *Nitrosomonas europaea*. Heme-heme interactions and ligand binding. *Journal of Biological Chemistry* 261:1126–1138.

Averill, B.A. and **Tiedje, J.M.** 1982. The chemical mechanism of microbial denitrification. *FEBS Letters* 138:8–12.

Baas Becking, L.G.M. and **Parks, G.S.** 1927. Energy relations in the metabolism of autotrophic bacteria. *Physiological Reviews* 7:85–106.

Blumer, M., Chase, T., and **Watson, S.W.** 1969. Fatty acids in the lipids of marine and terrestrial nitrifying bacteria. *Journal of Bacteriology* 99:366–370.

Bock, E. 1976. Growth of *Nitrobacter* in the presence of organic matter: chemoorganotrophic growth of *Nitrobacter agilis. Archives of Microbiology* 108:305–312.

Bock, E. 1978. Lithotrophic and chemoorganotrophic growth of nitrifying bacteria. pp. 310–314, in Schlessinger, D. (editor), *Microbiology,* American Society for Microbiology, Washington, DC.

Bock, E. and **Heinrich, G.** 1969. Morphologische und Physiologische Untersuchungen an Zellen von *Nitrobacter winogradskyi* Buch. *Archiv. für Mikrobiologie* 69:149–159.

Boogerd, F.C., Van Verseveld, H.W., and **Stouthamer, A.H.** 1983. Dissimilatory nitrate uptake in *Paracoccus denitrificans* via a $\Delta\mu$H-dependent system and a nitrate-nitrite antiport system. *Biochimica et Biophysica Acta* 723:415–427.

Burrows, K.J., Cornish, A., Scott, D., and **Higgins, I.J.** 1984. Substrate specificities of the soluble and particulate methane mono-oxygenases of *Methylosinus trichosporium* OB3b. *Journal of General Microbiology* 130:3327–3333.

Cobley, J.G. 1976a. Energy-conserving reactions in phosphorylating electron-transport particles from *Nitrobacter winogradskyi.* Activation of nitrite oxidation by the electrical component of the protonmotive force. *Biochemical Journal* 156:481–491.

Cobley, J.G. 1976b. Reduction of cytochromes by nitrite in electron-transport particles from *Nitrobacter winogradskyi*: Proposal of a mechanism for H^+ translocation. *Biochemical Journal* 156:493–498.

Cobley, J.G. 1984. Oxidation of nitrite and formate in *Nitrobacter* membrane preparations: Evidence that both reactions are catalyzed by the same enzyme, pp. 170–183, in Strohl, W.R. and Tuovinen, O.H. (editors), *Microbial Chemoautotrophy,* Ohio State University Press, Columbus.

Davies, S.L., and **Whittenbury, R.** 1970. Fine structure of methane and other hydrocarbon utilizing bacteria. *Journal of General Microbiology* 61:227–232.

DiSpirito, A.A. and **Hooper, A.B.** 1986. Oxygen exchange between nitrate molecules during nitrite oxidation by *Nitrobacter. Journal of Biological Chemistry* 261:10534–10537.

DiSpirito, A.A., Lipscomb, J.D., and **Hooper, A.B.** 1986. Cytochrome aa_3 from *Nitrosomonas europaea. Journal of Biological Chemistry* 261:17048–17056.

DiSpirito, A.A., Taaffe, L.R., and **Hooper, A.B.** 1985a. Localization and concentration of hydroxylamine oxidoreductase and cytochromes *c*-552, *c*-554, c_m-553, c_m-552 and *a* in *Nitrosomonas europaea. Biochimica et Biophysica Acta* 806:302–330.

DiSpirito, A.A., Taaffe, L.R., Lipscomb, J.D., and **Hooper, A.B.** 1985b. A 'blue' copper oxidase from *Nitrosomonas europaea. Biochimica et Biophysica Acta* 827:320–326.

Drozd, J.W. 1976. Energy coupling and respiration in *Nitrosomonas europaea. Archives of Microbiology* 110:257–262.

Drozd, J.W. 1980. Respiration in the ammonia oxidizing chemoautotrophic bacteria, pp. 81–111, in Knowles, C.J. (editor), *Diversity of Bacterial Respiratory Systems,* Vol. 2, CRC Press, Boca Raton.

Dua, R.D., Bhandari, B., and **Nicholas, D.J.D.** 1979. Stable isotope studies on the oxidation of ammonia to hydroxylamine by *Nitrosomonas europaea. FEBS Letters* 106:401–404.

Erickson, R.H. and **Hooper, A.B.** 1972. Preliminary characterization of a variant CO-binding heme protein from *Nitrosomonas. Biochimica et Biophysica Acta* 275:231–244.

Falcone, A.B., Shug, A.L., and **Nicholas, D.J.D.** 1963. Some properties of a hydroxylamine oxidase from *Nitrosomonas europaea. Biochimica et Biophysica Acta* 77:199–208.

Faull, K.F., Wallace, W., and **Nicholas, D.J.D.** 1969. Nitrate oxidase and nitrate reductase in *Nitrobacter agilis. Biochemical Journal* 113:448–455.

Ferguson, S.J. 1982. Is a proton-pumping cytochrome oxidase essential for energy conservation in *Nitrobacter? FEBS Letters* 146:239–243.

Feitag, A., Rudert, M., and **Bock, E.** 1987. Growth of *Nitrobacter* by dissimilatory nitrate reduction. *FEMS Microbiology Letters* 48:105–109.

Friedman, S.H., Massefski, W., and **Hollocher, T.C.** 1986. Catalysis of intermolecular oxygen atom transfer by nitrite dehydrogenase of *Nitrobacter agilis. Journal of Biological Chemistry* 261:10538–10543.

Fukuoka, M., Fukumori, Y., and **Yamanaka, T.** 1987. *Nitrobacter winogradskyi* cytochrome a_1c_1 is an iron-sulfur molybdoenzyme having hemes *a* and *c*. *Journal of Biochemistry* 102:525–530.

Goreau, T.J., Kaplan, W.A., Wofsy, S.C., McElroy, M.B., Valois, F.W., and **Watson, S.W.** 1980. Production of NO_2^- and N_2O by nitrifying bacteria at reduced concentrations of oxygen. *Applied and Environmental Microbiology* 40:526–532.

Hanssum, B., Dohnt, G., and **Renger, G.** 1985. On the mechanism of ADRY agent interaction with the photosystem II donor site. *Biochimica et Biophysica Acta* 806:210–220.

Herweijer, M.A., Berden, J.A., and **Slater, E.C.** 1986. Uncoupler-inhibitor titrations of ATP-driven reverse electron transfer in bovine submitochondrial particles provide evidence for direct interaction between ATPase and NADH:Q oxidoreductase. *Biochimica et Biophysica Acta* 849:276–287.

Hofman, T. and **Lees, H.** 1953. The biochemistry of the nitrifying organisms: 4. The respiration and intermediary metabolism of *Nitrosomonas. Biochemical Journal* 54:579–583.

Hollocher, T.C., Tate, M.E., and **Nicholas, D.J.D.** 1981. Oxidation of ammonia by *Nitrosomonas europaea*: Definitive ^{18}O-tracer evidence that hydroxylamine formation involves a monooxygenase. *Journal of Biological Chemistry*: 256:10834–10836.

Hollocher, T.C., Kumar, S., and **Nicholas, D.J.D.** 1982. Respiration-dependent proton translocation in *Nitrosomonas europaea* and its apparent absence in *Nitrobacter agilis* during inorganic oxidation. *Journal of Bacteriology* 149:1013–1020.

Hooper, A.B. 1968. A nitrite-reducing enzyme from *Nitrosomonas europaea.* Preliminary characterization with hydroxylamine as electron donor. *Biochimica et Biophysica Acta* 162:49–65.

Hooper, A.B. 1969. Biochemical basis of obligate autotrophy in *Nitrosomonas europaea. Journal of Bacteriology* 97:776–779.

Hooper, A.B. 1984. Ammonia oxidation and energy transduction in the nitrifying bacteria, pp. 133–167, in Strohl, W.R. and Tuovinen, O.H. (editors), *Microbial Chemoautotrophy,* Ohio State University Press, Columbus.

Hooper, A.B. and **Balny, C.** 1982. Reaction of oxygen with hydroxylamine oxidoreductase of *Nitrosomonas*. Fast kinetics. *FEBS Letters* 144:299–303.

Hooper, A.B. and **DiSpirito, A.A.** 1985. In bacteria which grow on simple reductants, generation of a proton gradient involves extracytoplasmic oxidation of substrate. *Microbiological Reviews* 49:140–157.

Hooper, A.B. and **Nason, A.** 1965. Characterization of hydroxylamine-cytochrome *c* reductase from the chemoautotrophs *Nitrosomonas europaea* and *Nitrosocystis oceanus. Journal of Biological Chemistry* 240:4044–4057.

Hooper, A.B. and **Terry, K.R.** 1977. Hydroxylamine oxidoreductase from *Nitrosomonas*: Inactivation by hydrogen peroxide. *Biochemistry* 16:455–459.

Hooper, A.B. and **Terry, K.R.** 1979. Hydroxylamine oxidoreductase of *Nitrosomonas*: Production of nitric oxide from hydroxylamine. *Biochimica et Biophysica Acta* 571:12–20.

Hooper, A.B., DiSpirito, A.A., Olson, T.C., Andersson, K.K., Cunningham, W., and **Taafe, L.R.** 1984. Generation of the proton gradient by a periplasmic dehydrogenase, pp. 53–58, in Crawford, R.L. and Hanson, R.S. (editors), *Microbial Growth on C_1 Compounds,* American Society for Microbiology, Washington, D.C.

Hooper, A.B., Hansen, J., and **Bell, R.** 1967. Characterization of glutamate dehydrogenase from the ammonia-oxidizing chemoautotroph *Nitrosomonas europaea. Journal of Biological Chemistry* 242:288–296.

Hooper, A.B., Terry, K.R., and **Maxwell, P.C.** 1977. Hydroxylamine oxidoreductase of *Nitrosomonas*: Oxidation of diethyldithiocarbamate concomitant with stimulation of nitrite synthesis. *Biochimica et Biophysica Acta* 462:141–152.

Hooper, A.B., Maxwell, P.C., and **Terry, K.R.** 1978. Hydroxylamine oxidoreductase from *Nitrosomas*: Absorption spectra and content of heme and metal. *Biochemistry* 17:2984–2989.

Hooper, A.B. 1978. Nitrogen oxidation and electron transport in ammonia-oxidizing bacteria, pp. 299–304, in Schlessinger, D. (editor), *Microbiology*, American Society for Microbiology, Washington, D.C.

Hooper, A.B. and **Terry, K.R.** 1973. Specific inhibitors of ammonia oxidation in *Nitrosomonas. Journal of Bacteriology* 115:480–485.

Hooper, A.B. and **Terry, K.R.** 1974. Photoinactivation of ammonia oxidation in *Nitrosomonas. Journal of Bacteriology* 119:899–906.

Hubley, J.H., Thomson, A.W., and **Wilkinson, J.F.** 1975. Specific inhibitors of methane oxidation in *Methylosinus trichosporium. Archives of Microbiology* 102:199–202.

Hyman, M.R. and **Wood, P.M.** 1983. Methane oxidation by *Nitrosomonas europaea. Biochemical Journal* 212:31–37.

Hyman, M.R. and **Wood, P.M.** 1984a. Ethylene oxidation by *Nitrosomonas europaea. Archives of Microbiology* 137:155–158.

Hyman, M.R. and **Wood, P.M.** 1984b. Bromocarbon oxidations by *Nitrosomonas europaea*, pp. 49–52, in Crawford, R.L. and Hanson, R.S. (editors), *Microbial Growth on C_1 Compounds*, American Society for Microbiology, Washington, D.C.

Hyman, M.R. and **Wood, P.M.** 1985. Suicidal inactivation and labeling of ammonia monooxygenase by acetylene. *Biochemical Journal* 227:779–725.

Hyman, M.R., Sansome-Smith, A.W., Shears, J.H., and **Wood, P.M.** 1985. A kinetic study of benzene oxidation to phenol by whole cells of *Nitrosomonas europaea* and evidence for the further oxidation of phenol to hydroquinone. *Archives of Microbiology* 143:302–306.

Hynes, R.K. and **Knowles, R.** 1983. Inhibition of chemoautotrophic nitrification by sodium chlorate and sodium chlorite: a reexamination. *Applied and Environmental Microbiology* 45:1178–1182.

Hynes, R.K. and **Knowles, R.** 1978. Inhibition by acetylene of ammonia oxidation in *Nitrosomonas europaea. FEMS Microbiology Letters* 4:319–321.

Ingledew, W.J. and **Halling, P.J.** 1976. Paramagnetic centers of the nitrite oxidizing bacterium *Nitrobacter. FEBS Letters* 67:90–93.

Ingledew, W.J., Cobley, J.G., and **Chappell, J.B.** 1974. Cytochromes of *Nitrobacter* respiratory chain. *Transactions of the Biochemical Society* 2:149–151.

Jahnke, L.S., Lyman, C., and **Hooper, A.B.** 1984. Carbonic anhydrase, carbon dioxide levels and growth of *Nitrosomonas. Archives of Microbiology* 140:291–293.

Jones, R.D. and **Morita, R.Y.** 1983. Methane oxidation by *Nitrosococcus oceanus* and *Nitrosomonas europaea. Applied and Environmental Microbiology* 45:401–410.

Kiesow, L. 1963. Uber die Reduktion von Diphosphopyridinnucleotid bei der Chemosynthese. *Biochemische Zeitschrift* 338:400–406.

Kiesow, L.A. 1967. Energy-linked reactions in chemoautotrophic organisms. *Current Topics in Bioenergetics* 2:195–233.

Krüger, B. and **Meyer, O.** 1986. The pterin (bactopterin) of carbon monoxide dehydrogenase from *Pseudomonas carboxydoflava. European Journal of Biochemistry* 157:121–128.

Kumar, S. and **Nicholas, D.J.D.** 1982a. A proton motive force-dependent adenosine-5′ triphosphate synthesis in spheroplasts of *Nitrosomonas europaea. FEMS Microbiology Letters* 14:21–25.

Kumar, S. and **Nicholas, D.J.D.** 1982b. Assimilation of inorganic nitrogen compounds by *Nitrobacter agilis. Journal of General Microbiology* 128:1795–1801.

Kumar, S., Nicholas, D.J.D., and **Williams, E.H.** 1983. Definitive ^{15}N NMR evidence that water serves as a source of 'O' during nitrite oxidation by *Nitrobacter agilis. FEBS Letters* 152:71–74.

Lees, H. 1952. The biochemistry of the nitrifying organisms: 1. The ammonia-oxidizing systems of *Nitrosomonas. Biochemical Journal* 52:134–139.

Lipschultz, F., Zafiriou, O.C., Wofsy, S.C., McElroy, M.G., Valois, F.W., and **Watson, S.W.** 1981. Production of NO and N_2O by soil nitrifying bacteria. *Nature* 294:641–643.

Lipscomb, J.D. and **Hooper, A.B.** 1982. Resolution of multiple heme centers of hydroxylamine oxidoreductase from *Nitrosomonas*. 1. Electron paramagnetic resonance spectroscopy. *Biochemistry* 21:3965–3972.

Makula, R.A. 1978. Phospholipid composition of methane-utilizing bacteria. *Journal of Bacteriology* 134:771–777.

Malavolta, E., Delwiche, C.C., and **Burge, W.D.** 1960. Carbon dioxide fixation and phosphorylation by *Nitrobacter agilis. Biochemical and Biophysical Research Communications* 2:445–449.

Miller, D.J. and **Nicholas, D.J.D.** 1985. Characterization of a soluble cytochrome oxidase/nitrate reductase from *Nitrosomonas europaea. Journal of General Microbiology* 131:2851–2854.

Miller, D.J. and **Wood, P.M.** 1983. The soluble cytochrome oxidase of *Nitrosomonas europaea. Journal of General Microbiology* 129:1645–1650.

Murray, R.G.E. and **Watson, S.W.** 1965. Structure of *Nitrosocystis oceanus* and comparison with *Nitrosomonas* and *Nitrobacter. Journal of Bacteriology* 89:1594–1609.

Nicholas, D.J.D. 1978. Intermediary metabolism of nitrifying bacteria, with particular reference to nitrogen, carbon, and sulfur compounds, pp. 305–309, in Schlessinger, D. (editor), *Microbiology*, American Society for Microbiology, Washington, D.C.

Nicholas, D.J.D. and **Jones, O.T.G.** 1960. Oxidation of hydroxylamine in cell-free extracts of *Nitrosomonas europaea. Nature* 185:512–514.

O'Kelley, J.C. and **Nason, A.** 1970. Particulate formate oxidase from *Nitrobacter agilis. Biochimica et Biophysica Acta* 205:426–436.

O'Neil, J.G. and **Wilkinson, J.F.** 1977. Oxidation of ammonia by methane-oxidizing bacteria and the effects of ammonia on methane oxidation. *Journal of General Microbiology* 100:407–412.

Olson, T.C. and **Hooper, A.B.** 1983. Energy coupling in the bacterial oxidation of small molecules: an extracytoplasmic dehydrogenase in *Nitrosomonas. FEMS Microbiology Letters* 19:47–50.

Poth, M. and **Focht, D.C.** 1985. [15]N Kinetic analysis of N_2O production by *Nitrosomonas europaea*: an examination of nitrifier denitrification. *Applied and Environmental Microbiology* 49:1134–1141.

Prince, R.C. and **Hooper, A.B.** 1987. Resolution of the hemes of hydroxylamine oxidoreductase by redox potentiometry and electron spin resonance spectroscopy. *Biochemistry* 26:970–974.

Prince, R.C., Larroque, C., and **Hooper, A.B.** 1983. Resolution of the hemes of hydroxylamine oxidoreductase by redox potentiometry and optical spectroscopy. *FEBS Letters* 163:25–27.

Rao, P.S. and **Nicholas, D.J.D.** 1966. Studies on the incorporation of CO_2 by cells and cell-free extracts of *Nitrosomonas europaea. Biochimica et Biophysica Acta* 124:221–232.

Ritchie, G.A.F. and **Nicholas, D.J.D.** 1972. Identification of the sources of nitrous oxide produced by oxidative and reductive processes in *Nitrosomonas europaea. Biochemical Journal* 126:1181–1191.

Sewell, D.L., Aleem, M.I.H., and **Wilson, D.H.** 1972. The oxidation-reduction potentials and rates of oxidation of the cytochromes of *Nitrobacter agilis. Archives of Biochemistry and Biophysics* 153:312–319.

Shears, J.H. and **Wood, P.M.** 1985. Spectroscopic evidence for a photosensitive oxygenated state of ammonia monooxygenase. *Biochemical Journal* 226:499–507.

Shears, J.H. and **Wood, P.M.** 1986. Tri- and tetra-methylhydroquinone as electron donors for ammonia monooxygenase in whole cells of *Nitrosomonas europaea. FEMS Microbiology Letters* 33:281–284.

Smith, A.J., London, J., and **Stanier, R.Y.** 1967. Biochemical basis of obligate autotrophy in blue-green algae and thiobacilli. *Journal of Bacteriology* 94:972–983.

Smith, U., Ribbons, D.W., and **Smith, D.S.** 1970. The fine structure of *Methylcococcus capsulatus. Tissue and Cell* 2:513–520.

Solioz, M., Carafoli, E., and **Ludwig, B.** 1982. The cytochrome *c* oxidase of *Paracoccus denitrificans* pumps protons in a reconstituted system. *Journal of Biological Chemistry* 257:1579–1582.

Sone, N., Yanagita, Y., Hon-Nami, K., Fukumori, Y., and **Yamanaka, T.** 1983. Proton-pump activity of *Nitrobacter agilis* and *Thermus thermophilus* cytochrome *c* oxidases. *FEBS Letters* 155:150–154.

Stanley, S.H., Prior, S.D., Leak, D.J., and **Dalton, H.** 1983. Copper stress underlies the fundamental change in intracellular location of methane monooxygenase in methane-oxidizing organisms: studies in batch and continuous cultures. *Biotechnology Letters* 5:487–492.

Sundermeyer-Klinger, H., Meyer, W., Warninghoff, B., and **Bock, E.** 1984. Membrane-bound nitrite oxidoreductase of *Nitrobacter*: evidence for a nitrate reductase system. *Archives of Microbiology* 140:153–158.

Suzuki, I. and **Kwok, S.C.** 1981. A partial resolution and reconstitution of the ammonia-oxidizing system of *Nitrosomonas europaea*: role of cytochrome C-554. *Canadian Journal Biochemistry* 59:484–488.

Suzuki, I., Kwok, S.C., Dular, U., and **Tsang, D.C.Y.** 1981. Cell-free ammonia-oxidizing system of *Nitrosomonas europaea*: general conditions and properties. *Canadian Journal of Biochemistry* 59:477–483.

Suzuki, I., Dular, U., and **Kwok, S.C.** 1974. Ammonia and ammonium ion as substrate for oxidation by *Nitrosomonas* cells and extracts. *Journal of Bacteriology* 120:556–558.

Suzuki, I., Kwok, S.C., and **Dular, U.** 1976. Competitive inhibition of ammonia oxidation in *Nitrosomonas europaea* by methane, carbon monoxide or methanol. *FEBS Letters* 72:117–120.

Tanaka, Y., Fukumori, Y., and **Yamanaka, T.** 1983. Purification of cytochrome a_1c_1 from *Nitrobacter agilis* and characterization of nitrite oxidation system of the bacterium. *Archives of Microbiology* 135:265–271.

Terry, K.R. and **Hooper, A.B.** 1981. Hydroxylamine oxidoreductase: A 20-heme, 200,000 molecular weight cytochrome *c* with unusual denaturation properties which forms a 63,000 molecular weight monomer after heme removal. *Biochemistry* 20:7026–7032.

Terry, K.R. and **Hooper, A.B.** 1970. Polyphosphate and orthophosphate content of *Nitrosomonas europaea* as a function of growth. *Journal of Bacteriology* 103:199–206.

Tsang, D.C.Y. and **Suzuki, I.** 1982. Cytochrome *c*-554 as a possible electron donor in the hydroxylation of ammonia and carbon monoxide in *Nitrosomonas europaea. Canadian Journal of Biochemistry* 60:1018–1024.

Van Gool, A. and **Laudelout, H.** 1967. Spectrophotometric and kinetic study of nitrite and formate oxidation in *Nitrobacter winogradskyi. Journal of Bacteriology* 93:215–220.

Van Gool, A.P., Tobback, P.P., and **Fischer, I.** 1971. Autotrophic growth and synthesis of reserve polymers in *Nitrobacter winogradskyi. Archiv für Mikrobiologie* 76:252–264.

Vickery, L.E. and **Hooper, A.B.** 1981. EPR of hydroxylamine oxidoreductase from *Nitrosomonas europaea. Biochimica et Biophysica Acta* 670:291–293.

Wallace, W. and **Nicholas, D.J.D.** 1968. Properties of some reductase enzymes in the nitrifying bacteria and their relationship to the oxidase systems. *Biochemical Journal* 109:763–773.

Walter, B., Sidransky, E., Kristjansson, J.K., and **Hollocher, T.C.** 1978. Inhibition of denitrification by uncouplers of oxidative phosphorylation. *Biochemistry* 17:3039–3045.

Weining, S. and **Nicholas, D.J.D.** 1986. Intermediary metabolism in *Nitrobacter hamburgensis* strain X14. *Biochemistry International* 12:549–556.

Wetzstein, H.-G. and **Ferguson, S.J.** 1985. Respiration-dependent proton translocation and the mechanism of protonmotive force generation in *Nitrobacter winogradskyi. FEMS Microbiology Letters* 30:87–92.

Wieghardt, K., Woeste, M., Roy, P.S., and **Chaudhuri, P.** 1985. Kinetics and mechanism or nitrate reduction by a dimeric aquamolybdenum (III) complex in aqueous solution. Direct evidence for an oxo-group transfer. *Journal of the American Chemical Society* 107:8276–8277.

Williams, P.J.L. and **Watson, S.W.** 1968. Autotrophy in *Nitrosocystis oceanus. Journal of Bacteriology* 96:1640–1648.

Wood, P.M. 1986. Nitrification as a bacterial energy source, pp. 39–62, in Prosser, J.I. (editor), *Nitrification*, Special Publications of the Society for General Microbiology, Vol. 20, IRL Press, Oxford.

Yamanaka, T. and **Shinra, M.** 1974. Cytochrome C-552 and cytochrome C-554 derived from *Nitrosomonas europaea*. Purification, properties and their function in hydroxylamine oxidation. *Biochemical Journal* 75:1265–1273.

Yamanaka, T., Kamita, Y., and **Fukumori, Y.** 1981. Molecular and enzymatic properties of cytochrome aa_3-type terminal oxidase derived from *Nitrobacter agilis. Biochemical Journal* 89:265–273.

Yamanaka, T., Tanaka, Y., and **Fukumori, Y.** 1982. *Nitrobacter agilis* cytochrome *c*-550: isolation, physicochemical and enzymatic properties, and primary structure. *Plant and Cell Physiology* 23:441–449.

Yamazaki, T., Fukumori, Y., and **Yamanaka, T.** 1985. Cytochrome a_1 of *Nitrosomonas europaea* resembles aa_3-type cytochrome *c* oxidase in many respects. *Biochimica et Biophysica Acta* 810:174–183.

14

Physiology and Biochemistry of Phototrophic Bacteria

Hans G. Trüper

Institute for Microbiology
University of Bonn
Bonn, Federal Republic of Germany

Phototrophy is a property occurring in several groups of procaryotes that are phylogenetically only very distantly related (Woese, 1987; Trüper, 1987). In eubacteria, phototrophy is based on the possession of bacteriochlorophylls, whereas in archaebacteria it is based on the possession of bacteriorhodopsin (Oesterhelt and Krippahl, 1983). This chapter deals with photolithoautotrophic bacteria in the "classical sense," i.e., the anaerobic anoxygenic photolithoautotrophic bacteria (see Chapter 6). Included are members of the families Chromatiaceae and Ectothiorhodospiraceae ("purple sulfur bacteria"), Chlorobiaceae ("green sulfur bacteria"), and Chloroflexaceae. In another major group, the Rhodospirillaceae ("purple nonsulfur bacteria"), only certain members are able to grow photolithoautotrophically either with molecular hydrogen or with reduced sulfur compounds as photosynthetic electron donors (Table 14.1). Excluded from the present discussion are the oxygenic cyanobacteria and Prochlorales, the aerobic bacteriochlorophyll-possessing bacteria such as *Erythrobacter* (Shiba and Harashima, 1986) and *Methylobacterium* (Trüper, 1987), and the Heliobacteriaceae, all of which so far have been found to be photoheterotrophic (Beer-Romero and Gest, 1987). Photolithoautotrophic growth requires light, CO_2 (bicarbonate), anoxic conditions, and suitable photosynthetic electron donors, i.e., reduced sulfur compounds or molecular hydrogen. The habitats and ecological niches of phototrophic bacteria have been discussed in Chapter 6, chemolithotrophy in Chapter 15, biophysics of the photosynthetic system in Chapter 24, and molecular biology of carbon dioxide fixation in Chapter 25. The present chapter deals with three major aspects of the physiology and biochemistry of the photolithoautotrophic bacteria: utilization of light (*photo*trophy), utilization of inorganic electron donor (*litho*trophy), and CO_2 fixation (*auto*trophy). As an addition, a short section on assimilatory nitrogen and sulfur metabolism of photolithoautotrophic eubacteria is included.

Table 14.1 Occurrence of photolithotrophic capacity in species of the Rhodospirillaceae

Species	Utilization of inorganic electron donors		
	H_2	S^{2-}	$S_2O_3^{2-}$
Rhodospirillum			
fulvum	+	ND	–
mediosalinum	ND	+	–
molischianum	ND	ND	–
photometricum	ND	ND	–
rubrum	+	+	–
salexigens	–	–	–
salinarum	ND	–	–
Rhodopila globiformis	ND	–	(+)
Rhodopseudomonas			
acidophila	+	–	–
blastica	+	–	–
marina	ND	(+)	–
palustris	+	+	+
rutila	ND	–	–
sulfoviridis	ND	(+)	(+)
viridis	–	–	–
Rhodomicrobium			
vannielii	+	+	–
Rhodobacter			
adriaticus	–	+	+
capsulatus	+	+	–
euryhalinus	ND	+	+
sphaeroides	+	+	–
sulfidophilus	+	+	+
veldkampii	–	+	+
Rhodocyclus			
gelatinosus	±	–	–
purpureus	+	–	–
tenuis	+	–	–

+, photolithotrophy proven
(+), photolithoheterotrophy proven
±, strain differences found
–, unable to grow photolithotrophically
ND, not determined

14.1 PHOTOTROPHIC ENERGY METABOLISM

When S. N. Winogradsky did his studies on purple sulfur bacteria (see Chapter 1), he did not realize that they were phototrophic organisms but rather considered them as chemolithotrophic and in their physiology related to *Beggiatoa*. Proof of the phototrophic nature of these bacteria slowly emerged from the work of Engelmann (1888), Buder (1919), and finally van Niel (1931).

Although the energy yield from lithotrophic metabolism is usually low, phototrophic organisms are able to produce large amounts of ATP as long as light is present. In their natural environments, however, this is only the case at daytime (see Chapter 6). Therefore, all phototrophic organisms must in addition possess other energy-yielding metabolic processes, i.e., respiration and/or fermentation, which function during periods of darkness. In the light, ATP is produced by cyclic photophosphorylation. This term characterizes a process during which electrons are circled between reaction-center bacteriochlorophyll, a primary acceptor, and several electron carriers such as quinones, nonheme iron-sulfur proteins, and cytochromes, all associated with the photosynthetic membrane system. Figure 14.1 gives a simplified scheme of the arrangement and functions of these components during ATP synthesis. Light is harvested by antenna pigments (bacteriochlorophylls and carotenoids bound to proteins); several different light-harvesting complexes are known. In Chlorobiaceae and Chloroflexaceae there exist special light-harvesting organelles, the *chlorosomes*, that are attached to the inner side of the cytoplasmic membrane in direct association with the membrane-contained reaction centers. From the light-harvesting pigments, the energy is transferred to the reaction centers. In purple bacteria, the reaction center consists of 4 molecules of bacteriochlorophyll, 2 of bacteriophaeophytin, and a protein consisting of several subunits. Closely associated with the reaction center is an iron-ubiquinone complex which bridges the gap between the inner membrane surface and the reaction center; the latter is located in the membrane toward the outer surface (Figure 14.1). Another biochemical complex stretching through the membrane contains a cytochrome b, a cytochrome c_1, and an iron-sulfur protein. Ubiquinone acts as a mobile carrier within the membrane, whereas cytochrome c_2 can change its site on the outside of the cytoplasmic membrane. When light energy is transferred from the antenna complex to the reaction center, one electron per light quantum is translocated from bacteriochlorophyll to bacteriopheophytin and

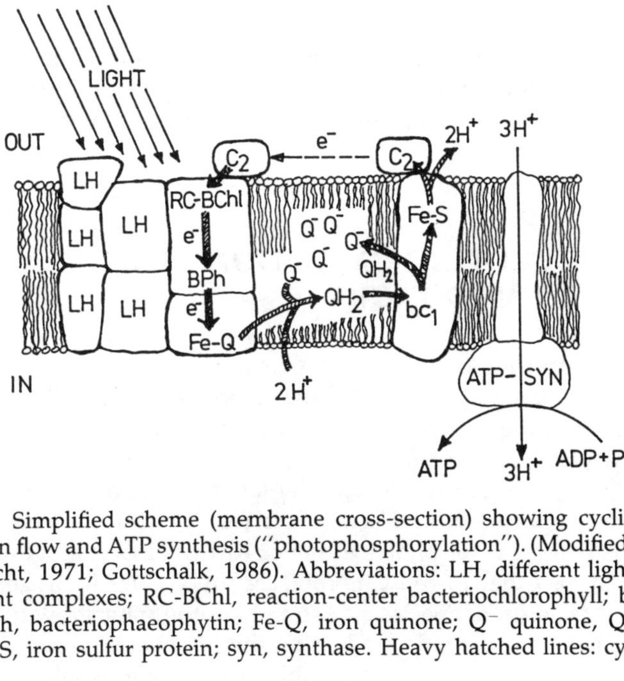

Figure 14.1 Simplified scheme (membrane cross-section) showing cyclic photosynthetic electron flow and ATP synthesis ("photophosphorylation"). (Modified after Drews and Giesbrecht, 1971; Gottschalk, 1986). Abbreviations: LH, different light-harvesting photopigment complexes; RC-BChl, reaction-center bacteriochlorophyll; bc_1, c_2, cytochromes; BPh, bacteriophaeophytin; Fe-Q, iron quinone; Q^- quinone, QH_2, reduced quinone; Fe-S, iron sulfur protein; syn, synthase. Heavy hatched lines: cyclic electron flow.

further to the ubiquinone iron complex. At the reaction center bacteriochlorophyll, the missing electron is immediately replaced by an electron from cytochrome c_2. From the reduced ubiquinone iron complex, the electron enters the intramembraneous ubiquinone pool together with two protons (per quinone molecule and electron) from the cytoplasm. The reduced ubiquinone travels to the bc_1 complex from where the electron is donated to cytochrome c_2, the two protons to the topological outside of the cell. Cytochrome c_2 acts as a mobile electron carrier between the bc_1 complex and the reaction center complex. The proton gradient created this way between out- and inside drives ATP synthase, thus forming ATP in the cytoplasm.

For the generation of NADH, anoxygenic lithoautotrophic bacteria require external electron donors such as H_2 or reduced sulfur compounds. Two different ways for NAD^+ reduction are known: 1) In the Chlorobiaceae, the primary photosynthetic electron acceptor is not a bacteriopheophytin but an iron-sulfur protein with a reduction potential of −540 mV. This allows the green sulfur bacteria to directly reduce NAD^+ via reduced ferredoxin. The electron removed by cyclic electron flow is made up by external electrons entering the system at a less electronegative site via cytochrome c_{555}, which again donates them to the reaction center bacteriochlorophyll. Cytochrome c_{555} behaves in analogy to c_2 in purple bacteria, whereas menaquinones replace ubiquinones as mobile intramembraneous hydrogen carriers. 2) In the Chromatiaceae, the Ectothiorhodaceae, and the facultatively photolithoautotrophic Rhodospirillaceae and Chloroflexaceae, NADH formation is not possible by direct reduction of NAD^+ via the primary acceptor but requires *reverse electron flow*. Reverse electron transfer is driven by the proton motive force provided by the light reaction mechanism described above, i.e., by cyclic photosynthetic electron flow. The electrons required for NAD^+ reduction are donated at less electronegative sites via c-type chrochromes. These electrons are derived from oxidation of H_2 or reduced sulfur compounds.

14.2 LITHOTROPHIC OXIDATION OF EXTERNAL ELECTRON DONORS

Molecular hydrogen Many phototrophic bacteria can grow with molecular hydrogen as the photosynthetic electron donor. This capability was detected more than 50 years ago in *Chromatium vinosum* strain D and other purple sulfur bacteria (Roelofsen, 1935; Gaffron, 1935), and since then *Rhodospirillum rubrum* (Ormerod and Gest, 1962), *Rhodobacter capsulatus* (Klemme and Schlegel, 1967), and many other purple nonsulfur bacteria (see Table 14.1) have been found to utilize hydrogen gas for photosynthetic growth. Lippert and Pfennig (1969) demonstrated also that in the Chlorobiaceae molecular hydrogen utilization occurs in the majority of species and strains. Autotrophic growth in the green sulfur bacteria is, however, dependent on the presence of a reduced source of sulfur, as these organisms are not capable of assimilatory sulfate reduction.

Hydrogen utilization depends upon the presence of the enzyme hydrogenase, and several "uptake" hydrogenases of phototrophic bacteria have been studied in much detail; for example, *Rhodospirillum rubrum* (Adams et al., 1980; Adams and Hall, 1979), *Rhodobacter capsulatus* (Kondratieva and Gogotov, 1983), *Chromatium vinosum* (Gitlitz and Krasna, 1975; Strekus et al., 1980), *Thiocapsa roseopersicina* (Kondratieva and Gogotov, 1981, 1983; see also Vignais et al., 1985; Gogotov, 1978, 1984, 1986; Schlegel and Schneider, 1978). Less well understood are the hydrogenases of *Ectothiorhodospira shaposhnikovii*, *Chlorobium limicola*, and *Chloroflexus aurantiacus* (Gogotov, 1986). In general, the enzymes appear to be membrane-bound and probably contain nickel and iron-sulfur clusters. The natural electron acceptors of these hydrogenases, as far as studied, are cytochromes of the c- or b-type in purple bacteria, and rubredoxin or cytochrome in *Chlorobium* (Gogotov, 1986).

Reduced sulfur compounds The use of reduced sulfur compounds as electron donors for anoxygenic photosynthesis has been found in several groups of phototrophic eubacteria; purple bacteria (families Rhodospirillaceae, Chromaticeae, Ectothiorhodospiraceae) and green bacteria (families Chlorobiaceae, Chloroflexaceae) (Trüper, 1981a, 1984; Trüper and Fischer, 1982). In the Chlorobiaceae and Chromatiaceae, all species oxidize sulfide and elemental sulfur, some also thiosulfate and sulfite. As shown in Table 14.1, not all species of Rhodospirillaceae are able to utilize reduced sulfur compounds during photosynthesis.

As first shown by van Niel (1931), photosynthetic carbon dioxide fixation and sulfide (or thiosulfate) oxidation are stoichiometrically linked (in equations 1–6 $<CH_2O>$ stands for organic matter):

$$CO_2 + 2H_2S \xrightarrow{\text{light}} <CH_2O> + H_2O + 2S^0 \quad (1)$$

$$3CO_2 + 2S^0 + 5H_2O \xrightarrow{\text{light}} 3<CH_2O> + 2H_2SO_4 \quad (2)$$

$$2CO_2 + H_2S + 2H_2O \xrightarrow{\text{light}} 2<CH_2O> + H_2SO_4 \qquad \text{Sum of 1 and 2} \quad (3)$$

$$CO_2 + 2Na_2S_2O_3 + H_2O \xrightarrow{\text{light}} <CH_2O> + 2S^0 + 2Na_2SO_4 \quad (4)$$

$$3CO_2 + 2S^0 + 5H_2O \xrightarrow{\text{light}} 3<CH_2O> + 2H_2SO_4 \quad (5)$$

$$2CO_2 + Na_2S_2O_3 + 3H_2O \xrightarrow{\text{light}} <2CH_2O> + H_2SO_4 + Na_2SO_4 \qquad \text{Sum of 4 and 5} \quad (6)$$

A great deal of information about the action of reduced sulfur compounds as photosynthetic electron donors has been derived from experiments with whole cell suspensions, in contrast to work with cell extracts or purified enzymes. Winogradsky from his experiments with enrichment cultures of Chromatiaceae stated in 1888: "These bacteria cannot live without sulfur which they store in their cells. They can only obtain the sulfur by oxidation of hydrogen sulfide, consequently also the latter is indispensable for their propagation. They oxidize their sulfur to sulfuric acid which they excrete from their cells in the form of a sulfate (mainly gypsum)," (Winogradsky, 1988; transl. from German).

Most species of phototrophic bacteria depend on or are at least capable of utilizing reduced sulfur compounds as electron donors. All organisms with this capacity utilize sulfide. Elemental sulfur is readily utilized by all Chromatiaceae, Ectothiorhodospiraceae, and Chlorobiaceae, whereas Rhodospirillaceae usually cannot oxidize it. The utilization of thiosulfate is more common in Chromatiaceae and Rhodospirillaceae than in Chlorobiaceae, and only very few species oxidize sulfite or tetrathionate (Trüper, 1981a).

During oxidation of sulfide, usually sulfur appears in the form of globules inside (Chromatiaceae) or outside (*Ectothiorhodospira*) the bacterial cells. The extracellular formation of elemental sulfur by alkaliphilic *Ectothiorhodospira* species is an interesting phenomenon. Because of the alkaline pH of their media, the chemical detection of polysulfides as intermediates of sulfide oxidation is possible. The average chain length of these polysulfides has been found to be 4 (Then and Trüper, 1983), but chain lengths between 2 and 7 may occur. Presumably the elemental sulfur originating from sulfide reacts immediately with surplus sulfide (at alkaline pH values) thus forming polysulfide. As soon as the sulfide is exhausted, the bacteria start to oxidize the polysulfides, and immediately elemental sulfur globules appear outside the cells (Then and Trüper, 1983; Then, 1984). Although the pH inside Chromatiaceae and the optimal pH of media for Chlorobiaceae and the Chromatiaceae is rather close to 7.0, so that polysulfides are much less stable, the formation of intra- and extracellular "sulfur globules" may follow the same scheme with different kinetics. The formation of elemental sulfur is discussed in detail in Chapter 16. Evidence is accumulating that these "sulfur globules" are not elemental sulfur but rather large globular agglomerates of long-chain polythionates. This idea is supported by activation energy measurements during cyanolysis (Then and Trüper, 1983, 1984; Then, 1984) the demonstration that only traces of sulfur ring molecules are present (Steudel, 1985), the hydrophilic properties of the globules (Then and Trüper, 1983, 1984), and density measurements (Guerrero et al., 1984). A detailed investigation of comparable "sulfur" globules of the chemolithoautotrophic *Thiobacillus ferrooxidans* by Steudel et al. (1987) revealed that these consisted of an inner core of S_8, S_7, S_9 and S_{12}, and an outer layer of long-chain polythionates (with more than 10 sulfur atoms) which render the globules hydrophilic, so that they are covered by a hydration shell. This structure has yet to be proven for the anaerobic phototrophic sulfur bacteria.

During growth of Chromatiaceae on thiosulfate, sulfur globules appear inside the cells, derived entirely from the sulfane group of thiosulfate (Smith, 1965; Trüper and Pfennig, 1966). Among the green thiosulfate-using bacteria, only *Chlorobium vibrioforme* f. *thiosulfatophilum* forms extracellular sulfur globules from thiosulfate (Steinmetz and Fischer, 1982).

Rhodopseudomonas sulfoviridis oxidizes sulfide via intermediate intracellular sulfur which, however, does not appear as microscopically visible globules inside the cells, although it has been chemically identified. The internal sulfur in this organism probably represents long-chain polysulfides or polythionates. In *Rhodobacter sulfidophilus,* measurable amounts of sulfite are excreted as an intermediate in sulfide and thiosulfate oxidation (Neutzling et al., 1985). *Rhodopseudomonas marina* produces thiosulfate from sulfide (Imhoff, 1983). Thiosulfate formation from sulfide was found to occur in thick suspensions of *Chlorobium limicola* f. *thiosulfatophilum* alongside the formation of elemental sulfur (Trüper et al., 1988). It was proven in this species that thiosulfate is not an intermediate of elemental sulfur oxidation and elemental sulfur is not an intermediate of thiosulfate oxidation. *Chlorobium vibrioforme* f. *thiosulfatophilum* also forms thiosulfate plus elemental sulfur from sulfide but the formation of sulfur continues at the expense of thiosulfate after sulfide is exhausted (Steinmetz and Fischer, 1982). In batch cultures of *Rhodomicrobium vannielii* (family Rhodospirillaceae), Hansen (1974) detected simultaneous thiosulfate and elemental sulfur formation during growth on sulfide. He proved, however, by chemostat cultures, that this was due to the formation of tetrathionate which chemically reacted with the residual sulfide forming thiosulfate plus sulfur. This chemical reaction mechanism does not account for the occurrence of thiosulfate plus sulfur in cultures of *Chlorobium limicola* f. *thiosulfatophilum,* because polythionates could not be detected (Trüper et al., 1988). Tetrathionate is formed from thiosulfate by *Rhodopila globiformis* (Then and Trüper, 1981), by *Chromatium vinosum* at pH values below 7.0 (Smith, 1966), and by *Rhodopseudomonas palustris* when concentrations of thiosulfate above 10 mM are present (Rodova and Pedan, 1980).

In Chromatiaceae and most other phototrophic bacteria, sulfate is the end product of sulfur oxidation, although the oxidation of extracellular sulfur by Chlorobiaceae and Ectothiorhodospiraceae is often incomplete. *Rhodospirillum rubrum, Rhodobacter capsulatus,* and *R. sphaeroides* cannot further oxidize elemental sulfur formed from sulfide (Trüper, 1981a).

The present state of knowledge on the enzymology of photolithotrophic sulfur metabolism in phototrophic bacteria is summarized in Figure 14.2. The enzymatic steps in Figure 14.2 are not common to all phototrophic bacteria but depict the different pathways for which evidence exists in different bacterial species. Figure 14.2 shows that all sulfur compounds oxidized to sulfate have to pass through sulfite. The scheme can be divided into three metabolic sections with a group of enzymes each: 1) oxidation of sulfide and elemental sulfur; 2) utilization of thiosulfate; 3) oxidation of sulfite.

Sulfide may be utilized by three different enzymes, leading to polysulfides, thiosulfate, or sulfite, respectively. The enzymatic step from sulfide to polysulfides and "sulfur" clearly involves cytochromes of the c type (Fischer and Trüper, 1977; Then and Trüper, 1983, 1984) as biocatalysts, and electron acceptors. The formation of thiosulfate from sulfide is catalyzed by flavocytochrome c as it occurs in *Chromatium* and *Chlorobium* species (Fischer, 1977). A "reverse" siroheme-containing sulfite reductase is responsible for sulfite formation from sulfide directly as well as from polysulfides and/or "elemental sulfur." This enzyme has been purified from *Chromatium vinosum* and cha-

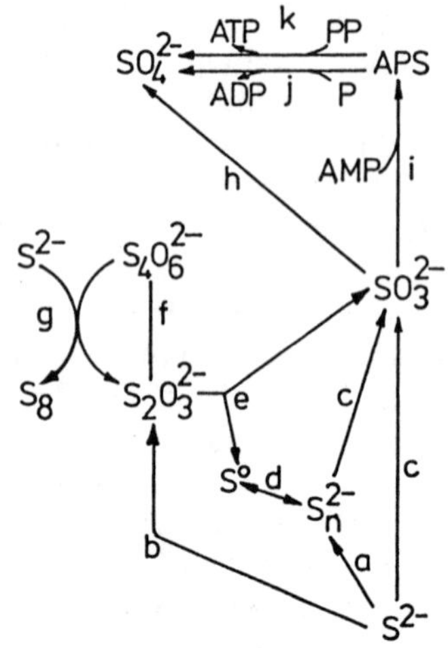

Figure 14.2 Pathways of anaerobic dissimilatory sulfur metabolism in phototrophic bacteria. S^0, globular "sulfur." a, cytochrome c; b, flavocytochrome c; c, reverse sulfite reductase; d, enzyme unknown (nonenzymatic?); e, thiosulfate sulfur transferase; f, thiosulfate:acceptor oxidoreductase; g, nonenzymatic reaction; h, sulfite:acceptor oxidoreductase; i, APS reductase; j, ADP sulfurylase; k, ATP sulfurylase.

racterized (Schedel et al., 1979). Thiosulfate is split to elemental sulfur and sulfite by an enzyme that may be measured by its thiosulfate-sulfur transferase (rhodanese) activity. *Chlorobium limicola* f. *thiosulfatophilum* (Kusai and Yamanaka, 1973), *Chromatium vinosum* (Smith, 1966), *Rhodopseudomonas palustris* (Knobloch et al., 1971), and *Rhodopila globiformis* (Then and Trüper, 1981) contain thiosulfate:acceptor oxidoreductase, earlier called tetrathionase, which forms tetrathionate from thiosulfate. An enzyme oxidizing or splitting tetrathionate has not as yet been found in phototrophic bacteria. Under natural environmental conditions, however, a chemical reduction of tetrathionate by sulfide may readily occur. The enzyme adenosine phosphosulfate (APS)* reductase, an essential enzyme in dissimilatory sulfate-reducing bacteria and two *Thiobacillus* species was found to occur also in Chromatiaceae and Chlorobiaceae (Thiele, 1968; Trüper and Peck, 1970). The enzyme was purified from *Thiocapsa roseopersicina* (Trüper and Rogers, 1971), and *Chlorobium limicola* f. *thiosulfatophilum* (Kirchhoff and Trüper, 1974). In *Chromatium vinosum* (Schwenn and Biere, 1979) and *Chromatium warmingii* (Leyendecker, 1983) it is firmly membrane bound and cannot be solubilized. Species of *Ectothiorhodospira* and the Rhodospirillaceae do not possess APS reductase (Trüper, 1981a; Trüper and Fischer, 1982). Also *Chromatium purpuratum* and *Chromatium gracile* lack this enzyme (Trüper and Fischer, 1982; Ulbricht, 1984). The intermediary formation of APS demands an additional enzyme to split off the sulfate moiety. This step allows conservation of the high-energy phosphate bond energy contained in APS. The enzyme ADP sulfurylase replaces the sulfate moiety of

*Adenosine phosphosulfate is a synonym for adenylylsulfate.

APS by inorganic phosphate, thus producing adenosine-diphosphate (equation 7):

$$APS + P_i \rightarrow ADP + SO_4^{2-} \quad (7)$$

ADP then can be disproportionated by the enzyme adenylate kinase, leading to the formation of 1 ATP and 1 AMP per 2 ADP. This pathway has been found in *Chlorobium vibrioforme* f. *thiosulfatophilum* (Trüper and Fischer, 1982; Bias and Trüper, 1987; Khanna and Nicholas, 1983) and in six species of the Chromatiaceae (Trüper and Fischer, 1982; Ulbricht, 1984) but not, however, in Rhodospirillaceae.

Another pathway that conserves the energy of APS is the replacement of its sulfate moiety by inorganic pyrophosphate, catalyzed by the action of ATP sulfurylase. Inorganic pyrophosphate is a common product of biosynthetic reactions, especially of protein biosynthesis, and therefore is readily available during the exponential growth phase of cells. ATP sulfurylase has been shown to exclusively catalyze this reaction in *Chlorobium limicola* f. *thiosulfatophilum* Bias and Trüper, 1987). An enzyme bypassing the formation of APS in phototrophic bacteria is sulfite:acceptor oxidoreductase (Trüper, 1981a; Trüper and Fischer, 1982):

$$SO_3^{2-} + H_2O \rightarrow SO_4^{2-} + 2e^- + 2H^+ \quad (8)$$

This enzyme occurs in most Chromatiaceae in addition to the APS pathway in a few cases instead of the latter (Ulbricht, 1984); it was also found in *Rhodobacter sulfidophilus, Rhodopseudomonas sulfoviridis* and *Rhodobacter adriaticus* (Neutzling et al., 1985). Sulfite:acceptor oxidoreductase has been partly purified and characterized from *Chromatium vinosum* (Ulbricht, 1984), *R. sulfidophilus, R. sulfoviridis,* and *R. adriaticus;* in the latter three species it is clearly a membrane-bound enzyme (Neutzling et al., 1985), whereas it is soluble in *C. vinosum* (Ulbricht, 1984). Dahl (1988) showed that the enzyme activity considered as sulfite:acceptor oxidoreductase in *Thiocapsa roseopersicina* by other authors is a nonenzymatic property of cell-free extracts, since *T. roseopersicina* is devoid of this enzyme.

The flow of electrons from reduced sulfur compounds to the photosynthetic reaction center bacteriochlorophyll is, in all species of the Chromatiaceae and Chlorobiaceae studied, mediated by c-type cytochromes in its final step. The nature of the direct electron acceptors (and, perhaps intermediate electron carriers) of the sulfur-oxidizing enzymes is far from being resolved in all cases. Besides c- and b-type cytochromes, flavocytochromes c, and also high-potential nonheme iron-sulfur proteins are being considered (Fischer, 1984). In *Rhodobacter sulfidophilus* there is evidence that in an initial dark reaction ubiquinone is reduced by sulfide, followed by light-driven reverse electron transport from ubiquinol to NAD mediated by NADH dehydrogenase (Brune and Trüper, 1986).

14.3 AUTOTROPHIC CO_2 FIXATION

In Chromatiaceae, Ectothiorhodospiraceae, and Rhodospirillaceae, autotrophic CO_2 fixation is performed by the reductive ribulose bisphosphate cycle (Calvin cycle) with ribulose bisphosphate carboxylase (see Chapters 19 and 25) and phosphoribulokinase as the key enzymes. Although not all species of phototrophic bacteria have been studied in detail, there is enzymological proof for the Calvin cycle in *Chromatium vinosum* (Fuller et al., 1961), *C. okenii* (Trüper, 1964), *Thiocapsa roseopersicina* (Kondratieva et al., 1976; Zhukov, 1976), *T. pfennigii* (Sahl and Trüper, 1977), *Ectothiorhodospira shaposhnikovii* (Firsov et al., 1974), *E. halophila* (Tabita and McFadden, 1974), *E. mobilis* (Sahl and Trüper, 1977), *Rhodospirillum rubrum* (Glover et al., 1952; Anderson and Fuller, 1967), *Rhodobacter capsulatus* (Stoppani et al., 1955), and *Rhodopseudomonas palustris* (Stokes and Hoare, 1969). Undoubtedly all species of the above-mentioned three families possess the Calvin cycle when they are growing photolithoautotrophically (for further details, see Kondratieva, 1979).

In the Chlorobiaceae, autotrophic CO_2 fixation occurs via a reverse tricarboxylic acid cycle (see Chapter 20) which was first proposed by Evans et al. (1966). Key enzymes of this cycle are the ferredoxin-dependent α-ketoglutarate and pyruvate synthases, and an ATP-dependent citrate lyase (Sintsov et al., 1980; Ivanovsky et al., 1970). The careful ^{14}C-labeling studies by Fuchs et al. (1980a, b) and respective ^{13}C-NMR studies by Paalme et al. (1982a, b), together with the characterization of the citrate lyase (Antranikian et al., 1982) have provided unequivocal final evidence for this then new mechanism of autotrophic CO_2 fixation.

The pathway of autotrophic CO_2 fixation in *Chloroflexus aurantiacus*, the only species of the Chloroflexaceae that has received a great deal of attention by physiological studies, has not yet been resolved. Holo and Sirevåg (1986) reported that in this organism they could not demonstrate ribulose bisphosphate carboxylase or phosphoribulokinase on one hand, nor citrate lyase or α-ketoglutarate synthase on the other hand. They made clear, however, that acetyl CoA is a central intermediate in the CO_2 fixation process.

The enzymatic equipment necessary to make use of the assimilated carbon in phototrophic bacteria does not substantially differ from that of other procaryotes. Versatility in carbon metabolism is highest in the Rhodospirillaceae, is less in the Ectothiorhodospiraceae and those Chromatiaceae with small cells (e.g., *C. vinosum, Thiocapsa roseopersicina*), still less in the Chromatiaceae with large cells (e.g., *C. okenii, Thiospirillum jenense*), and the least in the Chlorobiaceae (Trüper, 1981b). The latter two groups are obligately photolithoautotrophic, however, with a certain mixotrophic potential, i.e., they depend strictly on a supply of CO_2 and sulfide (or thiosulfate) even when they are also utilizing acetate or pyruvate.

14.4 ASSIMILATION OF INORGANIC NITROGEN AND SULFUR

Nitrogen assimilation In principle, assimilatory nitrogen and sulfur metabolism is not different from that of nonphototrophic bacteria. The majority of photolithoautotrophic bacteria are able to fix dinitrogen although in some species this capability is found only in certain strains. As dinitrogen fixation is linked with hydrogen (gas) production, this field has received much attention in research (for reviews see Vignais et al., 1985; Hallenbeck, 1987).

Ammonia is used as the preferred nitrogen source by all species of phototrophic bacteria. As in other bacteria, it is assimilated via glutamine synthetase and further distributed via transamination (the "GOGAT" system). Nitrate is a rather uncommon nitrogen source in phototrophic bacteria. Only a few species of the Rhodospirillaceae and Ectothiorhodospiraceae assimilate nitrate.

Sulfur assimilation In their natural environments, Chromatiaceae and Chlorobiaceae have access to abundant concentrations of reduced sulfur compounds—predominantly sulfide—for assimilation as well as for energy metabolism. Sulfur is bound to the carbon skeleton via O-acetylserine sulfhydrylase, forming cysteine (Hensel and Trüper, 1981, 1983). From cysteine, most other organic sulfur compounds needed can be synthesized. Assimilatory sulfate reduction is lacking in the large-cell species of Chromatiaceae, the Chlorobiaceae, and *Rhodopseudomonas sulfoviridis* (Neutzling and Trüper, 1982), whereas all other phototrophic bacteria are able to assimilate sulfate and reduce it to the sulfhydryl level by a carrier-bound process, along the same pathways as occur in nonphototrophic bacteria.

The author thanks the Fonds der Chemischen Industrie for support of his work.

REFERENCES

Adams, M.W.W. and **Hall, D.O.** 1979. Properties of the solubilized membrane-bound hydrogenase from the photosynthetic bacterium *Rhodospirillum rubrum. Archives of Biochemistry and Biophysics* 195:288–299.

Adams, M.W.W., Mortenson, L.E., and **Chen, J.S.** 1980. Hydrogenase. *Biochimica et Biophysica Acta* 594:105–176.

Anderson, L. and **Fuller, R.C.** 1967. Photosynthesis in *Rhodospirillum rubrum.* I. Autotrophic carbon dioxide fixation. *Plant Physiology* 42:487–490.

Antranikian, G., Herzberg, C., and **Gottschalk, G.** 1982. Characterization of ATP citrate lyase from *Chlorobium limicola. Journal of Bacteriology* 152:1284–1287.

Beer-Romero, P. and **Gest, H.** 1987. *Heliobacillus mobilis,* a peritrichously flagellated anoxyphototroph containing bacteriochlorophyll g. *FEMS Microbiology Letters* 41:109–114.

Bias, U. and **Trüper, H.G.** 1987. Species specific release of sulfate from adenylyl sulfate by ATP sulfurylase or ADP sulfurylase in the green sulfur bacteria *Chlorobium limicola* and *Chlorobium vibrioforme. Archives of Microbiology* 147:406–410.

Brune, D.C. and **Trüper, H.G.** 1986. Noncyclic electron transport in chromatophores from photolithotrophically grown *Rhodobacter sulfidophilus. Archives of Microbiology* 145:295–301.

Buder, J. 1919. Zur Bakteriologie des Bakteriopurpurins und der Purpurbakterien. *Jahrbuch für wissenschaftliche Botanik* 58:525–628.

Dahl, C. 1988. *Untersuchungen zur Enzymologie des Schwefelstoffwechsels in Thiocapsa roseopersicina unter chemo- und photolithotrophen Bedingungen.* Diploma thesis, University of Bonn.

Drews, G. and **Giesbrecht, P.** 1981. Die Strukturelemente der Prokaryotenzelle, pp. 471–541, in Metzner, H. (editor), *Die Zelle–Struktur und Funktion*, 3rd edition, Wissenschaftliche Verlagsgesellschaft mbH, Stuttgart.

Engelmann, T.W. 1888. Die Purpurbakterien und ihre Beziehungen zum Licht. *Botanische Zeitung* 46:661–669.

Evans, W.C.W., **Buchanan, B.B.**, and **Arnon, D.I.** 1966. A new ferredoxin-dependent carbon reduction cycle in a photosynthetic bacterium. *Proceedings of the National Academy of Sciences USA* 55:928–933.

Firsov, N.N., **Chernadiev, I.I.**, **Ivanovsky, R.N.**, **Kondratieva, E.N.**, **Udovina, N.V.**, and **Doman, N.G.** 1974. Pathways of assimilation of carbon dioxide by *Ectothiorhodospira shaposhnikovii*. *Mikrobiologiya* 43:214–219.

Fischer, U. 1977. *Die Rolle von Cytochromen in Schwefelstoffwechsel phototropher Schwefelbakterien*. Doctoral thesis, University of Bonn.

Fischer, U. 1984. Cytochromes and iron sulfur proteins in sulfur metabolism of phototrophic sulfur bacteria, pp. 383–407, in Müller, A. and Krebs, B. (editors), *Sulfur, its Significance for Chemistry, for the Geo-, Bio- and Cosmophere and Technology*. Elsevier, Amsterdam/Oxford/New York/Tokyo.

Fischer, U. and **Trüper, H.G.** 1977. Cytochrome c-550 of *Thiocapsa roseopersicina*: Properties and reduction of sulfide. *FEMS Microbiology Letters* 1:87–90.

Fuchs, G., **Stupperich, E.**, and **Jaenchen, R.** 1980a. Autotrophic CO_2 fixation in *Chlorobium limicola*. Evidence against the operation of the Calvin cycle in growing cells. *Archives of Microbiology* 128:56–64.

Fuchs, G., **Stupperich, E.**, and **Jaenchen, R.** 1980b. Autotrophic CO_2 fixation in *Chlorobium limicola*. Evidence for the operation of a reductive tricarboxylic acid cycle in growing cells. *Archives of Microbiology* 128:64–72.

Fuller, R.C., **Smillie, R.M.**, **Sisler, E.C.** and **Kornberg, H.L.** 1961. Carbon metabolism in *Chromatium*. *Journal of Biological Chemistry* 236:2140–2149.

Gaffron, H. 1935. Über die Kohlensäureassimilation der roten Schwefelbakterien II. *Biochemische Zeitschrift* 279:1–33.

Gitlitz, P.H. and **Krasna, A.J.** 1975. Structural and catalytical properties of hydrogenase from *Chromatium*. *Biochemistry* 14:2561–2568.

Glover, J., **Kamen, M.D.**, and **van Genderen, H.** 1952. Studies on the metabolism of photosynthetic bacteria. *Archives of Biochemistry and Biophysics* 35:384–408.

Gogotov, I.N. 1978. Relationships in hydrogen metabolism between hydrogenase and nitrogenase in phototrophic bacteria. *Biochimie* 60:267–275.

Gogotov, I.N. 1984. Hydrogenase of purple bacteria: properties and regulation of synthesis. *Archives of Microbiology* 140:86–90.

Gogotov, I.N. 1986. Hydrogenases of phototrophic microorganisms. *Biochimie* 68:181–187.

Gottschalk, G. 1986. *Bacterial Metabolism*. 2nd edition, Springer-Verlag New York/Berlin/Heidelberg/Tokyo. 359 pp.

Guerrero, R., **Mas, J.**, and **Pedros-Alió, C.** Buoyant density changes due to intracellular content of sulfur in *Chromatium warmingii* and *Chromatium vinosum*. *Archives of Microbiology* 137:350–356.

Hallenbeck, P.C. 1987. Molecular aspects of nitrogen fixation by photosynthetic prokaryotes. *Critical Reviews in Microbiology* 14:1–48.

Hansen, T.A. 1974. *Sulfide als electrondonor voor Rhodospirillaceae*. Doctoral thesis, University of Groningen.

Hensel, G. and **Trüper, H.G.** 1981. O-Acetylserine sulfhydrylase and S-sulfocysteine synthase activities of *Chromatium vinosum*. *Archives of Microbiology* 130:228–233.

Hensel, G. and **Trüper, H.G.** 1983. O-Acetylserine sulfhydrylase and S-sulfocysteine synthase activities of *Rhodospirillum tenue*. *Archives of Microbiology* 134:227–232.

Holo, H. and **Sirevåg, R.** 1986. Autotrophic growth and CO_2 fixation of *Chloroflexus aurantiacus*. *Archives of Microbiology* 145:173–180.

Imhoff, J.F. 1983. *Rhodopseudomonas marina* sp. nov., a new marine phototrophic purple bacterium. *Systematic and Applied Microbiology* 4:512–521.

Ivanovsky, R.N., Sintsov, N.V., and **Kondratieva, E.N.** 1980. ATP-linked citrate lyase activity in the green sulfur bacterium *Chlorobium limicola* forma *thiosulfatophilum. Archives of Microbiology* 128:239–242.

Kanna, S. and **Nicholas, D.J.D.** 1983. Substrate phosphorylation in *Chlorobium vibrioforme f. sp. thiosulfatophilum. Journal of General Microbiology* 129:1365–1370.

Kirchhoff, J. and **Trüper, H.G.** 1974. Adenylylsulfate reductase of *Chlorobium limicola. Archives of Microbiology* 100:115–120.

Klemme, J.H. and **Schlegel, H.G.** 1967. Photoreduktion von Pyridinnucleotid durch Chromatophoren aus *Rhodopseudomonas capsulata* mit molekularem Wasserstoff. *Archiv für Mikrobiologie* 59:185–196.

Knobloch, K., Eley, J.H., and **Aleem, M.I.H.** 1971. Thiosulfate-linked ATP-dependent NAD^+ reduction in *Rhodopseudomonas palustris. Archiv für Microbiologie* 80:97–114.

Kondratieva, E.N. 1979. Interrelation between modes of carbon assimilation and energy production in phototrophic purple and green bacteria, pp. 117–175, in Quayle, J.R. (editor), *International Review of Biochemistry, Microbial Biochemistry,* Vol. 21, University Park Press, Baltimore.

Kondratieva, E.N. and **Gogotov, I.N.** 1981. *Molecular Hydrogen in Microbial Metabolism.* Nauka, Moscow. 343 pp.

Kondratieva, E.N. and **Gogotov, I.N.** 1983. Production of molecular hydrogen in microorganisms. *Advances in Biochemical Engineering/Biotechnology* 28:139–191.

Kondratieva, E.N., Zhukov, V.G., Ivanovsky, R.N., Petushkova, Yu.P., and **Monosov, E.Z.** 1976. The capacity of phototrophic sulfur bacterium *Thiocapsa roseopersicina* for chemosynthesis. *Archives of Microbiology* 108:287–292.

Kusai, A. and **Yamanaka, T.** 1973. The oxidation mechanism of thiosulphate and sulphide in *Chlorobium thiosulfatophilum*: Roles of cytochrome c-551 and cytochrome c-553. *Biochimica et Biophysica Acta* 325:304–314.

Leyendecker, W. 1983. *Charakterisierung der partikelgebundenen APS-Reduktase aus Chromatium warmingii Stamm 6512.* Diploma thesis, University of Bonn.

Lippert, K.D. and **Pfennig, N.** 1969. Die Verwertung von molekularem Wasserstoff durch *Chlorobium thiosulfatophilum* Wachstum und CO_2-Fixierung. *Archiv für Mikrobiologie* 65:29–47.

Neutzling, O., Pfleiderer, C., and **Trüper, H.G.** 1985. Dissimilatory sulphur metabolism in phototrophic "nonsulphur" bacteria. *Journal of General Microbiology* 131:791–798.

Neutzling, O. and **Trüper, H.G.** 1982. Assimilatory sulfur metabolism in *Rhodopseudomonas sulfoviridis. Archives of Microbiology* 133:145–148.

Oesterhelt, D. and **Krippahl, G.** 1983. Phototrophic growth of halobacteria and its use for isolation of photosynthetically-deficient mutants. *Annales de Microbiologie* (Institut Pasteur) 134B:137–150.

Ormerod, J. and **Gest, H.** 1962. Hydrogen photosynthesis and alternative metabolic pathways in photosynthetic bacteria. *Bacteriological Reviews* 26:51–66.

Paalme, T., Olivson, A., and **Vilu, R.** 1982a. ^{13}C-NMR study of the glucose synthesis pathways in the bacterium *Chlorobium thiosulfatophilum. Biochimica et Biophysica Acta* 720:303–310.

Paalme, T., Olivson, A., and **Vilu, R.** 1982b. ^{13}C-NMR study of CO_2-fixation during the heterotrophic growth in *Chlorobium thiosulfatophilum. Biochimica et Biophysica Acta* 720:311–319.

Rodova, N.A. and **Pedan, L.V.** 1980. Metabolism of thiosulfate in *Rhodopseudomonas palustris. Mikrobiologiya* 49:221–226.

Roelofsen, P.A. 1935. On the metabolism of the purple sulfur bacteria. *Proceedings of the Royal Academy of Sciences, Amsterdam* 37:660–669.

Sahl, H.G. and **Trüper, H.G.** 1977. Enzymes of CO_2 fixation in Chromatiaceae. *FEMS Microbiology Letters* 2:129–132.

Shiba, T. and **Harashima, K.** 1986. Aerobic photosynthetic bacteria. *Microbiological Sciences* 3:376–378.

Sintsov, N.V., Ivanovskii, R.N., and **Kondratieva, E.N.** 1980. ATP-dependent citrate lyase in the green phototrophic bacterium *Chlorobium limicola. Mikrobiologiya* (English translation) 49:449–452.

Smith, A.J. 1965. The discriminative oxidation of the sulphur atoms of thiosulphate by a photosynthetic sulphur bacterium—Chromatium strain D. *Biochemical Journal* 94:27P.

Smith, A.J. 1966. The role of tetrathionate in the oxidation of thiosulfate by *Chromatium* sp. strain D. *Journal of General Microbiology* 42:371–380.

Schedel, M., Vanselow, M., and **Trüper, H.G.** 1979. Siroheme sulfite reductase isolated from *Chromatium vinosum. Archives of Microbiology* 121:29–36.

Schlegel, H.G. and **Schneider, K.** (editors) 1978. *Hydrogenases: Their Catalytic Activity, Structure and Function.* E. Goltze, Göttingen, 453 pp.

Schwenn, J.D. and **Biere, M.** 1979. APS-reductase activity in the chromatophores of *Chromatium vinosum* strain D. *FEMS Microbiology Letters* 6:19–22.

Steinmetz, M. and **Fischer, U.** 1982. Cytochromes of the green sulfur bacterium *Chlorobium vibrioforme* f. *thiosulfatophilum*, purification, characterization and sulfur metabolism. *Archives of Microbiology* 131:19–26.

Steudel, R. 1985. Neue Entwicklungen in der Chemie des Schwefels und des Selens. *Nova Acta Leopoldina,* Neue Folge, 59:231–246.

Steudel, R., Holdt, G., Göbel, T., and **Hazeu, W.** 1987. Chromatographische Trennung höherer Polythionate $S\mathring{O}_6^{2-}$ (n=3. . . 22) und deren Nachweis in Kulturen von *Thiobacillus ferrooxidans;* molekulare Zusammensetzung bakterieller Schwefelausscheidungen. *Angewandte Chemie* 99:143–146.

Stokes, J.E. and **Hoare, D.S.** 1969. Reductive pentose phosphate cycle and formate assimilation in *Rhodopseudomonas palustris. Journal of Bacteriology* 100:890–894.

Stoppani, A.O.M., Fuller, R.C. and **Calvin, M.** 1955. Carbon dioxide fixation by *Rhodopseudomonas capsulatus. Journal of Bacteriology* 69:491–501.

Strekus, T., Antanaitis, B.C., and **Krasna, A.J.** 1980. Characterization and stability of hydrogenase from *Chromatium. Biochimica et Biophysica Acta* 116:1–9.

Tabita, F.R. and **McFadden, B.A.** 1974. One-step isolation of microbial ribulose-1,5-diphosphate carboxylase. *Archives of Microbiology* 99:231–240.

Then, J. 1984. *Beiträge zur Sulfidoxidation durch Ectothiorhodospira abdelmalekii und Ectothiorhodospira halochloris.* Doctoral thesis, University of Bonn.

Then, J. and **Trüper, H.G.** 1981. The role of thiosulfate in sulfur metabolism in *Rhodopseudomonas globiformis. Archives of Microbiology* 130:143–146.

Then, J. and **Trüper, H.G.** 1983. Sulfide oxidation in *Ectothiorhodospira abdelmalekii.* Evidence for the catalytic role of cytochrome c-551. *Archives of Microbiology* 135:254–258.

Then, J. and **Trüper, H.G.** 1984. Utilization of sulfide and elemental sulfur by *Ectothiorhodospira halochloris. Archives of Microbiology* 139:295–298.

Thiele, H.H. 1968. Sulfur metabolism inThiorhodaceae. V. Enzymes of sulfur metabolism in *Thiocapsa floridana* and *Chromatium* species. *Antonie van Leeuwenhoek Journal of Microbiology and Serology* 34:350–356.

Trüper, H.G. 1964. CO_2-Fixierung und Intermediärstoffwechsel bei *Chromatium okenii* Perty. *Archiv für Mikrobiologie* 49:23–50.

Trüper, H.G. 1981a. Photolithotrophic sulfur oxidation, pp. 199–211, in Bothe, H. and Trebst, A. (editors), *Biology of Inorganic Nitrogen and Sulfur,* Springer-Verlag, Berlin/Heidelberg/New York.

Trüper, H.G. 1981b. Versatility of carbon metabolism in the phototrophic bacteria, pp. 116–121, in Dalton, H. (editor), *Microbial Growth on C_1 Compounds.* Heyden, London/Philadelphia/Rheine.

Trüper, H.G. 1984. Phototrophic bacteria and their sulfur metabolism, pp. 367–382, in Müller, A. and Krebs, B. (editors), *Sulfur, Its Significance for Chemistry, for the Geo-, Bio- and Cosmophere and Technology.* Elsevier, Amsterdam/Oxford/New York/Tokyo.

Trüper, H.G. 1987. Phototrophic bacteria (an incoherent group of prokaryotes). A taxonomic versus phylogenetic survey. *Microbiología SEM* 3:71–89.

Trüper, H.G. and **Fischer, U.** 1982. Anaerobic oxidation of sulphur compounds as electron donors for bacterial photosynthesis. *Philosophical Transactions of the Royal Society B* 298:529–542.

Trüper, H.G., Lorenz, C., Schedel, M., and **Steinmetz, M.** 1988. Metabolism of thiosulfate in *Chlorobium,* pp. 189–200, in Olson, J.M., Ormerod, J.G., Amesz, J., Stackebrandt, E., and Trüper, H.G. (editors), *Green Photosynthetic Bacteria,* Plenum Press, New York/London.

Trüper, H.G. and **Peck, H.D. Jr.** 1970. Formation of adenylylsulfate in photosynthetic bacteria. *Archiv für Mikrobiologie* 73:125–142.

Trüper, H.G. and **Pfennig, N.** 1966. Sulphur metabolism in Thiorhodaceae. III. Storage and turnover of thiosulphate sulphur in *Thiocapsa floridana* and *Chromatium* species. *Antonie van Leeuwenhoek Journal of Microbiology and Serology* 32:261–276.

Trüper, H.G. and **Rogers, L.A.** 1971. Purification and properties of adenylysulfate reductase from the phototrophic sulfur bacterium *Thiocapsa roseopersicina. Journal of Bacteriology* 108:1112–1121.

Ulbricht, H. 1984. *Aspekte des Energiegewinns durch Substratphosphorylierung im Zuge der Sulfitoxidation bei Chromaticeae und Thiobacillus denitrificans.* Doctoral thesis, University of Bonn.

van Niel, C.B. 1931. On the morphology and physiology of the purple and green sulphur bacteria. *Archiv für Mikrobiologie* 3:1–112.

Vignais, P.M., Colbeau, A., Willison, J.C., and Jouanneau, Y. 1985. Hydrogenase, nitrogenase, and hydrogen metabolism in photosynthetic bacteria. *Advances in Microbial Physiology* 26:155–234.

Winogradsky, S. 1888. *Beiträge zur Morphologie und Physiologie der Bacterien. Heft I. Zur Morphologie und Physiologie der Schwefelbacterien.* Verlag Arthur Felix, Leipzig. 120 pp.

Woese, C.R. 1987. Bacterial evolution. *Microbiological Reviews* 51:221–271.

Zhukov, V.G. 1976. Formation of ribulose-1,5-diphosphate carboxylase by *Thiocapsa roseopersicina* in different growth conditions. *Mikrobiologiya* 45:915–917.

15

Chemolithotrophy of Phototrophic Bacteria

Elena N. Kondratieva

Department of Microbiology
Moscow State University
Moscow, USSR

Some of Winogradsky's early work was on the purple bacteria, and he made a large contribution to the classification of these microorganisms on the basis of morphological studies (Winodgradsky, 1888). Among other things, Winogradsky suggested that these bacteria were capable of a chemosynthetic metabolism (i.e., chemolithoautotrophy), but this concept was forgotten after van Niel (1931) showed that the purple bacteria were able to use light energy.

It is now known, however, that many species of purple bacteria are able to grow not only in the presence of light but in the dark as well. Most purple nonsulfur bacteria and some purple sulfur bacteria are able to grow under dark heterotrophic conditions and to obtain energy for growth by the oxidation of organic compounds. This may be due either to anaerobic fermentation or—most often—to aerobic respiration (Kondratieva, 1979; Trüper and Pfennig, 1981).

It has also been discovered that several purple bacteria are able to grow under dark autotrophic conditions at the expense of energy provided by oxidation of some inorganic compounds. This capacity for a chemolithoautotrophic mode of life was found first in *Thiocapsa roseopersicina* (Bogorov, 1974) and then in some other species of purple sulfur (Gorlenko, 1974; Kämpf and Pfennig, 1980, 1986a, b; Kondratieva et al., 1981; Eichler and Pfennig, 1986), and purple nonsulfur (Siefert and Pfennig, 1979; Madigan and Gest, 1979) bacteria belonging to the families Chromatiaceae and Rhodospirillaceae (Table 15.1). However, the chemolithoautotrophic growth of some purple bacteria ceases after several cell generations. This phenomenon was observed in *Thiocystis violacea* and *Chromatium gracile* (Kämpf and Pfennig, 1986b). Also, some strains of *Rhodocyclus gelatinosus* are able to grow using carbon monoxide as the energy source for biosynthetic processes under dark anaerobic conditions (Uffen, 1981).

Other purple nonsulfur bacteria are able to grow under chemolithoautotrophic conditions in the presence of molecular hydrogen. Most purple sulfur

Table 15.1 Purple bacteria able to grow under chemolithoautotrophic conditions

Purple sulfur bacteria	Purple nonsulfur bacteria
Amoebobacter pedioformis	*Rhodobacter capsulatus*
Amoebobacter roseus	*Rhodobacter sulfidophilus*
Chromatium gracile	*Rhodocyclus gelatinosus*
Chromatium minus	*Rhodopseudomonas acidophila*
Chromatium vinosum	
Chromatium violascens	
Lamprobacter modestohalophilus	
Thiocapsa roseopersicina	
Thiocystis violacea	

bacteria are able to use sulfide and thiosulfate, and some of them are able to use H_2 as electron donors (see Table 14.1). Although molecular oxygen is required as a terminal electron acceptor, many strains of purple bacteria are able to grow in the dark only under microaerobic conditions (Kämpf and Pfennig, 1980, 1986b).

Some purple sulfur bacteria, grown under chemolithoautotrophic conditions, have only a restricted capacity or are unable to oxidize organic compounds and use them in addition to carbon dioxide as a source of carbon. Some strains of *T. roseopersicina* (Kondratieva et al., 1981) and *C. gracile* (Kämpf and Pfennig, 1986a) show this property.

The growth of these and some other purple sulfur bacteria in the light and in the dark is obligately dependent on the presence of inorganic sulfur compounds or molecular hydrogen. Only these hydrogen donors stimulate respiration of the cells (Table 15.2). The restricted capacity of some purple bacteria to utilize organic compounds seems to be correlated with the absence of a complete tricarboxylic acid cycle and a glyoxylate shunt (Kondratieva et al., 1981). In purple bacteria the metabolism of sulfur compounds in the light and in the dark is apparently similar. In both cases *T. roseopersicina* (Kondratieva et al., 1976) and *T. violacea* (Kämpf and Pfennig, 1980) oxidize thiosulfate to sulfate. But in the light, the yield of cell material formed per mol of thiosulfate consumed is lighter than in the dark.

Under various conditions *T. roseopersicina* oxidizes sulfide and thiosulfate to sulfite and then to sulfate via the formation of adenosine phosphosulfate (APS) This allows the synthesis of ATP by substrate-level phosphorylation:

$$SO_3^{2-} + AMP \rightarrow APS + 2e^-$$

$$APS + P_i \rightarrow ADP + SO_4^{2-}$$

$$2ADP \rightarrow AMP + ATP$$

Table 15.2 The rate of O_2 uptake by purple bacteria in the presence of various substrates

Substrate	*T. roseopersicina*	*E. shaposhnikovii*
Malate	n.d.	22.5
Succinate	0.1	8.8
Thiosulfate	25.0	1.7
Sulfide	34.5	24.0

Units, $nmol \cdot min^{-1} \cdot mg^{-1}$ protein
n.d., not detected

Also, the oxidation of inorganic sulfur compounds and H_2 in *T. roseopersicina* and other purple bacteria under chemolithoautotrophic growth conditions appears to supply the cells with energy as a result of the functioning of respiratory electron transport chains.

The components of these systems in different purple bacteria are cytochromes of b and c types and the terminal oxidase usually belongs to cytochrome o (Kondratieva et al., 1981; Wynn et al., 1985).

The rate of carbon dioxide fixation by purple bacteria in the dark is lower than in the light, but in both cases CO_2 is assimilated under autotrophic conditions via a Calvin cycle through the activity of the enzyme ribulose bisphosphate carboxylase (see Chapter 19). The functioning of the Calvin cycle in the dark is evident by the nature and the kinetics of labeled compounds produced during short-time exposure of the cells of *T. roseopersicina* to $^{14}CO_2$. The first labeled products of dark $^{14}CO_2$ fixation are 3-phosphoglycerate and sugar phosphates. The same labeled compounds were previously found in *Rhodospirillum rubrum* (Glover et al., 1957) and *Rhodobacter capsulatus* (Stoppani et al., 1955) after cells from light cultures had fixed $^{14}CO_2$ in the dark in the presence of H_2 and O_2.

It is also known that the level of ribulose bisphosphate carboxylase in the cells of *T. roseopersicina* under different conditions of growth is the same (Table 15.3). Other carboxylases are also present in *T. roseopersicina,* but their activity is lower than that of ribulose bisphosphate carboxylase (Kondratieva et al., 1981).

In contrast to *T. roseopersicina* strain BBS, the synthesis of ribulose bisphosphate carboxylase in the majority of purple bacteria under aerobic conditions is repressed (Table 15.3). In addition at high O_2 concentrations the ribulose bisphosphate carboxylase in purple bacteria as well as in other organisms may function as an oxygenase (see Chapter 15). This fact explains the production of glycolate by some purple bacteria.

Molecular oxygen represses the synthesis and inhibits the activity of hydrogenases in purple bacteria (Kondratieva and Gogotov, 1983) and possibly other enzymes required for chemolithoautotrophic growth. This is probably

Table 15.3 The activity of some enzymes involved in CO_2 fixation in purple bacteria grown under different conditions

	T. roseopersicina		*E. shaposhnikovii*	
Enzyme	Light	Dark	Light	Dark
Ribulose 1,5-bisphosphate carboxylase	26.6	25.7	46.0	1.2
Phosphoenolpyruvate carboxylase	2.3	3.5	0.5	1.1
Phosphoenolpyruvate carboxykinase	6.1	5.9	0.2	n.d.
Pyruvate carboxylase	2.9	2.4	2.8	1.9
Propionyl CoA carboxylase	6.3	1.8	43.6	22.3
Pyruvate synthase	9.6	12.1	16.5	17.8
2-Oxoglutarate synthase	2.2	2.3	15.4	11.7

Units, $nmol \cdot min^{-1} \cdot mg^{-1}$ protein
n.d., not detected

the reason that some purple bacteria reveal the capacity for chemolithoautotrophic growth only under microaerobic conditions.

Other species of purple bacteria exhibit the capacity for oxidation of inorganic sulfur compounds in the dark too. However, their growth under these conditions depends on the presence of organic compounds. These bacteria include *Ectothiorhodospira shaposhnikovii, E. mobilis, Rhodopseudomonas palustris* and some other species (Kondratieva, 1979; Kondratieva et al., 1981).

The cells of *E. shaposhnikovii* grown under dark, microaerobic conditions on media with acetate or other organic compounds contain ribulose bisphosphate carboxylase in sufficient quantity to autotrophically assimilate CO_2 when oxidizing sulfide or thiosulfate. This process, however, is only possible in the light and begins without a lag phase (Kondratieva et al., 1981). These results indicate that the oxidation of inorganic sulfur compounds does not provide *E. shaposhnikovii* and some other purple bacteria with sufficient energy for chemolithoautotrophic growth. This can probably be explained by the differences of the pathways of sulfur metabolism in different purple bacteria.

T. roseopersicina and some members of the family Chromatiaceae oxidize sulfide and thiosulfate via formation of adenosine phosphosulfate. In contrast, *E. shaposhnikovii* as well as other members of the Ectothiorhodospiraceae do not contain APS reductase (see Chapter 14). The oxidation of sulfur compounds in these bacteria may lead to ATP synthesis coupled with functioning of the respiratory electron transport systems but they cannot synthesize ATP as a result of adenosine phosphosulfate metabolism (Kondratieva et al., 1981).

It was also shown that *Rhodobacter capsulatus* can oxidize sulfide and thiosulfate in the dark. However, chemolithoautotrophic growth only occurred in the presence of H_2 as electron donor (Madigan and Gest, 1979).

Thus, among the phototrophic bacteria there are species able to grow under different conditions and able to adjust their metabolism accordingly. Of special interest is the capacity of some purple bacteria for chemolithoautotrophic mode of life. This fact as well as other data are in accordance with the phylogenetic relationships between the phototrophic bacteria and some chemolithoautotrophic procaryotes (Woese, 1987).

REFERENCES

Bogorov, L.G. 1974. About the properties of *Thiocapsa roseopersicina* strain BBS, isolated from the estuary of the White Sea. *Mikrobiologija* 43:326–332.

Eichler, B. and **Pfennig, N.** 1986. Characterization of a new platelet-forming purple sulfur bacterium, *Amoebobacter pedioformis* sp. nov. *Archives of Microbiology* 146:295–300.

Glover, J., Kamen, M.D., and **van Genderen, H.** 1952. Studies of the metabolism of photosynthetic bacteria. XII. Comparative light and dark metabolism of acetate and carbonate by *Rhodospirillum rubrum*. *Archives of Biochemistry and Biophysics* 35:384–408.

Gorlenko, V.M. 1974. The oxidation of thiosulfate of *Amoebobacter roseus* in the dark under microaerobic conditions. *Mikrobiologiya* 43:729–731.

Kämpf, C.N and Pfennig, N. 1980. Capacity of Chromaticeae for chemotrophic growth. Specific respiration rate of *Thiocystis violacea* and *Chromatium vinosum. Archives of Microbiology* 127:125–135.

Kämpf, C. and **Pfennig, N.** 1986a. Isolation and characterization of some chemoautotrophic Chromatiaceae. *Journal of Basic Microbiology* 26:507–515.

Kämpf, C. and **Pfennig, N.** 1986b. Chemoautotrophic growth of *Thiocystis violacea, Chromatium gracile,* and *C. vinosum* in the dark at various O_2 concentration. *Journal of Basic Microbiology* 26:517–531.

Kondratieva, E.N. 1979. Interrelation between modes of carbon assimilation and energy production in phototrophic purple and green bacteria, pp. 117–249, in Quayle, J.R. (editor), *Microbial Biochemistry,* Baltimore University Park Press.

Kondratieva, E.N. and **Gogotov, I.N.** 1983. Production of molecular hydrogen in microorganisms. *Advances in Biochemical Engineering and Biotechnology* 28:139–190.

Kondratieva, E.N., Ivanovsky, R.N., and **Krasilnikova, E.N.** 1981. Light and dark metabolism in purple sulfur bacteria, pp. 326–365, in Skulachev, V.P. (editor), *Soviet Science Reviews Section D, Biology Reviews.*

Kondratieva, E.N., Zhukov, V.G., Ivanovsky, R.N., and **Monosov, E.Z.** 1976. The capacity of phototrophic sulfur bacterium *Thiocapsa roseopersicina* for chemosynthesis. *Archives of Microbiology* 108:287–292.

Madigan, M.T. and **Gest, H.** 1979. Growth of the photosynthetic bacterium *Rhodopseudomonas capsulata* chemoautotrophically in the darkness with H_2 as energy source. *Journal of Bacteriology* 137:524–530.

Siefert, E. and **Pfennig, N.** 1979. Chemoautotrophic growth of *Rhodopseudomonas* species with hydrogen and chemotrophic utilization of methanol and formate. *Archives of Microbiology* 122:177–182.

Stoppani, A.O.M., Fuller, R.C., and **Calvin, M.** 1955. Carbon dioxide fixation by *Rhodopseudomonas capsulatus. Journal of Bacteriology* 69:491–501.

Trüper, H.G. and **Fischer, V.** 1982. Anaerobic oxidation of sulfur compounds as electron donors for photosynthesis. *Philosophical Transactions of the Royal Society London B* 292:529–549.

Trüper, H.G. and **Pfennig, N.** 1981. Characterization and identification of anoxygenic phototrophic bacteria, pp. 299–312, in Starr, M.P., Stolp, H., Trüper, H.G., Balows, A., and Schlegel, H.G. (editors), *The Prokaryotes,* Springer-Verlag, Berlin.

Uffen, R.L. 1981. Metabolism of carbon monoxide. *Enzyme and Microbial Technology* 3:197–206.

van Niel, C.B. 1931. On the morphology and physiology of the purple and green sulfur bacteria. *Archiv f:auur Mikrobiologie* 3:1–112.

Winogradsky, S. 1888. Beiträge zur Morphologie und Physiologie der Bacterien. Heft I. Zur Morphologie und Physiologie der Schwefelbacterien. Verlag von Arthur, Felix, Leipzig. 120 pp.

Woese, C.R. 1987. Bacterial evolution. *Microbiological Reviews* 51:221–271.

Wynn, R.M., Kämpf, C., Gaul, D.L., Choi, W.K., Shaw, K.W., and **Knaff, D.B.** 1985. The membrane-bound electron-transfer components of aerobically grown *Chromatium vinosum. Biochimica et Biophysica Acta* 808:85–93.

16

On the Nature of the "Elemental Sulfur" (S^0) Produced by Sulfur-oxidizing Bacteria—a Model for S^0 Globules

Ralf Steudel

Institute of Inorganic and Analytical Chemistry
Technical University of Berlin
Berlin, Federal Republic of Germany

16.1 INTRODUCTION

In 1887 Winogradsky proposed the name "sulfur bacteria" for organisms which seemed to accumulate elemental sulfur in their cells when living under suitable conditions, e.g., in water containing H_2S (see Chapter 1). He vividly described the rapid buildup and disappearance of these sulfur inclusions by *Beggiatoa* depending on whether or not the aqueous medium contained H_2S, and already addressed the question whether this sulfur may be a stored nutrient or an excretion (i.e., a decontamination) product. Many authors have since described the formation and the properties of this "elemental sulfur," which has also been designated by the chemical formula S^0. According to these reports, S^0 forms transparent droplets (globules) which are deposited inside or outside the corresponding cells. The droplets reach diameters of up to 1 μm, are of spherical or ellipsoidal shape, and dissolve at least partly in various organic solvents like acetone, chloroform, ethanol, and carbon disulfide. S^0 is hydrophilic and of white or pale-yellow color, shows a higher index of refraction than water, and has been described as liquid-like and amorphous on the basis of X-ray diffraction results (Hageage et al., 1970). The buoyant density of S^0 produced by *Chromatium* has been determined as 1.22 g cm^{-3} (Guerrero et al., 1984). When allowed to stand in the wet state or on drying, the sulfur globules eventually convert to crystalline S_8. However, it has never been demonstrated that the S^0 globules consist of 100 percent sulfur.

Surprisingly, many of the properties reported above do not match the properties of any known chemical allotrope of elemental sulfur, raising the question of whether S^0 is a new form of elemental sulfur or rather a sulfur-rich compound. In this chapter, the details of the chemistry of the S^0 globules of sulfur bacteria are discussed.

16.2 ELEMENTAL SULFUR

Sulfur is the element with the largest number of allotropes, most of which consist of cyclic nonplanar molecules. The following have been prepared as pure compounds: S_6, S_7, S_8, S_9, S_{10}, S_{11}, S_{12}, S_{13}, S_{15}, S_{18}, S_{20}. In addition, there exists polymeric sulfur (S_∞), both as amorphous and crystalline material (Steudel, 1982, 1984b).

Commercial elemental sulfur is usually a mixture of S_8 and S_∞ with traces of S_7 present also (Steudel and Holz, 1988). Only recrystallization from an organic solvent yields pure S_8 which exists as orthorhombic (α) and monoclinic (β and γ) allotropes. The best analytical technique to study such mixtures and materials is reversed-phase HPLC followed by laser Raman spectroscopy.

The cyclic compounds S_6 through S_{20} form yellow to orange-yellow crystals which are soluble in carbon disulfide and to a lesser degree in certain other organic solvents, but the solubility decreases with increasing ring size and with increasing symmetry of the molecule. The melting points are found in the range 39–148°C, with S_7 having the lowest and S_{12} the highest value. Since only α-S_8 is thermodynamically stable, all others tend to convert to α-S_8, either at room temperature (S_6, S_7, S_9, S_{10}, S_{15}) or on gentle heating. This holds also for S_∞, which is unsoluble in organic solvents but slowly reverts to S_8, S_7, and traces of other S_n molecules on contact with such solvents (Steudel et al., 1984a).

Regardless of the molecular size, all sulfur allotropes are hydrophobic, are not wetted by water, and barely dissolve in water: the solubility of α-S_8 in water of 25°C is only 5 μg/kg (Boulegue, 1978). The densities of all crystalline sulfur allotropes are in the range 1.9–2.2 g cm^{-3}.

Liquid sulfur at the triple point (115°C) has a density of 1.80 g cm^{-3} which decreases on heating; linear extrapolation to 25°C results in a value of about 1.85 g cm^{-3}, which is much higher than the value reported for bacterial S^0. At 116°C an equilibrated sulfur melt consists of 93.6 percent S_8, 3.1 percent S_7, 0.5 percent S_6, 0.3 percent S_9, 0.4 percent S_{12}, and other trace components including S_∞ (Steudel et al., 1985a). On cooling to 25°C it crystallizes as α- and β-S_8, while most of the nonS_8 components slowly convert to S_∞. Only tiny droplets on clean glass surfaces sometimes remain liquid at 25°C for some time; these supercooled droplets are of deep-yellow color (Das and Das Gupta, 1939), certainly hydrophobic, and of too high a density to be candidates for bacterial S^0. The reactivity of the various sulfur allotropes is very different, S_8 being the least and S_6 and S_7 the most reactive. Since liquid sulfur contains S_7 it is more reactive than S_8. The same holds for S_∞ which is commercially available and which is a constituent of "flower of sulfur."

16.3 SULFUR-RICH COMPOUNDS

If the bacterial sulfur globules have a sulfur content of <100 percent, their molecular composition may be discussed in terms of sulfur-rich compounds. Such compounds are usually chainlike and of the type $R{-}S_n{-}R$. They are known to form long homologous series, for instance, with R being H, Cl, Br, SO_3^-, or an organic group:

$H{-}S_n{-}H$ Polysulfanes

$Cl{-}S_n{-}Cl$ Dichlorosulfanes

$R{-}S_n{-}H$ Monoorganylsulfanes

$R{-}S_n{-}R$ Bisorganylsulfanes

$^-O_3S{-}S_n{-}SO_3^-$ Polythionates (Sulfanebissulfonates)

$^-S{-}S_n{-}S^-$ Polysulfides

The chain length n can reach values of up to 50 in most of these series. Only the ionic polysulfides are restricted to a chain length of 8. Of course, in this context only compounds should be considered which are stable at 25°C for at least 24 h and which do not react with water at pH 7 ± 2, the pH which can be assumed for the bacterial cytoplasm. This holds for the organic polysulfanes, R_2S_n, as well as for the polythionates, $S_n(SO_3)_2^{2-}$, but in addition the ionic polysulfides S_n^{2-} should be discussed which are known with $n = 2{-}8$.

Since chains of sulfur atoms, like rings, are hydrophobic, the required hydrophilic nature can only be achieved by suitable end groups R. The chemistry of surfactants demonstrates that R has to carry an ionic charge or to contain many polar groups (like the –OH in carbohydrates) to make an otherwise hydrophobic and large molecule at least partly hydrophilic and to guarantee a certain solubility in water which seems to be mandatory for the transport of the S^0 material through membranes, cell walls, etc.

16.4 IONIC POLYSULFIDES

Water soluble sulfides like Na_2S dissolve in H_2O according to equation (1):

$$S^{2-} + H_2O \rightleftharpoons HS^- + OH^- \qquad (1)$$

Due to the extremely low second dissociation constant of H_2S ($K_{D2} = 10^{-17}$ at 20°C; Giggenbach, 1971b), the sulfide ion S^{2-} exists at very high pH values only. Solutions containing HS^- or S^{2-} dissolve elemental sulfur by nucleophilic attack according to equations (2)–(4):

$$HS^- + OH^- + S_8 \rightleftharpoons S_9^{2-} + H_2O \qquad (2)$$

$$HS^- + OH^- + S_9^{2-} \rightleftharpoons 2S_5^{2-} + H_2O \qquad (3)$$

$$2S_5^{2-} \rightleftharpoons 2S_{5-x}^{2-} + S_{5+x}^{2-} \qquad (4)$$

The chain length n can assume all values from 2 to 7, and its maximum value strongly depends on the pH of the solution (equation 3). Saturation with elemental sulfur yields an average chain length of only $\bar{n} = 4.5$. Lowering of the pH results in formation of SH^- and long chains and finally in sulfur precipitation (equations 2 and 3). This sulfur is crystalline and consists of S_8; it neither forms liquid-like nor oily phases nor is it hydrophilic. There are also no ionic polysulfides known which are liquid at 25°C. The equilibria of equations (2)–(4) are rapidly established in water at 25°C (Giggenbach, 1972).

Under special conditions such as pouring a polysulfide solution into cold hydrochloric acid, the protonation of polysulfide anions produces polysulfanes, H_2S_n, which are always obtained as a mixture with $n = 2$–35; this mixture is yellow oily liquid, hydrophobic, and completely decomposes at 25°C within a few hours to yield crystalline S_8 and H_2S. The sulfanes H_2S_n can best be detected and determined by ^{1}H–NMR spectroscopy (Hahn, 1985).

No polysulfide anions larger than S_7^{2-} have ever been detected in aqueous solution, but in addition these solutions contain the radical anion S_3^- originating from the dissociation of S_6^{2-}, especially at higher temperatures (Giggenbach, 1971a; Chivers and Drummond, 1972):

$$S_6^{2-} \rightleftharpoons 2S_3^- \tag{5}$$

The blue S_3^- has been detected in hot polysulfide solutions and in melts by ultraviolet (UV) electron-spin resonance (ESR) and resonance Raman spectroscopy and may be responsible for the high reactivity of aqueous polysulfide which, for example, rapidly takes up molecular oxygen to produce thiosulfate and solid S_8 according to equation (6) (Steudel et al., 1986):

$$S_n^{2-} + \tfrac{3}{2}O_2 \rightarrow S_2O_3^{2-} + \tfrac{n-2}{8} S_8 \tag{6}$$

This reaction results in a discoloration of the formerly deep yellow or orange solution. Monosulfides are also subject to autoxidation; commercial sodium sulfide, usually formulated as $Na_2S \cdot 9H_2O$, therefore usually also contains polysulfides, sulfite, and thiosulfate, as can be demonstrated by ion chromatography.

Sulfite reacts with polysulfides transforming them into thiosulfate, and on strong heating aqueous polysulfide reversibly disproportionates to thiosulfate and monosulfide (Giggenbach, 1974).

It is obvious from this discussion that ionic polysulfides can exist in neutral and alkaline solutions only (pH $\geq$6; Teder, 1971; Giggenbach, 1972); they do not form polymeric unsoluble hydrophilic salts, and their properties show that they are not candidates for the components of bacterial S^0, although dissolved polysulfides have been observed in the media of *Ectothiorhodospira abdelmalekii* cultures (Then and Trüper, 1983; 1984).

16.5 ORGANIC POLYSULFANES

Organic polysulfanes, R_2S_n, are known as pure materials with up to 8 sulfur atoms in the chain (Nakabayashi et al., 1964), but in mixtures all members with up to 20 atoms have been detected by high-performance liquid chromatography (HPLC) (Möckel et al., 1983; Göbel, 1984). These mixtures usually form yellow oily liquids which are unsoluble in water and slowly decompose to a mixture of short-chain polysulfanes and elemental sulfur (Göbel, 1984). Cyclic polysulfanes with a very high sulfur content are also known (Steudel et al., 1985b). However, no hydrophilic polysulfanes have so far been produced which are soluble in water, despite a high sulfur content (>80 percent); carboxylic groups (–COOH) are not hydrophilic enough to give a polysulfane dicarbonic acid with hydrophilic properties. Organic polysulfanes of the type RS_nH with $n = 1–3$ and $R = (C_6H_5)_3C$ have also been prepared but these compounds are strongly hydrophobic and unstable at 20°C (Nakabayashi et al., 1964).

16.6 POLYTHIONATES

Polythionates, $S_n(SO_3)_2^{2-}$, are a well-known class of compounds and form a long homologous series with n having been claimed to reach values up to 140 (Weitz et al., 1956d). Only the first members of this series ($n = 3–6$) have been prepared in a pure state, as sodium and potassium salts. While tetra- and pentathionates are colorless, the sulfur-rich members are pale yellow (Weitz et al., 1956a–d). The solubility of the alkali metal salts in water decreases with increasing values of n; however, the salts $K_2S_nO_6$ ($n = 3–6$) are quite soluble in water but insoluble in pure ethanol. The aqueous solutions are decomposed by alkaline substances and they must be stabilized by addition of some acid. Polythionates react in neutral solution with sulfite to give thiosulfate and trithionate and with cyanide to give thiocyanate, thiosulfate, and sulfate; sulfide causes polythionates to decompose to thiosulfate and elemental sulfur.

Polythionates are best prepared from thiosulfate and dichlorosulfanes, S_mCl_2, according to equation (7):

$$2K_2S_2O_3 + S_mCl_2 \rightarrow K_2S_n(SO_3)_2 + 2KCl \quad (7)$$

where $n = m + 2$

The anions $S_n(SO_3)_2^{2-}$ are chain-like according to X-ray structural analyses for all members with $n = 1–4$ and can be separated and detected by ion-pair chromatography using a UV absorbance detector; the largest ion detected so far is $S_{22}O_6^{2-}$ (Steudel et al., 1987) but there is no doubt that much longer chains are possible.

Polythionates are also formed on acid decomposition of thiosulfate, and numerous papers on this reaction have been published. Elemental sulfur consisting of S_6, S_7, and S_8 is formed when aqueous sodium thiosulfate is decomposed by hydrochloric acid (Steudel and Mäusle, 1979):

$$Na_2S_2O_3 + 2HCl \rightarrow \tfrac{1}{n}S_n + 2NaCl + SO_2 + H_2O \quad (8)$$

The following reaction mechanism explaining the formation of these products has been proposed (Davis, 1958). In this mechanism, sulfane-monosulfonic acids, $H–S_m–SO_3H$, are the key intermediates:

$$S_2O_3^{2-} + H^+ \rightleftharpoons HS_2O_3^- \tag{9}$$

$$HS_2O_3^- + S_2O_3^{2-} \rightleftharpoons HS_3O_3^- + SO_3^{2-} \tag{10}$$

$$HS_3O_3^- + S_2O_3^{2-} \rightleftharpoons HS_4O_3^- + SO_3^{2-} \tag{11}$$

etc.

$$HS_nO_3^- + S_2O_3^{2-} \rightleftharpoons HS_{n+1}O_3^- + SO_3^{2-} \tag{12}$$

$$HS_{n+1}O_3^- \rightleftharpoons S_n + HSO_3^- \ (n = 6,7,8) \tag{13}$$

$$HSO_3^- + H^+ \rightleftharpoons H_2O + SO_2 \tag{14}$$

The escaping SO_2 removes sulfite from the solution and drives reactions (9)–(13) to the right. The additional formation of polythionates probably results from the reaction:

$$2HS_nO_3^- \rightleftharpoons H_2S + S_{n-1}O_6^{2-} \tag{15}$$

Since the formation of S_8 demonstrates the existence of sulfane-monosulfonic acids with up to 9 sulfur atoms, reaction (15) should then provide polythionates with at least up to 17 sulfur atoms. By ion chromatography, all members up to $S_{12}O_6^{2-}$ have been detected in this mixture (Steudel and Holdt, 1986). Still higher polythionates (up to $S_{22}O_6^{2-}$) have been detected in a reaction mixture from $K_2S_2O_3$, S_2Cl_2 and aqueous HCl (Steudel et al., 1987).

Sulfur-rich polythionates are also the main constituents of hydrophilic sulfur sols prepared from either aqueous thiosulfate and concentrated sulfuric acid (called "Raffo sol") or from aqueous sulfite or sulfur dioxide gas (SO_2) and sulfide or H_2S (called "Selmi sol"). These sols have been extensively investigated using ion chromatography, reversed-phase HPLC, infrared, and Raman spectroscopy, as well as with other techniques (Steudel et al., 1988a, and unpublished results). A freshly prepared Raffo sol after purification by repeated NaCl precipitation and repeptization in water showed the composition $Na_{1.64}S_{28.6}O_6 \cdot {}^{5.9}/_{n}S_n \cdot 1.0\ NaCl$. The elemental sulfur, S_n (17 percent of the total sulfur content), is present as S_8 and other rings ($n = 6–14$) but 83 percent of the sulfur is found as long-chain polythionates which in water form micelles and probably vesicles which keep the otherwise water-unsoluble elemental sulfur in solution. On aging at 25°C, long-chain polythionates slowly decompose according to equation (16):

$$S_mO_6^{2-} \rightarrow S_n + S_{m-n}O_6^{2-} \tag{16}$$

This reaction has been followed by HPLC and ion chromatography; the sulfur S_n finally precipitates from the aqueous phase in which the short-chain polythionates $S_{m-n}O_6^{2-}$ are dissolved (Steudel et al., 1988a). Freshly prepared diluted Raffo sols are stable for months at 25°C and can be heated to 150°C for a short time without precipitation of sulfur (Gmelin, 1953).

The behavior of long-chain polythionate anions in water must resemble that of surfactant molecules with hydrophilic groups at both chain ends, as in alkane disulfonates, $^{-}O_3S{-}(CH_2)_n{-}SO_3^{-}$. Compounds of this type (α, ω-bis-amphiphiles or "bolaamphiphiles") with n >10 are sparingly soluble in water at 25°C but form micelles and monolayer membranes which may result in stable spherical vesicles of up to 1 μm diameter (Fuhrhop and Mathieu, 1984; Fuhrhop and Fritsch, 1986).

The particles in freshly prepared Raffo sols are about 0.1–0.5 μm in diameter and are negatively charged (Donaldson and Johnston, 1969; Kerker et al., 1963; Verstraete, 1943). The average chain length of the polythionate anions in these sols follows from the analytical composition as about 30 sulfur atoms, but chains with up to 50 sulfur atoms may be present. A linear $S_{50}O_6^{2-}$ anion has a length of about 6 nm. A micelle of the type shown in Figure 16.1A and formed from such ions will therefore have a diameter of about 0.006 μm. Even if the micelle is formed from folded polythionate anions and has a nucleus consisting of elemental sulfur, as indicated in Figure 16.1B, the diameter would still be much too small to match the experimental value of 0.1–0.5 μm which is of the same order of magnitude as the diameter of bacterial S^0 globules. This latter model can also be rejected on grounds that its S_n content is too high (n = 6–12) compared to the experimental finding of only 17 percent S_n.

Micelles formed from bolaamphiphiles do not easily fuse together to form larger aggregates but monolayer membranes are spontaneously formed instead, especially when molecules of different length are available (Fuhrhop et al., 1986; Hauser et al., 1983). These membranes tend to form spherical unilamellar

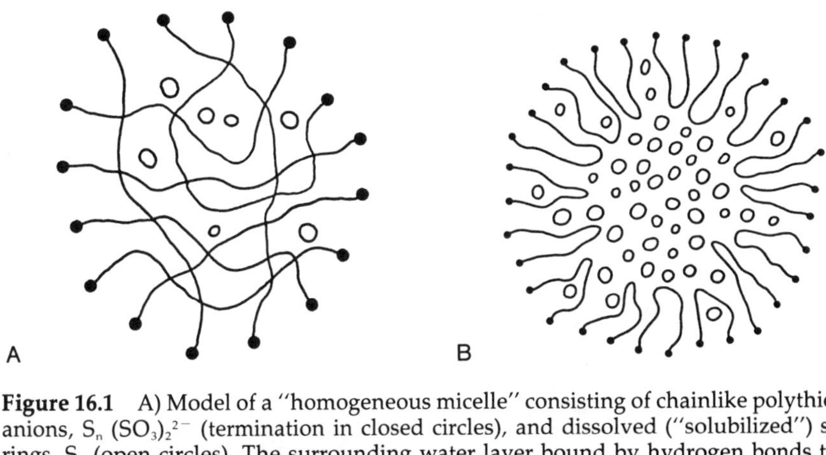

Figure 16.1 A) Model of a "homogeneous micelle" consisting of chainlike polythionate anions, $S_n\,(SO_3)_2^{2-}$ (termination in closed circles), and dissolved ("solubilized") sulfur rings, S_n (open circles). The surrounding water layer bound by hydrogen bonds to the hydrophilic sulfonate groups has been omitted. The number of chain molecules within the micelle may be as high as 50 and a number of cations will be linked to this negatively charged particle by electrostatic forces (after Steudel et al., 1987, 1988a). B) Model of a "heterogeneous micelle" with a core of densely packed sulfur ring molecules (open circles). The surface layer consists of folded long-chain polythionate anions which make this sulfur-rich particle hydrophilic. Some sulfur rings may also be dissolved in the anion layer.

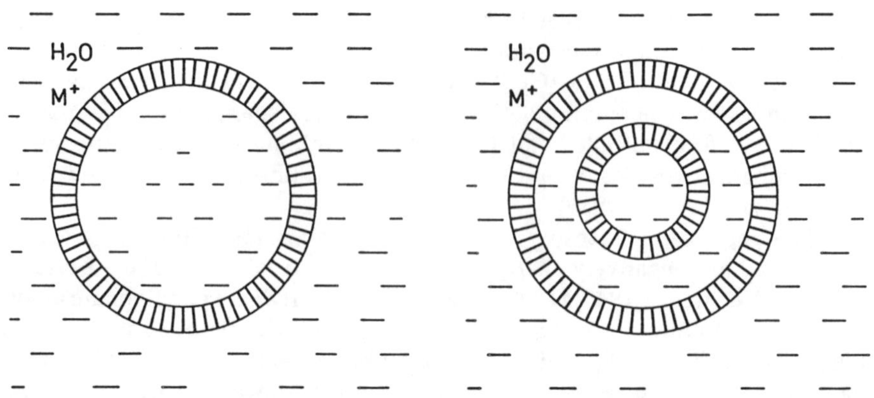

Figure 16.2 Schematic representation of a unilamellar vesicle (left) consisting of a membrane-like envelope of chainlike molecules of radial orientation and a water-filled cavity ("water pool") inside. The interior and surrounding water phases may contain cations (M^+) attracted to the negatively charged membrane surfaces. Large enough vesicles may contain one or more smaller vesicles in their cavity (multilamellar vesicle; right). Micelles may also be present in these cavities.

or multilamellar vesicles with an aqueous internal cavity (Figure 16.2). Such vesicles are more stable than micelles and are not in a thermodynamic equilibrium with the "monomeric constituents" dissolved in the surrounding aqueous phase; the concentration of these monomers is usually very small (Pfüller, 1986).

Obviously, vesicles are much larger than micelles and therefore the particles in hydrophilic sulfur sols most probably consist of polythionate vesicles which contain a certain amount of elemental sulfur dissolved in the hydrophobic membrane (Figure 16.3). The density of such particles is a function of the diameter and the number of polythionate layers. If only one layer is present consisting of polythionates of average composition $S_{40}O_6^{2-}$, the layer had a thickness of about 5 nm. A spherical particle of diameter 100 nm (0.1 μm) has

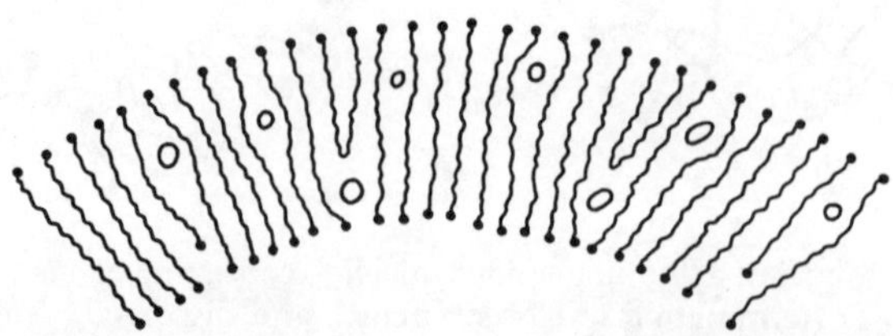

Figure 16.3 Probable structure of the polythionate membrane of the vesicle-type particles in hydrophobic sulfur sols. Anions of different length will be present and sulfur rings seem to be dissolved in the hydrophobic liquid-like part of the membrane. Since the helical sulfur chains are quite flexible as far as their length is concerned, chains with different number of sulfur atoms may adjust to each other. From the structure of polymeric sulfur (Tuinstra, 1967), the length of an undistorted helix can be estimated as 1.38 nm per 10 atoms and the van der Waals diameter as 0.46 nm.

a volume of $4.2 \times 10^{-21} m^3$ of which $1.1 \times 10^{-21} m^3$ or 26 percent would be occupied by the polythionate ions while 74 percent are filled by the water inside the membrane of the vesicle. Taking the densities of water as 1.00 g cm^{-3} and of liquid-like sulfur chains as 1.85 g cm^{-3} (see section 16.2), the overall density would be 1.22 g cm^{-3} (neglecting the exterior water sphere linked by hydrogen bonds to the sulfonate groups). Higher overall density values will be obtained for smaller or multilamellar vesicles; the spontaneous formation of the latter (Figure 16.2) has been observed by electron microscopy in the case of phosphatidylcholine/phosphatidic acid and similar mixtures (Hauser et al., 1983, 1985; Adler and Schiemann, 1985). The presence of polythionate micelles in the intravesicular water pool also would enhance the density.

16.7 S^0 GLOBULES IN SULFUR BACTERIA

The question now arises whether the polythionate vesicles discussed in the previous section can serve as models to understand the formation and properties of bacterial "sulfur globules." Polythionates have been detected in the media of many sulfur bacteria, particularly Thiobacilli, Chromatiaceae, and Chlorobiaceae. By the application of more sensitive, more specific, and more gentle detection techniques like ion-chromatography, which can detect as little as 5 μg/ml of $S_4O_6^{2-}$ (Steudel and Holdt, 1986), it may be found that polythionates are ubiquitous compounds in cultures of sulfur bacteria.

Sulfur-oxidizing bacteria are usually grown with either sulfide or thiosulfate as electron donors. Both compounds easily form polythionates:

$$2HS^- \xrightarrow[-2H^+]{\text{oxidation}} S_2^{2-} \xrightarrow{\text{oxidation}} S_2O_3^{2-} \xrightarrow{\text{oxidation}} \tfrac{1}{2}S_4O_6^{2-} \qquad (17)$$

Tetrathionate may then be hydrolyzed enzymatically at its weakest and most polar S–S bond:

$$S_4O_6^{2-} + H_2O \rightarrow HS_3O_3^- + HSO_4^- \qquad (18)$$

This process resembles the well-known ATP hydrolysis process. However, there may also be an enzymatically catalyzed direct pathway from thiosulfate to sulfanemonosulfonates. The sulfanemonosulfonate $HS_3O_3^-$ may enter reactions like that of equation (11), resulting in polythionates and elemental sulfur according to equations (12) through (15). Polythionates with up to 17 sulfur atoms have in fact been detected in cultures of *Thiobacillus ferrooxidans* (Steudel et al., 1987). This reaction mechanism can explain why bacterial S^0 is formed entirely from the terminal sulfur atom of thiosulfate (Schedel and Trüper, 1980).

A careful and critical evaluation of the published properties of bacterial sulfur globules shows that these particles must be either identical with or very similar to the polythionate vesicles described above. As outlined below, all properties of bacterial S^0 can be understood on the basis of a vesicle nature:

Color	Hydrophilic sulfur is colorless or very pale yellow as are bacterial S^0 globules. Liquid sulfur is deep yellow, even at 20°C.
Particle size	The particle diameters in hydrophilic sulfur sols (0.1–0.5 μm) are in the same range as the size of S^0 globules; of course, there will be a size distribution in both cases.
Hydrophilic nature:	S^0 is hydrophilic. The only known hydrophilic sulfur-rich compounds are long-chain polythionates.
Liquid state	The particles in dilute hydrophilic sulfur sols remain liquid for weeks, as do S^0 globules. Rapid crystallization of S_8 is induced by drying of sulfur sols (Steudel et al., 1988a) and of sulfur globules (Hageage et al., 1970).
Density	The average density of polythionate vesicles will be much lower than the density of elemental sulfur; as has been shown above, it matches the observed density of S^0 globules.
Water content	The S^0 globules have been described as "hydrated sulfur" (Guerrero et al., 1984), which fits the picture of a polythionate membrane enclosing a water-filled cavity.
Particle structure	S^0 globules are usually either spherical or ellipsoidal, as are vesicles. The content of a broken S^0 globule or "sulfur vacuole" (Nicolson and Schmidt, 1971) has been described as a liquid with "myelin-like figures." These features may be caused by pieces from the broken membrane of polythionates (Figure 16.3).
Reactivity	The sulfur of S^0 globules is more reactive toward cyanide than S_8, and the same holds for polythionates (Schedel, 1977; Schedel and Trüper, 1980; Then, 1984). Hydrophilic sulfur can easily be reactivated by the reverse reactions of equations (10)–(13) for instance (Donaldson and Johnston, 1969).
Solubility	Polythionates do not dissolve in carbon disulfide but are soluble in polar organic solvents and to a certain degree in water either as molecular ions or as mi-

celles. Only the S_n content ($n = 6$–12) of hydrophilic sulfur will rapidly dissolve in carbon disulfide (CS_2) (Steudel et al., 1988a). S^0 globules hardly dissolve inCS_2 also (only 5–10 percent), the major part of the sulfur is insoluble (Then, 1984; Moriarty and Nicholas, 1970) unless the long-chain polythionates have decomposed according to equation (16); this reaction occurs during drying, aging, and warming up of the sample; the S_n formed will then dissolve in CS_2.

UV absorption

Reliable UV spectra of S^0 in intact cells or in cell extracts are difficult to obtain because of the strong protein and nucleotide absorption at wavelengths shorter than 300 nm. The absorption spectra of dissolved elemental sulfur (Steudel et al., 1988b), of polythionates (Schmidt and Sand, 1964), and of inorganic polysulfides (Teder, 1971) are similar and characterized by an increase in absorption at shorter wavelengths. Unless the molar extinction coefficients are determined, UV spectra cannot be used to discriminate between the three types of compounds, especially since the chain length and ring size, respectively, also influences the spectra.

Sulfur content

Surprisingly little if any quantitative data about bacterial sulfur globules are available and this applies also to the sulfur content, which has been investigated by means of UV absorption after extraction with ethanol (Schmidt et al., 1971) and by X-ray microanalysis after air drying (Lawry et al., 1981; Vetter, 1985). A high sulfur content of the material in the globules has always been stated but the techniques applied did not provide an exact quantitative analysis of the undecomposed globule material; these techniques also do not discriminate between elemental sulfur, polythionates, or other forms of sulfane sulfur. Electron microscopy of several types of sulfur bacteria has demonstrated that the dried globules contain an electron-dense layer of sulfur attached to the protein envelope of the "sulfur vacuoles" (Strohl et al., 1981; Lawry et al., 1981). This picture is in perfect agreement with the vesicle model of S^0 globules proposed here. In some cases the electron-dense layer is 4–5 nm thick (Strohl et al., 1981), which can be interpreted as a unilamellar polythionate vesicle; in other cases this layer was found to be 40 nm thick (Strohl et al., 1981) probably as a result of a multilayer vesicle structure which on drying has collapsed in such a way that the single layers cannot be observed any more.

Birefringence

Liquid sulfur near the melting point is an isotropic fluid (Das and Das Gupta, 1939) and does not show any optical birefringence. However, sulfur globules from *Chromatium* and other bacteria when investigated with polarized light showed a maltese-cross pattern which is typical for liquid crystals (anisotropic fluids) which are birefringent (Vetter, 1985; Hageage et al., 1970). These observations are fully in accord with if not evidence for the vesicle structure with radially oriented polythionate chains proposed in this work. It is obvious that a lamellar structure such as that shown in Figure 16.3 must show highly anisotropic physical properties.

16.8 CONCLUSIONS

It has been shown that the "elemental sulfur" or "S^0" present in cultures of certain sulfur bacteria as an intracellular or extracellular storage product most likely consists of long-chain polythionates. These anions in water form micelles of about 5 nm diameter which are not visible in a light microscope. They also form vesicles (uni- or multilamellar) which form large globules of up to 1 μm diameter. All chemical and physical properties of bacterial S^0 reported in the literature are in agreement with this view. The term "elemental sulfur" should therefore not be applied any longer to bacterial S^0; it should be replaced by either "polythionates" or "hydrophilic sulfur," which at the present time are synonyms.

The author is grateful to his co-workers whose names appear in the references, and to Mrs. K. Thiel and Dr. E.-M. Strauss for assistance. This work was supported by the Deutsche Forschungsgemeinschaft.

REFERENCES

Adler, K. and **Schiemann, J.** 1985. Characterization of liposomes by scanning electron microscopy and the freeze-fracture technique. *Micron and Microscopica Acta* 16:109–113.

Boulegue, J. 1978. Solubility of elemental sulfur in water at 298 K. *Phosphorus and Sulfur* 5:127–128.

Chivers, T. and **Drummond, I.** 1972. Characterization of the trisulfur radical anion S_3^- in blue solutions of alkali polysulfides in hexamethylphosphoramide. *Inorganic Chemistry* 11:2525–2527.

Das, S.R. and **Das Gupta, K.** 1939. X-Ray diffraction by supercooled liquid sulphur. *Nature* 143:332.

Davis, R.E. 1958. Displacement reactions at the sulfur atom. I. An interpretation of the decomposition of acidified thiosulfate. *Journal of the American Chemical Society* 80:3565–3569.

Donaldson, G.W. and **Johnston, F.J.** 1969. The reaction of colloidal sulfur with sulfite. *The Journal of Physical Chemistry* 73:2064–2068.

Fuhrhop, J.-H. and **Mathieu, J.** 1984. Wege zu funktionalen Vesikelmembranen ohne Proteine. *Angewandte Chemie* 96:134–137.

Fuhrhop, J.-H. and **Fritsch, D.** 1986. Bolaamphiphiles form ultrathin, porous, and unsymmetric monolayer lipid membranes. *Accounts of Chemical Research* 19:130–137.

Fuhrhop, J.-H., David, H.-H., Mathieu, J., Liman, U., Winter, H.-J., and **Boekema, E.** 1986. Bolaamphiphiles and monolayer lipid membranes made from 1,6,19,24-tetraoxa-3,21-cyclohexatriacontadiene-2,5,20, 23-tetrone. *Journal of the American Chemical Society* 108:1785–1791.

Giggenbach, W. 1971a. The blue solutions of sulfur in water at elevated temperatures. *Inorganic Chemistry* 10:1306–1311.

Giggenbach, W. 1971b. Optical spectra of highly alkaline sulfide solutions and the second dissociation constant of hydrogen sulfide. *Inorganic Chemistry* 10:1333–1338.

Giggenbach, W. 1972. Optical spectra and equilibrium distribution of polysulfide ions in aqueous solution at 20°C. *Inorganic Chemistry* 11:1201–1207.

Giggenbach, W. 1974. Equilibria involving polysulfide ions in aqueous sulfide solutions up to 240°C. *Inorganic Chemistry* 13:1724–1730.

Gmelin Handbuch der Anorganischen Chemie. 1953. 8. Auflage. Schwefel, Teil A, Lieferung 2, p. 485–502, Verlag Chemie, Weinheim.

Göbel, T. 1984. Synthese und Eigenschaften von organylsubstituierten Polysulfanen R_2S_n (n = 2–9). Ein Beitrag zum Stoffwechsel der Schwefelbakterien. *Diplomarbeit, Technische Universität Berlin.*

Guerrero, R., Mas, J., and **Pedrós-Alió, C.** 1984. Buoyant density changes due to intracellular content of sulfur in *Chromatium warmingii* and *Chromatium vinosum*. *Archives of Microbiology* 137:350–356.

Hageage, G.J. Jr., Eanes, E.D., and **Gherna, R.L.** 1970. X-Ray diffraction studies of the sulfur globules accumulated by Chromatium species. *Journal of Bacteriology* 101:464–469.

Hahn, J. 1985. Zusammensetzung von Rohsulfan, Nachweis der Sulfane H_2S_9 bis H_2S_{35}. *Zeitschrift für Naturforschung* 40b:263–272.

Hauser, H., Gains, N., and **Müller, M.** 1983. Vesiculation of unsonicated phospholipid dispersions containing phosphatidic acid by pH adjustment: physiochemical properties of the resulting unilamellar vesicles. *Biochemistry* 22:4775–4781.

Kerker, M., Daby, E., Cohen, G.L., Kratohvil, J.P., and **Matijevic, E.** 1963. Particle size distribution in LaMer sulfur sols. *Journal of Physical Chemistry* 67:2105–2111.

Lawry, N.H., Jani, V., and **Jensen, T.E.** 1981. Identification of the sulfur inclusion body in Beggiatoa alba B18LD by energy-dispersive X-Ray microanalysis. *Current Microbiology* 6:71–74.

Möckel, H.J., Freyholdt, T., and **Weiss, J.** 1983. The HPLC of divalent sulfur. Elemental sulfur rings and nonionic sulfur compounds, pp. 160–185 in Molnar, I. (editor), *Practical Aspects of Modern HPLC*, Walter de Gruyter, Berlin.

Moriarty, D.J.W. and **Nicholas, D.J.D.** 1970. Products of sulphide oxidation in extracts of *Thiobacillus concretivorus. Biochimica et Biophysica Acta* 197:143–151.

Nakabayashi, T., Tsurgi, J., and **Yabuta, T.** 1964. Organic polysulfides. IV. Synthesis of bis(triphenylmethyl) polysulfides. *Journal of Organic Chemistry* 29:1236–1238.

Nicolson, G.L. and **Schmidt, G.L.** 1971. Structure of the Chromatium sulfur particle and its protein membrane. *Journal of Bacteriology* 105:1142–1148.

Pfüller, U. 1986. Mizellen-Vesikel-Mikroemulsionen, Tensidassoziate und ihre Anwendung in Analytik und Biochemie, Springer Verlag, Berlin.

Schedel, M. 1977. Untersuchungen zur Anaeroben Oxidation Reduzierter Schwefelverbindungen durch *Thiobacillus denitrificans, Chromatium vinosum,* und *Chlorobium limicola.* Dissertation, Universität Bonn.

Schedel, M. and **Trüper, G.** 1980. Anaerobic oxidation of thiosulfate and elemental sulfur in *Thiobacillus denitrificans. Archives of Microbiology* 124:205–210.

Schmidt, G.L., Nicolson, G.L., and **Kamen, M.D.** 1971. Composition of the sulfur particle of *Chromatium vinosum* strain D. *Journal of Bacteriology* 105:1137–1141.

Schmidt, M. and **Sand, T.** 1964. Physikalische Charakterisierung der Sulfan-Disulfonate $K_2S_3O_6$ bis $K_2S_6O_6$; UV Extinktionen wässriger Sulfan-Disulfonatlösungen. *Journal of Inorganic and Nuclear Chemistry* 26:1173–1177.

Steudel, R. and **Mäusle, H.-J.** 1979. Molecular composition of the sulfur from acidified sodium thiosulfate. *Zeitschrift für Allgemeine und Anorganische Chemie* 457:165–173.

Steudel, R. 1982. Homocyclic sulfur molecules. *Topics in Current Chemistry* 102:149–176.

Steudel, R., Paßlack-Stephan, S., and **Holdt, G.** 1984a. Thermal polymerization and depolymerization reactions of 10 sulfur allotropes studied by HPLC and DSC. *Zeitschrift für Allgemeine und Anorganische Chemie* 517:7–42.

Steudel, R. 1984b. Sulfur: the significance for chemistry, for the geo-, bio-, and cosmophere and technology, pp. 3–37, Müller, A. and Krebs, B. (editors), *Elemental Sulfur and Related Homocyclic Compounds and Ions,* Chapter 1, Elsevier Science Publ., Amsterdam.

Steudel, R., Strauss, R., and **Koch, L.** 1985a. Quantitative HPLC-analysis and thermodynamics of the sulfur melt. *Angewandte Chemie International Edition in English* 24:59–60.

Steudel, R., Strauss, R., and **Jensen, D.** 1985b. Qualitative and quantitative HPLC analysis of chlorosulfanes S_nCl_2 (n = 1–30) after derivatization. *Chemikerzeitung* 109:349–350.

Steudel, R., Holdt, G., and **Nagorka, R.** 1986. On the autoxidation of aqueous sodium polysulfide. *Zeitschrift für Naturforschung* 41b:1519–1522.

Steudel, R. and **Holdt, G.** 1986. Ion-pair chromatographic separation of polythionates $S_n O_6^{2-}$ with up to thirteen sulfur atoms. *Journal of Chromatography* 361:379–384.

Steudel, R., Holdt, G., Göbel, T., and **Hazeu, W.** 1987. Chromatographic separation of higher polythionates $S_nO_6^{2-}$ (n = 3–22) and their detection in cultures of *Thiobacillus ferrooxidans;* molecular composition of bacterial sulfur secretions. *Angewandte Chemie International Edition in English* 26:151–153.

Steudel, R. and **Holz, B.** 1988. Detection of reactive sulfur molecules (S_6, S_7, S_9, $S\infty$) in commercial sulfur, in sulfur minerals, and in sulfur melts slowly cooled to 20°C. *Zeitschrift für Naturforschung* 43b:581–589.

Steudel, R., Göbel, T., and **Holdt, G.** 1988a. The molecular composition of hydrophilic sulfur sols prepared by acid decomposition of thiosulfate. *Zeitschrift für Naturforschung* 43b:203–218.

Steudel, R., Jensen, D., Göbel, P., and **Hugo, P.** 1988b. Optical absorption spectra of the homocyclic sulfur molecules S_n (n = 6,7,8,9,10,12,15,20) in solution. *Berichte der Bunsengesellschaft für Physikalische Chemie* 92:118–122.

Strohl, W.R., Geffers, I., and **Larkin, J.M.** 1981. Structure of the sulfur inclusion envelopes from four Beggiatoas. *Current Microbiology* 6:75–79.

Teder, A. 1971. The equilibrium between elementary sulfur and aqueous polysulfide solutions. *Acta Chemica Scandinavia* 25:1722–1728.

Then, J. and **Trüper, G.** 1983. Sulfide oxidation in *Ectothiorhodospira abdelmalekii.* Evidence for the catalytic role of cytochrome c-551. *Archives of Microbiology* 135:254–258.

Then, J. 1984. Beiträge zur Sulfidoxidation durch *Ectothiorhodospira abdelmalekii* und *Ectothiorhodospira halochloris*. Dissertation, Universität Bonn.

Then, J., and **Trüper, G.** 1984. Utilization of sulfide and elemental sulfur by *Ectothiorhodospira halochloris*. *Archives of Microbiology* 139:295–298.

Tuinstra, F. 1967. *Structural Aspects of the Allotropy of Sulfur and the Other Divalent Elements*. Waltman, Delft.

Verstraete, E.O.K. 1943. Der Ionenaustausch an kolloidem Schwefel, Beitrag zur Kenntnis des Ionenaustausches in lyophoben kolloiden Systemen. *Kolloid-Zeitschrift* 102:251–267.

Vetter, R.D. 1985. Elemental sulfur in the gills of three species of clams containing chemoautotrophic symbiotic bacteria: a possible inorganic energy storage compound. *Marine Biology* 88:33–42.

Weitz, E. and **Spohn, K.** 1956a. Über höhere Polythionsäuren, II. Mitteilung: Polythionsäuren mit mehr als 6 Schwefelatomen. *Chemische Berichte* 10:2332–2345.

Weitz, E., Becker, F., and **Gieles, K.** 1956b. Über höhere Polythionsäuren, III. Mitteilung: Vergleich der Darstellungsmethoden für Kaliumhexathionat. *Chemische Berichte* 10:2345–2352.

Weitz, E., Becker, F., Gieles, K., and **Alt, B.** 1956c. Über höhere Polythionsäuren, IV. Mitteilung: Polythionsäuren mit 15–40 Schwefelatomen. *Chemische Berichte* 10:2353–2365.

Weitz, E., Gieles, K., Singer, J., and **Alt, B.** 1956d. Über höhere Polythionsäuren, V. Mitteilung: Über die Polythionat-Natur der hydrophilen Odenschen Schwefelsole. *Chemische Berichte* 10:2365–2375.

Winogradsky, S. 1887. Über Schwefelbakterien. *Botanische Zeitung* 31:490–507.

17

Aerobic Hydrogen-oxidizing (Knallgas) Bacteria

Hans Günter Schlegel

Institute for Microbiology
University of Göttingen
Göttingen, Federal Republic of Germany

17.1 INTRODUCTION

The aerobic chemolithoautotrophic hydrogen-oxidizing (hydrogenotrophic) bacteria are characterized by their ability to grow on a mineral medium under an atmosphere of gaseous hydrogen and oxygen* and carbon dioxide as carbon source. Since the isolation of the first representatives of this physiological group, they have been called "Knallgas bacteria" (Kaserer, 1906; Ruhland, 1924). This short designation emphasizes that these aerobic bacteria differ unmistakably from the anaerobic hydrogen-oxidizing bacteria such as the methanogenic, acetogenic, and sulfidogenic chemolithoautotrophic bacteria. For reasons of brevity and clearness the term Knallgas bacteria will therefore be used in this chapter.

Various aspects of the Knallgas bacteria have been reviewed: habitats, isolation, identification, and taxonomy (Aragno and Schlegel, 1981), physiology and biochemistry (Bowien and Schlegel, 1981), the regulation and genetics of enzymes (Schlegel, 1984), evolution (Friedrich, 1985), and biotechnology (Schlegel and Schneider, 1985; Sebo, 1986). Among the lithoautotrophs, the Knallgas bacteria are useful models for the study of problems of general biological interest. Furthermore, the structure, formation and regulation of the key enzymes of lithoautotrophy have been studied at the molecular level in these organisms (see Chapters 23 and 24).

17.2 HABITATS AND ECOLOGICAL NICHES OF KNALLGAS BACTERIA

Although the first Knallgas bacteria were isolated many years ago, in the past 30 years many new genera, species, and strains of Knallgas bacteria have been

*The German word "Knallgas" is often used to refer to this inflammable mixture of hydrogen and oxygen (*Knall* means *explosion* in German).

isolated (Table 17.1). With more extensive knowledge of the physiology, biochemistry, and genetics of bacteria and with a wealth of information on the physical and chemical parameters of various ecosystems, we may ask basic questions about the habitats (i.e., the dwelling places) and ecological niches (i.e., the functions) of microorganisms again. It, therefore, appears justified to begin this chapter on the Knallgas bacteria with a consideration of possible habitats and ecological niches (Schlegel and Jannasch, 1981).

Where in nature do ecosystems or microenvironments exist which are permanently or transiently supplied with gaseous hydrogen? Hydrogen plays an important role in the biosphere. It is a product of anaerobic fermentations and an important intermediate in the anaerobic food chain. Most of this anaerobically produced hydrogen is converted *in situ* to methane, acetic acid, and hydrogen sulfide, and only traces leave the anoxic zones and reach oxygen-containing environments. Only the hydrogen formed by N_2-fixing bacteria, especially by those in root nodules, is exclusively available to aerobic organisms.

Part of the hydrogen produced on land and in the oceans reaches the atmosphere. Hydrogen is also a product of various chemical reactions going on in the gaseous phase in the atmosphere, due to conversions of methane, carbon monoxide, and other trace gases. In fact, the most important reservoir of gaseous hydrogen in nature is the troposphere, which contains 2.2×10^9 tons H_2; it is, however, present at very high dilution; the mixing ratio is only 0.56×10^{-6} v/v or 0.56 ppmv. The major fraction of the atmospheric hydrogen is oxidized in the soil, which is considered to be the dominant sink of the atmospheric H_2 cycle (Conrad and Seiler, 1980a, b, c). It has thus been assumed that Knallgas bacteria of the soil are responsible for the decomposition of atmospheric H_2 in nature and that soil is the primary ecological niche of the Knallgas bacteria (Conrad, 1984).

However, the distribution and function of the Knallgas bacteria in nature and their participation in the geochemical cycles cannot be described as clearly as the habitats of other chemolithotrophs such as the nitrifying bacteria. For the majority of Knallgas bacteria, hydrogen autotrophy is primarily an alter-

Table 17.1 Genera and species of Knallgas bacteria

Gram-negative species:		**Gram-positive species:**
Alcaligenes eutrophus	*Pseudomonas facilis*	*Arthrobacter* strain 11/X
A. hydrogenophilus	*P. flava*	*Bacillus schlegelii*
A. ruhlandii	*P. pseudoflava*	*B. tusciae*
A. latus	*P. hydrogenovora*	*Mycobacterium gordonae*
A. paradoxus	*P. hydrogenothermophila*	*Nocardia autotrophica*
A. eutrophus CH34	*P. palleronii*	*N. opaca*
Aquaspirillum autotrophicum	*P. thermophila*	
Azospirillum lipoferum	*P. saccharophila*	
Calderobacterium hydrogenophilum	*Renobacter vacuolatum*	
Derxia gummosa	*Rhizobium japonicum*	
Flavobacterium autothermophilum	*Xanthobacter autotrophicus*	
Hydrogenobacter thermophilus	*X. flavus*	
Microcyclus aquaticus	Most carboxydobacteria (see	
Paracoccus denitrificans	Chapter 18)	

native mode of nutrition; Knallgas bacteria are facultative autotrophs and nutritionally very versatile. It is not known which of their nutritional capabilities confers the decisive selective advantage to them. Since they belong to many different genera, generalizations are not possible.

There are only a few places in nature where molecular hydrogen is released, converted, or consumed and will possibly be transiently or permanently available to aerobic microorganisms. We consider four such locations: 1) zones above or within anoxic ecosystems such as water bodies of ponds, swamps, and lakes, and the gaseous drainage from landfills and compost heaps; 2) the soil on top of root systems bearing nodules with N_2-fixing bacteria; 3) the top soil layer in contact with atmospheric hydrogen; and 4) geothermal areas such as hydrothermal vents and hot springs.

Molecular hydrogen escaping from anoxic ecosystems Two kinds of anoxic ecosystems where gaseous or dissolved molecular hydrogen escapes have been recognized: water bodies of ponds, ditches, and lakes, and soil and other solid material covering heaps of organic matter such as silage, organic waste (landfills), and litter (compost). In many lakes and ponds of the temperate climate zones a sediment is formed containing exogenous litter or endogenously produced plankton, i.e., organic material such as green algae, cyanobacteria, and other phototrophic bacteria. Starting within the sediment and continuing in the hypolimnion, the organic matter, mainly consisting of cell walls and cytoplasmic components of bacteria and algae, is decomposed by fermentative bacteria (Conrad et al., 1983b, 1985; Zeikus, 1983). The biodegradative anaerobic food chain involves polymer-degrading bacteria, H_2-producing acetogenic bacteria, homoacetogenic bacteria, methanogenic and sulfidogenic bacteria. Hydrogen plays an important role in the substrate transfer between the bacteria involved; H_2 is, however, *in situ* scavenged by the methanogenic, acetogenic, and sulfidogenic bacteria and does not apppear in the gaseous state. Dissolved hydrogen is not even detectable in the anoxic waters above the sediment and thus does not escape to the oxic water layers of the epilimnion (Schink and Zeikus, 1984). However, Knallgas bacteria are present on the sediment surface and in the epilimnion. This distribution may be due to the association of Knallgas bacteria with cyanobacteria which produce hydrogen as by-product of N_2 fixation. This conclusion is based on the correlation of bacterial cell counts and the distribution of H_2 in the water column (Figure 17.1). The presence of Knallgas bacteria at the sediment surface may be due to their association with cyanobacteria which sedimented. Although no studies have been done, Knallgas bacteria may inhabit the cell envelopes and slime sheaths of cyanobacteria.

Compost heaps and landfills lend special conditions to the decomposition and mineralization of organic matter such as litter and organic waste. Contrary to lake sediments and sewage digesters, these ecosystems are not permanently water-saturated and anoxic. Therefore, the biodegradative processes of the anaerobic food chain may not proceed undisturbed. The temporal course of decomposition in landfills and compost heaps is rather characterized by fluctua-

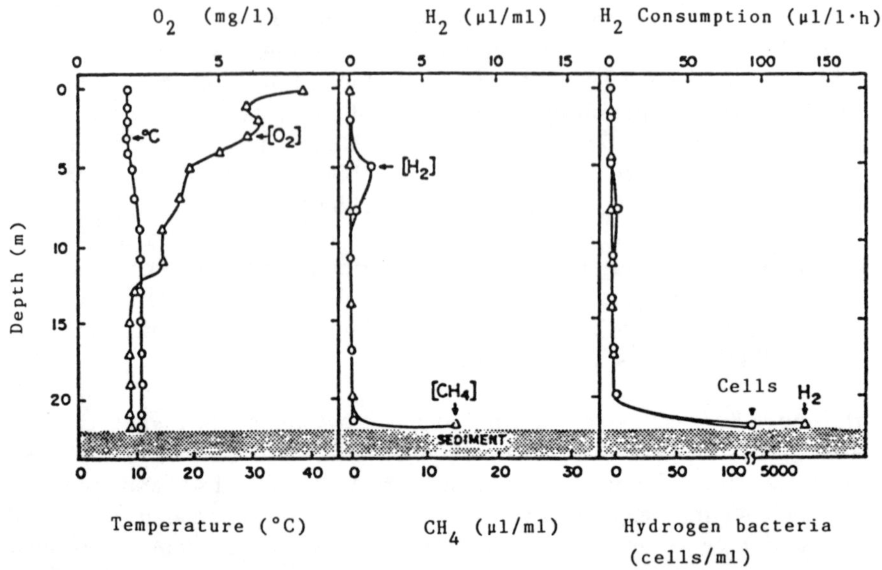

Figure 17.1 Relation of Lake Mendota water column depth to *in situ* temperature, oxygen, hydrogen, methane, and aerobic hydrogenase activity. From Schink and Zeikus (1984).

tions in water-saturation, drying and access of oxygen from the air, the productivity depending on rainfall and drought. The irregular periodic changes result in a partial interruption of fermentation processes, especially those involving interspecies hydrogen transfer. In consequence, a substantial portion of the molecular hydrogen may escape its primary production site in the gaseous state and enter the stream of gaseous decomposition products (CO_2 and CH_4) released from anoxic zones or microenvironments. Up to 30 percent (v/v) H_2 was measured in the drainage gas of landfills (Figure 17.2). The transient access of air in dry periods will provide conditions for the growth of Knallgas bacteria (and of course methanotrophic bacteria). Recent model experiments with soil columns flushed upward with a CO_2-containing Knallgas mixture showed that bacterial growth and H_2 consumption actually occurs. Most of the organisms were N_2-fixing Knallgas bacteria (Dugnani et al., 1986).

Molecular hydrogen escaping from root nodules Hydrogen escaping from nitrogen-fixing bacteria of endosymbiotic plant-bacterial systems should provide a selective advantage for those aerobic bacteria which are able to use hydrogen for energy generation and growth. The nitrogenase reaction results in the production of not only ammonia but also gaseous hydrogen: $N_2 + 8[H] \rightarrow 2NH_3 + H_2$. The production of hydrogen by the nitrogenases is an intrinsic property of the enzyme. Most nitrogen-fixing bacteria produce hydrogenase, and the H_2 produced by nitrogenase may have a protective function for the oxygen-sensitive nitrogenase by providing a reducing agent which *in situ* can

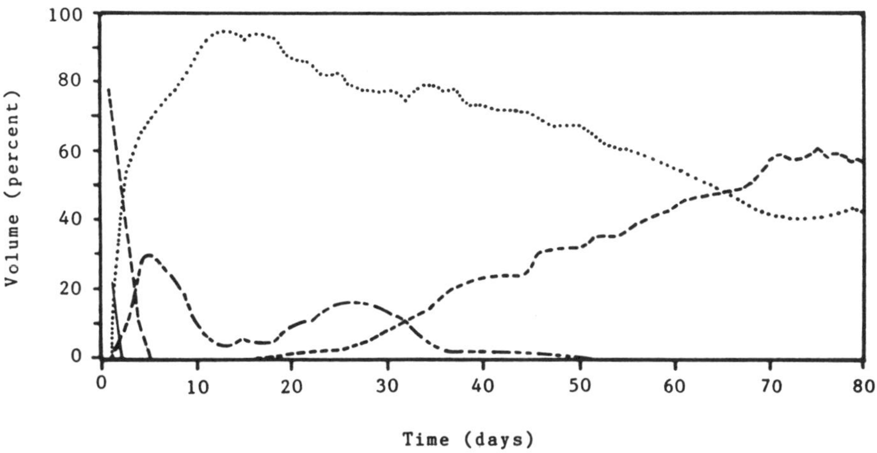

Figure 17.2 Evolution of gases from a mixture of organic wastes deposited under anoxic conditions kept at 33–35°C. Symbols: ·······, carbon dioxide; -------, methane; ————, gaseous hydrogen; ----, gaseous nitrogen; ———, oxygen. Redrawn from Dugnani et al. (1986).

trap diffusing O_2 by reduction. Only rhizobia are exceptions; the majority of strains and species lack hydrogenase. As a consequence root nodules generally emit H_2 into the environment.

As shown by Conrad and Seiler (1980b), hydrogen is released from legumes into the atmosphere in significant amounts, adding about 1 to 5 percent of the total H_2 production to the global hydrogen budget. Hydrogen evolution from clover fields reflected the growth season of clover *Trifolium hybridum* (Figure 17.3) and the N_2 fixation rates and ceased in darkness or was absent in ammonium-fertilized plots. Whether root nodules or rhizothamnia containing *Frankia* as symbiont as well as free-living N_2-fixing bacteria evolve H_2 *in situ* is not known. The seasonal measurements of H_2 production rates in clover fields indicated that the H_2 consumption (decomposition) rates followed the peak of H_2 production rates with a lag of several months (Figure 17.3). Whether the increase of the decomposition rates originated from the increase of populations of Knallgas bacteria was not examined. The role of nitrogenase-produced hydrogen as a source for growth of Knallgas bacteria remains so far obscure.

Atmospheric hydrogen The soil accounts for 95 percent of the total sink activity of the atmospheric hydrogen (Conrad and Seiler, 1979), the involvement of the Knallgas bacteria is highly improbable. The K_m values (H_2) for the H_2 decomposition activity of soils were extremely low (24–60 μl H_2/l, equivalent to H_2 concentrations in water of about 20–45 nM.) In contrast, the K_m values (H_2) for H_2 oxidation of all Knallgas bacteria tested are much higher, greater than 1 μM H_2 throughout (Conrad et al., 1983a). The K_m values (H_2) for cell-free hydrogenases were reported to range between 1 and 60 μM H_2

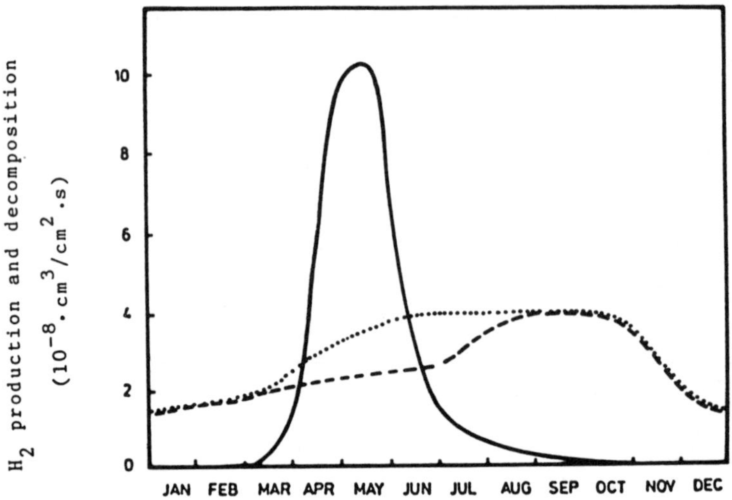

Figure 17.3 Seasonal variation of the gross H_2 production and decomposition rates observed on uncultivated grass and clover fields. Symbols: ———, H_2 production on clover; — — —, H_2 decomposition on clover; ·······, H_2 decomposition on grass. Redrawn from Conrad and Seiler (1980c).

(Schlegel and Schneider, 1978). Experiments were also conducted on the H_2 threshold, i.e., the H_2 concentration beneath which H_2 is not taken up and used by suspensions of intact cells of hydrogen bacteria (Figure 17.4 and Table 17.2). When suspensions of *Aquaspirillum autotrophicum* SA32, *Alcaligenes eutrophus* H16, or *Xanthobacter autotrophicus* 7c were incubated under air containing 47 to 200 ppmv H_2, the H_2 concentration decreased logarithmically with time until a particular H_2 concentration was reached. These threshold values varied between 1 and 8 ppmv H_2 among different species (Table 17.2) but never dropped below 1 ppmv H_2. As a control, experiments with moist soil and with lake water as catalysts were included. With those samples, H_2 concentrations as low as 0.01 ppmv H_2 were reached. These experiments conclusively show that the Knallgas bacteria tested are not able to utilize hydrogen at the concentrations found in the atmosphere because of low affinity for hydrogen (high K_m) and high H_2 threshold value. "The observed H_2 consumption by soils and waters may be due to H_2 consumption by unknown bacteria or to oxidation of H_2 by abiontic hydrogenases" (Conrad et al., 1983a).

In a discussion on aerobic microorganisms able to utilize and grow on atmospheric trace gases (H_2, CO, CH_4), Conrad (1984) concluded, "The presently described strains of hydrogen bacteria, carboxydobacteria, and methanotrophic bacteria cannot grow on atmospheric trace gases as sole energy source, because their affinities are much lower than those observed in soil and water ecosystems." Besides the well-known bacterial flora, which is at least sporadically supplied by the gases from endogenous sources, Conrad postulates the existence of a "high affinity" flora able to use the atmospheric trace gases. Thus, effort should be expended to enrich for and isolate more strains of Knall-

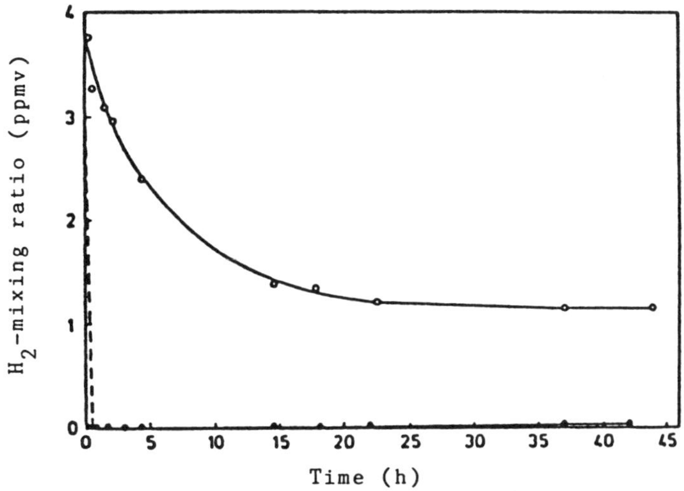

Figure 17.4 Change of the H_2 mixing ration in the headspace over a suspension of *Aquaspirillum autotrophicum* SA32 (•,○) and over a soil sample(— — —). The headsppace of flasks contained air with hydrogen (○), and air without hydrogen (•). Redrawn from Conrad et al. (1983a).

gas bacteria, to examine their enzymes, and to study their behavior under mixotrophic conditions.

Geothermal areas Further sources of molecular hydrogen and other electron donors for chemolithoautotrophic microorganisms are the spreading centers of the sea floor and volcanic areas. In the spreading centers, molten magma rises to the sea floor, cools, and forms the new surface of the ocean bottom. When seawater penetrates through fissures into the basaltic rock, it warms and reacts with minerals, thus dissolving salts and producing ions (HS^-, NH_4^+, Fe^{2+}, Mn^{2+}) and releasing volcanic gases (H_2, CO, CH_4). Several of these substances, when reaching the ocean floor at hydrothermal vents or cold seeps, are potential substrates of aerobic chemoautotrophic life (Karl, 1987). The macro- and microorganisms growing on the basis of bacterial primary production in the light-

Table 17.2 Apparent K_m values for hydrogen and threshold concentrations of H_2 of representative hydrogen bacteria

Strain or species	K_m (μM)	Threshold dissolved H_2 (nM)	Threshold of H_2 in gas phase (ppmv)
Alcaligenes eutrophus H16	>1	6.2	8.3
Alcaligenes paradoxus SA29	>1	4.2	5.5
Xanthobacter autotrophicus 7C	>1	5.0	6.7
Xanthobacter SA35	>1	0.9	1.3
Aquaspirillum autotrophicum SA32	>1	1.9	2.5
Strain PP5	>1	2.1	2.8
Strain PP18	>1	24	33
Strain PP19	>1	133	178

From Conrad et al. (1983a).

less depth of the oceans have been discovered by Jannasch and Wirsen (1979, 1981), and are discussed in Chapter 8. Regrettably, Knallgas bacteria have not yet been isolated from waters of the hydrothermal vents and their surroundings although several species of anaerobic H_2-utilizing bacteria have been discovered (see Chapter 9).

Geothermal areas on the land provide ecosystems for Knallgas bacteria, too. The volcanic gases contain up to 4–10 percent (v/v) H_2; hot springs deliver waters which are either not contaminated by organic material or contain only traces of organic substrates. In these ecosystems, thermophilic Knallgas bacteria such as *Bacillus tusciae* (Bonjour and Aragno, 1984), *Hydrogenobacter thermophilus* (Kawasumi et al., 1980; Kristjansson et al., 1985), and *Hydrogenobacter* spec. (Bonjour and Aragno, 1986) have been isolated.

17.3 CHARACTERISTICS OF KNALLGAS BACTERIA

As discussed above, our knowledge of the ecology of Knallgas bacteria is exceedingly fragmentary and their roles in aerobic food webs uncertain. Further work on the ecology of these organisms will be greatly benefitted by the extensive knowledge of the basic physiological properties of these organisms: nutritional requirements, growth rates, tolerance to environmental conditions, dormancy, regulation of biosynthesis, and genetic structure. The earlier work on these topics has been reviewed by Bowien and Schlegel (1981), and the following discussion emphasizes more recent work.

Hydrogenases are the key enzymes in the activation or formation of molecular hydrogen (Schlegel and Schneider, 1978; Adams et al., 1981); they catalyze the reversible reaction

$$H_2 \rightleftharpoons 2H^+ + 2e^-$$

In Knallgas bacteria, hydrogenases have two functions: 1) to channel electrons into the electron transport chain, and 2) to generate reducing power for CO_2 fixation. Two hydrogenases are known: one is membrane bound and does not reduce NAD *in vitro*; the other is a soluble, cytoplasmic NAD-reducing enzyme. Both enzymes can serve either of the two functions mentioned above. A few hydrogenases have been isolated from various Knallgas bacteria and their structures compared; all contained nickel and iron-sulfur clusters.

Most Knallgas bacteria contain only a membrane-bound hydrogenase. This type of enzyme delivers electrons in the middle of the electron transport chain, probably to ubiquinones and cytochrome *c*. Reduction of NAD requires reverse electron transport. There is an astonishing similarity between the membrane-bound hydrogenases of Knallgas bacteria, the nitrogen-fixing bacteria, and fermentative bacteria (Pinkwart et al., 1983; Arp et al., 1985; Podzuweit et al., 1986; Podzuweit et al., 1988; Knüttel et al., 1988).

By solubilizing the enzyme proteins from purified cytoplasmic membranes and by phase separation, chromatography on DEAE-sephacel, and Procion-

Red HE-3B-sepharose and related steps, enzyme preparations with specific activities of about 100 to 300 μmol $H_2 \cdot min^{-1} \cdot mg$ protein^{-1} (with methylene blue as acceptor) and molecular weights of about 100,000 were obtained. Sodium dodecyl sulfate (SDS) gel electrophoresis revealed two nonidentical subunits with relative molecular weights of 62,000 to 67,000 and 30,000 to 35,000. The studies were carried out on *Alcaligenes eutrophus* H16 (Schink and Schlegel, 1979, 1980; Podzuweit et al., 1987; Arp et al., 1985), *A. eutrophus* type strain (Schneider et al., 1983; Podzuweit et al., 1987), *A. latus* (Pinkwart et al., 1983; Arp et al., 1985), *Paracococcus denitrificans* (Knüttel et al., 1988), *Pseudomonas pseudoflava* (Schneider et al., 1983; Weiss et al., 1980) and *Rhizobium japonicum* (Harker et al., 1986; Stults et al., 1986; Arp and Burris, 1981; Arp et al., 1985; Tilak et al., 1984). Very similar data were reported on the hydrogenase from *Azotobacter vinelandii* (Arp et al., 1985; Seefeldt and Arp, 1986; Seefeldt et al., 1986).

The enzymes of the Knallgas and N_2-fixing bacteria differ from the hydrogenases isoenzyme I and II from *Escherichia coli* insofar as the latter enzymes are reported to be tetramers consisting of two large and two small subunits and to have a very low specific activity with methylene blue compared with benzyl viologen as acceptors (Sawers and Boxer, 1986; Ballantine and Boxer, 1986; Harker et al., 1986).

In summary, the membrane-bound hydrogenases of the Knallgas bacteria investigated appear to resemble each other to a high degree (Table 17.3). They even showed immunological relationships and similar electron-spin resonance spectra. Future work should include new well-characterized Knallgas species, other N_2-fixing bacteria, and the characteristic species of phototrophic bacteria to which high 16S rRNA homologies exist. We look forward to seeing data on amino acid and DNA sequencing of the subunits.

Table 17.3 Composition of membrane-bound hydrogenases

	Molecular weight		
Strain or species	Holoenzyme	Subunit	Reference
Paracoccus denitrificans	100,000	64,800 34,800	Knüttel et al., 1988
Alcaligenes eutrophus H16	95,000	62,000 31,000	Podzuweit et al., 1987
A. eutrophus type strain	95,000	62,000 31,000	Podzuweit et al., 1987
A. latus	101,000	67,000 34,000	Pinkwart et al., 1983
Pseudomonas pseudoflava	98,000	65,000 30,000	Weiss et al., 1980
Rhizobium japonicum	90–99,000	60–64,000 30–35,000	Harker et al., 1986 Seefeldt and Arp, 1986
Azotobacter vinelandii	98,000	67,000 31,000	Seefeld and Arp, 1986
Escherichia coli K-12			
Isoenzyme I	200,000	2 × 64,000 2 × 35,000	Sawers and Boxer, 1986
Isoenzyme II	180,000	2 × 61,000 2 × 30,000	Ballantine and Boxer, 1986

Soluble, cytoplasmic hydrogenases able to reduce NAD are formed by only a few Knallgas bacteria. In *Nocardia opaca* there is only this type of enzyme. Various *Alcaligenes eutrophus* strains, *A. hydrogenophilus,* and some newly isolated strains contain both types, the soluble and the membrane-bound hydrogenase (Figure 17.5).

The soluble hydrogenase of *A. eutrophus* H16 consists of four subunits with molecular weights of 63,000, 56,000, 30,000, and 26,000 (Schneider et al., 1984a, b; Hornhardt et al., 1986). One of the large subunits (SU1) contains flavin mononucleotide, the other (SU2) nickel, and both contain FeS clusters. The enzyme from *N. opaca* can be easily cleaved into two dimers each consisting of a large and small subunit (Figure 17.6). Acceptor studies showed that one dimer (Su1 and 3) exhibits diaphorase activity and is reactive with NAD and artificial acceptors, the other dimer (SU2 and 4) contains the H_2-activating site and reduces only methylene blue and some other acceptors. The holoenzyme can be reconstituted from both dimers and reduces NAD with H_2. The soluble hydrogenase is a unique enzyme and has so far only been detected in the bacteria mentioned above.

The soluble hydrogenase confers a high efficiency and rate of growth to its possessors. While bacteria having only a membrane-bound hydrogenase require approximately 10 to 12 mol H_2 to assimilate 1 mol of CO_2, the strains containing both hydrogenases require only 4.5 to 6 mol H_2 to assimilate 1 mol CO_2 (Schlegel and Bartha, 1961; Schink and Schlegel, 1978a, b). In the latter strains, the specific rate of H_2 uptake is strictly controlled by the presence of CO_2. If CO_2 is lacking, the H_2 uptake rate is less than a third compared with the rate in the presence of CO_2. The presence of both hydrogenases is thus

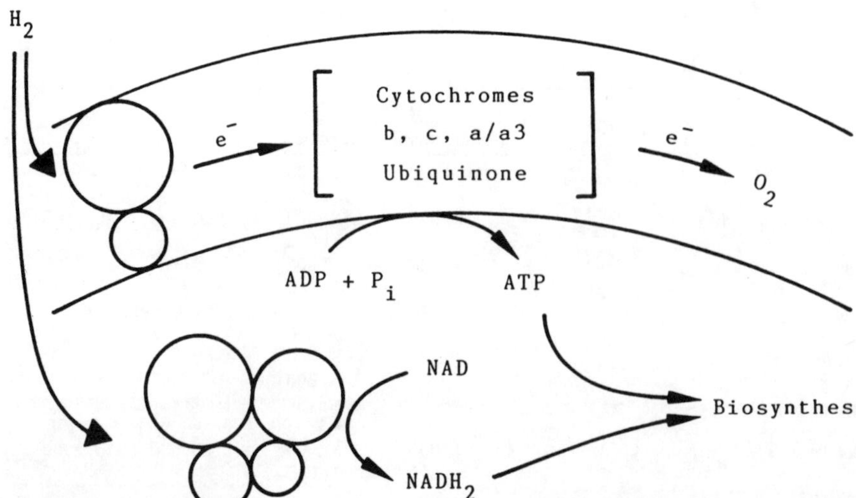

Figure 17.5 The function of the two hydrogenases present in *Alcaligenes eutrophus,* membrane-bound and soluble. The hydrogenase subunits are indicated by the circles. From Schlegel (1976).

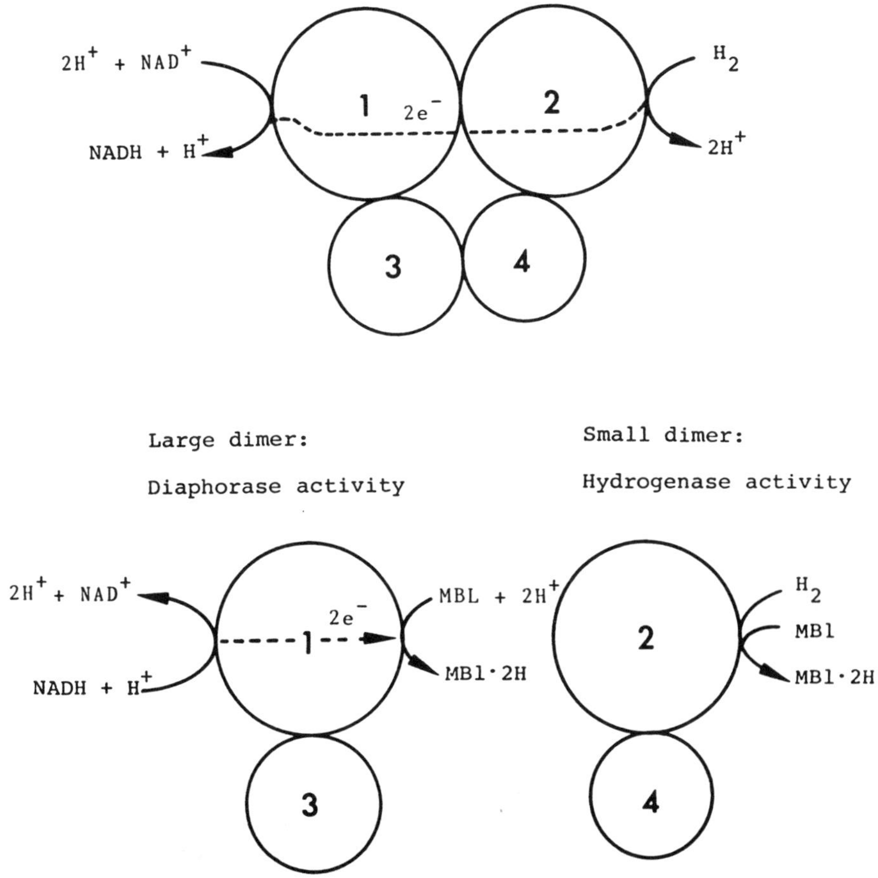

Figure 17.6 Enzyme model for the native hydrogenase of *Nocardia opaca* 1b. The enzyme dissociates into two dimer subunits at low ionic strength or in the absence of nickel salt. The subunit types are numbered. Redrawn from Hornhardt et al. (1986).

well correlated with high efficiencies of substrate consumption and high growth rates. These properties, in common with robustness, tolerance to temporal absence of oxygen, and longevity under adverse conditions are the reasons why *A. eutrophus* strains have been chosen for industrial production of single-cell protein or poly-β-hydroxybutyric acid (Schlegel et al., 1981).

Recently, the subunits of the soluble hydrogenases from *Nocardia opaca* 1b and *A. eutrophus* H16 were partially sequenced (C. Zaborosch and H. Kratzin, in preparation). The amino acid sequences of the amino-terminal ends of the respective subunits from *Nocardia* and from *Alcaligenes* are characterized by a high degree of homology. The homologies amount to the following percentages: SU1 53 percent, SU2 76 percent, SU3 80 percent, and SU4 65 percent. This is the more astonishing since the enzymes compared originate from a Gram-positive (*Nocardia*) and a Gram-negative bacterium (*A. eutrophus*). Clearly, these high homologies are of interest for conclusions about the evolution of soluble hydrogenases as well as ideas on the spreading of hydrogen lithoautotrophy by means of lateral (horizontal) gene transfer.

As the Knallgas bacteria are facultative lithoautotrophs and can grow as well with organic substrates, lithoautotrophy can be considered to be an alternate metabolic activity which is not obligately essential for the survival of the cell. There was thus reason to expect that the genetic information for the key enzymes or metabolic pathways of lithoautotrophy (*Aut*) is localized on disposable genetic elements such as plasmids. The *en bloc* transfer of *Aut* was first shown in *Nocardia* where Aut^+ can be conjugally transferred from *N. opaca* 1b to *Nocardia erythropolis* or *Corynebacterium hydrocarboclastus* (Reh and Schlegel, 1981; Ecker et al., 1986). The transfer is, however, not as simple as it appeared and poses problems (Sensfuss et al., 1986). In *A. eutrophus* a megaplasmid (pHG1; 460 kb) was discovered which codes for the hydrogenase and can be conjugally transferred (Friedrich et al., 1981). Many hydrogen bacteria contain plasmids (Gerstenberg et al., 1982) but so far only those of *A. eutrophus* H16 (pHG1), *A. hydrogenophilus* (pHG21a) and *Pseudomonas facilis* (pHG22a) have been studied. Details on genetic localization of autotrophy characters are presented in Chapter 23.

17.4 THE CO_2-FIXING SYSTEM

Three metabolic systems of CO_2 fixation are known to serve organisms that grow with CO_2 as the sole carbon source. These are the ribulose bisphosphate (Calvin) cycle (see Chapter 19), the reductive tricarboxylic acid cycle and the acetyl-CoA pathway (see Chapter 20). Carbon dioxide assimilation is a reductive process and is dependent on energy (ATP) and reduction power (NAD(P)H). These driving forces can be supplied by energy derived from light or from the oxidation of inorganic energy sources. As CO_2 fixation is a reductive process, we have strictly to differentiate between anaerobic and aerobic conditions. Under anaerobic conditions, reducing substances are generally present in the environment, whereas under aerobic conditions, reducing substances are generally absent and must be synthesized by the organism. Thus, CO_2 fixation in the presence of oxygen has to be linked to higher expenditure of energy than under strictly anaerobic conditions.

The three metabolic systems of CO_2 fixation in fact differ from each other by the expenditure of energy and by the involvement of irreversible enzymatic steps. For the reduction of 3 mol CO_2 to the level of triosephosphate in the Calvin cycle 9 mol of ATP are required, whereas in the reductive tricarboxylic acid cycle 5 mol of ATP are required and in the acetyl-CoA pathway only 3 or 4 mol ATP are required. While in the acetyl-CoA pathway irreversible steps are not involved, and apparently the same enzymes are involved in the synthesis of acetyl CoA and in its dissimilation (Fuchs, 1986; Chapter 20) in the other two CO_2-fixing sequences, several enzyme-catalyzed steps are irreversible.

Until recently there was no doubt that all aerobic chemolithoautotrophic bacteria fix CO_2 via the ribulosebisphosphate cycle. In all members of this physiological group so far studied, the key enzymes and/or the primary fixation products had been detected, and the cycle had been found operational. However, Knallgas bacteria which do not fix carbon dioxide via the Calvin cycle were discovered by Minoda and collaborators in hot springs of the Izu Peninsula and Kyushu, Japan (Kawasumi et al., 1980). The six isolates were described as a new genus and species, *Hydrogenobacter thermophilus,* with strain TK-6 as the type species (Kawasumi et al., 1984). Morphologically, physiologically, and biochemically the isolates represent a new unique type of bacterium. They are Gram-negative, nonmotile, nonspore-forming, long, straight rods with DNA base composition between 43.5 and 43.9 mol percent G + C, have their temperature optimum at 70 to 75°C, and do not use any of 48 organic compounds (or complex medium ingredients tested). The type strain contains a new type of quinone, a sulfur-containing 2-methylthio-1,4-naphthoquinone (Ishii et al., 1983, 1987).

Very similar strains assigned to *H. thermophilus* have also been isolated from Icelandic hot springs containing up to 10 percent (v/v) H_2 (Kristjansson et al., 1985). Although *Hydrogenobacter* does not grow on agar media with Knallgas and had to be purified by the dilution method, it has been found to grow well on thiosulfate agar (Alfredsson et al., 1986; Bonjour and Aragno, 1986). This finding greatly facilitates the isolation procedure.

As the *Hydrogenobacter* strains did not utilize any organic substrate, this trait of obligate autotrophy was studied in detail using *H. thermophilus* strain TK-6. The cells lack various enzymes of the Embden-Meyerhof and the Entner-Doudoroff pathway as well as 2-oxoglutarate dehydrogenase (Shiba et al., 1982). In the presence of H_2 and O_2, small amounts of glucose, acetate, and pyruvate were incorporated into intact cells, and pyruvate, malate, and acetate stimulated autotrophic growth, but no growth occurred in the absence of H_2 (Shiba et al., 1984).

The most striking metabolic trait of *Hydrogenobacter thermophilus* is the absence of the enzymes of the Calvin cycle. Incorporation of $^{14}CO_2$ carbon into intermediary metabolites during short-term CO_2 fixation experiments showed that aspartate and glutamate were the first labeled products, showing negative slopes of percent radioactivity within the first 30 seconds. These data and the presence of the key enzymes of the reductive tricarboxylic acid cycle, fumarate reductase and ATP citrate lyase, as well as evidence for the presence of 2-oxoglutarate synthase and pyruvate synthase, strongly support the conclusion that in *H. thermophilus* a reductive tricarboxylic acid cycle is operating (Shiba et al., 1985). This mode of autotrophic carbon dioxide fixation had been first proposed to operate in the phototrophic green sulfur bacterium *Chlorobium limicola* (Evans et al., 1966; see also Ivanowsky, 1980; Antranikian et al., 1982; Fuchs et al., 1980a and b). Before *Hydrogenobacter* became known the cycle was, however, not expected to support autotrophic CO_2 fixation in aerobic bacteria.

Obligate autotrophy has so far not been detected in Knallgas bacteria other than the thermophilic *Hydrogenobacter*. It is of interest whether obligately autotrophic mesophilic Knallgas bacteria exist and whether the reductive tricarboxylic acid cycle is operating in those Knallgas bacteria which are microaerophilic, slowly growing, and were so far almost completely neglected when isolating Knallgas bacteria. Increased devotion to the slow growers and ecologically important bacteria will certainly result in new fascinating insights on the comparative biochemistry of autotrophic CO_2 fixation.

According to Woese (1987), all Gram-negative aerobic bacteria are descendents of the purple phototrophic bacteria. As both the purple phototrophs as well as the majority of the aerobic lithoautotrophic bacteria fix CO_2 via the Calvin cycle, the descendence theory implies that the Calvin cycle was maintained in the course of the evolution of the lithoautotrophs. Consequently, the enzymes ribulose bisphosphate carboxylase/oxygenase and ribulose-5-phosphate kinase should reflect the evolutionary relationship by various degrees of homology. Research on the structure of these enzymes, both on the enzymic level and on the genetic level, which is being carried out by Bowien and his collaborators with *A. eutrophus* (see Chapter 24) needs therefore to be extended to representatives of phototrophic bacteria and their descendent lithoautotrophs.

The enzymes of the Calvin cycle catalyzing the reductive steps from 3-phosphoglycerate to fructose-6-phosphate are also involved in the gluconeogenetic and glycolytic pathways of metabolism. As already discussed (Bowien and Schlegel, 1981), some of these reactions must be regulated differently depending on their involvement in autotrophic or heterotrophic metabolism. It may, therefore, be postulated that in the facultative autotrophs isofunctional enzymes exist. As an example, fructose bisphosphatase in *Nocardia opaca* was studied, and two forms of this enzyme were found in autotrophically grown cells, whereas only one was detected in heterotrophically grown cells (Amachi and Bowien, 1979).

The problem of isoenzymes is of considerable interest from another point of view. The distribution of hydrogen autotrophy in various species of quite different genera, in Gram-positive and Gram-negative bacteria, in spore-forming bacteria, in both motile and nonmotile bacteria, suggests that lateral gene transfer mechanisms may have been involved in the spreading of genes coding for autotrophy through diverse taxonomical units. This hypothesis was supported by the discovery that in *N. opaca* the key enzymes of lithoautotrophy were transferred *en bloc* by conjugation (Reh and Schlegel, 1975; Ecker et al., 1986). A further argument was provided by the discovery that the hydrogenase in *A. eutrophus* (Anderson et al., 1981; Friedrich et al., 1981) as well as ribulose bisphosphate carboxylase and phosphoribulokinase (Bowien et al., 1984) are encoded on a conjugative plasmid. Comparative studies on the structure and distribution of enzymes both on the genetic and biochemical level are required to clarify relevant questions.

17.5 NITROGEN-FIXING KNALLGAS BACTERIA

In the early days of comparative studies on nitrogen-fixing bacteria (1949–1953) a rather strict correlation between the ability to fix nitrogen and the presence of hydrogenase activities was discovered, and P. W. Wilson made the prediction that all hydrogenase-containing organisms might also be able to fix nitrogen (Wilson, 1952). On the basis of this reasoning, nitrogen fixation was discovered in many phototrophic bacteria and in *Desulfovibrio desulfuricans*. The examination of other H_2-activating bacteria such as *Escherichia coli, Proteus vulgaris, Scenedesmus* and *Hydrogenomonas* (= *Alcaligenes eutrophus*) failed to detect N_2 fixation. Only after nitrogen fixation had been discovered in *Xanthobacter autotrophicus* (Gogotov and Schlegel, 1974), new screening experiments were started, resulting in the discovery of nitrogen fixation in such well-known Knallgas bacteria as *Alcaligenes latus, Renobacter vacuolatum, Microcyclus aquaticus*, as well as the detection of the ability to grow autotrophically among such nitrogen-fixing bacteria as *Azospirillum lipoferum, Derxia gummosa* and *Rhizobium japonicum* (reviewed by Bowien and Schlegel, 1981).

The most prominent metabolic type among the N_2-fixing Knallgas bacteria was found in the genus *Xanthobacter* (Wiegel et al., 1978; Wiegel and Schlegel, 1984). Like the majority of aerobic nitrogen-fixing bacteria, *Xanthobacter* is able to fix N_2 only under decreased oxygen partial pressures (p_{O2} around 0.01 bar), while in the presence of ammonium salt it grows very well at 0.21 bar O_2. The N_2-fixing system in *X. autotrophicus* and *X. flavus* was found to involve two new ferredoxins (Berndt et al., 1978; Schrautemeier, 1981; Yates et al., 1989).

The original isolates of *Xanthobacter*, strains 7C and 14g, originated from enrichments in the presence of ammonium salt (Schneider et al., 1973; Tunail and Schlegel, 1974), but subsequent enrichments for N_2-fixing Knallgas bacteria invariably resulted in the isolation of bacterial strains almost identical to strains 7C and 14g (Wiegel and Schlegel, 1976). On the basis of the comparison of these strains, the new genus *Xanthobacter* and the species *X. autotrophicus* were established (Wiegel et al., 1978). Also assigned to this species were well-characterized Knallgas bacteria, strains Y38 and N34 (Nakamura et al., 1985), strain H4–14, which is the methylotrophic partner of a methanotrophic coculture (Lidstrom-O'Connor et al., 1983; Weaver and Lidstrom, 1985; Murrell and Lidstrom, 1983), strain GJ10 able to degrade halogenated aliphatic compounds (Janssen et al., 1985), further strains able to grow on cyclohexane as carbon source (Magor et al., 1986) and strains enriched in Knallgas-flushed soil columns (Dugnani et al., 1986). *X. autotrophicus* appears to be a nutritionally very versatile organism and is even suited for genetic studies involving conjugational gene transfer (Wilke, 1980; Weaver and Lidstrom, 1985).

17.6 KNALLGAS BACTERIA AND THE ANAEROBIC ENVIRONMENT

The Knallgas bacteria are strictly respiratory bacteria; they synthesize ATP by oxidative phosphorylation via the respiratory chain, either aerobically with oxygen or anaerobically with nitrate, nitrite, or nitrous oxide as electron acceptors. Only a few Knallgas bacteria are able to grow with organic substrates on the basis of dissimilatory nitrate reduction, and with H_2 as the sole hydrogen donor only *Paracoccus denitrificans* is able to grow.

It is astonishing that no Knallgas bacterium has been reported to grow anaerobically via a fermentative ATP regeneration. For a long time, the absence of fermentation enzymes was taken for granted and strict aerobic Knallgas bacteria were considered to have lost the genetic information for fermentation enzymes. The excretion of fermentation products such as 2,3-butanediol, succinate, lactate, ethanol, acetate, CO_2, and H_2 during incubation of *A. eutrophus* H16 or N9A under conditions of oxygen deficiency (Vollbrecht et al., 1979) and the formation of fermentation enzymes (Figure 17.7) under these conditions (Schlegel and Vollbrecht, 1980; Steinbüchel et al., 1983; Steinbüchel and Schlegel, 1983a, b) indicated that *A. eutrophus* is cryptic and contains silent genes for fermentation enzymes. This finding raised the question why *A. eutrophus* cannot grow fermentatively under anaerobic conditions. *A. eutrophus* degrades

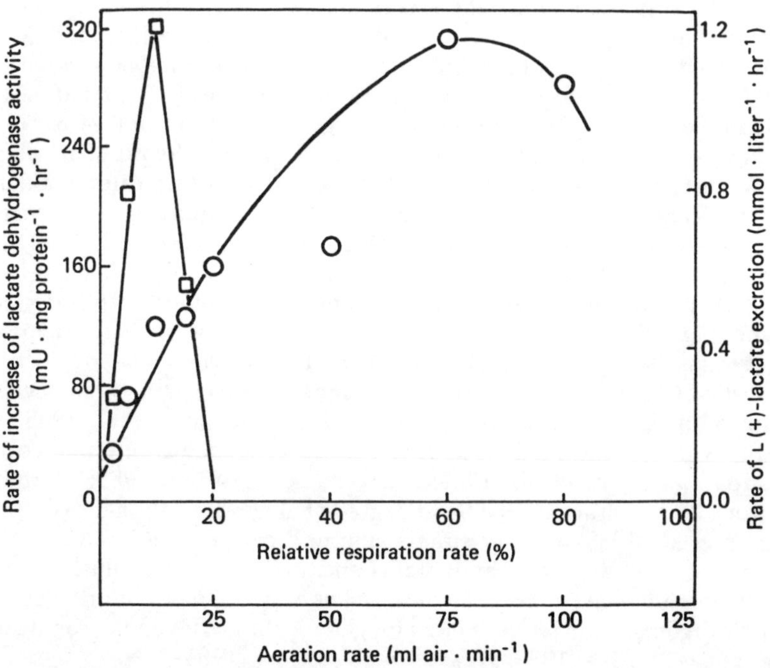

Figure 17.7 Dependence of the excretion of L(+)-lactate and the increase of lactate dehydrogenase activity on the relative respiration rate in *Alcaligenes eutrophus*. The initial rates of the increase of specific activity of lactate dehydrogenase (○) and of lactate excretion (□) following a shift from the maximal respiration rate to a low respiration rate were plotted against the low respiration rate calculated as relative respiration rate. The latter is defined as the actual enforced respiration rate of the cells over the maximum respiration rate observed under oxygen saturating conditions. From Steinbüchel and Schlegel (1983a).

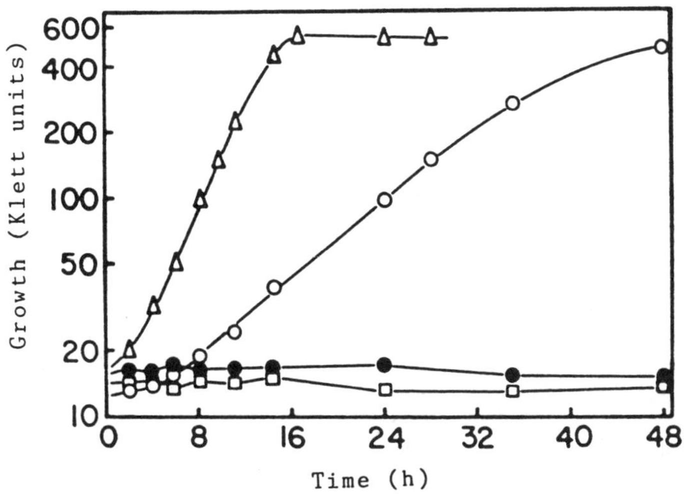

Figure 17.8 Growth of wild type and KDPG aldolase-deficient strains of *Alcaligenes eutrophus*. The strains were incubated aerobically in a fructose medium. Symbols: △, wild-type strain *A. eutrophus* H16; □, KDPG aldolase-deficient mutant C145; ○, transconjugant derived from C145 and containing plasmid pGE42 with the *pfk*A gene. From Steinbüchel (1986).

fructose via the Entner-Doudoroff (ED) pathway. Among the bacteria using the ED pathway, only *Zymomonas mobilis* is able to grow anaerobically by fermentation. The low yield of only 1 ATP per hexose may be the reason for the scarcity of fermenting bacteria among organisms degrading sugars via the ED pathway. This reasoning prompted A. Steinbüchel to try to convert *A. eutrophus* to a fermentative bacterium by establishing the Embden-Meyerhof pathway, which provides 2 ATP per hexose. When the phosphofructokinase (PFK) gene *pfk*A from *E. coli* was transferred to a mutant of *A. eutrophus*, which did not grow on hexose due to deficiency of 2-keto-3-desoxy-6-phosphoaldolase, the transconjugant had gained the ability to synthesize PFK and to grow on fructose (Figure 17.8; Steinbüchel, 1986). However, fructose degradation via the Embden-Meyerhof glycolytic pathway did not result in the ability to grow anaerobically by fermentation, so that the reason for the lack of fermentative growth remains a mystery.

Also lacking is the metabolic type which combines the capability to grow anaerobically as a fermentative bacterium and aerobically as a Knallgas bacterium, perhaps because such an organism has not been sought in enrichment cultures!

17.7 VERSATILITY OF KNALLGAS BACTERIA

It is certainly too early to make generalizations on the versatility of Knallgas bacteria. Only a small proportion of species and strains isolated from nature has been studied in detail. However, it is clear that nutritionally the Knallgas bacteria are a very diverse group. The ability to grow as an aerobic hydrogen-oxidizing bacterium is just a common denominator of single members belonging to many different physiological groups of bacteria such as the nitrogen-fixing bacteria, methylotrophs, denitrifiers, and carboxydotrophs.

With respect to the number of substrates utilized the Knallgas bacteria differ from each other widely. There are very versatile species which are able to grow on about 50 among 150 substrates tested (Davis et al., 1970), and others which are fastidious and can only grow on a few organic acids. At the end of the scale are obligate lithoautotrophs unable to utilize any organic substrate, such as the recently described species *Hydrogenobacter thermophilus* (Kawasumi et al., 1980, 1984), *Calderobacterium hydrogenophilum* (Kryukow et al., 1983), *Hydrogenobacter* sp. (Kristjansson et al., 1985), and other strains of *Hydrogenobacter* (Bonjour and Aragno, 1986). All these bacteria were isolated from geothermal areas and are thermophiles. The habitat, low in organic carbon and often rich in H_2 as well as hydrogen sulfide and sulfur, explains the obligate autotrophic nature of these strains as well as the ability to grow as sulfur oxidizers. The latter property overlaps with other lithoautotrophic groups which had already been detected in mesophiles such as *Paracoccus denitrificans, Aquaspirillum autotrophicum, Xanthobacter autotrophicus, Pseudomonas palleronii,* and *P. pseudoflava* (Friedrich and Mitrenga, 1981).

The correlation between the degree of nutritional versatility and the habitats of Knallgas bacteria can so far scarcely be discussed. Data exist only for extreme types of the just-mentioned obligately autotrophic species; these bacteria were isolated from geothermal wells where hydrogen is permanently present, and the content of organic matter is very low. As already mentioned, the actual habitat for most of the Knallgas bacteria is not known. Although many strains or species descriptions mention the location from which the soil or water sample was collected, detailed descriptions of the habitat (environmental characteristics, etc.), are lacking. In this context, an observation made by the author is worth mentioning. Attempts were made to isolate a variety of strains closely related to the metal-resistant bacterium CH34. This strain had been isolated by Mergeay et al. (1978) and tentatively assigned to *A. eutrophus* due to the presence of two hydrogenases (Mergeay et al., 1985). The strain differs, however, from the classical strains of *A. eutrophus* (H16, N9A, type strain, G27, G29, B19 and others) by a few nutritional markers and by the localization of the autotrophy characters on the chromosome and by the resistance to nickel and cobalt coded for by one plasmid (163 kb) and resistance to cadmium, zinc, and cobalt coded on another plasmid (238 kb). The resistances to nickel and cobalt are inducible (Siddiqui and Schlegel, 1987). Strain CH34 was isolated from a zinc decantation tank. We enriched for related strains using samples from municipal and industrial waste water plants of several cities and metal-processing industries. In spite of repeated attempts, a strain very closely related to CH34 was found nowhere except in the waste water plant of Göttingen, to which the waste water of the laboratory is released (Timotius and Schlegel, 1987; T. Schmidt and H. G. Schlegel, in preparation).

The observation indicates that strain CH34 may be unique with respect to its robustness and viability; it obviously draws advantage from its nutritional versatility and maintains its specific characters even in the absence of specific selective conditions.

17.8 CONCLUSIONS

Knallgas bacteria have been studied in quite some detail and are considered the model organisms for studying the chemolithoautotrophic mode of life. However, the present status is unsatisfactory since studies have concentrated on only a few strains, those having outstanding properties. The first choices for research are the fast growing, easy to handle representatives. Indeed, it has been stated that *Alcaligenes eutrophus* H16 is "a microorganism which can almost as easily be grown as bakers' yeast" (Seebach and Züger, 1982). It is the task of the microbiologists to provide such organisms to chemists, biochemists, and physicists. But the microbiologist should not worship heroes and stick to only unique organisms. The microbiologist ought to study the diversity of microorganisms, the foot soldiers included. Many slowly growing bacteria, difficult to grow, maintain, and study in the laboratory, may have more important functions in the biosphere than the outstanding representatives. Microbiologists should, therefore, be encouraged to study ecosystems, isolate new bacteria, and study even slow growers. Enrichment culture offers lots of opportunities: Knallgas bacteria associated with slime sheaths of H_2-releasing cyanobacteria or root nodules and other Knallgas bacteria which may be filamentous, gliding, magnetotactic, or facultatively fermenting. There may be members with unusual resistance properties or strains that do not tolerate the ingredients of nutrient media at the generally used concentrations. The Knallgas bacteria are a group of microorganisms whose function and dwelling place in nature can only be supposed, but not known in detail. Studies on the ecology, physiology, biochemistry, and genetics of Knallgas bacteria provide numerous opportunities for solving fundamental problems.

The author is aware of the excellent contributions of the many collaborators who worked with him in the course of 30 years and is grateful to them for sharing lasting devotion to the Knallgas bacteria with him. Thanks are also due to the Deutsche Forschungsgemeinschaft for continual financial support.

REFERENCES

Adams, M.W.W., Mortenson, L.E., and **Clan, J.S.** 1981. Hydrogenase. *Biochimica et Biophysica Acta* 594:105–176.

Alfredsson, G.A., Ingason, A., and **Kristjansson, J.K.** 1986. Growth of thermophilic obligately autotrophic hydrogen-oxidizing bacteria on thiosulfate. *Letters in Applied Microbiology* 2:21–23.

Amachi, T. and **Bowien, B.** 1979. Characterization of two fructose bisphosphatase isoenzymes from the hydrogen bacterium *Nocardia opaca* lb. *Journal of General Microbiology* 113:347–356.

Andersen, K., Tait, R.C., and **King, W.R.** 1981. Plasmids required for utilization of molecular hydrogen by *Alcaligenes eutrophus. Archives of Microbiology* 129:384–390.

Antranikian, G., Herzberg, C., and **Gottschalk, G.** 1982. Characterization of ATP citrate lyase from *Chlorobium limicola. Journal of Bacteriology* 152:1284–1287.

Aragno, M. and **Schlegel, H.G.** 1981. The hydrogen-oxidizing bacteria, pp. 865–893, in Starr, M.P., Stolp, H., Trüper, H.G., Balows, A., Schlegel, H.G. (editors), *The Prokaryotes. A Handbook on Habitats, Isolation and Identification of Bacteria,* Vol. I, Springer-Verlag, Berlin.

Arp, D.J. and **Burris, R.H.** 1981. Kinetic mechanism of the hydrogen-oxidizing hydrogenase from soybean nodule bacteroids. *Biochemistry* 20:2234–2240.

Arp, D.J., McCollum, L.C., and **Seefeldt, L.C.** 1985. Molecular and immunological comparison of membrane-bound, H_2-oxidizing hydrogenases of *Bradyryhizobium japonicum, Alcaligenes eutrophus, Alcaligenes latus,* and *Azotobacter vinelandii. Journal of Bacteriology* 163:15–20.

Ballantine, S.P. and **Boxer, D.H.** 1986. Isolation and characterization of a soluble active fragment of hydrogenase isoenzyme 2 from the membranes of anaerobically-grown *Escherichia coli. European Journal of Biochemistry* 156:277–284.

Berndt, H., Lowe, D.J., and **Yates, M.G.** 1978. The nitrogen-fixing system of *Corynebacterium autotrophicum.* Purification and properties of the nitrogenase components and two ferredoxins. *European Journal of Biochemistry* 86:133–142.

Bonjour, F. and **Aragno, M.** 1984. *Bacillus Atusciae,* a new species of thermoacidophilic, facultatively chemolithoautotrophic hydrogen oxidizing sporeformer from a geothermal area. *Archives of Microbiology* 139:397–401.

Bonjour, F. and **Aragno, M.** 1986. Growth of thermophilic, obligatory chemolithoautotrophic hydrogen-oxidizing bacteria related to *Hydrogenobacter* with thiosulfate and elemental sulfur as electron and energy source. *FEMS Microbiology Letters* 35:11–15.

Bowien, B. and **Schlegel, H.G.** 1981. Physiology and biochemistry of aerobic hydrogen-oxidizing bacteria. *Annual Review of Microbiology* 35:405–452.

Bowien, B., Friedrich, B., and **Friedrich, C.G.** 1984. Involvement of megaplasmids in heterotrophic derepression of the carbon dioxide assimilating enzyme system in *Alcaligenes* spp. *Archives of Microbiology* 139:305–310.

Conrad, R. 1984. Capacity of aerobic microorganisms to utilize and grow on atmospheric trace gases (H_2, CO, CH_4), pp. 461–467, in Klug, M.J. and Reddy, C.A. (editors), *Current Perspectives in Microbial Ecology,* American Society for Microbiology, Washington, D.C.

Conrad, R. and **Seiler, W.** 1979. The role of hydrogen bacteria during the decomposition of hydrogen by soil. *FEMS Microbiology Letters* 6:143–145.

Conrad, R. and **Seiler, W.** 1980a. Decomposition of atmospheric hydrogen by soil microorganisms and soil enzymes. *Soil Biology and Biochemistry* 13:43–49.

Conrad, R. and **Seiler, W.** 1980b. The role of microbiological processes for the atmospheric hydrogen cycle. *Forum Mikrobiologie* 3:219–225.

Conrad, R. and **Seiler, W.** 1980c. Contribution of hydrogen production by biological nitrogen fixation to the global hydrogen budget. *Journal of Geophysical Research* 85:5493–5498.

Conrad, R., Aragno, M., and **Seiler, W.** 1983a. The inability of hydrogen bacteria to utilize atmospheric hydrogen is due to threshold and affinity for hydrogen. *FEMS Microbiology Letters* 18:207–210.

Conrad, R., Aragno, M., and **Seiler, W.** 1983b. Production and consumption of hydrogen in a eutrophic lake. *Applied and Environmental Microbiology* 45:502–510.

Conrad, R., Weber, M., and **Seiler, W.** 1983. Kinetics and electron transport of soil hydrogenases catalyzing the oxidation of atmospheric hydrogen. *Soil Biology and Biochemistry* 15:167–173.

Conrad, R., Phelps, T.J., and **Zeikus, J.G.** 1985. Gas metabolism evidence in support of the juxtaposition of hydrogen-producing and methanogenic bacteria in sewage sludge and lake sediments. *Applied and Environmental Microbiology* 50:595–601.

Conrad, R., Lupton, F.S., and **Zeikus, J.G.** 1987. Hydrogen metabolism and sulfate-dependent inhibition of methanogenesis in a eutrophic lake sediment (Lake Mendota). *FEMS Microbiology Ecology* 45:107–115.

Davis, D.H., Stanier, R.Y., Doudoroff, M., and **Mandel, M.** 1970. Taxonomic studies on some Gram-negative polarly flagellated "hydrogen bacteria" and related species. *Archiv für Mikrobiologie* 70:1–13.

Dugnani, L., Wyrsch, I., Gandolla, M., and **Aragno, M.** 1986. Biological oxidation of hydrogen in soils flushed with a mixture of H_2, CO_2, O_2, and N_2. *FEMS Microbiology Ecology* 38:347–351.

Ecker, C., Reh, M., and **Schlegel, H.G.** 1986. Enzymes of the autotrophic pathway in mating partners and transconjugants of *Nocardia opaca* 1b and *Rhodococcus erythropolis*. *Archives of Microbiology* 145:280–286.

Evans, M.C.W., Buchanan, B.B., and **Arnon, D.I.** 1966. A new ferredoxin-dependent carbon reduction cycle in a photosynthetic bacterium. *Proceedings of the National Academy of Sciences of the United States of America* 55:928–934.

Fauque, G.D., Berlier, Y.M., Czechowski, M.H., Dimon, B., Lespinat, P.A., and **LeGall, J.** 1987. A proton-deuterium exchange study of three types of *Desulfovibrio* hydrogenases. *Journal of Industrial Microbiology* 2:15–23.

Friedrich, B. 1985. Evolution of chemolithoautotrophy, pp. 205–234, in Schleifer, K.H. and Stackebrandt, E. (editors), *Evolution of Prokaryotes*, Academic Press, Inc., London.

Friedrich, B., Hogrefe, C., and **Schlegel, H.G.** 1981. Naturally occurring genetic transfer of hydrogen-oxidizing ability between strains of *Alcaligenes eutrophus*. *Journal of Bacteriology* 147:198–205.

Friedrich, C.G. and **Mitrenga, G.** 1981. Oxidation of thiosulfate by *Paracoccus denitrificans* and other hydrogen bacteria. *FEMS Microbiology Letters* 10:209–212.

Fuchs, G., Stupperich, E., and **Jaenchen, R.** 1980a. Autotrophic CO_2 fixation in *Chlorobium limicola*. Evidence against the operation of the Calvin cycle in growing cells. *Archives of Microbiology* 128:56–63.

Fuchs, G., Stupperich, E., and **Eden, G.** 1980b. Autotrophic CO_2 fixation in *Chlorobium limicola*. Evidence for the operation of a reductive tricarboxylic acid cycle in growing cells. *Archives of Microbiology* 128:64–71.

Gerstenberg, C., Friedrich, B., and **Schlegel, H.G.** 1982. Physical evidence for plasmids in autotrophic, especially hydrogen-oxidizing bacteria. *Archives of Microbiology* 133:90–96.

Gogotov, J.N. and **Schlegel, H.G.** 1974. N_2 fixation by chemoautotrophic hydrogen bacteria. *Archives of Microbiology* 97:359–362.

Harker, A.R., Zuber, M., and **Evans, H.J.** 1986. Immunological homology between the membrane-bound uptake hydrogenases of *Rhizobium japonicum* and *Escherichia coli*. *Journal of Bacteriology* 165:579–584.

Hornhardt, S., Schneider, K., and **Schlegel, H.G.** 1986. Characterization of a native subunit of the NAD-linked hydrogenase isolated from a mutant of *Alcaligenes eutrophus* H16. *Biochimie* 68:15–24.

Ishii, M., Kawasumi, T., Igarashi, Y., Kodama, T., and **Minoda, Y.** 1983. 2-Methylthio-1,4-naphthoquinone, a new quinone from an extremely thermophilic hydrogen bacterium. *Agricultural and Biological Chemistry* 47:167–169.

Ishii, M., Kawasumi, T., Igarashi, Y., Kodama, T., and **Minoda, Y.** 1987. 2-Methylthio-1,4-naphthoquinone, a unique sulfur-containing quinone from a thermophilic hydrogen-oxidizium bacterium, *Hydrogenobacter thermophilus*. *Journal of Bacteriology* 169:2380–2384.

Ivanowsky, R.N., Sintsov, N.V., and **Kondratieva, E.N.** 1980. ATP-linked citrate lyase activity in the green sulfur bacterium *Chlorobium limicola* forma *thiosulfatophilum*. *Archives of Microbiology* 128:239–241.

Jannasch, H.W. and **Wirsen, C.O.** 1979. Chemosynthetic primary production at East Pacific sea floor spreading centers. *Bio Science* 29:592–598.

Jannasch, H.W. and **Wirsen, C.O.** 1981. Morphological survey of microbial mats near deep-sea thermal vents. *Applied and Environmental Microbiology* 41:528–538.

Janssen, D.B., Scheper, A., Dijkhuizen, L., and **Witholt, B.** 1985. Degradation of halogenated aliphatic compounds by *Xanthobacter autotrophicus* GJ10. *Applied and Environmental Microbiology* 49:673–677.

Karl, D.M. 1987. Bacterial production at deep-sea hydrothermal vents and cold seeps: evidence for chemosynthetic primary production, pp. 319–360, in Fletcher, M., Gray, T.R.G., Jones, J.G. (editors), *Ecology of Microbial Communities*, Symposia of the Society of General Microbiology, Vol. 41, Cambridge University Press, London.

Kaserer, H. 1906. Über die Oxidation des Wasserstoffs und des Methans durch Mikroorganismen. *Zentralblatt für Bakteriologie Abteilung II* 15:573–576.

Kawasumi, T., Igarashi, Y., Kodama, I., and **Minoda, Y.** 1980. Isolation of strictly thermophilic and obligately autotrophic hydrogen bacteria. *Agricultural and Biological Chemistry* 44:1985–1986.

Kawasumi, T., Igarashi, Y., Kodama, T., and **Minoda, Y.** 1984. *Hydrogenobacter thermophilus* gen. nov., sp. nov., an extremely thermophilic, aerobic, hydrogen-oxidizing bacterium. *International Journal of Systematic Bacteriology* 34:5–10.

Knüttel, K., Schneider, K., and **Schlegel, H.G.** 1988. The membrane-bound hydrogenase of *Paracoccus denitrificans*. Purification and molecular characterization *European Journal of Biochemistry* (in press).

Kristjansson, J.K., Ingason, A., and **Alfredsson, G.A.** 1985. Isolation of thermophilic obligately autotrophic hydrogen-oxidizing bacteria, similar to *Hydrogenobacter thermophilus,* from Icelandic hot springs. *Archives of Microbiology* 140:321–325.

Kryukov, V.R., Savelyeva, N.D., and **Pusheva, M.A.** 1983. *Calderobacterium hydrogenophilum* nov. gen. nov. sp., an extreme thermophilic hydrogen bacterium, and its hydrogenase activity. *Mikrobiologiya* 52:781–788.

Lidstrom-O'Connor, M., Fulton, G.L., and **Wopat, A.E.** 1983. *Methylobacterium ethanolicum*: a syntrophic association of two methylotrophic bacteria. *Journal of General Microbiology* 129:3139–3148.

Magor, A.M., Warburton, J., Trower, M.K., and **Griffin, M.** 1986. Comparative study of the ability of three *Xanthobacter* species to metabolize cycloalkanes. *Applied and Environmental Microbiology* 52:665–671.

Mergeay, M., Houba, C., and **Gerits, J.** 1978. Extrachromosomal inheritance controlling resistance to cadmium, cobalt, and zinc ions: evidence from curing in a Pseudomonas. *Archives Internationales de Physiologie et de Biochimie* 86:440–441.

Mergeay, M., Nies, D., Schlegel, H.G., Gerits, J., Charles, P., and **van Gijsegem, F.** 1985. *Alcaligenes eutrophus* CH34 is a facultative chemolithotroph with plasmid-bound resistance to heavy metals. *Journal of Bacteriology* 162:328–334.

Murrell, J.C. and **Lidstrom, M.E.** 1983. Nitrogen metabolism in *Xanthobacter* H4–14. *Archives of Microbiology* 136:219–221.

Nakamura, Y., Yamanobe, T., and **Ooyama, J.** 1985. Identification of nitrogen-fixing hydrogen bacterium strain N34 and its oxygen-resistant segregant strain Y38. *Agricultural and Biological Chemistry* 49:1703–1709.

Pinkwart, M., Schneider, K., and **Schlegel, H.G.** 1983. Purification and properties of the membrane-bound hydrogenase from N_2-fixing *Alcaligenes latus. Biochimica et Biophysica Acta* 745:267–278.

Podzuweit, H.G., Arp, D.J., Schlegel, H.G., and **Schneider, K.** 1986. Investigation of the H_2-oxidizing activities of *Alcaligenes eutrophus* H16 membranes with artificial electron acceptors, respiratory inhibitors and redox-spectroscopic procedures. *Biochimie* 68:103–111.

Podzuweit, H.G., Schneider, K., and **Knüttel, K.** 1987. Comparison of the membrane-bound hydrogenases from *Alcaligenes eutrophus* H16 and *Alcaligenes eutrophus* type strain. *Biochimica et Biophysica Acta* 905:435–446.

Reh, M. and **Schlegel, H.G.** 1975. Chemolithoautotrophie als eine übertragbare, autonome Eigenschaft von *Nocardia opaca* 1b. *Nachrichten der Akademie der Wissenschaften zu Göttingen, II. Math.-Physik. Klasse* 12:207–216.

Reh, M. and **Schlegel, H.G.** 1981. Hydrogen autotrophy as a transferable genetic character of *Nocardia opaca* 1b. *Journal of General Microbiology* 126:327–336.

Ruhland, W. 1924. Beiträge zur Physiologie der Knallgasbakterien. *Jahrbücher für Wissenschaftliche Botanik* 63:321–389.

Sawers, R.G. and **Boxer, D.H.** 1986. Purifications and properties of membrane-bound hydrogenase isoenzyme 1 from anaerobically grown *Escherichia coli* K12. *European Journal of Biochemistry* 156:265–275.

Schink, B. and **Schlegel, H.G.** 1978a. Hydrogen metabolism in aerobic hydrogen-oxidizing bacteria. *Biochimie* 60:297–305.

Schink, B. and **Schlegel, H.G.** 1978b. Mutants of *Alcaligenes eutrophus* defective in autotrophic metabolism. *Archives of Microbiology* 117:123–129.

Schink, B. and **Schlegel, H.G.** 1979. The membrane-bound hydrogenase of *Alcaligenes eutrophus*. I. Solubilization, purification, and biochemical properties. *Biochimica et Biophysica Acta* 567:351–324.

Schink, B. and **Schlegel, H.G.** 1980. The membrane-bound hydrogenase of *Alcaligenes eutrophus*: II. Localization and immunological comparison with other hydrogenase systems. *Antonie van Leeuwenhoek Journal of Microbiology and Serology* 46:1–14.

Schink, B. and **Zeikus, J.G.** 1984. Ecology of aerobic hydrogen-oxidizing bacteria in two freshwater lake ecosystems. *Canadian Journal of Microbiology* 30:260–265.

Schlegel, H.G. 1976. The Physiology of Hydrogen Bacteria. The fifth A.J. Kluyver memorial lecture. *Antonie van Leeuwenhoek Journal of Microbiology and Serology* 42:181–201.

Schlegel, H.G. 1984. Studies on the regulation and genetics of enzymes of *Alcaligenes eutrophus*, pp. 187–209, in Codd, G. (editor), *Aspects of Microbial Metabolism and Ecology*, Academic Press, London.

Schlegel, H.G. and **Bartha, R.** 1961. "Leerlauf"-H_2-Oxydation und "Rückkoppelung" bei Knallgasbakterien. *Naturwissenschaften* 48:414–415.

Schlegel, H.G. and **Jannasch, H.W.** 1981. Prokaryotes and their habitats, pp. 43–82, in Starr, M.P., Stolp, H., Trüper, H.G., Balows, A., Schlegel, H.G. (editors), *The Prokaryotes. A Handbook on Habitats, Isolation, and Identification of Bacteria*, Springer-Verlag, Berlin-Heidelberg.

Schlegel, H.G., Ibrahim, M.E.K., Wilde, E., Schneider, K., Schlesier, M., Friedrich, B., and **Malik, K.A.** 1981. Detrimental and beneficial effects of oxygen exerted on hydrogen-oxidizing bacteria, pp. 107–129, in Lyons, J.M., Valentine, R.C., Phillips, D.A., Rains, D.W., Huffaker, R.C. (editors), *Genetic Engineering of Symbiotic Nitrogen Fixation and Conservation of Fixed Nitrogen*, Plenum Press, New York.

Schlegel, H.G. and **Schneider, K.** 1985. Microbial metabolism of hydrogen, pp. 439–457, in Moo-Young, M. (editor), *Comprehensive Biotechnology, Vol. 1, Pergamon Press, Oxford.*

Schneider, K., Rudolph, V., and **Schlegel, H.G.** 1973. Description and physiological characterization of a coryneform hydrogen bacterium strain 14g. *Archiv für Mikrobiologie* 93:179–193.

Schneider, K., Patil, D.S., and **Cammack, R.** 1983. ESR properties of membrane-bound hydrogenases from aerobic hydrogen bacteria. *Biochimica et Biophysica Acta* 748:353–361.

Schneider, K., Schlegel, H.G., and **Jochim, K.** 1984a. Effect of nickel on activity and subunit composition of purified hydrogenase from *Nocardia opaca* 1b. *European Journal of Biochemistry* 138:533–541.

Schneider, K., Cammack, R., and **Schlegel, H.G.** 1984b. Content and localization of FMN, Fe-S clusters and nickel in the NAD-linked hydrogenases of *Nocardia opaca* 1b. *European Journal of Biochemistry* 142:75–84.

Schrautemeier, B. 1981. The role of ferredoxin in the nitrogen-fixing hydrogen bacterium *Xanthobacter autotrophicus*. *FEMS Microbiology Letters* 12:153–157.

Sebo, P. 1986. Hydrogen-oxidizing bacteria in biotechnology. *TIBTECH* 4:321–323.

Seebach, D. and **Züger, M.** 1982. Über die Depolymerisierung von Poly-(R)-3-hydroxybuttersäureester (PHB). *Helvetica Chimica Acta* 65:495–503.

Seefeldt, L.C. and **Arp, D.J.** 1986. Purification to homogeneity of *Azotobacter vinelandii* hydrogenase: a nickel and iron containing β dimer. *Biochimie* 68:25–34.

Seefeldt, L.C., Fox, C.A., and **Arp, D.J.** 1986. Reversible inactivation of the O_2-labile hydrogenases from *Azotobacter vinelandii* and *Rhizobium japonicum. Journal of Biological Chemistry* 261:10688–10694.

Sensfuß, C., Reh, M., and **Schlegel, H.G.** 1986. No correlation exists between the conjugative transfer of the autotrophic character and that of plasmids in *Nocardia opaca* strains. *Journal of General Microbiology* 132:997–1007.

Shiba, H., Kawasumi, T., Igarashi, Y., Kodama, T., and **Minoda, Y.** 1982. The deficient carbohydrate metabolic pathways and the incomplete tricarboxylic acid cycle in an obligately autotrophic hydrogen-oxidizing bacterium. *Agricultural and Biological Chemistry* 46:2341–2345.

Shiba, H., Kawasumi, T., Igarashi, Y., Kodama, T., and **Minoda, Y.** 1984. Effect of organic compounds on the growth of an obligately autotrophic hydrogen-oxidizing bacterium *Hydrogenobacter thermophilus* TK-6. *Agricultural and Biological Chemistry* 48:2809–2813.

Shiba, H., Kawasumi, T., Igarashi, Y., and **Minoda, Y.** 1985. The CO_2 assimilation via the reductive tricarboxylic acid cycle in an obligately autotrophic, aerobic hydrogen-oxidizing bacterium, *Hydrogenobacter thermophilus. Archives of Microbiology* 141:198–203.

Siddiqui, R.A. and **Schlegel, H.G.** 1987. Plasmid pMOL28-mediated inducible nickel resistance in *Alcaligenes eutrophus* strain CH34. *FEMS Microbiology Letters* 43:9–13.

Steinbüchel, A., Kuhn, M., Niedrig, M., and **Schlegel, H.G.** 1983. Fermentation enzymes in strictly aerobic bacteria: comparative studies on strains of the genus *Alcaligenes* and on *Nocardia opaca* and *Xanthobacter autotrophicus. Journal of General Microbiology* 129:2825–2835.

Steinbüchel, A. and **Schlegel, H.G.** 1983a. NAD-linked L(+)-lactate dehydrogenase from the strict aerobe *Alcaligenes eutrophus*. 1. Purification and properties. *European Journal of Biochemistry* 130:321–328.

Steinbüchel, A. and **Schlegel, H.G.** 1983b. NAD-linked L(+)-lactate dehydrogenase from the strict aerobe *Alcaligenes eutrophus*. 2. Kinetic properties and inhibition by oxaloacetate. *European Journal of Biochemistry* 130:219–334.

Steinbüchel, A. 1986. Expression of the *Escherichia coli* pfkA gene in *Alcaligenes eutrophus* and in other Gram-negative bacteria. *Journal of Bacteriology* 166:319–327.

Stults, L.W., Sray, W.A., and **Maier, R.J.** 1986. Regulation of hydrogenase biosynthesis by nickel in *Bradyrhizobium japonicum. Archives of Microbiology* 146:280–283.

Tilak, K.V.B.R., Schneider, K., and **Schlegel, H.G.** 1984. Autotrophic growth of strains of *Rhizobium* and properties of isolated hydrogenase. *Current Microbiology* 10:49–52.

Timotius, K. and **Schlegel, H.G.** 1987. Aus Abwässern isolierte nickel-resistente Bakterien. *Nachrichten der Akademie der Wissenschaften zu Göttingen, II. Math.-Physik. Klasse.* No. 3, pp.15–23.

Tunail, N. and **Schlegel, H.G.** 1974. A new coryneform hydrogen bacterium: *Corynebacterium autotrophicum* strain 7C. I. Characterization of the wild-type strain. *Archives of Microbiology* 100:341–350.

Vollbrecht, D., Schlegel, H.G., Stoschek, G., and **Janczikowski, A.** 1979. Excretion of metabolites by hydrogen bacteria. IV. Respiration rate-dependent formation of primary metabolites and of poly-3-hydroxybutanoate. *European Journal of Applied Microbiology and Biotechnology* 7:267–276.

Weaver, C.A. and **Lidstrom, M.E.** 1985. Methanol dissimilation in *Xanthobacter* H4–14: Activities, induction and comparison to *Pseudomonas* AM1 and *Paracoccus denitrificans. Journal of General Microbiology* 131:2183–2197.

Weiss, A.R., Schneider, K., and **Schlegel, H.G.** 1980. Purification and properties of the membrane-bound hydrogenase of *Pseudomonas pseudoflava* Ga3. *Current Microbiology* 3:317–320.

Wiegel, J. and **Schlegel, H.G.** 1976. Enrichment and isolation of nitrogen-fixing hydrogen bacteria. *Archives of Microbiology* 107:139–142.

Wiegel, J., Wilke, D., Baumgarten, J., Opitz, R., and **Schlegel, H.G.** 1978. Transfer of the nitrogen-fixing hydrogen bacterium *Corynebacterium autotrophicum* Baumgarten et al. to *Xanthobacter* gen. nov. *International Journal of Systematic Bacteriology* 28:573–581.

Wiegel, K.W. and **Schlegel, H.G.** 1984. Genus *Xanthobacter* Wiegel, Wilke, Baumgarten, Opitz, and Schlegel 1978, 573 AL. pp. 325–333, in N.R. Krieg, Holt, J.G. (editors), *Bergey's Manual of Systematic Bacteriology* Vol. 1, Williams and Wilkins, Baltimore.

Wilke, D. 1980. Conjugational gene transfer in*Xanthobacter autotrophicus* GZ29. *Journal of General Microbiology* 117:431–436.

Wilson, P.W. 1952. The comparative biochemistry of nitrogen fixation. *Advances in Enzymology* 13:345–375.

Woese, C.R. 1987. Bacterial evolution. *Microbiological Reviews* 51:221–271.

Yates, M.G., O'Donnell, J., Lowe, D.J., and **Bothe, H.** 1978. Ferredoxins from nitrogen-fixing bacteria. Physical and chemical characterization of two ferredoxins from *Mycobacterium flavum* 301. *European Journal of Biochemistry* 85:291–299.

Zeikus, J.G. 1983. Metabolic communication between biodegradative populations in nature, in Slater, J.G., Whittenbury, R., Wimpenny, J.W.T. (editors), *Society for General Microbiology, Symposium 34, Microbes in their natural environments,* Cambridge University Press, Cambridge.

18

Aerobic Carbon Monoxide-oxidizing Bacteria

Ortwin Meyer

Department of Microbiology
University of Bayreuth
Bayreuth, Federal Republic of Germany

Although Kistner (1953, 1954) had already reported on the aerobic oxidation of CO by an isolate which he named *Hydrogenomonas carboxydovorans* (see Chapter 1), microbiologists discredited the existence of such bacteria. For instance: "If, as we consider rather unlikely, further experimental work shows that such bacteria can obtain their energy autotrophically by the oxidation of carbon monoxide, they will provide a unique example of autotrophy in the higher bacteria." (Bisset and Grace, 1954.)

Indeed, the samples of *H. carboxydovorans* deposited by Kistner could not be revitalized (Davis, 1967; Davis et al., 1969, 1970; Meyer and Schlegel, 1978). Fortunately, however, new strains of CO-oxidizing bacteria became available through the work of Zavarzin and co-workers and through our own isolations (Table 18.1). These isolates proved the existence of bacteria capable of growing aerobically at the expense of CO. Since these new isolations, various aspects of carboxydotrophic bacteria have been covered in numerous reviews (Hegeman, 1984; Kim and Hegeman, 1983; Meyer, 1985, 1986; Meyer and Fiebig, 1985; Meyer and Rohde, 1984; Meyer and Schlegel, 1983; Meyer et al., 1986; Nozhevnikova and Yurganov, 1978; Park and Hegeman, 1984; Schlegel and Meyer, 1981; Williams and Colby, 1986; Zavarzin, 1978; Zavarzin and Nozhevnikova, 1976, 1977).

18.1 CHARACTERISTICS OF SELECTED CARBOXYDOTROPHIC BACTERIA

Several strains of *P. carboxydovorans* have been isolated in our laboratory (Table 18.1), and most of the work on the physiology and biochemistry of CO oxidation has been carried out with *P. carboxydovorans* strain OM5. Zavarzin and

Table 18.1 Characteristics of selected carboxydotrophic bacteria

Species	Gram type	DNA mol % G+C	Electron acceptor	Growth with H_2+CO_2[b]	N_2 fixation	Temperature class
Pseudomonas carboxydovorans	−	61–64	O_2	+	−	Mesophile
P. carboxydohydrogena	−	58–62	O_2	+	−	Mesophile
P. carboxydoflava	−	66–68	O_2	+	−	Mesophile
P. compransoris	−	63–66	O_2	+	−	Mesophile
P. gazotropha	−	63–67	O_2	+	−	Mesophile
Arthrobacter sp.	+	63–70	O_2	+	−	Mesophile
Bacillus schlegelii	+	64–69	O_2	+	−	Thermophile
Streptomyces G26	+	ND[a]	O_2	ND	−	Thermophile
P. thermocarboxydovorans	−	72	O_2	−	−	Thermophile
Alcaligenes carboxydus	−	63–64	O_2	−	−	Mesophile
Acinetobacter sp.	−	43–45	O_2	ND	−	Mesophile
Azotobacter sp.	−	66	O_2	−	+	Mesophile
Azomonas sp.	−	55	O_2	−,+[c]	+	Mesophile
Isolates S17 and A305	−	ND	O_2	+	+	Mesophile
Rhizobium japonicum	−	ND	Nitrate	+	+	Mesophile

[a]ND, not determined.
[b]H_2 as energy source and CO_2 as carbon source.
[c]Depends on the strain tested.

co-workers have obtained *P. carboxydohydrogena* (Meyer et al., 1980; Sanjieva and Zavarzin, 1971; Savelieva and Nozhevnikova, 1973; Zavarzin and Nozhevnikova, 1976, 1977), *P. carboxydoflava, P. compransoris, P. gazotropha,* and *Alcaligenes carboxydus* (Zavarzin and Nozhevnikova, 1976, 1977). All of these isolates are mesophiles (see Table 18.1). Krüger and Meyer (1984) succeeded in the isolation of strains of thermophilic carboxydotrophic *Bacilli.* Most of them could be assigned as new strains of *Bacillus schlegelii,* and the authors found that even the type strain of *B. schlegelii* could utilize CO. These thermophiles had a temperature optimum of 65 °C and did not grow at temperatures below 50°C. Subsequently, several strains of moderately thermophilic CO-oxidizers, designated *Pseudomonas thermocarboxydovorans* C2, S1, and FE, were isolated in Colby's laboratory (Lyons et al., 1984). Their temperature optimum was 50 °C, and no growth occurred below 37 °C. Another moderate thermophile was *Streptomyces* G26 (Bell et al., 1985). The capability of *Arthrobacter* 11/x, *Bacillus schlegelii,* and *Streptomyces* G26 to grow with CO indicates that this property is found in Gram-positive bacteria as well. Among the first carboxydotrophic bacteria which became known were the strains S17 and A305 of Ooyama (1976). These bacteria were capable of growing with CO under conditions of dinitrogen fixation. Later, there was a report showing growth of a strain of *Rhizobium japonicum* on CO, with nitrate serving as the terminal electron acceptor (Gunatilaka et al., 1983). CO-oxidizing strains of N_2-fixing *Azotobacter* and *Azomonas* are also known (Park and Hegeman, 1984).

The metabolism of carboxydotrophic bacteria is facultative, e.g., all isolates can grow heterotrophically on organic substrates as well. With only few exceptions (*Alcaligenes carboxydus, Pseudomonas thermocarboxydovorans*), most carboxydotrophic bacteria also can grow chemolithotrophically, using H_2 as energy source and CO_2 as carbon source. Further, *Pseudomonas carboxydoflava* can grow mixotrophically with pyruvate using CO or H_2 as additional energy sources (Kiessling and Meyer, 1982); CO or CO_2 derived from its oxidation are not assimilated under mixotrophic conditions. Whether the carboxydotrophic bacteria should be considered a subgroup of the hydrogen-oxidizing bacteria, having gained the ability of growing at the expense of CO, or the hydrogen bacteria represent a subgroup of the CO-oxidizing bacteria having lost the ability to utilize CO, is a matter of dispute and still remains to be elucidated. So far, the basis for the close relationship of the two physiological bacterial groupings is not understood.

Several of the carboxydotrophic bacteria mentioned in Table 18.1 can grow anaerobically using nitrate as the ultimate electron acceptor, reducing it to N_2 via nitrite. With *Pseudomonas carboxydoflava* and *Pseudomonas carboxydohydrogena,* suitable electron donors for denitrification were CO, H_2, or organic substrates. Denitrification by carboxydotrophic bacteria growing with CO represents a novel type of metabolism and is currently under investigation.

18.2 PLASMIDS OF CARBOXYDOTROPHIC BACTERIA

Particular strains of carboxydotrophic bacteria contain plasmids (Gerstenberg et al., 1982; Kwon and Kim, 1985; Park and Hegeman, 1984), and a recent survey revealed plasmids in 8 out of 20 different strains (Kraut and Meyer, 1988). The amount of plasmid DNA detected depended on the growth phase and was high in exponentially growing cells (Kraut and Meyer, 1988). The sizes of plasmids ranged from 3.5 to 558 kilobases (kb) (Table 18.2). Megaplasmids of 428 and 558 kb were resolved in *Alcaligenes carboxydus,* and small plasmids were found in *Azomonas* and *Azotobacter.* Restriction digest patterns of plasmids from different carboxydotrophic bacteria were dissimilar. The strains

Table 18.2 Plasmids detected in carboxydotrophic bacteria*

Strain		Plasmid designation	Approximate size (kb)
Pseudomonas carboxydovorans	OM2	pHCG4	128
	OM4	pHCG5-a	158
		pHCG5-b	128
	OM5	pHCG3	128
	OM5-24	Δ^1pHCG3	113
	OM5-29	Δ^2pHCG3	74
Pseudomonas carboxydoflava		pHCG2-a	110
		pHCG2-b	45
Alcaligenes carboxydus		pHCG1-a	558
		pHCG1-b	428
		pHCG1-c	129
Azomonas B1		pHCG6-a	254
		pHCG6-b	200
Azomonas C2		pHCG7	3.5
*Azotobacter*A1		pHCG8	3.5

*Gerstenberg et al., 1982; Kraut and Meyer, 1988; Kwon and Kim, 1985; Kwon et al., 1986; Meyer et al., 1986; Park and Hegeman, 1984.
ND, not determined.

OM2, OM4, and OM5 of *P. carboxydovorans* contained a 128 kb plasmid. Indistinguishable restriction patterns suggested that this plasmid is presumably the same in different strains. Strain OM4 harbored an additional 158 kb plasmid. Strain OM3 carried no plasmid. Nine cured mutants obtained from *P. carboxydovorans* OM5, as well as the deletion mutant OM5-29 (carrying a 54 kb deletion on plasmid pHCG3), could not grow chemolithotrophically with CO or H_2 plus CO_2, as they were devoid of CO dehydrogenase, hydrogenase, and ribulosebisphosphate carboxylase. The deletion mutant OM5-24 carried a 15 kb deletion on plasmid pHCG3 and retained the ability to grow with CO. As the mutant could not grow with H_2 plus CO_2 and was devoid of hydrogenase, it was concluded that structural and/or regulatory genes of CO dehydrogenase, hydrogenase, and ribulosebisphosphate carboxylase reside on plasmid pHCG3 of *P. carboxydovorans* (Kraut and Meyer, 1988). The deletion mutants suggested clustering of the genes mentioned. Three plasmids were resolved in *Alcaligenes carboxydus* and two in *Azomonas* B1 and in *P. carboxydoflava*. A 3.5 kb plasmid was found in each of the N_2-fixing carboxydotrophs *Azomonas* and *Azotobacter* (Kraut and Meyer, 1988; Kwon and Kim, 1985; Park and Hegeman, 1984). Another small plasmid of similar size was resolved in *P. carboxydovorans*. It coded for a protease which was constitutive and specific for the medium subunit of *P. carboxydovorans* CO dehydrogenase. The protease had no activity with the same subunit of the *P. carboxydohydrogena* enzyme (Kwon and Kim, 1985; Kwon et al., 1986).

We are currently investigating the N-terminal amino acid sequences of CO dehydrogenases purified from different carboxydotrophic bacteria (S. Herwig, personal communication). Although the CO dehydrogenases from *Pseudomonas carboxydovorans* and *Pseudomonas carboxydoflava* showed a high degree of similarity with respect to their physical and chemical properties, the N-terminal amino acid sequences of the small subunits were different:

Met–Ala–Lys–Ala–His–Ile–Glu–Leu–Thr–Ile–Asn–Gly–His–Pro–Val–Glu–Ala–Leu–Val–Glu–Pro (*P. carboxydovorans*)

Met–Ala–Lys–Lys–Ile–Ile–Thr–Val–Asn–Val–(Cys)–Gly–Lys–Ala–Gln–Glu–Lys–Ala–Val–(Leu or Glu)–Pro–(?)–Thr–Ser–(Gln or Leu)–Ile (*P. carboxydoflava*)

With *P. carboxydovorans* the analysis of the medium subunit revealed the following sequence:

Met–Met–Ile–Pro–Gly–Ser–Phe–Asp–Tyr–His–Arg–Pro–Lys–Ser–Ile–(Thr)–Asp–(?)–Val

On the basis of these sequences, mixtures of synthetic oligonucleotides can be used for screening recombinant DNA clones derived of carboxydotrophic bacteria for CO dehydrogenase structural genes.

18.3 CO OXIDATION AND ASSIMILATION OF CARBON

Cells of *P. carboxydovorans* growing aerobically oxidize CO as follows (Meyer and Schlegel, 1978; 1983):

$$1\ O_2 + 2.19\ CO \rightarrow 1.83\ CO_2 + 0.36\ \text{cell carbon}$$

Table 18.3 Percentages of CO-carbon assimilated by different carboxydotrophic bacteria

Strain	Percentage of CO incorporated into cell carbon	t_d with CO
Hydrogenomonas carboxydovorans	16	ND
Pseudomonas carboxydovorans	16	18–20 hr
Pseudomonas carboxydohydrogena	4	12
Pseudomonas carboxydoflava	4	12
Pseudomonas gazotropha	5	ND
Pseudomonas compransoris	4	18
Achromobacter carboxydus	4	42
Bacillus schlegelii type strain	3	6–7
Bacillus schlegelii OMT1	8	3
Bacillus schlegelii OMT4	1	3–4
Bacillus schlegelii OMT7	10	4
Bacillus OMT2	2	3–4

t_d, doubling time. ND, not determined.
Data from Kistner, 1953; Meyer and Schlegel, 1978, Zavarzin and Nozhevnikova, 1977; Krüger and Meyer, 1984; Lyons et al., 1984.

About one molecule of O_2 is required for the conversion of two molecules CO to about two molecules of CO_2. Similar stoichiometries have been found with other carboxydotrophic bacteria (Table 18.3). Although the free-energy change of CO oxidation (−514 kJ or −123 kcal) is even somewhat higher than that of H_2 oxidation (−474 kJ or −113 kcal), growth with CO is relatively slow with most mesophilic isolates (Table 18.3). On the other hand, the thermophilic strains grow exceptionally well.

Using *Pseudomonas carboxydovorans*, it has been established that gaseous CO_2, which rather than bicarbonate is the actual substrate of ribulosebisphosphate carboxylase, is the first product of the CO dehydrogenase reaction and that the bacteria are devoid of carbonic anhydrase activity (Futo and Meyer, 1986). About 1 to 16 percent of the CO_2 derived from CO was incorporated into cell carbon, depending on the species and on the growth conditions (Table 18.3). It is also obvious from these measurements that the major portion of the CO_2 formed is released into the environment, leading to a slight acidification of the growth medium. Carbon assimilation of carboxydotrophic bacteria occurs via the Calvin cycle. All species examined in that respect showed specific activities of ribulosebisphosphate carboxylase and phosphoribulokinase sufficient to allow the observed growth rates with CO. It has also been established that CO is fixed into cell carbon after its conversion to CO_2 (Meyer and Schlegel, 1983; Zavarzin and Nozhevnikova, 1977).

18.4 CO-INSENSITIVE ELECTRON TRANSPORT AND GENERATION OF ENERGY AND REDUCING POWER

The metabolism of carboxydotrophic bacteria is strictly respiratory. Suitable electron acceptors are O_2 and with several strains, nitrate and nitrite. Respiration involves an electron transport system with quinones and cytochromes of the b-, c-, and a-types (Figure 18.1), and the cytochrome composition of carboxydotrophic bacteria grown on CO, H_2 plus CO_2, or pyruvate has been studied by spectroscopy at room and liquid nitrogen temperatures (Cypionka and Meyer, 1983a; Kim and Hegeman, 1981). Although CO inhibited growth and respiration of most aerobic bacteria examined in that respect (Cypionka and Meyer, 1982), the carboxydotrophic bacteria tolerated concentrations of up to 90 percent CO in the culture head space. However, CO drastically decreased both growth and respiration of *Achromobacter carboxydus* and *Pseudomonas compransoris*, even at low concentrations. For example, the growth rate

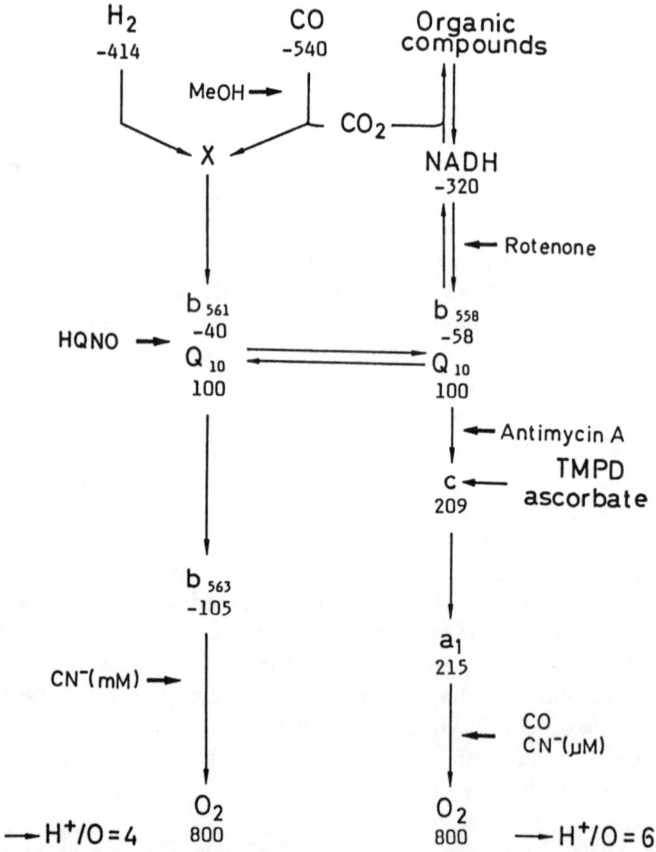

Figure 18.1 CO-sensitive and CO-insensitive electron transport in *Pseudomonas carboxydovorans* (Cypionka and Meyer, 1983b; Meyer et al., 1986) and the path of formation of NAD(P)H (Jacobitz and Meyer, 1986; Meyer et al., 1986). The numbers refer to midpoint potentials, which are in mV. The action points of inhibitors are indicated by arrows. X refers to the physiological electron acceptor, which possibly is cytochrome b_{561}. The cytochromes are indicated by their α-absorption maxima in room temperature redox-difference spectra. HQNO, 2-heptyl-4-quinolinol 1-oxide; TMPD, tetramethyl-*p*-phenylenediamine.

of *A. carboxydus* could be stimulated twofold by decreasing the CO concentration in the gas phase from 50 to 3 percent (Cypionka and Meyer, 1982). With the exception of *C. compransoris*, resistance toward CO is a constitutive property of carboxydotrophic bacteria since it is also evident under heterotrophic growth conditions. Type and concentration of most respiratory chain components were not significantly influenced by the presence or absence of CO, and it was established that CO dehydrogenase is not involved in the detoxification of CO (Cypionka and Meyer, 1982). It must be noted, however, that certain b-type cytochromes were detected in CO-grown but not in pyruvate-grown cells of *P. carboxydovorans* (cytochrome b_{558}—room temp. = b_{561}—low temp.), *P. carboxydohydrogena* (b_{558}), *P. carboxydoflava* (b_{563}), and *Arthrobacter* 11/x (b_{561}). Cytochrome b_{558} was specifically present in CO- as well as H_2-grown cells of *P. carboxydovorans*, but was not detectable in heterotrophically grown bacteria (Cypionka and Meyer, 1983a).

A specific function of cytochrome b_{558} in the reverse electron flow leading to the formation of NAD(P)H would not be surprising. The respiratory chain of *P. carboxydovorans* is branched at the level of b-type cytochromes or ubiquinone (Figure 18.1). The branch which uses cytochrome a_1 as the terminal oxidase is sensitive to CO, and thus cytochrome a_1 could be identified by photochemical action spectroscopy in the presence of CO (Cypionka and Meyer, 1983b). Resistance toward CO is due to the functioning of cytochrome b_{563} as a CO-insensitive terminal oxidase of the alternative branch. Cytochrome b_{563} has a low midpoint potential (Figure 18.1) and an exceptionally high affinity for O_2 ($K_m=0.5\ \mu M\ O_2$). These properties keep it oxidized and thus prevent binding of CO. During growth of carboxydotrophic bacteria with CO or H_2 plus CO_2, electrons are delivered by the action of CO dehydrogenase or hydrogenase, respectively. Because of the inhibition of cytochrome a_1 by CO, electrons are then transported via the CO-insensitive branch of the respiratory chain (Figure 18.1). In the absence of CO, electrons released from H_2 or organic substrates have a preference for the CO-sensitive branch.

Proton translocation measurements with intact cells of *P. carboxydovorans* revealed H^+/O ratios of 4 with CO (Cypionka et al., 1984). These were 6 with H_2 or pyruvate in the absence of CO (Figure 18.1). Obviously, in the presence of CO, when all electrons must use the CO-insensitive pathway, less energy is conserved than in the CO-sensitive pathway. This interpretation is in accord with the low midpoint redox potential of -11 to -215 mV of cytochrome b_{563} (Cypionka et al., 1985), suggesting that electrons derived from CO have to overcome a strong kinetic hindrance. This idea is also supported by the slower growth of *P. carboxydovorans* with CO compared to H_2 or pyruvate. The interpretation is challenged, however, by the observation that CO has no effect on the growth rate of *P. carboxydovorans* in the presence of organic substrates. What is clearly needed are determinations of growth yields under various conditions.

Formation of reduced pyridine nucleotides (NAD(P)H) was studied in cell suspensions of *P. carboxydovorans* pulsed with CO or H_2 in the presence of O_2 (Jacobitz and Meyer, 1986). The experiments indicated that in bacteria growing with CO or H_2, electrons and a proton motive force (pmf), generated by respiration, are required to drive reverse electron transfer for the formation of reduced pyridine nucleotides. The path of reverse electron flow was deduced from inhibitor studies (Figure 18.1). It involved CO dehydrogenase, the ubiquinone–cytochrome b region, and NADH dehydrogenase, acting in the reverse direction.

18.5 CO DEHYDROGENASE

CO oxidation in carboxydotrophic bacteria is mediated by CO dehydrogenase as follows:

$$CO + H_2O + A\ (\text{oxidized}) \rightarrow CO_2 + AH_2\ (\text{reduced})$$

The enzyme catalyzes the oxidation of CO to CO_2 using hydrogen ions from water as the oxidant. The two electrons and protons formed can be transferred to oxidized artificial two-electron acceptors (abbreviated A in the above reaction), which are thereby reduced. Up to now, all CO dehydrogenases studied from different bacterial sources had no activity with oxidized pyridine nucleotides, flavins, or any other soluble physiological electron acceptors (Bell et al., 1985; Cypionka et al., 1980; Meyer and Schlegel, 1979, 1980).

CO dehydrogenases have been purified to different extent from the cytoplasm of the bacteria listed in Table 18.4. Their cofactor composition is complex as they consistently contain FAD, two types of 2Fe–2S centers, molybdenum (Mo), and a pterin (Bray et al., 1983; Krüger and Meyer, 1986; Meyer, 1982; Meyer and Rajagopalan, 1984a). In addition, the enzyme from *Pseudomonas carboxydovorans* has the capacity to bind selenite and is thereby activated (Meyer and Rajagopalan, 1984b). As is discussed below, CO dehydrogenase solubilized from cytoplasmic membranes of *P. carboxydovorans* copurified with cytochrome b_{561} (S. Jacobitz, personal communication).

CO dehydrogenases are CO:acceptor oxidoreductases. Their anabolic function is to provide CO_2 for assimilation by ribulosebisphosphate carboxylase. Their catabolic function is to provide electrons and protons for the generation of a proton motive force. As studied in *P. carboxydovorans*, the enzyme is not involved in the detoxification of CO (Cypionka and Meyer, 1982).

Molecular masses ranging from 230 to 400 kDa have been reported for different CO dehydrogenases (Table 18.4). The values determined were found to depend considerably on the method applied. However, with a given method same molecular masses were found for different enzymes. It seems that CO dehydrogenases are consistently composed of three subunits of different size (Table 18.4).

CO dehydrogenases from *P. carboxydovorans* and *P. carboxydohydrogena* were examined by electron paramagnetic resonance (EPR) spectroscopy (Bray et al., 1983). In the temperature range of 8.5–80 K, both enzymes, after reduction by dithionite or by CO in the presence of methylene blue, displayed

Table 18.4 Properties of CO dehydrogenases purified from different sources

Organism	M_r (kDa)	Subunit composition (kDa)	Atoms of Mo	FAD	Iron sulfur centers	K_m (μM CO)	pI	Temp for max act (°C)	Electron acceptors
P. carboxydovorans (1)	230–300	70+33+17 85+25+14	2	2	+	53	4.7	63	MB, LV, PMS, TB, DCPIP
P. carboxydohydrogena (2)	230–300 400	70+33+17 85+28+14	2	2	+	63	4.6	ND	MB, LV, PMS, TB, DCPIP
P. carboxydoflava (3)	230–300	70+33+17	2	2	+	ND	ND	ND	MB, LV, PMS, TB, DCPIP
B. schlegelii (4)	230	ND	+	+	+	ND	ND	95	MV, MB, PMS, TB, DCPIP, FC
P. thermocarboxydovorans (5)	230–310	ND	0.7	1.8	+	0.5	ND	80	MB, LV, PMS, DCPIP, FC
Streptomyces G26 (6)	ND	ND	+	ND	ND	1.3	ND	ND	MV, BV, MB, LV, PMS, TB, DCPIP, FC

Abbreviations: BV, benzyl viologen; DCPIP, dichlorophenolindophenol; FAD, flavin adenine dinucleotide; FC, ferricyanide; kDa, kilodaltons; LV, Lauth's violet; MB, methylene blue; Mm, relative molecular weight; MV, methyl viologen; ND, not determined; pI, isoelectric pH; PMS, phenazine methosulfate; TB, toluylene blue.

(1) Kim and Kim, 1984; Kim and Lee, 1986; Meyer, 1982, 1985, 1986; Meyer and Fiebig, 1985; Meyer and Rohde, 1986; Meyer and Schlegel, 1983; Meyer et al., 1986. (2) Kim and Hegeman, 1981a, 1983; Kim and Lee, 1986; Meyer et al., 1986. (3) Meyer et al., 1986. (4) Krüger and Meyer, 1984. (5)(6) Bell et al., 1985; Williams and Colby, 1986.

EPR spectra due to 2 different types of Fe_2S_2 centers (designated Fe–S 1 and Fe–S 2). These signals were very similar to those of the Fe–S 1 and Fe–S 2 centers of milk xanthine oxidase. That each of these centers occurs in duplicate is indicated by the analysis of 8 nonheme iron atoms and 8 acid labile sulfur atoms per molecule of CO dehydrogenase. EPR spectroscopy also revealed signals from Mo(V) atoms in two different chemical environments. As was to be expected for catabolic enzymes, CO dehydrogenases had a high affinity for their substrate (Table 18.4). This is especially true for the enzymes from *P. thermocarboxydovorans* and *Streptomyces* G26.

The isoelectric points of the CO dehydrogenases from *P. carboxydovorans* and *P. carboxydohydrogena* (Table 18.4) agree with the observation that these enzymes show a high affinity toward different kinds of anion exchangers.

CO dehydrogenases have been shown to reduce several artificial electron acceptors using CO as electron donor (Table 18.4). In general, the viologen dyes or other dyes with a more negative redox potential were ineffective. Exceptions were the enzymes from *Streptomyces* G26, which displayed similar activities with phenazine methosulfate (PMS) as with benzyl viologen (Bell et al., 1985) and the enzyme from *B. schlegelii* which had some activity with methyl viologen (Krüger and Meyer, 1984). The latter two enzymes did not react with oxidized pyridine nucleotides, as is true for all other CO dehydrogenases studied thus far.

18.6 IMMUNOLOGICAL RELATIONSHIPS

Using polyclonal antibodies raised against CO dehydrogenase from *P. carboxydovorans* or *P. carboxydohydrogena*, Ouchterlony double immunodiffusion revealed that most enzymes, including those from Gram-positive sources, were immunologically related to each other showing reactions of partial identity (Table 18.5). However, the enzymes from the thermophilic *Bacilli, P. compransoris,* and *Acinetobacter* were immunologically unrelated to all others.

18.7 UTILIZATION OF CO REQUIRES MOLYBDENUM

As Mo is required for the formation of active CO dehydrogenase, the element must be present in the growth medium of carboxydotrophic bacteria utilizing CO. In contrast, growth of carboxydotrophic bacteria with H_2 plus CO_2 or organic substrates occurs in the absence of Mo. A culture of *Bacillus* OMT2 deprived of molybdate was shown to start growing again autotrophically on CO when molybdate was added (Krüger and Meyer, 1984). Similar observations have been made with other carboxydotrophic bacteria (Kalnowski, 1980). However, there is the interesting exception of an *Acinetobacter* strain capable of growing with CO in the absence of molybdate (Cho et al., 1985), suggesting

Table 18.5 Immunological reactivity of CO dehydrogenases from different carboxydotrophic bacteria

	Reaction with antibodies prepared against CO dehydrogenase from	
Organism	*Pseudomonas carboxydovorans*[a]	*Pseudomonas carboxydohydrogena*[b]
Pseudomonas carboxydovorans OM5	I[c]	PI[d]
P. carboxydovorans OM2, OM3, OM4	PI	ND[e]
Pseudomonas carboxydohydrogena	PI	I
Pseudomonas carboxydoflava	PI	ND
Alcaligenes carboxydus	PI	ND
Arthrobacter 11/x	PI	ND
Kalnowski's strain 5	PI	ND
Azomonas sp.1, sp.2, 3B	ND	PI
Pseudomonas compransoris	NI[f]	ND
Bacillus schlegelii type strain, OMT1, OMT4, OMT7	NI	ND
Bacillus OMT2	NI	ND
Acinetobacter JC1	ND	NI

[a] Meyer and Rohde, 1984; Krüger and Meyer, 1984; [b] Kim et al., 1982; Park and Hegeman, 1984; Cho et al., 1985; [c] I, identity; [d] PI, partial identity; [e] ND, not determined; [f] NI, nonidentity

a novel type of CO dehydrogenase. This enzyme has not been purified so far, and further studies are eagerly awaited.

18.8 PRESENCE OF A PTERIN

Molybdoenzymes, with the exception of nitrogenase, harbor a molybdenum cofactor (MoCo) which is a complex of a pterin, Mo and an oxo- or sulfido group attached to the metal (Figure 18.2). The Mo in MoCo is chelated by the 2 sulfhydryl groups at the side chain of the pterin. MoCo was originally characterized by its ability to complement NADPH-nitrate reductase activity in extracts of induced mycelia of the *Neurospora crassa nit-1* mutant (Ketchum et al., 1970; Nason et al., 1970, 1971; for a review see Johnson, 1980). That MoCo is a noncovalent constituent of CO dehydrogenase was evident from the ability of unfolded enzyme to reconstitute aponitrate reductase in *nit-1* extracts (Krüger and Meyer, 1986; Meyer, 1985, 1986; Meyer and Rajagopalan, 1984a; Meyer and Schlegel, 1983; for a review see Meyer et al., 1986). As was elaborated by the group at Duke (Johnson et al., 1980, 1984; Johnson and Rajagopalan, 1982; Rajagopalan et al., 1982; Rajagopalan, 1985) and later confirmed by others (Ishizuka et al., 1983; Krüger and Meyer, 1987), the pterin occurring in molybdoenzymes is substituted in the 6-position, containing the tetrahydroform of the ring and one phosphate group in its side chain. Molybdopterin released from protein linkage is extremely unstable and becomes rapidly inactivated, forming a number of fluorescent and relatively stable derivatives by unknown degradative pathways. Though molybdopterin is unstable in air, it is relatively stable when kept anaerobic in the presence of reductants (Hawkes and Bray, 1984a,b; Wahl et al., 1984) or when the sulfhydryl groups in its side chain are

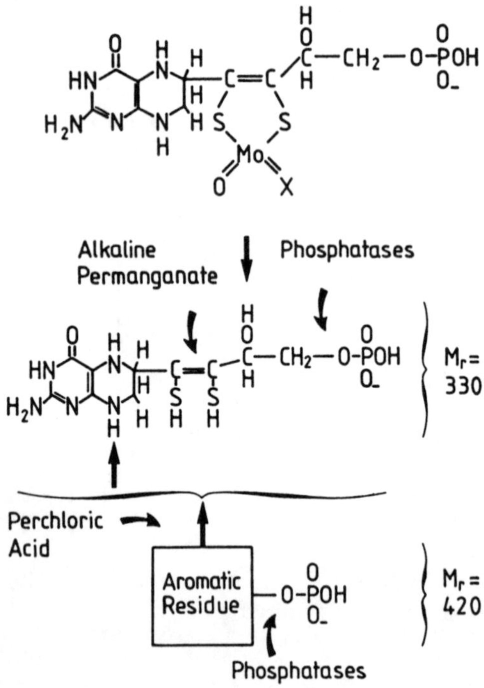

Figure 18.2 Structure of the molybdenum cofactor containing tetrahydromolybdopterin (top), and presumed structural elements of bactopterin (bottom). See text for references.

protected by thiol reagents (Mendel and Alikulov, 1983). A dicarboxyderivative of molybdopterin has been prepared under conditions favoring retention of the structural features of the molecule (K. V. Rajagopalan, personal communication). Specific radioactivity of (1-^{14}C)-iodoacetamide incorporated relative to the amount of phosphate indicated two alkylation sites per pterin. Energy dispersive X-ray analysis of the derivative showed the presence of two sulfurs. An exact mass corresponding to the molecular formula $C_{14}H_{18}N_7O_5S_2$ was obtained for the MH^+ ion of the alkylated dephosphorylated compound by fast atom bombardment mass spectroscopy. ^{1}H-NMR spectra of the phosphorylated forms of alkylated molybdopterin, in conjunction with other data, provided strong corroboration for the validity of the proposed structure of molybdopterin (Johnson and Rajagopalan, 1982), as a 6-alkylpterin with a 4-carbon side chain, containing an enedithiol on C1′ and C2′, a secondary alcohol on C3′ and a phosphorylated primary alcohol on C4′. As isolated the dicarboxyamidomethylmolybdopterin was found to be a 5,6,7,8-tetrahydropterin (Figure 18.2).

CO dehydrogenases purified from different carboxydotrophic bacteria were found to contain a pterin exhibiting characteristics not very different from molybdopterin (Meyer and Rajagopalan, 1984a). It has been postulated that CO dehydrogenases contain a modified molybdopterin (Figure 18.2) which has been termed bactopterin (Krüger and Meyer, 1986). The proposal is based on the following considerations: Bactopterin contains a single pterin ring and carries a side chain at carbon atom 6 of the pterin nucleus, as indicated by the formation of pterin-6-carboxylic acid upon alkaline permanganate oxidation (Krüger and Meyer, 1987), as has been determined for molybdopterin (Rajagopalan et al., 1982). Organic phosphate analyses revealed 8 mol of organic phosphate per mol of enzyme (Meyer and Rajagopalan, 1984a). This phosphate content has been confirmed by labeling CO dehydrogenase with ^{32}P-phosphate and was in conflict with the assumption of 2 FAD and 2 molybdopterins (yielding a total of 6 phosphate residues) per enzyme molecule (Krüger and Meyer, 1986). Upon gel filtration, ^{32}P-labeled bactopterin revealed an apparent M_r of about 730, a figure about twice that obtained with molybdopterin (Krüger and Meyer, 1986). Labeled bactopterin was cleaved by perchloric acid into two fragments with apparent M_r of about 420 and 330, respectively. As the large fragment was UV-absorbing and nonfluorescent, it was considered not to be a pterin. The small fragment was UV-absorbing, fluorescent, and its apparent M_r was indistinguishable from that of molybdopterin. Each fragment carried a single phosphate group (Krüger and Meyer, 1986). It was thus suggested that bactopterin is composed of molybdopterin or a very similar pterin, carrying an additional substituent which might be an organic compound, but is not a pterin (Krüger and Meyer, 1986, 1987). Studies employing phosphate-cleaving enzymes revealed the presence of two monophosphoric acid monoesters in bactopterin, and the affinity of the reduced compound for thiol-sepharose pointed to the presence of thiol(s) (Krüger and Meyer, 1987). In addition, bactopterin was less susceptible to air oxidation than molybdopterin (Krüger and Meyer, 1986). The exact chemical structure of bactopterin is currently under investigation. By estimating the apparent M_r (Krüger and Meyer, 1986; Krüger et al., 1987), bactopterin has been resolved in CO dehydrogenases (*Pseudomonas carboxydovorans, Pseudomonas carboxydohydrogena,* and *Pseudomonas carboxydoflava*), xanthine dehydrogenase (*Clostridium barkeri, Veillonella alcalescens*), nicotinic acid dehydrogenase (*Bacillus niacini*), nitrite oxidoreductase (*Nitrobacter hamburgensis*), and nitrate reductase (*Escherichia coli, Pseudomonas stutzeri*). Instead, molybdopterin was resolved in xanthine oxidase (bovine milk) and sulfite oxidase (chicken liver). It has been suspected from these findings that bactopterin might be a common cofactor of eubacterial molybdoenzymes and that the occurrence of molybdopterin is restricted to eucaryotic molybdoenzymes (Krüger and Meyer, 1986; Krüger et al., 1987). It is hoped that final proof for this hypothesis will be obtained in due course.

18.9 INTRACELLULAR LOCATION OF CO DEHYDROGENASE

The intracellular location of CO dehydrogenase in *P. carboxydovorans* has been examined by immunoelectronmicroscopy using the ferritin and the protein A-gold methods (Meyer and Rohde, 1984; Rohde et al., 1984, 1985). The intracellular location of the enzyme depended on the growth phase. In exponential cells nearly 90 percent of CO dehydrogenase were associated with the inner surface of the cytoplasmic membrane. However, this ratio decreased to 50 percent in the stationary growth phase. The actual amount of CO dehydrogenase associated with the cytoplasmic membrane was paralleled by the respiratory oxidation rate of CO (Rohde et al., 1985). In other words, a high percentage of membrane-bound CO dehydrogenase resulted in high respiration rates. *Vice versa,* respiration rates decreased upon dissociation of the enzyme from the membrane. The results explain the inability of CO dehydrogenase to act on soluble physiological electron acceptors and indicate that *in vivo* CO dehydrogenase transfers the electrons to a membrane-bound electron acceptor. As will be discussed below, there are indications that the physiological electron acceptor of CO dehydrogenase might be cytochrome b_{561}.

18.10 MEMBRANE-BOUND CO DEHYDROGENASE

Upon disintegration and ultracentrifugation of extracts obtained from *P. carboxydovorans* or other carboxydotrophic bacteria, there is always a 30 to 50 percent fraction of CO dehydrogenase which stays with the cytoplasmic membrane (Cypionka et al., 1980; Meyer and Schlegel, 1979). The membrane-bound enzyme could be solubilized with CHAPS or octylglucoside (S. Jacobitz, personal communication). After 2–3 treatments with these detergents, CO dehydrogenase was entirely removed from the membranes, whereas the electron transport chain components remained membrane-bound. Interestingly, di-or trivalent cations (e.g., Mg^{2+} or La^{3+}) caused the reattachment of CO dehydrogenase to membranes deprived of the enzyme. The activities of CO dehydrogenase with O_2 and with methylene blue very effectively complemented inactive cytoplasmic membranes in the presence of 15 mM $LaCl_3$. Although it seems unlikely that La^{3+} plays a role *in vivo,* this is a system which nicely mimics the process *in vitro* and demonstrates a dependence on the surrounding milieu of the affinity of CO dehydrogenase for the cytoplasmic membrane.

Another major achievement has been the isolation and purification of membrane-bound CO dehydrogenase (S. Jacobitz, personal communication). Cytoplasmic membranes of *P. carboxydovorans* were solubilized with dodecylmaltoside, and the enzyme was purified. The last step of the purification procedure was gel filtration on Sepharose-6B, resulting in two well-resolved peaks (Figure 18.3). The first material to be eluted was the cytochrome bc_1-complex as is evident from the redox-difference spectrum. The second peak contained CO dehydrogenase. However, the visible absorption spectrum of the enzyme was unusual in showing a strong absorption peak around 415 nm which is absent in the spectrum of the enzyme purified from the cytoplasmic fraction. The redox-difference spectra of CO dehydrogenase in the α-region at room

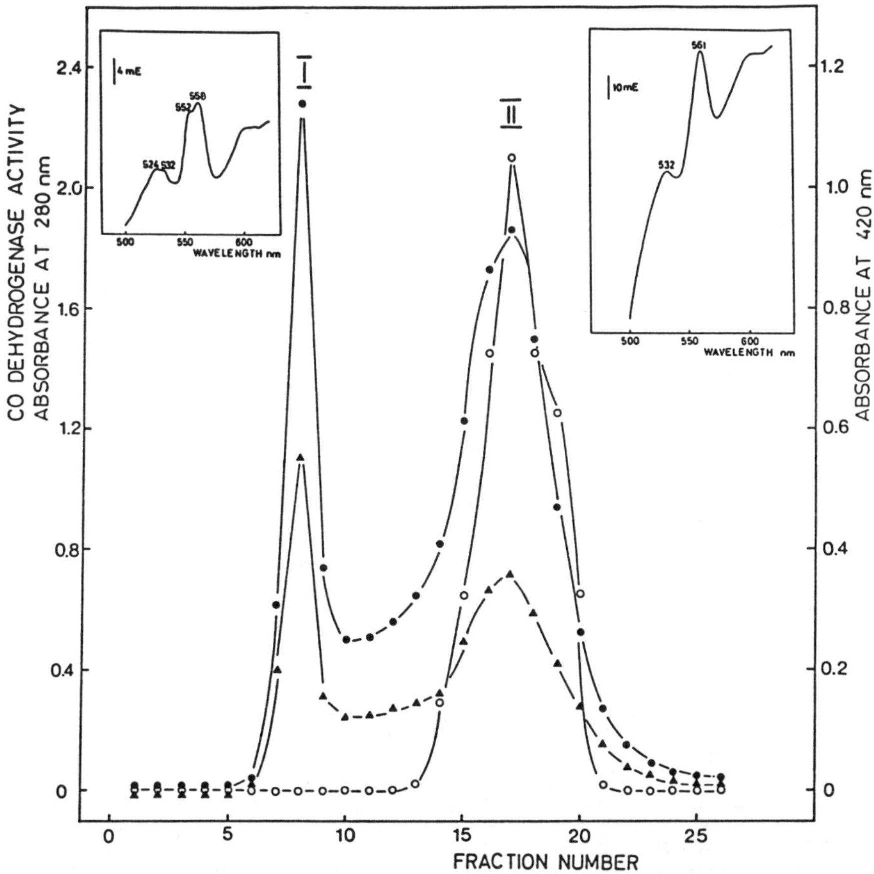

Figure 18.3 Gel filtration on Sepharose-6B of dodecylmaltoside-solubilized CO dehydrogenase. Symbols: CO→ methylene blue activity (o), protein absorbance, (A_{280}) (•) and cytochrome absorbance, (A_{420}) (▲). The insets show the dithionite-reduced minus oxidized difference spectra of pool I (left) and pool II (right) at room temperature.

and at liquid nitrogen temperature revealed the presence of cytochrome b_{561} (Figure 18.3, insets). The peak at 561 nm in the room temperature spectrum shifted to 558 nm in the low temperature spectra.

18.11 CONCLUDING REMARKS

Although CO is a rather strange and frequently toxic growth substrate, it is apparent from the data summarized in this review that its utilization by the carboxydotrophic bacteria under aerobic conditions is now firmly established. The presence of CO dehydrogenase, an electron transport system which is not inhibited by CO, and the ability to invoke a reverse electron flow for the generation of reduced pyridine nucleotides are indispensable prerequisites for the utilization of CO by these bacteria. CO dehydrogenase is a molybdo iron-sulfur flavoprotein. In addition, it contains the molybdenum cofactor which is a complex of Mo and a pterin. The proposed structure of molybdopterin from bovine milk xanthine oxidase and chicken liver sulfite oxidase has now been confirmed (Figure 18.2). Obviously, we are now in most acute need of chemical and spectroscopical data proving the structural differences between bactopterin and molybdopterin, suggested by the indirect results discussed. In addition, much more molybdoenzymes need to be examined for the occurrence of the two pterins. Now that the existence in *Pseudomonas carboxydovorans* of a branched respiratory chain, composed of a CO-insensitive and a sensitive branch, has been firmly established, the discovery of further branches and very complex respiratory systems in the facultatively denitrifying carboxydotrophs may not be so far away. Although substantial advances have been made in our understanding of the special physiological and biochemical features of carboxydotrophic bacteria, continued efforts are necessary to understand the genetic basis of CO oxidation and the function of plasmids, to dissect the components of the denitrifying system, to identify the physiological electron acceptor of CO dehydrogenase, to elucidate the intramolecular electron flow in the enzyme, and to determine the efficiency of the different branches of the electron transport system by estimating growth yields in the presence and absence of CO. In addition, the mechanism of the attachment of CO dehydrogenase to the cytoplasmic membrane and its regulation *in vivo* are scarcely understood. Also of interest is an understanding of the role of carboxydotrophic bacteria in nature, a question which seems to be completely unsolved up to now.

I wish to thank my colloborators and students for their continuing contributions to this field. I also thank the various authors who gave permission to quote their unpublished results, and I am particularly grateful to K. V. Rajagopalan for making available some of his unpublished results. Work in my laboratory is supported by grants from the Deutsche Forschungsgemeinschaft and the Fonds der Chemischen Industrie.

REFERENCES

Bell, J.M., Williams, E., and **Colby, J.** 1985. Carbon monoxide oxidoreductases from thermophilic carboxydobacteria, pp. 153–159, in Poole, R.K. and Dow, C.S. (editors), *Microbial gas metabolism: mechanistic, metabolic and biotechnological aspects,* Academic Press, London.

Bisset, K.A. and **Grace, J.B.** 1954. The nature and relationships of autotrophic bacteria, pp. 28–53, in Fry, B.A. and Peel, J.L. (editors), *Autotrophic microorganisms,* At the University Press, Cambridge.

Bray, R.C., George, G.N., Lange, R., and **Meyer, O.** 1983. Studies by e.p.r. spectroscopy of carbon monoxide oxidases from *Pseudomonas carboxydovorans* and *Pseudomonas carboxydohydrogena. Biochemical Journal* 211:687–694.

Cho, J.W., Yim, H.S., and **Kim, Y.M.** 1985. *Acinetobacter* isolates growing with carbon monoxide. *Korean Journal of Microbiology* 23:1–8.

Cypionka, H. and **Meyer, O.** 1982. Influence of carbon monoxide on growth and respiration of carboxydobacteria and other aerobic organisms. *FEMS Microbiology Letters* 15:209–214.

Cypionka, H. and **Meyer, O.** 1983a. The cytochrome composition of carboxydotrophic bacteria. *Archives of Microbiology* 135:293–298.

Cypionka, H. and **Meyer, O.** 1983b. Carbon monoxide-insensitive respiratory chain of *Pseudomonas carboxydovorans*. *Journal of Bacteriology* 156:1178–1187.

Cypionka, H., Meyer, O., and **Schlegel, H.G.**1980. Physiological characteristics of various species of strains of carboxydobacteria. *Archives of Microbiology* 127:301–307.

Cypionka, H., van Verseveld, H.W., and **Stouthamer, A.H.** 1984. Proton translocation coupled to carbon monoxide-insensitive and—sensitive electron transport in *Pseudomonas carboxydovorans*. *FEMS Microbiology Letters* 22:209–213.

Cypionka, H., Reijners, W.N.M., van Wielink, J.E., Oltmann, L.F., and **Stouthamer, A.H.** 1985. Half reduction potentials and oxygen affinity of the cytochromes of *Pseudomonas carboxydovorans*. *FEMS Microbiology Letters* 27:189–193.

Davis, D.H. 1967. *Studies on the Gram-negative hydrogen bacteria and related organisms*. Ph.D. thesis. University of California at Davis. 117 pp.

Davis, D.H., Doudoroff, M., and **Stanier, R.Y.** 1969. Proposal to reject the genus *Hydrogenomonas*: Taxonomic implications. *International Journal of Systematic Bacteriology* 19:375–390.

Davis, D.H., Stanier, R.Y., Doudoroff, M., and **Mandel, M.** 1970. Taxonomic studies on some Gram-negative polarly flagellated "hydrogen bacteria" and related species. *Archiv für Mikrobiologie* 70:1–13.

Futo, S. and **Meyer, O.** 1986. CO_2 is the first species formed upon CO oxidation by CO dehydrogenase from *Pseudomonas carboxydovorans*. *Archives of Microbiology* 145:358–360.

Gerstenberg, C., Friedrich, B., and **Schlegel, H.G.** 1982. Physical evidence for plasmids in autotrophic, especially hydrogen-oxidizing bacteria. *Archives of Microbiology* 133:90–96.

Gunatilaka, M.K., Allen, G.C., and **Neal, J.L.** 1983. Anaerobic nitrate-dependent growth of *Rhizobium japonicum* on carbon monoxide. Abstracts, *4th International Symposium on Microbial Growth on One Carbon Compounds*, Minneapolis, poster session II, 2–4.

Hawkes, T.R. and **Bray, R.C.** 1984a. Quantitative transfer of the molybdenum cofactor from xanthine oxidase and from sulphite oxidase to the deficient enzyme of the *nit-1* mutant of *Neurospora crassa* to yield active nitrate reductase. *Biochemical Journal* 219:481–493.

Hawkes, T.R. and **Bray, R.C.** 1984b. Studies by electron-paramagnetic-resonance spectroscopy of the environment of the metal in the molybdenum cofactor of molybdenum-containing enzymes. *Biochemical Journal* 222:587–600.

Hegeman, G.D. 1984. Oxidation of carbon monoxide by aerobic bacteria, pp. 21–25, in Crawford, R.L. and Hanson, R.S. (editors), *Microbial growth on C_1 compounds*, American Society for Microbiology, Washington, D.C.

Ishizuka, M., Ushio, K., Toraya, T., and **Fukui, S.** 1983. Formation of thieno(3,2-g)pterins from the molybdenum cofactor. *Biochemical and Biophysical Research Communications* 111:537–543.

Jacobitz, S. and **Meyer, O.** 1986. Reduced pyridine nucleotides in *Pseudomonas carboxydovorans* are formed by reverse electron transfer linked to proton motive force. *Archives of Microbiology* 145:372–377.

Johnson, J.L. 1980. The molybdenum cofactor common to nitrate reductase, xanthine dehydrogenase and sulfite oxidase, pp. 347–383, in Coughlan, P. (editor), *Molybdenum and Molybdenum-containing Enzymes*, Pergamon, Oxford.

Johnson, J.L. and **Rajagopalan, K.V.** 1982. Structural and metabolic relationship between the molybdenum cofactor and urothione. *Proceedings of the National Academy of Sciences USA* 79:6856–6860.

Johnson, J.L., Hainline, B.E., and **Rajagopalan, K.V.** 1980. Characterization of the molybdenum cofactor of sulfite oxidase, xanthine oxidase, and nitrate reductase. *Journal of Biological Chemistry* 255:1783–1786.

Johnson, J.L., Hainline, B.E., Rajagopalan, K.V., and **Arison, B.H.** 1984. The pterin component of the molybdenum cofactor. Structural characterization of two fluorescent derivatives. *Journal of Biological Chemistry* 259:5414–5422.

Kalnowski, G. 1980. *Physiologische Untersuchung der Kohlenmonoxid-Oxidation und des chemolithotrophen Wachstums an zwei neu isolierten Carboxidobakterien.* Ph.D. thesis, Technische Universität Braunschweig, F.R.G.

Ketchum, P.A., Cambier, H.Y., Frazier III, W.A., Madansky, C.H., and **Nason, A.** 1970. In vitro restoration of *Neurospora* assimilatory nitrate reductase from protein subunits of a *Neurospora* mutant and the xanthine-oxidizing or aldehyde oxidase systems of higher animals. *Proceedings of the National Academy of Sciences USA* 66:1016–1023.

Kiessling, M. and **Meyer, O.** 1982. Profitable oxidation of carbon monoxide or hydrogen during heterotrophic growth of *Pseudomonas carboxydoflava. FEMS Microbiology Letters* 13:333–338.

Kim, Y.M. and **Hegeman, G.D.** 1981a. Purification and some properties of carbon monoxide dehydrogenase from *Pseudomonas carboxydohydrogena. Journal of Bacteriology* 148:904–911.

Kim, Y.M. and **Hegeman, G.D.** 1981b. Electron transport system of an aerobic carbon monoxide-oxidizing bacterium. *Journal of Bacteriology* 148:991–994.

Kim, Y.M. and **Hegeman, G.D.** 1983. Oxidation of carbon monoxide by bacteria. *International Reviews of Cytology* 81:1–32.

Kim, E.S. and **Kim, Y.M.** 1984. Subunit structure of carbon monoxide oxidase from *Pseudomonas carboxydovorans. Korean Biochemical Journal* 17:141–147.

Kim, Y.M. and **Lee, W.H.** 1986. Stable subunit pattern of carbon monoxide dehydrogenases from *Pseudomonas carboxydohydrogena* and *Pseudomonas carboxydovorans. Korean Biochemical Journal* 19:75–80.

Kim, Y.M., Kirkconnel, S., and **Hegeman, G.** 1982. Immunological relationships among carbon monoxide dehydrogenases of carboxydobacteria. *FEMS Microbiology Letters* 13:219–223.

Kistner, A. 1953. On a bacterium oxidizing carbon monoxide. *Proceedings Koninklijke Nederlandse Akademie van Wetenschappen Series C,* 56:443–450.

Kistner, A. 1954. Conditions determining the oxidation of carbon monoxide and of hydrogen by *Hydrogenomonas carboxydovorans. Proceedings Koninklijke Nederlandse Akademie van Wetenschappen, Series C,* 57:186–195.

Kraut, M. and **Meyer, O.** 1988. Plasmids in carboxydotrophic bacteria: physical and restriction analysis. *Archives of Microbiology* 149:540–546.

Krüger, B. and **Meyer, O.** 1984. Thermophilic *Bacilli* growing with carbon monoxide. *Archives of Microbiology* 139:402–408.

Krüger, B. and **Meyer, O.** 1986. The pterin (bactopterin) of carbon monoxide dehydrogenase from *Pseudomonas carboxydoflava. European Journal of Biochemistry* 157:121–128.

Krüger, B. and **Meyer, O.** 1987. Structural elements of bactopterin from *Pseudomonas carboxydoflava* carbon monoxide dehydrogenase. *Biochimica et Biophysica Acta* 912:357–364.

Krüger, B., Meyer, O., Nagel, M., Andreesen, J.R., Meincke, M., Bock, E., Blümle, S., and **Zumft, W.G.** 1987. Evidence for the presence of bactopterin in the eubacterial molybdoenzymes nicotinic acid dehydrogenase, nitrite oxidoreductase, and respiratory nitrate reductase. *FEMS Microbiology Letters* 48:225–227.

Kwon, M.O. and **Kim, Y.M.** 1985. Relationship between carbon monoxide dehydrogenase and a small plasmid in *Pseudomonas carboxydovorans. FEMS Microbiology Letters* 29:155–159.

Kwon, M.O., Chung, I.K., and **Kim, Y.M.** 1986. Role of a small plasmid in the modification of carbon monoxide dehydrogenase in *Pseudomonas carboxydovorans. FEMS Microbiology Letters* 37:113–116.

Lyons, C.M., Justin, P., Colby, J., and **Williams, E.** 1984. Isolation, characterization, and autotrophic metabolism of a moderately thermophilic carboxydobacterium, *Pseudomonas thermocarboxydovorans* sp. nov. *Journal of General Microbiology* 130:1097–1105.

Mendel, R.R. and **Alikulov, Z.A.** 1983. Reversible immobilization of molybdenum cofactor on a gel matrix via sulphhydryl groups. *Journal of Chromatography* 267:409–413.

Meyer, O. 1982. Chemical and spectral properties of carbon monoxide:methylene blue oxidoreductase, the molybdenum-containing iron-sulfur flavoprotein from *Pseudomonas carboxydovorans. Journal of Biological Chemistry* 257:1333–1334.

Meyer, O. 1985. Metabolism of aerobic carbon monoxide-utilizing bacteria, pp. 131–151, in Poole, R.K. and Dow, C.S. (editors), *Microbial gas metabolism,* Academic Press, London.

Meyer, O. 1986. Biologie und Biotechnologie aerob Kohlenmonoxide-oxidierender Bakterien, pp. 3–31, in Crueger, W., Esser, K., Präve, P., Schlingmann, M., Thauer, R., and Wagner, F. (editors), *Jahrbuch Biotechnologie,* Carl Hanser, München.

Meyer, O. and **Fiebig, K.** 1985. Enzymes oxidizing carbon monoxide, pp. 147–168, in Degn, H., Cox, R.P., and Toftlund, H. (editors), *Gas enzymology,* Reidel, Dordrecht.

Meyer, O. and **Rajagopalan, K.V.** 1984a. Molybdopterin in carbon monoxide oxidase from carboxydotrophic bacteria. *Journal of Bacteriology* 157:843–848.

Meyer, O. and **Rajagopalan, K.V.** 1984b. Selenite-binding to carbon monoxide oxidase from *Pseudomonas carboxydovorans.* Selenium binds covalently to the protein and activates specifically the CO $\rightarrow$ methylene blue reaction. *Journal of Biological Chemistry* 259:5612–5617.

Meyer, O. and **Rohde, M.** 1984. Enzymology and bioenergetics of carbon monoxide-oxidizing bacteria, pp. 26–33, in Crawford, R.L. and Hanson, R.S. (editors), *Microbial growth on C_1 compounds,* American Society for Microbiology, Washington.

Meyer, O. and **Schlegel, H.G.** 1978. Reisolation of the carbon monoxide-utilizing bacterium *Pseudomonas carboxydovorans* (Kistner) comb. nov. *Archives of Microbiology* 118:35–43.

Meyer, O. and **Schlegel, H.G.** 1979. Oxidation of carbon monoxide in cell extracts of *Pseudomonas carboxydovorans. Journal of Bacteriology* 137:811–817.

Meyer, O. and **Schlegel, H.G.** 1980. Carbon monoxide:methylene blue oxidoreductase from *Pseudomonas carboxydovorans. Journal of Bacteriology* 141:74–80.

Meyer, O. and **Schlegel, H.G.** 1983. Biology of aerobic carbon monoxide-oxidizing bacteria. *Annual Review of Microbiology* 37:277–310.

Meyer, O., Lalucat, J. and **Schlegel, H.G.** 1980. *Pseudomonas carboxydohydrogena* (Sanjieva and Zavarzin) comb. nov., a monotrichous, nonbudding, strictly aerobic, carbon monoxide-utilizing hydrogen bacterium previously assigned to *Seliberia. International Journal of Systematic Bacteriology* 30:189–195.

Meyer, O., Jacobitz, S., and **Krüger, B.** 1986. Biochemistry and physiology of aerobic carbon monoxide-utilizing bacteria. *FEMS Microbiology Reviews* 39:161–179.

Nason, A., Antonie, A.D., Ketchum, P.A., Frazier III, W.A., and **Lee, D.K.** 1970. Formation of assimilatory nitrate reductase by in vivo inter-cistronic complementation in *Neurosspora crassa. Proceedings of the National Academy of Sciences USA* 65:137–144.

Nason, A., Lee, K.-Y., Pan, S.-S., Ketchum, P.A., Lamberti, A., and **DeVries, J.** 1971. In vitro formation of assimilatory reduced nicotinamide adenine dinucleotide phosphate:nitrate reductase from a *Neurospora* mutant and a component of molybdenum enzymes. *Proceedings of the National Academy of Sciences USA* 66:133–137.

Nozhevnikova, A.N. and **Yurganov, L.N.** 1978. Microbiological aspects of regulating the carbon monoxide content in the earth's atmosphere, pp. 203–244, in Alexander, M. (editor), *Advances in microbial ecology,* vol. 2, Plenum, New York.

Ooyama, J. 1976. Bacterial culture by CO_2 or CO and N_2 as the sole sources of C and N, pp. 237–245, in Schlegel, H.G., Gottschalk, G. and Pfennig, N. (editors), *Microbial production and utilization of gases,* Goltze, Göttingen.

Park, Y. and **Hegeman, G.** 1984. The oxidation of carbon monoxide by bacteria, pp. 211–218, in Strohl, W.R. and Tuovinen, O.H. (editors), *Microbial chemoautotrophy,* Ohio State University Press, Columbus.

Rajagopalan, K.V. 1985. Chemistry and biology of the molybdenum cofactor. 611th meeting, Galway, Ireland. *Biochemical Society Transactions* 13:401–403.

Rajagopalan, K.V., Johnson, J.L., and **Hainline, B.E.** 1982. The pterin of the molybdenum cofactor. *Federation Proceedings* 41:2608–2612.

Rohde, M., Mayer, F., and **Meyer, O.** 1984. Immunocytochemical localization of carbon monoxide oxidase in *Pseudomonas carboxydovorans.* The enzyme is attached to the inner aspect of the cytoplasmic membrane. *Journal of Biological Chemistry* 259:14788–14792.

Rohde, M., Mayer, F., Jacobitz, S., and **Meyer, O.** 1985. Attachment of CO dehydrogenase to the cytoplasmic membrane is limiting the respiratory rate of *Pseudomonas carboxydovorans. FEMS Microbiology Letters* 28:141–144.

Sanjieva, E.U. and **Zavarzin, G.A.** 1971. Oxidation of carbon monoxide by *Seliberia carboxydohydrogena. Doklady Akademii Nauk. SSSR* 196:956–958.

Savelieva, N.D. and **Nozhevnikova, A.N.** 1973. Autotrophic growth of *Seliberia carboxydohydrogena* during the oxidation of hydrogen and carbon monoxide. *Mikrobiologiya* 41: 813–817 (*Microbiology* 41:719–723).

Schlegel, H.G. and **Meyer, O.** 1981. Microbial growth on carbon monoxide, formate and hydrogen: a biochemical assessment, pp. 105–115, in Dalton, H. (editor), *Microbial growth on C_1 compounds,* Heyden, London.

Wahl, R.C., Hageman, R.V., and **Rajagopalan, K.V.** 1984. *Archives of Biochemistry and Biophysica* 230:264–273.

Williams, E. and **Colby, J.** 1986. Biotechnological applications of carboxydotrophic bacteria. *Microbiological Sciences* 3:149–153.

Zavarzin, G.A. 1978. *Hydrogen bacteria and carboxydobacteria.* Nauka Publ., USSR Academy of Sciences, Moscow. 207 pp.

Zavarzin, G.A. and **Nozhevnikova, A.N.** 1976. CO oxidizing bacteria, pp. 207–213, in Schlegel, H.G., Gottschalk, G. and Pfennig, N. (editors), *Microbial production and utilization of gases,* Goltze, Göttingen.

Zavarzin, G.A. and **Nozhevnikova, A.N.** 1977. Aerobic carboxydobacteria. *Microbial Ecology* 3:305–326.

19

The Ribulose Bisphosphate Pathway of CO_2 Fixation

Bruce A. McFadden

Biochemistry/Biophysics Program
Washington State University
Pullman, Washington, U.S.A.

The work leading to the discovery of the pathway of autotrophic CO_2 fixation was discussed in Chapter 3 and the key role of the enzyme ribulose bisphosphate carboxylase was emphasized. In the present chapter, evidence will be considered which implicates the ribulose bisphosphate (RuBP) or Calvin cycle in the assimilation of CO_2 by numerous chemo- and photolithotrophic bacteria and oxygenic photosynthetic species. An additional focus will be upon the key enzyme of that cycle ribulose bisphosphate carboxylase/oxygenase (RuBisCO). Additional pathways of autotrophic CO_2 fixation are discussed in Chapter 20.

19.1 CO_2 ASSIMILATION

Athough by 1957 numerous studies of chemolithotrophic bacteria had clearly established the dependence of CO_2 fixation upon exergonic chemical reactions, the mechanism of coupling was uncertain. This was clarified that year in research on *Hydrogenomonas* (now *Pseudomonas) facilis* in which CO_2 fixation that was dependent upon the oxidation of H_2 by O_2 (the so-called Knallgas reaction) was studied. The striking inhibition of $^{14}CO_2$ fixation at a concentration 2,4-dinitrophenol (0.7 mM) that actually stimulated the Knallgas reaction suggested that the coupling was mediated by ATP (McFadden and Atkinson, 1957). The effects of chelating compounds, sulfhydryl reagents, and arsenite established that proportional inhibition of the Knallgas reaction and CO_2 fixation occurred and similarities to CO_2 fixation by isolated chloroplasts were noted (McFadden and Atkinson, 1957). In parallel research, it had become evident that the pathway of CO_2 fixation in the chemolithotroph *Thiobacillus denitrificans* was identical (Aubert et al., 1956) to that which had been established (see Figure 19.1) for green plant photosynthesis by Calvin and co-workers (summarized by Calvin, 1962).

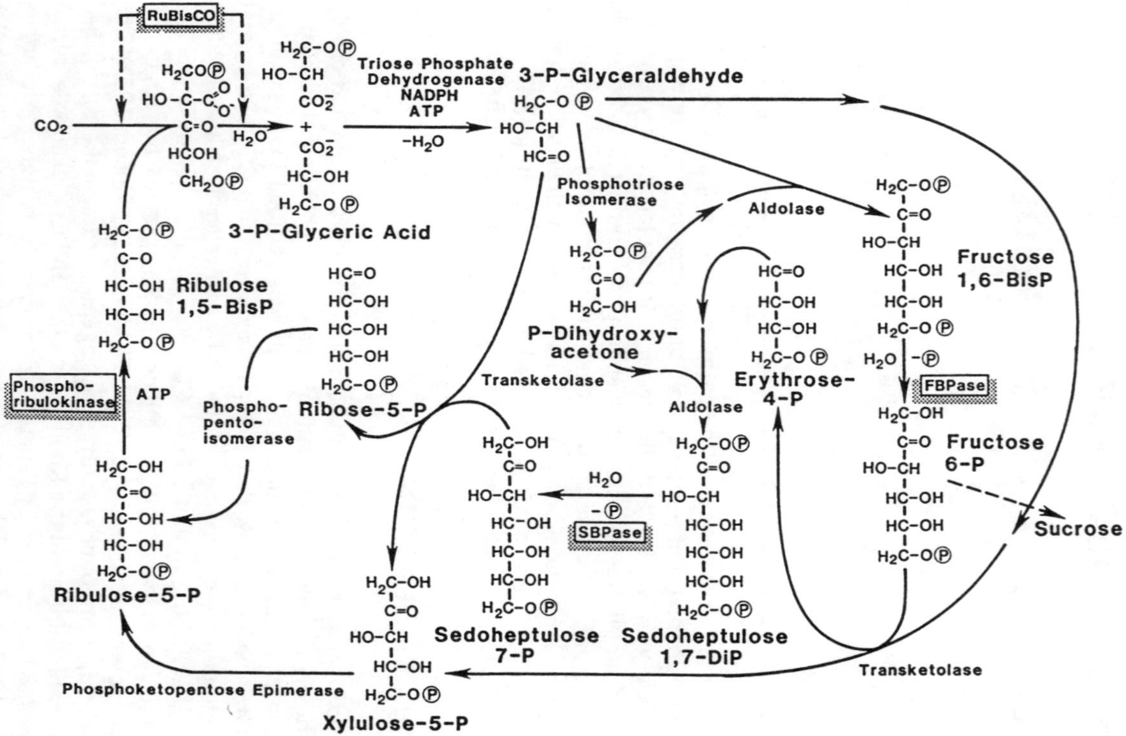

Figure 19.1 The RuBP (Calvin) cycle. The following enzymes, which function uniquely in the cycle, are highlighted by shading: RuBisCO (ribulose bisphosphate carboxylase/oxygenase), phosphoribulokinase, and SBPase (sedoheptulose 1,7-bisphosphatase). In addition, FBPase (fructose 1,6-bisphosphatase), is shaded to stress the fact that in some procaryotes this activity and SBPase may occur together in a bifunctional enzyme.

In that era, then, there were hints that diverse autotrophic forms of life utilize CO_2 by a common mechanism, reinforcing the unity of biochemistry. This was further indicated by the demonstration that extensive $^{14}CO_2$ fixation by extracts of *Pseudomonas facilis** was dependent upon NADH and ATP (McFadden and Tu, 1965, 1967). Subsequent research established that, indeed, the RuBP cycle (Figure 19.1) functions in diverse autotrophic bacteria. Basically, however, experiments reduced to the common approach of tracing the pathway of $^{14}CO_2$ fixation into intermediates, the first of which was 3-phosphoglycerate (Figure 19.1). By this criterion, function of the RuBP cycle has been firmly established in anoxygenic photolithotrophs belonging to the genera *Chromatium, Rhodobacter**, and *Rhodospirillum,* in the oxygenic cyanobacteria (blue-green algae) and in numerous chemolithotrophs (McFadden, 1978). The latter include *Alcaligenes eutrophus**, *P. facilis**, several thiobacilli, formate-grown *Pseudomonas oxalaticus, Paracoccus denitrificans*, Nitrobacter agilis,* and *N. winogradskyi.*

Although the function of the RuBP cycle is nearly ubiquitous in aerobic CO_2-grown chemolithotrophs (McFadden, 1973), this cycle may be uniquely absent in the thermophilic hydrogen-oxidizing obligate autotroph *Hydrogenobacter thermophilus* (Shiba et al., 1985). For example, kinetic studies of $^{14}CO_2$ utilization by cells suspended in buffer (i.e., in the absence of inorganic components required for growth) and enzyme data supported CO_2 fixation by a reductive tricarboxylic acid cycle (described in Chapter 20). Moreover, two unique enzymes of the RuBP cycle, ribulose bisphosphate carboxylase/oxygenase (RuBisCO) and phosphoribulokinase (Figure 19.1) were reported to be absent.

Among the photolithotrophic bacteria, function of the RuBP cycle remains controversial in *Chlorobium limicola* (McFadden, 1978). Pulse-labeling experiments with CO_2-grown cells clearly identified 3-phosphoglycerate as the principal labeled product arising from $^{14}CO_2$ after 5 seconds. Moreover, the percentage of radioactivity incorporated into phosphoglycerate revealed the steepest negative slope with time (Buchanan et al., 1972). Perhaps because RuBisCO could not be detected (Buchanan et al., 1972; Buchanan and Sirevåg, 1976), and in spite of its detection in other laboratories (Smillie et al., 1962; Tabita et al., 1974), it was proposed that *C. limicola* utilized a reductive carboxylic acid cycle in CO_2 fixation (Buchanan et al., 1972; see also Chapter 20). Indeed most enzymes of the proposed cycle were detectable in extracts of *C. limicola.* It should be emphasized that assays for phosphoribulokinase (Figure 19.1) were also negative in extracts of this organism (Buchanan and Sirevåg, 1976), in contrast to an earlier report (Smillie et al., 1962). Quandt et al. (1977) and, independently, Sirevåg et al. (1977), examined isotope discrimination in the assimilation of $^{13}CO_2$ and $^{12}CO_2$. Values of $\delta\ ^{13}C$

$$\left(\frac{^{13}C/^{12}C\ \text{sample}}{^{13}C/^{12}C\ \text{standard}} - 1\right) \times 10^3$$

---

*In the older literature some species of *Rhodobacter* were classified in the genus *Rhodopseudomonas.* Also, *Alcaligenes eutrophus* and *Pseudomonas facilis* were both given generic designations as *Hydrogenomonas,* and *Paracoccus denitrificans* as *Micrococcus.*

were first determined for cells and medium (in terms of per mille deviation from the standard). The δ ^{13}C value for the medium was then corrected as necessary for the fractionation resulting from conversion of HCO_3^- to CO_2. Values of $\Delta\delta$ ^{13}C (δ $^{13}C_{cells}$ $-$ δ $^{13}C_{medium, corrected}$) were found to cluster between -2.51 and -5.19 for five strains and/or species of *Chlorobium*. In contrast, much more negative values of -10.58 to -19.57 were obtained for *Rhodobacter capsulatus, Rhodospirillum rubrum* and *Chromatium vinosum* (Quandt et al., 1977), all of which utilize the RuBP cycle (Figure 19.1). These comparisons assume added significance when it is considered that prior research (Park and Epstein, 1960) had demonstrated that assimilation of $^{12}CO_2$ and $^{13}CO_2$ by C_3 plants* resulted in a discrimination against $^{13}CO_2$ by RuBisCO (Figure 19.1) which was similar to that seen by Quandt et al. (1977). In contrast, cellular carbon in C_4 plants* had a relatively higher proportion of ^{13}C (Bender, 1971; Smith and Epstein, 1971), reflecting the smaller discrimination against $^{13}CO_2$ by PEP carboxylase. Quandt et al. (1977) concluded that either RuBisCO was absent in *Chlorobium* or that another enzyme preceding RuBisCO in primary CO_2 fixation (as in C_4 plants) discriminated relatively less against $^{13}CO_2$. The latter caveat illustrates the difficulty in correlating $\Delta\delta$ ^{13}C values with the function of pathways of net CO_2 assimilation such as the RuBP cycle. In related research to assess the pathway of CO_2 fixation by *C. limicola*, it has been reported that the reductive tricarboxylic cycle functions in *C. limicola* grown in the presence of ^{14}C-pyruvate (Fuchs et al., 1980a) or ^{14}C-propionate (Fuchs et al., 1980b). These claims are irrelevant, however, to the potential of this organism to utilize the RuBP cycle. It is well established that the inclusion of organic compounds during the growth of some autotrophic species represses the level of RuBisCO and phosphoribulokinase (see next section). To recapitulate, in spite of conflicting data the total evidence available does suggest that the RuBP cycle may not function in CO_2 assimilation by *Chlorobium*.

In other phototrophic bacteria such as members of the genera *Chloroflexus* (Pierson and Castenholz, 1974) and *Heliobacterium* (Gest and Favinger, 1983), growth under strictly autotrophic conditions remains to be established. Consequently, the role of the RuBP cycle in CO_2 fixation has not been assessed.

In marked contrast to utilization of the Calvin cycle by most aerobic chemolithotrophic bacteria, numerous anaerobic chemolithotrophic bacteria appear to assimilate CO_2 in an acetyl CoA-synthesizing pathway. These comprise some sulfate-reducing, acetogenic, and methanogenic strains including the well-studied *Methanobacterium thermoautotrophicum* (Fuchs, 1986 and Chapter 20). In that organism, for example, exposure of cells to progressively shorter pulses with $^{14}CO_2$ (culminating in a two-second exposure) identified alanine as the principal earliest labeled product. 3-Phosphoglycerate was much less heavily labeled. The labeling pattern of alanine suggested that carbons 2 and 3 originated from acetyl CoA into which two CO_2 molecules had been reductively incorporated (Stupperich and Fuchs, 1981). Obviously, the use of brief pulse

*C_3 and C_4 plants incorporate CO_2 into phosphoglycerate and C_4 dicarboxylic acid(s) in the primary CO_2 fixing steps catalyzed by RuBisCO and phosphoenolpyruvate (PEP) carboxylase, respectively.

labeling is a far superior approach to prolonged exposure (for hours or days) of growing cells in radioisotopic CO_2. Such long exposures result in considerable recycling of metabolic intermediates accompanied by randomization of label and leave little prospect of identifying primary products of CO_2 fixation. A more comprehensive coverage of autotrophic assimilation of CO_2 involving acetyl CoA as the primary product will be given in Chapter 20.

In concluding this section, it should be emphasized that a valid examination of the pathway of autotrophic assimilation must elucidate the kinetics of $^{14}CO_2$ incorporation into early intermediates. Alternatively, cells in steady-state $^{14}CO_2$ fixation can be subjected to elimination of the energy source (e.g., light or an exergonic chemical reaction) or reduction of the activity of $^{14}CO_2$. Pool sizes of postulated intermediates can then be examined. For example, if the RuBP cycle were operative, the expectation would be that the pool sizes of labeled phosphoglycerate and RuBP would increase and decrease, respectively, with elimination of the energy source (Calvin, 1962). In addition CO_2 fixation studies must be conducted on cells which: a) *have been grown on CO_2 in the complete absence of organic compounds* and b) *are incubated under strictly autotrophic conditions of growth during the fixation period.* Only by meeting these criteria may the pathway of primary carbon assimilation be identified. Correct identification is of profound importance to our understanding of the relatedness of autotrophic organisms and, ultimately of their evolution.

19.2 UNIQUE ENZYMES IN THE RUBP CYCLE

Three enzymes, ribulose bisphosphate carboxylase/oxygenase (RuBisCO), phosphoribulokinase, and sedoheptulose 1,7-bisphosphatase (SBPase), function uniquely in the RuBP cycle (Figure 19.1). The SBPase activity, although separable from alkaline fructose 1,6-bisphosphatase (FBPase) in plant extracts, may be encoded by the same gene as that for FBPase in some procaryotes (McFadden, 1973). In 1973 it was postulated that genes for these three unique catalysts may comprise an operon in some autotrophic bacteria (McFadden, 1973). Indeed, this may be the case in *Alcaligenes eutrophus;* also, in this organism, a single bifunctional enzyme contains both SBPase and FBPase activities (B. Bowien, personal communication; see Chapter 24).

By implication, these enzymes should be coordinately regulated, but such is not the case for RuBisCO and phosphoribulokinase in *P. facilis* (McFadden and Tu, 1967) as shown in Table 19.1. Clearly, if the corresponding structural genes are part of an operon in *P. facilis,* additional factors must regulate the level of these two enzymes. Nevertheless, specific activities for RuBisCO and phosphoribulokinase are highest after autotrophic growth (Table 19.1). This suggests that one useful indicator of the utilization of the RuBP cycle by a test organism after growth on CO_2 may prove to be the detection in extracts of one or more of the following enzymes: RuBisCO, phosphoribulokinase, or SBPase. Additionally, the potential for expression of these activities may be examined

Table 19.1 Ribulose bisphosphate carboxylase/oxygenase (RuBisCO) and phosphoribulokinase in *Pseudomonas facilis*[a]

	Specific activity[b]	
Growth substrate	RuBisCO	Phosphoribulokinase
CO_2	67	20
Fructose	51	22
Glucose	51	21
Ribose	25	12
Glutamate	2	13
Lactate	10	13
Succinate	2	13
Acetate	1	2

[a]Taken from McFadden and Tu (1967).
[b]Expressed as nanomol of product per minute per milligram of protein.

by checking for the corresponding genes using suitable hybridization probes. In this connection, it is important to note that in *Rhodobacter sphaeroides* there are two genetic loci for RuBisCO. These encode form I and form II enzymes which differ significantly in structure (see section 19.4). Recent research suggests that each of these loci is part of a cluster which also contains phosphoribulokinase and FBPase genes, i.e., there may be two chromosomal operons encoding unique enzymes of the RuBP cycle (see Chapter 26).

Apart from considerations of the coordinate expression of genes for autotrophic CO_2 fixation, it is evident from Table 19.1 that levels of both RuBisCO and phosphoribulokinase are drastically reduced after growth of *P. facilis* on acetate. This observation underscores the importance of culturing test organisms under strictly autotrophic conditions prior to probing the pathway of CO_2 assimilation. Some organisms may be even more sensitive than *P. facilis* in terms of repression of the RuBP cycle by organic compounds.

19.3 EVOLUTION OF THE RUBP CYCLE—THE ESTABLISHMENT OF RUBISCO

It is reasonable to assume that the first organisms appeared in waters enriched in organic compounds and that they had limited biosynthetic capacity. Moreover, these heterotrophic organisms should have been anaerobic and had complex nutritional requirements as do many fermentative bacteria even today (McFadden, 1973; McFadden and Tabita, 1974). Presumably ancient organisms catalyzed substrate-level phosphorylations and their oxidation-reduction reactions were mediated by pyridine nucleotides. Thus ATP and NADH (or structurally similar primitive counterparts) would have been available to couple with newly acquired reductive biosynthetic steps which enabled the synthesis of an essential nutrient (McFadden, 1976). Horowitz was the first to emphasize the profound selective advantage accruing to organisms which acquired the capacity to synthesize essential nutrients that had been depleted from the environment. Logically, this led to the postulate that the evolution of biosynthetic sequences had occurred from their termini (as we know them today) toward earlier steps (Horowitz, 1945)—a process often referred to as retroevolution. A further proposal is that establishment of the Calvin cycle occurred at a time when the resynthesis of an intermediate by CO_2 fixation would have been selectively advantageous (McFadden, 1973). In those cells, the recruitment of phosphoribulokinase and SBPase (Figure 19.1) from preexisting phosphofructokinase and FBPase, respectively, has also been suggested. It is easy to imagine how genes for the latter two enzymes evolved by duplication and mutation to specify increasingly better catalysts. With this line of reasoning, it is plausible that the acquisition of RuBisCO triggered the establishment of the RuBP cycle and, accordingly, permitted the evolution of ancestors of many autotrophic bacteria.

An interest in the evolution of autotrophic forms has been the *raison d'être* of our deep interest in RuBisCO. In the remainder of this chapter, salient features of the structure and function of this enzyme, which catalyzes the rate-limiting step in the assimilation of CO_2 by many autotrophs, will be considered.

19.4 RIBULOSE BISPHOSPHATE CARBOXYLASE/ OXYGENASE

A dual-function enzyme A carboxylase activity catalyzing the carboxylation of ribulose 1,5-bisphosphate (RuBP) by $^{14}CO_2$ to form carboxyl-labeled phosphoglycerate was discovered in extracts of *Chlorella* in 1954 (Quayle et al., 1954). In 1973, an oxygenase activity was characterized in this carboxylase (Lorimer et al., 1973), leading to the recognition that the enzyme, which was designated ribulose bisphosphate carboxylase/oxygenase (RuBisCO), initiated photosynthesis as well as photorespiration (Andrews and Lorimer, 1978) in plants. It was reported that catalysis by the oxygenase results in oxygenolysis of RuBP to form phosphoglycerate (from carbons 3–5) and phosphoglycolate (from carbons 1–2), the photorespiratory substrate. Curiously, the oxygenase activity was found also in the carboxylase from *Rhodospirillum rubrum* (McFadden, 1974) and was found to correlate with light-dependent glycolate excretion (Storrø and McFadden, 1981). These and numerous other findings have led to the recognition that RuBP carboxylases from divergent sources have an oxygenase. An important enigma, then, is why the function of RuBisCO may contribute to both photosynthesis and photorespiration, which appear to be fundamentally opposed processes. Indeed, glycolate excretion by photosynthetic microorganisms is a poorly understood and apparently wasteful process. This dual function of a single enzyme contributing to both productive and ostensibly wasteful phenomena is unprecedented in biology. Inevitably, it raises intriguing questions about the prospects and consequences of controlling RuBP carboxylase:oxygenase ratios in plants (McFadden, 1980).

Activation and catalysis Much of the literature on RuBisCO dealing with activation, catalysis, structure, and function has been reviewed in detail elsewhere [see, for example, (Miziorko and Lorimer, 1983; McFadden et al., 1986)]. Only milestones in our understanding of these phenomena and properties will be considered in this and the following two sections.

Activation of both activities of RuBisCO is a two-step process shown below for the spinach enzyme (Lorimer et al., 1976; Badger and Lorimer, 1976):

$$\underset{(lys201)}{\text{E–NH}_2} + CO_2 \underset{slow}{\rightleftharpoons} \text{E–NH–CO}_2^- \underset{fast}{\overset{Mg^{2+}}{\rightleftharpoons}} \text{E–NH–CO}_2^- \cdot Mg^{2+}$$

Both activities in each of two bacterial enzymes respond similarly (Purohit and McFadden, 1979; Robison et al., 1979); presumably this is true of RuBisCO from all sources. Although the assumption is widespread that both activities decay (e.g., with lowered P_{CO_2}) at equivalent rates, there is evidence for a 3-fold longer decay half-time for the carboxylase than the oxygenase in toluene-permeabilized cells of *Rhodospirillum rubrum* (Storrø and McFadden, 1983). In general, however, the data suggest that both activities are similar in kinetics of appearance and decay. The activation process shown assumes added significance when the design of experiments to trace the pathway of $^{14}CO_2$ fixation in intact cells is considered (Section 19.1). Some studies have been conducted in the absence of exogenous Mg^{2+} and CO_2 under conditions in which RuBisCO is inactive. For example, the failure to detect ^{14}C-phosphoglycerate as an early labeled product using *Hydrogenobacter thermophilus* cells which had been suspended in phosphate buffer was, in itself, inconclusive with regard to function of the RuBP cycle (Shiba et al., 1985).

With regard to catalysis, both carboxylase and oxygenase activities occur in a common catalytic center (McFadden, 1980; Miziorko and Lorimer, 1983). Nevertheless, it is very likely that the oxygenolytic and carboxylative pathways catalyzed by RuBisCO diverge after a common early RuBP-binding step (McFadden, 1980). In fact, these pathways have markedly different pH activity profiles on the alkaline side in toluene-permeabilized *R. rubrum* (Storrø and McFadden, 1983). Moreover, enzyme from this source shows excellent oxygenase but no carboxylase activity with Co^{2+} (Robison et al., 1979) and, in general, the V_{max} ratios for carboxylase:oxygenase are around 10 and 1 in the presence of Mg^{2+} and Mn^{2+}, respectively (McFadden, 1980).

Research on the mechanism of catalysis has focused on the carboxylation reaction. Advances have been rapid, especially through use of the "transition-state analog," 2-C-carboxy-D-arabinitol 1,5-bisphosphate (CABP). In Figure 19.2, the structure of this compound is compared with that of the reaction intermediate (Miziorko and Lorimer, 1983; Lorimer et al., 1986).

The analog CABP is bound extremely tightly to active RuBisCO ($K_D \simeq 10^{-11}$ M) with a half-time of dissociation at 30°C of days (Miziorko and Lorimer, 1983). Moreover, the binding of CABP renders Me^{2+} and CO_2 exchange inert in active enzyme from both spinach and *R. rubrum* (Miziorko and Lorimer, 1983; Miziorko and Sealy, 1980) and strongly indicates that activator and substrate CO_2 molecules are different (Miziorko and Lorimer, 1983). The carbamate (derived from activator CO_2) is in the second sphere of Me^{2+} of activated spinach enzyme (McFadden, 1980) and with enzyme from this source (Styring

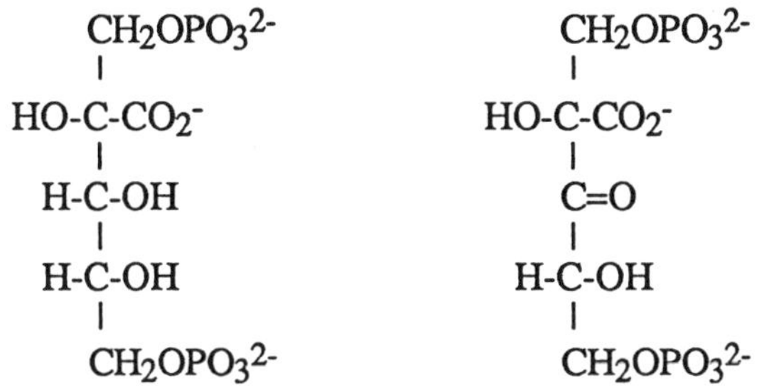

CABP   Reaction Intermediate

Figure 19.2 Comparison of the structure of 2-C-carboxy-D-arabinitol 1,5-bisphosphate (CABP) with the reaction intermediate of ribulose bisphosphate carboxylase.

and Brändèn, 1985) or *R. rubrum* (Miziorko and Sealy, 1984), the carboxylate of CABP is coordinated directly with Me^{2+}. In addition, the C2-hydroxyl of CABP (Miziorko and Sealy, 1984) or of one product molecule 3-phosphoglycerate (Styring and Brändèn, 1985) is coordinated directly with Me^{2+}.

In other research on the carboxylation mechanism, trapping experiments have suggested participation of an RuBP carbanion (Bhagwat and McFadden, 1983a). Elegant affinity labeling experiments by Hartman and co-workers have suggested that *lys*-175 and -334 are in the active-site domain of the spinach enzyme (Hartman et al., 1986). Finally, results with the histidine-modifying reagent, diethyl pyrocarbonate, have suggested that *his*-298 is in the same domain (Saluja and McFadden, 1982; Igarashi and McFadden, 1985).

Structure The dominant form of RuBisCO is composed of eight 55-kDa large (L) and eight 15-kDa small (S) subunits. Each L subunit contains one catalytic center and S subunits are of unknown function (McFadden, 1980). Recent evidence, obtained with a chemical crosslinker and directed mutagenesis, has established that the active-site domain of the *Rhodospirillum rubrum* enzyme, which is near the carboxy terminus, is in close contact with the amino terminus of the adjacent L subunit (Lee et al., 1987). In this procaryote the enzyme is uniquely an L_2 dimer. In contrast the L_8S_8 form of the enzyme is found in numerous bacteria and in all eucaryotes (McFadden, 1980; McFadden et al., 1986).

Elsewhere we have proposed that genes for L subunits arose from a common ancestral gene and that S subunits were added later in evolution but before the appearance of atmospheric oxygen (McFadden, 1973; McFadden and Tabita, 1974). Indeed, L subunits have a high degree of homology, especially in functional regions, whereas there is considerably greater sequence divergence in S subunits (McFadden et al., 1986). Presumably the L_2 dimer from *R. rubrum* represents a more primitive enzyme (McFadden et al., 1986). In another photolithotroph, *Rhodobacter sphaeroides*, there are two forms of the enzyme: L_8S_8 (I) and L_n (form II of uncertain aggregation state). On the basis of research in Tabita's laboratory (reviewed by McFadden et al., 1986), the form II enzyme resembles the L_2 dimer from *R. rubrum* whereas form I, encoded by a unique set of structural genes, is more like the eucaryotic form. Although there are indications that S subunits are required for activity of the L_8S_8 form from two cyanobacteria, the recent isolation of a fully active L_8 core from *Chromatium vinosum* (Torres-Ruiz and McFadden, 1985) establishes that aggregates of L subunits from some sources are catalytically competent. In this instance, the L_8 enzyme was derived from the L_8S_8 form by centrifugal fractionation (Torres-Ruiz and McFadden, 1987a). One catalytic feature of enzymes lacking small subunits is a considerably lower specificity factor, which is approximately proportional to the carboxylase:oxygenase ratio (Jordan and Ogren, 1981).

In green algae and plants, S and L subunits are encoded by nuclear and chloroplast genomes, respectively. S subunits are synthesized on cytoplasmic ribosomes as 20-kDa precursor molecules which are taken up by chloroplasts. A 5-kDa leader sequence is removed to yield mature S subunits which assemble with L subunits to yield L_8S_8 RuBisCO. A 700-kDa protein, thought to bind to L subunits, may participate in the assembly process in higher plants (Musgrove and Ellis, 1986). In procaryotes, mature S subunits are synthesized directly, presumably because transit sequences are not required (McFadden et al., 1986). In those organisms, the L and S genes in an operon are expressed as a single transcript (McFadden and Small, 1988).

As stressed earlier, S subunits of RuBisCO are of unknown function. However, there is evidence for the strong attachment of these subunits to the membrane of *C. vinosum* (Torres-Ruiz and McFadden, 1987b). Indeed the L_8S_8 enzyme from that source may be anchored to the cytoplasmic membrane via a 700-kDa binding protein that is closely analogous to the "large subunit binding protein" from higher plants (Torres-Ruiz and McFadden, 1988). The biological significance of this association remains to be established.

Directed mutagenesis of RuBisCO genes The directed mutagenesis of various clones of the L subunit gene of RuBisCO has been reviewed (McFadden and Small, 1988). The most significant findings are those of Hartman and colleagues at the Oak Ridge National Laboratory. Using a clone of the *R. rubrum* gene, which is expressed in *E. coli* as a native L_2 structure, they have established that replacement of *his*-291 or *lys*-166 (corresponding to 298 and 175 in the spinach L subunit) results in 60 percent and >99 percent loss in activity, respectively. Thus, they have concluded that *his*-291 is nonessential but that *lys*-166 is essential, perhaps in deprotonating RuBP (McFadden and Small, 1988). This is of interest because of prior research implying that *his*-291 is at the active center (Igarashi and McFadden, 1985). Indeed, crystallography suggests that this residue is on the rim of the active site in the L_2 enzyme from *R. rubrum* (C. Brändèn, personal communication). One other residue is of special interest. Research using butane-2,3-dione and phenylglyoxyal has indicated that there is an essential arginine residue in RuBisCO from higher plants, as well as in *Pseudomonas oxalaticus* and *R. rubrum* (Lawlis and McFadden, 1978; Schloss et al., 1978). The most highly conserved *arg* residue is that at position 288 in the *R. rubrum* enzyme (corresponding to *arg*-295 in spinach). Because chemical modifications of *arg* are relatively unstable, the most direct approach in probing the function of *arg*-288 will be through directed mutagenesis. It can be hypothesized that such experiments will disclose a binding function for this residue.

Mutagenesis has been directed at the middle conserved sequence (Voordouw et al., 1987) in the S subunit gene (*tyr*-54 through *phe*-63) for RuBisCO from *Anacystis nidulans*. Replacement of *trp* residues at 55 and 58 (incorrectly designated as 54 and 57) by *phe* residues, singly and in combination, was accomplished but there was no loss in activity of RuBisCO after expression of the mutated S subunits in the presence of wild-type L subunits. In contrast, the first conserved region (residues 9–21) of this gene has been mutagenized and the following pairs of changes have been constructed: *ser tyr* → *ala asp; thr phe ser* → *ala phe ala; pro pro* → *ala ala*; and *arg arg* → *gly gly*. Each of the first four mutated small subunit genes has been religated with the L subunit gene and expressed in *E. coli* with 29–94 percent loss in RuBisCO activity (McFadden and Small, 1988). This general approach appears to be of considerable promise in elucidating the function of the small subunit of RuBisCO.

19.5 CONCLUSION

Considering the global importance of photosynthesis and the fact that RuBisCO is the single most abundant protein in the world (Ellis, 1979), it is not surprising that the ribulose bisphosphate (Calvin) cycle is dominant among autotrophic species. In fact, tracer experiments utilizing $^{14}CO_2$ support its function as do the detection of two key enzymes, ribulose bisphosphate carboxylase/oxygenase (RuBisCO) and phosphoribulokinase, in numerous procaryotic and eucaryotic autotrophs. Nevertheless, an alternative fixation pathway certainly functions in various anaerobic autotrophs such as the methanogens (see Chapter 20). An assessment of the natural distribution of the potential for Calvin cycle fixation must await the development of suitable DNA hybridization probes. In this way, the existence of genes for RuBisCO, phosphoribulokinase and sedoheptulose bisphosphatase, which are unique catalysts in the Calvin cycle, can be tested directly.

The author acknowledges support by NIH grant GM-19972, a John Simon Guggenheim Fellowship, and Alexander von Humboldt American Senior Scientist Award (from W. Germany) for much of the research done in this laboratory and cited in this chapter.

REFERENCES

Andrews, T.J. and **Lorimer, G.H.** 1978. Photorespiration—still unavoidable. *FEBS Letters* 90:1–9.

Aubert, J.P., Milhaud, G., and **Millet, J.** 1956. Metabolisme du carbone dans la chimioautotrophie. Mode d'incorporation de l'anhydride carbonique. *Comptes Rendus de l'Académie des Sciences* 243:2059–2062.

Badger, M.R. and **Lorimer, G.** 1976. Activation of ribulose-1,5-bisphosphate oxygenase—the role of Mg^{2+}, CO_2 and pH. *Archives of Biochemistry and Biophysica* 175:723–729.

Bender, M.M. 1971. Variations in the $^{13}C/^{12}C$ ratios of plants in relation to the pathway of photosynthetic carbon dioxide fixation. *Phytochemistry* 10:1239–1244.

Bhagwat, A. and **McFadden, B.A.** 1983a. Evidence for a carbanion intermediate in catalysis by spinach ribulose bisphosphate carboxylase/oxygenase. *Archives of Biochemistry and Biophysica* 223:604–609.

Bhagwat, A. and **McFadden, B.A.** 1983a. Pyridoxal phosphate as a probe in the active site of ribulose bisphosphate carboxylase/oxygenase. *Archives of Biochemistry and Biophysics* 223:610–617.

Buchanan, B.B., Schürmann, P., and **Shanmugam, K.T.** 1972. Role of the reductive carboxylic acid cycle in a photosynthetic bacterium lacking ribulose-1,5-diphosphate carboxylase. *Biochimica et Biophysica Acta* 283:136–145.

Buchanan, B.B. and **Sirevåg, R.** 1976. Ribulose 1,5-diphosphate carboxylase and *Chlorobium thiosulfatophilum. Archives of Microbiology* 109:15–19.

Calvin, M. 1962. The path of carbon in photosynthesis. *Science* 135:879–889.

Ellis, R.J. 1979. The most abundant protein in the world. *Trends in Biochemical Sciences* 4:241–244.

Fuchs, G., Stupperich, E., and **Jaenchen, R.** 1980a. Autotrophic CO_2 fixation in *Chlorobium limicola.* Evidence against the operation of the Calvin cycle in growing cells. *Archives of Microbiology* 128:56–63.

Fuchs, G., Stupperich, E., and Eden, G. 1980b. Autotrophic CO_2 fixation in *Chlorobium limicola.* Evidence for the operation of a reductive tricarboxylic acid cycle in growing cells. *Archives of Microbiology* 128:64–71.

Fuchs, G. 1986. CO_2 fixation in acetogenic bacteria: variations on a theme. *FEMS Microbiology Reviews* 39:181–213.

Gest, H. and **Favinger, J.**. 1983. *Heliobacterium chlorum,* an anoxygenic brownish-green photosynthetic bacterium containing a new form of bacteriochlorophyll. *Archives of Microbiology* 136:11–16.

Hartman, F.C., Stringer, C.D., Milanez, S., and **Lee, E.H.** 1986. The active site of Rubisco. *Philosophical Transactions of the Royal Society, London B* 313:379–395.

Horowitz, N.H. 1945. On the evolution of biochemical syntheses. *Proceedings of the National Academy of Sciences U.S.A.* 31:153–157.

Igarashi, Y. and **McFadden, B.A.** 1985. Active-site histidine in ribulose bisphosphate carboxylase/oxygenase modified by diethyl-pyrocarbonate. *Biochemistry* 24:3957–3962.

Jordan, D.B. and **Ogren, W.L.** 1981. Species variation in the specificity of ribulose bisphosphate carboxylase/oxygenase. *Nature* 291:513–515.

Lawlis, V.B. and **McFadden, B.A.** 1978. Modification of ribulose bisphosphate carboxylase by 2,3-butadione. *Biochemical and Biophysical Research Communications* 80:580–585.

Lee, E.H., Soper, T.S., Mural, R.J., Stringer, C.D., and **Hartman, F.C.** 1987. An intersubunit interaction at the active site of ribulose bisphosphate carboxylase/oxygenase as revealed by cross-linking and site-directed mutagenesis. *Biochemistry* 26:4599–4604.

Lorimer, G.H., Andrews, T.J., and **Tolbert, N.E.** 1973. Ribulose diphosphate oxygenase II. Further proof of reaction products and mechanism of action. *Biochemistry* 12:18–23.

Lorimer, G.H., Badger, M.R., and **Andrews, T.J.** 1976. The activation of ribulose-1,5-bisphosphate carboxylase by carbon dioxide and magnesium ions. Equilibria, kinetics, a suggested mechanism and physiological implications. *Biochemistry* 15:529–536.

Lorimer, G.H., Andrews, T.J., Pierce, J., and **J.V. Schloss.** 1986. 2′-Carboxy-3-keto-D-arabinitol 1,5-bisphosphate, the six-carbon intermediate of the ribulose bisphosphate carboxylase reaction. *Philosophical Transactions of the Royal Society, London B* 313:397–407.

McFadden, B.A. and **Atkinson, D.E.** 1957. The biochemistry of *Hydrogenomonas.* V. Factors affecting autotrophic fixation of carbon dioxide. *Archives of Biochemistry and Biophysics* 66:16–22.

McFadden, B.A. and **Tu, C.L.** 1967. Regulation of autotrophic and heterotrophic carbon dioxide fixation in *Hydrogenomonas facilis. Journal of Bacteriology* 93:886–893.

McFadden, B.A. 1974. The oxygenase activity of ribulose diphosphate carboxylase from *Rhodospirillum rubrum. Biochemical and Biophysical Research Communications* 60:312–317.

McFadden, B.A. and **Tabita, F.R.** 1974. D-ribulose-1,5-diphosphate carboxylase and the evolution of autotrophy. *Biosystems* 6:93–112.

McFadden, B.A. 1976. Evolution of C_1 metabolism with a focus on CO_2 utilization, pp. 267–280, in Schlegel, H.G., Gottschalk, G., and Pfennig, N. (editors), *Microbial Production and Utilization of Gases (H_2, CH_4, CO),* E. Goltze K. G., Göttingen, W. Germany.

McFadden, B.A. 1978. Assimilation of one-carbon compounds, pp. 219–304, in Ornston, N.O. and Sokatch, J.R. (editors), *The Bacteria,* Vol. 6, Academic Press, Inc., New York.

McFadden, B.A. 1980. A perspective of ribulose bisphosphate carboxylase/oxygenase, the key catalyst in photosynthesis and photorespiration. *Accounts of Chemical Research* 13:394–399.

McFadden, B.A., Torres-Ruiz, J., Daniell, H., and **Sarojini, G.** 1986. Interaction, functional relations and evolution of large and small subunits in RuBisCO from prokaryota and eukaryota. *Philosophical Transactions of the Royal Society, London B* 313:347–358.

McFadden, B.A. and **Small, C.L.** 1988. Cloning, expression and directed mutagenesis of the genes of ribulose bisphosphate carboxylase/oxygenase. *Photosynthesis Research,* Molecular Biology 3:(in press).

Miziorko, H.M. and **Sealy, R.C.** 1980. Characterization of the ribulosebisphosphate carboxylase-carbon dioxide-divalent cation-carboxypenitol bisphosphate complex. *Biochemistry* 19:1167–1171.

Miziorko, H.M. and **Lorimer, G.H.** 1983. Ribulose-1,5-bisphosphate carboxylase-oxygenase. *Annual Reviews of Biochemistry* 52:507–535.

Miziorko, H.M. and **Sealy, R.C.** 1984. Electron spin resonance studies of ribulosebisphosphate carboxylase:identification of activator cation ligands. *Biochemistry* 23:479–485.

Musgrove, J.E. and **Ellis, R.J.** 1986. The Rubisco large subunit binding protein. *Philosophical Transactions of the Royal Society, London B* 313:419–429.

Park, R. and **Epstein, S.** 1960. Carbon isotope fractionation during photosynthesis. *Geochimica et Cosmochimica Acta* 21:110–126.

Pierson, B.K. and **Castenholz, R.W.** 1974. A phototrophic gliding filamentous bacterium of hot springs, *Chloroflexus aurantiacus,* gen. and sp. nov. *Archives of Microbiology* 100:5–24.

Purohit, K. and **McFadden, B.A.** 1979. Ribulose 1,5-bisphosphate carboxylase and oxygenase from *Thiocapsa roseopersicina*: activation and catalysis. *Archives of Biochemistry and Biophysics* 194:101–106.

Quandt, L., Gottschalk, G., Ziegler, H., and **Stichler, W.** 1977. Isotope discrimination by photosynthetic bacteria. *FEMS Microbiology Letters* 1:125–128.

Quayle, J.R., Fuller, R.C., Benson, A.A., and **Calvin, M.** 1954. Pentose phosphate and CO_2 fixation with spinach extracts. *Journal of the American Chemical Society* 76:3610–3611.

Robison, P.D., Martin, M.N., and **Tabita, F.R.** 1979. Differential effects of metal ions on *Rhodospirillum rubrum* ribulose bisphosphate carboxylase/oxygenase and stoichiometric incorporation of HCO_3^- into a cobalt (III)-enzyme complex. *Biochemistry* 18:4453–4458.

Saluja, A. and **McFadden, B.A.** 1982. Modification of active site histidine in ribulose bisphosphate carboxylase/oxygenase. *Biochemistry* 21:89–95.

Schloss, J.A., Norton, I.L., Stringer, C.D., and **Hartman, F.C.** 1978. Inactivation of ribulosebisphosphate carboxylase by modification of arginyl residues with phenylglyoxal. *Biochemistry* 17:5626–5631.

Shiba, H., Kawasumi, T., Igarashi, Y., Kodama, T., and **Minoda, Y.** 1985. The CO_2 assimilation via the reductive tricarboxylic acid cycle in an obligately autotrophic, aerobic hydrogen-oxidizing bacterium, *Hydrogenobacter thermophilus. Archives of Microbiology* 141:198–203.

Sirevåg, R., Buchanan, B.B., Berry, J.A., and **Troughton, J.H.** 1977. Mechanisms of CO_2 fixation in bacterial photosynthesis studied by the carbon isotope fractionation technique. *Archives of Microbiology* 112:35–38.

Smillie, R.M., Rigopoulos, N., and **Kelly, H.** 1962. Enzymes of the reductive pentose phosphate cycle in the purple and in the green photosynthetic sulphur bacteria. *Biochimica et Biophysica Acta* 56:612–614.

Smith, B.N. and **Epstein, S.** 1971. Two categories of $^{13}C/^{12}C$ ratios of higher plants. *Plant Physiology* 47:380–384.

Storrø, I. and **McFadden, B.A.** 1983. Ribulose bisphosphate carboxylase/oxygenase in toluene-permeabilized *Rhodospirillum rubrum. Biochemical Journal* 212:45–54.

Stupperich, E. and **Ruchs, G.** 1981. Products of CO_2 fixation and ^{14}C-labeling pattern of alanine in *Methanobacterium thermoautotrophicum* pulse-labeled with $^{14}CO_2$. *Archives of Microbiology* 130:294–300.

Styring, S. and **Brändén, R.** 1985. Identification of ligands to the metal ion in Cu(II)-activated ribulose 1,5-bisphosphate carboxylase/oxygenase by the use of electron paramagnetic resonance spectroscopy and ^{17}O-labeled ligands. *Biochemistry* 24:6011–6019.

Tabita, F.R., McFadden, B.A., and **Pfennig, N.** 1974. D-ribulose-1,5-diphosphate carboxylase in *Chlorobium thiosulfatophilum* Tassajara. *Biochimica et Biophysica Acta* 341:187–194.

Torres-Ruiz, J. and **McFadden, B.A.** 1985. Isolation of L_8 and L_8S_8 forms of ribulose bisphosphate carboxylase/oxygenase from *Chromatium vinosum. Archives of Biochemistry and Biophysics* 142:55–60.

Torres-Ruiz, J. and **McFadden, B.A.** 1987b. Association of RuBisCO small subunits with a membrane fraction from *Chromatium vinosum*, pp. 5.419–5.422, in Biggins, J. (editor) *Progress in Photosynthesis Research*, Martinus Nijhoff, Boston, Massachusetts.

Torres-Ruiz, J.A. and **McFadden, B.A.** 1988. A homolog of ribulose bisphosphate carboxylase/oxygenase-binding protein in *Chromatium vinosum. Archives Biochemistry and Biophysica* 261: 196–204.

Voordouw, G., De Vries, P.A., van den Berg, W.A.M., and **De Clerck, E.P.J.** 1987. Site-directed mutagenesis of the small subunit of ribulose-1,5-bisphosphate carboxylase/oxygenase from *Anacystis nidulans. European Journal of Biochemistry* 163:591–598.

20

Alternative Pathways of Autotrophic CO_2 Fixation

Georg Fuchs

Department of Applied Microbiology
University of Ulm
Ulm, Federal Republic of Germany

Bacterial energy metabolism is characterized by a great diversity of different pathways for the breakdown of organic compounds. This is due to the diversity of chemical structures produced by living organisms which microbes can degrade. In contrast, the major building blocks of all cells are very similar and so are the central biosynthetic pathways. One such pathway supposed to be universal is the reductive pentose phosphate cycle (Calvin cycle) for CO_2 assimilation by autotrophic organisms. It was assumed to operate in all autotrophic plants and microorganisms (see Chapter 19). It is now clear, however, that at least two other pathways function in strict anaerobic or microaerophilic eubacteria as well as in archaebacteria which enable these organisms to grow with CO_2 as the sole source of cell carbon. Interestingly, the two pathways have in common a synthesis of acetyl CoA, a central biosynthetic intermediate, from 2 CO_2 and 8 reducing equivalents.

20.1 AUTOTROPHY VERSUS HETEROTROPHY

Before describing these alternative pathways of autotrophic CO_2 fixation, it may be helpful to discuss the biochemical potential of anaerobic bacteria to perform reversible carboxylation or decarboxylation reactions. Table 20.1 shows some common, main CO_2 fixation reactions which occur in heterotrophic and in autotrophic anaerobes, besides the many additional CO_2 fixation reactions of minor importance. Using these carboxylations, one could design a dozen or so of cyclic mechanisms which result in net CO_2 fixation and regeneration of one of the CO_2 acceptor molecules, as in the Calvin cycle. Surprisingly, so far there is evidence for only one such alternative cyclic mechanism of autotrophic CO_2 fixation, the reductive citric acid cycle. In this alternative pathway, the

Table 20.1 Important CO_2 fixation reactions in chemotrophic anaerobic bacteria[a]

Reaction	$\Delta G^{0\prime}$ (2[H] = H_2) (kJ/mol CO_2)	CO_2 acceptor	Enzyme
1. $CO_2 + 2[H] \rightleftharpoons HCO^- + H^+$	+3.4	Mo or W enzyme	Formate dehydrogenase
2. $CO_2 + 2[H] \rightleftharpoons [CO] + H_2O$	< +20.1	Ni-enzyme	CO dehydrogenase
3. CO_2 + Methanofuran $\rightleftharpoons$ Carboxy-methanofuran	?	Methanofuran	Not yet characterized
4. $CH_3\text{–CO–SCoA} + CO_2 + 2[H] \rightleftharpoons CH_3\text{–CO–COO}^- + H^+ + \text{CoASH}$	+16.3	Acetyl coenzyme A	Pyruvate synthase
5. $^-\text{OOC–CH}_2\text{–CH}_2\text{–CO–SCoA} + CO_2 + 2[H] \rightleftharpoons$ $^-\text{OOC–CH}_2\text{–CH}_2\text{–CO–COO}^- + H^+ + \text{CoASH}$	+14.8	Succinyl coenzyme A	2-Oxoglutarate synthase
6. $^-\text{OOC–CH}_2\text{–CH}_2\text{–CO–COO}^- + CO_2 + 2[H] \rightleftharpoons$ $^-\text{OOC–CH}_2\text{–CH(COO}^-\text{)–CHOH–COO}^- + H^+$	−9.6	2-Oxoglutarate	Isocitrate dehydrogenase
7. $CH_3\text{–CO–COO}^- + CO_2 + \text{ATP} \rightarrow {}^-\text{OOC–CH}_2\text{–CO–COO}^- + H^+ +$ $\text{ADP} + P_i$	+0.1	Pyruvate	Pyruvate carboxylase
8. $CH_2\text{=C(O–PO}_3^{2-}\text{)–COO}^- + HCO_3^- \rightarrow$ $^-\text{OOC–CH}_2\text{–CO–COO}^- + P_i$	−19.7	Phosphoenolpyruvate (PEP)	PEP carboxylase
9. $\text{PEP} + HCO_3^- + P_i \rightleftharpoons {}^-\text{OOC–CH}_2\text{–CO–COO}^- + PP_i$	+2.3	Phosphoenolpyruvate (PEP)	PEP carboxykinase (PP)

[a]Note that not all of these reactions are present in one given organism. The electron donors [H] of the enzymes catalyzing reductive carboxylations are not specified. They may vary in different organisms.

citric acid cycle is reductively reversed which allows acetyl-CoA formation from $2CO_2$ and 8[H]. A second autotrophic pathway, the reductive acetyl coenzyme A pathway, is noncyclic: two CO_2 molecules are reduced independently by a series of different enzymes, one with 2[H] to the oxidation state of CO, the other with 6[H] to the oxidation state of methanol; these C_1 compounds are enzyme- or coenzyme-bound and are finally condensed to acetyl CoA. The overall reaction of both alternate CO_2 fixation pathways is the same,

$$2CO_2 + 8[H] \rightarrow (CH_2O)_2 + 2H_2O$$

where $(CH_2O)_2$ is activated acetate, acetyl CoA. Both pathways may be reversible in some of these organisms, in contrast to the irreversible Calvin cycle.

Traditionally, organisms are divided into two major categories, heterotrophs, which oxidize organic matter to CO_2, and autotrophs, which fix and reduce CO_2 to organic compounds (see Chapter 1). The biochemical mechanisms underlying these fundamental processes are considered to be totally different. However, in anaerobes in fact such a complete separation of the metabolic principles used in building up and degrading central metabolites does not exist. As will be discussed in Chapter 22, three different major metabolic pathways for the complete oxidation of organic compounds are recognized, the oxidative citric acid cycle, the oxidative acetyl-CoA pathway, and the oxidative pentose phosphate cycle (Figure 20.1).

Similarly, three different autotrophic CO_2 fixation mechanisms are known, the reductive citric acid cycle, the reductive acetyl-CoA pathway, and the reductive pentose phosphate cycle (see Figure 20.1).

20.2 DISTRIBUTION OF AUTOTROPHIC PATHWAYS IN BACTERIA

The distribution of the autotrophic pathways in procaryotes is summarized in Table 20.2, which also presents the best studied organisms of the respective

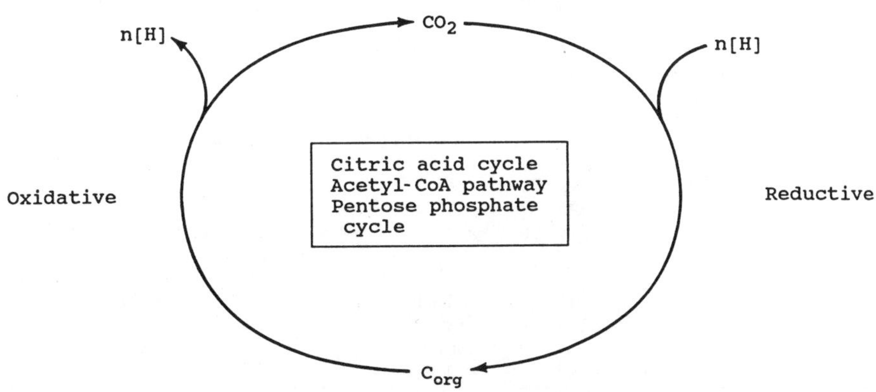

Figure 20.1 The three major metabolic pathways for the complete oxidation of organic compounds and for autotrophic CO_2 fixation.

groups. The Calvin cycle is confined to eubacteria, a remarkable rule to which so far no exception is known. Within the eubacteria, the Calvin cycle does not occur in strict anaerobes. In strict anaerobic eubacteria and in archaebacteria (both aerobic, anaerobic, and facultative ones) (for archaebacteria, see Kandler and Zillig, 1986) one of the alternate pathways is present. There is no clear rule about which eubacterial group uses which of these two processes. Within the archaebacteria, there appears to exist a separation into the methanogens including possibly the sulfate-reducing *Archaeoglobus* sp., which use the reductive acetyl-CoA pathway, and the sulfur-dependent aerobic and anaerobic (mostly thermophilic and acidophilic) archaebacteria, which may use a yet to be defined reductive citric acid cycle.

Energy sources The anaerobes derive energy by photosynthetic electron transport phosphorylation (green sulfur bacteria), sulfur reduction to H_2S (anaerobic sulfur reducers; *Thermoproteus*), sulfate reduction to H_2S (sulfate reducers), CO_2 reduction to CH_4 (methanogens), and CO_2 reduction to acetate (acetogens), respectively. The aerobic sulfur-oxidizing archaebacteria (*Sulfolobus*) gain energy by sulfur oxidation coupled to oxygen respiration. The thermophilic, microaerophilic, strictly autotrophic hydrogen-oxidizing (Knallgas) bacteria of unknown affiliation (*Hydrogenobacter, Calderobacterium*) oxidize H_2 with O_2.

Reducing power These chemolithoautotrophic organisms use H_2, formate, or H_2S (only phototrophs) as reductant, and the reduced compounds are oxidized via the enzymes hydrogenase, formate dehydrogenase, and H_2S dehydrogen-

Table 20.2 Distribution of CO_2 fixation pathways in autotrophic bacteria[a]

Calvin cycle	Reductive acetyl CoA pathway	Reductive citric acid cycle
Strict aerobic or facultatively aerobic eubacteria	Strict anaerobic (or microaerophilic) eubacteria; all archaebacteria	
	Homoacetogenic bacteria *Clostridium thermoaceticum* (Gram$^+$ type I) *Acetobacterium woodii* *Sporomusa* sp. (Gram$^+$ type II)	Green sulfur bacteria *Chlorobium limicola* Thermophilic hydrogen-oxidizing (Knallgas) bacteria (Gram$^-$) *Hydrogenobacter thermophilus*
	Most sulfate reducers (δ Purple bacteria) *Desulfobacterium autotrophicum* *Desulfovibrio baarsii*	Some sulfate reducers (δ Purple bacteria) *Desulfobacter hydrogenophilus*
	Methanogens (Archaebacteria I) *Methanobacterium thermoautotrophicum* *Methanosarcina barkeri*	

[a]Only the best studied autotrophs are included. The CO_2 fixation pathways in the sulfate-reducing archaebacteria (*Archaeoglobus* sp.), in the sulfur-dependent archaebacteria (*Thermoproteus* sp., *Sulfolobus* sp.), and in the gliding green sulfur bacteria (*Chloroflexus* sp.) remain to be studied.

ase, respectively. By action of these enzymes (in the case of H_2S, reversed electron transport is required) ferredoxin can be reduced ($E_0' \approx -410$ mV) which is a more potent reductant than NAD(P)H ($E_0' \approx -320$ mV).

Variations Many variations of these two common principles exist in nature, as will be discussed later. Many of the autotrophs prefer acetate as carbon source when available, in addition to CO_2; therefore, they are facultative autotrophs. Some organisms do not have a Calvin cycle, but details of their carbon metabolism are unknown. More work needs to be done to unravel their carbon metabolism and to discover new autotrophs and, possibly, new pathways.

20.3 REDUCTIVE CITRIC ACID CYCLE FOR AUTOTROPHIC CO_2 FIXATION

The well-known oxidative citric acid cycle is used by most (though not all) heterotrophs for the oxidation of acetyl CoA via citrate to 2 CO_2; the 8 reducing equivalents [H] formed can be consumed by different types of respiration. A reductive version of this pathway was first proposed to function in the fixation of 2 CO_2 with 8 [H] into acetate or acetyl CoA in the green sulfur bacterium *Chlorobium limicola* in 1966 (Evans et al., 1966; Sirevåg, 1975; Buchanan, 1979; Fuchs et al., 1980). Only by 1980 was the cycle completed by showing that an ATP citrate lyase (which replaces citrate synthase) is responsible for citrate cleavage (Ivanovsky et al., 1980). A simplified scheme of the pathway is given in Figure 20.2; the reactions, thermodynamic values under standard conditions (Thauer et al., 1977), and the equilibrium constants are given in Table 20.3. Its distribution is indicated in Table 20.2. For *Desulfobacter* sp. see Schauder et al., 1987; for *Hydrogenobacter* sp. see Shiba et al., 1985.

Outline of the reductive citric acid cycle pathway Reverse reactions of the citric acid cycle are used to form citrate from oxaloacetate, 2 CO_2, and 8 [H] (Figure 20.2). Oxaloacetate is reduced to succinate via malate dehydrogenase, fumarase, and fumarate reductase. Succinate becomes activated to succinyl CoA by action of succinyl-CoA synthetase which requires the expenditure of 1 ATP. Two consecutive CO_2 fixation reactions are reversible reductive carboxylations: 1) of succinyl CoA to 2-oxoglutarate (α-ketoglutarate), catalyzed by ferredoxin-dependent 2-oxoglutarate synthase; and 2) of 2-oxoglutarate to isocitrate, catalyzed by normal NADPH-dependent isocitrate dehydrogenase. 2-Oxoglutarate synthase differs from 2-oxoglutarate dehydrogenase in composition, electron donor, and important catalytic properties. This enzyme requires reduced ferredoxin rather than a reduced pyridine nucleotide, otherwise this reaction would be thermodynamically unfavorable ($E_0' = -490$ mV). The enzyme catalyzes a coenzyme A-dependent, ferredoxin-independent exchange of CO_2 into the C-1 carboxyl position of 2-oxoglutarate (for recent literature see Kerscher and Oesterheldt, 1982). Isocitrate is converted into citrate via aconitase. Finally,

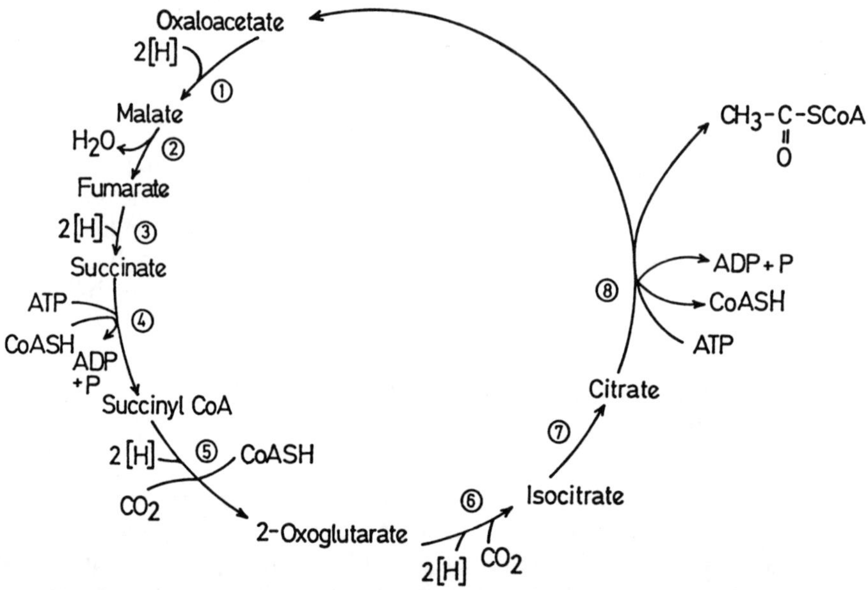

Figure 20.2 Reactions of the reductive citric acid cycle. The numbers refer to the enzyme reactions of Table 20.3. [H], reducing equivalent.

citrate is cleaved via ATP citrate lyase (with consumption of another ATP) into acetyl CoA, the CO_2 fixation product, and into oxaloacetate, the starting molecule of the cycle. The only virtually irreversible reaction of this pathway is fumarate reduction to succinate (however, see below).

Key enzymes and energy demand As key enzymes one may consider 2-oxoglutarate synthase in combination with ATP citrate lyase. Acetyl CoA assimilation into central intermediates requires further carboxylations. Acetyl CoA formation by this pathway requires 2 ATP when soluble fumarate reductase is involved. However, when fumarate reduction to succinate is catalyzed by a membrane-bound fumarate reductase using H_2 or formate as electron donor, approximately 1 ATP may be regained by electron transport phosphorylation; in this case only $\approx$ 1 ATP per acetyl CoA would be required. Such a modified reductive citric acid cycle should then be reversible and may be used in both directions even in a single organism (e.g., *Desulfobacter hydrogenophilus*). In other words, it may be used both for CO_2 fixation under autotrophic growth conditions and for acetate oxidation under heterotrophic growth conditions (Schauder et al., 1987; Möller et al., 1987; see also Chapter 22).

Modifications There are interesting modifications of this basic metabolic principle. The electron donors of malate dehydrogenase, fumarate reductase, and 2-oxoglutarate synthase may be different. The most interesting modifications are found in the sulfate-reducing *Desulfobacter hydrogenophilus*: Instead of a

Table 20.3 Reactions and enzymes of the reductive citric acid cycle and their thermodynamic values[a]

Reaction	$\Delta G^{0\prime}$ (kJ/mol) for 2[H] = H_2	k_{eq}	$\Delta G^{0\prime}$ (kJ/mol) for 2[H] = NAD(P)H+H^+	k_{eq}	E_0' [ox/red] (mV)	Enzyme (E.C. number)
1. Oxaloacetate^{2-} + 2[H] ⇌ Malate^{2-}	−47.9	4×10^{-9}	−29.8	6×10^{-6}	−166	Malate dehydrogenase (1.1.1.37)
2. Malate^{2-} ⇌ Fumarate^{2-} + H_2O	+3.7	0.22	+3.7	0.22	–	Fumarate hydratase (4.2.1.2)
3. Fumarate^{2-} + 2[H] ⇌ Succinate^{2-}	−86.0	10^{15}	−67.9	8×10^{11}	−32	Fumarate reductase (1.3.99.1)
4. Succinate^{2-} + ATP^{4-} + CoASH ⇌ Succinyl CoA$^-$ + ADP^{3-} + P^{2-}	+3.3	0.26	+3.3	0.26	–	Succinyl CoA synthetase (6.2.1.5)
5. Succinyl CoA$^-$ + 2[H] + CO_2 ⇌ 2-Oxoglutarate^{2-} + CoASH + H^+	+14.8	2.5×10^{-3}	+32.9	1.7×10^{-6}	−491	2-Oxoglutarate synthase (1.2.7.3)
6. 2-Oxoglutarate^{2-} + 2[H] + CO_2 ⇌ Isocitrate^{3-} + H^+	−9.6	4.8×10^{1}	+8.5	3.2×10^{-2}	−364	Isocitrate dehydrogenase ($NADP^+$) (1.1.1.42)
7. Isocitrate^{3-} ⇌ cis-Aconitate^{3-} ⇌ Citrate^{3-}	−6.7	1.2	−6.7	1.2	–	Aconitate hydratase (4.2.1.3)
8. Citrate^{3-} + ATP^{4-} + CoASH ⇌ Acetyl CoA + Oxaloacetate^{2-} + ADP^{3-} + P^{2-}	+5.6	10^{-1}	+5.6	10^{-1}	–	ATP citrate lyase (4.1.3.8)
8a. Citrate^{3-} ⇌ Acetate$^-$ + Oxaloacetate^{2-}	+1.7	0.5	+1.7	0.5	–	Theoretical: Citrate lyase (4.1.3.6)
8b. Citrate^{3-} + CoASH + H^+ ⇌ Acetyl CoA + Oxaloacetate^{2-} + H_2O	+37.4	2.8×10^{-7}	+37.4	2.8×10^{-7}	–	Theoretical: Citrate synthase (4.1.3.7)
1–8. $2CO_2$ + 8[H] + CoASH + $2ATP^{4-}$ ⇌ $CH_3COSCoA$ + H_2O + $2ADP^{3-}$ + $2P^{2-}$ + $2H^+$	−122.8	3×10^{21}	−50.4	6.8×10^{8}	−255	Reductive citric acid cycle

[a]The table includes some theoretical partial reactions, 8a, 8b. For details and variations of this pathway see the text. For calculations see Thauer et al. (1977).

soluble, NAD-linked malate dehydrogenase a possibly membrane-associated, NAD(P)-independent enzyme is found, which *in vitro* couples to artificial redox dyes; its physiological electron donor is unknown. Instead of the soluble fumarate reductase, a membrane-bound, perhaps menaquinone-dependent fumarate reductase/succinate dehydrogenase is present which may even generate ATP by electron transport phosphorylation coupled to fumarate reduction with H_2. Succinate appears to be activated by CoA transfer from acetyl CoA, rather than by succinyl CoA synthetase reaction (Schauder et al., 1987). This would require acetate to be reactivated by means of e.g. acetyl-CoA synthetase.

20.4 REDUCTIVE ACETYL-COA PATHWAY FOR AUTOTROPHIC CO_2 FIXATION

In the past few years a new autotrophic pathway, the reductive acetyl-CoA pathway, has emerged which may be used also in the oxidative direction for the complete oxidation of acetyl CoA in those heterotrophic anaerobes which do not have the enzymes of a complete oxidative citric acid cycle (see Chapter 22). The reductive version of this noncyclic pathway is used by acetogens, methanogens, and most autotrophic sulfate reducers for the synthesis of acetyl CoA. This two-carbon compound is formed by one-carbon to one-carbon addition of an enzyme-bound carbon monoxide and a tetrahydropterin-bound methanol. The general outline of the pathway is given in Figure 20.3; the reactions and their thermodynamic values under standard conditions are given in Table 20.4. Its distribution is indicated in Table 20.2. The reader is referred to Chapter 3 and to several papers dealing with various aspects of this pathway

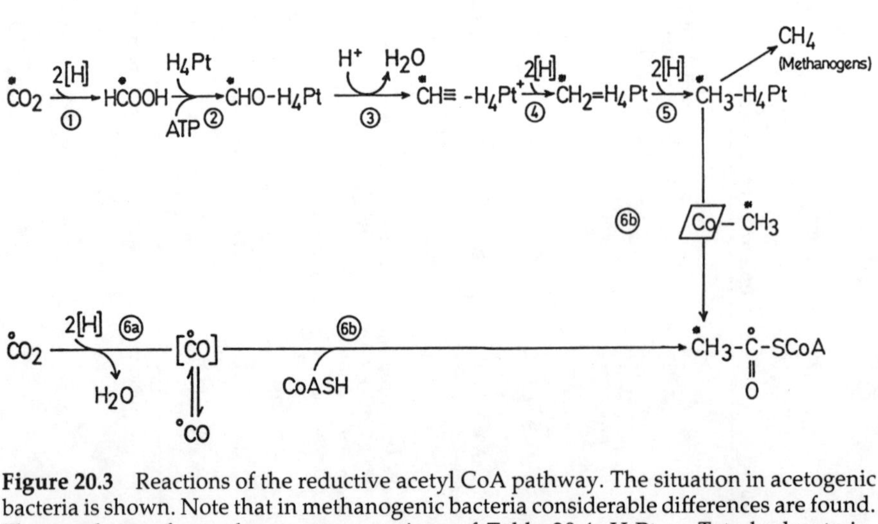

Figure 20.3 Reactions of the reductive acetyl CoA pathway. The situation in acetogenic bacteria is shown. Note that in methanogenic bacteria considerable differences are found. The numbers refer to the enzyme reactions of Table 20.4. H_4Pt = Tetrahydropterin, e.g., tetrahydrofolate. CHO —, formyl; CH ≡, methenyl; CH_2 =, methylene; CH_3 —, methyl; [H], reducing equivalent; [CO], bound CO; Co , corrinoid protein.

Table 20.4 Reactions and enzymes of the reductive acetyl CoA pathway and their thermodynamic values[a]

Reaction	$\Delta G^{0\prime}$ (kJ/reaction) for 2[H] = H_2	k_{eq}	$\Delta G^{0\prime}$ (kJ/reaction) for 2[H] = NAD(P)H+H^+	k_{eq}	E_0' [ox/red] (mV)	Enzyme (E.C. number)
1. $CO_2 + 2[H] \rightleftharpoons HCO_2^- + H^+$	+3.4	0.2	+21.5	1.7×10^{-4}	−421	Formate dehydrogenase (1.2.1.43)
2. $HCO_2^- + H_4$ folate + $ATP^{4-} \rightleftharpoons N^{10}$-formyl H_4 folate + $ADP^{3-} + P^{2-}$	−8.4	3×10^1	−8.4	3.10^1		Formyl-H_4 pterin synthetase (6.3.4.3)
3. N^{10}-formyl H_4 folate + $H^+ \rightleftharpoons N^5$, N^{10}-methenyl $FH_4^+ + H_2O$	−4.0	5	−4.0	5		Methenyl-H_4 pterin cyclohydrolase (3.5.4.9)
4. N^5,N^{10}-methenyl H_4 folate$^+$ + 2[H] $\rightleftharpoons N^5$,N^{10}-methylene H_4 folate + H^+	−23.0	10^4	−4.9	7	−295	Methylene-H_4 pterin dehydrogenase (1.5.1.15)
5. N^5,N^{10}-methylene H_4 folate + 2[H] $\rightleftharpoons N^5$-methyl H_4 folate	−57.3 to −37	10^{10} to 10^6	−39.2 to −19		−117 to −210	Methylene-H_4 pterin reductase (1.1.99.15)
6a. $CO_2 + 2[H] \rightleftharpoons CO + H_2O$	+20.1	3×10^{-4}	+38.2	2×10^{-7}	−524	CO dehydrogenase (1.2.99.2)
6b. N^5-methyl H_4 folate + CO + CoASH $\rightleftharpoons$ $CH_3COSCoA + H_4$ folate	−21.8 to −2	7×10^3 to 2	−21.8 to −2	7×10^3 to 2		Acetyl CoA synthase complex, including corrinoid enzyme, methyltransferase and CO dehydrogenase
6. N^5-methyl H_4 folate + CO_2 + 2[H] + CoASH $\rightleftharpoons CH_3$ COSCoA + H_4 folate + H_2O	−1.7 to −21	2 to 5 × 10^3	+16.4 to −4	10^{-3} to 5	−405 to −305	
1–6. $2CO_2$ + 8[H] + CoASH + $ATP^{4-} \rightleftharpoons$ CH_3 COSCoA + $2H_2O$ + $ADP^{3-} + P^{2-} + H^+$	−91.0	10^{16}	−18.6	2×10^3	−296	Reductive acetyl CoA pathway

[a]The reactions occurring in homoacetogenic bacteria are described. The table includes some partial reactions (6a, 6b). For details and variations of this pathway see the text. For calculations see Fuchs (1986).

(for general reviews see Wood and Utter, 1965; Ljungdahl and Wood, 1969, 1982; Wood et al., 1982; Zeikus, 1983; Diekert et al., 1985; Wood, 1985; Fuchs and Stupperich, 1986; Fuchs, 1986; Ljungdahl, 1986; Wood et al., 1986; for CO_2 fixation in methanogens see Taylor et al., 1976; Zeikus et al., 1977; Fuchs et al., 1978; Daniels and Zeikus, 1978; Fuchs and Stupperich, 1980, 1986; Kenealy and Zeikus, 1982; Zeikus, 1983; Stupperich and Fuchs, 1984a, b; Holder et al., 1985; Zeikus et al., 1985; Länge and Fuchs, 1987; for CO_2 fixation in acetogens, see Eden and Fuchs, 1982, 1983; Fuchs 1986; for CO_2 fixation in sulfate-reducing bacteria, see Jansen et al., 1984, 1985; Schauder et al., 1986; Brysch et al., 1987; and unpublished results).

Outline of the pathway It should be stressed that many modifications of this pathway exist. For historical reasons the pathway in acetogenic bacteria (*Clostridium thermoaceticum*) is considered here as representative. All the other organisms are less studied.

Origin of the methyl of acetate The pathway brings about a reduction of CO_2 to the level of methanol bound as N-methyl group to a tetrahydropterin (e.g., tetrahydrofolate); this gives rise to the methyl of acetate in acetyl CoA. In acetogenic and sulfate-reducing bacteria, formate is a free intermediate in CO_2 reduction via formate dehydrogenase; it is subsequently activated to a formyl tetrahydropterin at the expenditure of 1 ATP. The formyl is then reduced by several enzymes to the methyl tetrahydropterin. In methanogenic bacteria, even in those which can use formate, formate is not a free intermediate. Rather, CO_2 appears at first to be bound as a carbamate to the coenzyme methanofuran, then to be reduced to the formyl state, and then the formyl seems to be transferred to tetrahydromethanopterin (a tetrahydrofolate counterpart in methanogens) (for coenzymes in methanogens see Wolfe, 1985). Up to here also the equivalent of 1 ATP must be spent, but details are totally unknown. The formyl is then reduced to methyl tetrahydromethanopterin. These reactions are shared by CO_2 reduction to methane (Stupperich and Fuchs, 1984a, b; Länge and Fuchs, 1987).

Origin of the carboxyl of acetate and C_1 condensation The carboxyl group of acetate is synthesized from a bound carbon monoxide which is formed in a second CO_2 reduction catalyzed by the nickel-enzyme carbon monoxide dehydrogenase (Wood et al., 1986; Diekert et al., 1985). This enzyme is part of a complex which also brings about the condensation of the two C_1 precursors and coenzyme A. This requires that the methyl group from methyl tetrahydropterin is transferred by a methyltransferase to a corrinoid protein to form a methyl cobamide. CO dehydrogenase takes over this methyl and condenses it with the bound CO to an enzyme-bound energy-rich acetyl residue which becomes thiolyzed with CoA to yield acetyl CoA (see Chapter 3). The exact order of these events, the enzymes involved, their spatial organization, and other details remain to be studied. Again, many differences may exist in these autotrophs. The most interesting enzyme of this pathway is the nickel enzyme

CO dehydrogenase. Its composition and the reactions which it catalyzes are unique. *In vitro* it is easily tested as CO-oxidizing activity catalyzing $CO + H_2O + 2$ viologen $dye^{2+} \rightarrow CO_2 + 2$ viologen $dye^{+} + 2 H^{+}$. It catalyzes an isotope exchange between $^{14}CO_2$ and CO, as well as an isotope exchange between $^{14}CO_2$ or ^{14}CO, and the carboxyl of acetyl CoA. This implies that acetyl CoA is cleaved in bound methyl, CoA, and carbonyl, the carbonyl is exchanged with ^{14}CO (or with $^{14}CO_2$ which has to be reduced to a bound ^{14}C carbonyl), and acetyl CoA is reformed from its parts. When in growing cells acetyl-CoA synthesis is blocked, e.g., by propylation of the corrinoid protein, the cells produce large amounts of CO gas (Eikmanns et al., 1985). In addition, the metallo-organic compound methyl-cobamide of the corrinoid protein may become reduced to methane. The organisms with CO dehydrogenase therefore are also mini-methane producers (Schauder et al., 1986).

Assimilation and oxidation of reduced one-carbon compounds In contrast to the other CO_2 fixation mechanisms, the acetyl-CoA pathway allows the assimilation or oxidation of different reduced one-carbon compounds. CO can directly be incorporated into the carboxyl of acetate. Formate can serve as methyl precursor. Formaldehyde chemically reacts with tetrahydropterins to give methylene tetrahydropterin, which serves as precursor of the methyl of acetate. Methanol and methyl groups in methylamines appear to be incorporated into the methyl of acetate by different corrinoid enzymes. These compounds can also be oxidized. This does not necessarily mean that each individual species has the capability to perform all these reactions.

Key enzymes and energy demand It should be pointed out that the acetyl-CoA pathway shows considerable variations, depending on the organism. In addition it may be used in both degradative and biosynthetic directions even in a single organism (e.g., *Desulfobacterium autotrophicum*). As with the citric acid cycle, it is unclear what decides whether it serves for acetyl-CoA oxidation or for CO_2 fixation into acetyl CoA; there may be isoenzymes and other regulatory mechanisms involved. The nickel-enzyme carbon monoxide dehydrogenase, which is part of the yet to be defined multienzyme complex involved in the synthesis of acetyl CoA, is unique to this pathway (for aerobic Mo-enzyme CO dehydrogenase see Chapter 18). In addition, exceedingly high amounts of cobamides (100–1000 nmol/g dry cells) and of tetrahydropterins (2000 nmol/g dry cells) are required and have been found in these organisms (for cobamides, see Dolphin, 1982; Stupperich and Kräutler, 1987; for pterins, see Blakley and Benkovic, 1984).

The energy demand of the acetyl-CoA pathway cannot be specified yet, but it appears to be rather low. The synthesis of acetyl CoA from 2 CO_2, 4 H_2, and coenzyme A is almost energy neutral, even under cellular conditions. It may require only 1 ATP or less: One energy-requiring reaction is CO_2 reduction to the formyl tetrahydropterin level which requires 1 ATP; in acetogenic and sulfate-reducing bacteria ATP is consumed in the ATP-dependent formyl tetrahydropterin synthetase reaction. Another energy-dependent reaction (requiring less than 1 ATP equivalent) is CO_2 reduction to CO catalyzed by CO dehydrogenase, which is driven by an electrochemical potential; this fact may indicate that CO dehydrogenase is membrane linked (Eikmanns et al., 1985). The reduction of methylene tetrahydropterin to methyl tetrahydropterin with H_2 would be exergonic enough to be associated with energy conservation by electron transport phosphorylation; this remains to be shown. The condensation of bound CO and tetrahydropterin-bound methyl with CoA to give acetyl CoA does not require energy, nor does it probably yield energy (for discussion of energetic aspects of the acetyl-CoA pathway, see Fuchs, 1986). Hence, approximately 1 ATP may be required for autotrophic acetyl-CoA synthesis via the acetyl-CoA pathway.

Modifications There are remarkable modifications on the coenzyme and cosubstrate level: The electron donors of the four reducing steps catalyzed by CO dehydrogenase, formate dehydrogenase, methylene tetrahydropterin dehydrogenase, and methylene tetrahydropterin reductase, respectively, are unknown in many cases and may be quite different, including pyridine nucleotides, deazaflavins, ferredoxins, flavodoxins, quinones, and others. The tetrahydropterins and cobamides, which serve as C_1 carriers, are also different: methanopterin, sarcinapterin (van Beelen et al., 1984), folate, pteroyl tetraglutamate, and other folate derivatives were found. The methanopterin carries the formyl and methyl at N-5, whereas the folate and folate derivatives carry the formyl at N-10, the methyl at N-5 (Wolfe, 1985). In addition, in methanogens and possibly in *Archaeoglobus*, methanofuran may act as a formyl carrier (K. O. Stetter, personal communication). Similarly, as many as seven different cobamides were discovered in these organisms which differ in the base of the lower axial, fifth ligand to the cobalt (Stupperich and Kräutler, 1987; E. Stupperich, personal communication). On the enzyme level also large differences exist. CO_2 reduction to the formyl tetrahydropterin level differs in methanogens and is still a mystery (Wolfe, 1985). The enzymes involved in reduction of formyl to methyl tetrahydropterin may differ. Finally, the CO dehydrogenase/acetyl CoA synthase complex is not sufficiently studied in different organisms to allow conclusions as to its common features.

20.5 OTHER POSSIBLE PATHWAYS OF AUTOTROPHY

Most of the anaerobic autotrophic bacteria seem to use one of these alternate pathways of autotrophic CO_2 fixation, instead of the Calvin cycle. Many reactions, enzymes, and coenzymes remain to be studied, and more variations of the basic principles may emerge. This is especially true for the sulfur-dependent and sulfate-reducing archaebacteria (see Chapter 9). The evidence available suggests a yet to be defined reductive carboxylic acid cycle including intermediates of the citric acid cycle (Kandler and Stetter, 1981; Schäfer et al., 1986; unpublished results). So far there is no evidence for any of the known autotrophic pathways in the gliding green sulfur bacterium *Chloroflexus aurantiacus*: when acetate labeled at C-1 and at C-2, respectively, is metabolized, polyglucose is formed in which C-1 plus C-6 and C-2 plus C-5, respectively, becomes labeled. Clearly, this redistribution of label cannot be explained by the known pathways (Holo and Sirevåg, 1986; H. Holo, personal communication).

20.6 ASSIMILATION OF ACETYL COA FORMED BY THE ALTERNATIVE AUTOTROPHIC PATHWAYS

The incorporation of acetyl CoA into major cellular compounds does not proceed via a glyoxylic acid cycle. Rather acetyl CoA is reductively carboxylated to pyruvate, catalyzed by pyruvate synthase (Figure 20.4). This enzyme, which normally is ferredoxin dependent, is quite different from the NAD-dependent pyruvate dehydrogenase complex which irreversibly oxidizes pyruvate. Analogous to 2-oxoglutarate synthase, pyruvate synthase catalyzes a coenzyme A-dependent, ferredoxin-independent exchange of $^{14}CO_2$ and the carboxyl group of pyruvate. Further assimilation of pyruvate is different (see Figure 20.4) (for literature, see the work quoted in the sections on the pathways). Anaplerotic CO_2 fixation is catalyzed by PEP carboxylase, PEP carboxykinase (PP_i), or pyruvate carboxylase. PEP synthesis is either via pyruvate, water dikinase, or via pyruvate, phosphate dikinase. 2-Oxoglutarate synthesis in methanogens differs. Most methanogens synthesize 2-oxoglutarate via oxaloacetate reduction to succinyl CoA and via reductive carboxylation of succinyl CoA to 2-oxoglu-

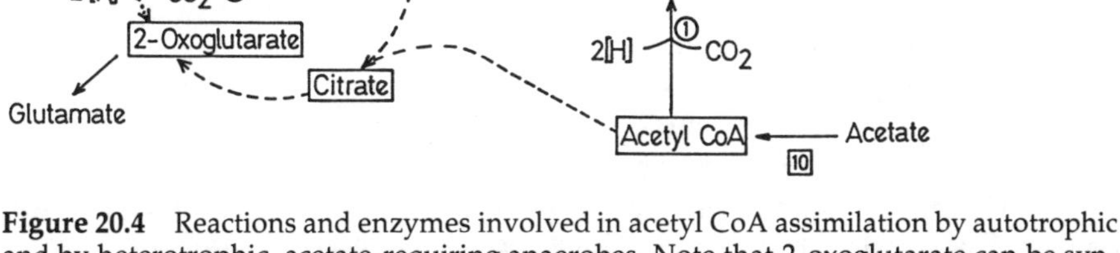

Figure 20.4 Reactions and enzymes involved in acetyl CoA assimilation by autotrophic and by heterotrophic, acetate-requiring anaerobes. Note that 2-oxoglutarate can be synthesized either from oxaloacetate via succinyl CoA carboxylation (. . .), or from citrate (---). For details and variations see the text. CO_2 fixation reactions (○) are: 1) pyruvate synthase, 2) PEP carboxylase, 3) PEP carboxykinase (PP), 4) pyruvate carboxylase, 5) 2-oxoglutarate synthase. Energy consuming reactions (□) are: 6) pyruvate, water dikinase, 7) pyruvate, phosphate dikinase, 8) succinyl CoA synthetase, 9) 3-phosphoglycerate kinase, 10) acetyl-CoA synthetase (required for heterotrophic acetate assimilation).

tarate. *Methanosarcina* sp. (Zeikus, 1983) and *Methanothrix* sp. (Ekiel et al., 1985) form citrate from acetyl CoA and oxaloacetate, and 2-oxoglutarate is formed from citrate. In any case acetyl CoA assimilation requires several additional main CO_2 fixation reactions, in addition to the two involved in acetyl CoA formation. The formation of one triose phosphate from acetyl CoA and CO_2 requires the expenditure of three ATP equivalents, two for pyruvate conversion into phosphoenolpyruvate, and one for phosphoglycerate activation prior to its reduction. Hence triose phosphate formation from 3 CO_2 via the reductive citric acid cycle requires 4–5 ATP, via the acetyl CoA pathway possibly only 3–4 ATP. In contrast, via the reactions of the Calvin cycle, 9 ATP are required. This may have an important ecological and evolutionary impact. The acetyl CoA assimilation by most heterotrophic strict anaerobes, which depend on acetate as a carbon source in addition to CO_2, proceeds in a similar way. Also facultative autotrophic anaerobes often preferentially assimilate acetate instead of CO_2. Acetate then is normally activated by acetyl CoA synthetase.

20.7 REGULATORY, ECOLOGICAL, AND EVOLUTIONARY ASPECTS OF ALTERNATIVE CO_2 FIXATION PATHWAYS

Besides the many unsolved biochemical problems, there are other interesting features of autotrophs which deserve to be studied. One problem is how facultative autotrophs regulate the autotrophic pathways when organic compounds such as acetate are available. Some bacteria, e.g., the autotrophic sulfate reducers, are able to oxidize organic compounds or to fix CO_2 via the same pathway. So far, no differences can be seen in the enzyme complement of heterotrophically and autotrophically grown *Desulfobacterium* and *Desulfobacter* species, respectively. What decides whether a compound is preferentially assimilated or oxidized and whether CO_2 or acetate is the preferred carbon source and what are the regulatory mechanisms? Which reduced substrates are preferred as electron donor when offered simultaneously? The answers to these and other questions are significant in view of the ecology of these organisms. It is surprising that within one physiological group of the sulfate reducers, which are also phylogenetically closely related, there are two different autotrophic pathways and two different pathways for acetyl CoA oxidation, namely the acetyl-CoA pathway in *Desulfobacterium autotrophicum* and related species, and a modified citric acid cycle in *Desulfobacter hydrogenophilus*. What are the physiological differences between the two pathways and their bearers? There are also some interesting evolutionary aspects of autotrophy. The isotopic carbon composition of sedimentary carbon, in which the ^{13}C deviates from inorganic carbonates by approximately −25 per mille (i.e., ^{13}C is present in less relative amounts than in inorganic carbonate) is being taken as indication for biological CO_2 fixation by autotrophs. According to these data, autotrophic

Table 20.5 $^{13}C/^{12}C$-Carbon isotope fractionation by different autotrophic bacteria[a]

Autotroph	$\Delta\delta^{13}C_{cell}$ (at pH 7)
1. *Alcaligenes eutrophus*	−26‰
Chromatium vinosum	−26‰[b]
(Calvin cycle)	
2. *Chlorobium limicola*	−12‰[b]
Desulfobacter hydrogenophilus	−10‰
Thermoproteus neutrophilus	− 8‰
(Reductive citric acid cycle)	
3. *Methanobacterium thermoautotrophicum*	−34‰[c]
Desulfobacterium autotrophicum	−36‰
Acetate formed by *Acetobacterium woodii*	−40‰
(Reductive acetyl CoA pathway)	

[a]The $\delta^{13}C_{cell}$ value represents the per mille deviation of the ^{13}C content of cell carbon compared with CO_2, with which the culture was grown at a high excess. The negative sign indicates that ^{13}C is depleted relative to CO_2 (unpublished results).
[b]From Quandt et al. (1977).
[c]From Fuchs et al. (1979).

ecosystems became a dominant component of the terrestrial carbon cycle as early as 3.5×10^9, if not 3.8×10^9 years ago. The $^{13}C/^{12}C$ carbon isotope discrimination by the 3 different autotrophic pathways is remarkably different, as shown in Table 20.5. It seems unlikely that the different carbon assimilation paths can be derived from one common ancient mechanism of CO_2 fixation. Therefore, they have developed independently at different times to meet different needs. If the autotrophic way of life has early emerged in evolution (see Chapter 27) and if some of the ancient mechanisms have persisted, one could speculate about evolution of the autotrophic pathways. The following characters of the acetyl-CoA pathway seem to be conservative (Table 20.6). These features relate to the following considerations or likely assumptions under which conditions carbon autotrophy may have arisen: 1) the early atmosphere was anaerobic, organic compounds were limiting, and volcanic activities were common; 2) archaebacteria and eubacteria might not have been separated yet; 3) the primitive autotrophs probably were energy-limited chemolithotrophic anaerobes; 4) CO_2 as well as volatile reduced one-carbon compounds (CO, CH_2O, etc.) were available as carbon sources as a consequence of volcanic activities; 5) common coenzymes and enzymes of intermediary metabolism were already existing. In the light of evolution it will be interesting to determine the degree of homology between the key enzymes of the different pathways in the different organisms. This is one impetus for further research. Also, biochemists and chemists may be stimulated by learning how easily bacteria

Table 20.6 Some characters of the acetyl-CoA pathway which are considered conservative

1. Occurrence of the pathway in anaerobes and in extreme thermophiles
2. Occurrence of the pathway in eubacteria ($Gram^+$ and $Gram^-$) and in archaebacteria
3. Minimal energy requirement of the pathway
4. Concomitant utilization of reduced one-carbon compounds such as CO, formate, formaldehyde, methanol
5. Strong requirement for different ubiquitous, yet varied coenzymic structures
6. Importance of metals in the catalysis of CO_2 fixation and reduction

reduce and fix CO_2 autotrophically, and chemical processes based on CO_2 reduction may become of industrial importance.

REFERENCES

Blakley, R.L. and **Benkovic, J.** (editors). 1984. *Folates and pterins*, Vol. I, II. John Wiley Sons, New York.

Brysch, K., Schneider, C., Fuchs, G., and **Widdel, F.** 1987. Lithoautotrophic growth of sulfate-reducing bacteria, and description of *Desulfobacterium autotrophicum* gen. nov., sp. nov. *Archives of Microbiology* 148:264–274.

Buchanan, B.B. 1979. Ferredoxin-linked carbon dioxide fixation in photosynthetic bacteria, pp. 416–424, in Gibbs, M. and Latzko, E. (editors), *Photosynthesis. II. Photosynthetic carbon metabolism and related processes.* Springer, Berlin, Heidelberg, New York.

Daniels, L. and **Zeikus, J.G.** 1978. One-carbon metabolism in methanogenic bacteria: analysis of short-term fixation products of $^{14}CO_2$ and $^{14}CH_3OH$ incorporated in whole cells. *Journal of Bacteriology* 136:75–84.

Diekert, G., Fuchs, G., and **Thauer, R.K.** 1985. Properties and function of carbon monoxide dehydrogenase from anaerobic bacteria, pp. 115–130, in Poole, R.K. and Dow, C.S. (editors), *Microbial gas metabolism: mechanistic, metabolic and biotechnological aspects,* Academic Press, London.

Dolphin, D. (editor). 1982. B_{12}, Vol. I, II, John Wiley Sons, New York, Toronto, Singapore.

Eden, G. and **Fuchs, G.** 1982. Total synthesis of acetyl coenzyme A involved in autotrophic CO_2 fixation in *Acetobacterium woodii. Archives of Microbiology* 133:66–74.

Eden, G. and **Fuchs, G.** 1983. Autotrophic CO_2 fixation in *Acetobacterium woodii*. II. Demonstration of enzymes involved. *Archives of Microbiology* 135:68–73.

Eikmanns, B., Fuchs, G., and **Thauer, R.K.** 1985. Formation of carbon monoxide from CO_2 and H_2 by *Methanobacterium thermoautotrophicum. European Journal of Biochemistry* 146:149–154.

Ekiel, J., Sprott, G.D., and **Patel, G.B.** 1985. Acetate and CO_2 assimilation by *Methanothrix concilii. Journal of Bacteriology* 162:905–908.

Evans, M.C.W., Buchanan, B.B., and **Arnon, D.I.** 1966. A new ferredoxin-dependent carbon reduction cycle in a photosynthetic bacterium. *Proceedings of the National Academy of Science (USA)* 55:928–934.

Fuchs, G. 1986. CO_2 fixation in acetogenic bacteria: Variations on a theme. *FEMS Microbiology Reviews* 39:181–213.

Fuchs, G., Stupperich, E., and **Thauer, R.K.** 1978. Acetate assimilation and the synthesis of alanine, aspartate and glutamate in *Methanobacterium thermoautotrophicum. Archives of Microbiology* 117:61–66.

Fuchs, G., Thauer, R.K., Ziegler, H., and **Stichler, W.** 1979. Carbon isotope fractionation by *Methanobacterium thermoautotrophicum. Archives of Microbiology* 120:135–139.

Fuchs, G., Stupperich, E., and **Eden, G.** 1980. Autotrophic CO_2 fixation in *Chlorobium limicola*. Evidence for the operation of a reductive tricarboxylic acid cycle in growing cells. *Archives of Microbiology* 128:64–71.

Fuchs, G. and **Stupperich, E.** 1980. Acetyl CoA, a central intermediate of autotrophic CO_2 fixation in *Methanobacterium thermoautotrophicum. Archives of Microbiology* 127:267–272.

Fuchs, G. and **Stupperich, E.** 1986. Carbon assimilation pathways in archaebacteria. *Systematic and Applied Microbiology* 7:364–369.

Holder, U., Schmidt, D.E., Stupperich, E., and **Fuchs, G.** 1985. Autotrophic synthesis of activated acetic acid from two CO_2 in *Methanobacterium thermoautotrophicum*. III.

Evidence for common one-carbon precursor pool and the role of corrinoid. *Archives of Microbiology* 141:229–238.

Holo, H. and **Sirevåg, R.** 1986. Autotrophic growth and CO_2 fixation of *Chloroflexus aurantiacus. Archives of Microbiology* 145:173–180.

Ivanovsky, R.N., Sintsov, N.V., and **Kondratieva, E.N.** 1980. ATP-linked citrate lyase activity in the green sulfur bacterium *Chlorobium limicola* forma *thiosulfatophilum. Archives of Microbiology* 128:239–241.

Jansen, K., Thauer, R.K., Widdel, F., and **Fuchs, G.** 1984. Carbon assimilation pathways in sulfate-reducing bacteria. Formate, carbon dioxide, carbon monoxide, and acetate assimilation by *Desulfovibrio baarsii. Archives of Microbiology* 138:257–262.

Jansen, K., Fuchs, G., and **Thauer, R.K.** 1985. Autotrophic CO_2 fixation by *Desulfovibrio baarsii*: demonstration of enzyme activities characteristic for the acetyl-CoA pathway. *FEMS Microbiology Letters* 28:311–315.

Kandler, O. and **Zillig, W.** (editors) 1986. *Archaebacteria '85*. Fischer, Stuttgart, New York.

Kandler, O.K. and **Stetter, K.O.** 1981. Evidence for autotrophic CO_2 assimilation in *Sulfolobus brierleyi* via a reductive carboxylic acid pathway. *Zentralblatt für Bakteriologie und Hygiene, I. Abteilung, Originale* C 2:111–121.

Kenealy, W.R. and **Zeikus, J.G.** 1982. One-carbon metabolism in methanogens: Evidence for synthesis of a two-carbon cellular intermediate and unification of catabolism and anabolism in *Methanosarcina barkeri. Journal of Bacteriology* 151:932–941.

Kerscher, L. and **Oesterhelt, D.** 1982. Pyruvate: ferredoxin oxidoreductase—new findings on an ancient enzyme. *Trends in Biochemical Sciences* 7:371–374.

Länge, S. and **Fuchs, G.** 1987. Autotrophic synthesis of activated acetic acid from CO_2 in *Methanobacterium thermoautotrophicum*. Synthesis from tetrahydromethanopterin-bound C_1 units and carbon monoxide. *European Journal of Biochemistry* 163:147–154.

Ljungdahl, L.G. 1986. The autotrophic pathway of acetate synthesis in acetogenic bacteria. *Annual Review of Microbiology* 40:415–450.

Ljungdahl, L.G. and **Wood, H.G.** 1969. Total synthesis of acetate from CO_2 by heterotrophic bacteria. *Annual Review of Microbiology* 23:515–538.

Ljungdahl, L.G. and **Wood, H.G.** 1982. Acetate biosynthesis, pp. 165–202, in Dolphin, D. (editor), B_{12}. John Wiley Sons, New York, Toronto, Singapore.

Möller, D., Schauder, R., Fuchs, G., and **Thauer, R.K.** 1987. Acetate oxidation to CO_2 via a citric acid cycle involving an ATP-citrate lyase: mechanism for the synthesis of ATP via substrate level phosphorylation in *Desulfobacter postgatei* growing on acetate and sulfate. *Archives of Microbiology* 148:202–207.

Quandt, L., Gottschalk, G., Ziegler, H., and **Stichler, W.** 1977. Isotope discrimination by photosynthetic bacteria. *FEMS Microbiology Letters* 1:125–128.

Schäfer, S., Barkowski, C., and **Fuchs, G.** 1986. Carbon assimilation by the autotrophic thermophilic archaebacterium *Thermoproteus neutrophilus. Archives of Microbiology* 146:301–308.

Schauder, R., Eikmanns, B., Thauer, R.K., Widdel, F., and **Fuchs, G.** 1986. Acetate oxidation to CO_2 in anaerobic bacteria via a novel pathway not involving reactions of the citric acid cycle. *Archives of Microbiology* 145:162–172.

Schauder, R., Widdel, F., and **Fuchs, G.** 1987. Carbon assimilation pathways in sulfate-reducing bacteria. II. Enzymes of a reductive citric acid cycle in the autotrophic *Desulfobacter hydrogenophilus. Archives of Microbiology* 148:218–225.

Shiba, H., Kawasumi, T., Igarashi, Y., Kodoma, T., and **Minoda, Y.** 1985. The CO_2 assimilation via the reductive tricarboxylic acid cycle in an obligately autotrophic aerobic hydrogen-oxidizing bacterium, *Hydrogenobacter thermophilus. Archives of Microbiology* 142:198–203.

Sirevåg, R. 1975. Photoassimilation of acetate and metabolism of carbohydrate in *Chlorobium thiosulfatophilum. Archives of Microbiology* 104:105–111.

Stupperich, E. and **Fuchs, G.** 1984a. Autotrophic synthesis of activated acetic acid from two CO_2 in *Methanobacterium thermoautotrophicum*. I. Properties of *in vitro* system. *Archives of Microbiology* 139:8–13.

Stupperich, E. and **Fuchs, G.** 1984b. Autotrophic synthesis of activated acetic acid from two CO_2 in *Methanobacterium thermoautotrophicum*. II. Evidence for different origins of acetate carbon atoms. *Archives of Microbiology* 139:14–20.

Stupperich, E. and **Kräutler, B.** 1987. Pseudo vitamin B_{12} or 5-hydroxybenzimidazolylcobamide are the corrinoids found in methanogenic bacteria. *Archives of Microbiology* 149:268–277.

Taylor, G.T., Kelly, D.P., and **Pirt, S.J.** 1976. Intermediary metabolism in methanogenic bacteria (*Methanobacterium*), pp. 173–180, in Schlegel, H.G., Gottschalk, G., and Pfennig, N. (editors), *Microbial production and utilization of gases (H_2, CH_4, CO)*, Goltze Verlag, Göttingen.

Thauer, R.K., Jungermann, K., and **Decker, K.** 1977. Energy conservation in chemotrophic anaerobic bacteria. *Bacteriological Reviews* 41:100–180.

Van Beelen, P., Labro, J.F.A., Keltjens, J.T., Geerts, W.J., Vogels, G.D., Laarhoven, W.H., Guijt, W., and **Haasnoot, C.A.G.** 1984. Derivatives of methanopterin, a coenzyme involved in methanogenesis. *European Journal of Biochemistry* 139:359–365.

Wolfe, R.S. 1985. Unusual coenzymes of methanogenesis. *Trends in Biochemical Sciences* 10:396–399.

Wood, H.G. 1985. Then and now. *Annual Review in Biochemistry* 54:1–41.

Wood, H.G. and **Utter, M.F.** 1965. The role of CO_2 fixation in metabolism, pp. 1–27, in Campbell, P.N. and Greville, G.D. (editors), Vol. 1, *Essays in Biochemistry*, Academic Press, London.

Wood, H.G., Drake, H.L., and **Hu, S.** 1982. Studies with *Clostridium thermoaceticum* and the resolution of the pathway used by acetogenic bacteria that grow on carbon monoxide or carbon dioxide and hydrogen. *Proceeding of Biochemistry Symposium* 29–56. Palo Alto, Calif. Annual Reviews.

Wood, H.G., Ragsdale, S.W., and **Pezacka, E.** 1986. The acetyl-CoA pathway of autotrophic growth. *FEMS Microbiology Review* 39:345–362.

Zeikus, J.G., Fuchs, G., Kenealy, W., and **Thauer, R.K.** 1977. Oxidoreductases involved in cell carbon synthesis of *Methanobacterium thermoautotrophicum. Journal of Bacteriology* 132:604–613.

Zeikus, J.G. 1983. Metabolism of one-carbon compounds by chemotrophic anaerobes. *Advances in Microbial Physiology* 24:215–299.

Zeikus, J.G., Kerby, R., and **Krzycki, J.A.** 1985. Single-carbon chemistry of acetogenic and methanogenic bacteria. *Science* 227:1167–1173.

21

Bioenergetics of Methanogenic and Acetogenic Bacteria

Gerhard Gottschalk

Institute for Microbiology
Georg-August University
Göttingen, Federal Republic of Germany

21.1 INTRODUCTION

Both acetogenic and methanogenic bacteria are able to gain energy from the reduction of carbon dioxide by molecular hydrogen:

$$4H_2 + CO_2 \rightarrow CH_4 + 2H_2O \qquad \Delta G^{0\prime} = -131 \text{ kJ/reaction}$$

$$4H_2 + 2CO_2 \rightarrow CH_3COO^- + H^+ + 2H_2O \qquad \Delta G^{0\prime} = -95 \text{ kJ/reaction}$$

However, the organisms carrying out these two reactions are very different (Table 21.1). **Acetogens** are typical eubacteria whereas the methanogens represent the largest and best known group of the archaebacteria (Balch et al., 1979; Jones et al., 1987). Chemolithotrophic acetogens are found in various genera of anaerobes (Fuchs, 1986). Among them are mesophiles and thermophiles, sporeformers and nonsporeformers. The known species are facultative chemolithotrophs. Sugars are generally the preferred substrates; some strains are unable to utilize sugars but grow on methanol, other alcohols, or N-methyl compounds such as betaine, e.g., *Sporomusa* species, (Möller et al., 1984). **Methanogenic bacteria** are also very diverse. The temperature range inhabited by them as a group is very large and reaches temperatures as high as 97°C (*Methanothermus sociabilis,* Lauerer et al., 1986). The shape and cell wall composition of methanogens varies considerably. They use a much more restricted range of substrates than the acetogens. Many of them are obligate chemolithotrophic (e.g., *Methanobacterium thermoautotrophicum*), whereas others may utilize formate, methanol, methylamines, or acetate (Mah et al., 1981; Hippe et al., 1979).

Table 21.1. Some species and properties of acetogenic and methanogenic bacteria

Acetogens	Methanogens
All eubacteria	All archaebacteria
Representative species: *Clostridium aceticum* *C. thermoaceticum* *C. thermoautotrophicum* *Sporomusa sphaeroides* *Acetobacterium woodii* *Acetogenium kivui* *Eubacterium limosum* *Peptostreptococcus productus*	**Representative species:** *Methanobacterium ruminantium* *M. thermoautotrophicum* *Methanobrevibacter arboriphilus* *Methanococcus voltae* *Methanospirillum hungatei* *Methanothermus fervidus* *Methanosarcina barkeri* *Methanothrix soehngenii*
Nutrition: Facultative chemolithotrophs, many acetogenic species not listed above are obligate chemoorganotrophs (not able to grow with H_2 + CO_2 alone)	**Nutrition:** Obligate chemolithotrophs (first 6 species) or facultative chemolithotrophs (*M. barkeri*); *M. soehngenii* is an obligate chemoorganotroph
Organic substrates utilized: Various sugars, organic acids, alcohols, and N-methyl compounds	**Organic substrates utilized:** Methanol, methylamines, or acetate
Coenzymes and carriers: Tetrahydrofolate is C_1-carrier in reduction of CO_2; cytochromes present in first 4 species, cytochromes absent in *A. woodii* and *E. limosum*; B_{12}-proteins involved in fermentation	**Coenzymes and carriers:** C_1-carriers in reduction of CO_2 are methanofuran, tetrahydromethanopterin, and coenzyme M; coenzyme F_{420} is the characteristic H-carrier; cytochromes absent in obligate chemolithotrophs, but present in *M. barkeri* and others; factors F_{430} (a nickel-tetrapyrrol) and B involved in terminal step of fermentation; B_{12}-proteins involved in fermentation

The stepwise reduction of CO_2 to acetate and methane, respectively, proceeds with the C_1-unit attached to certain coenzymes. This coenzyme function is fulfilled in acetogens by a well-known compound: tetrahydrofolate (Ljungdahl and Wood, 1982). Unique coenzymes, however, are involved as carriers in CO_2 reduction by methanogens (Wolfe, 1985; van Beelen et al., 1984): methanofuran, tetrahydromethanopterin, and coenzyme M (Figure 21.1). Additional unique compounds participate in methanogenesis: coenzyme F_{420}, a deazaflavin functioning as a hydrogen carrier, factor B (L-N-7mercaptoheptanolythreonine-phosphate, HS-HTP), and factor F_{430} (a nickel-tetrapyrrol), both involved in the terminal fermentation step, the methyl-coenzyme M methylreductase reaction (Pfaltz et al., 1982; Noll et al., 1986; Ankel-Fuchs et al., 1987). The pathways leading from CO_2 to acetate and methane, respectively, are depicted in Figures 21.2 and 21.3. In acetate synthesis, two molecules of CO_2 are first reduced by two enzyme systems to formate and carbon monoxide, respectively. Formate is then linked to tetrahydrofolate in a reaction associated with ATP hydrolysis. Three subsequent reduction steps lead to 5-methyltetrahydrofolate from which the methyl group is transferred to CO and coenzyme A in a very complex reaction (Wood et al., 1986). The acetyl CoA formed is finally converted to acetate via acetyl phosphate. ATP is formed in this terminal reaction.

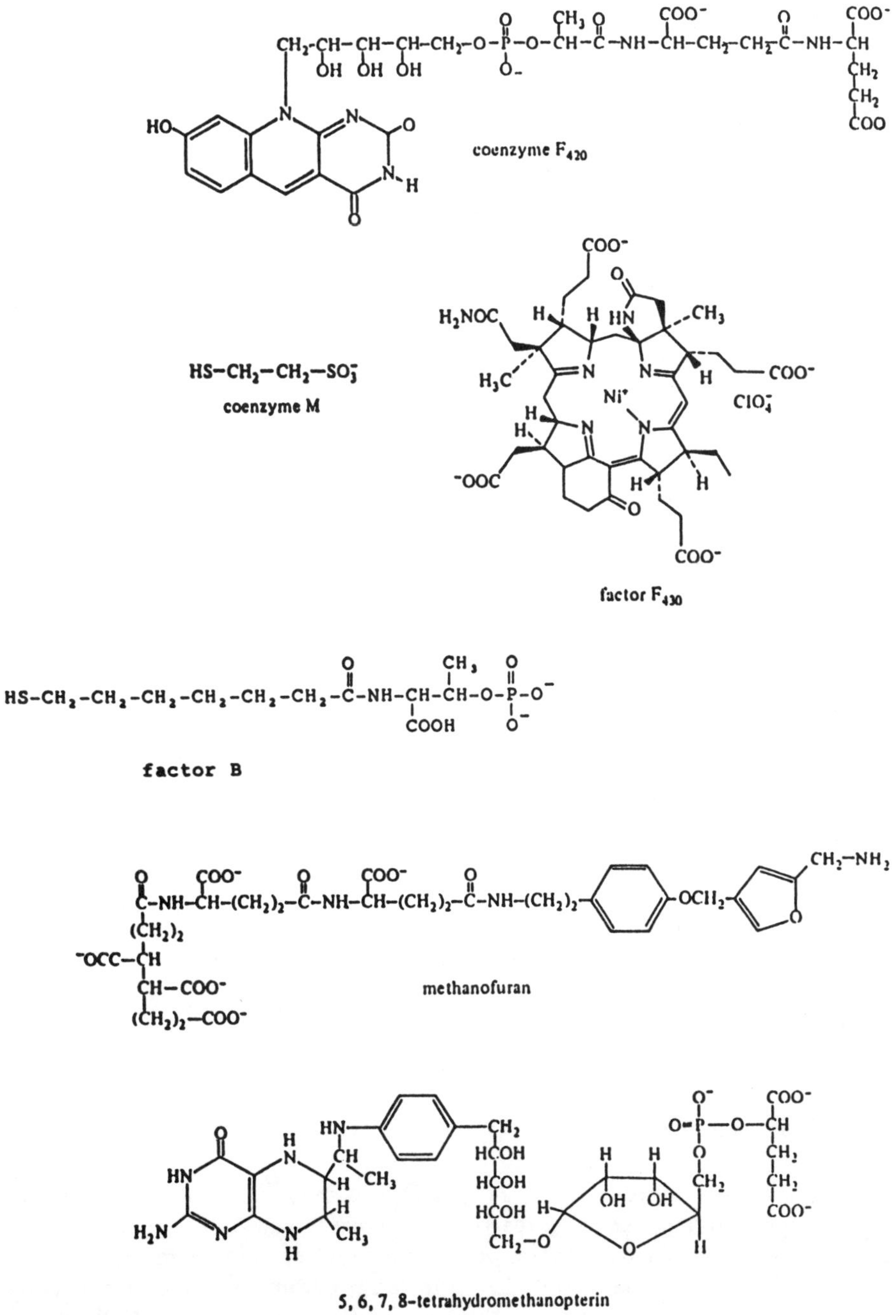

Figure 21.1 Coenzymes and factors involved in methanogenesis.

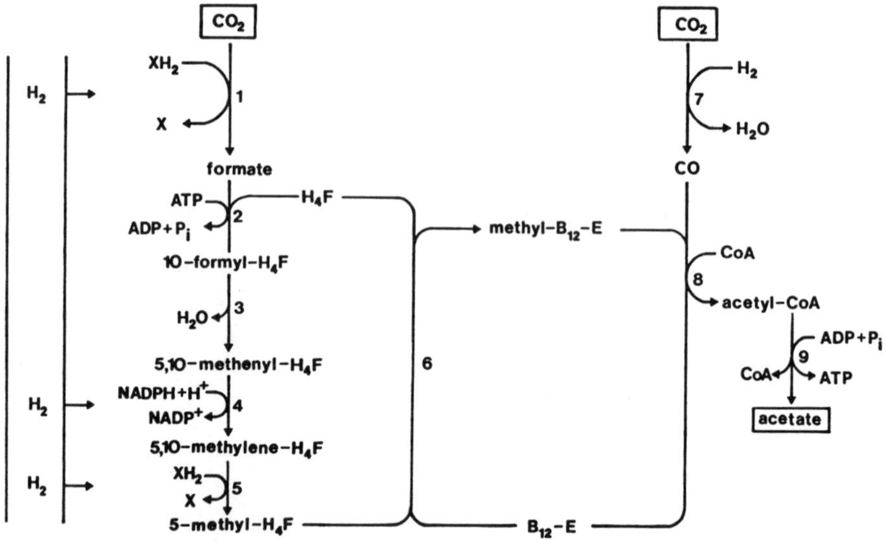

Figure 21.2 Path of acetate synthesis from $H_2 + CO_2$.

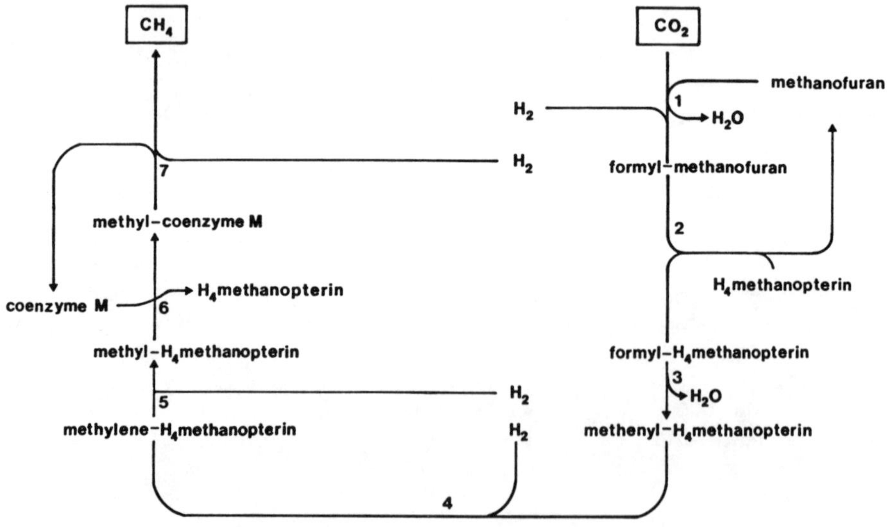

Figure 21.3 Path of methane synthesis from $H_2 + CO_2$.

In methane synthesis (Figure 21.3), CO_2 and methanofuran are converted into formyl-methanofuran in a reaction sequence which, *in vitro* has not yet been resolved into a carboxylation and a reduction reaction. Here, ATP is not required. Transfer of the formyl group to tetrahydromethanopterin and two further reduction steps lead to methyl-tetrahydromethanopterin (Noll et al., 1987). A methyl group transfer gives methyl-coenzyme M (Sauer, 1986), which is the substrate of the terminal reaction in which methane is released.

21.2 ENERGY METABOLISM OF ACETOGENS

Not much attention has been paid to the energy metabolism of acetogenic bacteria. When these organisms are grown chemoorganotrophically, comparatively high growth yields are observed, in the range of 50 g cells/mol glucose (*C. thermoaceticum*) to 68 g cells/mol fructose (*A. woodii*). With $H_2 + CO_2$ as energy source, yields amount to 5 g cells/mol acetate produced, indicating that approximately 0.5 ATP becomes available to the organisms in the conversion of $4H_2 + 2CO_2$ to 1 acetate (Fuchs, 1986).

How is this ATP synthesized? The thermodynamics of the reactions involved in acetate synthesis are summarized in Table 21.2. Following the conversion of CO_2 to formate (step 1), ATP is used in the formyl-tetrahydrofolate synthetase reaction (step 2). This ATP is gained back in the reaction catalyzed by acetate kinase (step 9). There are no further reactions involving ADP or ATP so that a net synthesis of ATP by substrate-level phosphorylation cannot occur. Thus, electron transport phosphorylation might be a source of ATP. A further inspection of the reactions depicted in Table 21.2 shows that three additional reactions are of bioenergetic significance: 1) The CO dehydrogenase reaction (step 6). It is endergonic in the direction $CO_2 \rightarrow CO$, and a requirement for energy has been demonstrated (Diekert et al., 1986); 2) The methylene-H_4 folate dehydrogenase reaction (step 4) is exergonic and could be operated with energy conservation; 3) The methylene-H_4 folate reductase (step 5) has the largest free energy change in the pathway (but see footnote b in the table), and ATP should be generated in this reaction (but nothing is known about the mechanism).

One finding is interesting in this connection. It was reported by Terraciano et al. (1987) that a Na^+/H^+ antiporter is present in *C. thermoaceticum*. Since Na^+ gradients established by such an antiporter have been shown to drive certain thermodynamically unfavorable reactions in methanogenic bacteria (see below) one may speculate that they are similarly involved in acetogens. Such an antiporter could couple one of the exergonic reactions with the endergonic reduction of CO_2 to CO. The involvement of the three mentioned reactions in

Table 21.2 Thermodynamics of acetate synthesis from $H_2 + CO_2$

Step	Reaction			$\Delta G^{0\prime a}$ (kJ/reaction)
1	CO_2 (gas) + H_2	⇌	$HCOO^- + H^+$	+ 3.4
2	$HCOO^-$ + H_4 folate + ATP^{4-}	⇌	N^{10}-formyl-H_4 folate + ADP^{3-} + P_i^{2-}	− 8.4
3	N^{10}-formyl-H_4 folate + H^+	⇌	N^5,N^{10}-methenyl-H_4 folate$^+$ + H_2O	− 4.0
4	N^5,N^{10}-methenyl-H_4 folate$^+$ + H_2	⇌	N^5,N^{10}-methylene-H_4 folate + H^+	−23.0
5	N^5,N^{10}-methylene-H_4 folate + H_2	⇌	N^5-methyl-H_4 folate	−57.3[b]
6	CO_2 (gas) + H_2	⇌	CO (gas) + H_2O	+20.1
7	N^5-methyl-H_4 folate + CO + CoA	⇌	acetyl CoA + H_4 folate	−21.8
8	Acetyl CoA + P_i^{2-}	⇌	acetyl phosphate$^-$ + CoA$^-$	+ 9.0
9	Acetyl phosphate$^-$ + ADP^{3-}	⇌	acetate + ATP^{4-}	−13.0

[a]$\Delta G^{0\prime}$ values taken from Thauer et al. (1977) and Fuchs (1986).
[b]A considerably lower value (−16.6 kJ/reaction) was calculated by Keltjens and van der Drift (1986).

the bioenergetics of the acetogens will be an interesting subject for future research.

21.3 ENERGY METABOLISM OF METHANOGENS

The thermodynamics of the reactions involved in methanogenesis from CO_2 + H_2 are summarized in Table 21.3. It is apparent an energy barrier is involved in this process because the conversion of CO_2, H_2, and methanofuran to formyl-methanofuran is associated with a positive free-energy change. Following a number of reactions with comparatively small free-energy changes, the terminal reactions, notably the last one, have large negative values. These facts make studies on the bioenergetics of methanogenesis from H_2 + CO_2 difficult because metabolic energy in a certain form has to be provided for the initial reaction. Therefore, it is understandable that the addition of uncouplers leads to abolishment of the whole CO_2 reduction process (Sauer et al., 1979). Consequently, results obtained in studies with CO_2 + H_2 have not led to any conclusions as to the mechanism of ATP synthesis in methanogenic bacteria (Daniels et al., 1984).

Some sort of a break-through was achieved when not the whole sequence of reactions was studied, but only the terminal one. This was possible, because *Methanosarcina barkeri* is able to convert methanol + H_2 to methane via methyl-coenzyme M and couples this conversion with ATP synthesis (Blaut and Gottschalk, 1984):

$$CH_3OH + HS\text{-}CoM \rightarrow CH_3\text{-}CoM + H_2O \qquad \Delta G^{0\prime} = 27.5 \text{ kJ/reaction}$$

$$CH_3\text{-}S\text{-}CoM + H_2 \rightarrow CH_4 + HS\text{-}CoM \qquad \Delta G^{0\prime} = -85 \text{ kJ/reaction}$$

It should be mentioned in this connection that the methanogenic organism *Methanosphaera stadtmaniae* uses exclusively methanol + H_2 as substrates (Miller and Wolin, 1985), and that the enzymes involved in the reactions given above have been extensively studied (van der Meijden et al., 1983; Wolfe, 1985).

Table 21.3 Thermodynamics of methane synthesis from H_2 + CO_2

Step	Reaction			$\Delta G^{0\prime a}$ (kJ/reaction)
1	$H_2 + CO_2 + MF$	⇌	formyl-MF + H_2O	+16
2	Formyl-MF + H_4MPT	⇌	formyl-H_4MPT + MF	− 5
3	Formyl-H_4MPT + H^+	⇌	methenyl-H_4MPT^+ + H_2O	− 2
4	Methenyl-H_4MPT^+ + H_2	⇌	methylene-H_4MPT + H^+	− 5
5	Methylene-H_4MPT + H_2	⇌	methyl-H_4MPT	−20[b]
6	Methyl-H_4MPT + HS-CoM	⇌	methyl-SCoM + H_4MPT	−29
7	Methyl-SCoM + H_2	⇌	CH_4 + HS-CoM	−85

[a]$\Delta G^{0\prime}$ values taken from Keltjens and van der Drift (1986).
[b]See analogous reaction in Table 21.2.
MF, methanofuran; H_4MPT, tetrahydromethanopterin

When the bioenergetics of methane formation from methanol + H_2 was studied using cell suspensions of *M. barkeri,* it could be concluded that ATP is synthesized by a chemiosmotic mechanism in the course of this fermentation. This conclusion is derived from the following evidence (Blaut and Gottschalk, 1985): 1)Protonophores such as tetrachlorosalicylanilide (TCS) behave "classically": The transmembrane electrochemical H^+ gradient is dissipated and the intracellular ATP content is decreased; methane formation, however, continues at a slightly increased rate; 2) N,N-dicyclohexylcarbodiimide (DCCD), an inhibitor of the ATP synthase, prevents ATP formation, inhibits methanogenesis, but leaves the electrochemical proton potential intact; the latter, however, is dissipated upon the addition of TCS (Figure 21.4).

These results are only in accordance with a sequence of events in which methanogenesis from methanol + H_2 gives rise to the generation of an electrochemical H^+ potential (proton motive force) which subsequently is taken advantage of by the ATP synthase for ATP formation. If ATP was synthesized by substrate-level phosphorylation it would not be understandable that the ATP level decreased in the presence of DCCD, that methanogenesis was largely inhibited under these conditions, and that the energized state of the membrane remained intact. A F_1F_0 type ATPase was identified in *M. barkeri* (Inatomi, 1986) and in the methanogenic bacterium, strain Gö 1 (Mayer et al., 1987). For the latter organism, results were obtained with methanol + H_2 analogous to those described above.

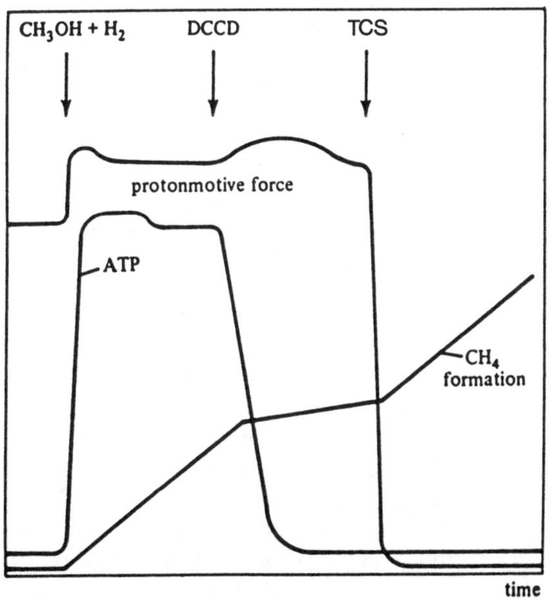

Figure 21.4 Inhibition of methane and ATP synthesis by N,N-dicyclohexylcarbodiimide (DCCD) and inability of DCCD and ability of tetrachlorosalicylanilide (TCS) to dissipate the proton motive force.

It is of interest to measure not only ATP formation as a result of methanogenesis from methanol + H_2, but also to follow the primary event in a chemiosmotic mechanism, that is proton extrusion. However, methanogenic bacteria have a very active Na^+/H^+ antiporter (Schönheit and Beimborn, 1985; Müller et al., 1986), the activity of which would interfere with the accumulation of protons outside the cell. For experimentation, this antiporter has to be switched off; this is possible in the methanol + H_2 system since here sodium is not required for methanogenesis and ATP synthesis (Blaut et al., 1985). Thus, any interference of the antiporter could be prevented by keeping the concentration of sodium in the medium as low as possible and by adding potassium ions which function as a competitive inhibitor of the antiporter. Using cell suspensions of *M. barkeri* incubated under a hydrogen atmosphere it was then possible to record pH changes when methanol pulses were applied. Values between 3–4 protons extruded per 1 methane formed from methanol were measured (Blaut et al., 1987). Proton extrusion was not observed when the protonophore TCS was added or when the cells were preincubated with the inhibitor of methanogenesis, bromoethane-sulfonic acid (BES). This is evidence that the electron transfer from H_2 to methyl-coenzyme M drives the translocation of protons across the cytoplasmic membrane of *M. barkeri*.

With these findings, the terminal enzyme of the methanogenic pathway, the methylreductase system, comes into focus. At least in one part of the catalytic cycle it must be in close contact with the cytoplasmic membrane. Therefore, an immunoelectron microscopic approach was taken to localize one of the components of the methylreductase in cells of methanogenic bacteria. *Methanococcus voltae* was studied with the help of antibodies raised against the 3 types of subunits of component C of the methylreductase system. Applying the protein A-gold technique, a localization of component C in the vicinity of the cytoplasmic membrane could be observed (Ossmer et al., 1986). Rarely, a comparable labeling pattern could also be seen using *Methanobacterium thermoautotrophicum*. In most cases, however, a considerable percentage of the label was also present in the cytoplasm. Using cells grown under conditions under which the synthesis of the methylreductase was limited, further evidence for a membrane-associated methylreductase system in *M. thermoautotrophicum* was obtained (Aldrich et al., 1987). In ultrathin sections of *Methanosarcina barkeri*, however, no indications of a membrane-associated enzyme could be obtained; enzyme molecules were always randomly distributed in the cell (Ossmer, 1986). The situation can be depicted as in Figure 21.5; the methylreductase might be loosely associated with the cytoplasmic membrane or a redox carrier might mediate between soluble and membrane-integrated components of the system. Such a redox carrier could be factor B (mercaptoheptanoylthreonine-phosphate, now designated HS-HTP), which is absolutely required for methylreductase activity. On the basis of its structure (see Figure 21.2) the following catalytic cycle of the methylreductase is apparent (Noll and Wolfe, 1987):

$$CH_3\text{-S-CoM} + \text{HS-HTP} \rightarrow CH_4 + \text{CoM-SS-HTP}$$

$$\text{CoM-SS-HTP} + H_2 \rightarrow \text{HS-CoM} + \text{HS-HTP}$$

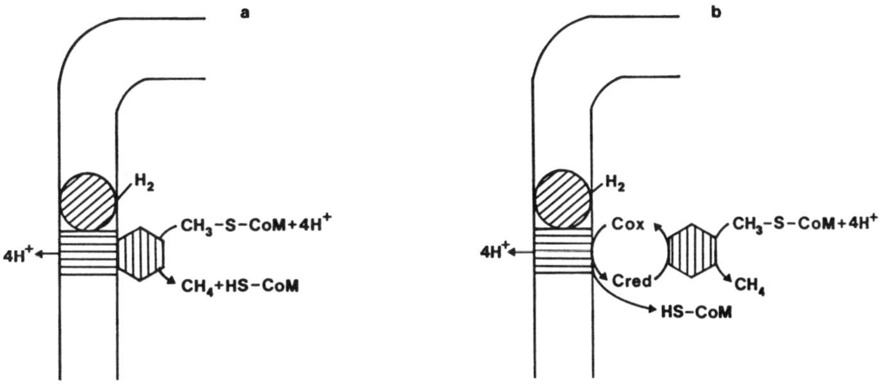

Figure 21.5 Models for the generation of an electrochemical transmembrane gradient. *a*, methylreductase attached to membrane-embedded component; *b*, methylreductase resides in the cytoplasm, and a redox carrier (factor B) mediates between the methylreductase and the membrane-embedded component. The whole entity (methylreductase plus membrane-embedded component) is referred to as the methylreductase system.

The mixed disulfide (CoM-SS-HTP) would then function as the actual electron acceptor at the membrane.

Methyl-coenzyme M is the central intermediate of methane formation from all known methanogenic substrates (Taylor and Wolfe, 1974; Shapiro and Wolfe, 1980; Lovley et al., 1984; Naumann et al., 1984); the composition of component C of the methylreductase system from various methanogenic species is very similar (Hartzell and Wolfe, 1986). Therefore, it is tempting to speculate that the mechanism of ATP synthesis is the same in all methanogens. However, on the basis of results obtained with *Methanococcus voltae* it was concluded that ATP synthesis occurred by an electron transfer-driven ATP synthesis not involving transmembrane ion gradients (Crider et al., 1985). Furthermore, a vanadate-sensitive E_1E_2-ATPase was described in *M. voltae* (Dharmavaram and Konisky, 1987). Future work has to show how these results fit into the picture outlined here for *M. barkeri*.

Acetate is the most difficult substrate for methanogenic bacteria and only a few species are able to utilize it (e.g., *M. barkeri* and *Methanothrix soehngenii*). Results from several laboratories have shown that acetate degradation by methanogens is in principal a reversal of acetogenesis from two C_1 precursors (see Figure 21.2) (Krzycki et al., 1985). Acetate is converted to acetyl coenzyme A which is cleaved into CO and methyl-coenzyme M. Subsequent reactions then yield CH_4 + CO_2 (Figure 21.6). It is apparent that in this reaction ATP is required in stoichiometric amounts for substrate activation. If one ATP is formed in the final step of methanogenesis, net ATP does not become available to the organisms. In this respect it was important that a coupling of CO oxidation to CO_2 plus H_2 with the phosphorylation of ADP could be demonstrated in acetate-grown *M. barkeri* (Bott et al., 1986).

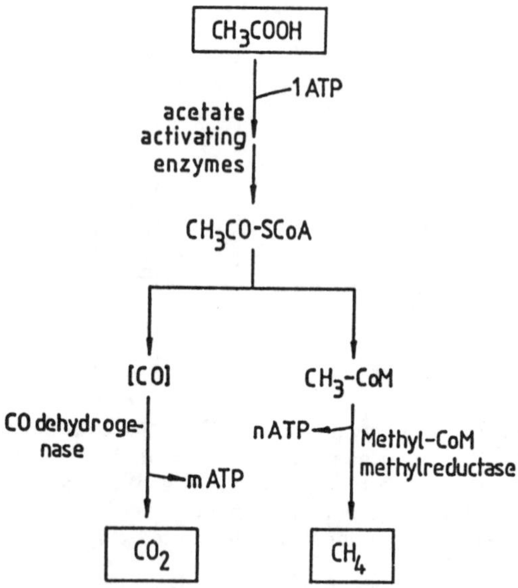

Figure 21.6 Tentative scheme of acetate conversion to methane and carbon dioxide.

21.4 ROLE OF SODIUM IONS IN METHANOGENESIS

Growth of all methanogenic bacteria depends on the presence of sodium ions in the growth medium (Perski et al., 1982). However, when just methanogenesis (not growth) from certain substrates was studied, it turned out that not all processes showed a requirement for Na^+. Methane could be produced from methanol + H_2 or trimethylamine + H_2 by cell suspensions of *M. barkeri* in the absence of Na^+. In addition, ATP could be synthesized under these conditions (Blaut et al., 1985; Müller et al., 1987), excluding a possible involvement of Na^+ in ATP synthesis. However, methane formation from CO_2 + H_2, methanol, trimethylamine, or acetate has an absolute requirement for Na^+. What Na^+ is doing in these fermentations may be deduced from the role it plays during methanogenesis from methanol. As depicted in Figure 21.7, the proton gradient generated in the methanogenic fermentation is not only used for ATP synthesis but also for the generation of a sodium gradient across the cytoplasmic membrane (Müller et al., 1986). This sodium gradient is then used to drive the oxidation of methanol to the level of formaldehyde, a reaction which is endergonic (Thauer et al., 1977):

$$CH_3OH \rightarrow CH_2O + H_2 \qquad \Delta G^{0\prime} = -44.8 \text{ kJ/reaction}$$

Methanol oxidation in *M. barkeri* is associated with Na^+ influx (Müller et al., 1987, 1988). However, nothing is known about the nature of the coupling mechanism. Similarly, Na^+ influx may provide the energy for overcoming energy barriers in methanogenesis from CO_2 + H_2 or acetate.

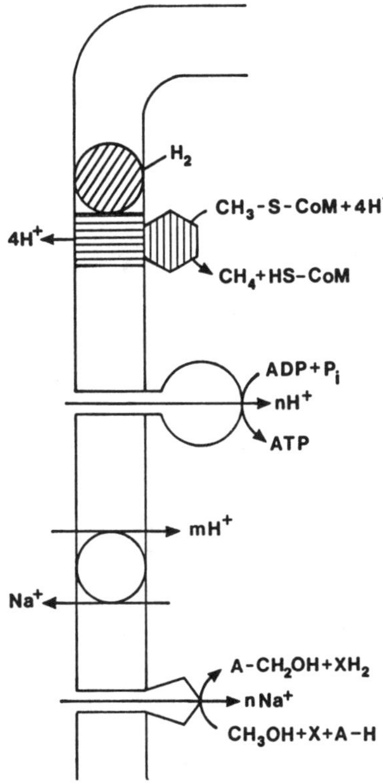

Figure 21.7 Generation and use of transmembrane H^+ and Na^+ gradients during methanogenesis from methanol by *M. barkeri*. AH stands for the C_1-carrier (H_4MPT?) in methanol oxidation.

Membrane-dependent redox reactions apparently play an important role in the energy metabolism of methanogenic bacteria. In this respect it is important to mention that the presence of membrane-bound iron-sulfur centers (Lancaster, 1980), cobamides (Dangel et al., 1987), and—in the methylotrophic methanogens—cytochromes (Kühn et al., 1983) has been demonstrated. How these carriers are involved in redox reactions is unknown.

The author wants to thank his co-workers for their excellent contributions to the results which came out of his laboratory. Thanks are also due to the Deutsche Forschungsgemeinschaft for financial support.

REFERENCES

Aldrich, H.C., Beimborn, D.B., Bokranz, M., and **Schönheit, P.** 1987. Immunocytochemical localization of methylcoenzyme M reductase in *Methanobacterium thermoautotrophicum*. *Archives of Microbiology* 147:190–194.

Ankel-Fuchs, D., Böcher, R., Thauer, R.K., Noll, K.M., and **Wolfe, R.S.** 1987. 7-mercaptoheptanoylthreonine phosphate functions as component B in ATP-independent methane formation from methyl-CoM with reduced cobalamin as electron donor. *FEBS Letters* 213:123–127.

Balch, W.E., Fox, G.E., Magrum, L.J., Woese, C.R., and **Wolfe, R.S.** 1979. Methanogens: reevaluation of a unique biological group. *Microbiological Reviews* 43:260–296.

Blaut, M. and **Gottschalk, G.** 1984. Coupling of ATP synthesis and methane formation from methanol and molecular hydrogen in *Methanosarcina barkeri. European Journal of Biochemistry* 141:217–222.

Blaut, M. and **Gottschalk, G.** 1985. Evidence for a chemiosmotic mechanism of ATP synthesis in methanogenic bacteria. *Trends in Biochemical Sciences* 10:486–489.

Blaut, M., Müller, V., Fiebig, K., and **Gottschalk, G.** 1985. Sodium ions and an energized membrane required by *Methanosarcina barkeri* for the oxidation of methanol to the level of formaldehyde. *Journal of Bacteriology* 164:95–101.

Blaut, M., Müller, V., and **Gottschalk, G.** 1987. Proton translocation coupled to methanogenesis from methanol + hydrogen in *Methanosarcina barkeri. FEBS Letters* 215:53–57.

Bott, M., Eikmanns, B., and **Thauer, R.K.** 1986. Coupling of carbon monoxide oxidation to CO_2 and H_2 with the phosphorylation of ADP in acetate-grown *Methanosarcina barkeri. European Journal of Biochemistry* 159:393–398.

Crider, B.P., Carper, S.W., and **Lancaster, J.R.** 1985. Electron transfer-driven ATP synthesis in *Methanococcus voltae* is not dependent on a proton electrochemical gradient. *Proceedings of the National Academy of Sciences, USA* 82:6793–6796.

Dangel, W., Schulz, H., Diekert, G., König, H., and **Fuchs, G.** 1987. Occurrence of corrinoid-containing membrane proteins in anaerobic bacteria. *Archives of Microbiology* 148:52–56.

Daniels, L., Sparling, R., and **Sprott, G.D.** 1984. The bioenergetics of methanogenesis. *Biochimica et Biophysica Acta* 768:113–163.

Dharmavaram, R.M. and **Konisky, J.** 1987. Identification of a vanadate-sensitive, membrane-bound ATPase in the archaebacterium *Methanococcus voltae. Journal of Bacteriology* 169:3921–3925.

Diekert, G., Schrader, E., and **Harder, W.** 1986. Energetics of CO formation and CO oxidation in cell suspensions of *Acetobacterium woodii. Archives of Microbiology* 144:386–392.

Fuchs, G. 1986. CO_2 fixation in acetogenic bacteria: variations on a theme. *FEMS Microbiology Reviews* 39:181–213.

Hartzell, P.L. and **Wolfe, R.S.** 1986. Comparative studies of component C from the methylreductase system of different methanogens. *Systematic and Applied Microbiology* 7:376–382.

Hippe, H., Caspari, D., Fiebig, K., and **Gottschalk, G.** 1979. Utilization of trimethylamine and other N-methyl compounds for growth and methane formation by *Methanosarcina barkeri. Proceedings of the National Academy of Sciences, USA* 76:494–498.

Inatomi, K.-J. 1986. Characterization and purification of the membrane-bound ATPase of the archaebacterium *Methanosarcina barkeri. Journal of Bacteriology* 167:837–841.

Jones, W.J., Nagle Jr., D.P., and **Whitman, W.B.** 1987. Methanogens and the diversity of archaebacteria. *Microbiological Reviews* 51:135–177.

Keltjens, J.T. and **van der Drift, C.** 1986. Electron transfer reactions in methanogens. *FEMS Microbiology Reviews* 39:259–303.

Krzycki, J.A., Lehmann, L.J., and **Zeikus, J.G.** 1985. Acetate catabolism by *Methanosarcina barkeri*: evidence for involvement of carbon monoxide dehydrogenase, methyl-coenzyme M and methylreductase. *Journal of Bacteriology* 163:1000–1006.

Kühn, W., Fiebig, K., Hippe, H., Mah, R.A., Huser, B.A., and **Gottschalk, G.** 1983. Distribution of cytochromes in methanogenic bacteria. *FEMS Microbiology Letters* 20:407–410.

Lancaster Jr., J.R. 1980. Soluble and membrane-bound paramagnetic centers in *Methanobacterium bryantii. FEBS Letters* 115:285–288.

Lauerer, G., Kristjansson, J.K., Langworthy, T.A., König, H., and **Stetter, K.O.** 1986. *Methanothermus sociabilis* sp. nov., a second species within the Methanothermaceae growing at 97°C. *Systematic and Applied Microbiology* 8:100–105.

Ljungdahl, L.G. and **Wood, H.G.** 1982. Acetate biosynthesis, pp. 165–202 in Dolphin, D. (editor), *Vitamin B_{12}*, Wiley, New York.

Lovley, D.R., White, R.H., and **Ferry, J.G.** 1984. Identification of methyl coenzyme M as an intermediate in methanogenesis from acetate in *Methanosarcina* spp. *Journal of Bacteriology* 160:521–525.

Mah, R.A., Smith, M.R., Ferguson, T., and **Zinder, S.** 1981. Methanogenesis from H_2–CO_2, methanol, and acetate by *Methanosarcina*, pp. 131–145, in Dalton, H. (editor), *Microbiological Growth on C_1 Compounds*, Heyden, London.

Mayer, F., Jussofie, A., Salzmann, M., Lübben, M., Rohde, M., and **Gottschalk, G.** 1987. Immunoelectron microscopic demonstration of ATPase on the cytoplasmic membrane of the methanogenic bacterium strain Gö 1. *Journal of Bacteriology* 169:2307–2309.

Miller, T.L., and **Wolin, M.J.** 1985. *Methanosphaera stadtmaniae* gen. nov., sp. nov.: a species that forms methane by reducing methanol with hydrogen. *Archives of Microbiology* 141:116–122.

Möller, B., Ossmer, R., Howard, B.H., Gottschalk, G., and **Hippe, H.** 1984. *Sporomusa*, a new genus of Gram-negative anaerobic bacteria including *Sporomusa sphaeroides* spec. nov. and *Sporomusa ovata* spec. nov. *Archives of Microbiology* 139:388–396.

Müller, V., Blaut, M., and **Gottschalk, G.** 1986. Generation of a transmembrane gradient of Na^+ in *Methanosarcina barkeri. European Journal of Biochemistry* 162:461–466.

Müller, V., Blaut, M., and **Gottschalk, G.** 1988. The transmembrane electrochemical gradient of Na^+ as driving force for methanol oxidation in *Methanosarcina barkeri. European Journal of Biochemistry* 172:601–606.

Müller, V., Kozianowski, G., Blaut, M., and **Gottschalk, G.** 1987. Methanogenesis from trimethylamine + H_2 by *Methanosarcina barkeri* is coupled to ATP formation by a chemiosmotic mechanism. *Biochimica et Biophysica Acta* 892:207–212.

Naumann, E., Fahlbusch, K., and **Gottschalk, G.** 1984. Presence of a trimethylamine: HS-coenzyme M methyltransferase in *Methanosarcina barkeri. Archives of Microbiology* 138:79–83.

Noll, K.M., Rinehart Jr., K.L., Tanner, R.S., and **Wolfe, R.S.** 1986. Structure of component B (7-mercaptoheptanoylthreonine phosphate) of the methylcoenzyme M methylreductase system of *Methanobacterium thermoautotrophicum. Proceedings of the National Academy of Sciences, USA* 83:4238–4242.

Noll, K.M., Donnelly, M.J., and **Wolfe, R.S.** 1987. Biochemical aspects of methane formation in *Methanobacterium thermoautotrophicum*, pp. 55–61, in van Verseveld, H.W. and Duine, J.A. (editors), *Microbial Growth on C_1 Compounds*, Martinus Nijhoff Publishers, Dordrecht.

Noll, K.M. and **Wolfe, R.S.** 1987. The role of 7-mercaptoheptanoylthreonine phosphate in the methylcoenzyme M methylreductase system from *Methanobacterium thermoautotrophicum. Biochemical and Biophysical Research Communications* 145:204–210.

Ossmer, R. 1986. Lokalisierung der Komponente C des Methylreduktase-Systems in *Methanosarcina barkeri* und in *Methanococcus voltae*. PhD Thesis, Göttingen.

Ossmer, R., Mund, T., Hartzell, P.L., Konheiser, U., Kohring, G.W., Klein, A., Wolfe, R.S., Gottschalk, G., and **Mayer, F.** 1986. Immunocytochemical localization of component C of the methylreductase system in *Methanococcus voltae* and *Methanobacterium autotrophicum. Proceedings of the National Academy of Sciences, USA* 83:5789–5792.

Perski, H.J., Schönheit, P., and **Thauer, R.K.** 1982. Sodium dependence of methane formation in methanogenic bacteria. *FEBS Letters* 143:323–326.

Pfaltz, A., Jaun, B., Fässler, A., Eschenmoser, A., Jaenchen, R., Gilles, H.-H., Diekert, G., and **Thauer, R.K.** 1982. Zur Kenntnis des Faktors F_{430} aus methanogenen Bakterien: Struktur des porphinoiden Ligandsystems. *Helvetica Chimica Acta* 65:828–865.

Sauer, F.D. 1986. Tetrahydromethanopterin, a component of the methane synthesizing complex of *Methanobacterium thermoautotrophicum. Biochemical and Biophysical Research Communications* 136:542–547.

Sauer, F.D., Erfle, J.D., and **Mahadevan, S.** 1979. Methane synthesis without the addition of adenosine triphosphate by cell membranes isolated from *Methanobacterium ruminantium. Biochemical Journal* 178:165–172.

Schönheit, P. and **Beimborn, D.B.** 1985. Presence of a Na^+/H^+ antiporter in *Methanobacterium thermoautotrophicum* and its role in Na^+ dependent methanogenesis. *Archives of Microbiology* 142:354–361.

Shapiro, S. and **Wolfe, R.S.** 1980. Methyl-coenzyme M, an intermediate in methanogenic dissimilation of C_1 compounds by *Methanosarcina barkeri. Journal of Bacteriology* 141:728–734.

Taylor, C.D. and **Wolfe, R.S.** 1974. Structure and methylation of coenzyme M. *Journal of Biological Chemistry* 249:4879–4885.

Terracciano, J.S., Schreurs, W.J.A., and **Kashket, E.R.** 1987. Membrane H^+ conductance of *Clostridium thermoaceticum* and *Clostridium acetobutylicum*: evidence for electrogenic Na^+/H^+ antiport in *Clostridium thermoaceticum. Applied and Environmental Microbiology* 53:782–786.

Thauer, R.K., Jungermann, K., and **Decker, K.** 1977. Energy conservation in chemolithotrophic anaerobic bacteria. *Bacteriological Reviews* 41:100–180.

Van Beelen, P., Stassen, A.P.M., Bosch, J.W.G., Vogels, G.D., Gruijt, W., and **Hassnoot, C.S.G.** 1984. Elucidation of the structure of methanopterin, a coenzyme from *Methanobacterium thermoautotrophicum,* using two-dimensional nuclear-magnetic-resonance techniques. *European Journal of Biochemistry* 138:563–571.

Van der Meijden, P., Heythuysen, H.J., Pouwels, A., Houwen, F., van der Drift, C., and **Vogels, G.D.** 1983. Methyltransferases involved in methanol conversion by *Methanosarcina barkeri. Archives of Microbiology* 134:238–242.

Wood, H.G., Ragsdale, S.W., and **Pezaka, E.** 1986. The acetyl-CoA pathway of autotrophic growth. *FEMS Microbiology Reviews* 39:345–362.

Wolfe, R.S. 1985. Unusual coenzymes of methanogenesis. *Trends in Biochemical Sciences* 10:396–399.

22

Energy Metabolism of Sulfate-reducing Bacteria

Rudolf K. Thauer

Department of Microbiology
University of Marburg
Marburg, Federal Republic of Germany

The capacity to use sulfate as electron acceptor for catabolic oxidation processes is phylogenetically widespread among anaerobic bacteria (Fowler et al., 1986) and has probably evolved early (Ohmoto and Felder, 1987). It is found in Gram-negative flagellated and nonflagellated eubacteria (*Desulfovibrio, Desulfobacter, Desulfobacterium, Desulfobulbus, Desulfococcus, Desulfosarcina, Desulfomonas, Thermosulfobacterium*), in gliding Gram-negative eubacteria (*Desulfonema*), in Gram-positive, endospore-forming eubacteria (*Desulfotomaculum*), and in a group of archaebacteria (*Archaeoglobus*) (Pfennig et al., 1981; Widdel and Pfennig, 1984; Postgate, 1984a, b; Widdel, 1988; Achenbach-Richter et al., 1987; Stetter et al., 1987).

Sulfate-reducing bacteria typically use H_2, lactate, and/or acetate (acetyl CoA) as electron donors. Some sulfate-reducing bacteria can also live at the expense of sulfite or thiosulfate disproportionation to sulfate and H_2S. Others can grow on aromatic compounds or alkanes and sulfate as sole energy source. With each of these substrates, the organisms have to cope with different energetic problems which are imposed on them by the thermodynamics of dissimilatory sulfate reduction.

The following discussion is restricted to energetic aspects of dissimilatory sulfate reduction and should be viewed as being exemplary and not exhaustive. The supporting bibliography considers mainly work published after the review article on the subject by Odom and Peck in 1984. For a comprehensive coverage of the attendant literature, the reader is referred to the book on sulfate-reducing bacteria by Postgate (1984a) and to the review on the biology and ecology of sulfate-reducing bacteria by Widdel (1988).

Thermodynamic data, unless otherwise noted, have been taken from Thauer et al. (1977). For a treatise of the mechanisms of energy conservation in anaerobic bacteria allowing fractional stoichiometries of ATP formation, see Thauer and Morris (1984) and Kröger et al. (1986).

22.1 CHEMOLITHOTROPHIC GROWTH ON H_2 AND SULFATE

Most sulfate-reducing bacteria can grow on H_2 and sulfate as sole energy source (see reaction 1) below (Sorokin, 1966a–c; Badziong et al., 1978; Brandis and Thauer, 1981; Widdel, 1980, 1988). Exceptions are *Desulfobacter postgatei, D. latus, Desulfovibrio baarsii, D. sapovorans, Desulfotomaculum acetoxidans, Desulfococcus multivorans, Desulfonema magnum, Desulfobacterium indolicum,* and *D. phenolicum* (Pfennig et al., 1981; Widdel and Pfennig, 1984; Postgate 1984a, b; Widdel, 1988).

Growth on H_2 and sulfate indicates that sulfate reduction to H_2S is coupled with a net synthesis of ATP from ADP and inorganic phosphate (P_i) and that ATP synthesis is by the mechanism of electron-transport-linked phosphorylation. Substrate-level phosphorylation, which is associated with the oxidation of substrates, is ruled out by the fact that H_2 is the electron donor (Thauer et al., 1977).

$$SO_4^{2-} + 4H_2 + 2H^+ \rightarrow H_2S + 4H_2O \qquad \Delta G^{0\prime} = -158 \text{ kJ/mol} \qquad (1)$$

For this reaction, $\Delta G^{0\prime}$ was calculated from Gibbs free energies of formation; H_2S in the gaseous state at 1 atmosphere pressure; $[SO_4^{2-}] = 1$ M; pH = 7.* At the H_2 concentrations prevailing in the natural habitats of these bacteria ($H_2 \approx 10$ Pa) (see Conrad et al., 1986) the free energy change is only -67 kJ/mol. The phosphorylation potential ($\Delta G'$) of the ATP/ADP + P_i system in bacteria is between -40 and -50 kJ/mol (Thauer et al., 1977). Since the thermodynamic efficiency of energy conversion is never 100 percent, it can be predicted that for each mole of sulfate reduced to H_2S with H_2, approximately 1 mol of ATP can be formed. Growth yields on H_2 and sulfate are consistent with these thermodynamic considerations (Nethe-Jaenchen and Thauer, 1984).

Sulfate transport The reactions leading from sulfate to H_2S (reaction 1) are catalyzed by enzymes (ATP sulfurylase, APS reductase, sulfite reductase) all located in the cytoplasm. Therefore, the first step in dissimilatory sulfate reduction must be the transport of the sulfate dianion into the cell through the cytoplasmic membrane against the electrical membrane potential which is inside negative (Figure 22.1). The available evidence (Cypionka, 1987, 1988) indicates that sulfate ($pK_2 = 1.96$) is actively taken up by the cells (sulfate can accumulate over 100-fold in the cells) and that active transport proceeds by electrogenic symport of SO_4^{2-} with 3 protons (a ΔpH is not essentially required for accumulation). Of the three protons, only one has to be actively translocated back out again since H_2S, the end product of sulfate reduction, is a weak acid ($pK_1 = 6.9$) which can leave the cell simply by diffusion. If in sulfate-reducing bacteria the $\rightarrow H^+$/ATP stoichiometry of the H^+-translocating ATP synthase/ATPase is 3 as it is in other organisms (see Thauer and Morris, 1984), then

*$\Delta G' = \Delta G^{0\prime} + RT\ 2.3 \log [H_2S]/[SO_4^{2-}] \times [H_2]^4$; R = 8.3143 J mol^{-1} $^0K^{-1}$; T = 298^{0}K.

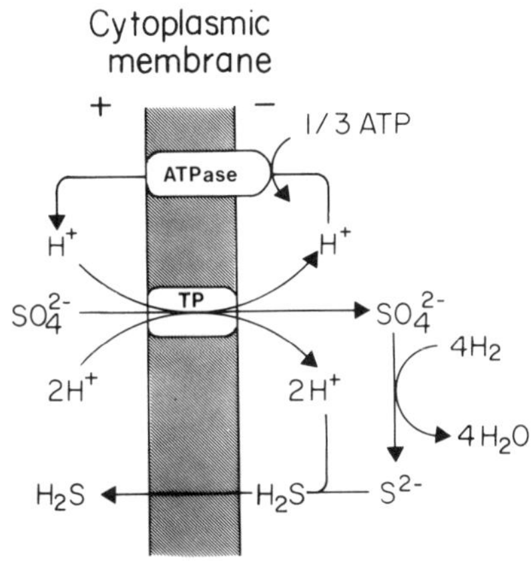

Figure 22.1 Sulfate transport via electrogenic symport with $3H^+$ in *Desulfovibrio desulfuricans* (proposed mechanism). In the scheme, ATP hydrolysis via a H^+-translocating ATPase drives the uptake of sulfate. Per mol of sulfate transported into the cells, 1/3 mol ATP is required. During dissimilatory sulfate reduction, however, the build up of the electrochemical proton potential is coupled to the reduction of SO_3^{2-} to H_2S rather than to the hydrolysis of ATP (see Figure 22.3). TP, transport protein.

the transport of sulfate into the cells requires an energy equivalent of 1/3 mol ATP (Figure 22.1).

The SO_4^{2-} transport system of sulfate-reducing bacteria is highly regulated. Deenergized cells are impermeable to SO_4^{2-} (Littlewood and Postgate, 1957; Varma et al., 1983) and transport is feedback inhibited by the intracellular sulfate concentration. The apparent K_s for sulfate in chemostat cultures is $<$ 10 μM (Nethe-Jaenchen and Thauer, 1984; Ingvorsen and Jørgensen, 1984; Ingvorsen et al., 1984). Ecologically this stringent control makes sense. In the natural habitats of these organisms (anaerobic sediments), the SO_4^{2-} concentration ranges from 28 mM (concentration in aerobic sea water) to $<$10 μM (concentration in the methanogenic zone). Symport of sulfate with protons into the cells would lead to an acidification of the cytoplasm if at high extracellular sulfate concentration, transport was not inhibited.

Sulfate reduction to sulfite Transport of sulfate into the cells is followed by its reduction to sulfite ($pK_2 = 6.8$) (reaction 2).

$$SO_4^{2-} + H_2 \rightarrow SO_3^{2-} + H_2O \qquad \Delta G^{0\prime} = +21 \text{ kJ/mol} \qquad (2)$$

The reduction of SO_4^{2-} to SO_3^{2-} ($E_0' = -520$ mV) with H_2 ($E_0' = -420$ mV) is an endergonic reaction which is driven by the exergonic hydrolysis of ATP. Coupling is achieved by the intermediary formation of adenosine phosphosulfate (APS) (Odom and Peck, 1984). Sulfate reacts with ATP to form APS and inorganic pyrophosphate (PP_i) (reaction 3) catalyzed by ATP sulfurylase. APS is then reduced to SO_3^{2-} and AMP (reaction 4) catalyzed by APS reductase (Stille and Trüper, 1984). Sulfate-reducing bacteria contain high specific activities of inorganic pyrophosphatase and adenlylate kinase. By their catalytic action the pyrophosphate formed in reaction 3 is hydrolyzed to 2 P_i (reaction 5) and the AMP formed in reaction (4) converted back to ADP (reaction 6).

$$SO_4^{2-} + ATP \rightarrow APS + PP_i \qquad \Delta G^{0\prime} = +46 \text{ kJ/mol/} \quad (3)$$

$$APS + 2[H] \rightarrow SO_3^{2-} + 2H^+ + AMP \qquad \Delta G^{0\prime} = -66.7 \text{ kJ/mol} \quad (4)$$

$$PP_i + H_2O \rightarrow 2P_i \qquad \Delta G^{0\prime} = -21.9 \text{ kJ/mol} \quad (5)$$

$$AMP + ATP \rightarrow 2ADP \qquad \Delta G^{0\prime} = 0 \text{ kJ/mol} \quad (6)$$

Reactions (3) to (6) (which can be combined to yield reaction 7) indicate that the reduction of sulfate to sulfite requires the input of 2 mol ATP.

$$SO_4^{2-} + 2ATP + H_2 \rightarrow SO_3^{2-} + 2H^+ + 2ADP + 2P_i \qquad \Delta G^{0\prime} = -42.6 \text{ kJ/mol} \quad (7)$$

There are arguments and findings, however, that the net ATP requirement could be somewhat lower: 1) Even at low H_2 concentration (H_2 = 10 Pa) reaction 7 is still exergonic by −19.8 kJ/mol. This energy amount would be "wasted" if 2 mol of ATP were required; 2) in all living cells several biosynthetic reactions exist which are driven by the hydrolysis of PP_i and not by ATP (Reeves, 1976; Kulaev, 1979), which indirectly reduces the amount of ATP consumed in reaction (7); 3) in *Desulfotomaculum nigrificans* an enzyme has been found which catalyzes the phosphorylation of acetate with PP_i yielding acetyl phosphate which, via acetate kinase, could generate ATP (Liu et al., 1982; Peck and LeGall, 1982); 4) reduction of APS ($E_0{}'$ = −70 mV) with H_2 ($E_0{}'$ = −420 mV) could be coupled with ATP synthesis via electron-transport-linked phosphorylation.

Growth yields of *Desulfovibrio vulgaris* on sulfate plus H_2 [$Y_{max}(SO_4^{2-})$ = 12.7 g/mol] and on SO_3^{2-} plus H_2 [$Y_{max}(SO_3^{2-})$ = 35.3 g/mol] indicate that per mol SO_3^{2-}, 2.8 times more ATP is formed than per mol SO_4^{2-} reduced to H_2S (Nethe-Jaenchen and Thauer, 1984). This finding suggests that per mol SO_4^{2-} reduced to SO_3^{2-}, 1.8 mol ATP is consumed (although this is an overinterpretation of growth yield data). Growth yields of *Desulfotomaculum nigrificans* on lactate plus sulfate and on lactate plus sulfite have been interpreted to indicate a 1:1 stoichiometry. The interpretation was based on the assumption that in this organism sulfite reduction to H_2S is not coupled with ATP synthesis via electron-transport-linked phosphorylation. In the meantime it has been determined, however, that *Desulfotomaculum* spp. can grow on H_2 and sulfate as sole energy source, clearly demonstrating that the latter assumption was wrong (Klemps et al., 1985; Cypionka and Pfennig, 1986).

Sulfite reduction to H_2S Reduction to APS to SO_3^{2-} is followed by the reduction of SO_3^{2-} to H_2S (E_0 = −120 mV), an exergonic reaction coupled with electron-transport-linked phosphorylation, as evidenced by growth of sulfate-reducing bacteria on H_2 and SO_3^{2-} (reaction 8).

$$SO_3^{2-} + 3H_2 + 2H^+ \rightarrow H_2S + 3H_2O \qquad \Delta G^{0\prime} = -179 \text{ kJ/mol} \quad (8)$$

The free energy change is sufficient to drive the synthesis of 2 to 3 mol ATP. Growth yields on SO_3^{2-} plus H_2 indicate that this is indeed the amount of ATP probably generated (Nethe-Jaenchen and Thauer, 1984). The mechanism of electron-transport-linked phosphorylation allows such fractional stoichiometries of ATP formation (Thauer and Morris, 1984).

The reduction of sulfite with H_2 in *Desulfovibrio* spp. (the only organism investigated in detail) involves a periplasmic hydrogenase, a cytoplasmic sulfite reductase, and several electron carriers: menaquinone (Collins and Widdel, 1986), cytochromes, ferredoxin, rubredoxin, and flavodoxin (Odom and Peck, 1984; LeGall and Peck, 1987; LeGall and Fauque, 1988). The function of most of these carriers has not yet been ascertained and the physiological electron donor of sulfite reductase is still unknown (this is also true for APS reductase). One of the major problems encountered in studying the enzymology of sulfite reduction with H_2 is that, upon breakage of the cells, the specific activity drops to a few percent of the *in vivo* value (Thauer and Badziong, 1980).

It has been disputed whether sulfite reduction to H_2S involves thiosulfate and trithionate as free intermediates. This possibility, although presently no longer favored, has not yet been completely ruled out (see e.g., Kim and Akagi, 1985).

The energy requiring and yielding steps in dissimilatory sulfate reduction with H_2 are summarized in Figure 22.2.

Mechanism of energy conservation Protonophores, which dissipate the electrochemical proton potential, and N, N'dicyclohexylcarbodiimide (DCCD), which blocks the proton translocating ATPase, inhibit sulfate reduction but not sulfite reduction with H_2 in intact cells of sulfate-reducing bacteria (Odom and Peck, 1984). This finding indicates that the ATP required to drive reaction (2) (sulfate reduction to sulfite) is generated in reaction (8) (sulfite reduction to H_2S) by a chemiosmotic mechanism. Attempts to prove this mechanism have, until now, failed due to problems in measuring the electrochemical proton

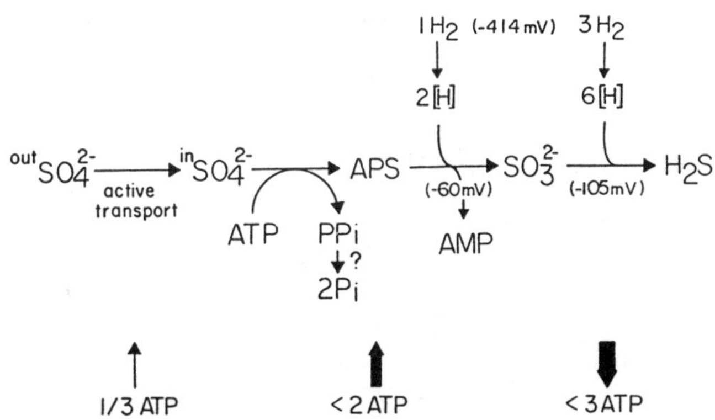

Figure 22.2 ATP-consuming and ATP-generating steps in dissimilatory sulfate reduction with H_2 (for explanations see text).

potential (very high unspecific binding to sulfate-reducing bacteria of the probes tetraphenylphosphonium bromide, benzoic acid, and methylamine used for $\Delta\psi$ and for ΔpH determination).

A study of the mechanism of coupling has also been hampered by the fact that until now it has not been possible to prepare from sulfate-reducing bacteria inside-out-vesicles capable of coupling the reduction of sulfite to H_2S with the phosphorylation of ADP; always both the soluble and the membrane fraction were required for activity (Peck, 1966; Herr, 1982; Paulsen, 1982).

For *Desulfovibrio*, a mechanism has been proposed which is based on the finding that bacteria of this genus contain a periplasmic hydrogenase and a cytoplasmic sulfite reductase, the latter enzyme probably being a peripheral membrane protein located at the inner aspect of the cytoplasmic membrane (Kremer et al., 1988) (Figure 22.3). Based on this topology it is assumed that the electron transport chain from H_2 to sulfite begins at the periplasmic site of the membrane and ends at the cytoplasmic site and is constructed such that only the electrons from H_2 are transported across the membrane, leaving behind the protons (Badziong and Thauer, 1980). By this mechanism of charge separation (and maybe by additional proton pumping) (Kobayashi et al., 1982), the free energy change associated with the reduction of sulfite is conserved in an electrochemical proton potential, which can be used to drive the phosphorylation of ADP via the proton-translocating ATP synthase present in the cytoplasmic membrane of these bacteria.

In *Desulfotomaculum orientis* the hydrogenase is cytoplasmic rather than periplasmic (Cypionka and Dilling, 1986), indicating that the proposed mechanism (Figure 22.3) cannot be generalized for Gram-positive sulfate-reducing bacteria.

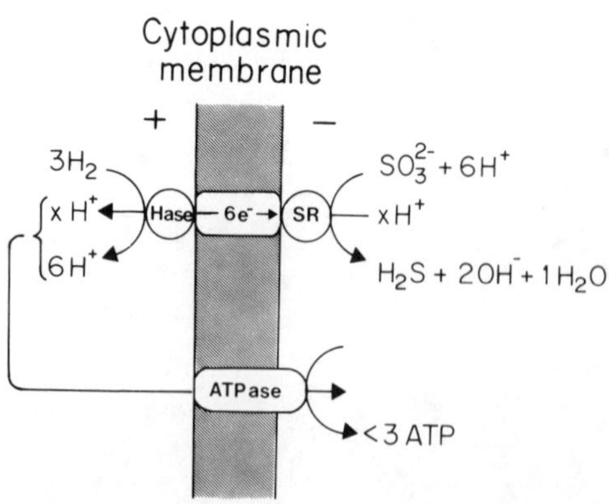

Figure 22.3 Coupling of dissimilatory sulfite reduction with the phosphorylation of ADP in *Desulfovibrio*, using H_2 as electron donor. The proposed mechanism is based on the topology of the electron transport components involved (see text). SR, sulfite reductase.

22.2 GROWTH ON LACTATE PLUS SULFATE

Many sulfate-reducing bacteria can grow on lactate and sulfate. Exceptions are most *Desulfobacter* species, *Desulfovibrio baarsii, Desulfotomaculum acetoxidans, Desulfococcus niacini, Desulfobacterium indolicum, D. phenolicum,* and *Desulfonema magnum.* Most *Desulfovibrio* species, all *Desulfobulbus* species, *Desulfotomaculum nigrificans, D. orientis, D. ruminis, Desulfomonas pigra,* and *Thermosulfobacterium commune* oxidize lactate only to the level of acetate and CO_2 (reaction 9), whereas most *Desulfobacterium* species, *Desulfococcus multivorans, Desulfosarcina variabilis,* and *Desulfonema limicola* are capable of complete oxidation of lactate to 3 CO_2 (reaction 10) (Widdel, 1980; Pfennig et al., 1981; Postgate, 1984a, b; Widdel and Pfennig, 1984; Widdel, 1988).

$$2\ \text{lactate}^- + SO_4^{2-} + 2H^+ \rightarrow 2\ \text{acetate}^- + 2CO_2 + H_2S + 2H_2O \qquad \Delta G^{0\prime} = -175.6\ \text{kJ/mol}\ H_2S \quad (9)$$

$$2\ \text{lactate}^- + 3SO_4^{2-} + 8H^+ \rightarrow 6CO_2 + 3H_2S + 6H_2O \qquad \Delta G^{0\prime} = -100.5\ \text{kJ/mol}\ H_2S \quad (10)$$

Lactate oxidation proceeds via pyruvate and acetyl CoA as intermediates. The formation of pyruvate from lactate is catalyzed by a membrane-associated, pyridine nucleotide-independent lactate dehydrogenase, and the dehydrogenation of pyruvate to acetyl CoA and CO_2 is carried out by a cytoplasmic pyruvate:ferredoxin oxidoreductase. Acetyl CoA is then either converted to acetate via phosphotransacetylase and acetate kinase (incomplete lactate oxidation) or further metabolized to CO_2 (complete lactate oxidation). It is still not clear how these reactions are coupled with dissimilatory sulfate reduction.

One theory of the mechanism of dissimilatory sulfate reduction is that the electrons from lactate and from pyruvate are transferred to protons via a cytoplasmic hydrogenase and the H_2 thus formed diffuses through the cytoplasmic membrane into the periplasm (Odom and Peck, 1981, 1984). There, H_2 enters the electron transport chain starting with periplasmic hydrogenase (Figure 22.3). Experimental evidence for this H_2 cycling mechanism has been provided for pyruvate as electron donor (Peck et al., 1987). The redox potential of the acetyl CoA + CO_2/pyruvate couple is sufficiently negative ($E_0' = -500$ mV) to allow the reduction of H^+ ($E_0' = -420$ mV). However, lactate oxidation with H^+ as electron acceptor is thermodynamically difficult to achieve. Even at very low H_2 partial pressures (H_2 = 10 Pa), the redox potential of the H^+/H_2 couple ($E_0' = -300$ mV) is still more negative than that of the pyruvate/lactate couple ($E_0' = -190$ mV). The finding that H_2, at 10^5 Pa pressure, has no effect on lactate oxidation with sulfate actually rules out free H_2 as an electron carrier between lactate and sulfate (Pankhania et al., 1986). Also, *Desulfovibrio sapovorans,* which grows on lactate plus sulfate, does not contain hydrogenase (Pfennig et al., 1981; Postgate, 1984b).

Some sulfate-reducing bacteria (for example, *Desulfobrio vulgaris, D. gigas, D. desulfuricans*) ferment lactate to acetate, CO_2 and H_2 when growing, in the absence of sulfate, in syntrophic association with H_2-consuming methanogenic bacteria, which keep the H_2 pressure low (Bryant et al., 1977; McInerney and Bryant, 1981; Traore et al., 1983; Chartrain and Zeikus, 1986; Conrad et al., 1986). These bacteria are capable of lactate oxidation to pyruvate and H_2 (reaction 11), a highly endergonic reaction which can proceed only if coupled to the exergonic oxidation of pyruvate to acetate, CO_2 and H_2 (reaction 12). The summation of these two reactions, the fermentation of lactate to acetate, is given in reaction (13).

$$\text{lactate}^- \rightarrow \text{pyruvate}^- + H_2 \qquad \Delta G^{0\prime} = +43.2 \text{ kJ/mol} \quad (11)$$

$$\text{pyruvate}^- + H_2O \rightarrow \text{acetate}^- + CO_2 + H_2 \qquad \Delta G^{0\prime} = -52 \text{ kJ/mol} \quad (12)$$

$$\text{lactate}^- + H_2O \rightarrow \text{acetate}^- + CO_2 + 2H_2 \qquad \Delta G^{0\prime} = -8.8 \text{ kJ/mol} \quad (13)$$

Experimental evidence for the mechanism of coupling was recently provided by the finding that H_2 formation from lactate in cell suspensions of *Desulfovibrio vulgaris* is inhibited by protonophores, by DCCD, and by arsenate, whereas H_2 formation from pyruvate is not (Pankhania et al., 1988). The three inhibitors have in common that they decrease the electrochemical proton potential, protonophores by directly dissipating the potential, DCCD by blocking the H^+-translocating ATPase present in these bacteria, and arsenate by hydrolyzing acetyl CoA and thus inhibiting ATP synthesis in the acetate kinase reaction. The data indicate that reaction (11) is coupled to reaction (12) via the electrochemical proton potential (Figure 22.4).

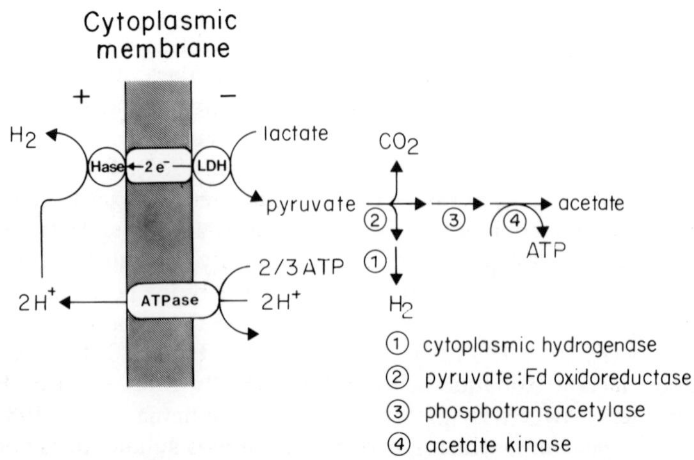

Figure 22.4 Proposed mechanism for the coupling of endergonic lactate conversion to pyruvate and H_2 to the hydrolysis of ATP in *Desulfovibrio vulgaris* fermenting 1 lactate to 1 acetate + $1CO_2$ + $2H_2$ (reactions 11–13). LDH, lactate dehydrogenase; Hase, hydrogenase.

22.3 GROWTH ON ACETATE PLUS SULFATE

For a long time it was believed that sulfate-reducing bacteria were not capable of acetate (acetyl CoA) oxidation to CO_2 (Postgate, 1979). However, Widdel and Pfennig succeeded in isolating numerous bacteria from marine sediments which can grow with sulfate on acetate (reaction 14) as sole energy source (Widdel, 1980; Pfennig et al., 1981; Widdel and Pfennig, 1984; Widdel, 1988).

$$\text{acetate}^- + SO_4^{2-} + 3H^+ \rightarrow 2CO_2 + H_2S + 2H_2O \qquad \Delta G^{0\prime} = -63 \text{ kJ/mol} \quad (14)$$

Two mechanisms were found to exist by which the sulfate-reducing bacteria can oxidize acetate to CO_2: a modified citric acid cycle and a novel C_1 pathway referred to as the oxidative acetyl CoA/carbon monoxide dehydrogenase pathway (Thauer, 1988).

Acetate oxidation via the citric acid cycle The first organism to be investigated in detail with respect to the mechanism of acetate oxidation was *Desulfobacter postgatei* (Widdel and Pfennig, 1981). By enzymatic studies and long-term labeling experiments, it was shown that acetate oxidation to CO_2 in this organism proceeds via a modified citric acid cycle (Figure 22.5). Interesting features are the involvement of a succinyl CoA: acetate CoA transferase for the activation of acetate, of a membrane-bound pyridine nucleotide-independent malate dehydrogenase, and of an ATP-citrate lyase (Thauer, 1982; Brandis-Heep et al., 1983; Gebhardt et al., 1983; Möller et al., 1987; Möller-Zinkhan and Thauer, 1988).

Only 4 of the 8 reducing equivalents generated during acetate oxidation in the citric acid cycle have a redox potential negative enough to allow the synthesis of ATP during electron transport to sulfite ($E_0{}' = -105$ mV) or APS

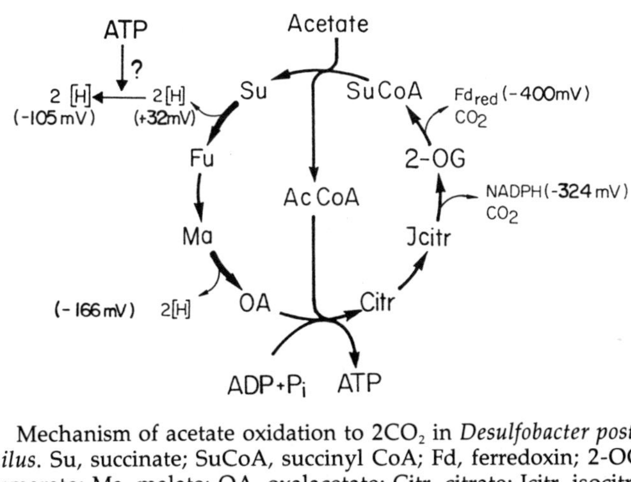

Figure 22.5 Mechanism of acetate oxidation to $2CO_2$ in *Desulfobacter postgatei* and *D. hydrogenophilus*. Su, succinate; SuCoA, succinyl CoA; Fd, ferredoxin; 2-OG, 2-oxoglutarate; Fu, fumarate; Ma, malate; OA, oxalacetate; Citr, citrate; Icitr, isocitrate; AcCoA, acetyl CoA.

(E_0' = −70 mV). These are the electrons accepted by NADP (E_0' = −324 mV) in the isocitrate dehydrogenase reaction and by ferredoxin (E_0' = −400 mV) in the 2-oxoglutarate:ferredoxin oxidoreductase reaction. The redox potential (E_0') of the oxaloacetate/malate couple is −166 mV and that of the fumarate/succinate couple is +32 mV. The reduction of sulfite or APS with succinate or malate can therefore not be coupled with the phosphorylation of ADP. The oxidation of succinate is probably even an energy-consuming reaction (Paulsen et al., 1986). Thus, during sulfite and APS reduction with acetate, probably less ATP is formed than required for the activation of SO_4^{2-}. To compensate for this, *D. postgatei* uses an ATP citrate lyase rather than a citrate synthase to catalyze the synthesis of citrate. The use of this enzyme enables this organism to couple the oxidation of acetate to 2 CO_2 with the formation of 1 ATP by the mechanism of substrate-level-phosphorylation and thus to gain a net amount of ATP from the oxidation of acetate with sulfate.

Interesting in this respect is the finding that in *Desulfuromonas acetoxidans*, which oxidizes acetate to CO_2 with S^0 as electron acceptor, the synthesis of citrate is catalyzed by citrate synthase (Gebhardt et al., 1985). S^0 reduction to H_2S (E_0' = −270 mV) with acetate is associated with a net synthesis of ATP by electron-transport-linked phosphorylation (Paulsen et al., 1986; Kröger et al., 1988).

Desulfobacter hydrogenophilus uses the same pathway of acetate oxidation as *D. postgatei* (Schauder et al., 1986).

Acetate oxidation via the acetyl CoA/carbon monoxide dehydrogenase pathway All other sulfate-reducing bacteria investigated so far use a totally different mechanism for the oxidation of acetate (*Desulfobacterium vacuolatum; Desulfosarcina variabilis; Desulfococcus niacini; Desulfococcus multivorans; Desulfovibrio baarsii; Desulfobacterium autotrophicum*; and *Desulfotomaculum acetoxidans*). This pathway, shown in Figure 22.6, is based on enzymatic studies, and on short-term and on long-term labeling experiments. Cell suspensions catalyze an isotopic exchange between CO_2 and the carboxyl group of acetate, which permits a direct demonstration of the reversible C–C-bound cleavage reaction centrally involved (Schauder et al., 1986; Spormann and Thauer, 1988).

The sulfate-reducing bacteria using this novel pathway characteristically contain high activities of carbon monoxide dehydrogenase. This enzyme has been shown to be involved in acetyl CoA formation with methyltetrahydrofolate and CO in acetogenic bacteria (see Chapter 3) and in autotrophic methanogens (see Chapter 20). It is also involved in acetyl CoA cleavage in *Methanosarcina barkeri* (see Chapter 21). The sulfate-reducing bacteria that have carbon monoxide dehydrogenase also contain high specific activities of acetate kinase and of phosphotransacetylase, indicating that, as in the above-mentioned acetogenic and methanogenic bacteria, acetyl CoA rather than acetate is the substrate of the reversible C–C cleavage reaction.

The methyl group of acetyl CoA is oxidized via C_1-tetrahydrofolate intermediates (some of the sulfate-reducing bacteria contain modified folates) to CO_2, 1 mol of ATP being formed by substrate-level phosphorylation in the

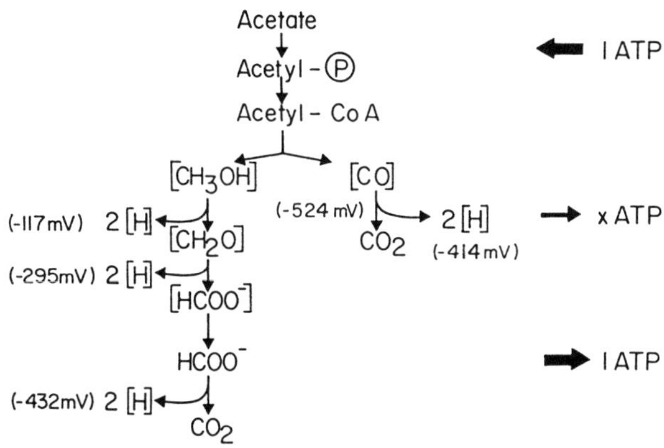

Figure 22.6 Proposed mechanism of acetate (acetyl CoA) oxidation to 2 CO_2 in *Desulfobacterium vacuolatum, Desulfosarcina variabilis, Desulfococcus niacini, Desulfococcus multivorans, Desulfovibrio baarsii, Desulfobacterium autotrophicum,* and *Desulfotomaculum acetoxidans.* [CH_3OH] = methyltetrahydrofolate; [CH_2O] = methylenetetrahydrofolate; [$HCOO^-$] = methenyltetrahydrofolate and formyltetrahydrofolate.

formyl tetrahydrofolate synthetase reaction (Figure 22.6). The redox potential of the methenyltetrahydrofolate/methylenetetrahydrofolate couple is −295 mV, that of the methylenetetrahydrofolate/methyltetrahydrofolate couple is −117 mV, and that of the CO_2/formate couple is −432 mV (Fuchs, 1986). Thus, with sulfite ($E_0{}'$ = −120 mV) or APS ($E_0{}'$ = −70 mV) as electron acceptors, only 4 of the 6 reducing equivalents generated during methyl group oxidation to CO_2 have redox potentials negative enough to drive the synthesis of ATP via electron-transport-linked phosphorylation.

The carboxyl group of acetyl CoA yields a C_1 unit at the oxidation level of CO, which is oxidized to CO_2. Electron transport from CO ($E_0{}'$ = −524 mV) to the redox level of the hydrogen electrode ($E_0{}'$ = −414 mV) is probably coupled with the phospohorylation of ADP, as this is the case in *Methanosarcina barkeri* (Bott et al., 1986; Bott and Thauer, 1987).

Interrelation between the mechanism of acetate oxidation and that of autotrophic CO_2 fixation Many of the acetate-oxidizing sulfate-reducing bacteria (or phylogenetically closely related species) can also live chemolithoautotrophically. As has been outlined in Chapter 20, autotrophic CO_2 fixation in these bacteria proceeds either via the reductive citric acid cycle or via a reductive acetyl CoA/carbon monoxide dehydrogenase pathway. It is interesting that these organisms always use essentially the same pathway for autotrophic CO_2 fixation that they use in reversal for acetyl CoA oxidation to CO_2 (Table 22.1).

22.4 CHEMOLITHOTROPHIC GROWTH ON THIOSULFATE OR SULFITE

Bak and Pfennig (1987) have isolated a new *Desulfovibrio, D. sulfodismutans* (strain ThAc 01), which conserves energy for growth from the disproportion-

Table 22.1 Pathways of acetate (acetyl CoA) oxidation to CO_2 and of autotrophic CO_2 fixation used by sulfate-reducing bacteria

	Pathway used for	
Genus	Acetyl CoA oxidation	Autotrophic CO_2 fixation
Desulfobacter	Oxidative TCA cycle*	Reductive TCA cycle
Desulfobacterium, *Desulfosarcina*, *Desulfococcus*, *Desulfovibrio baarsii*, *Desulfotomaculum*	Oxidative acetyl CoA/CO dehydrogenase pathway	Reductive acetyl CoA/CO dehydrogenase pathway

*TCA, tricarboxylic acid.

ation of thiosulfate (reaction 15) or of sulfite (reaction 16) to H_2S and sulfate (see also Bak and Cypionka, 1987).

$$S_2O_3^{2-} + H_2O \rightarrow SO_4^{2-} + H_2S \qquad \Delta G^{0\prime} = -27.6 \text{ kJ/mol } H_2S \quad (15)$$

$$4SO_3^{2-} + 2H^+ \rightarrow 3SO_4^{2-} + H_2S \qquad \Delta G^{0\prime} = -241.3 \text{ kJ/mol } H_2S \quad (16)$$

For growth, *D. sulfodismutans* requires not only CO_2 but also acetate as carbon source and is thus a chemolithoheterotroph. Another isolate (strain NTA 3), however, did not require acetate, indicating the capacity for autotrophic growth.

Desulfovibrio sulfodismutans can also grow by dissimilatory sulfate reduction with ethanol or lactate as electron donor. Out of 10 known species of sulfate-reducing bacteria tested, only *Desulfobacter curvatus* was able to disproportionate thiosulfate and sulfite. *Desulfovibrio vulgaris* (strain Marburg) slowly disproportionates sulfite but not thiosulfate.

Disproportionation of $S_2O_3^{2-}$ (reaction 15) probably proceeds via sulfite as intermediate. (E_0' of the $S_2O_3^{2-}/H_2S + SO_3^{2-}$ couple is -380 mV while that of the SO_4^{2-}/SO_3^{2-} couple is -520 mV. Electron transfer from sulfite to thiosulfate is thus energetically comparable with the transfer of electron from CO ($E_0' = -520$ mV) to protons ($E_0' = -420$ mV) a reaction which in *Methanosarcina barkeri* has been shown to be coupled with the synthesis of ATP via electron-transport-linked phosphorylation (Bott et al., 1986; Bott and Thauer, 1987).

The coupling of sulfite oxidation to sulfate ($E_0' = -520$ mV) with the reduction of sulfite to H_2S ($E_0' = -120$ mV) (reaction 16) is energetically much more favorable than the reduction of thiosulfate with sulfite (reaction 15). The reduction of sulfite in reaction 16 is probably associated with ATP synthesis via electron-transport-linked phosphorylation as this is the case when H_2 ($E_0' = -420$ mV) is the electron donor (see section 22.1). The thermodynamics would allow for the formation of 3 to 4 mol ATP per mol of sulfite reduced to H_2S.

In some *Thiobacillus* species and some phototrophic bacteria, sulfite is oxidized to sulfate via APS as intermediate (see Chapters 11 and 14). The formation of APS from sulfite and AMP is catalyzed by APS reductase. APS is then phosphorylytically cleaved to ADP and sulfate by the action of ADP sulfurylase. The ADP thus generated is converted to AMP and ATP with adenylate kinase. By this mechanism, 1 mol of ATP is formed by substrate-level phosphorylation per mol of sulfite oxidized to sulfate. It therefore has to be considered that during sulfite disproportionation, ATP could be formed exclusively by substrate-level phosphorylation rather than by electron-transport-linked phosphorylation.

22.5 DEGRADATION OF AROMATIC COMPOUNDS AND OF ALKANES

Sulfate-reducing bacteria have been isolated that can grow on benzoate (*Desulfococcus* sp., Widdel, 1980; *Desulfosarcina variabilis*, Widdel, 1988; *Desulfonema magnum*, Widdel et al., 1983; *Desulfotomaculum sapomandens*, Cord-Ruwisch and Garcia, 1985), on phenol derivatives (*Desulfobacterium phenolicum*, Bak and Widdel, 1986b), on catechols (*Desulfobacterium catecholicum*, Szewzyk and Pfennig, 1987), or on indole derivatives (*Desulfobacterium indolicum*, Bak and Widdel, 1986a) as electron donors for dissimilatory sulfate reduction. The aromatic compounds are oxidized to CO_2. How these bacteria, in the complete absence of molecular oxygen, degrade the aromatic ring system has not yet been elucidated (see Thauer and Morris, 1984).

There are indications that sulfate-reducing bacteria can also degrade alkanes (Widdel, 1988). In the absence of molecular oxygen, the oxidation of alkanes is mechanistically even more difficult to envisage than the anaerobic degradation of aromatic compounds.

22.6 GROWTH ON METHANOL AND SULFATE

Desulfovibrio carbinolicus (Nanninga and Gottschal, 1986, 1987), a *Desulfovibrio* sp. capable of growth on fructose (Cord-Ruwisch et al., 1986), *Desulfotomaculum orientis* (Klemps et al., 1985), and *Desulfobacterium catecholicum* (Szewzyk and Pfennig, 1987) can grow on methanol and sulfate as sole energy source (see also Braun and Stolp, 1985).

$$4CH_3OH + 3SO_4^{2-} + 6H^+ \rightarrow 4CO_2 + 3H_2S + 8H_2O \qquad \Delta G^{0\prime} = -133.6 \text{ kJ/mol } H_2S \qquad (17)$$

This is mentioned here, because earlier attempts to isolate sulfate-reducing bacteria with this capacity were not successful (Postgate, 1984a).

This work was supported by the Deutsche Forschungsgemeinschaft and by the Fonds der Chemischen Industrie. I thank H. Friedrich Widdel for helpful discussions.

REFERENCES

Achenbach-Richter, L., **Stetter, K.O.**, and **Woese, C.R.** 1987. A possible biochemical missing link among archaebacteria. *Nature* 327:348–349.

Badziong, W., **Thauer, R.K.**, and **Zeikus, J.G.** 1978. Isolation and characterization of *Desulfovibrio* growing on hydrogen plus sulfate as the sole energy source. *Archives of Microbiology* 116:41–49.

Badziong, W. and **Thauer, R.K.** 1980. Vectorial electron transport in *Desulfovibrio vulgaris* (Marburg) growing on hydrogen plus sulfate as sole energy source. *Archives of Microbiology* 125:167–174.

Bak, F. and **Cypionka, H.** 1987. A novel type of energy metabolism involving fermentation of inorganic sulphur compounds. *Nature* 326:891–892.

Bak, F. and **Pfennig, N.** 1987. Chemolithotrophic growth of *Desulfovibrio sulfodismutans* sp. nov. by disproportionation of inorganic sulfur compounds. *Archives of Microbiology* 147:184–189.

Bak, F. and **Widdel, F.** 1986a. Anaerobic degradation of indolic compounds by sulfate-reducing enrichment cultures, and description of *Desulfobacterium indolicum* gen. nov., sp. nov. *Archives of Microbiology* 146:170–176.

Bak, F. and **Widdel, F.** 1986b. Anaerobic degradation of phenol and phenol derivatives by *Desulfobacterium phenolicum* sp. nov. *Archives of Microbiology* 146:177–180.

Bott, M. and **Thauer, R.K.** 1987. Proton motive force-driven formation of CO from CO_2 and H_2 in methanogenic bacteria. *European Journal of Biochemistry* 168:407–412.

Bott, M., **Eikmanns, B.**, and **Thauer, R.K.** 1986. Coupling of carbon monoxide oxidation to CO_2 and H_2 with the phosphorylation of ADP in acetate-grown *Methanosarcina barkeri*. *European Journal of Biochemistry* 159:393–398.

Brandis, A. and **Thauer, R.K.** 1981. Growth of *Desulfovibrio* species on hydrogen and sulphate as sole energy source. *Journal of General Microbiology* 126:249–252.

Brandis-Heep, A., **Gebhardt, N.A.**, **Thauer, R.K.**, **Widdel, F.**, and **Pfennig, N.** 1983. Anaerobic acetate oxidation to CO_2 by *Desulfobacter postgatei*. 1. Demonstration of all enzymes required for the operation of the citric acid cycle. *Archives of Microbiology* 136:222–229.

Braun, M. and **Stolp, H.** 1985. Degradation of methanol by a sulfate-reducing bacterium. *Archives of Microbiology* 142:77–80.

Bryant, M.P, **Campbell, L.L.**, **Reddy, C.A.**, and **Crabill, M.R.** 1977. Growth of *Desulfovibrio* in lactate or ethanol media low in sulfate in association with H_2-utilizing methanogenic bacteria. *Applied and Environmental Microbiology* 33:1162–1169.

Chartrain, M. and **Zeikus, J.G.** 1986. Microbial ecophysiology of whey biomethanation: Intermediary metabolism of lactose degradation in continuous culture. *Applied and Environmental Microbiology* 51:180–187.

Collins, M.D. and **Widdel, F.** 1986. Respiratory quinones of sulphate-reducing and sulphur-reducing bacteria: A systematic investigation. *Systematic and Applied Microbiology* 8:8–18.

Conrad, R., **Schink, B.**, and **Phelps, T.J.** 1986. Thermodynamics of H_2-consuming and H_2-producing metabolic reactions in diverse methanogenic environments under in situ conditions. *FEMS Microbiology Ecology* 38:353–360.

Cord-Ruwisch, R. and **Garcia, J.-L.** 1985. Isolation and characterization of an anaerobic benzoate-degrading spore-forming sulfate-reducing bacterium, *Desulfotomaculum sapomandens* sp. nov. *FEMS Microbiology Letters* 29:325–330.

Cord-Ruwisch, R., **Ollivier, B.**, and **Garcia, J.-L.** 1986. Fructose degradation by *Desulfovibrio* sp. in pure culture and in coculture with *Methanospirillum hungatei*. *Current Microbiology* 13:285–289.

Cypionka, H. 1987. Uptake of sulfate, sulfite, and thiosulfate by proton-anion symport in *Desulfovibrio desulfuricans*. *Archives of Microbiology* 148:144–149.

Cypionka, H. and **Pfennig, N.** 1986. Growth yields of *Desulfotomaculum orientis* with hydrogen in chemostat culture. *Archives of Microbiology* 143:396–399.

Cypionka, H. and **Dilling, W.** 1986. Intracellular localization of the hydrogenase in *Desulfotomaculum orientis. FEMS Microbiology Letters* 36:257–260.

Fowler, V.J., Widdel, F., Pfennig, N., Woese, C.R., and **Stackebrandt, E.** 1986. Phylogenetic relationships of sulfate- and sulfur-reducing eubacteria. *Systematic and Applied Microbiology* 8:32–41.

Fuchs, G. 1986. CO_2 fixation in acetogenic bacteria: variations on a theme. *FEMS Microbiology Reviews* 39:181–213.

Gebhardt, N.A., Linder, D., and **Thauer, R.K.** 1983. Anaerobic acetate oxidation to CO_2 by *Desulfobacter postgatei*. 2. Evidence from ^{14}C-labelling studies for the operation of the citric acid cycle. *Archives of Microbiology* 136:230–233.

Gebhardt, N.A., Thauer, R.K., Linder, D., Kaulfers, P-M., and **Pfennig, N.** 1985. Mechanism of acetate oxidation to CO_2 with elemental sulfur in *Desulfuromonas acetoxidans. Archives of Microbiology* 141:392–398.

Herr, G. 1982. Untersuchungen zur Elektronentransport-Phosphorylierung von *Desulfovibrio vulgaris* (Marburg) im zellfreien System. *Diploma thesis,* Fachbereich Biologie, Philipps-Universität Marburg.

Ingvorsen, K. and **Jørgensen, B.B.** 1984. Kinetics of sulfate uptake by freshwater and marine species of *Desulfovibrio. Archives of Microbiology* 139:61–66.

Ingvorsen, K., Zehnder, A.J.B., and **Jørgensen, B.B.** 1984. Kinetics of sulfate and acetate uptake by *Desulfobacter postgatei. Applied and Environmental Microbiology* 47:403–408.

Kim, J.-H. and **Akagi, J.M.** 1985. Characterization of a trithionate reductase system from *Desulfovibrio vulgaris. Journal of Bacteriology* 163:472–475.

Klemps, R., Cypionka, H., Widdel, F., and **Pfennig, N.** 1985. Growth with hydrogen, and further physiological characteristics of *Desulfotomaculum* species. *Archives of Microbiology* 143:203–208.

Kobayashi, K., Hasegawa, H., Takagi, M., and **Ishimoto, M.** 1982. Proton translocation associated with sulfite reduction in a sulfate-reducing bacterium *Desulfovibrio vulgaris. FEBS Letters* 142:235–237.

Kremer, D.R., Veenhuis, M., Fauque, G., Peck, H.D., Jr., LeGall, J., Lampreia, J., and **Moura, J.J.G.** 1988. Immunocytochemical localization of APS reductase and bisulfite reductase in three *Desulfovibrio* species. *Archives of Microbiology* 150:296–301.

Kröger, A., Schröder, I., and **Paulsen, J.** 1986. Direct and reversed electron transport in anaerobic bacteria, pp. 93–104, in Dubourguier, H.C. et al. (editors), *Biology of Anaerobic Bacteria,* Elsevier Science Publishers, Amsterdam.

Kröger, A., Schröder, J., Paulsen, J., and **Beilmann, A.** 1988. Acetate oxidation with sulphur and sulphate as terminal electron acceptors, pp. 133–145, in Cole, J.A., and Ferguson, S. (editors), *The Nitrogen and Sulphur Cycles,* Society for General Microbiology Symposium.

Kulaev, I.S. 1979. *The Biochemistry of Inorganic Polyphosphates,* John Wiley & Sons, Chichester, New York, Brisbane, Toronto.

LeGall, J. and **Fauque, G.** 1988. Dissimilatory reduction of sulfur compounds, Chapter 11, in Zehnder, A.J.B. (editor), *Biology of Anaerobic Microorganisms,* John Wiley & Sons, New York (in press).

LeGall, J. and **Peck, Jr., H.D.** 1987. Amino-terminal amino acid sequences of electron transfer proteins from Gram-negative bacteria as indicators of their cellular localization: the sulfate-reducing bacteria. *FEMS Microbiology Reviews* 46:35–40.

Littlewood, D. and **Postgate, J.R.** 1957. On the osmotic behavior of *Desulphovibrio desulphuricans. Journal of General Microbiology* 16:596–603.

Liu, C.L., Hart, N., and **Peck Jr., H.D.** 1982. Inorganic pyrophosphate: energy source for sulfate-reducing bacteria of the genus *Desulfotomaculum. Science* 217:363–364.

McInerney, M.J. and **Bryant, M.P.** 1981. Anaerobic degradation of lactate by syntrophic associations of *Methanosarcina barkeri* and *Desulfovibrio* species and effect of H_2 on acetate degradation. *Applied and Environmental Microbiology* 41:346–354.

Möller, D., Schauder, R., Fuchs, G., and **Thauer, R.K.** 1987. Acetate oxidation to CO_2 via a citric acid cycle involving an ATP-citrate lyase: a mechanism for the synthesis of ATP via substrate level phosphorylation in *Desulfobacter postgatei* growing on acetate and sulfate. *Archives of Microbiology* 148:202–207.

Möller-Zinkhan, D., and **Thauer, R.K.** 1988. Membrane-bound NADPH dehydrogenase and ferredoxin: NADP oxidoreductase activity involved in electron transport during acetate oxidation to CO_2 in *Desulfobacter postgatei. Archives of Microbiology* 150:145–154.

Nanninga, H.J. and **Gottschal, J.C.** 1986. Isolation of a sulfate-reducing bacterium growing with methanol. *FEMS Microbiology Ecology* 38:125–130.

Nanninga, H.J. and **Gottschal, J.C.** 1987. Properties of *Desulfovibrio carbinolicus* sp. nov. and other sulfate-reducing bacteria isolated from an anaerobic-purification plant. *Applied and Environmental Microbiology* 53:802–809.

Nethe-Jaenchen, R. and **Thauer, R.K.** 1984. Growth yields and saturation constant of *Desulfovibrio vulgaris* in chemostat culture. *Archives of Microbiology* 137:236–240.

Odom, J.M. and **Peck Jr., H.D.** 1981. Hydrogen cycling as a general mechanism for energy coupling in the sulfate-reducing bacteria *Desulfovibrio* sp. *FEMS Microbiology Letters* 12:47–50.

Odom, J.M. and **Peck Jr., H.D.** 1984. Hydrogenase, electron-transfer proteins, and energy coupling in the sulfate-reducing bacteria *Desulfovibrio. Annual Review of Microbiology* 38:551–592.

Ohmoto, H. and **Felder, R.P.** 1987. Bacterial activity in the warmer, sulphate-bearing, Archaean oceans. *Nature* 328:244–246.

Pankhania, I.P., Gow, L.A., and **Hamilton, W.A.** 1986. The effect of hydrogen on the growth of *Desulfovibrio vulgaris* (Hildenborough) on lactate. *Journal of General Microbiology* 132:3349–3356.

Pankhania, I.P., Spormann, A.M., Hamilton, W.A., and **Thauer, R.K.** 1988. Lactate conversion to acetate, CO_2, and H_2 in cell suspensions of *Desulfovibrio vulgaris* (Marburg): Indications for the involvement of an energy driven reaction. *Archives of Microbiology* 150:26–31.

Paulsen, J. 1982. Vergleichende Untersuchungen zur Elektronentransport-Phosphorylierung von *Desulfovibrio gigas* und *Desulfovibrio vulgaris* (Marburg). *Diploma thesis,* Fachbereich Biologie, Philipps-Universität Marburg.

Paulsen, J., Kröger, A., and **Thauer, R.K.** 1986. ATP-driven succinate oxidation in the catabolism of *Desulfuromonas acetoxidans. Archives of Microbiology* 144:78–83.

Peck Jr., H.D. 1966. Phosphorylation coupled with electron transfer in extracts of the sulfate-reducing bacterium *Desulfovibrio gigas. Biochemical and Biophysical Research Communications* 22:112–118.

Peck Jr., H.D. and **LeGall, J.** 1982. Biochemistry of dissimilatory sulphate reduction. *Philosophical Transactions of the Royal Society of London, Series B* 298:443–466.

Peck Jr., H.D., LeGall, J., Lespinat, P.A., Berlier, Y., and **Fauque, G.** 1987. A direct demonstration of hydrogen cycling by *Desulfovibrio vulgaris* employing membrane-inlet mass spectrometry. *FEMS Microbiology Letters* 40:295–299.

Pfennig, N., Widdel, F., and **Trüper, H.G.** 1981. The dissimilatory sulfate-reducing bacteria, pp. 926–940, in Starr, M.P. et al., (editors), *The Prokaryotes,* Springer-Verlag, Berlin, Heidelberg, New York.

Postgate, J.R. 1979. *The sulphate-reducing bacteria.* 1st ed. Cambridge University Press, Cambridge, London, New York, Melbourne.

Postgate, J.R. 1984a. *The sulphate-reducing bacteria,* 2nd ed. Cambridge University Press, Cambridge, London, New York, Melbourne.

Postgate, J.R. 1984b. Genus *Desulfovibrio* Kluyver and van Niel 1936, 397[AL], pp. 666–672, in Krieg, N.R. and Holt, J.G. (editors), *Bergey's Manual of Systematic Bacteriology, Vol. 1,* Williams & Wilkins, Baltimore/London.

Reeves, R.E. 1976. How useful is the energy in inorganic pyrophosphate? *Trends in Biochemical Sciences* 1:53–55.

Schauder, R., Eikmanns, B., Thauer, R.K., Widdel, F., and **Fuchs, G.** 1986. Acetate oxidation to CO_2 in anaerobic bacteria via a novel pathway not involving reactions of the citric acid cycle. *Archives of Microbiology* 145:162–172.

Schauder, R., Widdel, F., and **Fuchs, G.** 1987. Carbon assimilation pathways in sulfate reducing bacteria. II. Enzymes of a reductive citric acid cycle in the autotrophic *Desulfobacter hydrogenophilus. Archives of Microbiology* 148:248–225.

Sorokin, Yu.I. 1966a. Sources of energy and carbon for biosynthesis in sulfate-reducing bacteria. *Microbiology (USSR) (Eng. Trans.)* 35:643–647.

Sorokin, Yu.I. 1966b. Investigations of the structural metabolism of sulfate-reducing bacteria with ^{14}C. *Microbiology (USSR) (Engl. Transl.)* 35:806–814.

Sorokin, Yu.I. 1966c. Role of carbon dioxide and acetate in the biosynthesis by sulphate-reducing bacteria. *Nature* 210:551–552.

Spormann, A.M., and **Thauer, R.K.** 1988. Anaerobic acetate oxidation to CO_2 by *Desulfotomaculum acetoxidans.* Demonstration of enzymes required for the operation of an oxidative acetyl-CoA/carbon monoxide dehydrogenase pathway. *Archives of Microbiology (in press).*

Stetter, K.O., Lauerer, G., Thomm, M., and **Neuner, A.** 1987. Isolation of extremely thermophilic sulfate reducers: Evidence for a novel branch of archaebacteria. *Science* 236:822–824.

Stille, W. and **Trüper, H.G.** 1984. Adenylylsulfate reductase in some new sulfate-reducing bacteria. *Archives of Microbiology* 137:145–150.

Szewzyk, R. and **Pfennig, N.** 1987. Complete oxidation of catechol by the strictly anaerobic sulfate-reducing *Desulfobacterium catecholicum* sp. nov. *Archives of Microbiology* 147:163–168.

Thauer, R.K. 1982. Dissimilatory sulphate reduction with acetate as electron donor. *Philosophical Transactions of the Royal Society of London, Series B* 298:467–471.

Thauer, R.K. 1988. Citric-acid cycle, 50 years of. Modifications and an alternative pathway in anaerobic bacteria. *Eur. J. Biochem. (in press).*

Thauer, R.K. and **Badziong, W.** 1980. Respiration with sulfate as electron acceptor, pp. 65–85, in Knowles, Ch.J. (editor), *Diversity of Bacterial Respiratory Systems, Vol. II,* CRC Press, Inc., Boca Raton, Florida.

Thauer, R.K. and **Morris, J.G.** 1984. Metabolism of chemotrophic anaerobes: old views and new aspects, pp. 123–168, in Kelly, D.P. and Carr, N.G. (editors), *The Microbe 1984: Part II Prokaryotes and Eukaryotes,* Society for General Microbiology Symposium 36, Cambridge University Press, Cambridge.

Thauer, R.K., Jungermann, K., and **Decker, K.** 1977. Energy conservation in chemotrophic anaerobic bacteria. *Bacteriological Reviews* 41:100–180.

Traore, A.S., Fardeau, M-L., Hatchikian, C.E., LeGall, J., and **Belaich, J-P.** 1983. Energetics of growth of a defined mixed culture of *Desulfovibrio vulgaris* and *Methanosarcina barkeri*: interspecies hydrogen transfer in batch and continuous cultures. *Applied and Environmental Microbiology* 46:1152–1156.

Varma, A., Schönheit, P., and **Thauer, R.K.** 1983. Electrogenic sodium ion/proton antiport in *Desulfovibrio vulgaris. Archives in Microbiology* 136:69–73.

Widdel, F. 1980. Anaerober Abbau von Fettsäuren und Benzoesäure durch neu isolierte Arten Sulfat-reduzierender Bakterien. *Doctoral thesis,* Mathematisch-Naturwissenschaftliche Fakultät der Georg-August-Universität zu Göttingen.

Widdel, F. 1988. Microbial ecology of sulfate and sulfur-reducing bacteria, Chapter 10, pp. 469–584, in Zehnder, A.J.B. (editor), *Biology of Anaerobic Microorganisms,* John Wiley & Sons, New York.

Widdel, F. and **Pfennig, N.** 1981. Studies on dissimilatory sulfate-reducing bacteria that decompose fatty acids. I. Isolation of new sulfate-reducing bacteria enriched with acetate from saline environments. Description of *Desulfobacter postgatei* gen. nov. sp. nov. *Archives of Microbiology* 129:395–400.

Widdel, F. and **Pfennig, N.** 1984. Dissimilatory sulfate- or sulfur-reducing bacteria, pp. 663–679, in Krieg, N.R. and Holt, J.G. (editors), *Bergey's Manual of Systematic Bacteriology, Vol. 1,* Williams & Wilkins, Baltimore/London.

Widdel, F., Kohring, G.-W., and **Mayer, F.** 1983. Studies on dissimilatory sulfate-reducing bacteria that decompose fatty acids. III. Characterization of the filamentous gliding *Desulfonema limicola* gen. nov. sp. nov., and *Desulfonema magnum* sp. nov. *Archives of Microbiology* 134:286–294.

23

Genetics of Energy Converting Systems in Aerobic Chemolithotrophs

Bärbel Friedrich

Institute of Plant Physiology,
Cell Biology, and Microbiology
Free University of Berlin
Berlin, Federal Republic of Germany

Although our current knowledge of the genetics of lithoautotrophic organisms is still in its infancy, it is the purpose of this contribution to document the enormous progress that has been achieved lately in this particular field of research. The survey is limited to energy-generating systems in aerobic lithotrophs: those organisms which oxidize hydrogen, carbon monoxide, ammonia, nitrite, ferrous iron, and sulfur and its reduced compounds. It will focus on the more recent developments in *Alcaligenes eutrophus*, one of the best known and most intensively studied lithoautotrophic organisms. Some of the aspects have been reviewed previously (Bowien and Schlegel, 1981; Schlegel and Schneider, 1985; Maier, 1986; Meyer et al., 1986; Kelly et al., 1987; Friedrich, 1987).

23.1 CHEMOLITHOTROPHIC MUTANTS

A comprehensive genetic investigation is based ideally on three criteria: the availability of mutants, a gene transfer system, and the access to recombinant DNA techniques and molecular analyses. The fact that the nitrifying, and many iron- and sulfur-oxidizing bacteria are obligately bound to the lithoautotrophic mode of life means that a mutation in a lithotrophic function is generally lethal. Hence, the availability of mutants is restricted to a few sulfur oxidizers, and carboxydotrophic and hydrogenotrophic bacteria which are facultatively lithoautotrophic. Three major groups of mutants can be distinguished (Table 23.1). Sox mutants, impaired in the oxidation of sulfur compounds, were obtained with *Thiobacillus novellus* (Davidson et al., 1985) and *Thiosphaera pantotropha* (Robertson and Kuenen, 1983; Chandra and Friedrich, 1986). Cox mutants, unable to grow with carbon monoxide, were isolated from *Pseudomonas carboxydovorans* (Meyer, 1985). Hox mutants, with deficiencies in hydrogen oxi-

Table 23.1 Lithotrophic mutants

Phenotype	Species employed	Mutagen	Defect[a]	Reference
Sox	*Thiobacillus novellus*	Tn*501*(Hg^r)	Sox^-	1
	Thiosphaera pantotropha	Tn*5*	Sox^-	
			$Sox^-Hox^-Nir^-$	2
			$Sox^-Fox^-Nar^-$ Xan^-	
Cox	*Pseudomonas carboxydovorans*	Chemicals	Cox^- Nar^-	3
			$Cox^-Hox^-Nar^-$	
Hox	Specified in the text	Chemicals	Hox^-, Hno^-	4
		Tn*5*	Hos^-,Hop^-	

(1) Davidson et al., 1985; (2) Chandra and Friedrich, 1986; (3) Meyer, 1985; (4) reviewed by Friedrich, 1987.
[a]Sox, sulfur oxidation; Hox, hydrogen oxidation; Nir, nitrite reduction; Fox, formate oxidation; Nar, nitrate reduction; Xan, xanthine degradation; Cox, carbon monoxide oxidation; Hno, pleiotropic phenotype described in the text; Hos, soluble hydrogenase; Hop, particulate hydrogenase.

dation (also designated in the literature as Hup mutants), were isolated from a variety of species including *Alcaligenes eutrophus* (reviewed by Friedrich, 1987), *Bradyrhizobium japonicum* (reviewed by Maier, 1986), *Pseudomonas facilis* (Warrelmann and Friedrich, 1986), *Rhodobacter capsulatus* (Takakuwa et al., 1983; Colbeau et al., 1986), *Xanthobacter autotrophicus* (Lehmicke and Lidstrom, 1985), and *Nocardia opaca* (Reh and Schlegel, 1981). The mutants were obtained after treatment of the cells with either chemical mutagens or transposons (Table 23.1). Transpositional insertion mutagenesis has been applied successfully to a wide range of Gram-negative lithotrophic bacteria, e.g., *T. novellus, Thiobacillus versutus* (Davidson et al., 1985), *T. pantotropha* (Chandra and Friedrich, 1986), *A. eutrophus* (Srivastava et al., 1982), *P. facilis* (Warrelmann and Friedrich, 1986), and *B. japonicum* (Maier, 1986). Screening for Hox mutants was improved by the development of rapid colony staining assays based on hydrogen-dependent reduction of either methylene blue (Haugland et al., 1983) or triphenyltetrazolium chloride (Schlegel and Meyer, 1985; Warrelmann and Friedrich, 1986). The identification of Sox mutants was based on the thiosulfate-related color change of the pH indicator phenol red (Chandra and Friedrich, 1986).

It is noteworthy that pleiotropic mutants occurred relatively frequently, irrespective of the mutagen and screening method used (Table 23.1). Molybdenum is required for growth with thiosulfate (Friedrich et al., 1986), thus Sox mutants were identified with simultaneous defects in other molybdenum-dependent activities like formate, nitrate, and xanthine metabolism (Chandra and Friedrich, 1986). Similar pleiotropic phenotypes were observed among Cox mutants of *P. carboxydovorans*. Some of these mutants were missing bactopterin, an essential cofactor of carbon monoxide dehydrogenase (Meyer, 1985 and Chapter 18). A simultaneous loss of two lithotrophic abilities, namely Sox and Hox or Cox and Hox, has been reported (Table 23.1). Moreover, Hox mutants with pleiotropic defects in nitrogenase were isolated from nitrogen-fixing hydrogen-oxidizing bacteria (Moshiri et al., 1983). Extremely pleiotropic mutants were identified among Hox mutants of *A. eutrophus*. These so-called Hno mutants (hydrogen, nitrate, and others) were altered in at least seven additional metabolic properties, including nitrate assimilation, denitrification, autotrophic carbon dioxide fixation, urease formation, the uptake of nickel ions, and C_4 dicarboxylic acids utilization (Römermann et al., 1988). These data point to a rather complex genetic structure and regulation which is obviously needed to establish lithotrophic metabolism.

23.2 GENETIC STABILITY AND NATURALLY OCCURRING GENE TRANSFER

The lithoautotrophic abilities of carboxydotrophic and hydrogenotrophic bacteria can be considered as a bypath of normal organotrophic energy conversion, advantageous during transient exhaustion of organic nutrients (Schlegel, 1975 and Chapter 17). In their natural environments these bacteria encounter inorganic and organic substrates and probably use them concomitantly, thus growing mixotrophically. Since the lithotrophic functions in these organisms are dispensable, questions are raised concerning the genetic stability and the intra- and interspecific transmission of lithotrophic traits.

Reh and Schlegel (1975) first reported a spontaneous loss of hydrogen autotrophy in strains of *N. opaca*. This deficiency could not be restored by reversion (back mutation) but was restored by mating with the hydrogen-autotrophic parent strain. Moreover, it was possible to transfer hydrogen autotrophy to nonlithoautotrophic recipients such as *Rhodococcus erythropolis* (Table 23.2; Reh and Schlegel, 1981). A frequent occurrence of nonrevertible hydrogenase mutants in the presence of plasmid-curing agents and sublethal temperature was observed with cells of *P. facilis* strain ATCC 15376 (Pootjes, 1977). Nevertheless, it seems as if genetic instability in this bacterium is strain-specific. Lithotrophic mutants derived from *P. facilis* strain ATCC 17695 stably expressed hydrogen autotrophy (Warrelmann and Friedrich, 1986), whereas strain DSM620 exhibited a substantial loss of the lithoautotrophic marker after exposing the cells to chemicals or as a result of a spontaneous event. In this respect it is of interest that *P. facilis* DSM620 was successfully used in conjugations as a donor to complement Hox deficiency in mutants derived from

Table 23.2 Naturally occurring gene transfer of lithoautotrophic traits

Donor	Mode of transfer	Host range	Plasmid involvement	Reference
Nocardia opaca	Cell contact	Intra-, interspecific, intergeneric	ND	1
Pseudomonas facilis DSM620	Conjugation	Intra-, interspecific	RP4-mobilization	2
Alcaligenes hydrogenophilus	Conjugation	Intra-, interspecific	pHG21-a	3
Xanthobacter autotrophicus	Cell contact Phage	Intrastrain-specific	ND	4 5
A. eutrophus H16	Conjugation	Intra-, interspecific	pHG1	6

ND, no data.
(1) Reh and Schlegel, 1975; Sensfuss et al., 1986; (2) J. Warrelmann and B. Friedrich, unpublished result; (3) Friedrich et al., 1984a; Umeda et al., 1986; (4) Wilke, 1980; (5) Wilke and Schlegel, 1979; (6) Friedrich et al., 1981c; Friedrich et al., 1984a.

strain ATCC 17695. Furthermore, hydrogen autotrophy was transferred from *P. facilis* DSM620 to the nonlithoautotrophic bacterium *Pseudomonas delafieldii* (Table 23.2). This transfer required the presence of the conjugative IncP1 plasmid RP4 (J. Warrelmann and B. Friedrich; unpublished result). A conjugal transfer of hydrogen-oxidizing ability from *Alcaligenes hydrogenophilus* to *Pseudomonas oxalaticus* has also been achieved. The transconjugants gained the capacity to grow autotrophically with hydrogen as energy source (Umeda et al., 1986). A gene transfer based on cell contact, permitting the transfer of large pieces of chromosomal DNA, was detected in *X. autotrophicus* GZ29 (Table 23.2). This system was applicable to chromosomal mapping of genes involved in lithoautotrophy, nitrogen fixation, and methanol consumption (Wilke, 1980). A generalized transducing bacteriophage provides a second mode of gene transfer in *X. autotrophicus* (Table 23.2). Single chromosomal characters were transducible at a frequency of 10^{-4} per marker and phage particle (Wilke and Schlegel, 1979). The transduction appears to be useful for fine structure DNA analyses.

Growth of various *A. eutrophus* wild-type strains in the presence of mitomycin C resulted in the loss of ability to grow lithoautotrophically at frequencies up to 43 percent (Lim et al., 1980; Andersen et al., 1981). Of the four wild-type strains tested, *A. eutrophus* ATCC 17697 exhibited a high degree of genetic instability, whereas in most other isolates hydrogen autotrophy was rather stably maintained. Occasionally nonrevertible hydrogenase mutants (Hox^-) were found which however did recover hydrogen autotrophy after mating with a Hox^+ donor (Table 23.2). The conjugal transfer of Hox between different strains and species of *Alcaligenes* occurred at frequencies up to 10^{-2} per donor (Friedrich et al., 1981c; Friedrich et al., 1984a).

The sulfur-oxidizing organism *T. pantotropha* has been shown to be lysogenic for a bacteriophage that has not been characterized so far. It has been suggested that phage formation may cause genetic instability in this strain (Mayer and Friedrich, 1986). The size of the *T. ferrooxidans* genome was estimated to be approximately two-thirds that of *E. coli*, and 6 percent of this genomic DNA consists of moderately repeated DNA sequences characterized as two distinct families. It is not known yet whether these repeated sequences are involved in gene rearrangements or amplification (Yates and Holmes, 1987).

23.3 PLASMIDS AND THEIR ROLE IN LITHOAUTOTROPHY

Naturally occurring conjugal transfer of hydrogen autotrophy and the spontaneous loss of this character suggested a location of genes necessary for hydrogen-dependent autotrophic growth on transmissible plasmids. In fact, extrachromosomal DNA was shown to be abundant in lithoautotrophic bacteria. Its presence has been demonstrated in sulfur-, iron-, nitrite-, and carbon monoxide-oxidizing strains (Table 23.3) as well as in hydrogenotrophs (Table 23.4), and the collection of data listed herein is not complete. However, all of the

Table 23.3 Plasmids in thiobacteria, nitrifiers, and carboxydobacteria

Bacterial strain	Plasmid Number	Plasmid Size (kb)	Reference
Thiosphaera pantotropha	2	110, >450	Chandra and Friedrich, 1986
Thiobacillus versutus	2	106, >450	Gerstenberg et al., 1982
T. acidophilus ATCC 27807	4	18, 27, 76, >90	Mao et al., 1980
T. intermedius	1	36	Davidson and Summers, 1983
T. ferrooxidans ATCC 13728	1	54	Martin et al., 1981
T. ferrooxidans ATCC 33020	3	6.7, 16, >23	Holmes et al., 1984
T. ferrooxidans FC	3	4.5, 12.4, 27.6	Rawlings et al., 1984
Nitrobacter hamburgensis	1	114	Kraft and Bock, 1984
Achromobacter carboxydus	3	150, 405, 458	Gerstenberg et al., 1982
Pseudomonas carboxydoflava	2	34, 116	Gerstenberg et al., 1982
P. carboxydovorans OM5	1	128	Gerstenberg et al., 1982
P. carboxydovorans DSM1227	3	3.5*	Kwon and Kim, 1985

*Three plasmids, but only one plasmid characterized.

Table 23.4 Plasmids and their physiological function in hydrogen bacteria

Strain	Plasmid Designation	Plasmid Size (kb)	Function[a]	Reference
Nocardia opaca MR11	pHG32-a	140	Thallium^r	1
	pHG32-b	17	Cryptic	
N. opaca MR22	pHG33	110	Thallium^r	1
Pseudomonas facilis ATCC 17695	pHG20	270	Cryptic	2
P. facilis DSM620	pHG22-a	440	Hox, Cfx	3
	pHG22-b	200	Cryptic	
Alcaligenes eutrophus H16	pHG1	450	Hox, Cfx, Nir, Fhp, Tra	4
A. eutrophus ATCC 17707	pHG12-a	>550	Cryptic	5
	pHG12-b	500	Hox, Ose, Tra	6
A. hydrogenophilus	pHG21-a	419	Hox, Tra	7
	pHG21-b	357	Cryptic	
A. eutrophus CH34	pMOL28	163	Ni^r, Co^r, Hg^r	8
	pMOL30	238	Zn^r, Cd^r, Co^r, Hg^r	
A. eutrophus ATCC 17706	pRMB1	ND	Hox, Cfx, Tra$^-$	9
Pseudomonas denitrificans ATCC 17741	pHG16-a	75	Cryptic	10
	pHG16-b	>450	Cryptic	
Bradyrhizobium japonicum SR-1	pRJa-SR1a	160	Cryptic	11
	pRJa-SR1b	420	Cryptic	
Rhodobacter capsulatus B10	Undesignated	129	Cryptic	12

[a]Phenotype designations are explained in the text.
ND, no data.
(1) Sensfuss et al., 1986; (2) Warrelmann and Friedrich, 1986; (3) J. Warrelmann and B. Friedrich, unpublished result; (4) reviewed by Friedrich, 1987; (5) C. Hogrefe and B. Friedrich, unpublished result; (6) Cangelosi and Wheelis, 1984; (7) Friedrich et al., 1984a; (8) Mergeay et al., 1985; (9) Behki et al., 1983; (10) Gerstenberg et al., 1982; (11) Cantrell et al., 1982; (12) Willison et al., 1987.

previously reported plasmids in thiobacteria, nitrifying strains, and carboxydobacteria are cryptic. It has been proposed that plasmids present in metal-leaching acidophiles may encode heavy-metal resistances (Summers and Silver, 1978) but a causal relationship has not yet been established. The plasmid patterns of chemolithotrophic thiobacilli are rather heterogenous with respect to plasmid number and size (Table 23.3). The 6.7 kb plasmid pTf1 from *T. ferrooxidans* ATCC 33020 was cloned into the *E. coli* specific plasmid pBR322 to construct potential cloning vectors for exchanging genetic information between *E. coli* and *T. ferrooxidans* (Holmes et al., 1984). A recombinant plasmid containing an origin of replication from a cryptic plasmid of *T. ferrooxidans* FC was shown to be able to replicate in *E. coli* (Rawlings et al., 1984). Moreover, nonconjugative endogenous plasmids of *T. ferrooxidans* were mobilized at high frequencies among *E. coli* strains (Rawlings and Woods, 1985). Broad-host-range plasmids of the incompatibility group P1 can be transferred and expressed in a variety of thiobacilli (Kulpa et al., 1983; Davidson and Summers, 1983). These achievements are encouraging for future developments of genetic systems in these organisms. No correlation seems to exist between thiosulfate-oxidizing ability and the presence of two large plasmids in *T. pantotropha* (Table 23.3). Although one Sox^- mutant was cured of the indigenous megaplasmid, it was still able to revert to the parental phenotype (Chandra and Friedrich, 1986). Furthermore, the loss of the 110 kb plasmid did not affect the lithotrophic phenotype of its host, and the Tn*5* insertions identified so far were found to map on the chromosomal DNA (C. G. Friedrich, personal communication).

The functions of plasmids in nitrifying and carboxydotrophic bacteria are still hypothetical. A relationship between the small 3.5 kb plasmid of *P. carboxydovorans* (Table 23.3) and carbon monoxide dehydrogenase has been discussed by Kwon and Kim (1985). They suggested an involvement of this plasmid in the production of carbon monoxide dehydrogenase and presented evidence for a plasmid-mediated proteolytic modification of the enzyme (Kwon et al., 1986). Some of the Cox mutants introduced in Table 23.1 exhibited physical alterations in the large plasmid indigenous to *P. carboxydovorans* (M. Kraut and O. Meyer, personal communication). More experimental data are required to substantiate a function of plasmids in carbon monoxide metabolism.

A survey on plasmids in hydrogen bacteria (Table 23.4) shows the presence of remarkably large plasmids, so-called megaplasmids. Although physical evidence for the existence of plasmids in *N. opaca* was presented, this extrachromosomal DNA does not seem to be involved in the conjugative transfer of hydrogen autotrophy. Either putative plasmids are too large or unstable to be detected or the lithoautotrophic character in *N. opaca* resides on an autonomous element which is integrated into the chromosome (Sensfuss et al., 1986). In this respect it is of interest that it was not possible to show the presence of plasmid DNA in *X. autotrophicus* (Gerstenberg et al., 1982), where conjugative gene transfer has been elucidated like in *N. opaca* (Table 23.2; Wilke, 1980). Nevertheless, resistance to 50 μM thallium acetate was assigned to the *N. opaca* plasmids pHG31-a and pHG33. Moreover, one of these plasmids appears to be associated with a restriction-modification system (Sensfuss et al., 1986).

Pootjes (1977) proposed a linkage between hydrogen oxidation and plasmid DNA in *P. facilis*. The location of Tn*5* insertions in Hox^- mutants of *P. facilis* ATCC 17695 did not confirm this assumption, although the strain was shown to be host of a 270 kb plasmid (Warrelmann and Friedrich, 1986). Recent experimental results indicate that the 440 kb plasmid pHG22-a detected in *P. facilis* DSM620 carries essential genes of lithoautotrophic metabolism (J. Warrelmann and B. Friedrich, unpublished result). This observation suggests that plasmid pattern and function may differ significantly even among closely related strains. Similar data were obtained with *A. eutrophus*. Although the majority of wild-type strains were shown to contain a single megaplasmid of approximately 450 kb, there are isolates such as *A. hydrogenophilus* (Ohi et al., 1979), *A. eutrophus* CH34 (Mergeay et al., 1985), and *A. eutrophus* ATCC 17707 which harbor two plasmids (Table 23.4). In order to detect plasmid molecules of such a large size, specific lysis procedures must be employed. The technique is mainly based on alkaline sodium dodecyl sulfate lysis of the cells, followed by heat treatment, salt precipitation, and phenol extraction; it was adapted for preparative purposes (Hogrefe and Friedrich, 1984; Nies et al., 1987). The sizes of the plasmids were routinely estimated by agarose gel electrophoresis in reference to standard plasmids of known molecular weight. Although the error within this size range is rather large, the value of 450 kb calculated for plasmid pHG1 of *A. eutrophus* H16 (Table 23.4) is most reliable since it is in good agreement with the contour length of the plasmid DNA measured by electron microscopy (Rohde et al., 1986).

A comparison of restriction patterns of pHG plasmids isolated from closely related *A. eutrophus* strains revealed a high degree of fragment homology, whereas plasmid pHG21-a from *A. hydrogenophilus* (Table 23.4) exhibited marked differences in restriction sites. No similarities were recognized between plasmids from *A. eutrophus* H16 and those of *P. facilis*, *P. denitrificans* (Hogrefe and Friedrich, 1984), and *A. eutrophus* CH34 (Nies et al., 1987). The latter observation is not surprising since the pMOL plasmids (Table 23.4) code for multiple resistances to heavy metal ions and not for lithoautotrophic genes (Mergeay et al., 1985). Plasmid-borne hydrogenase genes have been conclusively demonstrated only in various strains of *A. eutrophus*, in *A. hydrogenophilus*, possibly in *P. facilis* DSM620 (Table 23.4), and the nonautotroph *Rhizobium leguminosarum* (Brewin et al., 1980). Previously it was shown that plasmids pHG1 (Bowien et al., 1984; Klintworth et al., 1985), pAE7 (Andersen and Wilke-Douglas, 1984), and pRMB1 (Behki et al., 1983) are associated with autotrophic carbon dioxide fixation. Moreover, plasmid pHG1 also codes for determinants of denitrification (Römermann and Friedrich, 1985) and a soluble flavohemoprotein (K. Schmidt, personal communication) which is capable of binding oxygen reversibly but whose physiological role is unknown (Bowien and Schlegel, 1981). As discussed above, Hox encoding plasmids of *A. eutrophus* are generally self-transmissible by conjugation and hence must bear Tra and Mob functions (Friedrich et al., 1981c). Plasmid pRMB1 appears to be the only exception to this rule, lacking the conjugative ability (Behki et al., 1983).

Extrachromosomal DNA was also identified in nitrogen-fixing hydrogen bacteria. Although large plasmids were detected in *Bradyrhizobium japonicum* (Table 23.4), correlations of Hox and plasmid content could not be established. Two plasmids were found in nonrevertible hydrogenase mutants but unexpectedly the corresponding parent strain was free of discernible plasmid DNA (Cantrell et al., 1982). Mutants of *Rhodobacter capsulatus* B10 cured of the endogenous plasmid (Table 23.4) were unaffected in autotrophic growth, indicating no plasmid function in lithoautotrophy (Willison et al., 1987).

23.4 STRUCTURE OF HYDROGENASE

Hydrogenases catalyze the reaction $H_2 \rightleftharpoons 2H^+ + 2e^-$ and are thus involved in the oxidation as well as the formation of the simplest molecular metabolite. They are widespread in nature and are present in various microorganisms, where they are found in the cytoplasm, in the periplasmic space, and as integral membrane components. Hydrogenases have been subject of several review articles (Schlegel and Schneider, 1978; Adams et al., 1981; Schlegel and Schneider, 1985; see also Chapter 17). The enzymes that are most pertinent to this communication are the two hydrogenases present in *A. eutrophus*.

Despite a remarkable heterogeneity concerning the physical and catalytic properties, two distinct groups of hydrogenases have been recognized. Their distribution and subunit composition in aerobic lithoautotrophs is shown in Table 23.5. The largest group, represented by the so-called particulate hydrogenase (Hop), has in *A. eutrophus* an $\alpha\beta$-dimeric structure, a molecular weight near 100 kDa, and is characterized as a membrane-bound protein that is linked to the respiratory chain (Schink and Schlegel, 1979). As all hydrogenases examined so far, it contains iron-sulfur centers as prosthetic groups (Schneider et al., 1983). Nickel plays an important role in the formation of active hydrogenase in *A. eutrophus* (Friedrich et al., 1981b; Friedrich et al., 1982), and many, but not all, other hydrogenases also contain nickel as an essential constituent (reviewed by Hausinger, 1987). The hydrogenases from aerobic lithoautotrophs characterized to date belong exclusively to the class of nickel-iron-sulfur proteins. In most cases, the natural electron acceptor that interacts primarily with hydrogenase is still unknown. One of the few exceptions to this is the soluble hydrogenase present in *A. eutrophus* H16 and *N. opaca* 1b. The heterotetrameric protein (Table 23.5) reduces NAD^+ to NADH at the expense of dihydrogen (Schneider and Schlegel, 1976). The enzyme of *N. opaca* has been separated into two dimeric species, one with hydrogenase activity and the other with diaphorase activity (NADH:electron acceptor oxidoreductase; Schneider et al., 1984b). Although the nickel-containing dimer with subunits of 56 kDa and 27 kDa (Table 23.5) shares some structural homology with the particulate hydro-

Table 23.5 Structures of hydrogenases from aerobic lithoautotrophs

	Subunit composition (kDa)[a]						
	Hop[b]		Hos[b]				
Bacterial species	α	β	α	β	γ	δ	Reference
Alcaligenes eutrophus H16	61	30	56	26	63	30	1
Alcaligenes latus	67	34	Not present				2
Nocardia opaca 1b	Not present		56	27	64	31	3
Paracoccus denitrificans	64	35	Not present				4
Bradyrhizobium japonicum	65	35	Not present				5
Rhodobacter capsulatus	60	35	Not present				6

[a]Molecular weight, kilodaltons.
[b]Hop, particulate hydrogenase; Hos, soluble hydrogenase.
(1) Schink and Schlegel, 1979; Schneider et al., 1983; Schneider and Schlegel, 1976; Hornhardt et al., 1986; (2) Pinkwart et al., 1983; (3) Schneider et al., 1984a; Schneider et al., 1984b; (4) Knüttel et al., 1988; (5) Harker et al., 1985; Arp, 1985; Maier, 1986; (6) Seefeldt et al., 1987.

genase, there is no immunological relationship between the two proteins (K. Schneider, personal communication) as proposed before (Schneider and Piechulla, 1986). On the other hand, Arp et al. (1985) demonstrated cross reactivity between the particulate hydrogenases from *A. eutrophus, Alcaligenes latus,* and *B. japonicum*. This result indicates an extensive homology between the Hop proteins of various strains.

From a mutant of *A. eutrophus* H16 a single polypeptide of aproximately 57 kDa was isolated which corresponds to the second largest subunit of the soluble hydrogenase. It contained nickel, did not interact with NAD^+ as electron acceptor, but mediated weak catalysis with methylene blue, ferricyanide, and cytochrome *c*. The result suggests that this specific peptide contains the catalytic center of the hydrogenase protein (Hornhardt et al., 1986). Reconstitution experiments with the enzyme from *Nocardia* led to the conclusion that the NAD-reducing capacity of the soluble hydrogenase depends on the presence of the diaphorase dimer to which FMN is bound as a cofactor (Schneider et al., 1984a).

23.5 REGULATION OF HYDROGENASE SYNTHESIS

A number of physiological factors have been defined as regulatory signals of hydrogenase synthesis (Bowien and Schlegel, 1981). It seems as if the regulation of hydrogenase is as complex and diverse as its structure. The two hydrogenases in *A. eutrophus* were shown to be coordinately expressed under conditions of reductant limitation in the absence of hydrogen (Friedrich, 1982), an ambient temperature below 33°C (Friedrich and Friedrich, 1983), and in some wild-type strains at a low oxygen tension (Cangelosi and Wheelis, 1984). Organic substrates that support fast growth led to hydrogenase repression (Friedrich et al., 1981a). Based on these data, a heterotrophic system for the cultivation of hydrogenase-derepressed cells was designed (Friedrich et al., 1981b) which proved to be extremely useful for mutant characterizations and for the production of hydrogenase protein for biochemical purposes. As illustrated in Figure 23.1A, cells of *A. eutrophus* H16 were grown in a medium containing two substrates. During the substrate down-shift from fructose to glycerol, the formation of soluble and particulate hydrogenase was highly derepressed, reaching a fourfold higher level than that obtained under lithoautotrophic conditions. Up to 5 percent of the cell protein consisted of hydrogenase. Panel B of Figure 23.1 illustrates the temperature sensitivity of hydrogenase synthesis. Only thermoresistant mutants are capable of forming hydrogenase at a temperature above 33°C (Friedrich and Friedrich, 1983). Several other proteins of unknown physiological function were shown to be under the same control circuitry in *A. eutrophus* (Kärst and Friedrich, 1987). One of the most prominent proteins, a dimer formed at a molar ratio to soluble hydrogenase of 4:1, was found to be present in all hydrogen bacteria that do contain the NAD-dependent hydrogenase, including *N. opaca* (Kärst et al., 1987).

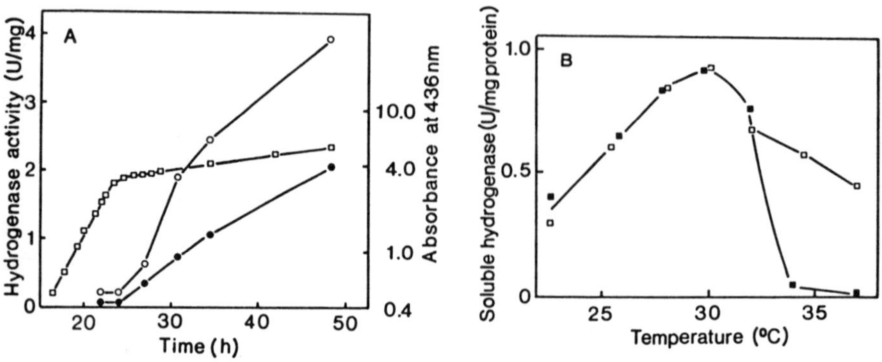

Figure 23.1 Derepression of hydrogenase in *Alcaligenes eutrophus.* (A) Cells were grown with fructose and glycerol as carbon sources. Growth [□–□], soluble hydrogenase [o–o], and particulate hydrogenase [•–•] activities. The data are taken from Friedrich et al., 1981b. (B) Soluble hydrogenase activity of the wild type H16 [■–■] and the thermoresistant mutant HF116 [□–□] in a succinate-limited chemostat. The data are taken from Friedrich and Friedrich, 1983.

Although the regulation of hydrogenase in other lithoautotrophs has not been as extensively studied as in *A. eutrophus,* the available data indicate that hydrogenase synthesis in most, if not all, hydrogen bacteria, is subject to a transcriptional control. The regulatory signals, however, may differ remarkably. In *A. hydrogenophilus,* whose hydrogen-oxidizing enzyme system is similar to that of *A. eutrophus,* molecular hydrogen is absolutely needed for hydrogenase induction (Friedrich et al., 1984). This has also been shown for *Paracoccus denitrificans* ATCC 17741 (Bowien and Schlegel, 1981). In other organisms the role of hydrogen as inducer of hydrogenase synthesis is not as clearly pronounced. In many cases hydrogenase expression is affected by the organic carbon source. For example, hydrogenase synthesis was inhibited but never completely repressed in *P. facilis* (Warrelmann and Friedrich, 1986) and *A. latus* (Doyle and Arp, 1987). In *B. japonicum* its level was shown to be dependent on carbon substrates and oxygen tension (Maier et al., 1978; Maier, 1986). Hydrogenase protein synthesis was induced by hydrogen during heterotrophic growth (van Berkum, 1987), and according to a recent observation DNA gyrase activity seems to be specifically involved in the expression of hydrogenase in *B. japonicum* (Novak and Maier, 1987).

23.6 CLONING OF HYDROGENASE GENES AND THEIR GENETIC STRUCTURE

Cosmids containing hydrogenase genes (*hox*) have first been isolated from gene banks of *B. japonicum* DNA (Cantrell et al., 1983). Plasmid pHU1 contains around 15 kb of hydrogenase-specific DNA organized in at least two transcriptional units. A similar plasmid (pHU52) with an extra 5.5 kb *Eco*RI fragment is supposed to encode the entire genetic information required for hydrogenase expression in *B. japonicum* in the free-living state (Lambert et al., 1985). Since then a number of *hox* genes have been isolated from autotrophic as well as heterotrophic organisms. The *hox* genes in *B. japonicum* are most probably chromosomally bound or present on large undetected plasmids (Cantrell et al., 1982). The plasmid location of *hox* genes in *A. eutrophus* raised the following questions: 1) Does the megaplasmid encode structural and/or regulatory genes? 2) Are chromosomal genes required in addition to the plasmid DNA? A straightforward answer was deduced from results of transfer experiments using strains with mutations in structural or regulatory *hox* genes as donors in matings with plasmid-free Hox$^-$ recipients. If the mutation mapped on the megaplasmid the transconjugants should inherit the mutant phenotype. The results of this study are summarized in Table 23.6. They indicate that *hox* genes coding for soluble hydrogenase (Hos) and particulate hydrogenase (Hop) are plasmid-borne (Hogrefe et al., 1984). With the aid of thermoresistant mutants (Figure 23.1B), in combination with interspecific plasmid exchange between *A. eutrophus* and *A. hydrogenophilus* (Friedrich et al., 1984a), we could further demonstrate a linkage between the Hox-specific regulation and the corresponding megaplasmid DNA. Moreover, the oxygen-sensitive Hox control in *A. eutrophus* ATCC 17707 was also reported to be associated with a plasmid (Cangelosi and Wheelis, 1984). The pleiotropic Hno mutants, described in section 23.1, however, were unimpaired in the plasmid DNA, indicating a requirement for at least one chromosomal gene for active hydrogenase expression (Hogrefe et al., 1984; Römermann et al., 1988). Chromosomal mutations causing the reduction or loss of both hydrogenases have also been detected in mutants of *A. eutrophus* ATCC 17707 (Cangelosi and Wheelis, 1984).

Another experimental strategy focused on the identification and precise mapping of *hox* genes on the genomic DNA of *A. eutrophus*. This was achieved by physical analysis of mutant DNA containing plasmid deletions or insertions and the construction of gene banks using megaplasmid DNA and chromosomal DNA. Both steps involved the preparation of pure plasmid DNA as outlined in section 23.3. A few examples of the restriction pattern obtained with the endonuclease *Eco*RI are documented in Figure 23.2. All Hox$^-$ deletion mutants exhibited the loss of an 8.6 kb fragment (Figure 23.2, panel A). This particular

Table 23.6 Transfer of plasmid-linked *hox* mutations into a plasmid-free recipient of *A. eutrophus*

Relevant phenotype[a]	
Donor	Transconjugant
Hos$^+$Hop$^+$HoxTs	Hos$^+$Hop$^+$HoxTs
Hos$^-$Hop$^+$	Hos$^-$Hop$^+$
Hos$^+$Hop$^-$	Hos$^+$Hop$^-$
HoxTr	HoxTr
Hoxi	Hoxi
Hno$^-$	Hno$^+$

[a]Hos, soluble hydrogenase; Hop, particulate hydrogenase; HoxTr, temperature-resistant hydrogenase formation; Hoxi, H_2-inducible hydrogenase synthesis; Hno, pleiotropic as described in section 23.1

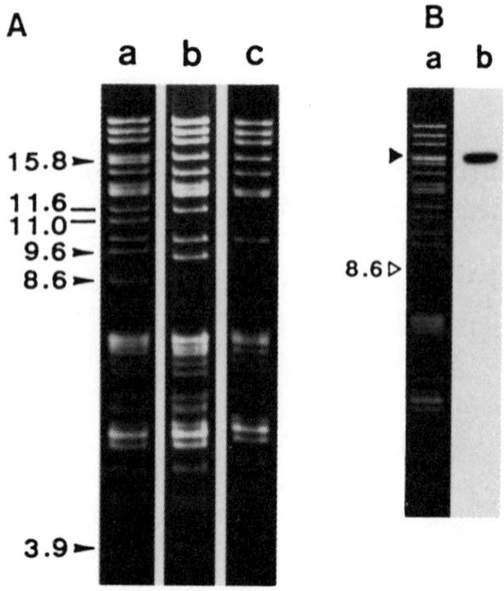

Figure 23.2 Physical analysis of megaplasmid DNA. (A) *Eco*RI restriction pattern of megaplasmids with deletions. Lane a, *A. eutrophus* H16 wild type; lane b, HF47, lane c, HF184. (B) *Eco*RI restriction pattern of a megaplasmid with a Tn*5* insertion (a) and Southern hybridization analysis of plasmid DNA of a Tn*5* insertion mutant hybridized with ^{32}P-labeled probes of the Tn*5*-containing vector pSUP5011 (b). Numbers indicate the sizes of the fragments in kilobase pairs.

fragment was size-enlarged in a Hox$^-$ insertion mutant due to the integration of transposon Tn*5* as also shown by DNA-DNA hybridization (Figure 23.2, panel B). From a series of experiments it was concluded that the mutations which rendered the cells Hos$^-$ or Hop$^-$, or even affected both activities (Hox$^-$), were clustered on megaplasmid pHG1 within a DNA sequence of around 100 kb (Kortlüke et al., 1987). A preliminary map of this region is illustrated in Figure 23.3.

This map was confirmed by results obtained from molecular cloning experiments of structural and regulatory *hox* genes (Eberz et al., 1986). The gene bank was constructed using the *E. coli* specific Mob site-containing vector pSUP202 (Simon et al., 1983). Hybrid plasmids with *hox* genes were identified by their ability to restore the activity of Hos$^-$, Hop$^-$, and Hox$^-$ mutants via homologous recombination. From these results we obtained evidence that genes coding for Hos activity (*hox*S) are located on an 11.6 kb *Eco*RI fragment, whereas the genes coding for Hop activity (*hox*P) are present on a distantly located 11.0 kb *Eco*RI fragment (Eberz et al., 1986). The latter fragment has previously been identified as the locus of megaplasmid-linked genes involved in autotrophic CO_2 fixation (Klintworth et al., 1985; see also Chapter 24). A DNA sequence of two contiguously arranged *Eco*RI fragments on 8.6 kb and 2.0 kb was shown to be essential for the expression of active Hos and Hop proteins (Eberz et al., 1986). The cloned fragments, indicated by the bold lines in Figure 23.3, are consistent with the sites of mutations mapped on plasmid pHG1. The *hox* DNA was subcloned into the broad-host-range vector pVK101 (Knauf and Nester,

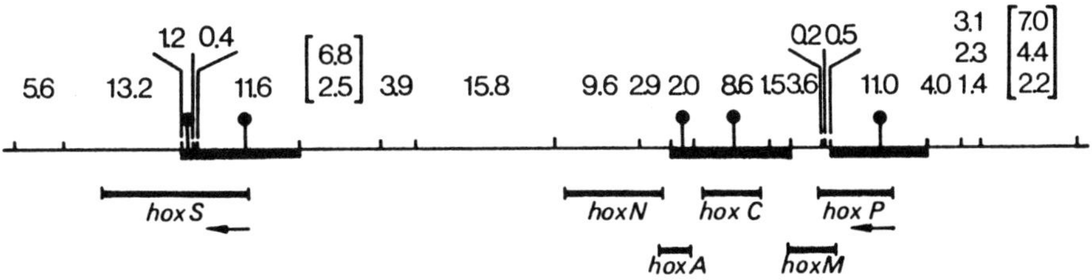

Figure 23.3 *Eco*RI restriction map of megaplasmid DNA from *Alcaligenes eutrophus*. The size of fragments is given in kilobase pairs. Thick lines indicate cloned genes; insertions are symbolized as ⚲. *hox* gene clusters that have been cloned and identified recently are listed below the map. Arrows indicate the orientation of transcription. The genes are specified in the text.

1982), and the resulting hybrids were used to complement the three classes of hydrogenase mutants (Hos$^-$, Hop$^-$, Hox$^-$). These studies clearly revealed that the cloned DNA consisted of incomplete structural *hox* gene sequences whereas the 10.6 kb fragment complemented all of the Hox$^-$ mutants impaired in both hydrogenases (Eberz et al., 1986). Cosmids with *hox* genes of *A. hydrogenophilus* have similarly been isolated from a gene bank of megaplasmid pHG21-a. The genes were identified by transferring the cosmids into *Pseudomonas oxalaticus* OX1 and screening transconjugants for the ability to grow autotrophically with hydrogen. One of the transconjugants expressed the membrane-bound Hop activity of *A. hydrogenophilus* (Yagi et al., 1986). The soluble hydrogenase genes of this organism have been cloned on a 13.5 kb *Hin*dIII fragment and could be expressed in *A. eutrophus* (A. Tran-Betcke and B. Friedrich, unpublished result). *hox* genes coding for the membrane-bound hydrogenase have also been cloned from the photosynthetic lithoautotroph *Rhodobacter capsulatus* (Colbeau et al., 1986) and the hydrogen bacterium *P. facilis* (J. Warrelmann and B. Friedrich, unpublished result).

A gene that was able to restore the pleiotropic Hno$^-$ phenotype in mutants of *A. eutrophus* was detected in a chromosomal DNA bank of this organism (D. Römermann and B. Friedrich, unpublished result). Although Hno was initially identified on the basis of its role in hydrogenase gene expression (Hogrefe et al., 1984), succeeding studies of Römermann et al. (1988) revealed a much broader physiological significance than originally envisaged. This was also supported by recent results: 1) a functionally similar gene was detected in *P. facilis* which was able to complement Hno$^-$ mutants of *A. eutrophus* (J. Warrelmann and B. Friedrich, unpublished result); and 2) by transferring the *hno* gene from *A. eutrophus* into an *ntrA* mutant of *Klebsiella aerogenes* it was possible to restore prototrophy (D. Römermann and B. Friedrich, unpublished result). The *ntrA* gene is known to be involved in nitrogen-controlled gene expression, e.g., nitrate assimilation, nitrogen fixation, and in C_4-dicarboxylate transport of *Rhizobium meliloti* (Ronson et al., 1987). The product of *ntrA* was identified as a sigma factor of an RNA polymerase which recognizes NtrA-specific promoter sequences (Hirschmann et al., 1985; Hunt and Magasanik, 1985). The complementation data suggest that the *hno* gene detected in *A. eutrophus* and *P. facilis* codes for an NtrA-like product.

23.7 THE CURRENT STATE OF HOX GENETICS IN *A. EUTROPHUS*

The cloning of the four structural *hoxS* and the two structural *hoxP* genes from *A. eutrophus* has been completed. The separate lines of Figure 23.3 indicate the location of the respective sequences within the lithotrophic gene cluster. To study further *hox* gene expression in *A. eutrophus,* a broad-host-range *lacZ* protein fusion vector (Rossen et al., 1985) was used to construct a hybrid containing a 5.0 kb *Bam*Hl fragment of the *hoxS* region fused to the *lacZ* gene from *E. coli.* Introducing this hybrid into the naturally Lac$^-$ bacterium *A. eutrophus* it was possible to study the regulation of the *hoxS* promoter by measuring the level of β-galactosidase activity. The expression of the *hoxS-lacZ* hybrid was repressed in mutants bearing lesions either in the plasmid gene *hoxA* or the chromosomal gene *hno* (Figure 23.4) whereas mutations in the structural genes *hoxS* and *hoxP* did not affect β-galactosidase expression (B. Friedrich, U. Warnecke, and A. Nies, unpublished result). Further experiments using the *hoxS-lacZ* hybrid and mutants carrying different insertions in the 10.6 kb control region of plasmid pHG1 revealed that the DNA sequence adjacent to *hoxA,* designated *hoxC* (Figure 23.4) is required for the synthesis of active Hos and Hop proteins but dispensible for the expression of β-galactosidase activity (G. Eberz and B. Friedrich, unpublished results). From these results we

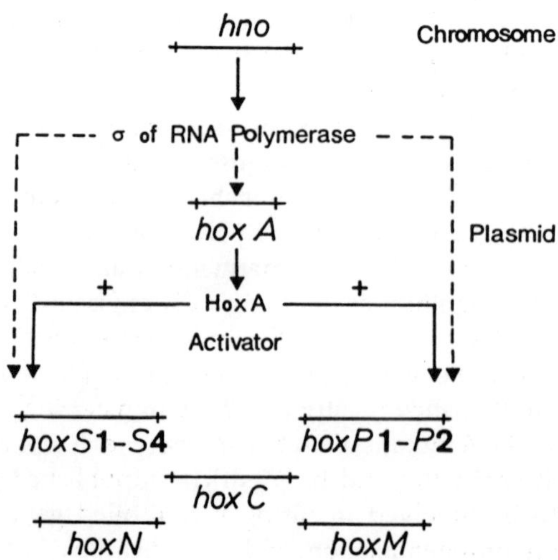

Figure 23.4 Current model of hydrogenase (*hox*) genetics of *Alcaligenes eutrophus* H16. Details are given in the text.

conclude that the *hoxC* product(s) are either involved in the processing of the two hydrogenases or in the incorporation of nickel or iron into these proteins. It was shown before that both processes require protein synthesis (Friedrich et al., 1984b).

A plasmid-bound nickel-related marker has been discovered in the course of deletion analyses. Although the precise location and function of *hoxN* (Figure 23.4) are still not known, experimental data suggest that *hoxN* maps between the *hoxA* gene and the *hoxS* sequence, near a site which was reported to contain an inverted repeat of about 7.2 kb (Rohde et al., 1986). Mutants bearing deletions in this part of the megaplasmid were only able to grow autotrophically with hydrogen in the presence of a high concentration of nickel ions. This nickel requirement correlated with the intracellular level of hydrogenase activity. It could be restored by introducing a hybrid plasmid containing a DNA sequence specified as *hoxN* (Figure 23.4). The nickel requirement of the mutants was severely enhanced by elevating the concentration of magnesium ions in the medium. It was shown before that one of the two systems that mediate the uptake of nickel ions in *A. eutrophus* is strongly inhibited by Mg^{2+} whereas high affinity nickel transport does not seem to be affected by divalent cations (Lohmeyer and Friedrich, 1987). These results suggest that one of the products of *hoxN* may be involved in nickel metabolism.

The membrane fraction of *A. eutrophus* H20 (ATCC 17700) was reported to be devoid of H_2-dependent methylene-blue-reducing activity. Immunological investigations revealed the presence of particulate antigen in the soluble fraction of the extract (Friedrich et al., 1986). The transfer of plasmid pHG1 from *A. eutrophus* H16 into a plasmid-cured derivative of strain H20 resulted in transconjugants that contained Hop-specific catalytic and immunological activity in the membrane fraction of the extract. Moreover, a 2.1 kb *Xho* fragment was identified in the gene bank of plasmid pHG1 which caused the Hop protein to attach to the membrane when introduced into strain H20 (C. Kortlüke and B. Friedrich, unpublished result). This result suggests that the gene product of this particular sequence, designated *hoxM* (Figure 23.4), may constitute an anchor to the membrane. It is interesting to note that *hoxM* represents a separate transcript although the sequence is closely linked to the structural *hoxP* genes.

Figure 23.4 summarizes the current model of *hox* genetics in *A. eutrophus* H16. It shows that the expression of the plasmid-linked genes *hoxS* and *hoxP* is under control of the chromosomal gene *hno* whose product is assumed to be a sigma factor of RNA polymerase. *hoxS* and *hoxP* gene expression also depend on the presence of the plasmid gene *hoxA*, which positively regulates structural *hox* gene transcription. It is not known whether *hoxA* itself is under control of *hno* or even subject to a separate cascade-like regulation process. However, we do know that at least three additional *hox* transcripts are essential for the expression of catalytically active Hos and Hop proteins: 1) *hoxN* seems to play a role in the supply of nickel; 2) *hoxC* is apparently involved in hydrogenase processing; and 3) *hoxM* mediates membrane association. It is not known yet whether these genes are part of the *hox* control circuitry.

23.8 CONCLUSIONS AND FUTURE ASPECTS

The techniques described in this paper should be generally applicable to the production of specific mutations in a variety of facultatively lithoautotrophic bacteria. These techniques should also be useful in the initiation of genetic studies. DNA fragments identified and cloned on the basis of transposon insertional inactivation can be further mutagenized to rapidly construct mutations on cloned DNA and then recombine these fragments into the genome of a lithotrophic host to scan for other mutant phenotypes.

Mutant studies with obligate lithoautotrophs will mostly be limited by the necessity to isolate the genes of interest first, to mutagenize them, and to determine the defective phenotype in a nonobligate lithotrophic host. A major limiting problem is that the respective genes have to be expressed therein, and we do not know in detail how many functions are actually involved in the establishment of lithoautotrophy. The advanced studies on the genetics of hydrogen oxidation indicate a system which seems to be as complex as bacterial nitrogen fixation. Despite this reservation it seems to be a pragmatic approach to clone the genes from obligate lithoautotrophs in facultative hosts instead of using the heterotrophic *E. coli* as the sole cloning organism. The successful development of thiobacilli-specific shuttle vectors looks promising for future genetic research with this group of bacteria. It has been shown that several facultative species are capable of using multiple inorganic energy sources, and that it is even possible to convert heterotrophic bacteria to facultative lithotrophs by transfer of endogenous plasmids or genetically manipulated recombinant DNA.

While we are at present unable to contemplate extensive genetic studies with obligately lithotrophic bacteria, it is likely that most of the immediate problems with facultative species will be resolved over the next few years. DNA sequence analyses of structural and regulatory hydrogenase genes as initiated with *Desulfovibrio* species (Voordouw and Brenner, 1985; Menon et al., 1987), the search for promoter regions, and the understanding of their function, will elucidate the extent of homology between hydrogenases from aerobic and anaerobic eubacteria and the corresponding enzymes from archaebacteria. Since hydrogenases are widespread in the world of procaryotes, they are excellently suited to study evolution. Finally, by evaluating interspecies homology and the organization of lithotrophic gene transcripts we may raise questions concerning the multivarious plasmids, their ecological importance, and their role in intergenomic rearrangements, cointegrate formation, and recombination with the chromosomal DNA.

I am grateful to my co-workers: G. Eberz, C. Kortlüke, D. Römermann, A. Tran-Betcke, U. Warnecke, and J. Warrelmann for providing unpublished data. A part of this work was supported by Deutsche Forschungsgemeinschaft grants awarded to B.F.

REFERENCES

Adams, M.W.W., Mortenson, L.E., and **Chen, J.S.** 1981. Hydrogenase. *Biochimica et Biophysica Acta* 594:105–176.

Andersen, K., Tait, R.C., and **King, W.R.** 1981. Plasmids required for utilization of molecular hydrogen by *Alcaligenes eutrophus. Archives of Microbiology* 129:384–390.

Andersen, K. and **Wilke-Douglas, M.** 1984. Construction and use of a gene bank of *Alcaligenes eutrophus* in analysis of ribulose bisphosphate carboxylase genes. *Journal of Bacteriology* 159:973–978.

Arp, D.J. 1985. *Rhizobium japonicum* hydrogenase: purification to homogeneity from soybean nodules, and molecular characterization. *Archives of Biochemistry and Biophysics* 237:504–512.

Arp, D.J., McCollum, L.C., and **Seefeldt, L.C.** 1985. Molecular and immunological comparison of membrane-bound, H_2-oxidizing hydrogenases of *Bradyrhizobium japonicum, Alcaligenes eutrophus, Alcaligenes latus,* and *Azotobacter vinelandii. Journal of Bacteriology* 163:15–20.

Behki, R.M., Selvaray, G., and **Iyer, V.N.** 1983. Hydrogenase and ribulose-1,5 bisphosphate carboxylase activities of *Alcaligenes eutrophus* ATCC 17706 associated with an indigenous plasmid. *Canadian Journal of Microbiology* 29:767–774.

Bowien, B., Friedrich, B., and **Friedrich, C.G.** 1984. Involvement of megaplasmids in heterotrophic derepression of the carbon dioxide assimilating enzyme system in *Alcaligenes* spp. *Archives of Microbiology* 139:305–310.

Bowien, B. and **Schlegel, H.G.** 1981. Physiology and biochemistry of aerobic hydrogen-oxidizing bacteria. *Annual Review of Microbiology* 35:405–452.

Brewin, N.J., DeJong, T.M., Phillips, D.A., and **Johnston, A.W.B.** 1980. Co-transfer of determinants for hydrogenase activity and nodulation ability in *Rhizobium leguminosarum. Nature* 288:77–79.

Cangelosi, G.A. and **Wheelis, M.L.** 1984. Regulation by molecular oxygen and organic substrates of hydrogenase synthesis in *Alcaligenes eutrophus. Journal of Bacteriology* 159:138–144.

Cantrell, M.A., Hickok, R.E., and **Evans, H.J.** 1982. Identification and characterization of plasmids in hydrogen uptake positive and hydrogen uptake negative strains of *Rhizobium japonicum. Archives of Microbiology* 131:102–106.

Cantrell, M.A., Haugland, R.A., and **Evans, H.J.** 1983. Construction of a *Rhizobium japonicum* gene bank and use in isolation of a hydrogen uptake gene. *Proceedings of the National Academy of Sciences USA* 80:181–185.

Chandra, T.S. and **Friedrich, C.G.** 1986. Tn5-induced mutations affecting sulfur-oxidizing ability (Sox) of *Thiosphaera pantotropha. Journal of Bacteriology* 166:446–452.

Colbeau, A., Godfroy, A., and **Vignais, P.M.** 1986. Cloning of DNA fragments carrying hydrogenase genes of *Rhodopseudomonas capsulata. Biochimie* 68:147–155.

Davidson, M.S. and **Summers, A.S.** 1983. Wide-host-range plasmids function in the genus *Thiobacillus. Applied and Environmental Microbiology* 46:565–572.

Davidson, M.S., Roy, P., and **Summers, A.O.** 1985. Transpositional mutagenesis of *Thiobacillus novellus* and *Thiobacillus versutus. Applied and Environmental Microbiology* 49:1436–1441.

Doyle, C. and **Arp, D.J.** 1987. Regulation of H_2 oxidation activity and hydrogenase protein levels by H_2, O_2, and carbon substrates in *Alcaligenes latus. Journal of Bacteriology* 169:4463–4468.

Eberz, G., Hogrefe, C., Kortlüke, C., Kamienski, A., and **Friedrich, B.** 1986. Molecular cloning of structural and regulatory hydrogenase (*hox*) genes of *Alcaligenes eutrophus* H16. *Journal of Bacteriology* 168:636–641.

Friedrich, B. 1987. Genetics of aerobic lithoautotrophs, pp. 230–237, in van Verseveld, H.W., Duine, J.A. (editors), *Microbial Growth on C_1 Compounds,* Martinus Nijhoff Publishers, Dordrecht.

Friedrich, C.G., Friedrich, B., and **Bowien, B.** 1981a. Formation of enzymes of autotrophic metabolism during heterotrophic growth of *Alcaligenes eutrophus. Journal of General Microbiology* 122:69–78.

Friedrich, B., Heine, E., Finck, A., and **Friedrich, C.G.** 1981b. Nickel requirement for active hydrogenase formation in *Acaligenes eutrophus. Journal of Bacteriology* 145:1144–1149.

Friedrich, B., Hogrefe, C., and **Schlegel, H.G.** 1981c. Naturally occurring genetic transfer of hydrogen-oxidizing ability between strains of *Alcaligenes eutrophus. Journal of Bacteriology* 147:198–205.

Friedrich, C.G. 1982. Derepression of hydrogenase during limitation of electron donors and derepression of ribulosebisphosphate carboxylase during carbon limitation of *Alcaligenes eutrophus. Journal of Bacteriology* 149:203–210.

Friedrich, B., Friedrich, C.G., Meyer, M., and **Schlegel, H.G.** 1984a. Expression of hydrogenase in *Alcaligenes* spp. is altered by interspecific plasmid exchange. *Journal of Bacteriology* 158:331–333.

Friedrich, C.G., Schneider, K., and **Friedrich, B.** 1982. Nickel in the catalytically active hydrogenase of *Alcaligenes eutrophus. Journal of Bacteriology* 152:42–48.

Friedrich, C.G. and **Friedrich, B.** 1983. Regulation of hydrogenase formation is temperature sensitive and plasmid coded in *Alcaligenes eutrophus. Journal of Bacteriology* 153:176–181.

Friedrich, C.G., Suetin, S., and **Lohmeyer, M.** 1984b. Nickel and iron incorporation into soluble hydrogenase of *Alcaligenes eutrophus. Archives of Microbiology* 140:206–211.

Friedrich, C.G., Meyer, O., and **Chandra, T.S.** 1986. Molybdenum-dependent sulfur oxidation in facultatively lithoautotrophic thiobacteria. *FEMS Microbiology Letters* 37:105–108.

Gerstenberg, C., Friedrich, B., and **Schlegel, H.G.** 1982. Physical evidence for plasmids in autotrophic, especially hydrogen-oxidizing bacteria. *Archives of Microbiology* 133:90–96.

Harker, A.R., Lambert, G.R., Hanus, F.J., and **Evans, H.J.** 1985. Further evidence that two unique subunits are essential for expression of hydrogenase activity in *Rhizobium japonicum. Journal of Bacteriology* 164:187–191.

Haugland, R.A., Hanus, F.J., Cantrell, M.A., and **Evans, H.J.** 1983. Rapid colony screening method for identifying hydrogenase activity in *Rhizobium japonicum. Applied and Environmental Microbiology* 45:892–897.

Hausinger, R.P. 1987. Nickel utilization by microorganisms. *Microbiological Reviews* 51:22–42.

Hirschmann, J., Wong, P.K., Sei, K., Keener, J., and **Kustu, S.** 1985. Products of nitrogen regulatory genes *ntrA* and *ntrC* of enteric bacteria activate *glnA* transcription *in vitro*: evidence that the *ntrA* product is a sigma factor. *Proceedings of the National Academy of Sciences USA* 82:7525–7529.

Hogrefe, C. and **Friedrich, B.** 1984. Isolation and characterization of megaplasmid DNA from lithoautotrophic bacteria. *Plasmid* 12:161–169.

Hogrefe, C., Römermann, D., and **Friedrich, B.** 1984. *Alcaligenes eutrophus* hydrogenase genes (Hox). *Journal of Bacteriology* 158:43–48.

Holmes, D.S., Lobos, J.H., Bopp, L.H., and **Welch, G.G.** 1984. Cloning of a *Thiobacillus ferrooxidans* plasmid in *Escherichia coli. Journal of Bacteriology* 157:324–326.

Hornhardt, S., Schneider, K., and **Schlegel, H.G.** 1986. Characterization of a native subunit of the NAD-linked hydrogenase isolated from a mutant of *Alcaligenes eutrophus* H16. *Biochimie* 68:15–24.

Hunt, T.P. and **Magasanik, B.** 1985. Transcription of *glnA* by purified *Escherichia coli* components: Core RNA polymerase and the products of *glnF, glnG,* and *glnL. Proceedings of the National Academy of Sciences USA* 82:8453–8457.

Kärst, U. and **Friedrich, C.G.** 1987. Identification of new peptides synthesized under the hydrogenase control system of *Alcaligenes eutrophus. Archives of Microbiology* 147:346–353.

Kärst, U., Suetin, S., and **Friedrich, C.G.** 1987. Purification and properties of a protein linked to the soluble hydrogenase of hydrogen-oxidizing bacteria. *Journal of Bacteriology* 169:2079–2085.

Kelly, D.P., Mason, J., and **Wood, A.P.** 1987. Energy metabolism in chemolithotrophs, pp. 186–194, in van Verseveld, H.W., Duine, J.A. (editors), *Microbial Growth on C_1 Compounds*, Martinus Nijhoff Publishers, Dordrecht.

Knauf, V.C. and **Nester, E.W.** 1982. Wide host range cloning vectors: a cosmid clone bank of an *Agrobacterium* Ti plasmid. *Plasmid* 8:45–54.

Knüttel, K., Schneider, K., and **Schlegel, H.G.** 1988. The membrane bound hydrogenase from *Paracoccus denitrificans*. Purification and molecular characterization. *European Journal of Biochemistry* (in press).

Kortlüke, C., Hogrefe, C., Eberz, G., Pühler, A., and **Friedrich, B.** 1987. Genes of lithoautotrophic metabolism are clustered on the megaplasmid pHG1 in *Alcaligenes eutrophus. Molecular and General Genetics* 210:122–128.

Kraft, I. and **Bock, E.** 1984. Plasmids in *Nitrobacter. Archives of Microbiology* 140:79–82.

Kulpa, C.F., Roskey, M.T., and **Travis, M.T.** 1983. Transfer of plasmid RP1 into chemolithotrophic *Thiobacillus neapolitanus. Journal of Bacteriology* 156:434–436.

Kwon, M.O. and **Kim, Y.M.** 1985. Relationship between carbon monoxide dehydrogenase and a small plasmid in *Pseudomonas carboxydovorans. FEMS Microbiology Letters* 29:155–159.

Kwon, M.O., Chung, I.K., and **Kim, Y.M.** 1986. Role of a small plasmid in the modification of carbon monoxide dehydrogenase in *Pseudomonas carboxydovorans. FEMS Microbiology Letters* 37:113–116.

Lambert, G.R., Hanus, F.J., Harker, A.R., Cantrell, M.A., Russel, S.A., and **Evans, H.J.** 1985. Intra- and interspecies transfer and expression of *Rhizobium japonicum* hydrogen uptake genes and autotrophic growth capability. *Proceedings of the National Academy of Sciences USA* 82:3232–3236.

Lehmicke, L.G. and **Lidstrom, M.E.** 1985. Organization of genes necessary for growth of the hydrogen-methanol autotroph *Xanthobacter* sp. strain H4–14 on hydrogen and carbon dioxide. *Journal of Bacteriology* 162:1244–1249.

Lim, S.T., Andersen, K., Tait, R., and **Valentine, R.C.** 1980. Genetic engineering in agriculture; hydrogen uptake (*hup*) genes. *Trends in Biochemical Sciences* 5:167–170.

Lohmeyer, M. and **Friedrich, C.G.** 1987. Nickel transport in *Alcaligenes eutrophus. Archives of Microbiology* 149:130–135.

Maier, R.J. 1986. Biochemistry, regulation, and genetics of hydrogen oxidation in *Rhizobium. CRC Critical Reviews in Biotechnology* 3:17–38; CRC Press.

Maier, R.J., Campbell, N.E.R., Hanus, F.J., Simpson, F.B., Simpson, S.A., and **Evans, H.J.** 1978. Expression of hydrogenase in free-living *Rhizobium japonicum. Proceedings of the National Academy of Sciences USA* 75:3258–3262.

Mao, M.W.H., Dugan, P.R., Martin, P.A.W., and **Tuovinen, O.H.** 1980. Plasmid DNA in chemoorganotrophic *Thiobacillus ferrooxidans* and *T. acidophilus. FEMS Microbiology Letters* 8:121–125.

Martin, P.A.W., Dugan, P.R., and **Tuovinen, O.H.** 1981. Plasmid DNA in acidophilic, chemolithotrophic thiobacilli. *Canadian Journal of Microbiology* 27:850–853.

Mayer, F. and **Friedrich, C.G.** 1986. Higher order structural organization of the nucleoid in the thiobacterium *Thiosphaera pantotropha. FEMS Microbiology Letters* 37:109–112.

Menon, N.K., Peck, H.D. Jr., Le Gall, J., and **Przybyla, A.E.** 1987. Cloning and sequencing of the genes encoding the large and small subunits of the periplasmic (NiFeSe) hydrogenase of *Desulfovibrio baculatus. Journal of Bacteriology* 169:5401–5407.

Mergeay, M., Nies, D., Schlegel, H.G., Gerits, J., Charles, P., and **van Gijsegem, F.** 1985. *Alcaligenes eutrophus* CH34 is a facultative chemolithoautotroph with plasmid-bound resistance to heavy metals. *Journal of Bacteriology* 162:328–334.

Meyer, O. 1985. Metabolism of aerobic carbon monoxide-utilizing bacteria, pp. 131–152, in Poole, R.K. and Dow, C.S. (editors), *Microbial Gas Metabolism*, Academic Press, Orlando, Florida.

Meyer, O., Jacobitz, S., and **Krüger, B.** 1987. Biochemistry and physiology of aerobic carbon monoxide-utilizing bacteria. *FEMS Microbiological Reviews* 39:161–179.

Moshiri, F.M., Stults, F., Novak, P., and **Maier, R.J.** 1983. Nif^-Hup^- mutants of *Rhizobium japonicum*. *Journal of Bacteriology* 155:926–929.

Nies, D., Mergeay, M., Friedrich, B., and **Schlegel, H.G.** 1987. Cloning of plasmid genes encoding resistance to cadmium, zinc, and cobalt in *Alcaligenes eutrophus* CH34. *Journal of Bacteriology* 169:4865–4868.

Novak, P.D. and **Maier, R.J.** 1987. Inhibition of hydrogenase synthesis by DNA gyrase inhibitors in *Bradyrhizobium japonicum*. *Journal of Bacteriology* 169:2708–2712.

Ohi, K., Takada, N., Komemushi, S., Okazaki, M., and **Miura, Y.** 1979. A new species of hydrogen-utilizing bacterium. *Journal of General and Applied Microbiology* 25:53–58.

Pinkwart, M., Schneider, K., and **Schlegel, H.G.** 1983. Purification and properties of the membrane-bound hydrogenase from N_2-fixing *Alcaligenes latus*. *Biochimica et Biophysica Acta* 745:267–278.

Pootjes, C.F. 1977. Evidence for plasmid coding of the ability to utilize hydrogen gas by *Pseudomonas facilis*. *Biochemical and Biophysical Research Communications* 76:1002–1006.

Rawlings, D.E., Pretorius, I., and **Woods, D.R.** 1984. Expression of a *Thiobacillus ferrooxidans* origin of replication in *Escherichia coli*. *Journal of Bacteriology* 158:737–738.

Rawlings, D.E. and **Woods, D.R.** 1985. Mobilization of *Thiobacillus ferrooxidans* plasmids among *Escherichia coli* strains. *Applied and Environmental Microbiology* 49:1323–1325.

Reh, M. and **Schlegel, H.G.** 1975. Chemolithoautotrophie als eine übertragbare, autonome Eigenschaft von *Nocardia opaca* 1b. *Nachrichten der Akademie der Wissenschaften in Göttingen II*. Mathematisch-Physikalische Klasse 12, 207–216.

Reh, M. and **Schlegel, H.G.** 1981. Hydrogen autotrophy as a transferable genetic character of *Nocardia opaca* 1b. *Journal of General Microbiology* 126:327–336.

Robertson, L.A. and **Kuenen, J.G.** 1983. *Thiosphaera pantotropha* gen. nov. sp. nov., a facultatively anaerobic, facultatively autotrophic sulphur bacterium. *Journal of General Microbiology* 129:2847–2855.

Römermann, D., Lohmeyer, M., Friedrich, C.G., and **Friedrich, B.** 1988. Pleiotropic mutants from *Alcaligenes eutrophus* defective in the metabolism of hydrogen, nitrate, urea, and fumarate. *Archives of Microbiology* 149:471–475.

Römermann, D. and **Friedrich, B.** 1985. Denitrification by *Alcaligenes eutrophus* is plasmid dependent. *Journal of Bacteriology* 162:852–854.

Rohde, C.W., Johannssen, W., and **Mayer, F.** 1986. Electron microscopic localization of hydrogenase genes on the megaplasmid pHG1 in *Alcaligenes eutrophus*. *Molecular and General Genetics* 202:476–480.

Ronson, C.W., Nixon, B.T., Albright, L.M., and **Ausubel, F.M.** 1987. The *Rhizobium meliloti ntrA (rpoN)* gene is required for diverse metabolic functions. *Journal of Bacteriology* 168:2424–2431.

Rossen, L., Shearman, C.A., Johnston, A.W.B., and **Downie, J.A.** 1985. The *nodD* gene of *Rhizobium leguminosarum* is autoregulatory and in the presence of plant exudates induces the *nodA, B, C* genes. *EMBO Journal* 4:3369–3373.

Schink, B. and **Schlegel, H.G.** 1979. The membrane-bound hydrogenase of *Alcaligenes eutrophus*. 1. Solubilization, purification, and biochemical properties. *Biochimica et Biophysica Acta* 567:315–324.

Schlegel, H.G. 1975. Mechanisms of chemoautotrophy, pp. 9–60, in Kinne, O. (editor), *Marine Ecology II*. Part 1, John Wiley & Sons, London.

Schlegel, H.G. and **Meyer, M.** 1985. Isolation of hydrogenase regulatory mutants of hydrogen-oxidizing bacteria by a colony-screening method. *Archives of Microbiology* 141:377–383.

Schlegel, H.G. and **Schneider, K.** 1978. Introductory report: distribution and physiological role of hydrogenases in microorganisms, pp. 15–44, in Schlegel, H.G. and Schneider, K. (editors), *Hydrogenases: Their Catalytic Activity, Structure and Function,* Goltze KG, Göttingen.

Schlegel, H.G. and **Schneider, K.** 1985. Microbial metabolism of hydrogen, pp. 439–457, in Moo-Young, M. (editor), *Comprehensive Biotechnology 1,* Pergamon Press, Oxford.

Schneider, K., Patil, D.S., and **Cammack, R.** 1983. ESR properties of membrane-bound hydrogenases from aerobic hydrogen bacteria. *Biochimica et Biophysica Acta* 748:353–361.

Schneider, K. and **Schlegel, H.G.** 1976. Purification and properties of soluble hydrogenase from *Alcaligenes eutrophus* H16. *Biochimica et Biophysica Acta* 452:66–80.

Schneider, K., Cammack, R., and **Schlegel, H.G.** 1984a. Content and localization of FMN, Fe-S clusters and nickel in the NAD-linked hydrogenase of *Nocardia opaca* 1b. *European Journal of Biochemistry* 142:75–84.

Schneider, K., Schlegel, H.G., and **Jochim, K.** 1984b. Effect of nickel on activity and subunit composition of purified hydrogenase from *Nocardia opaca* 1b. *European Journal of Biochemistry* 138:533–541.

Schneider, K. and **Piechulla, B.** 1986. Isolation and immunological characterization of the four nonidentical subunits of the soluble NAD-linked hydrogenase from *Alcaligenes eutrophus* H16. *Biochimie* 68:5–13.

Seefeldt, L.G., Mc Collum, L.C., Doyle, C.M., and **Arp, D.J.** . 1987. Immunological and molecular evidence for membrane bound dimeric hydrogenase in *Rhodopseudomonas capsulata. Biochimica et Biophysica Acta* 914:299–303.

Sensfuss, C., Reh, M., and **Schlegel, H.G.** 1986. No correlation exists between the conjugative transfer of the autotrophic character and that of plasmids in *Nocardia opaca* strains. *Journal of General Microbiology* 132:997–1007.

Simon, R., Priefer, U., and **Pühler, A.** 1983. A broad host range mobilization system for *in vivo* genetic engineering: transposon mutagenesis in Gram-negative bacteria. *Biotechnology* 1:784–791.

Srivastava, V.S., Urban, S.M., and **Friedrich, B.** 1982. Mutagenesis of *Alcaligenes eutrophus* by insertion of the drug-resistance transposon Tn*5*. *Archives of Microbiology* 131:203–207.

Summers, A.O. and **Silver, S.** 1978. Microbial transformations of metals. *Annual Review of Microbiology* 32:637–672.

Takakuwa, S., Odom, J.M., and **Wall, J.D.** 1983. Hydrogen uptake deficient mutants of *Rhodopseudomonas capsulata. Archives of Microbiology* 136:20–25.

Umeda, F., Min, H., Urushira, M., Okazaki, M., and **Miura, Y.** 1986. Conjugal transfer of hydrogen-oxidizing ability of *Alcaligenes hydrogenophilus* to *Pseudomonas oxalaticus. Biochemical and Biophysical Research Communications* 137:108–113.

van Berkum, D. 1987. Expression of uptake hydrogenase and hydrogen oxidation during heterotrophic growth of *Bradyrhizobium japonicum. Journal of Bacteriology* 169:4565–4569.

Voordouw, G. and **Brenner, S.** 1985. Nucleotide sequence of the gene encoding the hydrogenase from *Desulfovibrio vulgaris* (Hildenborough). *European Journal of Biochemistry* 148:515–520.

Warrelmann, J. and **Friedrich, B.** 1986. Mutants of *Pseudomonas facilis* defective in lithoautotrophy. *Journal of General Microbiology* 132:91–96.

Wilke, D. 1980. Conjugational gene transfer in *Xanthobacter autotrophicus* GZ29. *Journal of General Microbiology* 117:431–436.

Wilke, D. and **Schlegel, H.G.** 1979. A defective generalized transducing bacteriophage in *Xanthobacter autotrophicus* GZ29. *Journal of General Microbiology* 115:403–410.

Willison, J.C., Magnin, J.P., and **Vignais, P.M.** 1987. Isolation and characterization of *Rhodobacter capsulatus* strains lacking endogenous plasmids. *Archives of Microbiology* 147:134–142.

Yagi, K., Min, H., Urushihara, M., Manabe, Y., Umeda, F., and **Miura, Y.** 1986. Isolation of hydrogen-oxidation gene from *Alcaligenes hydrogenophilus* and its expression in *Pseudomonas oxalaticus. Biochemical and Biophysical Research Communications* 137:114–119.

Yates, J.R. and **Holmes, O.S.** 1987. Two families of repeated DNA sequences in *Thiobacillus ferrooxidans. Journal of Bacteriology* 169:1861–1870.

24

Molecular Biology of Carbon Dioxide Assimilation in Aerobic Chemolithotrophs

Botho Bowien

Institute of Microbiology
Georg-August University
Göttingen, Federal Republic of Germany

Evidence for the operation of the reductive pentose phosphate cycle (Calvin cycle; Bassham and Calvin, 1957) as the pathway of CO_2 assimilation in the majority of autotrophic bacteria is discussed in Chapter 19. A solid body of information on the biochemistry and control of this primary biosynthetic process has accumulated during the past three decades after the detection of the Calvin cycle in chemo- and photolithotrophic bacteria. However, since generally little is known about the genetics of these organisms, it is not surprising that knowledge of the genetics of the CO_2-assimilating enzyme systems remained rather restricted. In fact, it was not until 1975 that experiments concerned with the genetics of the system in a chemolithotroph were reported (Reh and Schegel, 1975). The availability and application of the powerful methods of molecular biology and genetic engineering in recent years has considerably stimulated research on this most interesting topic. This chapter deals mainly with the molecular and cellular regulation of Calvin cycle enzymes (referred to as Cfx enzymes) as well as their genetic organization in chemolithotrophs. Relevant biochemical features of Cfx enzymes are also considered.

24.1 SOME MOLECULAR AND CATALYTIC PROPERTIES OF CALVIN CYCLE ENZYMES

The Calvin cycle is a complex autocatalytic pathway comprising thirteen reactions catalyzed by eleven enzymes (see also Chapter 19). While nine reactions also occur in the central carbon metabolism of heterotrophs, the remaining four are specific for carbon flow within the cycle. The four unique enzymes are phosphoribulokinase (PRK) and ribulose-1,5-bisphosphate carboxylase/oxygenase (Rubisco) in one section of the cycle and fructose-1,6-bisphosphate/

sedoheptulose-1,7-bisphosphate aldolase (ALD) and fructose-1,6-bisphosphatase/sedoheptulose-1,7-bisphosphate (FBPase/SBPase) in another section. It is of interest to note that the enzymes of each pair function in successive reactions. PRK and Rubisco are generally considered as the key enzymes whose presence indicates the potential operation of the Calvin cycle in an organism. Although information on bacterial ALDs and FBPase/SBPases involved in the cycle are rather scarce, the available evidence suggests that these enzymes are bifunctional. They convert both fructose and sedoheptulose bisphosphates.

Biochemical investigations on Cfx enzymes in chemolithotrophs have largely focused on the actual CO_2-fixing enzyme Rubisco and PRK. As in plants and most other autotrophs (Miziorko and Lorimer, 1983), the Rubisco of chemolithotrophs is a hexadecameric protein with a relative molecular mass (M_r) of about 540,000 that is composed of eight large catalytic subunits (L, $M_r \approx$ 53,000) and eight small subunits (S, $M_r \approx$ 14,000) of unknown function. Rubisco is a bifunctional enzyme which not only catalyzes the carboxylation of ribulose 1,5-bisphosphate (RuBP) to yield two molecules 3-phosphoglycerate, but also the oxygenolytic cleavage of RuBP, resulting in the formation of 2-phosphoglycolate and 3-phosphoglycerate. However, the affinity of Rubisco to CO_2 is at least one order of magnitude higher than that for oxygen. The enzyme has been characterized from various hydrogen-oxidizing bacteria (Bowien and Schlegel, 1981) including the symbiotic nitrogen fixer *Bradyrhizobium japonicum* (Purohit et al., 1982), the nitrifier *Nitrobacter agilis* (Harrison et al., 1979), and several sulfur oxidizers (McCarthy and Charles, 1975; Charles and White, 1976; Snead and Shively, 1978; Bowman and Chollet, 1980). The facultative chemoorganoautotroph *Pseudomonas oxalaticus* was reported to contain an unusual Rubisco with an L_6S_6 structure (Lawlis et al., 1979), but this finding requires independent confirmation.

The Rubisco from the hydrogen bacterium *Alcaligenes eutrophus* has been most thoroughly studied. A quaternary structure model for the enzyme from strain H16 was proposed on the basis of electron and immunoelectron microscopic analyses (Bowien et al., 1976; Bowien and Mayer, 1978). It is a four-layered structure with fourfold rotational symmetry, consisting of two central eclipsed layers, each of four L subunits. The S subunits are arranged in two outer layers, each made up of four polypeptides. X-ray diffraction and small-angle X-ray scattering studies performed with the crystallized and dissolved enzyme, respectively, supported the principal features of the model (Bowien et al., 1980; Meisenberger et al., 1984; Pal et al., 1985; Holzenburg et al., 1987). The enzyme changes its sedimentation behavior (Bowien and Gottschalk, 1982) and undergoes a configurational change (Meisenberger et al., 1984) when its activation state is altered by varying the concentrations of HCO_3^- and Mg^{2+} ions. The amino acid sequences of both subunits of the chromosomally encoded enzyme from strain ATCC 17707 (see section 24.4) has been deduced from the nucleotide sequences of the cloned genes (Andersen and Caton, 1987). Known sequences of the L subunit from other hexadecameric Rubiscos exhibit at least 80 percent homology. The L subunit from ATCC 17707 is only about 57 percent homologous with the corresponding proteins. However, large segments of the polypeptide are strongly conserved and many sequence differences represent conservative amino acid substitutions. The overall structure of the *A. eutrophus* enzyme is thus expected to closely resemble the other plant-type Rubiscos.

The wealth of information on Rubisco is contrasted by a relative lack of knowledge about PRK despite the crucial importance of the enzyme in the operation of the Calvin cycle. Among PRKs studied from chemolithotrophs, the enzyme from *A. eutrophus* is the only one so far investigated with respect to its molecular properties (Abdelal and Schlegel, 1974a; Siebert et al., 1981). In fact, it is the best characterized procaryotic PRK. The native enzyme has a M_r of 256,000 and was shown to consist of eight identical subunits of M_r 33,000 (Siebert et al., 1981; Siebert and Bowien, 1984). Similarities in the M_r values of procaryotic PRKs are evident from the data compiled in Table 24.1. An octomeric structure may be a more general property of bacterial PRKs. The enzyme from the phototrophic nonsulfur purple bacterium *Rhodopseudomonas acidophila* resembles the *A. eutrophus* PRK in its native and subunit sizes (Rippel and Bowien, 1984). This is not the case for the enzyme from the cyanobacterium *Chlorogloeopsis fritschii*, which is a hexamer of 40 kilodalton subunits (Marsden and Codd, 1984). A hexameric structure has also been proposed for the PRK from the nonsulfur purple bacterium *Rhodobacter capsulatus* (Tabita, 1980). Eucaryotic PRKs, on the other hand, are dimers consisting of 41–46 kilodalton subunits (Table 24.1). It appears that the structure of the enzyme in phylogenetically distinct groups of autotrophs is more diverse than that of Rubisco. From an evolutionary point of view, it will be of considerable interest to study the question of whether or not the PRK genes in organisms belonging to different phylogenetic groups arose from a common ancestral sequence, as can be envisaged to have occurred at least with the L subunit gene (McFadden and

Table 24.1 Molecular properties of phosphoribulokinase (PRK) from various autotrophs

Organism	Relative molecular mass (M_r) Holoenzyme	Subunits	Number of subunits	Reference
Bacteria:				
Alcaligenes eutrophus	256,000	33,000	8	Siebert et al. (1981)
Rhodopseudomonas acidophila	248,000	32,000	8	Rippel and Bowien (1984)
Rhodobacter capsulatus	220,000	36,000	6	Tabita (1980)
Chromatium vinosum	240,000	ND	—	Hart and Gibson (1971)
Chlorogloeopsis fritschii	230,000	40,000	6	Marsden and Codd (1984)
Algae and plants:				
Scenedesmus obliquus	84,000	42,000	2	Lazaro et al. (1986)
Bryopsis maxima	90,000	41,000	2	Satoh et al. (1985)
Chlamydomonas reinhardtii	ND	42,000	—	Salvucci and Ogren (1985)
Tobacco	90,000	46,000	2	Kagawa (1982)
Spinach	90,000	44,000	2	Porter et al. (1986)
Wheat	83,000	42,000	2	Surek et al. (1985)

ND, no data.

Tabita, 1974; Takabe and Akazawa, 1975) and possibly also with the S subunit gene of Rubisco.

Surprisingly little is known about the FBPase/SBPases from autotrophic bacteria. Among the chemoautotrophs, the enzymes from *A. eutrophus* (Abdelal and Schlegel, 1974b) and another hydrogen bacterium, *Nocardia opaca* 1b (Amachi and Bowien, 1979), have been studied in some detail. Their molecular sizes fall into the same range (M_r = 130,000–190,000) as those of the enzymes from *Rhodopseudomonas palustris* (Springgate and Stachow, 1972a), the cyanobacterium *Synechococcus leopoliensis (Anacystis nidulans)* (Gerbling et al., 1986), mammalian tissues (Tejwari, 1983), and the cytosol of plant cells (Zimmermann et al., 1978). The latter enzyme is a specific FBPase. Plant chloroplasts contain two specific bisphosphatases, a tetrameric FBPase (M_r = 160,000; Zimmermann et al., 1976) and a dimeric SBPase (M_r = 66,000; Cadet et al., 1987). FBPase/SBPases are generally tetramers of identical subunits, although the quaternary structure of the *A. eutrophus* and *N. opaca* enzymes was not examined. The bacterial FBPase/SBPases functioning in the Calvin cycle share the property of dual specificity with the mammalian enzyme. This is rather unexpected since the enzyme from the heterotrophic bacterium *Escherichia coli* only converts fructose bisphosphate (Fraenkel et al., 1966), and SBPase has no obvious function in heterotrophic carbon metabolism. A particularly interesting constellation exists in *N. opaca*. Heterotrophically grown cells of the organism contain a specific FBPase, whereas both activities are present in autotrophic cells. The presence of these activities are correlated with the synthesis of a bifunctional "autotrophic" FBPase/SBPase under autotrophic conditions in addition to a "heterotrophic" FBPase (Amachi and Bowien, 1979). The occurrence and significance of isoenzymes operating in the metabolic section at the interface between autotrophic and heterotrophic carbon metabolism in facultatively autotrophic bacteria will be discussed later in this chapter. In the nonsulfur purple bacterium *Rhodospirillum rubrum*, FBPase forms a complex with PRK that has been isolated but not further characterized regarding its physical nature (Joint et al., 1972a). Such an association of two regulatory enzymes could be of specific importance for the control of carbon flow through the Calvin cycle. So far, definite evidence for the existence of complexes of Calvin cycle enzymes in other autotrophic bacteria is still lacking, but there was one report on a potential association of PRK and Rubisco in the sulfur oxidizers *Thiobacillus thioparus* and *Thiobacillus neapolitanus* (MacElroy et al., 1968a). Research along these lines could, however, prove to be rewarding in view of the recent isolation of a multimeric complex from the green alga *Scenedesmus obliquus* composed of PRK and glyceraldehyde-3-phosphate dehydrogenase (GAPD) (Nicholson et al., 1987).

24.2 ENZYME REGULATION BY METABOLITES

The control of CO_2 assimilation in chemolithotrophs has been investigated to some extent by examining regulatory properties of Cfx enzymes. Similar to the chloroplast enzymes of higher plants and algae and to the cyanobacterial enzymes, PRK and FBPase/SBPase are main targets of regulation at the level of enzyme activity. It is firmly established that light is involved in the regulation of these and another Cfx enzyme, GAPD, in oxygenic photosynthetic organisms, exerting its action by the ferredoxin/thioredoxin system (Buchanan, 1980; Crawford et al., 1984). Anoxygenic phototrophic bacteria like the nonsulfur and sulfur purple bacteria do not possess this system but instead modulate the enzymes by metabolic effectors (Crawford et al., 1984) as do the chemolithotrophs.

The primary metabolic signals for this regulation are provided by the energetic state ("energy charge") of the cell and the availability of reducing power. This reflects the high demand of ATP and NADH (3 mol ATP and 2 mol NADH per mol CO_2 fixed into triose phosphate) for CO_2 assimilation via the Calvin cycle. Consequently, elevated intracellular concentrations of AMP, indicating a shortage of ATP supply furnished by the energy-converting system, should result in diminished CO_2 assimilation rates. Inhibition of ATP-dependent CO_2 fixation by AMP was first observed in cell-free extracts of *T. thioparus* (Johnson and Peck, 1965) and the hydrogen bacterium *Pseudomonas facilis* (McFadden and Tu, 1965, 1967). The site of inhibition was later identified to be PRK (MacElroy et al., 1968b, 1969). The enzyme from *Thiobacillus ferrooxidans* (Gale and Beck, 1966) and *A. eutrophus* (Abdelal and Schlegel, 1974a) is also subject to AMP inhibition like that from the purple bacteria *Rhodobacter sphaeroides* (Rindt and Ohmann, 1969), *R. rubrum* (Joint et al., 1972b), and *Chromatium vinosum* (Hart and Gibson, 1971). Among the PRKs tested, only the enzymes from *T. neapolitanus* (MacElroy et al., 1972) and *R. acidophila* (Rippel and Bowien, 1984) do not significantly respond to AMP. Phosphoenolpyruvate can be considered as another general inhibitor of bacterial PRK (Ballard and MacElroy, 1971; MacElroy et al., 1972; Abdelal and Schlegel, 1974a; Hart and Gibson, 1971; Rippel and Bowien, 1984). This central intermediate of carbon metabolism may serve as a dual regulatory signal: indirect indication of the energy state and, probably more important, function as a feedback inhibitor since it is a closely derived product of the Calvin cycle.

NADH, the reductant of CO_2 in lithotrophic bacteria (NADPH is used by plants and cyanobacteria), is a specific activator of PRK from most of these organisms studied so far, including all tested hydrogen bacteria (MacElroy et al., 1969; Abdelal and Schlegel, 1974a; Bowien and Schlegel, 1981), *Nitrobacter winogradskyi* (Kiesow et al., 1977), and the nonsulfur purple bacteria (Rindt and Ohmann, 1969; Joint et al., 1972b; Tabita, 1980; Rippel and Bowien, 1984). It was not found to be an effector of the enzyme from *T. neapolitanus* (MacElroy et al., 1972) and *C. vinosum* (Hart and Gibson, 1971). *A. eutrophus* PRK is regulated by still another mechanism. Partial inactivation of the enzyme was observed upon transition of autotrophically growing cells of the organism to heterotrophic or mixotrophic growth conditions with pyruvate as the most effective organic substrate tested (Leadbeater and Bowien, 1984; Bowien and Leadbeater, 1984). The inactivation of PRK was accompanied by a decrease in the CO_2 fixation rate of the cells, and reactivation of the enzyme occurred upon exhaustion of the pyruvate from the medium of the mixotrophic culture. Although the molecular mechanism of these regulatory processes is still unknown, they appear to contribute to the control of CO_2 assimilation in *A. eutrophus*.

Relatively few data are available on the regulation of FBPase/SBPase from chemolithotrophs. In contrast to the enzyme from most heterotrophs that plays a key role in gluconeogenesis (Tejwari, 1983), the Calvin cycle FBPase/SBPase from chemolithotrophic bacteria is not inhibited by AMP (Abdelal and Schlegel, 1974b; Amachi and Bowien, 1979). The enzyme from *T. neapolitanus* may be an exception in this respect (Johnson and MacElroy, 1973). Instead, inhibition by ATP, that was also reported for the enzyme from some nonsulfur purple bacteria (Joint et al., 1972b; Springgate and Stachow, 1972b), is a more typical regulatory feature of "autotrophic" FBPase/SBPase. The metabolic significance of ATP inhibition is unclear since an elevated intracellular concentration of ATP is expected to stimulate rather than retard the highly energy-demanding CO_2 assimilation. Feedback-type inhibitions of the enzyme by ribulose 5-phosphate (Abdelal and Schlegel, 1974b), RuBP (Amachi and Bowien, 1979), and phosphoenolpyruvate (Johnson and MacElroy, 1973; Amachi and Bowien, 1979) were also reported. *N. opaca* is so far the only autotrophic bacterium in which two FBPase(/SBPase) isoenzymes have been recognized (see also section 24.1). The bifunctional "autotrophic" enzyme exhibits the characteristic regulatory properties just mentioned—inhibition by ATP, RuBP, and phosphoenolpyruvate—whereas the specific "heterotrophic" FBPase is inhibited by AMP and activated by phosphoenolpyruvate (Amachi and Bowien, 1979), underscoring the different physiological roles for the two isoenzymes in autotrophic and heterotrophic carbon metabolism.

Reports on the regulation of the remaining Cfx enzymes by metabolites are rare. The bacterial Rubisco, like the plant enzyme, requires activation by its substrate CO_2 in the presence of Mg^{2+} ions for full catalytic competence in both the carboxylase and the oxygenase function (see Chapter 19). Active L_8S_8 enzymes, but not the low-molecular-mass Rubiscos, are inhibited by 6-phosphogluconate, the initial intermediate of the oxidative pentose phosphate and Entner-Doudoroff pathways of hexose degradation (Tabita and McFadden, 1972; Bowien, 1977; Lawlis et al., 1979). On the contrary, the activation of inactive Rubisco can be stimulated by this compound and other metabolites such as fructose 1,6-bisphosphate, NADPH, and 2-phosphoglycolate (Lawlis et al., 1978; Tabita and Coletti, 1979). All these effectors possibly interact with the catalytic RuBP-binding site on the L subunit and thereby may stabilize the active or inactive forms of the enzyme (Lorimer, 1981). Whether or not these regulations observed *in vitro* are of metabolic importance cannot be conclusively answered yet.

Oxygen is a competitive inhibitor of Rubisco with respect to CO_2 and serves as substrate of the enzyme's oxygenase function. The oxygenation of RuBP results in the primary formation of 2-phosphoglycolate that is converted into glycolate. Glycolate is the substrate for photorespiration, a process opposing photosynthetic CO_2 assimilation in oxygenic phototrophs (Lorimer, 1981). Formation of this compound and its excretion has also been observed with chemolithotrophs. It was shown that in autotrophically growing *A. eutrophus* (Codd et al., 1976; King and Andersen, 1980) and *T. neapolitanus* (Cohen et al., 1979) low CO_2 and high oxygen tensions cause a substantial excretion of glycolate. High CO_2 suppresses excretion almost completely. Thus, there is a mutually antagonistic effect of CO_2 and oxygen on glycolate production, as is to be expected because of the bifunctionality of Rubisco.

24.3 CONTROL OF ENZYME SYNTHESIS

Knowledge of the control patterns in the synthesis of Cfx enzymes in autotrophic bacteria is indispensable for the study of the regulatory processes on the genetic level. This is, of course, most significant for organisms which are able to adjust the rates of enzyme synthesis to the requirements of given growth conditions. There are considerable differences as to the extent of this control among chemolithotrophs. Obligate chemolithotrophs, such as the majority of nitrifiers and sulfur oxidizers, are specialists relying on autotrophic CO_2 fixation as their sole mode of carbon assimilation. Hence, they generally regulate synthesis of Cfx enzymes to a limited degree (Smith and Hoare, 1977). Facultative chemolithotrophs, on the other hand, also have the ability to grow heterotrophically or mixotrophically, using organic compounds as energy and/or carbon sources. Transitions between these different nutritional regimes require the organisms to tightly control their lithoautotrophic enzyme systems on both the activity and the synthesis level. Metabolic economy dictates that the functions of these systems be drastically reduced during heterotrophic growth, which is generally observed.

In most investigations on the control of Cfx enzyme synthesis, only Rubisco activities have been determined and assumed to represent all Calvin cycle enzymes. Another shortcoming of these studies is the lack of necessary information on the correlation between enzyme activity and enzyme content in the bacteria grown on different substrates. Assuming that there is such a general correlation, partial or complete repression of the synthesis of Rubisco (and PRK when determined) occurs in sulfur bacteria (Tabita and Lundgren, 1971; McCarthy and Charles, 1974; Smith et al., 1980), the nitrifier *N. agilis* (Smith and Hoare, 1968), and in hydrogen bacteria (Bowien and Schlegel, 1981) during heterotrophic growth. The extent of repression apparently depends on the organic substrate utilized. Rubisco and PRK synthesis is also repressed in the facultative organoautotroph *P. oxalaticus* when grown on acetate or fructose (Gordon and McFadden, 1980; Dijkhuizen and Harder, 1984). During growth on mixtures on fructose and formate, the latter substrate relieves the repression, resulting in mixotrophic growth with simultaneous assimilation of organic carbon and CO_2 (Dijkhuizen and Harder, 1984). Like *P. oxalaticus*, many lithotrophs are able to utilize formate by oxidizing it first to CO_2 which is then assimilated via the Calvin cycle. *A. eutrophus* also has mixotrophic capabilities. Succinate strongly represses Rubisco synthesis in this organism. However, in the presence of hydrogen as additional energy source, Rubisco is derepressed and the CO_2 derived from the dissimilation of succinate is completely refixed via the Calvin cycle (Kärst and Friedrich, 1984).

Regulation of Cfx enzyme synthesis in *A. eutrophus* H16 has been intensely studied with the aim of recognizing signal metabolites directing the control processes. Growth substrates such as pyruvate, succinate, and acetate cause strong repression of Rubisco and PRK whereas fructose and gluconate are only moderately repressing substrates (Friedrich et al., 1981c; Leadbeater et al., 1982). As seen in Table 24.2, the synthesis of the two key Cfx enzymes is coordinately regulated. This mode of regulation is expected to be a general feature of autotrophs, although a noncoordinate regulation of the enzymes was reported for *P. facilis* (McFadden and Tu, 1967; see also Chapter 19). In *A. eutrophus*, Rubisco is produced in large excess over PRK (Leadbeater et al., 1982; Table 24.2), emphasizing the necessity of relating the mechanisms governing the expression of the *cfx* genes and the organization of these genes. It is of interest to note that the activities of FBPase/SBPase, ALD, and GAPD follow basically the same regulatory pattern as Rubisco and PRK (Friedrich et al., 1981c), suggesting a common control for these Cfx enzymes. Lithoautotrophic growth under limiting carbon (CO_2) supply leads to a strong derepression of Rubisco (Friedrich, 1982) and PRK (Table 24.2) beyond the levels present in autotrophic cells grown under nonlimiting conditions. A similar response of Rubisco levels to CO_2 limitation has also been found in the facultative autotroph *Thiobacillus versutus* (Smith et al., 1980) and even in the obligate autotroph *T. neapolitanus* (Beudeker et al., 1980). The same type of regulation is observed in phototrophic bacteria (Slater and Morris, 1973; Sarles and Tabita, 1983). This indicates that versatile and specialist lithotrophs share some common principles in the control of Cfx enzyme synthesis. End-product repression, characteristic of biosynthetic pathways, is a plausible mechanism to explain this regulation. Direct or indirect end products of the Calvin cycle may serve as signal metabolites. Evidence has been obtained that phosphoenolpyruvate may play such a role in *A. eutrophus* (Im and Friedrich, 1983). Studies with *P. oxalaticus* implicated 3-phosphogly-

Table 24.2 Activities and protein ratios of Rubisco and PRK in *A. eutrophus* H16 grown under autotrophic and heterotrophic conditions

	Specific activity (units/mg protein)		Enzyme content (% soluble protein)		
Substrate	Rubisco	PRK	Rubisco	PRK	Molar ratio Rubisco/PRK[a]
Autotrophic:					
H_2/CO_2[b]	0.127	0.230	6.8	0.7	4.6
H_2/CO_2, limiting CO_2[c]	0.475	0.475	28.1	2.0	6.7
Formate	0.139	0.257	7.6	0.9	4.0
Heterotrophic:					
Fructose	0.026	0.035	1.4	0.1	6.7
Pyruvate	<0.001	<0.001	n.d.[d]	n.d.	—

[a]Rubisco: M_r = 534,000 (Bowien and Mayer, 1978); PRK: M_r = 256,000 (Siebert et al., 1981).
[b]Lithoautotrophic growth under 80% H_2/10% CO_2/10% O_2 (v/v).
[c]Lithoautotrophic growth with CO_2 lowered to 1% (v/v).
[d]n.d., not detectable.

cerate or a closely related metabolite (perhaps phosphoenolpyruvate?) in this repression/derepression mechanism (Dijkhuizen and Harden, 1984).

A question that has been mostly neglected when discussing the enzyme inventory of facultative chemolithotrophs relates to the reactions in the reducing and regenerating segments of the Calvin cycle. All these reactions (except formation and hydrolysis of sedoheptulose 1,7-bisphosphate) also occur in the glycolytic and/or gluconeogenic pathways of heterotrophic carbon metabolism. It is not known whether monoenzymes are sufficient or whether isoenzymes have to be formed during autotrophic or heterotrophic growth. The different metabolic situations may require enzymes with suitable catalytic and regulatory properties. Also, an organism may be unable to adjust the rate of synthesis of these enzymes to the necessary rate of carbon flow. Formation of isoenzymes would avoid such complications. The enzymes catalyzing the reaction sequence leading from 3-phosphoglycerate to fructose 6-phosphate are of special interest since most of them (for instance, 3-phosphoglycerate kinase, GAPD, and FBPase/SBPase) are (potential) targets of activity control. So far, there is only one reported case of such a Cfx isoenzyme in autotrophic bacteria. Autotrophically grown cells of *N. opaca* synthesize two FBPases/(SBPases) with apparently different metabolic functions (Amachi and Bowien, 1979; see also sections 24.1 and 24.2). Preliminary evidence suggests that this bacterium also forms ALD isoenzymes belonging to class I and class II aldolases (B. Bowien, unpublished results). These findings are in accord with the current knowledge about the genetic organization of the lithoautotrophic enzyme systems in *N. opaca*. A genetic study performed with *A. eutrophus* indicated that this organism also synthesizes isoenzymes for this part of autotrophic carbon metabolism (discussed below).

24.4 GENETICS OF CALVIN CYCLE ENZYMES

The refractoriness of most autotrophic bacteria to genetic analysis by conventional methods has long prevented any progress in the elucidation of the genetics of Cfx enzyme systems. Significant advances in the field have only been made during the past decade and especially within the last five years. Because of their metabolic versatility and relatively easy accessibility to manipulation by the modern techniques of molecular genetics, the phototrophic nonsulfur purple bacteria (see Chapter 26) and, among chemolithotrophs, the hydrogen bacteria have been favorite objects of study. The investigations of the genetics of Cfx systems focused mainly on *A. eutrophus*, although the first contribution dealt with *N. opaca* (Reh and Schlegel, 1975). *P. facilis*, strains of *Xanthobacter*, and *B. japonicum* have also received some attention in this respect. However, as will become evident from the subsequent discussion, we are just beginning to understand some principles of *cfx* gene organization and function in these organisms.

Organization and function of cfx* genes in *A. eutrophus Initial attempts to gain information on the organization of *cfx* genes in *A. eutrophus* were concerned with the isolation and characterization of conventionally induced mutants deficient in autotrophic CO_2 fixation (Cfx^-). All mutants derived from strain H16 (ATCC 17699) were devoid of both Rubisco and PRK activities and the corresponding enzyme proteins (Schink and Schlegel, 1978; Friedrich et al., 1979). This suggested the existence of a regulatory locus in the genome of the organism that controlled the expression of the respective genes and confirmed the coordinate regulation of the key Cfx enzymes. Many additional trials in the author's laboratory to isolate Cfx^- mutants of strain H16 carrying defects in structural *cfx* genes failed. In contrast, Andersen (1979) succeeded in isolating such mutants from the type strain of *A. eutrophus* (ATCC 17697) that produced catalytically inactive Rubisco. As became evident in subsequent studies, the type strain differs in an important organizational aspect of its *cfx* genes from other strains of the species.

Several *A. eutrophus* strains were found to harbor self-transmissible megaplasmids of about 450 kilobase pairs (kb) in size that carry the information for hydrogen-oxidizing ability (Hox; Andersen et al., 1981; Friedrich et al., 1981b; see Chapter 23). In strain H16, Hox is encoded on megaplasmid pHG1. Although loss of pHG1 abolishes the ability of the organism to grow lithoautotrophically with hydrogen and CO_2, the potential for organoautotrophic growth with formate is not affected. Thus, all *cfx* genes required for the expression of the autotrophic phenotype are probably present on the chromosome. The plasmid is, however, necessary for derepression of Rubisco and PRK during heterotrophic growth of H16 on fructose or gluconate. The megaplasmids from other *A. eutrophus* strains confer the same phenotype with respect to *cfx* expression (Behki et al., 1983; Bowien et al., 1984). This applies even to pHG21-a from a related hydrogen bacterium, *Alcaligenes hydrogenophilus* M50, but not to pHG2 from *A. eutrophus* type strain, which can be considered as indirect evidence that pHG2 lacks information involved in *cfx* expression (see below).

First proof for the presence of functional Rubisco genes (*cfxL* and *cfxS*) on plasmid pAE7 (=pHG12-a) of *A. eutropphus* ATCC 17707 came from complementation experiments with a Rubisco-deficient mutant derived from the type strain (Andersen and Wilke-Douglas, 1984). In that study, both the plasmid and chromosomally encoded Rubisco genes were cloned, establishing a reiteration of these genes on the two genetic entities of strain ATCC 17707. This finding, together with the demonstrated involvement of megaplasmids in heterotrophic *cfx* gene derepression in *A. eutrophus*, suggested that this type of reiteration of Rubisco genes may occur in most strains of the species and comprise additional *cfx* genes such as the PRK gene (*cfxP*). In fact, evidence for reiteration of *cfxP* in strain H16 was obtained. A chromosomally ($cfxP_c$) and a pHG1-encoded copy ($cfxP_p$) of the gene were localized by hybridizing restriction-digested genomic DNA of the organism with a synthetic oligodeoxynucleotide probe complementary to a portion of the amino-terminal sequence of the PRK subunit (Klintworth et al., 1985). The genes are highly homologous and located on a 12-kb ($cfxP_c$) and an 11-kb ($cfxP_p$) *Eco*RI restriction fragment

(Figure 24.1). Heterologous expression of the individual genes under the control of the *lac* promoter of pUC vectors resulted in the production of active recombinant PRKs in *E. coli*. The experiment revealed that two types of PRK subunits are formed in *A. eutrophus*. The product from *cfxP*$_c$ has a slightly larger size (M_r = 33,000) than that from *cfxP*$_p$ (M_r = 32,500). The sizes of the recombinant proteins formed in *E. coli* coincide with those of the corresponding authentic subunits (Klintworth et al., 1988; Husemann et al., 1988). Expression of both *cfxP* copies in strain H16 results in the synthesis of two PRK isoenzymes with very similar structural properties.

Strain H16 was also shown to contain chromosomally and pHG1-encoded Rubisco genes (Husemann et al., 1988). In Southern hybridizations, the Rubisco gene from *R. rubrum* (Somerville and Somerville, 1984) was used as a heterologous probe to detect the L subunit gene (*cfxL*). A synthetic oligodeoxynucleotide complementary to a portion of the amino terminus of the S subunit served as a homologous probe for *cfxS*. The corresponding Rubisco genes are highly homologous and located on the same *Eco*RI fragments as the respective *cfxP* genes (Figure 24.1). There are two very similarly organized clusters of *cfx* genes with closely neighboring Rubisco genes and *cfxP* located about 3.5 kb downstream of *cfxS*. The regions between the Rubisco and PRK genes are also homologous, so the two *cfx* clusters appear to be duplicate regions in the genome of strain H16. This notion is emphasized by the fact that the genes in both sets exhibit the same transcriptional directions (Figure 24.1, arrows), as could be concluded from expression experiments performed in *E. coli* with the genes cloned in pUC vectors.

Expression of active recombinant Rubiscos was possible not only under the direction of the *lac* promoters from the vectors but also under the control of the promoters belonging to the Rubisco genes. In contrast, no promoter activity detectable in *E. coli* is associated with the PRK genes (Husemann et al., 1988). Analogous to what was observed with the PRK isoenzymes, the

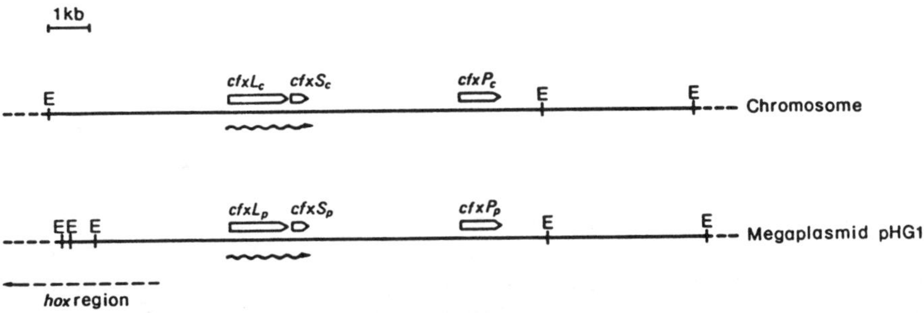

Figure 24.1 DNA segments from the chromosome and megaplasmid pHG1 of *A. eutrophus* H16, carrying *cfx* genes. Cleavage sites of restriction endonuclease *Eco*RI are indicated (E). The positions and transcriptional directions of the genes encoding the Rubisco large (*cfxL*) or small (*cfxS*) subunits and the PRK subunits (*cfxP*) are shown by thick arrows. The wavy arrows depict the *cfxL*-*cfxS* transcripts. Indices c and p refer to chromosomal and plasmid genes, respectively. Sizes are given in kilobases.

sizes of the L and S subunits from the recombinant Rubiscos were identical to the authentic polypeptides, but unlike the PRKs, there were no size differences between the corresponding chromosomally and pHG1-encoded Rubisco subunits, indicating an even closer structural similarity between the two Rubisco isoenzymes. Two conclusions can be drawn from these findings: 1) the ribosome-binding sites of the *cfx* genes are recognized in *E. coli;* and 2) at least the Rubisco genes have promoters active in the foreign host. The chromosomal Rubisco genes from strain ATCC 17707 that are also very closely linked (Andersen and Caton, 1987) were expressed from their own promoter in another heterologous host, *Pseudomonas aeruginosa* (Andersen and Wilke-Douglas, 1987). The heterologous expression of the pHG1-encoded Rubisco genes of strain H16 from their own promoter is a strong argument for their functionality. This conclusion is further supported by the proved functionality of $cfxP_p$ and the coordinate formation of Rubisco and PRK in *A. eutrophus.* Strains H16 and ATCC 17707 thus possess active megaplasmid-encoded *cfx* genes although the existence of $cfxP_p$ has not yet been demonstrated for the latter strain. The *cfx* gene cluster on pHG1 is encoded adjacent to the large region comprising the *hox* genes (Figure 24.1; Kortlüke et al., 1987). Part of the Hox information is in fact contained in the 11-kb *Eco*RI fragment about 1.5 kb upstream of $cfxL_p$. The proximity of the gene clusters for lithoautotrophic metabolism raises the question if this is a structural manifestation of a regulatory relationship between the two plasmid-borne gene systems. The reiteration of functional *cfx* gene clusters on two separate genetic entities in *A. eutrophus* strains is a novel type of gene organization. However, reiterated clusters of Rubisco and PRK genes also occur on the chromosome of *R. sphaeroides* (Gibson and Tabita, 1987; see Chapter 26).

Transcription of the Rubisco genes in strain ATCC 17707 gives rise to the production of a predominant mRNA about 2.1 kb long, indicating that the genes constitute an operon (Andersen and Wilke-Douglas, 1987). The *cfxL-cfxS* cotranscript formed in strain H16 is of the same length (Figure 24.1). Its abundance correlates with the Rubisco contents of the cells. The primary expression control of the operon is thus exerted via regulation of the transcription rate (Husemann et al., 1988). In view of the coordinate regulation of the Cfx enzymes, this type of control is also expected to govern expression of *cfxP*. A *cfxP*-specific transcript could not yet be detected in H16, so the intriguing question of whether or not the three *cfx* genes belong to a single operon cannot be answered on this basis. Since a 2.1-kb transcript is also synthesized in pHG1-cured mutants of H16, it is reasonable to assume that the transcription of both the chromosomal and plasmid *cfxL-cfxS* operons result in the formation of identically sized mRNAs. Close linkage of the L and S genes and the transcriptional order L gene-S gene seems to be the rule in autotrophic bacteria. Linkage was shown in various cyanobacteria (Shinozaki and Sugiura, 1983; Nierzwicki-Bauer et al., 1984; Tiboni et al., 1984), in *Cyanophora* cyanelles (Heinhorst and Shively, 1983), and in phototrophic bacteria (Viale et al., 1985; Gibson and Tabita, 1986). Cotranscription has been demonstrated for the cyanobacteria *A. nidulans* (Shinozaki and Sugiura, 1985) and *Anabaena* 7120

(Nierzwicki-Bauer et al., 1984) and the cyanelles (Starnes et al., 1985). The spacer regions between the genes in the various organisms have different lengths and contain no promoter but a ribosome-binding site immediately upstream of the S gene. The chromosomal spacer in *A. eutrophus* ATCC 17707 is very short (47 basepairs; Andersen and Caton, 1987), suggesting an effective translational coupling between *cfxL* and *cfxS*.

Several strains of *A. eutrophus* and *A. hydrogenophilus* have been tested for the location of their *cfx* genes, with the aim of establishing if reiteration of the genes on their chromosomes and megaplasmids is a common property of these bacteria. Southern hybridizations with *cfxL* and *cfxP* probes obtained from *A. eutrophus* H16 indeed revealed that plasmid pHG2 from *A. eutrophus* type strain (TF93) does not encode for *cfxL* and *cfxP* genes, whereas pHG3 and pHG7 from strains N9A and H20, respectively, like pHG1 from H16, contain the genes (Figure 24.2; J. Rietschel and B. Bowien, unpublished results). Plasmid pHG21-a from *A. hydrogenophilus* M50 also encodes *cfx* genes. *A. eutrophus* CH34, which harbors two plasmids conferring heavy metal resistance but not a pHG-type megaplasmid (Mergeay et al., 1985), has only chromosomal *cfx* genes. The nonautotrophic strain *A. eutrophus* JMP134 carrying the pesticide degradation plasmid pJP4 (Don and Pemberton, 1981) does not contain *cfx* genes detectable with the probes employed. The lack of heterotrophic derepression of Rubisco and PRK in *A. eutrophus* type strain (Bowien et al., 1984) is apparently a reflection of the missing *cfx* genes on pHG2. This fact provides a plausible explanation for the different outcome of selections for Cfx^- mutants derived from the type strain and from strain H16 (see above). Since in H16 both sets of *cfx* genes are functional mutations affecting structural genes of either set are phenotypically complemented by the activity of the second gene

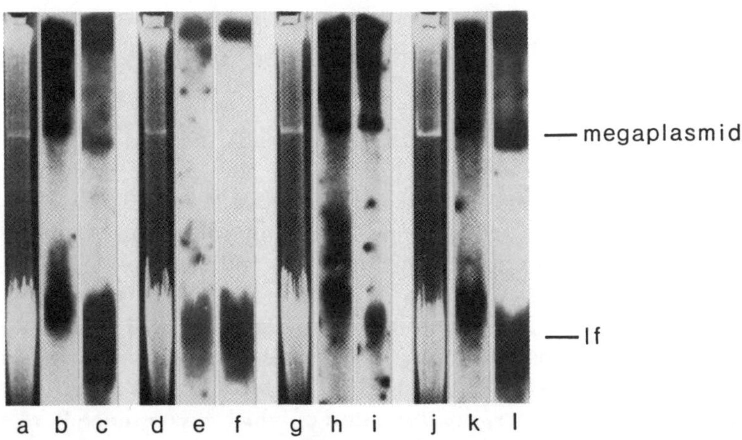

Figure 24.2 Detection of *cfx* genes on the megaplasmids from several *A. eutrophus* strains by Southern hybridization. Agarose gel electrophoresis (lanes a, d, g, and j) and autoradiographs are shown. a–c, DNA from strain H16 (pHG1); d–f, type strain TF93 (pHG2); g–i, strain N9A (pHG3) and j–l, strain H20 (pHG7). ^{32}P-labeled *cfxL* (b, e, h, and k) and *cfxP* (c, f, i, and l) were used. *lf*, linear fragments.

copy. Consequently, the only type of Cfx$^-$ mutants isolated from H16 were regulatory mutants which fail to express both sets of *cfx* genes. Such a complementation is not possible in the type strain consistent with the isolation of a specifically Rubisco-deficient mutant. An analogous approach for the isolation of corresponding mutants from H16 would require the use of a pHG1-cured derivative as the parental strain. This approach has actually been made in the author's laboratory (U. Windhövel and B. Bowien, unpublished results). Transposon Tn*5*-induced Cfx$^-$ mutants derived from pHG1-cured strain HF210 had insertions within the *cfx* region of the chromosomal 12-kb *Eco*RI fragment (see Figure 24.1). The strong polar effect of Tn*5* on the expression of the *cfx* genes distal to the site of insertion can be considered as one line of evidence for the existence of a large chromosomal *cfx* operon comprising the Rubisco and PRK genes within a region of at least 7 kb. Assuming an equivalent arrangement of the plasmid *cfx* genes into a single transcriptional unit, the wild-type strain H16 possesses two very similarly organized *cfx* operon. Their expression is controlled by a common regulatory locus, as suggested by the properties of the Cfx$^-$ mutants derived from H16 (see above). Thus, the *cfx* operons form a regulon. The control region was tagged by Tn*5* insertional mutagenesis of H16. The insertion in the resulting Cfx$^-$ mutant also mapped on the chromosomal 12-kb *Eco*RI fragment (Bowien et al., 1987). Since this mutation exerts its effect in *cis* and *trans*, the region thus codes for a diffusible gene product required for expression of both *cfx* gene sets. Further analyses of the *cfx* regions by Southern hybridizations with heterologous probes and by complementation experiments with appropriate mutants of *E. coli* indicated that genes of other Cfx enzymes, FBPase/SBPase and GAPD, are encoded in both clusters. These genes are under the same control as the Rubisco and PRK genes. Formation of FBPase/SBPase and GAPD isoenzymes in H16, which remains to be formally shown, is the logical consequence of this gene arrangement. It will certainly be of great interest to find out the definite number of Cfx enzymes encoded in the clusters.

***cfx* Genes in other chemolithotrophs** As already mentioned, *N. opaca* was the first chemolithotroph with which studies on the genetics of its lithoautotrophic enzyme systems were initiated. The original observation has been that the lithoautotrophic character can be transferred by conjugation from the wild-type strain to mutants deficient in this trait (Reh and Schlegel, 1975). Transfer to related heterotrophic bacteria like *Rhodococcus erythropolis* and *Corynebacterium hydrocarboclastus* is also possible (Reh and Schlegel, 1981). Mutants of *N. opaca* deficient in lithoautotrophy lack the three key enzymes of hydrogen autotrophy, hydrogenase, Rubisco and PRK. Transconjugants regain the ability to synthesize these enzymes, the formation of which is coordinately regulated (Ecker et al., 1986). An initial hypothesis was that the relevant genes are located on a conjugative plasmid, but subsequent studies convincingly ruled out this possibility. The current view is that the lithoautotrophic character resides on the chromosome of the organism, functioing as an independent self-transmissible genetic element of still unknown size and structure (Sensfuss et al., 1986).

Appart from the Rubisco and PRK genes, other Cfx enzyme genes may as well be encoded on the putative lithoautotrophy genetic element. This is suggested by the properties of lithoautotrophically negative mutants carrying additional gluconeogenic lesions in ALD or GAPD. Such mutants regain the lithoautotrophic capability upon retransfer of the genetic element (M. Reh, personal communication). ALD and GAPD genes seem to be reiterated on the genetic element, and the formation of "autotrophic" and "heterotrophic" ALD and GAPD isoenzymes in *N. opaca* must be postulated. There is in fact preliminary evidence for ALD isoenzymes. Two FBPase/SBPase isoenzymes are synthesized by the organism, and the way their synthesis is regulated suggests that the isoenzyme induced in autotrophically growing cells may also be encoded on the lithoautotrophy element (see sections 24.1 and 24.3).

The Rubisco L gene of *N. opaca* was localized by Southern hybridization, using the L gene from *R. rubrum* as a heterologous probe, on a chromosomal 5.1-kb *Eco*RI restriction fragment. For the detection of the PRK gene, the $cfxP_p$ gene from *A. eutrophus* H16 was employed as a probe. The gene was localized on a chromosomal 20-kb *Eco*RI fragment. These two *cfx* genes of *N. opaca* are not as closely clustered as in *A. eutrophus,* and there is no indication for a reiteration. Moreover, the *hox* genes are located in a region of the lithoautotrophy element that is not adjacent to the known *cfx* genes (M. Reh, personal communication). It is tempting to speculate that the genes of the lithoautotrophic enzyme system in *N. opaca* are organized in several operons the expression of which is coordinately controlled. A regulatory linkage between the hydrogenase system and Rubisco also exists in *P. facilis* DSM550 (ATCC 17695), as was concluded from mutant studies (Warrelmann and Friedrich, 1986). This *P. facilis* strain carries only chromosomal *hox* and *cfx* genes. However, in strain DSM620 the genes are clustered on a nonconjugative megaplasmid, pHG22-a, and do not occur on the chromosome (J. Warrelmann and B. Friedrich, personal communication). Plasmid locations of lithoautotrophy gene clusters are obviously not a unique property of *A. eutrophus* strains.

Some initial investigations into the genetics of the Cfx enzyme system have been reported for the nitrogen-fixing hydrogen bacteria *Xanthobacter autotrophicus* GZ29 and *Xanthobacter* sp. strain H4–14. Strains carrying single mutations that were defective in both Rubisco and PRK but had normal hydrogenase activity were derived from X. *autotrophicus* (Wilke, 1980). This phenotype may indicate that the expression of the respective *cfx* genes in this organism is under a control similar to that operating in *A. eutrophus.* In *Xanthobacter* strain H4–14, a presumably chromosomal region containing *cfx* genes was marked by Tn5 insertional mutagenesis. It was isolated after complementation of the corresponding Cfx^- mutants with a gene library of the organism (Lehmicke and Lidstrom, 1985). The size of the region could be delimited to about 3.2 kb. Although the definite nature of the *cfx* genes encoded on the region was not identified, the data suggest that the structural Rubisco genes might have been cloned. It is clear, however, that additional *cfx* genes are present in this DNA segment. There is no close linkage between *cfx* and *hox* genes in *Xanthobacter* strain H4–14.

Synthesis of hydrogenase or formate dehydrogenase and Cfx enzymes is coordinately regulated in free-living cells of the symbiotic nitrogen fixer *B. japonicum*. A regulatory relationship between the systems is indicated by the properties of mutants affected in these activities (Manian and O'Gara, 1982; Manian et al., 1984; Merberg and Maier, 1984). McClung and Chelm (1987) identified a chromosomal region of about 18 kb, isolated from a gene bank of strain BJ110 by complementation of mutants unable to grow on formate, that could contain structural genes for formate dehydrogenase, Rubisco and PRK and/or regulatory gene(s) essential for expression of the structural genes. This is an indication of a relatively close clustering of *cfx* genes and genes of one of the energy-converting enzyme systems in these facultative chemoautotrophs. The presence of Rubisco L genes has also been demonstrated in the genomes of the hydrogen bacterium *Paracoccus denitrificans*, of *P. oxalaticus*, the nitrifier *Nitrobacter hamburgensis*, and the sulfur bacteria *Thiobacillus intermedius*, *T. neapolitanus*, and *Thiothrix nivea*, using Southern hybridization with suitable heterologous probes (Shively et al., 1986). In *N. hamburgensis*, reiterated L genes were detected with one gene located on the chromosome and the second on a megaplasmid. The isolation of two hexadecameric Rubisco isoenzymes from the organism supports this finding (Harris et al., 1988).

24.5 CONCLUSIONS AND OUTLOOK

The newly available methods in molecular biology and genetics and the rapid development in these fields open the way for a comprehensive characterization of the Calvin cycle (Cfx) enzyme systems in autotrophic bacteria. A prerequisite for achieving this goal is a combination of physiological, biochemical, and genetic approaches. The first successful steps in this direction have been taken and future research on Cfx systems should continue to concentrate on two questions of major interest: 1) number and organization of *cfx* genes, and 2) genetic regulation of the systems. Although the facultative autotrophs are the more attractive objects for study because of their pronounced regulatory responses to nutritional changes, obligate autotrophs may also prove to be of interest since some types of control (e.g., derepression of Cfx enzymes under carbon limitation) are found in members of both groups. Special attention should be paid to the regulatory interrelationships between the energy-converting and Cfx enzyme systems. Metabolic signals must function to establish a basic coordination in the expression of the systems. However, this coordination is not necessarily strict under all conditions, as exemplified by the very different rates of hydrogenase and Rubisco synthesis in *A. eutrophus*. Cells growing on glycerol, which is an extremely poor substrate for the organism, derepress their two hydrogenases to levels several times higher than in autotrophic cells, whereas Rubisco synthesis remains low at about one-sixth of the rate observed under autotrophic conditions (Friedrich et al., 1981a). The identification of metabolites functioning as signals and effectors in these regulatory processes must certainly be one of the aims of this research.

Most progress in the elucidation of the genetics of a Cfx enzyme system has been made with *A. eutrophus*. This bacterium may serve as a model organism for the study of this topic. Yet, the reiteration of *cfx* genes on the chromosomes and megaplasmids of most *A. eutrophus* strains is probably an exception rather than the rule among chemolithotrophs. An exclusively chromosomal location is generally to be expected. The two *cfx* gene clusters in strain H16 are highly homologous and have a virtually identical arrangement of their individual genes. It is thus justified to consider them as duplicate *cfx* clusters. With *hox* and *cfx* genes clustered on pHG1 within a region of about 100 kb (Kortlüke et al., 1987), the megaplasmid carries information conferring the lithoautotrophic character on the organism. However, the expression of both systems depends on chromosomally encoded regulatory functions (see also Chapter 23).

Questions regarding the origin of the presumptive *cfx* gene duplication and the metabolic advantage of such a gene organization for the organism cannot be answered at the present state of knowledge of the system. Further characterization of the gene clusters with emphasis on identification of their component genes and transcriptional organization together with extensive sequence analyses will be required. Comparisons with the *hox* and *cfx* clusters of *P. facilis* or, if present, of other organism could provide valuable clues on whether or not DNA regions containing this information are similarly organized and may have been spread among different bacteria. Horizontal transfer of lithoautotrophy by plasmids or other transferable elements is a reasonable assumption.

The recent finding that not only Rubisco and PRK but also FBPase/SBPase and GAPD genes are encoded within the *cfx* regions of *A. eutrophus* H16 considerably expands our concept of the size of the system. Like *N. opaca*, the organism seems to synthesize isoenzymes functioning in central carbon metabolism. The identification of these isoenzymes and their genes and the determination of their functional and genetic regulation should be a most rewarding task. The occurrence of Cfx isoenzymes may well be a widespread feature among autotrophic bacteria. Even obligate autotrophs should be studied in this respect to uncover the potential diversity of Cfx systems. In any case, continued research efforts will undoubtedly bring about new exciting insights into the cellular and molecular control of autotrophic CO_2 fixation in bacteria.

The author thanks his co-workers and students for their dedicated efforts to obtain the results reported from his laboratory. The financial support of the Deutsche Forschungsgemeinschaft is also gratefully acknowledged.

REFERENCES

Abdelal, A.T.H. and **Schlegel, H.G.** 1974a. Purification and regulatory properties of phosphoribulokinase from *Hydrogenomonas eutropha* H16. *Biochemical Journal* 139:481–489.

Abdelal, A.T. and **Schlegel, H.G.** 1974b. Purification and regulatory properties of fructose 1,6-bisphosphatase from *Hydrogenomonas eutropha. Journal of Bacteriology* 120:304–310.

Amachi, T. and **Bowien, B.** 1979. Characterization of two fructose bisphosphatase isoenzymes from the hydrogen bacterium *Nocardia opaca* 1b. *The Journal of General Microbiology* 113:347–356.

Andersen, K. 1979. Mutations altering the catalytic activity of a plant-type ribulose bisphosphate carboxylase/oxygenase in *Alcaligenes eutrophus. Biochimica et Biophysica Acta* 585:1–11.

Andersen, K. and **Caton, J.** 1987. Sequence analysis of the *Alcaligenes eutrophus* chromosomally encoded ribulose bisphosphate carboxylase large and small subunit genes and their gene products. *Journal of Bacteriology* 169:4547–4558.

Andersen, K., Tait, R.C., and **King, W.R.** 1981. Plasmids required for utilization of molecular hydrogen by *Alcaligenes eutrophus. Archives of Microbiology* 129:384–390.

Andersen, K. and **Wilke-Douglas, M.** 1984. Construction and use of a gene bank of *Alcaligenes eutrophus* in the analysis of ribulose bisphosphate carboxylase genes. *Journal of Bacteriology* 159:973–978.

Andersen, K. and **Wilke-Douglas, M.** 1987. Genetic and physical mapping and expression in *Pseudomonas aeruginosa* of the chromosomally encoded ribulose bisphosphate carboxylase genes of *Alcaligenes eutrophus. Journal of Bacteriology* 169:1997–2004.

Ballard, R. and **MacElroy, R.D.** 1971. Phosphoenolpyruvate, a new inhibitor of phosphoribulokinase in *Pseudomonas facilis. Biochemical and Biophysical Research Communications* 44:614–618.

Bassham, J.A. and **Calvin, M.** 1957. *The Path of Carbon in Photosynthesis.* Prentice-Hall, Englewood Cliffs, New Jersey. 104 pp.

Behki, R.M., Selvaraj, G., and **Iyer, V.N.** 1983. Hydrogenase and ribulose-1,5-bisphosphate carboxylase activities of *Alcaligenes eutrophus* ATCC 17706 associated with an indigenous plasmid. *Canadian Journal of Microbiology* 29:767–774.

Beudeker, R.F., Cannon, G.C., Kuenen, J.G., and **Shively, J.M.** 1980. Relation between D-ribulose-1,5-bisphosphate carboxylase, carboxysomes and CO_2 fixing capacity in the obligate chemolithotroph *Thiobacillus neapolitanus* grown under different limitations in the chemostat. *Archives of Microbiology* 124:185–189.

Bowien, B. 1977. D-Ribulose 1,5-bisphosphate carboxylase from *Paracoccus denitrificans. FEMS Microbiology Letters* 2:263–266.

Bowien, B. and **Gottschalk, E.-M.** 1982. Influence of the activation state of the sedimentation properties of ribulose bisphosphate carboxylase from *Alcaligenes eutrophus. The Journal of Biological Chemistry* 257:11845–11847.

Bowien, B. and **Leadbeater, L.** 1984. Molecular and cellular regulation of carbon dioxide assimilation in bacteria, pp. 9–13, in Crawford, R.L. and Hanson, R.S. (editors), *Microbial Growth on C_1 Compounds,* American Society for Microbiology, Washington, D.C.

Bowien, B. and **Schlegel, H.G.** 1981. Physiology and biochemistry of aerobic hydrogen-oxidizing bacteria. *Annual Review of Microbiology* 35:405–452.

Bowien, B., Friedrich, B., and **Friedrich, C.G.** 1984. Involvement of megaplasmids in heterotrophic derepression of the carbon-dioxide assimilating enzyme system in *Alcaligenes* spp. *Archives of Microbiology* 139:305–310.

Bowien, B., Husemann, M., Klintworth, R., and **Windhövel, U.** 1987. Metabolic and molecular regulation of the CO_2-assimilating enzyme system in aerobic chemoautotrophs, pp. 21–27, in van Verseveld, H.W. and Duine, J.A. (editors), *Microbial Growth on C_1 Compounds,* Martinus Nijhoff Publishers, Dordrecht.

Bowien, B., Mayer, F., Codd, G.A., and **Schlegel, H.G.** 1976. Purification, some properties and quaternary structure of the D-ribulose 1,5-diphosphate carboxylase of *Alcaligenes eutrophus. Archives of Microbiology* 110:157–166.

Bowien, B., Mayer, F., Spiess, E., Pähler, A., Englisch, U., and **Saenger, W.** 1980. On the structure of crystalline ribulosebisphosphate carboxylase from *Alcaligenes eutrophus. European Journal of Biochemistry* 106:405–410.

Bowman, L.H. and **Chollet, R.** 1980. Presence of two subunit types in ribulose 1,5-bisphosphate carboxylase from *Thiobacillus intermedius. Journal of Bacteriology* 141:652–657.

Buchanan, B.B. 1980. Role of light in the regulation of chloroplast enzymes. *Annual Review of Plant Physiology* 31:341–374.

Cadet, F., Meunier, J.-C., and **Ferté, N.** 1987. Isolation and purification of chloroplastic spinach (*Spinacia oleracea*) sedoheptulose-1,7-bisphosphatase. *The Biochemical Journal* 241:71–74.

Charles, A.M. and **White, B.** 1976. Physical properties and metabolite regulation of ribulose bisphosphate carboxylase from *Thiobacillus* A2. *Archives of Microbiology* 108:203–209.

Codd, G.A., Bowien, B., and **Schlegel, H.G.** 1976. Glycollate production and excretion by *Alcaligenes eutrophus. Archives of Microbiology* 110:167–171.

Cohen, Y., de Jonge, T., and **Kuenen, J.G.** 1979. Excretion of glycollate by *Thiobacillus neapolitanus* grown in continuous culture. *Archives of Microbiology* 122:189–194.

Crawford, N.A., Sutton, C.W., Yee, B.C., Johnson, T.C., Carlson, D.C., and **Buchanan, B.B.** 1984. Contrasting modes of photosynthetic enzyme regulation in oxygenic and anoxygenic prokaryotes. *Archives of Microbiology* 139:124–129.

Dijkhuizen, L. and **Harder, W.** 1984. Regulation of autotrophic and heterotrophic metabolism in *Pseudomonas oxalaticus* OX1. Growth on fructose and on mixtures of fructose and formate in batch and continuous cultures. *The Journal of General Microbiology* 130:447–457.

Don, R.H. and **Pemberton, J.M.** 1981. Properties of six pesticide degradation plasmids isolated from *Alcaligenes paradoxus* and *Alcaligenes eutrophus. Journal of Bacteriology* 145:681–686.

Ecker, C., Reh, M., and **Schlegel, H.G.** 1986. Enzymes of the autotrophic pathway in mating partners and transconjugants of *Nocardia opaca* 1b and *Rhodococcus erythropolis. Archives of Microbiology* 145:280–286.

Fraenkel, D.G., Pontremoli, S., and **Horecker, B.L.** 1966. The specific fructose diphosphatase of *Escherichia coli*: properties and partial purification. *Archives of Biochemistry and Biophysics* 114:4–12.

Friedrich, B., Heine, E., Finck, A., and **Friedrich, C.G.** 1981a. Nickel requirement for active hydrogenase formation in *Alcaligenes eutrophus. Journal of Bacteriology* 145:1144–1149.

Friedrich, B., Hogrefe, C., and **Schlegel, H.G.** 1981b. Naturally occurring genetic transfer of hydrogen-oxidizing ability between strains of *Alcaligenes eutrophus. Journal of Bacteriology* 147:198–205.

Friedrich, C.G. 1982. Derepression of hydrogenase during limitation of electron donors and derepression of ribulosebisphosphate carboxylase during carbon limitation of *Alcaligenes eutrophus. Journal of Bacteriology* 149:203–210.

Friedrich, C.G., Bowien, B., and **Friedrich, B.** 1979. Formate and oxalate metabolism in *Alcaligenes eutrophus. The Journal of General Microbiology* 115:185–192.

Friedrich, C.G., Friedrich, B., and **Bowien, B.** 1981c. Formation of enzymes of autotrophic metabolism during heterotrophic growth of *Alcaligenes eutrophus. The Journal of General Microbiology* 122:69–78.

Gale, N.L. and **Beck, J.V.** 1966. Competitive inhibition of phosphoribulokinase by AMP. *Biochemical and Biophysical Research Communications* 24:792–796.

Gerbling, K.-P., Steup, M., and **Latzko, E.** 1986. Fructose 1,6-bisphosphatase form B from *Synechococcus leopoliensis* hydrolyzes both fructose and sedoheptulose bisphosphate. *Plant Physiology* 80:716–720.

Gibson, J.L. and **Tabita, F.R.** 1986. Isolation of the *Rhodopseudomonas sphaeroides* form I ribulose 1,5-bisphosphate carboxylase/oxygenase large and small subunit genes and expression of the active hexadecameric enzyme in *Escherichia coli. Gene* 44:271–278.

Gibson, J.L. and **Tabita, F.R.** 1987. Organization of phosphoribulokinase and ribulose bisphosphate carboxylase/oxygenase genes in *Rhodopseudomonas (Rhodobacter) sphaeroides. Journal of Bacteriology* 169:3685–3690.

Gordon, G.L.R. and **McFadden, B.A.** 1980. Enzymes in *Pseudomonas oxalaticus* grown in formate, acetate, and fructose. *Current Microbiology* 3:377–380.

Harris, S., Ebert, A., Schütze, E., Diercks, M., Bock, E., and **Shively, J.M.** 1988. Two different genes and gene products for the large subunit of ribulose-1,5-bisphosphate carboxylase/oxygenase (RuBisCOase) in *Nitrobacter hamburgensis. FEMS Microbiology Letters* 49:267–271.

Harrison, D., Rogers, L.J., and **Smith, A.J.** 1979. D-Ribulose 1,5-bisphosphate carboxylase of the nitrifying bacterium, *Nitrobacter agilis. FEMS Microbiology Letters* 6:47–51.

Hart, B.A. and **Gibson, J.** 1971. Ribulose 5-phosphate kinase from *Chromatium* sp. strain D. *Archives of Biochemistry and Biophysics* 144:308–321.

Heinhorst, S. and **Shively, J.M.** 1983. Encoding of both subunits of ribulose-1,5-bisphosphate carboxylase by organelle genome of *Cyanophora paradoxa. Nature* 304:373–374.

Holzenburg, A., Mayer, F., Harauz, G., van Heel, M., Tokuoka, R., Ishida, T., Harata, K., Pal, G.P., and **Saenger, W.** 1987. Structure of D-ribulose-1,5-bisphosphate carboxylase/oxygenase from *Alcaligenes eutrophus* H16. *Nature* 325:730–732.

Husemann, M., Klintworth, R., Büttcher, V., Salnikow, J., Weissenborn, C., and **Bowien, B.** 1988. Chromosomally and plasmid-encoded gene clusters for CO_2 fixation (*cfx* genes) in *Alcaligenes eutrophus. Molecular and General Genetics* 214:112–120.

Im, D.-S. and **Friedrich, C.G.** 1983. Fluoride, hydrogen, and formate activate ribulosebisphosphate carboxylase formation in *Alcaligenes eutrophus. Journal of Bacteriology* 154:803–808.

Johnson, E.J. and **MacElroy, R.D.** 1973. Regulation in the chemolithotroph *Thiobacillus neapolitanus*: fructose-1,6-diphosphatase. *Archiv für Mikrobiologie* 93:23–28.

Johnson, E.J. and **Peck, H.D., Jr.** 1965. Coupling of phosphorylation and carbon dioxide fixation in extracts of *Thiobacillus thioparus. Journal of Bacteriology* 89:1041–1050.

Joint, I.R., Morris, I., and **Fuller, R.C.** 1972a. Purification of a complex of alkaline fructose 1,6-bisphosphatase and phosphoribulokinase from *Rhodospirillum rubrum. The Journal of Biological Chemistry* 247:4833–4838.

Joint, I.R., Morris, I., and **Fuller, R.C.** 1972b. Possible regulatory characteristics of the fructose diphosphatase-phosphoribulokinase complex from *Rhodospirillum rubrum. Biochimica et Biophysica Acta* 276:333–337.

Kagawa, T. 1982. Isolation and purification of ribulose-5-phosphate kinase from *Nicotiana glutinosa,* pp. 695–705, in Edelman, M., Hallick, R.B., and Chua, N.H. (editors), *Methods in Chloroplast and Molecular Biology,* Elsevier, Amsterdam.

Kärst, U. and **Friedrich, C.G.** 1984. Mixotrophic capabilities of *Alcaligenes eutrophus. The Journal of General Microbiology* 130:1987–1994.

Kiesow, L.A., Lindsley, B.F., and **Bless, J.W.** 1977. Phosphoribulokinase from *Nitrobacter winogradskyi*: activation by reduced nicotinamide adenine dinucleotide and inhibition by pyridoxal phosphate. *Journal of Bacteriology* 130:20–25.

King, W.R. and **Andersen, K.** 1980. Efficiency of CO_2 fixation in a glycolate oxidoreductase mutant of *Alcaligenes eutrophus* which exports fixed carbon as glycolate. *Archives of Microbiology* 128:84–90.

Klintworth, R., Husemann, M., Salnikow, J., and **Bowien, B.** 1985. Chromosomal and plasmid locations for phosphoribulokinase genes in *Alcaligenes eutrophus. Journal of Bacteriology* 164:954–956.

Klintworth, R., Husemann, M., Weissenborn, C., and **Bowien, B.** 1988. Expression of the plasmid-encoded phosphoribulokinase gene from *Alcaligenes eutrophus. FEMS Microbiology Letters* 49:1–6.

Kortlüke, C., Hogrefe, C., Eberz, G., Pühler, A., and **Friedrich, B.** 1987. Genes of lithoautotrophic metabolism are clustered on the megaplasmid pHG1 in *Alcaligenes eutrophus. Molecular and General Genetics* 210:122–128.

Lawlis, V.B., Gordon, G.L.R., and **McFadden, B.A.** 1978. Regulation of activation of ribulose bisphosphate carboxylase from *Pseudomonas oxalaticus. Biochemical and Biophysical Research Communications* 84:699–705.

Lawlis, V.B. Jr., Gordon, G.L.R., and **McFadden, B.A.** 1979. Ribulose 1,5-bisphosphate carboxylase/oxygenase from *Pseudomonas oxalaticus. Journal of Bacteriology* 139:287–298.

Lazaro, J.J., Sutton, C.W., Nicholson, S., and **Powls, R.** 1986. Characterisation of two forms of phosphoribulokinase isolated from the green alga, *Scenedesmus obliquus. European Journal of Biochemistry* 156:423–429.

Leadbeater, L. and **Bowien, B.** 1984. Control of autotrophic carbon assimilation in *Alcaligenes eutrophus* by inactivation and reactivation of phosphoribulokinase. *Journal of Bacteriology* 157:95–99.

Leadbeater, L., Siebert, K., Schobert, P., and **Bowien, B.** 1982. Relationship between activities and protein levels of ribulosebisphosphate carboxylase and phosphoribulokinase in *Alcaligenes eutrophus. FEMS Microbiology Letters* 14:263–266.

Lehmicke, L.G. and **Lidstrom, M.E.** 1985. Organization of genes necessary for growth of the hydrogen-methanol autotroph *Xanthobacter* sp. strain H4–14 on hydrogen and carbon dioxide. *Journal of Bacteriology* 162:1244–1249.

Lorimer, G.H. 1981. The carboxylation and oxygenation of ribulose 1,5-bisphosphate: the primary events in photosynthesis and photorespiration. *Annual Review of Plant Physiology* 32:349–383.

MacElroy, R.D., Johnson, E.J., and **Johnson, M.K.** 1968a. Characterization of ribulose disphosphate carboxylase and phosphoribulokinase from *Thiobacillus thioparus* and *Thiobacillus neapolitanus. Archives of Biochemistry and Biophysics* 127:310–316.

MacElroy, R.D., Johnson, E.J., and **Johnson, M.K.** 1968b. Allosteric regulation of phosphoribulokinase activity. *Biochemical and Biophysical Research Communications* 30:678–682.

MacElroy, R.D., Johnson, E.J., and **Johnson, M.K.** 1969. Control of ATP-dependent CO_2 fixation in extracts of *Hydrogenomonas facilis*: NADH regulation of phosphoribulokinase. *Archives of Biochemistry and Biophysics* 131:272–275.

MacElroy, R.D., Mack, H.M., and **Johnson, E.J.** 1972. Properties of phosphoribulokinase from *Thiobacillus neapolitanus. Journal of Bacteriology* 112:532–538.

Manian, S.S. and **O'Gara, F.** 1982. Induction and regulation of ribulose bisphosphate carboxylase activity in *Rhizobium japonicum* during formate-dependent growth. *Archives of Microbiology* 131:51–54.

Manian, S.S., Gumbleton, R., Buckley, A.M., and **O'Gara, F.** 1984. Nitrogen fixation and carbon dioxide assimilation in *Rhizobium japonicum. Applied and Environmental Microbiology* 48:276–279.

Marsden, W.J.N. and **Codd, G.A.** 1984. Purification and molecular and catalytic properties of phosphoribulokinase from the cyanobacterium *Chlorogloeopsis fritschii. The Journal of General Microbiology* 130:999–1006.

McCarthy, J.T. and **Charles, A.M.** 1974. CO_2 fixation by the facultative autotroph *Thiobacillus novellus* during autotrophy-heterotrophy interconversions. *Canadian Journal of Microbiology* 20:1577–1584.

McCarthy, J.T. and **Charles, A.M.** 1975. Properties and regulation of ribulose diphosphate carboxylase from *Thiobacillus novellus. Archives of Microbiology* 105:51–59.

McClung, C.R. and **Chelm, B.K.** 1987. A genetic locus essential for formate-dependent growth of *Bradyrhizobium japonicum. Journal of Bacteriology* 169:3260–3267.

McFadden, B.A. and **Tabita, F.R.** 1974. D-Ribulose-1,5-diphosphate carboxylase and the evolution of autotrophy. *BioSystems* 6:93–112.

McFadden, B.A. and **Tu, C.L.** 1965. Ribulose diphosphate carboxylase and CO_2 incorporation in extracts of *Hydrogenomonas facilis. Biochemical and Biophysical Research Communications* 19:728–733.

McFadden, B.A. and **Tu, C.L.** 1967. Regulation of autotrophic and heterotrophic carbon dioxide fixation in *Hydrogenomonas facilis. Journal of Bacteriology* 93:886–893.

Meisenberger, O., Pilz, I., Bowien, B., Pal, G.P., and **Saenger, W.** 1984. Small angle X-ray study on the structure of active and inactive ribulose bisphosphate carboxylase from *Alcaligenes eutrophus. The Journal of Biological Chemistry* 259:4463–4465.

Merberg, D. and **Maier, R.J.** 1984. Coordinate expression of hydrogenase and ribulose bisphosphate carboxylase in *Rhizobium japonicum* Hup[c] mutants. *Journal of Bacteriology* 160:448–450.

Mergeay, M., Nies, D., Schlegel, H.G., Gerits, J., Charles, P., and **van Gijsegem, F.** 1985. *Alcaligenes eutrophus* CH34 is a facultative chemolithotroph with plasmid-bound resistance to heavy metals. *Journal of Bacteriology* 162:328–334.

Miziorko, H.M. and **Lorimer, G.H.** 1983. Ribulose-1,5-bisphosphate carboxylase/oxygenase. *Annual Review of Biochemistry* 52:507–535.

Nicholson, S., Easterby, J.S., and **Powls, R.** 1987. Properties of a multimeric protein complex from chloroplasts possessing potential activities of NADP-dependent glyceraldehyde-3-phosphate dehydrogenase and phosphoribulokinase. *European Journal of Biochemistry* 162:423–431.

Nierzwicki-Bauer, S.A., Curtis, S.E., and **Haselkorn, R.** 1984. Cotranscription of genes encoding the small and large subunits of ribulose-1,5-bisphosphate carboxylase in the cyanobacterium *Anabaena* 7120. *Proceedings of the National Academy of Sciences USA* 81:5961–5965.

Pal, G.P., Jakob, R., Hahn, U., Bowien, B., and **Saenger, W.** 1985. Single and twinned crystals of ribulose-1,5-bisphosphate carboxylase-oxygenase from *Alcaligenes eutrophus. The Journal of Biological Chemistry* 260:10768–10770.

Porter, M.A., Milanez, S., Stringer, C.D., and **Hartman, F.C.** 1986. Purification and characterization of ribulose-5-phosphate kinase from spinach. *Archives of Biochemistry and Biophysics* 245:14–23.

Purohit, K., Becker, R.R., and **Evans, H.J.** 1982. D-Ribulose-1,5-bisphosphate carboxylase/oxygenase from chemolithotrophically grown *Rhizobium japonicum. Biochimica et Biophysica Acta* 715:230–239.

Reh, M. and **Schlegel, H.G.** 1975. Chemolithoautotrophie als übertragbare, autonome Eigenschaft von *Nocardia opaca* 1b. *Nachrichten der Akademie der Wissenschaften zu Göttingen. II. Mathematisch-Physikalische Klasse* 12:207–216.

Reh, M. and **Schlegel, H.G.** 1981. Hydrogen autotrophy as a transferable genetic character of *Nocardia opaca* 1b. *The Journal of General Microbiology* 126:327–336.

Rindt, K.-P. and **Ohmann, E.** 1969. NADH and AMP as allosteric effectors of ribulose-5-phosphate kinase in *Rhodopseudomonas sphaeroides. Biochemical and Biophysical Research Communications* 36:357–364.

Rippel, S. and **Bowien, B.** 1984. Phosphoribulokinase from *Rhodopseudomonas acidophila. Archives of Microbiology* 139:207–212.

Salvucci, M.E. and **Ogren, W.L.** 1985. A *Chlamydomonas reinhardtii* mutant with catalytically and structurally altered ribulose-5-phosphate kinase. *Planta* 165:340–347.

Sarles, L.S. and **Tabita, F.R.** 1983. Derepression of the synthesis of D-ribulose 1,5-bisphosphate carboxylase/oxygenase from *Rhodospirillum rubrum. Journal of Bacteriology* 153:458–464.

Satoh, H., Okada, M., Nakayama, K., and **Murata, T.** 1985. Purification of ribulose 5-phosphate kinase and minor polypeptides of pyrenoid from the green alga *Bryopsis maxima. Plant and Cell Physiology* 26:931–940.

Schink, B. and **Schlegel, H.G.** 1978. Mutants of *Alcaligenes eutrophus* defective in autotrophic metabolism. *Archives of Microbiology* 117:123–129.

Sensfuss, C., Reh, M., and **Schlegel, H.G.** 1986. No correlation exists between the conjugative transfer of the autotrophic character and that of plasmids in *Nocardia opaca* strains. *The Journal of General Microbiology* 132:997–1007.

Shinozaki, K. and **Sugiura, M.** 1983. The gene for the small subunit of ribulose-1,5-bisphosphate carboxylase/oxygenase is located close to the gene for the large subunit in the cyanobacterium *Anacystis nidulans* 6301. *Nucleic Acids Research* 11:6957–6964.

Shinozaki, K. and **Sugiura, M.** 1985. Genes for the large and small subunits of ribulose-1,5-bisphosphate carboxylase/oxygenase constitute a single operon in a cyanobacterium, *Anacystis nidulans* 6301. *Molecular and General Genetics* 200:27–32.

Shively, J.M., Devore, W., Stratford, L., Porter, L., Medlin, L., and **Stevens, S.E. Jr.** 1986. Molecular evolution of the large subunit of ribulose-1,5-bisphosphate carboxylase/oxygenase (RuBisCO). *FEMS Microbiology Letters* 37:251–257.

Siebert, K. and **Bowien, B.** 1984. Evidence for an octameric structure of phosphoribulokinase from *Alcaligenes eutrophus. Biochimica et Biophysica Acta* 787:208–214.

Siebert, K., Schobert, P., and **Bowien, B.** 1981. Purification, some catalytic and molecular properties of phosphoribulokinase from *Alcaligenes eutrophus. Biochimica et Biophysica Acta* 658:35–44.

Slater, J.H. and **Morris, I.** 1973. Photosynthetic carbon dioxide assimilation by *Rhodospirillum rubrum. Archiv für Mikrobiologie* 88:213–223.

Smith, A.J. and **Hoare, D.S.** 1968. Acetate assimilation by *Nitrobacter agilis* in relation to its "obligate autotropphy." *Journal of Bacteriology* 95:844–855.

Smith, A.J. and **Hoare, D.S.** 1977. Specialist phototrophs, lithotrophs, and methylotrophs: a unity among a diversity of procaryotes? *Bacteriological Reviews* 41:419–448.

Smith, A.L., Kelly, D.P., and **Wood, A.P.** 1980. Metabolism of *Thiobacillus* A2 grown under autotrophic, mixotrophic and heterotrophic conditions in chemostat culture. *The Journal of General Microbiology* 121:127–138.

Snead, R.M. and **Shively, J.M.** 1978. D-Ribulose-1,5-bisphospphate carboxylase from *Thiobacillus neapolitanus. Current Microbiology* 1:309–314.

Somerville, C.R. and **Somerville, S.C.** 1984. Cloning and expression of the *Rhodospirillum rubrum* ribulosebisphosphate carboxylase gene in *E. coli. Molecular and General Genetics* 193:214–219.

Springgate, C.F. and **Stachow, C.S.** 1972a. Fructose 1,6-diphosphatase from *Rhodopseudomonas palustris.* I. Purification and properties. *Archives of Biochemistry and Biophysics* 152:1–12.

Springgate, C.F. and **Stachow, C.S.** 1972b. Fructose 1,6-diphosphatase from *Rhodopseudomonas palustris.* II. Regulatory properties. *Archives of Biochemistry and Biophysics* 152:13–20.

Starnes, S.M., Lambert, D.H., Maxwell, E.S., Stevens, S.E., Porter, R.D., and **Shively, J.M.** 1985. Cotranscription of the large and small subunit genes of ribulose bisphosphate carboxylase/oxygenase in *Cyanophora paradoxa. FEMS Microbiology Letters* 28:165–169.

Surek, B., Heilbronn, A., Austen, A., and **Latzko, E.** 1985. Purification and characterization of phosphoribulokinase from wheat leaves. *Planta* 165:507–512.

Tabita, F.R. 1980. Pyridine nucleotide control and subunit structure of phosphoribulokinase from photosynthetic bacteria. *Journal of Bacteriology* 143:1275–1280.

Tabita, F.R. and **Coletti, C.** 1979. Carbon dioxide assimilation in cyanobacteria: regulation of ribulose 1,5-bisphosphate carboxylase. *Journal of Bacteriology* 140:452–458.

Tabita, R. and **Lundgren, D.G.** 1971. Utilization of glucose and the effect of organic compounds on the chemolithotroph *Thiobacillus ferrooxidans. Journal of Bacteriology* 108:328–333.

Tabita, F.R. and **McFadden, B.A.** 1972. Regulation of ribulose 1,5-diphosphate carboxylase by 6-phospho-D-gluconate. *Biochemical and Biophysical Research Communications* 48:1153–1159.

Takabe, T. and **Akazawa, T.** 1975. Molecular evolution of ribulose-1,5-bisphosphate carboxylase. *Plant and Cell Physiology* 16:1049–1060.

Tejwari, G.A. 1983. Regulation of fructose-bisphosphatase activity. *Advances in Enzymology* 54:121–194.

Tiboni, O., di Pasquale, G., and **Ciferri, O.** 1984. Cloning and expression of the genes for ribulose-1,5-bisphosphate carboxylase from *Spirulina platensis. Biochimica et Biophysica Acta* 783:258–264.

Viale, A.M., Kobayashi, H., Takabe, T., and **Akazawa, T.** 1985. Expression of genes for subunits of plant-type RuBisCO from *Chromatium* and production of the enzymically active molecule in *Escherichia coli. FEBS Letters* 192:283–288.

Warrelmann, J. and **Friedrich, B.** 1986. Mutants of *Pseudomonas facilis* defective in lithoautotrophy. *The Journal of General Microbiology* 132:91–96.

Wilke, D. 1980. Conjugational gene transfer in *Xanthobacter autotrophicus* GZ29. *The Journal of General Microbiology* 117:431–436.

Zimmermann, G., Kelly, G.J., and **Latzko, E.** 1976. Efficient purification and molecular properties of spinach chloroplast fructose 1,6-bisphosphatase. *European Journal of Biochemistry* 70:361–367.

Zimmermann, G., Kelly, G.J., and **Latzko, E.** 1978. Purification and properties of spinach leaf cytoplasmic fructose-1,6-bisphophatase. *The Journal of Biological Chemistry* 253:5952–5956.

25

Energy Transduction in Phototrophic Bacteria

Gerhart Drews

Institute for Biology
University of Freiburg
Freiburg, Federal Republic of Germany

25.1 INTRODUCTION

The phylogeny of the phototrophic bacteria and their evolutionary relationships with the cyanobacteria and green plants are discussed in Chapter 27. Phototrophic bacteria transform light energy via a membrane-bound photosynthetic apparatus into an electrochemical gradient of protons, which can be used for generation of ATP, reduction of NAD^+, active transport across the membrane, or other energy-consuming processes. In contrast to the cyanobacteria, phototrophic bacteria are anoxygenic, i.e., they do not produce oxygen by a water-splitting system in the photosynthetic apparatus. Formation and function of the photosynthetic apparatus in the phototrophic bacteria depends on very low oxygen tension in the surroundings. Phototrophic bacteria have adapted to various ecological habitats by formation of different metabolic pathways (Trüper, 1987; see also Chapter 14).

On the basis of sequence homology of 16S rRNA the evolutionary relationships of these bacteria have been studied and a phylogenetic hierarchical system has been constructed in which the photosynthetic bacteria are found in several divisions: division 7 (green sulfur bacteria), 8 (green nonsulfur bacteria), and 9 (purple bacteria) of the eubacterial kingdom (Stackebrandt, 1985; Woese et al., 1984; see also Chapter 27). The purple bacteria form the largest and most heterogeneous group of phototrophic bacteria, mostly belonging to the alpha subdivision but also to the beta and gamma subdivisions (Fox et al., 1980; Gibson et al., 1985; Stackebrandt, 1985; Woese et al., 1984, 1985a, b). Some species are more closely related to chemotrophic bacteria than to other phototrophic bacteria. Thus, the facultative phototrophic bacterium *Rhodopseudomonas palustris* and the lithotrophic ammonia oxidizer *Nitrobacter winogradskyi* are more homologous by 16S rRNA than the facultative phototrophic bacterium *Rhodobacter capsulatus* and *Rhodopseudomonas palustris.*

If this system really reflects the phylogeny of photosynthetic bacteria, it is questionable whether the photosynthetic apparatus was a constituent of all ancestors of present photosynthetic bacteria from the very early beginning of evolution. The genetic information for building the photosynthetic apparatus could possibly have been transferred from an archetype of a photosynthetic bacterium to different chemotrophic bacteria during later stages of evolution. However, the strong similarity in the organization and molecular structure of the photosynthetic apparatus suggests an evolutionary relationship of the photosynthetic genes of different photosynthetic bacteria, at least in purple bacteria. The data, however, are not sufficient to support any attempt to construct an evolutionary tree of the bacterial photosynthetic apparatus. In this article the organization of the energy-converting system of phototrophic bacteria and its genetic basis will be described. Cyanobacteria, which have an oxygenic type of photosynthesis (van Liere and Walsby, 1982; Ho and Krogmann, 1982), and halobacteria, which transform light energy via a bacteriorhodopsin proton pump (Lanyi, 1978), are not discussed.

25.2 PRINCIPLES OF ORGANIZATION AND FUNCTION OF THE BACTERIAL PHOTOSYNTHETIC APPARATUS

The photosynthetic apparatus of all purple bacteria (Rhodospirillaceae, Chromatiaceae, and Ectothiorhodospiraceae) is bound to the intracytoplasmic membrane system or, in few cases, exclusively to the cytoplasmic membrane. It consists of the photochemical reaction center, antenna light-harvesting pigment-protein complexes, mobile electron carriers (cytochromes, quinones), the cytochrome b/c_1 complex and the H^+/ATPase (Figure 25.1). The intracytoplasmic membrane is structured as a characteristic and species-specific net of vesicles, tubuli, or flat membrane aggregates (Oelze and Drews, 1972; Sprague and Varga, 1986).

The photosynthetic apparatus of green bacteria (Chlorobiaceae, Chloroflexaceae) contains chlorosomes. These are oblong vesicular structures attached to the inner side of the cytoplasmic membrane by a baseplate (Figure 25.2). Chlorosomes contain most of the antenna pigments (bacteriochlorophyll c and carotenoids) bound to small polypeptides localized in the lipid-rich interior of the chlorosomes. The reaction center, part of the light-harvesting bacteriochlorophyll *a* (Bchl) (B808–868 in *Chloroflexus*) and the electron transport system are localized in the cytoplasmic membrane. A second light-harvesting Bchl *a* (B795) seems to be localized in the baseplate of chlorosomes (Figure 25.2; Blankenship and Fuller, 1986).

The major or exclusive function of the bacterial photosynthetic apparatus is the formation of an electrochemical proton potential across the membrane, created by cyclic electron transport and coupled proton translocation after charge separation has occurred in the reaction center. ATP is produced at the proton

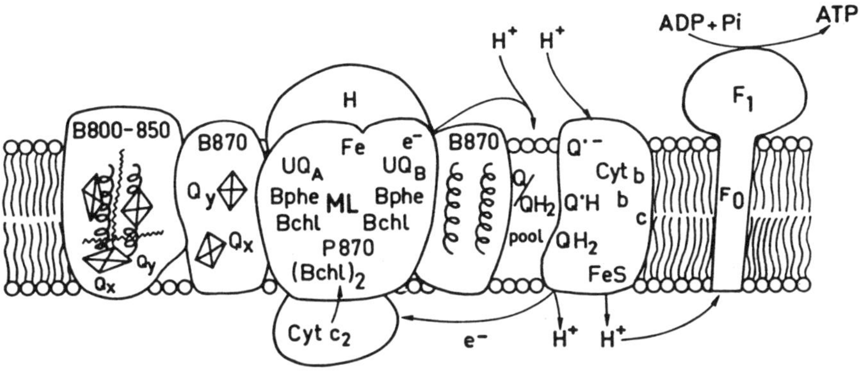

Figure 25.1 Scheme of the photosynthetic apparatus of facultative phototrophic bacteria. The photochemical reaction center, consisting of subunits H, M, and L, is surrounded by antenna pigment-protein complexes B870 and B800–850. Other membrane-bound complexes are ubiquinone-cytochrome b/c_1 oxidoreductase and proton (H^+) ATPase (consisting of membrane-bound "Proton channel" F_0 and enzymatic unit F_1). The helices drawn in antenna complexes symbolize the alpha-helical portion of the pigment-binding polypeptides. The rectangles symbolize tetrapyrrole rings and the zigzag line carotenoid molecules; Q_x and Q_y are the transition dipoles indicating the vectorial quantity of the molecular axes of the tetrapyrrol ring; Bchl, bacteriochlorophyll; Bphe, bacteriopheophytin; UQ, ubiquinone A and B; P870, $(Bchl)_2$ primary donor, special pair of Bchl molecules; Cyt c_2, cytochrome c_2, moves freely between cytochrome c/c_1 oxidoreductase and reaction center shuttling electrons; Q/QH_2, pool of ubiquinones shuttling electrons and protons in the membrane UQ_B and Q in the cyt b/c_1 oxidoreductase; FeS, Rieske iron center; Q'H, QH, Q', cyt b, cyt c are redox components of the cytochrome b/c oxidoreductase.

(H^+) ATPase using the electrochemical proton gradient as driving force. NAD(P) reduction is not coupled to the photochemical reaction, but to a noncyclic electron transport driven by ATP or the proton motive force of the membrane (Dutton, 1986).

25.3 THE PIGMENT-PROTEIN COMPLEXES OF THE PHOTOSYNTHETIC APPARATUS

The light-harvesting or antenna pigment protein system serves to absorb light quanta and to convert them to singlet excited states, which migrate in a random walk over the antenna system to the photochemical reaction center where the light energy is trapped and converted into an electrochemical potential of free energy (a membrane potential and redox potential difference (ΔEh) between an oxidant and a reductant) (Knox, 1986; Dutton, 1986; Sauer, 1986). Most of the absorbed light energy, which cannot be used by the reaction center for charge separation, is delivered from the antenna system closest to the reaction center by fluorescence emission. The high efficiency of light energy transduction depends on an exact orientation of the axes of the pigment molecules (Q_x and

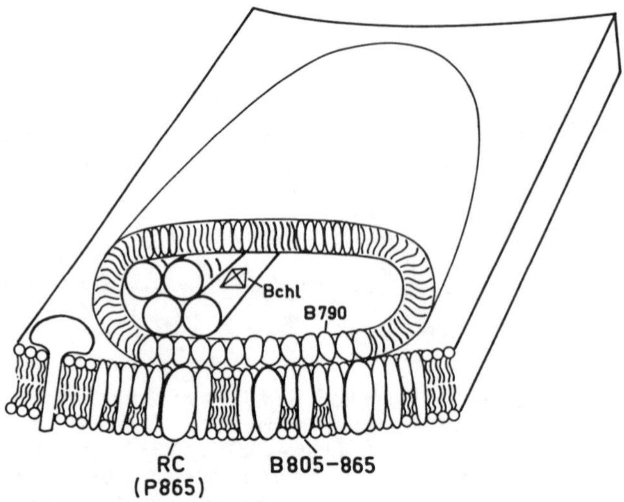

Figure 25.2 Scheme of organization of chlorosomes. The oblong, vesicle-like structures are equivalents of the light-harvesting phycobilisomes. They are delimited by a lipid layer containing polypeptides. Bchl (bacteriochlorophyll) *c* is bound to rod-like proteins in the chlorosomes. The reaction center (RC, P865) and additional light-harvesting pigments (B808–865) are localized in the cytoplasmic membrane.

Q_y transition moments) relative to the plane of the membrane and optimization of the distance of the pigment molecules (Knox, 1986; Breton, 1986). The specific positioning of the pigment molecules is a function of pigment-binding integral membrane proteins. The pigments Bchl and carotenoids (crt) are bound noncovalently to the antenna and reaction center polypeptides (Liaen-Jensen, 1978; Drews, 1985; Oelze, 1985). The polypeptides of the light-harvesting complexes in purple bacteria have molecular weights below 10,000. They are amphipathic molecules which span the membrane once by an alpha-helical domain of hydrophobic amino acids. The N-terminal regions are exposed to the cytoplasmic side of the membrane (Tadros et al., 1987) and contain a characteristic pattern of amphiphilic and charged amino acids. The C-terminal portions of the light-harvesting proteins contain also charged and hydrophilic amino acid residues (α) or they are very short and dominated by hydrophobic amino acyl residues (β). In most phototrophic bacteria two different antenna systems occur. Invariably, two different polypeptides (alpha and beta) bind two or three molecules of Bchl and one or two molecules of crt (Cogdell, 1986). A histidine residue in the central alpha-helical domain and some conserved amino acyl residues close to this histidine are ligands for magnesium and other groups of the tetrapyrrol ring of Bchl. The binding of Bchls to proteins shifts the near infrared absorption bands from 770 nm (monomeric Bchl *a* in acetone-methanol) to 800–880 nm (Drews, 1985; see also Figure 25.7 later). The overlapping of the long wavelength absorption bands of the different pigment-protein complexes is a prerequisite for efficient energy transfer. The antenna complexes in the membrane are oligomers of the basic subunits (Drews, 1985).

The antenna proteins of the green bacteria contain different components and a different organization (Blankenship and Fuller, 1986). The antenna proteins bind Bchls *a, c, d,* or *e* (Oelze, 1985). The light-harvesting Bchl *a* proteins are partially localized in the baseplate of chlorosomes. The structure of two Bchl *a*-binding proteins has been determined (Fenna et al., 1974; Daurat-Larroque et al., 1986). The Bchl *c* is believed to be bound to the surface of rodlike proteins embedded in the chlorosomes (Wechsler et al., 1984).

The reaction center of purple bacteria contains two pigment-binding (M, L; molecular weight [M_r] 34,000, 31,000) and one nonpigment-binding polypeptide (H, M_r 28,000). The M and L subunits bind 4 mol Bchl *a* or *b*, 2 mol bacteriopheophytin, 2 mol quinone, and 1 atom Fe (Figure 25.3; Thornber, 1986). The molecular structure of the reaction center of *Rhodopseudomonas viridis* (Bchl *b*) has been analyzed at 3-Angstrom resolution by X-ray spectroscopy and the exact protein-pigment interactions elucidated (Michel and Deisenhofer, 1986; Michel et al., 1986). These results give a good basis for learning more about the process of charge separation and electron tunneling between the special pair of Bchl and bacteriopheophytin molecules, as well as electron transport to the quinones and an understanding of the kinetics of the

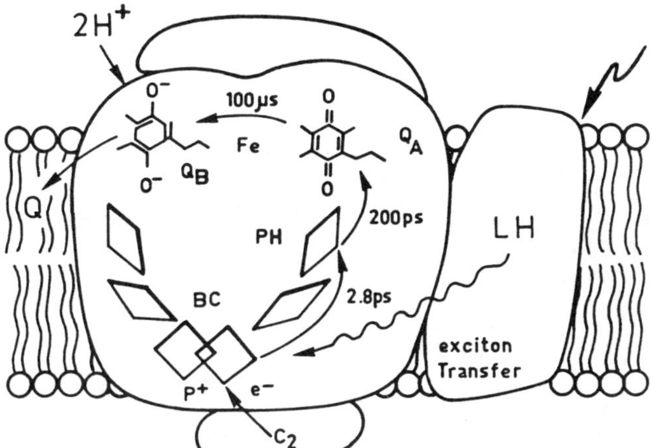

Figure 25.3 Photochemical reaction center, organization, and primary events in the purple bacteria. The special pair of two bacteriochlorophyll molecules (P^+), the single bacteriochlorophyll (BC) and bacteriopheophytin molecule (PH), and the ubiquinones A and B are bound to the M and L subunits of the reaction center. The H subunit covers the integral membrane polypeptides M and L on the cytoplasmic surface of the membrane. Absorbed light energy creates excited states in light-harvesting (LH) pigment-protein molecules. The energy of excited states is trapped by the P^+ in the reaction center. One electron is ejected from the special pair after excitation and tunneled in about 3 picoseconds (ps) to the primary acceptor bacteriopheophytin. The electron transfer from there to ubiquinone A (Q_A) takes about 200 ps and the reduction of ubiquinone B 100 μs. By these processes a charge separation across the membrane (P^+Q^-) is created. P^+ is reduced by an electron from cytochrome c_2 (see Figure 25.1). Q_H donates electrons and protons to the quinone pool in the membrane.

formation of short-lived intermediates and the high efficiency of energy conservation in the reaction center (Norris and van Brakel, 1986; Parson and Ke, 1982; Figure 25.3). The reaction centers of purple bacteria have structural and functional features in common with the photosystem II of higher plants, such as the presence of herbicide-binding proteins, pheophytin as primary acceptor, two quinones as secondary electron acceptors, and 1 atom of iron per reaction center.

The reaction center of *Chloroflexus aurantiacus* contains two polypeptides of M_r 24,000 and 24,500 which bind a special pair of Bchl *a*, molecules, as well as a single molecule of Bchl *a*, three molecules of bacteriopheophytin, and two molecules of menaquinone (Vasmel and Amesz, 1983; Blankenship et al., 1983; Shiozawa et al., 1987). The primary photochemistry seems to be functionally analogous to that of the reaction center of the Rhodospirillaceae (Kirmaier et al., 1983; 1986; Shuvalov et al., 1986).

The reaction center of Chlorobiaceae appears to be different from that of the purple bacteria and Chloroflexaceae, being more similar to the photosystem I reaction center of higher plants and that of the Bchl *g*-containing reaction center of *Heliobacter chlorum*, but it differs from the photosystem II reaction center in purple bacteria. The primary acceptor in the reaction center of Chlorobiaceae seems to be Bchl *c*. The core complex containing reaction center and light-harvesting Bchls is similar to the core complex of photosystem I of plants.

25.4 THE RESPIRATORY SYSTEM IN PURPLE BACTERIA

Many representatives of the nonsulfur purple bacteria (Rhodospirillaceae) are able to grow under chemotrophic conditions aerobically in the dark by means of electron transport (oxidative) phosphorylation. The cells of these species contain a branched respiratory chain which is bound to the cytoplasmic membrane but is also present in the intracytoplasmic membrane under conditions of phototrophic growth (Oelze and Drews, 1972; Smith and Pinder, 1987; Figure 25.4). Cytochrome b/c_1 oxidoreductase and H^+/ATPase serve both the photosynthetic and the respiratory energy transducing system (Baccarini-Melandri and Melandri, 1978; Zannoni and Melandri, 1980). A third mode of energy metabolism is active under anaerobic dark conditions. The system catalyzes electron transport between organic molecules like pyruvate and external electron acceptors in the periplasmic space like trimethylamine N-oxide or dimethyl sulfoxide under anaerobic dark conditions (Yen and Marrs, 1977; McEwan et al., 1985). Light intensity and oxygen tension regulate the photosynthetic versus the respiratory electron transport in cells which have developed both systems.

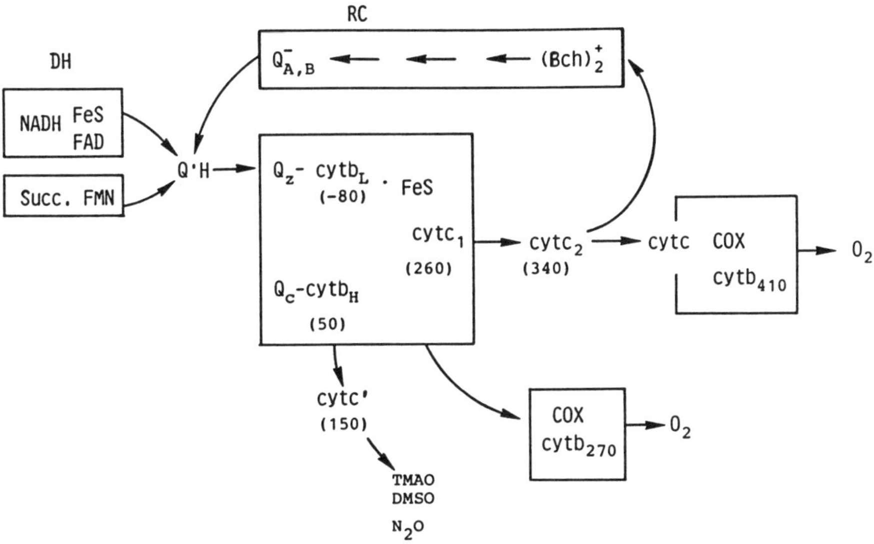

Figure 25.4 The branched electron transport system (ETS) of *Rhodobacter capsulatus*, consisting of the photochemical cyclic ETS via reaction center (RC), the aerobic, respiratory ETS with the high and the low potential cytochrome oxidases (COX), and the anaerobic dark ETS which passes electrons to the external acceptors trimethyl amine oxide (TMAO), dimethyl sulfoxide (DMSO), and N_2O. A central enzyme complex for photochemical and respiratory electron transport is the ubiquinone-cytochrome b/c_1 oxidoreductase which catalyzes a vectorial H^+ transport from the cytoplasmic to the extracytoplasmic space and an electron transport to cytochrome c_2 (Q cycle). This enzyme contributes to the formation of the electrochemical proton gradient. It contains high and low potential cytochromes b_H and b_L, cytochrome c_1 and Rieske iron center. This enzyme receives electrons from the quinone pool which is reduced either by light-driven electron transport (via the reaction center) or respiratory-driven electron transport (via NADH or succinate dehydrogenases (DH)). The Q-Cyt b/c_1 oxidoreductase donates electrons to the high potential (COX, cytochrome b_{410}) or the low potential cytochrome oxidase (COX cyt b_{270}). The high-potential oxidase (cyanide- but not CO-sensitive) is a proton pump and couples to oxidative phosphorylation. Under anaerobic dark conditions, electrons can move possibly via cytochrome c′ to electron acceptors like TMAO or DMSO or N_2O, which have the function of an electron sink. It is unknown whether this electron transport pathway contributes to energy conservation.

25.5 GENETIC RESEARCH WITH PHOTOSYNTHETIC BACTERIA

Rhodobacter capsulatus was the first phototrophic bacterium for which a useful genetic exchange system was discovered (Marrs, 1974). The gene carrier is a phage-like particle (gene transfer agent, GTA) which can transfer about 2 kb of any *R. capsulatus* chromosomal DNA to another strain of this species. This process, called "capsduction," was used for mapping of pigment markers (Yen and Marrs, 1976) and is also useful for transposon mutagenesis and other genetic transfer within *R. capsulatus* (Scolnik and Marrs, 1987). A number of broad host-range plasmids, such as RP1 (Pemberton and Bowen, 1981) and pRK290 (Ditta et al., 1980), are able to transfer conjugative markers as a shuttle vector between a large number of Gram-negative bacteria.

For production, isolation, and mapping of mutated genes, transposon mutagenesis is a very powerful tool. The plasmid pSUP201::Tn5 (Simon et al., 1983) transfers the Tn5 transposon from *E. coli* to *R. capsulatus* cells, where the Tn5 kanamycin resistance gene is randomly inserted into the genome and expressed, but the plasmid does not replicate. By selection for kanamycin resistant exconjugants and selection for markers such as deficiency of Bchl synthesis, a good yield of mutants is obtained (Kaufmann et al., 1984).

A transformation system is at present time only known for *R. sphaeroides* (Fornari and Kaplan, 1982). An updated review of all genetic tools and methods used for genetic research with photosynthetic bacteria has been published recently (Scolnik and Marrs, 1987).

25.6 LOCALIZATION OF PHOTOSYNTHETIC AND AUTOTROPHIC GENES ON THE CHROMOSOMES OF PHOTOSYNTHETIC BACTERIA

The genes for the synthesis of Bchl and carotenoids and for the pigment-binding polypeptides were found to be clustered on the genome of *R. capsulatus* (Marrs, 1981; Figure 25.5). These genes were mapped on a 46 kb fragment of the *R. capsulatus* chromosome and mobilized on the R′-plasmid pRPS404 (Marrs, 1981; Thaylor et al., 1983; Zsebo and Hearst, 1984). The *puf* and the *puh* operons, coding for the reaction center and the light-harvesting I proteins, have been cloned and sequenced (Table 25.1). The structure of the pigment genes is unknown. The *puc* operon (Figure 25.5), coding for the light-harvesting II proteins (B800–850), is localized outside of the pRPS404 gene cluster (Drews et al., 1985). The photosynthetic genes of *Rhodobacter sphaeroides* are organized similarly. Photosynthetic genes of some other phototrophic bacteria have been cloned but not mapped (Table 25.1). Some preliminary results suggest that regulatory genes for the expression of the photosynthetic genes are mapped outside of the pigment, reaction center, and light-harvesting gene clusters.

The genes coding for b and c′ cytochromes are located on a 30 kbp DNA region of the chromosome of *R. capsulatus* (H. Hüdig, unpublished, Figure 25.6). The structure of the *fbc* operon (genes for the cytochrome b/c_1 oxidoreductase) and of the genes for cytochrome c_2 and c′ have been determined (Table 25.1). It is, however, unknown whether all genes for the respiratory system are clustered.

The genes for the F_1 part of the ATPase form the *atp* operon. They have been sequenced (Table 25.1). ATPase, cyt b/c_1 oxidoreductase and cytochrome c_2 serve for both photosynthetic and the respiratory electron transport (Figure 25.4). Future work will be needed to discern the functional and evolutionary relationships between the photosynthetic and respiratory apparatuses.

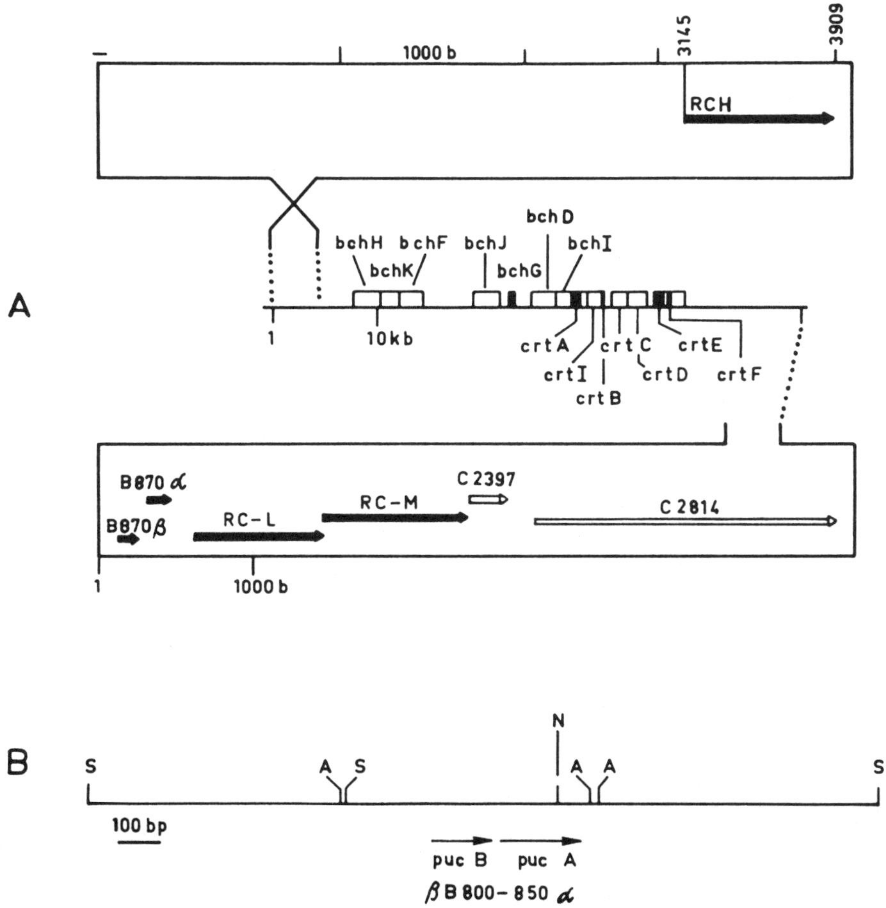

Figure 25.5 Genetic map of the photosynthetic apparatus of *Rhodobacter capsulatus* (Marrs, 1981; Thaylor et al., 1983; Zsebo and Hearst, 1984; Zhu and Hearst, 1986). A. The gene cluster carried by the R′ plasmid pRPS404 contains about 46 kb DNA. In the central part are genes for the biosynthesis of bacteriochlorophyll (bch) and carotenoids (crt). The portions of DNA containing *puh* (H subunit of reaction center, RCH) and *puf* operons (B870 β, B870 α antenna pigment binding polypeptides, RC-LM subunits of RC) are drawn enlarged (see mapping distances on the base line). There are other open reading frames on the cloned DNA fragments which have been sequenced. Since their function is unknown, they have been mostly omitted in the drawings for reasons of clarity. The restriction pattern has been omitted. B. The DNA coding for genes of the *puc* operon (B800–850 α, β, LH polypeptides) is localized separately from photosynthetic gene cluster (A) on the chromosome. The mapping distances are given in nucleotide bases (bp). Restriction fragments, S, *SmaI*; A, *ApaI*; N, *NruI*.

The structural genes for the nitrogenase enzyme form an operon while the regulatory genes for nitrogenase and nitrogen metabolism are dispersed in several groups on the chromosome of *R. capsulatus* (Haselkorn, 1986).

Table 25.1 Genes cloned from photosynthetic bacteria[a]

Operon	Genes	Products	Species	Sequence determined	Reference
atp	*unc*H, *unc*A, *unc*G, *unc*D, *unc*C	$\delta, \alpha, \gamma, \beta\ \epsilon$	*Rps. blastica*	+	1
atp	*unc*H, *unc*A, *unc*G, *unc*D, *unc*C	Subunits of ATPase	*Rs. rubrum*	+	2
	bch A–L	Bchl biosynthesis	*Rb. capsulatus*	–	3
	crt A–F, J	Crt biosynthesis	*Rb. capsulatus*	–	3
	crt A–F	Crt biosynthesis	*Rb. sphaeroides*	–	5
	crt I	Phytoene dehydrogenase	*Rb. capsulatus*	–	4
	cyc A	Cytochrome c_2	*Rb. capsulatus*	+	6
	cyc A	Cytochrome c_2	*Rb. sphaeroides*	+	7
fbc (pet)	*fbc* F, B, C	FeS-Rieske protein Cytochrome b Cytochrome c_1	*Rb. capsulatus*	+	8, 9, 10, 11
	gln A	Glutamine synthetase	*Rb. capsulatus*	–	12
	hup A, B	Hydrogenase	*Rb. capsulatus*	–	13
nif	*nif* H, D, K	Nitrogenase	*Rb. capsulatus*	+	14, 15
	nif A	Nitrogenase regulation	*Rb. capsulatus*	–	16
	nif R1 (ntrC)	Regulatory	*Rb. capsulatus*	–	17
	ntr 27–29	Regulatory	*Rb. capsulatus*	–	18
puc	*puc* A, B	LHII α, β	*Rb. capsulatus*	+	19, 20
puc	*puc* A, B	LHII	*Rb. sphaeroides*	+	21
puf	*puf* A, B, L, M	LHI α, β, RC-M, L	*Rb. capsulatus*	+	22
puf	*puf* A, B, L, M	LHI, RC-M, L	*Rps. viridis*	+	34
	puf A, B	LHI α, β	*Rs. rubrum*	+	23
puf	*puf* B, A, L, M	LHI α, β, RC-L, M	*Rb. sphaeroides*	–	24
puf	*puf* A, B	LHI α, β	*Rb. sphaeroides*	+	33

Table 25.1 Continued

Operon	Genes	Products	Species	Sequence determined	Reference
puh	*puh* A	RC-H	*Rb. capsulatus*	+	22
	puh A	RC-H	*Rb. sphaeroides*	−	25
	rbc L	Rubisco, large subunit	*Rb. sphaeroides*	−	26
	rbc R	Rubisco, form II	*Rb. sphaeroides*	−	27, 28
	rbc S	Rubisco, small subunit	*Rb. sphaeroides*		26
	rbc	Rubisco	*Rs. rubrum*	+	29, 30
	rrn	16S, 23S, 5S rRNA	*Rb. capsulatus*	−	31
	rrn	5S rRNA	*Rb. capsulatus*	+	32
	rrn	23S rRNA	*Rb. capsulatus*	+	35

References: 1)Tybulewicz, V.L.J., Falk, G., Walker, J.E. (1984) J. Mol. Biol. 179:185–214; 2)Falk, G., Hampe, A., Walker, J.E. (1985) Biochem. J. 228:391–407; 3)Taylor, D.P., Cohen, S.N., Clark, W.G., Marrs, B.L. (1983) J. Bacteriol. 154:580–590; 4)Giuliano, G., Pollock, D., Scolnick, P.A. (1986) J. Biol. Chem. 261:12925–12929; 5)Pemberton, J.M., Harding, C.M. (1986) Curr. Microbiol. 14:25–29; 6)Daldal, F., Cheng, S., Applebaum, J., Davidson, E., Prince, R.C. (1986) Proc. Natl. Acad. Sci. USA 83:2012–2016; 7)Donohue, T.J., McEwan, A.G., Kaplan, S. (1986) J. Bacteriol. 168:962–972; 8)Gabellini, N., Harnisch, U., McCarthy, J.E.G., Hauska, G., Sebald, W. (1985) EMBO Jour. 4:549–553; 9)Gabellini, N., Sebald, W. (1986). Eur. J. Biochem. 154:569–579; 10)Daldal, F., Davidson, E., Cheng, S. (1987) J. Mol. Biol. 195:1–12; 11)Davidson, E., Daldal, F. (1987) J. Mol. Biol. 195:13–24; 12)Scolnik, P.A., Virosco, J., Haselkorn, R. (1983) J. Bacteriol. 155:180–185; 13)Colbeau, A., Godfroy, A., Vignais, P.M. (1986) Biochimie 68:147–155; 14)Artges, P., Scolnik, P.A., Haselkorn, R. (1983) J. Bacteriol. 156:251–256; 15)Haselkorn, R. (1986) Ann. Rev. Microbiol. 40:525–547; 16)Ahombo, G., Willison, J.C., Vignais, P.M. (1986) Mol. Gen. Genet. 205:442–445; 17)Kranz, R.G., Haselkorn, R. (1985) Gene 40:203–215; 18)Willison, J.C., Ahombo, G., Chabert, J., Magnin, J.P., Vignais, P.M. (1985) J. Gen. Microbiol. 131:3001–3015; 19)Drews, G., Kaufmann, N., Klug, G. (1985) in *Molecular Biology of the Photosynthetic Apparatus*, pp. 211–222, K.E. Steinback, S. Bonitz, C.J. Arntzen, L. Bogorad (eds.), Cold Spring Harbor Lab.; 20)Youvan, D.C., Ismail, S. (1985) Proc. Natl. Acad. Sci. USA 82:58–62; 21)Ashby, M.K., Coomber, S.A., Hunter, C.N. (1987) FEBS Lett. 213:245–248; 22)Youvan, D.C., Bylina, E.J., Alberti, M., Begusch, H., Hearst, J. (1984) Cell 37:949–959; 23)Bérard, J., Bélanger, G., Corriveau, P., Gingras, G. (1986) J. Biol. Chem. 261:82–87; 24)Zhu, Y.S., Kiley, P.J., Donohue, T.J., Kaplan, S. (1986) J. Biol. Chem. 261:10366–10374; 25)Donohue, T.J., Hoger, J.H., Kaplan, S. (1986) J. Bacteriol. 168:953–961; 26)Gibson, J., Tabita, F.R. (1986) Gene 44:271–278; 27)Quivey, R.G., Tabita, F.R. (1984) Gene 31:91–101; 28)Müller, E.D., Chory, J., Kaplan, S. (1985) J. Bacteriol. 161:469–472; 29)Somerville, C.R., Somerville, S.C. (1984) Mol. Gen. Genet. 193:214–219; 30)Nargang, F., McIntosh, L., Somerville, C. (1984) Mol. Gen. Genet. 193:220–224; 31)Yu, P.-L., Hohn, B., Falk, H., Drews, G. (1982) Mol. Gen. Genet. 188:392–398; 32)Van den Eynde, H., Vandenberghe, A., De Wachter, R. (1986) Nucl. Acids Res. 14:8688; 33)Kiley, P.J., Donohue, T.J., Havelka, W.A., Kaplan, S. (1987) J. Bact. 169:742–750; 34)Michel, H., Weyer, K.A., Gruenberg, H., Dunger, I., Oesterhelt, D., Lottspeich, F. (1986) EMBO Jour. 5:1149–1158; 35)K.H. Schleifer et al. unpublished.

[a]Abbreviations: *Rps., Rhodopseudomonas; Rb., Rhodobacter; Rs., Rhodospirillum*; crt, carotenoids; LH, light harvesting, RC, reaction center; Bchl, bacteriochlorophyll; Rubisco, ribulose bisphosphate carboxylase; +, nucleotide sequence is known; Kaplan, S., Marrs, B.L. (1986) Proposed nomenclature for photosynthetic procaryotes, ASM News 52:242.

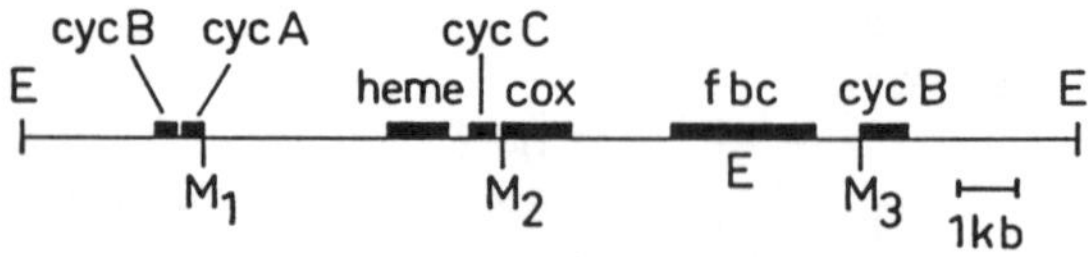

Figure 25.6 Cytochrome gene cluster of *Rhodobacter capsulatus* (H. Hüdig, unpublished). E, *EcoRI* restriction cut. Mapped genes: *cox*, cytochrome oxidase (b410) (see Figure 25.4); *cycA*, cytochrome c_2; *cycB*, cytochrome c'; *cycC*, cytochrome c_{234} (membrane bound); *fbc*, operon for cytochrome b/c oxidoreductase; M1, cycB-minus phenotype; M2, O_2-dependent regulation of oxidase; M3, incorporation of heme in cytochrome oxidase inhibited.

Genes for photosynthesis and autotrophic growth are localized on the chromosome of *R. capsulatus* and not on plasmids. This has been shown by curing *R. capsulatus* strains of plasmids and the demonstration that strains lacking endogenous plasmids still grow photosynthetically and autotrophically (Willison et al., 1987). The autotrophic genes *hup* (membrane-bound hydrogenase), *rbc* (ribulose bisphosphate carboxylase), and *aut* (regulatory) genes, which have been cloned (Table 25.1), are dispersed in three groups on the chromosome (J.-P. Magnin, P.M. Vignais, unpublished).

In addition to *R. capsulatus* and *R. sphaeroides, Rhodopseudomonas viridis, Rhodospirillum rubrum* and *Chloroflexus aurantiacus* have also been investigated genetically. It is not surprising that DNA probes taken from conservative regions of structural genes for pigment-binding proteins, nitrogenase, or ribulose bisphosphate carboxylase hybridize with DNA of different photosynthetic bacteria (Beatty and Cohen, 1983; see also references in Table 25.1). It is, however, too early to speculate on evolution of the bacterial photosynthetic apparatus on the basis of conservative homologous DNA structures. The diversity in structure, function, and composition of the components of the photosynthetic and respiratory systems in members of the various taxonomic groups of phototsynthetic bacteria indicates that the energy-transducing systems as well underwent an evolution in different ecological niches over long periods. Of special interest will be comparative genetic studies on the photosynthetic apparatus of the obligate aerobic marine bacterium *Erythrobacter* and related species which have a bacterial type of photosynthesis (anoxygenic) (Harashima et al., 1980, 1982; Shiba, 1984).

25.7 FORMATION OF THE PHOTOSYNTHETIC APPARATUS AND ITS REGULATION

Bacterial photosynthesis is anoxygenic and is only active under anaerobic conditions. Facultative phototrophic bacteria are able to adapt to a chemotrophic (aerobic, dark) or phototrophic (anaerobic, light) mode of metabolism. The oxygen partial pressure is the major external signal triggering a process of functional and structural membrane change from the respiratory to the photosynthetic mode of energy transduction. Other factors like light intensity, substrate concentration, and growth rate influence the process of adaptation as well. Here we will consider only the influence of oxygen tension on the formation of the photosynthetic and the respiratory apparatus in the membrane system of facultative phototrophic bacteria.

In natural habitats, oxygen tension can differ to a great extent in time and space. For example, upper layers of a lake, which are populated with oxygenic cyanobacteria and algae, are oversaturated with oxygen during the day while deeper in the lake the oxygen tension is low because oxygenic photosynthesis is absent and the dissolved oxygen is consumed by respiration and often not replenished due to restricted vertical mixing. Facultative phototrophic bacteria are free of photosynthetic pigments and other constituents of the photosynthetic apparatus when grown in the dark at high oxygen tensions (Drews and Oelze, 1981). Lowering of the oxygen tension below a species-specific threshold value induces a rapid synthesis of bacteriochlorophyll (Bchl), pigment-binding proteins, and other components of the photosynthetic apparatus (Figure 25.7). The pigment-protein complexes of the photosynthetic apparatus are assembled at specific invagination sites of the cytoplasmic membrane (*Rhodospirillum rubrum, Rhodobacter capsulatus, R. sphaeroides*; Figure 25.8), which are the starting

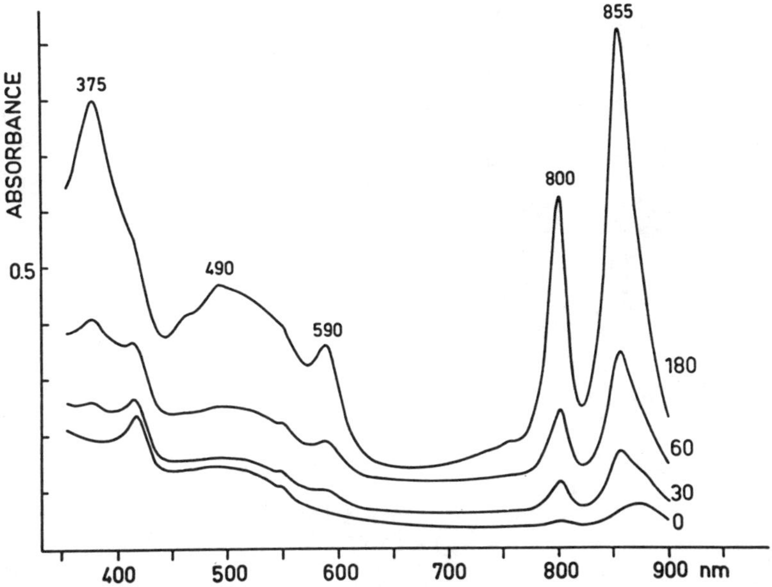

Figure 25.7 Induction of formation of pigment-protein complexes in strain 37b4 of *Rhodobacter capsulatus* by lowering of oxygen partial pressure in chemotrophic cultures (G. Klug, unpublished). Absorption spectra taken at 0, 30, 60, and 180 minutes of incubation. 490 nm, absorption maximum of carotenoids; 375, 590, 800, 855 nm absorption maxima of bacteriochlorophyll. The absorption maxima of the light-harvesting complex B800–850 predominates in most of the spectra. After 30 minutes of incubation, the ratio B800–850/B870 is smaller. The B870 component is seen as a shoulder at 870 nm.

points for the formation of intracytoplasmic membranes. In a later stage of differentiation, most of the photosynthetic apparatus is localized on the intracytoplasmic membrane system (Oelze and Drews, 1972). When phototrophically-grown cells are shifted to high oxygen tension under dark growth conditions, the synthesis of Bchl and pigment-binding polypeptides is stopped immediately. These components of the photosynthetic apparatus are, however, not degraded but diluted out in growing cells. Components of the respiratory chain, especially the terminal oxidase, are preferentially synthesized and incorporated into the membrane. The area of intracytoplasmic membrane decreased. Therefore, the process of membrane differentiation is determined by quantitative and qualitative changes in the patterns of incorporation and not by different turnover rates of membrane constituents. The phospholipids play an important role in membrane differentiation (Donohue and Kaplan, 1986).

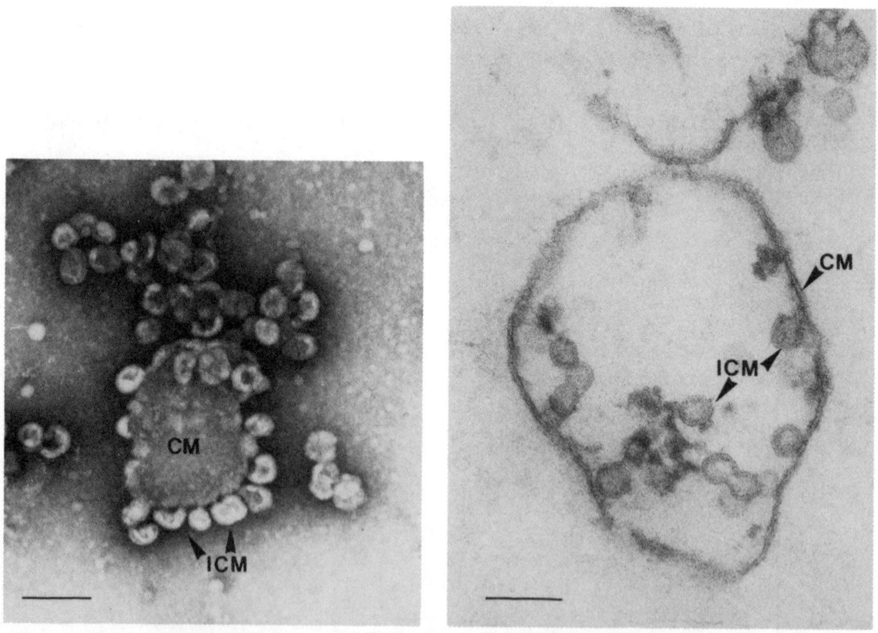

A B

Figure 25.8 A. Photosynthetic membranes isolated from *Chromatium vinosum*. The membrane fraction was prepared from spheroplasts and separated by sucrose gradient centrifugation (Hurlbert et al., 1974). The electron micrograph is taken from a negatively stained preparation. Intracytoplasmic membrane vesicles, bearing the photosynthetic apparatus, are bound to each other or to the cytoplasmic membrane. It has been proposed (Oelze and Drews, 1972; Reilly and Niederman, 1986) that the invagination sites at the cytoplasmic membrane are the sites for assembling of new photosynthetic units. B. Thin section of spheroplasts of *Rhodobacter capsulatus* prepared from cells induced to form the photosynthetic apparatus. The vesicle-like intracytoplasmic membrane structures were formed after insertion of bacteriochlorophyll and pigment-binding proteins into the membrane. A, Hurlbert et al., 1974; B, Dierstein et al., 1981.

The process of oxygen sensing and signal-transfer in the cell is unknown. Redox- or energy charge-controlled molecules have been proposed but not proved to be involved in regulation (Drews and Oelze, 1981; Drews, 1987).

It has also been shown that the synthesis of pigment-binding polypeptides is controlled at the level of transcription by derepression or activation of genes (Clark et al., 1984; Klug et al., 1985). After lowering of oxygen tension in the dark, the level of messenger RNA (mRNA) for pigment-binding polypeptides of reaction center and light-harvesting complexes increased strongly, followed by the synthesis of these proteins and their assembly in the membrane with pigments (Klug et al., 1985). It has been shown that the *puf* operon (coding for RC-M, RC-L, and light-harvesting I α, β polypeptides) in *R. capsulatus* is transcribed from more than one promoter. The most active of these promoters is located about 700 bp upstream from the *pufB* gene. This promoter is regulated by oxygen tension. In the region between *pufB* and the second promoter, a new gene called *pufQ* has been found, which is required for an early stage of Bchl synthesis (C. E. Bauer, M. Eleuterio, D. A. Young, and B. C. Marrs, unpublished). The synthesis of Bchl is regulated at the level of the activity of key enzymes for tetrapyrrole synthesis and by activation of genes for Bchl-synthesizing enzymes (Lascelles, 1978; Biel and Marrs, 1983). The activity of the first enzyme, the δ-aminolevulinate synthetase, was proposed to be regulated by a feedback mechanism (Lascelles, 1987) or a switch off/on mechanism (Viale et al., 1987).

Inhibition of Bchl synthesis by the competitive inhibitor levulinate immediately stops the synthesis of pigment-binding polypeptides but not the induced synthesis of mRNA for these polypeptides. Mutant strains, blocked in Bchl synthesis, express the genes for reaction center and light-harvesting polypeptides at a reduced rate (Klug et al., 1986). It is unknown at which level the synthesis and insertion of these proteins is inhibited. It has been suggested that polypeptides which cannot correctly assemble with pigments in the membrane are subject to a fast turnover (Dierstein, 1983). In a first step, the pigment-binding polypeptides are bound to the membrane, possibly at specific recognition sites. At this stage they are sensitive to proteolytic attack. The next step, the translocation of the polypeptides through the lipid bilayer, is dependent from the membrane potential or electrochemical proton gradient (Dierstein and Drews, 1986). The stable assembly of the proteins depends from the pigments, that is Bchl for reaction center and light-harvesting I complex and additionally carotenoids for the light-harvesting II complex (Drews et al., 1976; Zsebo and Hearst, 1984).

The regulation of synthesis and assembly of the pigment-proteins is a complex process. Variations of oxygen tension determine the transcription of photosynthetic genes and the activity of enzymes for pigment synthesis. There is a mutual influence of Bchl and Bchl-binding proteins on different steps of synthesis and assembly. Variations of oxygen tension and light intensity seem to effect the expression of genes coding not only for enzymes of pigment synthesis and pigment-binding proteins but also for regulatory proteins. Possibly a cascade of events determines the fine tuning in the formation of the photosynthetic apparatus similarly as in other complex regulatory processes such as nitrogenase.

The work described in this article was supported by Deutsche Forschungsgemeinschaft and Fonds der Chemischen Industrie.

REFERENCES

Baccarini-Melandri, A. and **Melandri, B.A.** 1978. Coupling factors, pp. 615–628, in Clayton, R.K. and Sistrom, W.R. (editors), *The Photosynthetic Bacteria*, Plenum Press, New York.

Beatty, J.T. and **Cohen, S.N.** 1983. Hybridization of cloned *Rhodopseudomonas capsulata* photosynthesis genes with DNA from other photosynthetic bacteria. *Journal of Bacteriology* 154:1440–1445.

Biel, A.J. and **Marrs, B.L.** 1983. Transcriptional regulation of several genes for bacteriochlorophyll biosynthesis in *Rhodopseudomonas capsulata* in response to oxygen. *Journal of Bacteriology* 156:686–694.

Blankenship, R.E., Feick, R., Bruce, B.D., Kirmaier, C., Holten, D., and **Fuller, R.C.** 1983. Primary photochemistry in the facultative green photosynthetic bacterium *Chloroflexus aurantiacus. Journal of Cellular Biochemistry* 22:251–261.

Blankenship, R.E. and **Fuller, R.C.** 1986. Membrane topology and photochemistry of the green photosynthetic bacterium *Chloroflexus aurantiacus*. pp. 390–399, in Staehelin, L.A. and Arntzen, C.J. (editors), *Photosynthesis III*, Springer-Verlag, Berlin, Heidelberg, New York.

Breton, J. 1986. Molecular orientation of the pigments and the problem of energy trapping in photosynthesis, pp. 319–326, in Staehelin, L.A. and Arntzen, C.J. (editors), *Photosynthesis III*, Springer-Verlag, Berlin, Heidelberg, New York.

Clark, W.G., Davidson, E., and **Marrs, B.L.** 1984. Variation of levels of mRNA coding for antenna and reaction center polypeptides in *Rhodopseudomonas capsulata* in response to changes in oxygen concentration. *Journal of Bacteriology* 157:945–948.

Cogdell, R.J. 1986. Light-harvesting complexes in the purple photosynthetic bacteria, pp. 252–259, in Staehelin, L.A. and Arntzen, C.J. (editors), *Photosynthesis III*, Springer-Verlag, Berlin, Heidelberg, New York.

Daurat-Larroque, S.T., Brew, K., and **Fenna, R.E.** 1986. The complete amino acid sequence of a bacteriochlorophyll *a* protein from *Prosthecochloris aestuarii. Journal of Biological Chemistry* 261:3607–3615.

Dierstein, R. 1983. Biosynthesis of pigment-protein complex polypeptides in bacteriochlorophyll-less mutant cells of *Rhodopseudomonas capsulata* YS. *FEBS Letters* 160:281–286.

Dierstein, R., Schumacher, A., and **Drews, G.** 1981. On insertion of pigment-associated polypeptides during membrane biogenesis in *Rhodopseudomonas capsulata. Archives of Microbiology* 128:376–383.

Dierstein, R. and **Drews, G.** 1986. Effect of uncoupler on assembly pathway for pigment-binding protein of bacterial photosynthetic membranes. *Journal of Bacteriology* 168:167–172.

Ditta, G., Stanfield, S., Corbin, D., and **Helsinki, D.R.** 1980. Broad host range cloning system for Gram-negative bacteria. *Proceedings of the National Academy of Sciences USA* 77:7347–7351.

Donohue, T.J. and **Kaplan, S.** 1986. Synthesis and assembly of bacterial photosynthetic membranes, pp. 632–639, in Staehelin, L.A. and Arntzen, C.J. (editors), *Photosynthesis III*, Springer-Verlag, Berlin, Heidelberg, New York.

Drews, G. 1985. Structure and functional organization of light-harvesting complexes and photochemical reaction centers in membranes of phototrophic bacteria. *Microbiological Reviews* 49:59–70.

Drews, G. 1987. The effect of oxygen partial pressure on formation of the bacterial photosynthetic apparatus, pp. 3–11, in *Oxygen Sensing in Tissues*, H. Acker (editor), Springer-Verlag, Heidelberg, Berlin.

Drews, G. and **Oelze, J.** 1981. Organization and differentiation of membranes of phototrophic bacteria. *Advances in Microbial Physiology* 22:1–92.

Drews, G., Dierstein, R., and **Schumacher, A.** 1976. Genetic transfer of the capacity to form bacteriochlorophyll protein complexes in *Rhodopseudomonas capsulata. FEBS Letters* 68:132–136.

Drews, G., Kaufmann, N., and **Klug, G.** 1985. The bacterial photosynthetic apparatus, pp. 211–222, in, Steinback, K.E., Bonitz, S., Arntzen, D.J., Bogorad, L. (editors), *Molecular Biology of the Photosynthetic Apparatus*, Cold Spring Harbor Laboratory, Cold Spring Harbor, New York.

Dutton, P.L. 1986. Energy transduction in anoxygenic photosynthesis, pp. 197–237, in, Staehelin, L.A. and Arntzen, C.J. (editors), *Photosynthesis III*, Springer-Verlag, Berlin, Heidelberg, New York.

Fenna, R.E., Matthews, B.W., Olson, J.M., and **Shaw, E.K.** 1974. Structure of a bacteriochlorophyll-protein from the green photosynthetic bacterium *Chlorobium limicola*: crystallographic evidence for a trimer. *Journal of Molecular Biology* 84:231–240.

Fornari, C.S. and **Kaplan, S.** 1982. Genetic transformation of *Rhodopseudomonas sphaeroides* by plasmid DNA. *Journal of Bacteriology* 152:89–97.

Fox, G.E., Stackebrandt, E., Hespell, R.B., Gibson, J., Maniloff, J., Dyer, T., Wolfe, R.S., Balch, W., Tanner, R., Magrum, L., Zablen, L.B., Blakemore, R., Gupta, R., Luehrsen, K.R., Bonen, L., Lewis, K.N., Chen, B.J-K.N., and **Woese, C.R.** 1980. The phylogeny of prokaryotes. *Science* 209:457–463.

Gibson, J., Ludwig, J.W., and **Stackebrandt, E.** 1985. The phylogeny of the green photosynthetic bacteria: Absence of a close relationship between *Chlorobium* and *Chloroflexus*. *Systematic and Applied Microbiology* 6:152–156.

Hale, M.B., Blankenship, R.E., and **Fuller, R.C.** 1983. Menaquinone is the sole quinone in the facultative aerobic green photosynthetic bacterium *Chloroflexus aurantiacus*. *Biochimica et Biophysica Acta* 723:376–382.

Harashima, K., Hayashi, J., Ikari, T., and **Shiby, T.** 1980. Oxygen-stimulated synthesis of bacteriochlorophyll and carotenoids in marine bacteria. *Plant and Cell Physiology* 21:1281–1294.

Harashima, K., Nakagawa, J., Ikari, T., and **Shiba, T.** 1982. Photochemical activities of bacteriochlorophyll in aerobic heterotrophs. *Plant and Cell Physiology* 23:185–193.

Haselkorn, R. 1986. Organization of the genes for nitrogen fixation in photosynthetic bacteria and cyanobacteria. *Annual Review of Microbiology* 40:525–547.

Ho, K.K., and **Krogmann, D.W.** 1982. Photosynthesis, pp. 191–214, in, Carr, N.G. and Whitton, B.A. (editors), *The Biology of Cyanobacteria*, Blackwell Scientific Publications, Oxford.

Hurlbert, R.E., Golecki, J.R., and **Drews, G.** 1974. Isolation and characterization of *Chromatium vinosum* membranes. *Archives of Microbiology* 101:169–186.

Kaufmann, N., Hüdig, H., and **Drews, G.** 1984. Transposon Tn5 mutagenesis of genes for the photosynthetic apparatus in *Rhodopseudomonas capsulatus*. *Molecular and General Genetics* 198:153–158.

Kirmaier, C., Holten, D., Feick, R., and **Blankenship, R.E.** 1983. Picosecond measurements of the primary photochemical events in reaction centers isolated from the facultative green photosynthetic bacterium *Chloroflexus aurantiacus*. *FEBS Letters* 158:73–78.

Kirmaier, C., Blankenship, R.E., and **Holten, D.** 1986. Formation and decay of radical-pairs state $P+I^-$ in *Chloroflexus* reaction centers. *Biochimica et Biophysica Acta* 850:275–285.

Klug, G., Kaufmann, N., and **Drews, G.** 1985. Gene expression of pigment-binding proteins of the bacterial photosynthetic apparatus: transcription and assembly in the membrane of *Rhodopseudomonas capsulata. Proceedings of the National Academy of Sciences USA* 82:6485–6489.

Klug, G., Liebetanz, R., and **Drews, G.** 1986. The influence of bacteriochlorophyll biosynthesis on formation of pigment-binding proteins and assembly of pigment protein complexes in *Rhodopseudomonas capsulata. Archives of Microbiology* 146:284–291.

Knox, R.S. 1986. Trapping events in light-harvesting assemblies, pp. 286–298, in Staehelin, L.A. and Arntzen, C.J. (editors), *Photosynthesis III*, Springer-Verlag, Berlin, Heidelberg, New York.

Lanyi, J.K. 1978. Light energy conversion in *Halobacterium halobium. Microbiological Reviews* 42:682–707.

Lascelles, J. 1978. Regulation of pyrrole synthesis, pp. 795–808, in, Clayton, R.K. and Sistrom, W.S. (editors), *The Photosynthetic Bacteria*, Plenum Press, New York.

Liaanen-Jensen, S. 1978. Chemistry of carotenoid pigments, pp. 233–247, in, Clayton, R.K. and Sistrom, W.R. (editors), *The Photosynthetic Bacteria*, Plenum Press, New York.

Marrs, B. 1974. Genetic recombination in *Rhodopseudomonas capsulata. Proceedings of the National Academy of Sciences USA* 71:971–973.

Marrs, B.L. 1981. Mobilization of the genes for photosynthesis from *Rhodopseudomonas capsulata* by a promiscuous plasmid. *Journal of Bacteriology* 146:1003–1012.

McEwan, A.G., Wetzstein, H.G., Ferguson, S.J., and **Jackson, J.B.** 1985. Periplasmic location of the terminal reductase in trimethylamine N-oxide and dimethylsulphoxide respiration in the photosynthetic bacterium *Rhodopseudomonas capsulata. Biochimica et Biophysica Acta* 806:410–417.

Michel, H. and **Deisenhofer, J.** 1986. X-ray diffraction studies on a crystalline bacterial photosynthetic reaction center, pp. 371–381, in, Staehelin, L.A. and Arntzen, C.J. (editors), *Photosynthesis III*, Springer-Verlag, Berlin, Heidelberg, New York.

Michel, H., Epp, O., and **Deisenhofer, J.** 1986. Pigment-protein interactions in the photosynthetic reaction center from *Rhodopseudomonas viridis. The EMBO Journal* 5:2445–2451.

Norris, J.R. and **van Brakel, G.** 1986. Energy trapping in photosynthesis as probed by the magnetic properties of reaction centers, pp. 353–370, in Staehelin, L.A. and Arntzen, C.J. (editors), *Photosynthesis III*, Springer-Verlag, Berlin, Heidelberg, New York.

Oelze, J. 1985. Analysis of bacteriochlorophylls, pp. 257–284, in Gottschalk, G. (editor), *Methods in Microbiology*, Vol. 18, Academic Press, New York, London.

Oelze, J. and **Drews, G.** 1972. Membranes of photosynthetic bacteria. *Biochimica et Biophysica Acta* 265:209–239.

Parson, W.W. and **Ke, B.** 1982. Primary photochemical reactions, pp. 331–386, in Govindjee (editor), *Photosynthesis Vol. I*, Academic Press, New York, London.

Pemberton, J.M. and **Bowen, A.R.S.G.** 1981. High-frequency chromosome transfer in *Rhodopseudomonas sphaeroides* promoted by broad-host range plasmid RP1 carrying mercury transposon. *Journal of Bacteriology* 147:110–117.

Reilly, P.A. and **Niederman, R.A.** 1986. Role of apparent membrane growth initiation sites during photosynthetic membrane development in synchronously dividing *Rhodopseudomonas sphaeroides. Journal of Bacteriology* 167:153–159.

Sauer, K. 1986. Photosynthetic light reactions, pp. 85–97, in, Staehelin, L.A. and Arntzen, C.J. (editors), *Photosynthesis III*, Springer-Verlag, Berlin, Heidelberg, New York.

Shiba, T. 1984. Utilization of light energy by the strictly aerobic bacterium *Erythrobacter* sp. *Journal of General and Applied Microbiology* 30:239–244.

Scolnik, P.A. and **Marrs, B.L.** 1987. Genetic research with photosynthetic bacteria. *Annual Review of Microbiology* 41:703–726.

Shiozawa, J.A., Lottspeich, F., and **Feick, R.** 1987. The photochemical reaction center of *Chloroflexus aurantiacus* is composed of two structurally similar polypeptides. *European Journal of Biochemistry* 167:595–600.

Shuvalov, V.A., Vasmel, H., Amesz, J., and **Duysens, L.N.M.** 1986. Picosecond spectroscopy of the charge separation in reaction centers of *Chloroflexus aurantiacus* with selective excitation of the primary donor. *Biochimica et Biophysica Acta* 851:361–368.

Smith, L. and **Pinder, P.B.** 1978. Oxygen-linked electron transport and energy conservation, pp. 641–656, in, Clayton, R.K. and Sistrom, W.R. (editors), *The Photosynthetic Bacteria,* Plenum Press, New York.

Sprague, S.G. and **Varga, A.R.** 1986. Membrane architecture of anoxygenic photosynthetic bacteria, pp. 603–619, in, Staehelin, L.A. and Arntzen, C.J. (editors), *Photosynthesis III, Photosynthetic Membranes and Light-harvesting Systems,* Springer-Verlag, Berlin, Heidelberg, New York.

Stackebrandt, E. 1985. Phylogeny and phylogenetic classification of prokaryotes, pp. 309–334, in, Schleifer, K.H. and Stackebrandt, E. (editors), *Evolution of Prokaryotes,* FEMS Symposium 29, Academic Press, London.

Tadros, M.H., Frank, R., Dörge, B., Gad'on, N., Takemoto, J.Y., and **Drews, G.** 1987. Orientation of the B800–850, B870 and reaction center polypeptides on the cytoplasmic and periplasmic surfaces of *Rhodobacter capsulatus* membranes. *Biochemistry* 26:7680–7687.

Thaylor, D.P., Cohen, S.N., Clark, W.G., and **Marrs, B.L.** 1983. Alignment of genetic and restriction maps of the photosynthetic region of the *Rhodopseudomonas capsulata* chromosome by a conjugation-mediated marker rescue technique. *Journal of Bacteriology* 154:580–590.

Thornber, J.P. 1986. Biochemical characterization and structure of pigment-proteins of photosynthetic organisms, pp. 98–142, in, Staehelin, L.A. and Arntzen, C.J. (editors), *Photosynthesis III,* Springer-Verlag, Berlin, Heidelberg, New York.

Trüper, H.G. 1987. Phototrophic bacteria an incoherent group of prokaryotes. A taxonomic versus phylogenetic survey. *Microbiologia SEM* 3:71–89.

van Liere, L. and **Walsby, A.E.** 1982. Interaction of cyanobacteria with light, pp. 9–46, in, Carr, N.G. and Whitton, B.A. (editors), *The Biology of Cyanobacteria,* Blackwell Sci. Publication, Oxford.

Vasmel, H. and **Amesz, J.** 1983. Photoreduction of menaquinone in the reaction center of the green photosynthetic bacterium *Chloroflexus aurantiacus. Biochimica et Biophysica Acta* 724:118–122.

Viale, A.A., Wider, A.E., and **del C. Battle, A.M.** 1987. Porphyrin biosynthesis in *Rhodopseudomonas palustris*—XII. *International Journal of Biochemistry* 19:379–383.

Wechsler, T., Suter, F., Fuller, R.C., and **Zuber, H.** 1984. The complete amino acid sequence of the bacteriochlorophyll *c*-binding polypeptide from chlorosomes of the green photosynthetic bacterium *Chloroflexus aurantiacus. FEBS Letters* 181:173–178.

Willison, J.C., Magnin, J.P., and **Vignais, P.M.** 1987. Isolation and characterization of *Rhodobacter capsulatus* strains lacking endogenous plasmids. *Archives of Microbiology* 147:134–142.

Woese, C.R., Stackebrandt, E., Macke, T., and **Fox, G.E.** 1985a. A phylogenetic definition of the major eubacterial taxa. *Systematic and Applied Microbiology* 6:143–151.

Woese, C.R., Weisburg, W.G., Paster, G.J., Hahn, C.M., Tanner, R.S., Krieg, N.R., Koops, H.-P., Harms, H., and **Stackebrandt, E.** 1984. The phylogeny of purple bacteria—the beta subdivision. *Systematic and Applied Microbiology* 5:327–336.

Woese, C.R., Weisburg, W.G., Hahn, C.M., Paster, B.J., Zablen, L.B., Lewis, B.J., Macke, T.J., Ludwig, W., and **Stackebrandt, E.** 1985b. The phylogeny of purple bacteria—the gamma subdivision. *Systematic and Applied Microbiology* 6:25–33.

Woese, C.R., Stackebrandt, E., Weisburg, W.G., Paster, B.J., Madigan, M.T., Fowler, J., Hahn, C.M., Blanz, P., Gupta, R., Nealson, K.H., and **Fox, G.E.** 1985c. The phylogeny of purple bacteria—the alpha subdivision. *Systematic and Applied Microbiology* 5:315–326.

Yen, H.C. and **Marrs, B.** 1977. Growth of *Rhodopseudomonas capsulata* under anaerobic dark conditions with dimethyl sulfoxide. *Archives of Biochemistry and Biophysics* 181:411–418.

Zannoni, D. and **Baccarini-Melandri, A.** 1980. Respiratory electron flow in facultative photosynthetic bacteria, pp. 184–202, in, Knowles, C.J. (editor), *Diversity of Bacterial Respiratory Systems,* CRC Press, Boca Raton.

Zhu, Y.S. and **Hearst, J.E.** 1986. Regulation of expression of genes for light-harvesting antenna proteins, reaction center polypeptides and enzymes of bacteriochlorophyll and carotenoid biosynthesis. *Proceedings of the National Academy of Sciences USA* 83:7613–7617.

Zsebo, K.M. and **Hearst, J.E.** 1984. Genetic-physical mapping of a photosynthetic gene cluster from *R. capsulata. Cell* 37:937–947.

26

Molecular Biology of Carbon Dioxide Fixation in Photosynthetic Bacteria

F. Robert Tabita

Department of Microbiology
University of Texas at Austin
Austin, Texas, U.S.A.

With the advent of molecular biology, the molecular control of CO_2 fixation may be approached and examined with considerable acuity. It will be the purpose of this chapter to review some of the salient features of CO_2 fixation in photosynthetic procaryotes and to examine how molecular biological studies are pointing the way toward unraveling the details of this complex and key biochemical life process.

26.1 GENETIC AND MOLECULAR BIOLOGICAL STUDIES IN PHOTOSYNTHETIC BACTERIA

Over the last few years, both the Rhodospirillaceae and the cyanobacteria have been employed to study important metabolic processes. The Rhodospirillaceae, all of which can grow heterotrophically, are conveniently employed in molecular biological research where transformation, conjugation, and capsduction are routinely and effectively used (for a review, see Scolnik and Marrs, 1987). In the cyanobacteria, selected organisms have been shown to be transformable (Porter, 1986) and shuttle vectors have been used to transfer genes from *Escherichia coli* to unicellular (Buzby et al., 1983; Gendel et al., 1983; Golden and Sherman, 1983; Kuhlemeier et al., 1981) as well as filamentous, heterocystous strains (Wolk et al., 1984). The ability to use the tools of molecular biology has allowed for detailed probing of the regulation of fundamental processes such as photochemical energy generation, membrane biogenesis, nitrogen fixation, nitrogen metabolism, and hydrogen metabolism (Haselkorn, 1986; Scolnik and Marrs, 1987), and now CO_2 fixation.

26.2 PATHWAYS OF CO_2 FIXATION IN PHOTOSYNTHETIC BACTERIA

There are several routes by which CO_2 may be converted to organic carbon in photosynthetic procaryotes. These include the reductive pentose phosphate pathway or Calvin cycle (Chapters 3 and 19), the reductive tricarboxylic acid cycle (Chapter 20), and the ancillary routes involving C_4 organic acids and citrulline (Tabita, 1987). The citrulline route accounts for up to 20 percent of all the organic carbon produced by various cyanobacteria and involves the enzymes carbamyl phosphate synthetase and ornithine transcarbamylase (Chen et al., 1987a, b). However, even under conditions where the carbamyl phosphate pathway is active, the bulk of the carbon synthesized is via the Calvin cycle. From a global view, it is clear that the Calvin cycle is the most significant autotrophic carbon assimilatory scheme.

26.3 KEY METABOLIC STEPS OF THE CALVIN CYCLE

Ribulose 1,5-bisphosphate carboxylase/oxygenase (RuBisCO) catalyzes the reaction in which the enediol of RuBP is attacked by one of the gaseous substrates CO_2 or O_2 (see Chapter 19). This enzyme is thus a bifunctional protein and is capable of catalyzing the first reaction of the competing pathways of carbon assimilation and photo/chemorespiration. By virtue of being the key reaction of both metabolic routes, considerable emphasis on the structure, function, and mechanistic details of this protein has occurred over the years (reviewed by Miziorko and Lorimer, 1983), not to mention that agricultural productivity is directly dependent on the action of this enzyme. Thus far, it is solely among the photosynthetic bacteria that variations from the usual L_8S_8 large (native M_r of about 550,000) hexadecameric structure of RuBisCO are found (Tabita, 1981). It is well established that the small subunits ($M_r \sim 15{,}000$) are required for the activity of the large subunits of L_8S_8 enzymes (Andrews and Ballment, 1983; Jordan and Chollet, 1985; Takabe et al., 1984). Exactly how the small subunits influence activity, beyond exerting conformational effects on the large subunits, is not clear at this time; however, a mutagenesis study suggests a direct role in catalysis (Voordouw et al., 1987). Still unexplained is how the dimeric (L_2) *Rhodospirillum rubrum* and *Rhodobacter sphaeroides* form II enzyme (which is probably a tetramer) functions in the absence of small subunits.

Phosphoribulokinase (PRK), the other key and unique enzyme of the Calvin cycle, has been isolated from photosynthetic bacteria (Gibson and Tabita, 1987; Hallenbeck and Kaplan, 1987; Marsden and Codd, 1984; Rippel and Bowien, 1984; Tabita, 1980). In all cases, PRK from autotrophic bacteria, including chemolithoautotrophic hydrogen bacteria (Siebert et al., 1981), is comprised of a single 32–36,000 molecular weight polypeptide and the protein has a native molecular weight of 200–256,000. A distinguishing feature of many bacterial PRK enzymes is the nearly absolute requirement of NADH for activity, a feature not shared by enzymes from some photo- and chemolithoautotrophic bacteria and all oxygen-evolving photosynthetic procaryotes and eucaryotes (Tabita, 1981).

Other enzymes, such as glyceraldehyde-3-phosphate dehydrogenase (GAP), 3-phosphoglyceric acid kinase (PGA), and fructose 1,6-bisphosphatase (FBP), catalyze key reactions involved in the reduction of the first CO_2 fixation product, 3-phosphoglyceric acid, and its subsequent conversion to ribulose-5-phosphate. It is interesting that a complex of PRK-FBP was isolated from *R. rubrum* (Joint et al., 1972) and a complex of PRK-GAP has been found in extracts of the green alga *Scenedesmus* (Nicholson et al., 1987). These studies suggest that a coordination of key enzymes of the Calvin cycle may occur *in vivo* to maximize efficient CO_2 fixation; however, the precise significance of such complexes has not been established as yet. The recent finding that the FBP and PRK genes (*fbp* and *prk*, respectively) are adjacent on the *R. sphaeroides* chromosome and are cotranscribed when placed behind a *lac* promoter in *E. coli* expression vectors (Gibson and Tabita, 1988) suggests that PRK-FBP complex formation in photosynthetic bacteria may be a consequence of the cotranscription of these genes.

26.4 METABOLIC CONTROL OF CO_2 FIXATION IN PHOTOSYNTHETIC BACTERIA

With the exception of the green sulfur bacteria, all photosynthetic procaryotes fix CO_2 by the Calvin reductive pentose phosphate pathway. Evidence for the utilization of this route to reduce CO_2 is overwhelming, and includes label incorporation (Glover et al., 1952; Stoppani et al., 1955), enzymatic (Fuller and Gibbs, 1959), carbon isotope fractionation (Quandt et al., 1977), and genetic/molecular (Tabita et al., 1987) studies. Organisms belonging to the Rhodospirillaceae are probably the most versatile organisms found in nature since they are capable of five distinct modes of metabolism, including growth under conditions where CO_2 is not required as a carbon source (Madigan and Gest, 1979). Early investigations by van Niel (1936), Gaffron (1935), and Muller (1933) indicated that these photosynthetic bacteria incorporated CO_2 when the electron donor was more reduced than the cell protoplasm, but released CO_2 into the medium (from the organic carbon supplied) when the cells were supplied with a more oxidized electron donor. It is interesting that early short-term label incorporation studies which established the presence of the Calvin cycle were performed with cells that were grown under conditions where there is a net increase in the amount of CO_2 found in the culture. Thus, even when CO_2 is not a prime source of carbon, the Rhodospirillaceae use this pathway, most likely as an anaplerotic source of carbon.

In 1960, Lascelles showed that CO_2 fixation in *Rhodopseudomonas* (now called *Rhodobacter*) *sphaeroides*, is under metabolic control since RuBisCO synthesis immediately ceased when photosynthetically grown cells were exposed to oxygen. In addition, there was an instantaneous increase in the differential rate of RuBisCO synthesis when dark, aerobically grown cells (containing bacteriochlorophyll) were transferred to photosynthetic conditions (Lascelles, 1960). In *R. rubrum*, the levels of RuBisCO were found to vary substantially depending on the growth conditions (Anderson and Fuller, 1967). These studies, along with the early results using electron donors of different oxidation state, led Lascelles to hypothesize that CO_2 may act as an electron sink when the electron donor is reduced, and the oxidation state of some intracellular metabolite derived from the electron donor and carbon source might serve to regulate RuBisCO synthesis. This theory has also been proposed to explain regulation in facultative chemolithoautotrophic bacteria as well (Dijkhuizen and Harder, 1984; Friedrich et al., 1979).

From the above studies, it was not surprising that *R. rubrum* (Tabita and McFadden, 1974), *R. sphaeroides* (Gibson and Tabita, 1977a), *R. capsulatus* (Gibson and Tabita, 1977b; Shively et al., 1984), and *Rhodopseudomonas blastica* (Dow, 1987) apparently derepress RuBisCO synthesis when cells are cultured in a butyrate-bicarbonate medium. Physiological studies with *R. rubrum* indicated that $NaH^{14}CO_3$ incorporation paralleled the growth of the culture; however, only when the concentration of bicarbonate in the medium decreased to 2 mM or less was there an increase in the level of RuBisCO, which occurred well into stationary phase (Sarles and Tabita, 1983). Subsequent studies indicated that butyrate and bicarbonate were taken up simultaneously (Figure 26.1), yet the free concentration of butyrate in the medium remained at approximately 20 mM. Since 20 mM butyrate is not limiting for growth, these results suggested that HCO_3^- (CO_2) limitation significantly affects the repression/derepression of RuBisCO synthesis, a result substantiated by growing cells photolithotrophically at low and high levels of CO_2 with H_2 as the electron donor (Sarles and Tabita, 1983). The rather dramatic change in the level of RuBisCO synthesis, dependent on the environmental growth conditions and availability of CO_2, is illustrated by two-dimensional gel electrophoresis (Figure 26.2) and illustrates the extremely powerful regulation attained with these organisms, where the levels of RuBisCO vary from nearly zero to 50 percent of the soluble protein.

Two different RuBisCO enzymes were found in *R. sphaeroides* (Gibson and Tabita, 1977a), form I (L_8S_8) and form II (L_x); the synthesis of each was found to be derepressed in the butyrate-bicarbonate growth medium. However, subsequent studies indicated that both the rate and extent of form I and form II RuBisCO synthesis differed, indicating that synthesis of the two enzymes is independently regulated (Jouanneau and Tabita, 1986). Our early experiments indicated that *R. capsulatus* also contains two distinct RuBisCO enzymes (Gibson and Tabita, 1977b), and this prompted the examination of several different representatives of the Rhodospirillaceae to determine if other organisms produce duplicate enzymes, particularly L_x forms. Using the polyacrylamide gel assay (Gibson and Tabita, 1977a, b) developed by Whitman (1978), it was shown that *Rhodospirillum tenue, Rhodospirillum molischianum, Rhodospirillum fulvum, Rhodomicrobium vannielii*, and *Rhodopseudomonas palustris* all synthe-

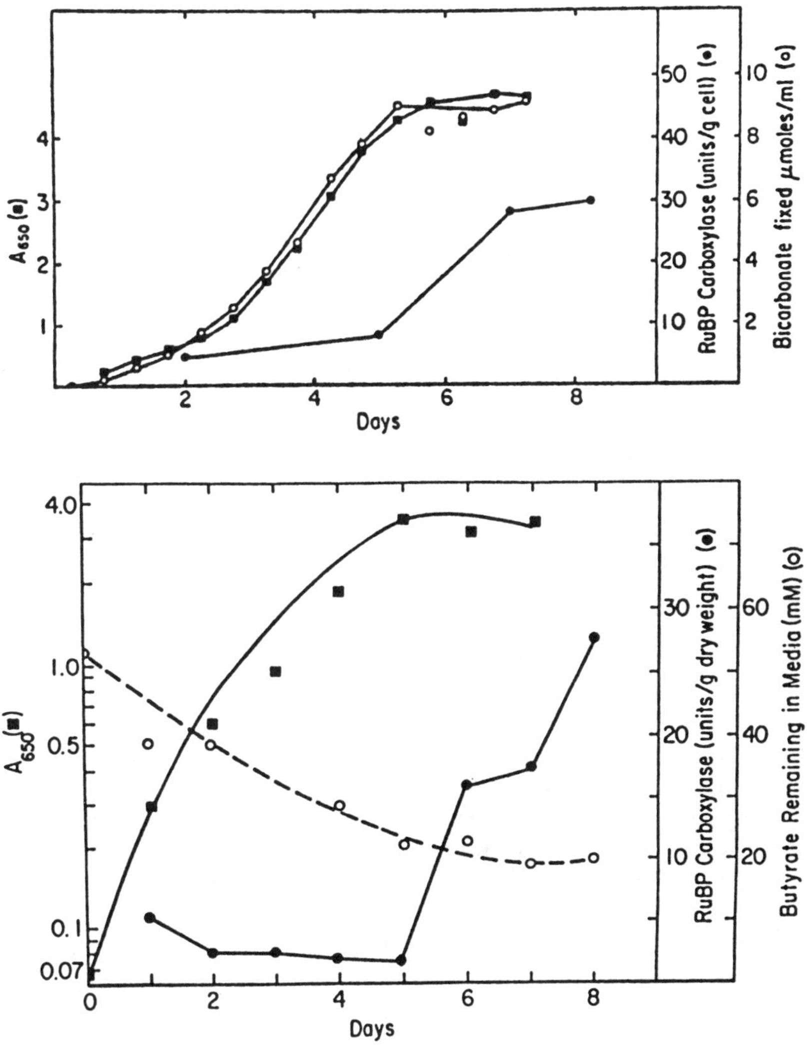

Figure 26.1 Simultaneous uptake of butyrate and $H^{14}CO_3^-$ by cultures of *Rhodospirillum rubrum.* Growth is measured by absorbance of 650 nm (A_{650}).

sized large molecular weight RuBisCO, presumably L_8S_8 structures, based on the cross-reactivity to antisera to the *R. sphaeroides* form I enzyme (Figure 26.3). These enzymes all co-electrophoresed and co-eluted from gel filtration columns with the spinach and *R. sphaeroides* form I RuBisCO, enzymes of known L_8S_8 structure. No dimeric form of RuBisCO was found in other *Rhodospirillum* species; antisera to the form II *R. sphaeroides* and *R. rubrum* enzymes failed to cross-react with extracts from these organisms, with the exception of *R. palustris.* Here it was found that both L_8S_8 and L_x enzymes were present, based on the antibody data and the specific elution of a presumptive L_8 enzyme

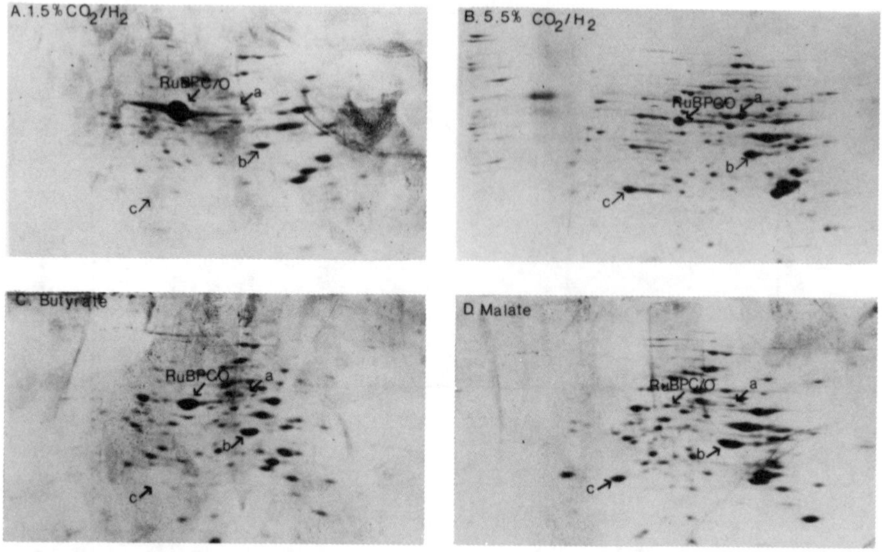

Figure 26.2 Two-dimensional polyacrylamide gel electrophoresis of crude extracts of phototrophically grown cells of *R. rubrum* cultured in malate, butyrate-bicarbonate, 5.5% CO_2/94.5% H_2, or 1.5% CO_2/98.5% H_2 atmosphere. Lower case letters represent proteins that are synthesized according to the indicated growth conditions. RuBPC/O, ribulose bisphosphate carboxylase/oxygenase.

separate from a L_8S_8 enzyme from dye-ligand columns (J. L. Gibson and F. R. Tabita, unpublished results). Presumably the failure to note this L_x enzyme with the polyacrylamide gel assay is due to the close molecular size of the L_8 enzyme compared to the L_8S_8 RuBisCO. In any case, there is more and more information accumulating to indicate that several organisms contain duplicate and distinct RuBisCO enzymes, particularly in the photosynthetic bacteria. At least in *R. sphaeroides*, we have found these genes to be chromosomally encoded. In the chemolithoautotrophic hydrogen bacterium *Alcaligenes eutrophus*, there are plasmid-encoded and chromosomally encoded RuBisCO genes (Andersen and Wilke-Douglas, 1987; Bowien et al., 1987), apparently the result of a gene duplication that results in the production of identical L_8S_8 enzymes. Clearly, such is not the case in *R. sphaeroides*, where completely different forms of enzyme are produced that are immunologically (Gibson and Tabita, 1977a, c), catalytically (Gibson, 1980; Gibson and Tabita, 1977a; Jordan and Ogren, 1981), and structurally (Gibson and Tabita, 1985) distinct. Recent hybridization and sequence analyses of the form I and form II structural genes also indicate that these are distinct entities (as discussed in subsequent paragraphs).

Although RuBisCO levels are drastically affected by the environmental growth conditions, recent immunological and molecular studies from our laboratory indicate that the synthesis of PRK does not precisely parallel this response. This was somewhat surprising since these enzymes catalyze the two unique and consecutive reactions of the Calvin cycle. Similar results were ob-

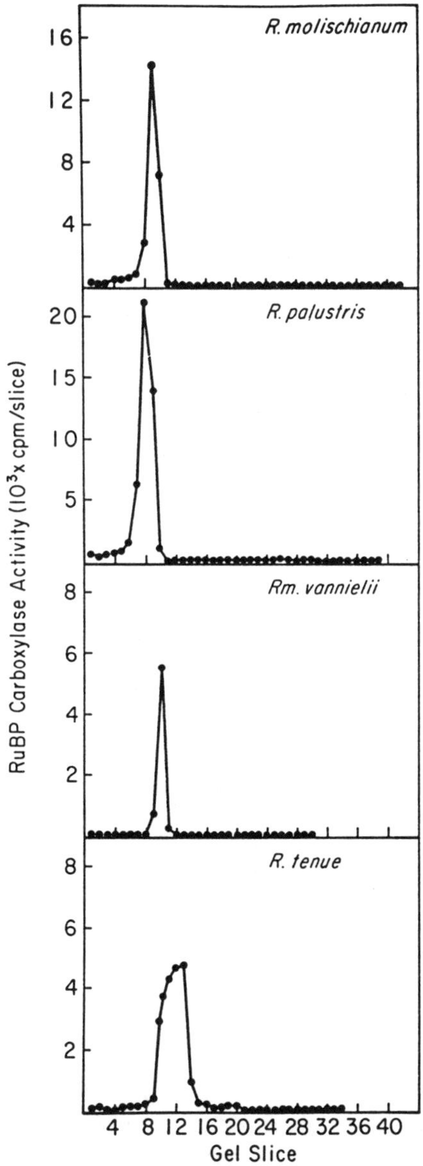

Figure 26.3 Polyacrylamide gel assay of ribulose bisphosphate carboxylase/oxygenase activity of crude extracts of various members of the Rhodospirillaceae.

tained for FBP and GAP. These physiological studies may suggest that the Calvin cycle structural genes are not localized in a single operon or alternatively specific portions of the relevant transcript may be selectively degraded.

Aside from possible control of the activity of key enzymes by various intracellular metabolites (Tabita, 1987), there is now compelling evidence to indicate that RuBisCO in photosynthetic bacteria may be regulated by posttranslational modification. Addition of organic carbon to photolithoautotrophically grown *R. sphaeroides* resulted in the loss of enzyme units from the cells. Furthermore, the decrease in specific activity greatly exceeded that which could be accounted for by the repression of enzyme synthesis. Initial attempts to purify the inactivated RuBisCO failed, since the enzyme became reactivated during the purification regimen. However, a series of kinetic experiments, using specific antisera and based on the catalytic idiosyncrasies of each enzyme, allowed for an unambiguous determination that the form I enzyme was specifically affected (Jouanneau and Tabita, 1987). Moreover, rapid separation and purification of the form I and form II enzymes was achieved by dye-ligand chromatography on Green A agarose columns, confirming that the form I enzyme was somehow modified. Most importantly, inactivated form I RuBisCO was easily separated from small amounts of active enzyme when extracts were prepared from inactivated cells, indicative of a change in the physicochemical properties of the enzyme. Subsequent work showed that under normal conditions the form I enzyme is found in a complex or aggregated state and is associated with one or more phosphorylated compounds. After the addition of organic substrates to the cells, active metabolism involving protein synthesis results in the modification or alteration of the form I enzyme such that there is a loss of activity, removal of the phosphorylated material, and shift in the conformation of the enzyme (Jouanneau and Tabita, 1987). The modification of the enzyme and nature of the phosphorylated compound that associates with the form I protein is under scrutiny. Clearly the enzyme is not covalently modified with phosphate since the label is rapidly removed upon heating or the addition of detergents, weak acid, or alkali. Other results from our laboratory indicate that the *R. rubrum* enzyme may be subject to another type of posttranscriptional inactivation, an oxidative modification that precedes the proteolytic degradation of the enzyme (Cook, 1986).

26.5 GENETICS OF CARBON DIOXIDE FIXATION IN PHOTOSYNTHETIC BACTERIA

Isolation and cloning of the structural genes The RuBisCO gene from *R. rubrum* was the first structural gene of the Calvin cycle to be cloned (Somerville and Somerville, 1984). This was accomplished after the gene was isolated from a genomic library of *R. rubrum* DNA constructed in bacteriophage λ1059. The gene was sequenced and the deduced amino acid sequence was found to show only 25 percent homology to the large subunit protein from higher plants (Nargang et al., 1984). Due to the structural and catalytic similarity of the *R. sphaeroides* form II RuBisCO to the *R. rubrum* enzyme (Gibson and Tabita, 1979), it was assumed that the *R. rubrum* gene might be employed as a heterologous probe to isolate the *R. sphaeroides* gene. Hybridization to total genomic DNA from *R. sphaeroides* resulted in the identification of a 3 kb *Eco*RI fragment that exhibited strong homology. This fragment was subsequently shown to contain the entire coding sequence for the form II gene (Quivey and Tabita, 1984; Muller et al., 1985; Tabita et al., 1986), designated *rbpL* to distinguish it from the large subunit gene of L_8S_8 enzymes, *rbcL* (Gibson and Tabita, 1987). Comparison of the deduced amino acid sequence of the *rbpL* gene product from *R. sphaeroides* and *R. rubrum* indicated there was about 75 percent homology (Wagner et al., 1988). In addition, several highly conserved regions, corresponding to the active site and activation site regions of large subunits of higher plants and *R. rubrum* were found in the *R. sphaeroides* form II sequence. Thus, despite the lack of small subunits, and a low overall homology, the L_x enzymes perform the same function and catalyze the same reactions as L_8S_8 enzymes. Inspection of the sequences of the two L_x enzymes now available indicates no domains which might be homologous to the three conserved regions of the small subunit of L_8S_8 enzymes. The L_x *R. rubrum* and *R. sphaeroides* form II enzymes are relatively insensitive to feedback effectors, such as 6-phosphogluconate (Schloss et al., 1979; Tabita and McFadden, 1972; Whitman et al., 1979), and the unactivated form of the L_x enzyme does not bind RuBisCO tightly (Gibson and Tabita, 1979; Tabita et al., 1986). Perhaps comparisons of the L_x sequences to the large subunits of L_8S_8 enzymes might point to residues important for these effects (and subsequent site-specific mutagenesis), as well as providing a molecular basis for the low affinity of L_x enzymes for CO_2 (Jordan and Ogren, 1981).

Attempts to isolate the *R. sphaeroides* form I RuBisCO genes, using the form II gene (*rbpL*) as a probe, met with limited success, no doubt due to a low homology of the primary structure of the large subunits of these proteins (Gibson and Tabita, 1985). However, the form I *rbcLrbcS* genes were highly reactive to an *Anacystis nidulans* probe, indicating the *R. sphaeroides* form I genes were more reactive to probes from unrelated organisms than to the *rbpL* gene from the same organism (Gibson and Tabita, 1986). The form I genes were subsequently isolated and cloned and shown to be cotranscribed, similar to other procaryotic *rbcLrbcS* genes (Andersen and Wilke-Douglas, 1987; Bowien et al., 1987; Nierzicki-Bauer et al., 1984; Shinozaki and Sugiura, 1985; Tiboni et al., 1984; Vakeria et al., 1986; Viale et al., 1985). Thus far the *rbcLrbcS* genes have been found to be cotranscribed in four cyanobacteria (*Anacystis nidulans, Anabaena* 7120, *Spirulina platensis, Chlorogloeopsis fritschii*), two photosynthetic bacteria (*R. sphaeroides* and *Chromatium vinosum*), and one chemolithoautotrophic bacterium (*Alcaligenes eutrophus*), suggesting that this operon structure may be universal in all procaryotic organisms that contain L_8S_8 RuBisCO enzymes. A reasonable hypothesis is that the regulation of transcription might also be universal, but this remains to be determined, as discussed later.

Recently, two distinct PRK genes were isolated from *R. sphaeroides* (Gibson and Tabita, 1987). These genes were found to be closely associated with the form I and form II RuBisCO genes. Unlike the situation in *A. eutrophus* (Klintworth et al., 1985; see Chapter 24), both of the PRK genes of *R. sphaeroides* were found on the chromosome. The form I PRK gene (*prkA*) was found to be about 1 kb upstream from the *rbcLrbcS* genes (Gibson and Tabita, 1987; Hallenbeck and Kaplan, 1987), a significantly different arrangement from *A. eutrophus,* where both the plasmid and chromosomal PRK genes are about 4 kb downstream from the *rbcLrbcS* gene cluster (Klintworth et al., 1985). Even more unique was the location of *prkB,* which was 3.5–4.0 kb upstream from *rbpL.*

Using plasmids from a pVK102 cosmid library, other Calvin cycle structural genes were localized in the form I and form II PRK-RuBisCO gene clusters. Plasmids pJG336 and pJG106 contain the form I and form II carboxylase-kinase genes, respectively. Heterologous probes containing FBP and GAP coding sequences, two other regulatory enzymes of the Calvin cycle (Fuller and Gibbs, 1959; Springgate and Stachow, 1972), were used to localize the corresponding genes in *R. sphaeroides.* Several lines of evidence indicate that there are two FBP genes (*fbpA* and *fbpB*) immediately upstream from *prkA* and *prkB,* respectively (Gibson and Tabita, 1988). The *fbpAprkA* genes are cotranscribed as, presumably, are the *fbpBprkB* genes. A GAP structural gene (*gapB*) was found solely in the form II CO_2 fixation (*cfx*) cluster, but not in the form I *cfx* cluster. These studies and the arrangement of the key Calvin cycle structural genes are summarized (Figure 26.4). It is interesting that a "common sequence," indicated by the black box immediately upstream from *rbcL* and *rbpL,* is found in both *cfx* clusters. A 37,000 molecular weight protein is encoded by the form I common sequence, which was reported to be transcribed in the opposite direction of the *prkA* gene (Hallenbeck and Kaplan, 1987). However, studies from our

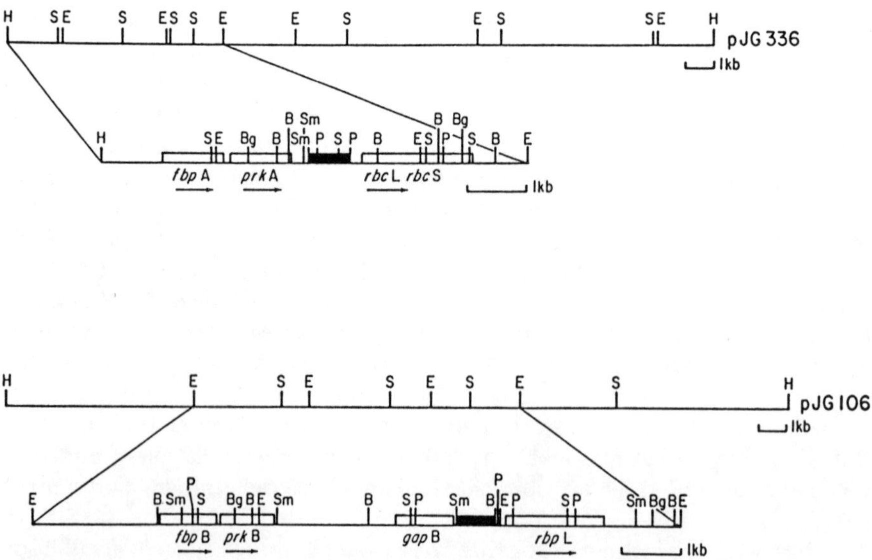

Figure 26.4 Organization of Calvin cycle structural genes in *Rhodobacter sphaeroides.* The form I and form II *cfx* clusters are shown to be organized in two regions of the *R. sphaeroides* chromosome. The organization of the genes in cosmids pJG336 (top, form I cluster) and pJG106 (bottom, form II cluster) is illustrated.

laboratory indicate that all the structural genes and the two common sequences of *cfx*I and *cfx*II are transcribed in the same direction (Gibson and Tabita, 1988). It is tempting to suggest that the two common sequences immediately upstream from the RuBisCO genes represent regulatory elements that encode proteins which affect the expression of both form I and form II RuBisCO. The apparent independent regulation of the expression of the two distinct RuBisCO enzymes of *R. sphaeroides* to similar environmental growth conditions suggests that the putative regulatory sequences might interact in a complex manner.

The significance of two *cfx* clusters is under intense study in our laboratory. We have approached this problem by either inactivating or deleting one of the two copies of the relevant structural genes. In the case of RuBisCO, inactivation of either the form I or form II RuBisCO genes by Tn*5* transpositional inactivation or insertion of an antibiotic resistance cartridge within the coding sequence does not prevent *R. sphaeroides* from growing with CO_2 as the carbon source (Falcone et al., 1988). Thus, either of the enzymes will support autotrophic growth, even at low levels of atmospheric CO_2 (1.5 percent). Since the affinity of the form I RuBisCO for CO_2 is somewhat higher than the form II enzyme, we expected that the form I^- mutant strain would be incapable of growth at low levels of CO_2; however, it is entirely possible that at still lower levels of atmospheric CO_2, this effect will become manifested.

All of the structural genes in Figure 26.4 have been expressed and most of the gene products have been purified and characterized (Chory et al., 1985; Gibson and Tabita, 1986, 1987, 1988; Hallenbeck and Kaplan, 1987; Muller et al., 1985; Quivey and Tabita, 1984; Tabita et al., 1986). The complete nucleotide and deduced amino acid sequence is available for the form II RuBisCO (Wagner et al., 1988) and sequencing of the other structural genes will soon be completed. Thus, there is no doubt as to the arrangement of these *cfx* genes in *R. sphaeroides*. Southern hybridization by itself provides much useful information but should always be confirmed by expression in *E. coli* (or other organism) and/or sequence data. A summary of the expression and available sequence information on the cloned Calvin cycle structural genes from photosynthetic procaryotes is shown in Table 26.1. Expression and facile purification of recombinant proteins suggests that mutagenesis studies will yield much information relative to the structure, function, and catalytic mechanism of these enzymes, studies which have already begun in earnest with the *R. rubrum* RuBisCO (Hartman et al., 1987a, b).

Regulation of structural genes In addition to transcriptional regulation of form II RuBisCO gene expression (Zhu and Kaplan, 1985), and other structural genes of the Calvin cycle, there may be other posttranscriptional factors that impinge on regulation, as recently described in eucaryotic photosynthetic systems (Berry et al., 1985, 1986; Inamine et al., 1985; Prioul and Reyss, 1987). One approach to identify regulatory determinants is to isolate mutants incapable of derepressing the *cfx* system. With the photosynthetic bacteria (Rhodospirillaceae), one can take advantage of their unique physiology to isolate such mutants; i.e., strains that grow photoheterotrophically on oxidized organic

Table 26.1 Expression of Calvin cycle structural genes from photosynthetic bacteria

Vector	Gene	Enzyme	Source	Active recombinant enzyme purified	Sequence available	References
pXG21	*rbp*L	RuBisCO[b]	*Rhodospirillum rubrum*	Yes	Yes	Somerville and Somerville, 1984
pUC8	*rbp*L	RuBisCO form II	*Rhodobacter sphaeroides*	Yes	Yes	Tabita et al., 1986
pUC8	*rbc*L*rbc*S	RuBisCO form I	*Rhodobacter sphaeroides*	Yes	Partial	Gibson and Tabita, 1986
pKK223-3	*rbc*L*rbc*S	RuBisCO	*Chromatium vinosum*	Yes	Yes	Viale et al., 1985
pL, pUC9, pUC13, pUC18[a]	*rbc*L*rbc*S	RuBisCO	*Anacystic nidulans*	Yes	Yes	Gatenby et al., 1985; Tabita and Small, 1985; Voordouw et al., 1987; Lee et al.[a]
pUC19	*rbc*L*rbc*S	RuBisCO	*Anabaena* 7120	Yes	Yes	Gurevitz et al., 1985
λCharon 4A	*rbc*L*rbc*S	RuBisCO	*Chlorogloeopsis fritschii*	No	No	Vakeria et al., 1986
	*rbc*L*rbc*S	RuBisCO	*Spirulina platensis*	No	No	Tiboni et al., 1984
pUC8, pUC19	*prk*A	PRK form I	*Rhodobacter sphaeroides*	Yes	Partial	Gibson and Tabita, 1987; Hallenbeck and Kaplan, 1987
pUC8	*prk*B	PRK form II	*Rhodobacter sphaeroides*	Yes	Partial	Gibson and Tabita, 1987
pUC9	*fbp*A	FBP form I	*Rhodobacter sphaeroides*	No	No	Gibson and Tabita, 1988
pUC18	*fbp*B	FBP form II	*Rhodobacter sphaeroides*	No	No	Gibson and Tabita, 1988
pUC18	*gap*B	GAP	*Rhodobacter sphaeroides*	No	No	Gibson and Tabita, 1988

[a]B. Lee, A.K. Rai, and F.R. Tabita, unpublished results.
[b]RuBisCO, ribulose bisphosphate carboxylase/oxygenase; PRK, phosphoribulokinase; GAP, glyceraldehyde 3-phosphate dehydrogenase.

electron donors but do not grow with reduced organic (butyrate) or inorganic (hydrogen gas) electron donors. Such strains contain low basal levels of RuBisCO when grown on malate since the Calvin cycle assumes an anaplerotic function under these conditions (Tabita, 1984). Several mutants of this type were isolated (A. M. Rainey and F. R. Tabita, unpublished results; Weaver and Tabita, 1983) and at least three different genetic loci were found to complement these mutations (A. M. Rainey and F. R. Tabita, unpublished results; Weaver and Tabita, 1985). Complementation involved screening transformants for reacquisition of the ability to grow photoautotrophically and derepression of RuBisCO synthesis. Successful complementation is striking and results in the synthesis of high levels of active RuBisCO protein. These regulatory elements are most likely *trans*-acting factors since, in most cases, *rec*A negative parental strains were initially mutagenized. It should be noted that there is an imperfect palindrome upstream from the ATG initiation site of the *R. sphaeroides rbpL* gene which could function as a *cis*-acting regulatory site to bind to a *trans*-acting regulatory factor of dyad symmetry (Wagner et al., 1988).

26.6 CONCLUSIONS

It is obvious from this discussion that we are just beginning to understand the factors involved in regulating CO_2 fixation in photosynthetic procaryotes. However, the system is now well defined, the molecular and genetic tools are available, and a good physiological/biochemical emphasis will continue to point out the proper direction for further work.

Original research was supported by United States Public Health Service Grant GM 24497 from the National Institutes of Health, U.S. Department of Agriculture Grants 87-CRCR-1-2330 and 83-CRCR-1-1344, and Robert A. Welch Grant F-691. I am pleased to acknowledge the contributions of my colleagues, associates, and students whose dedicated work makes this review possible.

REFERENCES

Andersen, K. and **Wilke-Douglas, M.** 1987. Genetic and physical mapping and expression in *Pseudomonas aeruginosa* of the chromosomally encoded ribulose bisphosphate carboxylase genes of *Alcaligenes eutrophus. Journal of Bacteriology* 169:1997–2004.

Anderson, L. and **Fuller, R.C.** 1967. Photosynthesis in *Rhodospirillum rubrum*. III. Metabolic control of reductive pentose phosphate and tricarboxylic acid cycle enzymes. *Plant Physiology* 42:497–502.

Andrews, T.J. and **Ballment, B.** 1983. The function of the small subunits of ribulose bisphosphate carboxylase-oxygenase. *The Journal of Biological Chemistry* 258:7514–7518.

Berry, J.O., Nikolau, B.J., Carr, J.P., and **Klessig, D.F.** 1985. Transcriptional and post-transcriptional regulation of ribulose 1,5-bisphosphate carboxylase gene expression in light-and dark-grown amaranth cotyledons. *Molecular and Cellular Biology* 5:2238–2246.

Berry, J.O., Nikolau, B.J., Carr, J.P., and **Klessig, D.F.** 1986. Translation regulation of light-induced ribulose 1,5-bisphosphate carboxylase gene expression in Amaranth. *Molecular and Cellular Biology* 6:2347–2353.

Bowien, B., Husemann, M., Klintworth, R., and **Windhövel, U.** 1987. Metabolic and molecular regulation of the CO_2-assimilating enzyme system in aerobic chemoautotrophs, pp. 21–27, in Van Verseveld, H.W. and Duine, J.A. (editors), *Microbial Growth on C_1 Compounds,* Martinus Nijhoff, Dordrecht.

Buzby, J.S., Porter, R.D., and **Stevens, S.E., Jr.** 1983. Plasmid transformation in *Agmenellum quadruplicatum* PR-6: construction of biphasic plasmids and characterization of their transformation properties. *Journal of Bacteriology* 154:1446–1450.

Chen, C., van Baalen, C., and **Tabita, F.R.** 1987a. Nitrogen starvation mediated by DL-7-azatryptophan in the cyanobacterium *Anabaena* sp. strain CA. *Journal of Bacteriology* 169:1107–1113.

Chen, C., van Baalen, C., and **Tabita, F.R.** 1987b. DL-7-Azatryptophan and citrulline metabolism in the cyanobacterium *Anabaena* sp. strain IF. *Journal of Bacteriology* 169:1114–1119.

Chory, J., Muller, E.D., and **Kaplan, S.** 1985. DNA-directed *in vitro* synthesis and assembly of the form II D-ribulose-1,5-bisphosphate carboxylase oxygenase from *Rhodopseudomonas sphaeroides. Journal of Bacteriology* 161:307–313.

Cook, L.S. 1986. Physiological regulation of ribulose 1,5-bisphosphate carboxylase/oxygenase *Rhodospirillum rubrum.* Ph.D. Dissertation, The University of Texas at Austin.

Dijkuizen, L. and **Harder, W.** 1984. Current views on the regulation of autotrophic carbon dioxide fixation via the Calvin cycle in bacteria. *Antonie van Leeuwenhoek: Journal of Microbiology and Serology* 50:473–487.

Dow, C.S. 1987. CO_2 fixation in *Rhodopseudomonas blastica,* pp. 28–37, in Van Verseveld, H.W. and Duine, J.A. (editors), *Microbial Growth on C_1 Compounds,* Martinus Nijhoff, Dordrecht.

Falcone, D.F., Quivey, R.G., Jr., and **Tabita, F.R.** 1988. Transposon mutagenesis and physiological analysis of strains containing inactivated form I and form II ribulose bisphosphate carboxylase/oxygenase genes in *Rhodobacter sphaeroides. Journal of Bacteriology* 170:5–11.

Friedrich, C.G., Bowien, B., and **Friedrich, B.** 1979. Formate and oxalate metabolism in *Alcaligenes eutrophus. The Journal of General Microbiology* 115:185–192.

Fuller, R.C. and **Gibbs, M.** 1959. Intracellular and phylogenetic distribution of ribulose 1,5-diphosphate carboxylase and D-glyceraldehyde 3-phosphate dehydrogenases. *Plant Physiology* 34:324–329.

Gaffron, H. 1935. Über den Stoffwechsel der Purpurbakterien. *Biochemische Zeitschrift* 275:301–310.

Gatenby, A.A., van der Vies, S.M., and **Bradley, D.** 1985. Assembly in *E. coli* of a functional multi-subunit ribulose bisphosphate carboxylase from a blue-green alga. *Nature* 314:617–620.

Gendel, S., Strauss, N., Pulleyblank, D., and **Williams, J.** 1983. A novel shuttle vector for the cyanobacterium *Anacystis nidulans. FEMS Microbiology Letters* 19:291–294.

Gibson, J.L. 1980. Different forms of ribulose 1,5-bisphosphate carboxylase from *Rhodopseudomonas sphaeroides.* Ph.D. Dissertation, The University of Texas at Austin.

Gibson, J.L. and **Tabita, F.R.** 1977a. Different molecular forms of D-ribulose-1,5-bisphosphate carboxylase from *Rhodopseudomonas sphaeroides. The Journal of Biological Chemistry* 252:943–949.

Gibson, J.L. and **Tabita, F.R.** 1977b. Isolation and preliminary characterization of two forms of ribulose 1,5-bisphosphate carboxylase from *Rhodopseudomonas capsulata. Journal of Bacteriology* 132:818–823.

Gibson, J.L. and **Tabita, F.R.** 1977c. Characterization of antiserum directed against form II ribulose 1,5-bisphosphate carboxylase from *Rhodopseudomonas sphaeroides. Journal of Bacteriology* 131:1020–1022.

Gibson, J.L. and **Tabita, F.R.** 1979. Activation of ribulose 1,5-bisphosphate carboxylase from *Rhodopseudomonas sphaeroides*: probable role of the small subunit. *Journal of Bacteriology* 140:1023–1027.

Gibson, J.L. and **Tabita, F.R.** 1985. Structural differences in the catalytic subunits of form I and form II ribulose 1,5-bisphosphate carboxylase/oxygenase from *Rhodopseudomonas sphaeroides. Journal of Bacteriology* 164:1188–1193.

Gibson, J.L. and **Tabita, F.R.** 1986. Isolation of the *Rhodopseudomonas sphaeroides* form I ribulose 1,5-bisphosphate carboxylase/oxygenase large and small subunit genes and expression of the active hexadecameric enzyme in *Escherichia coli. Gene* 44:271–278.

Gibson, J.L. and **Tabita, F.R.** 1987. Organization of phosphoribulokinase and ribulose bisphosphate carboxylase/oxygenase genes in *Rhodopseudomonas (Rhodobacter) sphaeroides. Journal of Bacteriology* 169:3685–3690.

Gibson, J.L. and **Tabita, F.R.** 1988. Localization and mapping of CO_2 fixation genes within two gene clusters in *Rhodobacter sphaeroides. Journal of Bacteriology* 170:2153–2158.

Glover, J., Kamen, M.D., and **van Gemerden, H.** 1952. Studies on the metabolism of photosynthetic bacteria. XII. Comparative light and dark metabolism of acetate and carbonate by *Rhodospirillum rubrum. Archives of Biochemistry and Biophysics* 35:343–408.

Golden, S.S. and **Sherman, L.A.** 1983. A hybrid plasmid is a stable cloning vector for the cyanobacterium *Anacystis nidulans* R2. *Journal of Bacteriology* 155:966–972.

Gurevitz, M., Somerville, C.R., and **McIntosh, L.** 1985. Pathway of assembly of ribulosebisphosphate carboxylase/oxygenase from *Anabaena* 7120 expressed in *Escherichia coli. Proceedings of the National Academy of Sciences U.S.A.* 82:6546–6550.

Hallenbeck, P. and **Kaplan, S.** 1987. Cloning of the gene for phosphoribulokinase activity from *Rhodobacter sphaeroides* and its expression in *Escherichia coli. Journal of Bacteriology* 169:3669–3678.

Hartman, F.C., Larimer, F.W., Mural, R.J., Machanoff, R., and **Soper, T.S.** 1987a. Essentiality of glu-48 of ribulose bisphosphate carboxylase/oxygenase as demonstrated by site-directed mutagenesis. *Biochemical and Biophysical Research Communications* 145:1158–1163.

Hartman, F.C., Soper, T.S., Niyogi, S.K., Mural, R.J., Foote, R.S., Mitra, S., Lee, E.H., Machanoff, R., and **Larimer, F.W.** 1987b. Function of lys-166 of *Rhodospirillum rubrum* ribulosebisphosphate carboxylase/oxygenase as examined by site-directed mutagenesis. *The Journal of Biological Chemistry* 262:3496–3501.

Haselkorn, R. 1986. Organization of the genes for nitrogen fixation in photosynthetic bacteria and cyanobacteria. *Annual Review of Microbiology* 40:525–547.

Inamine, G., Nash, B., Weissbach, H., and **Brot, N.** 1985. Light regulation of the synthesis of the large subunit of ribulose-1,5-bisphosphate carboxylase in peas: evidence for translational control. *Proceedings of the National Academy of Sciences U.S.A.* 82:5690–5694.

Joint, I.R., Morris, I., and **Fuller, R.C.** 1972. Purification of a complex of alkaline fructose 1,6-bisphosphatase and phosphoribulokinase from *Rhodospirillum rubrum. The Journal of Biological Chemistry* 247:4833–4838.

Jordan, D.B. and **Chollet, R.** 1985. Subunit dissociation and reconstitution of ribulose-1,5-bisphosphate carboxylase from *Chromatium vinosum. Archives of Biochemistry and Biophysics* 236:487–496.

Jordan, D.B. and **Ogren, W.L.** 1981. Species variation in the specificity of ribulose bisphosphate carboxylase/oxygenase. *Nature* 291:513–515.

Jouanneau, Y. and **Tabita, F.R.** 1986. Independent regulation of synthesis of form I and form II ribulose bisphosphate carboxylase-oxygenase in *Rhodopseudomonas sphaeroides. Journal of Bacteriology* 165:620–624.

Jouanneau, Y. and **Tabita, F.R.** 1987. *In vivo* regulation of form I ribulose 1,5-bisphosphate carboxylase/oxygenase from *Rhodopseudomonas sphaeroides. Archives of Biochemistry and Biophysics* 254:290–303.

Klintworth, R., Husemann, M., Salnikow, J., and **Bowien, B.** 1985. Chromosomal and plasmid locations for phosphoribulokinase genes in *Alcaligenes eutrophus. Journal of Bacteriology* 164:954–956.

Kuhlemeier, C.J., Borrias, W.E., van den Hondel, C.A.M.J.J., and **Van Arkel, G.A.** 1981. Vectors for cloning in cyanobacteria: construction and characterization of two recombinant plasmids capable of transformation to *Anacystis nidulans* R2 and *Escherichia coli* K12. *Molecular and General Genetics* 184:247–252.

Lascelles, J. 1960. The formation of ribulose 1,5-diphosphate carboxylase by growing cultures of *Athiorhodaceae. The Journal of General Microbiology* 23:499–510.

Madigan, M.T. and **Gest, H.** 1979. Growth of the photosynthetic bacterium *Rhodopseudomonas capsulata* chemoautotrophically in darkness with H_2 as the energy source. *Journal of Bacteriology* 137:524–530.

Marsden, W.J.N. and **Codd, G.A.** 1984. Purification and molecular and catalytic properties of phosphoribulokinase from the cyanobacterium *Chlorogloeopsis fritschii. The Journal of General Microbiology* 130:999–1006.

Miziorko, H.M. and **Lorimer, G.H.** 1983. Ribulose-1,5-bisphosphate carboxylase-oxygenase. *Annual Review of Biochemistry* 52:507–535.

Muller, E.D., Chory, J., and **Kaplan, S.** 1985. Cloning and characterization of the gene product of the form II ribulose-1,5-bisphosphate carboxylase gene of *Rhodopseudomonas sphaeroides. Journal of Bacteriology* 161:469–472.

Muller, F.M. 1933. On the metabolism of the purple sulfur bacteria in organic media. *Archiv für Mikrobiologie* 4:131–166.

Nargang, F., McIntosh, L., and **Somerville, C.** 1984. Nucleotide sequence of the ribulose bisphosphate carboxylase gene from *Rhodospirillum rubrum. Molecular and General Genetics* 193:220–224.

Nicholson, S., Easterby, J.S., and **Powls, R.** 1987. Properties of a multimeric protein complex from chloroplasts possessing potential activities of NADPH-dependent glyceraldehyde-3-phosphate dehydrogenase and phosphoribulokinase. *European Journal of Biochemistry* 162:423–431.

Nierzwicki-Bauer, S.A., Curtis, S.E., and **Haselkorn, R.** 1984. Cotranscription of genes encoding the small and large subunits of ribulose-1,5-bisphosphate carboxylase in the cyanobacterium *Anabaena* 7120. *Proceedings of the National Academy of Sciences U.S.A.* 81:5961–5965.

Porter, R.D. 1986. Transformation in cyanobacteria. *CRC Critical Reviews in Microbiology* 13:111–132.

Prioul, J.-L. and **Reyss, A.** 1987. Acclimation of ribulose bisphosphate carboxylase and mRNAs to changing irradiance in adult tobacco leaves. *Plant Physiology* 84:1238–1243.

Quandt, L., Gottschalk, G., Ziegler, H., and **Stichler, W.** 1977. Isotope discrimination by photosynthetic bacteria. *FEMS Microbiology Letters* 1:125–128.

Quivey, R.G., Jr. and **Tabita, F.R.** 1984. Cloning and expression in *Escherichia coli* of the form II ribulose 1,5-bisphosphate carboxylase/oxygenase from *Rhodopseudomonas sphaeroides. Gene* 31:91–101.

Rippel, S. and **Bowien, B.** 1984. Phosphoribulokinase from *Rhodopseudomonas acidophila. Archives of Microbiology* 139:207–212.

Sarles, L.E. and **Tabita, F.R.** 1983. Derepression of the synthesis of D-ribulose 1,5-bisphosphate carboxylase/oxygenase from *Rhodospirillum rubrum. Journal of Bacteriology* 153:458–464.

Schloss, J.V., Phares, E.F., Long, M.V., Norton, I.L., Stringer, C.D., and **Hartman, F.C.** 1979. Isolation, characterization, and crystallization of ribulosebisphosphate carboxylase from autotrophically grown *Rhodospirillum rubrum. Journal of Bacteriology* 137:490–501.

Scolnik, P.A. and **Marrs, B.L.** 1987. Genetic research with photosynthetic bacteria. *Annual Review of Microbiology* 41:703–726.

Shinozaki, K. and **Sugiura, M.** 1985. Genes from the large and small subunits of ribulose-1,5-bisphosphate carboxylase/oxygenase constitute a single operon on a cyanobacterium *Anacystis nidulans. Molecular and General Genetics* 200:27–32.

Shively, J.M., Davidson, E., and **Marrs, B.L.** 1984. Derepression of the synthesis of the intermediate and large forms of ribulose-1,5-bisphosphate carboxylase/oxygenase in *Rhodopseudomonas capsulata. Archives of Microbiology* 138:233–236.

Siebert, K., Schobert, P., and **Bowien, B.** 1981. Purification and some catalytic and molecular properties of phosphoribulokinase from *Alcaligenes eutrophus. Biochimica et Biophysica Acta* 658:35–44.

Somerville, C.R. and **Somerville, S.C.** 1984. Cloning and expression of the *Rhodospirillum rubrum* ribulose bisphosphate carboxylase gene in *E. coli. Molecular and General Genetics* 193:214–219.

Springgate, C.F. and **Stachow, C.S.** 1972. Fructose 1,6-diphosphatase from *Rhodopseudomonas palustris.* I. Purification and properties. *Archives of Biochemistry and Biophysics* 152:1–12.

Stoppani, A.O.M., Fuller, R.C., and **Calvin, M.** 1955. Carbon dioxide fixation by *Rhodopseudomonas capsulatus. Journal of Bacteriology* 69:491–501.

Tabita, F.R. 1980. Pyridine nucleotide control and subunit structure of phosphoribulokinase from photosynthetic bacteria. *Journal of Bacteriology* 143:1275–1280.

Tabita, F.R. 1981. Molecular regulation of carbon dioxide assimilation in autotrophic microorganisms, pp. 70–81, in Dalton, H. (editor), *Microbial Growth on C_1 Compounds,* Heyden and Son, London.

Tabita, F.R. 1987. Carbon dioxide fixation and its regulation in cyanobacteria, pp. 96–117, in Fay, P. and van Baalen, C. (editors), *The Cyanobacteria,* Elsevier, Amsterdam.

Tabita, F.R., Gibson, J.L., Jouanneau, Y., Falcone, D.L., and **Rainey, A.M.** 1987. Molecular biology and regulation of microbial CO_2 fixation, pp. 238–235, in Van Verseveld, H.W. and Duine, J.A. (editors), *Microbial Growth on C_1 Compounds,* Martinus Nijhoff, Dordrecht.

Tabita, F.R., Gibson, J.L., Mandy, W.J., and **Quivey, R.G., Jr.** 1986. Synthesis and assembly of a novel recombinant ribulose bisphosphate carboxylase/oxygenase. *Bio/Technology* 4:138–141.

Tabita, F.R. and **McFadden, B.A.** 1972. Regulation of ribulose-1,5-diphosphate carboxylase by 6-phospho-D-gluconate. *Biochemical and Biophysical Research Communications* 48:1153–1159.

Tabita, F.R. and **McFadden, B.A.** 1974. D-Ribulose 1,5-diphosphate carboxylase from *Rhodospirillum rubrum.* I. Levels, purification, and effect of metallic ions. *The Journal of Biological Chemistry* 249:3453–3458.

Tabita, F.R. and **Small, C.L.** 1985. Expression and assembly of active cyanobacterial ribulose-1,5-bisphosphate carboxylase/oxygenase in *Escherichia coli* containing stoichiometric amounts of large and small subunits. *Proceedings of the National Academy of Sciences U.S.A.* 82:6100–6103.

Takabe, T., Rai, A.K., and **Akazawa, T.** 1984. Interaction of constituent subunits in ribulose 1,5-bisphosphate carboxylase from *Aphanothece halophytica. Archives of Biochemistry and Biophysics* 229:202–221.

Tiboni, O., di Pasquale, G., and **Ciferri, O.** 1984. Cloning and expression of the genes for ribulose-1,5-bisphosphate carboxylase from *Spirulina platensis. Biochimica et Biophysica Acta* 782:258–264.

Vakeria, D., Codd, G.A., Hawthornthwaite, A.M., and **Stewart, W.D.P.** 1986. Construction of a gene bank and identification of the ribulose bisphosphate carboxylase/oxygenase genes from the nutritionally versatile cyanobacterium *Chlorogloeopsis fritschii. Archives of Microbiology* 145:228–233.

van Niel, C.B. 1936. On the metabolism of Thiorhodaceae. *Archiv für Mikrobiologie* 7:323–358.

Viale, A.M., Kobayashi, H., Takabe, T., and **Akazawa, T.** 1985. Expression of genes for subunits of plant-type RubisCO from *Chromatium* and production of the enzymatically active molecule in *Escherichia coli. FEBS Letters* 192:283–288.

Voordouw, G., DeVries, P.A., Van den Berg, W.A.M., and **DeClerk, E.P.J.** 1987. Site-directed mutagenesis of the small subunit of ribulose-1,5-bisphosphate carboxylase/oxygenase from *Anacystis nidulans. European Journal of Biochemistry* 163:591–598.

Wagner, S.J., Stevens, S.E., Jr., Nixon, B.T., Lambert, D.H., Quivey, R.G., Jr., and **Tabita, F.R.** 1988. Nucleotide and deduced amino acid sequence of the *Rhodobacter sphaeroides* gene encoding form II ribulose-1,5-bisphosphate carboxylase/oxygenase and comparison with other deduced form I and form II sequences. *FEMS Microbial. Lett.*, (in press).

Weaver, K.E. and **Tabita, F.R.** 1983. Isolation and partial characterization of *Rhodopseudomonas sphaeroides* mutants defective in the regulation of ribulose bisphosphate carboxylase/oxygenase. *Journal of Bacteriology* 156:507–515.

Weaver, K.E. and **Tabita, F.R.** 1985. Complementation of a *Rhodopseudomonas sphaeroides* ribulose bisphosphate carboxylase-oxygenase regulatory mutant from a genomic library. *Journal of Bacteriology* 164:147–154.

Whitman, W.B. 1978. Ribulose 1,5-bisphosphate carboxylase from *Rhodospirillum rubrum.* Ph.D. Dissertation, The University of Texas at Austin.

Whitman, W.B., Martin, M.N., and **Tabita, F.R.** 1979. Activation and regulation of ribulose bisphosphate carboxylase-oxygenase in the absence of small subunits. *The Journal of Biological Chemistry* 254:10184–10189.

Wolk, C.P., Vonshak, A., Kehoe, P., and **Elhai, J.** 1984. Construction of shuttle vectors capable of conjugative transfer from *E. coli* to nitrogen-fixing filamentous cyanobacteria. *Proceedings of the National Academy of Sciences U.S.A.* 81:1561–1565.

Zhu, Y.S. and **Kaplan, S.** 1985. Effects of light, oxygen, and substrates on steady-state levels of mRNA coding for ribulose 1,5-bisphosphate carboxylase and light-harvesting and reaction center polypeptides on *Rhodopseudomonas sphaeroides. Journal of Bacteriology* 162:925–932.

27

Evolutionary Relationships of the Autotrophic Bacteria

Thomas D. Brock

Department of Bacteriology
University of Wisconsin-Madison
Madison, Wisconsin, U.S.A.

Autotrophs are defined as organisms that use carbon dioxide (CO_2) as their immediate and major source of cellular carbon (see Chapter 1). It seems likely that bacteria were the first autotrophs to arise on earth. Although all organisms, even heterotrophs, fix some CO_2, autotrophs obtain all or virtually all of their cellular carbon from this inorganic source. Although a classification based on carbon nutrition is relatively straightforward, confusion arises when one attempts to combine terms relating to carbon nutrition with those relating to energy metabolism (see Section 1.1). In relation to energy, two broad groups of organisms are recognized, those using light, called *phototrophs*, and those using chemical energy, called *chemotrophs*. Many phototrophs and chemotrophs are autotrophs, but there are also many organisms in each energy grouping which are heterotrophs. Thus, we distinguish photoautotrophs and photoheterotrophs, as well as chemoautotrophs and chemoheterotrophs. Another term, *lithotroph*, referring to an organism which uses an inorganic energy source, finds considerable utility when discussing bacterial autotrophs, although if a lithotroph is an autotroph it is no different than a chemoautotroph.

27.1 DIVERSITY OF BACTERIAL AUTOTROPHS

As has been discussed extensively in this book, a wide variety of bacterial autotrophs are known. The history of their discovery was outlined briefly in Section 1.4 and a brief summary of the bacterial autotrophs is presented in Table 27.1. It is convenient, for the present discussion, to consider separately the two bacterial kingdoms, Archaebacteria and Eubacteria (Woese, 1987). As seen in Table 27.1, among the archaebacteria are two major types of autotrophs, the methanogens and the sulfur-dependent bacteria. Although not all orga-

Table 27.1 Diversity of the autotrophic bacteria

Anaerobic bacteria	Aerobic bacteria
Archaebacteria	**Archaebacteria**
Methanogenic	Sulfur-dependent (some)
Sulfur-dependent (some)	
Eubacteria	**Eubacteria**
Anoxygenic phototrophs	*Oxygenic phototrophs*
Purple bacteria	Cyanobacteria
Green bacteria	Prochlorophytes
Lithotrophs	*Lithotrophs*
Acetogenic	Sulfur-oxidizing
Sulfate-reducing	Iron-oxidizing
	Nitrifying
	Hydrogen-oxidizing
	Carboxydobacteria

nisms in these two groups are autotrophs, autotrophy is so widely distributed among these bacteria as to be considered an important primary characteristic. As we will note later, several distinct CO_2 fixation pathways are known but these are found in both the autotrophic archaebacteria as well as the eubacteria. The Calvin cycle of CO_2 fixation (see Chapter 19) seems to be restricted to the eubacteria.

Among the eubacterial autotrophs, we find two major subgroups, the *phototrophs* and the *lithotrophs*. Although not all phototrophs and lithotrophs are autotrophs, the autotrophic character is distributed so widely among these organisms that it must again appear as a primary character in their evolution. Finally, we must divide the phototrophs into two separate groups, the anoxygenic and the oxygenic phototrophs. Anoxygenic phototrophic bacteria include various groups of purple and green bacteria, whereas the oxygenic phototrophs are the cyanobacteria. The anoxygenic phototrophics were probably the earliest phototrophs on earth and are considered by most workers to be the forerunners of the cyanobacteria. The latter group was of major importance in the evolution of life on earth, as they were the first net producers of molecular oxygen (O_2). The archaebacterial halobacteria are able to use light energy but they are not strictly phototrophs, as they are unable to grow with light as sole energy source.

27.2 APPROACHES TO THE STUDY OF BACTERIAL EVOLUTION

Three major approaches to the study of bacterial evolution are available. 1) *Paleomicrobiological and biogeochemical*. The search for remnants of ancient life in sedimentary rocks has become a widely developed field, and one that provides significant information regarding the origin of bacterial autotrophy (Schopf, 1983). 2) *Molecular phylogeny*. This approach is best expressed in the seminal studies of Woese (1987) using 16S ribosomal RNA molecules. 3) *Comparative biochemistry*. With this approach, pathways and mechanisms for CO_2 fixation (see Chapter 3) and energy metabolism in various bacterial autotrophs are compared. In the following, we will briefly consider the evidence from each of these areas and then attempt to combine the various observations in the development of a general model of autotrophic evolution. For convenience, we discuss comparative biochemistry first.

27.3 PRINCIPAL PATHWAYS OF BIOLOGICAL CARBON FIXATION

The various pathways of CO_2 fixation have been presented in earlier chapters in this book (see especially Chapters 3, 20, 22). An analysis of these pathways is interesting in terms of the comparative biochemical approach to evolution, but it is also relevant biogeochemically.

A comparison of the principal biochemical pathways of CO_2 fixation is shown in Table 27.2, which also includes the names of representative biological groups that possess these pathways. The most widely distributed pathway, and the one found in all oxygenic phototrophs, as well as the purple bacteria and lithotrophic autotrophs, is the Calvin cycle, in which the initial stage of CO_2 fixation involves the enzyme *ribulose-bisphosphate carboxylase (RuBP carboxylase)* (see Chapters 19, 26). Finally, two pathways apparently unique to anaerobic bacteria are the *reverse tricarboxylic acid cycle,* found in the green bacteria, and the *acetyl-CoA pathway,* found in *Acetobacterium* as well as those sulfate-reducing bacteria and methanogens which are autotrophs (see Chapter 20). The diversity of pathways suggests that autotrophic CO_2 fixation is a property which has arisen more than once during evolution.

27.4 PALEOMICROBIOLOGY AND BIOGEOCHEMISTRY

Isotope fractionation in CO_2 fixation Organic carbon in ancient sediments shows various levels of isotopic fractionation between ^{12}C and ^{13}C, and an

Table 27.2 Principal biochemical pathways of CO_2 fixation

RuBP carboxylase. Calvin cycle
Algae
Cyanobacteria
Purple bacteria
Lithotrophic eubacteria
Thiobacillus
Nitrifying bacteria
Reductive tricarboxylic acid cycle
Green sulfur bacteria
Sulfur-dependent archaebacteria (?)
Acetyl-CoA pathway
Acetobacterium
Sulfate-reducing autotrophs
Methanogens

analysis of organic matter in rocks can be used to provide clues as to the time scale of autotrophic CO_2 fixation. $^{12}C/^{13}C$ fractionation is mainly due to kinetic isotope effects of the CO_2 fixing enzymes, with preference for the lighter isotope. A summary of isotope fractionation data from laboratory and field studies is given in Figure 27.1.

By looking at $^{12}C/^{13}C$ ratios of organic matter in autotrophs, evidence for the activity of various autotrophic organisms can be obtained. As discussed by Schidlowski et al. (1983), the sedimentary carbon record suggests that autotrophic activity has existed on Earth since at least 3.5 billion years before the present. ". . .the uniformity of [$^{12}C/^{13}C$ fractionation] suggests an extreme degree of evolutionary conservatism for the basic biochemical mechanisms of carbon fixation." Note that the "lightest" $^{12}C/^{13}C$ fractionation is associated with the methanogens, and there is some evidence to suggest that these organisms have been present since the beginning of life on earth.

Paleomicrobiology Study of fossil microorganisms naturally focuses on events that took place in the Precambrian period, since it was during that period that

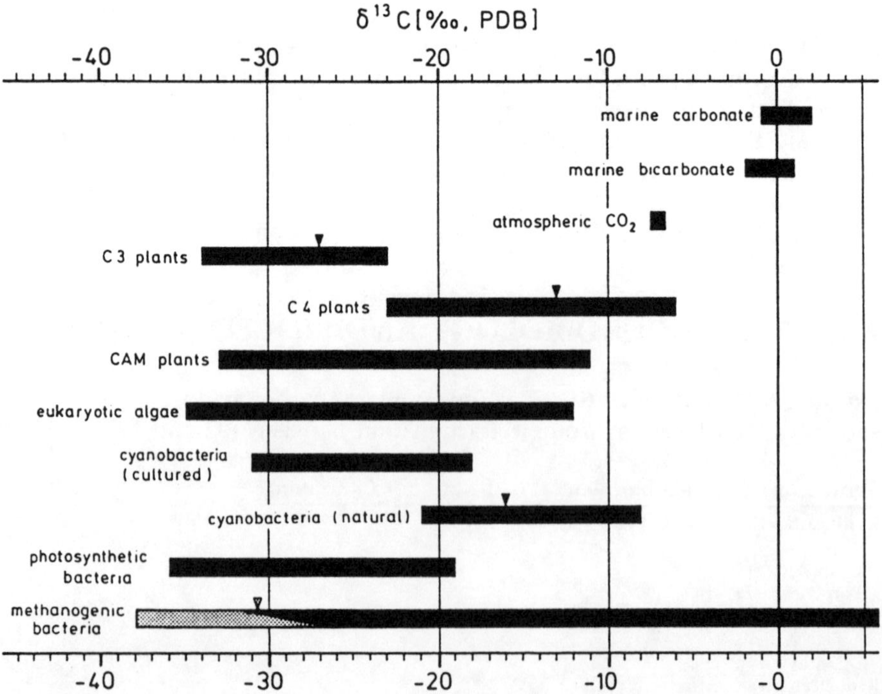

Figure 27.1 Carbon isotope composition of extant autotrophic organisms compared with the inorganic carbon of the environment. Variability is due to differences in growth conditions during the metabolic period that isotope fractionation is occurring. The numbers represent depletion of ^{13}C over ^{12}C as compared to a standard (Peedee belemnite, PDB). Data from Schidlowski et al. (1983).

virtually all of the evolutionary diversity in the bacterial autotrophs developed. Studies in this area have been well summarized in the book edited by Schopf (Schopf, 1983). It is convenient to divide the Precambrian into two major eras, the Archean and the Proterozoic. The Archean period extends from the time when the earth was formed, about 4.5-4.6 billion years before the present, until about 2.5 billion years, the time when the transition of the earth's atmosphere from anaerobic to aerobic was occurring. The Proterozoic period extends from 2.5 billion years to 0.5 billion years before the present, at the beginning of the Paleozoic, when the first traces of invertebrate animals can be seen in the rocks. Extensive evidence for ancient life, including evidence for bacterial autotrophs, has been found in sedimentary rocks from the Archean.

Two types of evidence exist: the microfossils themselves, and the general arrangement of the fossil remains in structures called *stromatolites* (see below), which were formed by the massive development of phototrophic microorganisms (Walter, 1983). Although one cannot make detailed conclusions about physiology from morphological observations, the general arrangement and chemical environment of various fossils provide clues as to the physiological nature of the organisms.

Stromatolites The strongest evidence for the evolution of autotrophs comes from a study of sedimentary rock formations containing stromatolites (Walter, 1976; Walter, 1983). Comparison between ancient stromatolites and modern analogs has provided a solid basis for the conclusion that stromatolitic deposits in Archean and Proterozoic rocks were formed by filamentous phototrophic microorganisms, either oxygenic (cyanobacteria) or anoxygenic (green filamentous bacteria). Undisputed stromatolites have been found in rocks as old as 3.5 billion years before the present, and a number of stromatolitic deposits are seen throughout the Archean. What is the evidence that these stromatolites were formed by autotrophs? First, geological evidence shows that the carbonates in these rocks is biogenic, precipitated in place by a biological mechanism. (CO_2 fixation results in an alkalinization of marine environments and a concomitant precipitation of carbonates.) Second, there is strong evidence that the microorganisms which formed these ancient stromatolites were phototactic, and, by inference, phototrophic (Walter, 1983). Third, where microfossils exist, many filamentous structures can be seen that resemble modern phototrophic microbes. Fourth, the overall arrangements of these stromatolites (nodules, laminae, etc.) are strikingly similar to the kinds of structures formed by modern phototrophic bacteria. Finally, fractionation isotope analysis indicates that the organic carbon associated with these stromatolites was probably formed by CO_2 fixation processes. Counted by major stratigraphic units, at least eleven occurrences of stromatolites have been recognized in the Archean (2.5 to 3.6 billion years before the present), with stromatolites abundant by 2.8 billion years. Archean stromatolites were formed by filamentous procaryotes, among which were autotrophs. The geological evidence is therefore strong that autotrophs were among the earliest forms of life on earth.

Unfortunately, nonphototrophic autotrophs do not provide good records in ancient rocks. However, sulfur isotope and sulfur biogeochemical studies summarized in Schopf (1983) provide some evidence that sulfate-reducing bacteria were present early in the Archean. Although most sulfate-reducing bacteria are anaerobic heterotrophs, several autotrophic sulfate reducers are known. Further, autotrophic *sulfur* reducers, and autotrophic methanogens, both members of the kingdom Archaebacteria, are known to exist. As discussed by Schopf et al. (1983), carbon isotopic evidence of sediments from 2.5–2.8 billion years with anomalously light values similar to values only measured by methanogens (see Figure 27.1), suggests the presence of methanogens very early in earth history.

Schopf et al. (1983) have constructed a "best-guess" scenario that is illustrated in Figure 27.2. As indicated, the earth became a habitable planet at least 3.8 billion years ago, perhaps much earlier. Chemical evolution led to the formation of a wide variety of organic molecules, including many biochemical molecules. The origin of life occurred among this random and rich mixture of chemicals. Physiologically, the first organisms must have been anaerobes, since no molecular oxygen was present in the environment. These early organisms were presumably chemoheterotrophs, relying on abiotically produced foodstuffs as their source of carbon and energy. Because of the absence of inorganic electron acceptors, these initial organisms must have been fermentative anaerobes. Whether they were archaebacteria or eubacteria, or members of the hypothetical *progenote* (Woese, 1987) will remain an open question until the evolutionary tree is rooted.

From this primary biota evolved a secondary array of microorganisms, still anaerobic, among which may have been the first autotrophs. Schopf et al. (1983) believe that these first autotrophs were the methanogens, but phototrophic autotrophs were also early forms, presumably deriving their chlorophyll pigments from the metalloporphyrin-containing compounds of methanogens or other chemotrophic anaerobes. The earliest evolving photo-responsive microorganisms were probably facultative, capable of using light energy when available but switching to dark metabolism. Many representatives of such facultative phototrophs are found among the modern phototrophic bacteria (see Chapters 6, 14, and 15. Because of the great increase in energy available once light-driven energy metabolism had evolved, there should have been a selective advantage for phototrophic organisms. However, the presence of ultraviolet radiation (UV) at the earth's surface would have restricted these organisms to deep marine habitats, where they would have been protected from harmful UV. Subsequently, mechanisms of UV protection may have evolved, but it was not until oxygenic photosynthesis led to the development of the ozone shield that widespread colonization of the earth by phototrophs could have occurred.

Oxygen-evolving phototrophs probably arose sometime during the Late Archean (see Figure 27.2). These organisms, now extinct, would be classified as cyanobacteria. They were probably facultative phototrophs, and also facultative oxygen-evolvers, using H_2S as electron donor (like the purple and green

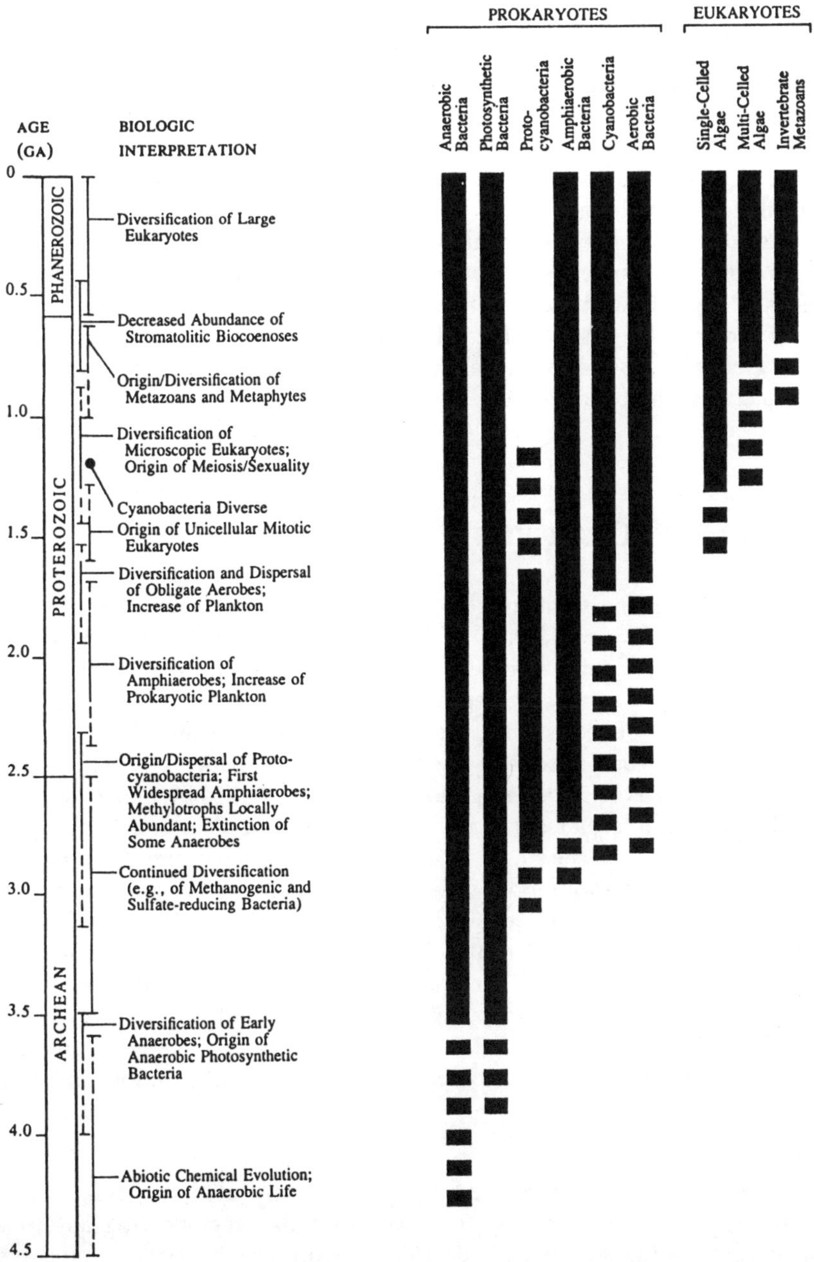

Figure 27.2 Time scale of evolutionary events during the Precambrian, showing the approximate times at which various biological groups are thought to have arisen. Schopf et al. (1983).

bacteria) when available, and switching to H_2O as electron donor when H_2S became depleted.

The earliest oxygen-respiring heterotrophs would undoubtedly have been facultative, using oxygen when available and switching to an anaerobic mode of energy metabolism when oxygen became depleted (as many modern bacteria do). It is not clear when lithotrophic autotrophs such as the sulfur and nitrifying bacteria arose, but since they are predominantly aerobes, they most likely arose after the cyanobacteria.

To summarize the geological evidence, as far as autotrophs are concerned: the first autotrophs were probably the methanogens and sulfur-dependent archaebacteria, followed by the anaerobic phototrophs (green and purple bacteria), and then by the cyanobacteria. We will discuss below the phylogeny of these groups and see how closely the geological ideas correlate with studies on molecular phylogeny.

27.5 MOLECULAR PHYLOGENY

Nucleic acid sequencing techniques have revolutionized the study of bacterial phylogeny and have profoundly changed our ideas about bacterial evolution. As Woese (1987) has discussed, the cell retains within it the history of its own evolution, and by sequencing genes, it is possible to read some of this history. Until the advent of sequencing, ideas about bacterial evolution were derived exclusively from comparisons of phenotypic characteristics, but sequence comparisons permit us to compare genotypes. Ribosomal RNA molecules are especially useful molecular chronometers, and the comparison of molecular sequences of the RNA of the small ribosomal subunit has permitted major insights into the origin and evolutionary divergence of different bacterial groups. Although this work is still in relatively early stages, it has already led to major realignments in the bacterial groups. The available data are of particular interest in an analysis of evolution of bacterial autotrophy. Adopting these new groupings based on genetics, we can examine phenotypes and see how they correlate with the underlying genotypes. In the present analysis, we are concerned with the various autotrophic phenotypes that relate to CO_2 fixation and energy generation.

The universal phylogenetic tree Figure 27.3 illustrates in a broad way the relationships of all extant life. The groups on the tree showing autotrophic phenotypes are highlighted. The fact that the autotrophic phenotype is found in so many groups suggests that autotrophy was an early phenotype, and has spread throughout the various biological phyla as evolution has proceeded. The early evolution of autotrophy is consistent with the geological evidence already discussed above.

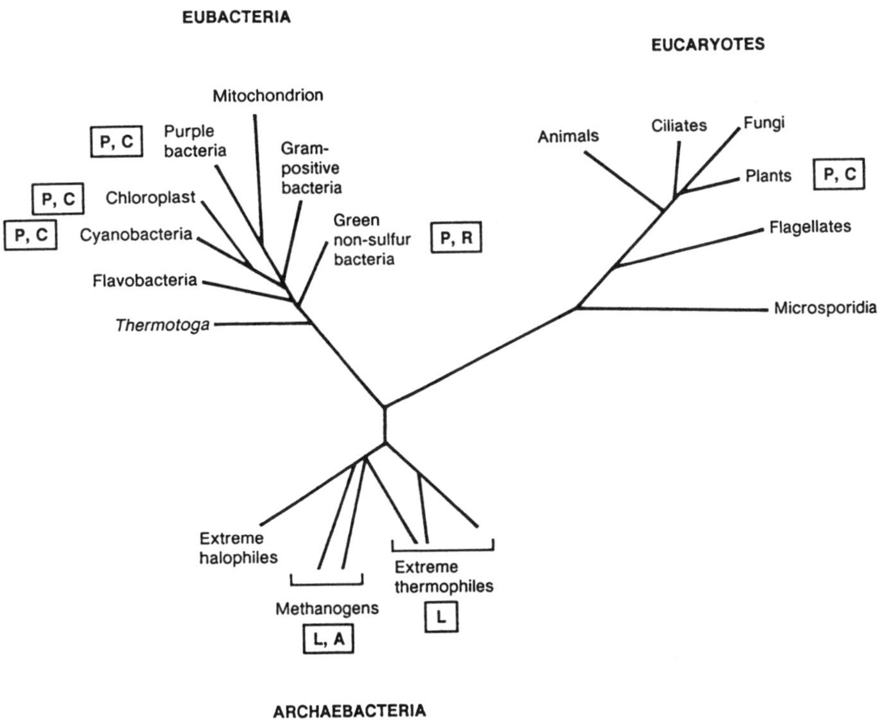

Figure 27.3 Universal phylogenetic tree determined from ribosomal RNA sequence comparisons. The autotrophic phenotypes are highlighted. P, phototroph; L, lithotroph; C, Calvin cycle; A, acetyl-CoA pathway; R, reverse tricarboxylic acid cycle. The tree is based on Woese (1987).

27.6 AUTOTROPHY AND PHOTOTROPHY

As we have seen, there are several quite distinct biochemical pathways of autotrophic CO_2 fixation (see Table 27.2). These various pathways are highlighted in the groupings listed in Figure 27.3. The fact that the Calvin cycle and oxygenic photosynthesis based on chlorophyll *a* are present in such distantly related groups as green plants and cyanobacteria suggests that this mode of photosynthesis evolved early, before the two major kingdoms, Eubacteria and Eucaryotes, diverged. Indeed, Woese et al. (1985a) have postulated that the ancestor common to *all* eubacteria was itself photosynthetic.

27.7 LITHOTROPHY AND THE "PURPLE BACTERIA"

If we turn to the lithotrophic phenotypes, we find that in the Eubacteria these all fall within the phylum which Woese et al. (1985b) have called the "purple bacteria." An outline of the tree for this group is shown in Figure 27.4, and a

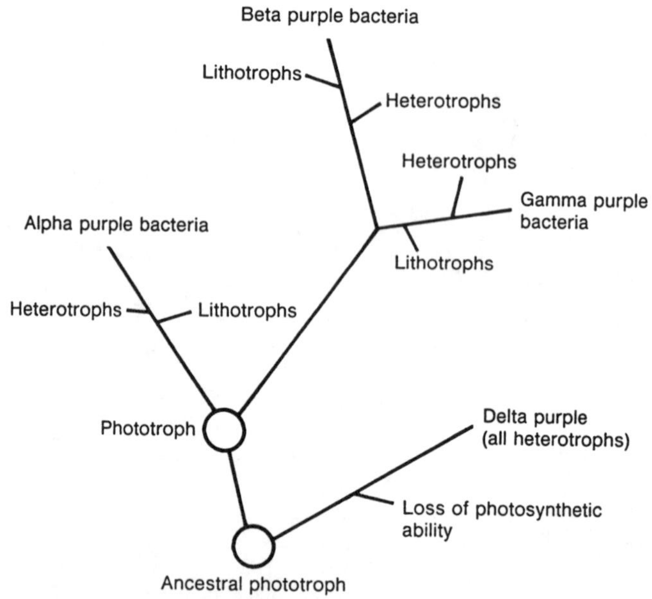

Figure 27.4 Overall phylogenetic tree of the "purple" bacterial lineage, showing the relationships of phototrophic and lithotrophic groups. Based on Woese (1987).

listing of the genera in the various subgroups is given in Table 27.3. As seen, lithotrophic representatives are found in all the major groups. Wherever studied, the CO_2 fixation pathway in the purple bacteria follows the Calvin cycle.

The purple bacterial group is the largest and most physiologically diverse of all the groups of eubacteria. As seen in Figure 27.4, the purple bacterial group can be divided into four distinct subdivisions, the *alpha, beta, gamma,* and *delta* groups. Phototrophic purple bacteria form the heart of three of these subdivisions, but heterotrophic and lithotrophic representatives are distributed across the groups as well.

Table 27.3 Major autotrophic genera of the purple bacteria group*

Alpha group	**Beta group**	**Gamma group**
Rhodospirillum	*Rhodocyclus*	*Chromatium*
Rhodopseudomonas	*Nitrosomonas*	*Thiospirillum*
Rhodobacter	*Thiobacillus*	*Beggiatoa*
Rhodomicrobium	*Derxia*	*Thiothrix* (?)
Rhodopila	*Nitrosococcus*	*Ectothiorhodospira*
Nitrobacter	*Nitrosolobus*	*Thiocapsa*
Bradyrhizobium	*Nitrosospira*	*Thiocystis*
Pseudomonas	*Nitrosoribrio*	*Thiodictyon*
Xanthobacter		*Thiomicrospira*
		Delta group
		Sulfate-reducing bacteria

*See Woese (1987) for details.

It is unlikely that the complex membrane systems and photopigments required for photosynthesis arose in so broad a phylogenetic group more than once, making it likely that the common ancestor of all bacteria in the purple bacteria group was photosynthetic. The genetic evidence for the spread of the photosynthetic character among bacteria is discussed by Drews in Chapter 25. The physiological diversity characteristic of the group probably arose as the result of the exchange of photosynthetic capacity for other physiological processes such as lithotrophy better adapted to new and expanding ecological niches. Such a hypothesis nicely accounts for the similar internal membrane structures found in such seemingly unrelated organisms as *Rhodopseudomonas* and *Nitrobacter*, both of which have polar stacks of lamellar membranes, or the morphological relatedness between phototrophic spirilla such as *Thiospirillum* and heterotrophic spirilla such as *Spirillum volutans*.

The nonsulfur purple bacteria are found in both the *alpha* and *beta* subgroups (see Table 27.3), whereas the purple sulfur bacteria are found in the *gamma* subgroup. It is interesting that certain lithotrophic sulfur bacteria which store sulfur internally are also found within the *gamma* subgroup. If the storage of sulfur is a unique biochemical property, than one could imagine how the process of phototrophic sulfur storage could have been recruited as a mechanism to store sulfur for use as an energy source lithotrophically by *Beggiatoa* and its relatives. Since many phototrophic sulfur bacteria are capable of lithotrophic growth in the dark under microaerobic conditions, the relationship between these two kinds of organisms as revealed by molecular phylogeny seems quite reasonable.

As summarized in Table 27.3, lithotrophic bacteria found in the purple bacteria group include *Nitrobacter* in the *alpha* group, *Nitrosomonas* in the *beta* group, *Beggiatoa* in the *gamma* group, and sulfate-reducing bacteria in the *delta* group. As far as is known, all of these organisms use the Calvin cycle for CO_2 fixation. Such wide dispersion of the lithotrophic character would suggest that it arose early during evolution. One interpretation is that autotrophic CO_2 fixation via the Calvin cycle arose early, and that various lithotrophic energy generation mechanisms were acquired later. Because lithotrophic energy metabolism in all these organisms involves aerobic metabolism, these organisms could not have arisen until significant oxygen was present in the earth's atmosphere, around 2.5 billion years before the present. Genetic evidence for the widespread distribution of the Calvin cycle is discussed by Tabita in Chapter 26.

27.8 GREEN BACTERIA

Phenotypically, the green phototrophic bacteria have always been considered to represent a unique group, and molecular phylogenetic studies confirm this. There are two separate kinds of green bacteria, the green sulfur bacteria (represented by the genus *Chlorobium*), and a group that Woese has called the green nonsulfur bacteria (represented by the genus *Chloroflexus*). Studies on physiology, pigments, and fine structure had previously suggested a close relationship between *Chlorobium* and *Chloroflexus*, but phylogenetically these organisms appear only distantly related (Figure 27.3). These organisms are now separated into two phyla, the green sulfur bacteria and the green nonsulfur bacteria (the latter somewhat anomolously, since *Chloroflexus* definitely metabolizes sulfur). The green nonsulfur bacteria are an evolutionary puzzle because of their similarity in phenotype to the green sulfur bacteria. Because most of the members of the green nonsulfur bacteria are thermophiles, it is possible that the thermophilic property is of more significance in defining the phylogeny of this group than its phototrophic character.

27.9 HELIOBACTERIUM AND RELATED ORGANISMS

These photosynthetic organisms present another puzzle, in that they fit phylogenetically with the Gram-positive bacteria (Woese et al., 1985), yet are phototrophic. Heliobacteria are an entirely separate group of anoxygenic phototrophs which contain a structurally distinct photopigment (bacteriochlorophyll *g*). The group consists of the genera *Heliobacterium, Heliospirillum,* and *Heliobacillus*. Although these organisms appear physiologically related to the nonsulfur purple bacteria, they are strictly anaerobic and are unable to grow by respiratory means as do the other nonsulfur purple bacteria. The structure of bacteriochlorophyll *g* (bchl *g*) is unique. Unlike all other bacteriochlorophylls but like chlorophyll *a* (chl *a*) of green plants, bchl *g* has a vinyl group on ring I of the tetrapyrrole molecule. The only structural difference between bchl *g* and chl *a* is on ring II, which is reduced in bchl *g*. Under conditions of aerobic growth, the double bond on ring II becomes oxidized, leading to the conversion of bchl *g* to chl *a*. This remarkable process has led to the suggestion that heliobacteria represent a bridge between anoxygenic and oxygenic phototrophs. However, because the molecular phylogeny studies place the heliobacteria within the Gram-positive bacteria (specifically, close to the genus *Clostridium*), the relationship between bchl *g* and chl *a* may simply be a fortuitous one.

27.10 CONCLUSION

The phylogenetic data and comparative physiology summarized in Figures 27.3 and 27.4 reveal that the autotrophic character is an ancient one. However, the data show clearly that there are several modes of autotrophy, each of which probably arose independently. Among the archaebacteria, phototrophic autotrophy is absent, as is the Calvin cycle of CO_2 fixation. Autotrophy in the archaebacteria is strictly lithotrophic, primarily anaerobic. Not all of the methanogens are autotrophic, since some grew only with acetate or a methyl compound. CO_2 is fixed by the acetyl-CoA pathway. Among the sulfur-dependent archaebacteria, energy generation involves the oxidation of either H_2 (with elemental sulfur as electron acceptor) or reduced sulfur compounds (with O_2 or ferric iron as electron acceptor), and CO_2 is fixed (probably) by the reverse tricarboxylic acid cycle.

Although the halobacteria, members of the archaebacteria, are able to use light energy, they are not capable of phototrophic *growth*. Phototrophic growth seems to be confined to the Kingdoms Eubacteria and Eucaryotes, and was probably an early evolutionary event. The first phototrophic organisms were probably anoxygenic, related to the purple or green bacteria, but oxygenic photosynthesis arose quite early, probably around 2.5 billion years before the present. The wide phylogenetic diversion between the cyanobacteria and the green plants, both of which share this common photosynthesis, is consistent with an early origin of oxygenic photosynthesis. The Calvin cycle of CO_2 fixation must also have been an early event, probably occurring first within the purple bacteria, but soon spreading to the lithotrophs and oxygenic phototrophs. The other modes of CO_2 fixation, the reverse TCA cycle and the acetyl-CoA pathway, could well have arisen more than once, since both of these pathways involve enzymes that also play roles in general catabolism.

Although the picture sketched above is reasonably satisfying, its provisional nature should be emphasized. Only a relatively small number of organisms have been characterized by molecular phylogeny, and further work may well lead to modifications. However, it seems unlikely that further studies will alter the conclusion that photosynthesis was an early event in the history of life on earth.

Although further molecular phylogenetic studies will clearly be of value, it should be emphasized that they should not be done to the exclusion of studies on the physiology and biochemistry of autotrophy. We urgently need to know the details of energy generation and CO_2 fixation in those little-studied groups among the archaebacteria and eubacteria where current data have already shown that the conventional pathways do not occur. Detailed comparisons between green sulfur and green nonsulfur eubacteria, and sulfur-dependent archaebacteria, will also be of great value.

Finally, one should not overlook the desirability of searching for more new organisms. Even without organized effort, a large number of completely new organisms have been uncovered among the autotrophic bacteria over the past decade or two. Conventional wisdom would suggest that classification of the organisms already isolated permits only a rough outline of the diversity of autotrophic bacteria. When more organisms have been studied, quite new insights may arise.

REFERENCES

Chapman, D.J. and **Gest, H.** 1983. Terms used to describe biological energy conversions, electron transport processes, interactions of cellular systems with molecular oxygen, and carbon nutrition. pp. 459–463, in Schopf, J.W. (editor), *Earth's Earliest Biosphere,* Princeton University Press, Princeton, N.J.

Schidlowski, M., Hayes, J.M. and **Kaplan, I.R.** 1983. Isotopic inferences of ancient biochemistries: carbon, sulfur, hydrogen, and nitrogen, pp. 149–186, in Schopf, J. W. (editor), *Earth's Earliest Biosphere, Princeton University Press, Princeton, N.J.*

Schopf, J.W. (editor). 1983. *Earth's Earliest Biosphere. Its Origin and Evolution.* Princeton University Press, Princeton, N.J. 543 pages.

Schopf, J.W., Hayes, J.M. and **Walter, M.R.** 1983. Evolution of earth's earliest ecosystems: recent progress and unsolved problems. pp. 361–384, in Schopf, J.W. (editor), *Earth's Earliest Biosphere,* Princeton University Press, Princeton, N.J.

Walter, M.R. (editor). 1976. *Stromatolites.* Elsevier, Amsterdam. 790 pages.

Walter, M.R. 1983. Archean stromatolites: evidence of the earth's earliest benthos. pp. 187–213, in Schopf, J.W. (editor), *Earth's Earliest Biosphere,* Princeton University Press, Princeton, N.J.

Woese, C.R., Debrunner-Vossbrinck, B.A., Oyaizu, H., Stackebrandt, E., and **Ludwig, W.** 1985a. Gram-positive bacteria: possible photosynthetic ancestry. *Science* 229:762–765.

Woese, C.R., Stackebrandt, E., Macke, T.J., and Fox, G.E. 1985b. A phylogenetic definition of the major eubacterial taxa. *Systematic and Applied Microbiology* 6:143–151.

Woese, C.R. 1987. Bacterial Evolution. *Microbiological Reviews* 51:221–271.

Index

Brock/Springer Series in Contemporary Bioscience

Series Editor: Thomas D. Brock
University of Wisconsin–Madison

Tom Fenchel
ECOLOGY OF PROTOZOA: The Biology of Free-living Phagotrophic Protists

Johanna Döbereiner and Fábio O. Pedrosa
NITROGEN-FIXING BACTERIA IN NONLEGUMINOUS CROP PLANTS

Tsutomu Hattori
THE VIABLE COUNT: Quantitative and Environmental Aspects

Roman Saliwanchik
PROTECTING BIOTECHNOLOGY INVENTIONS: A Guide for Scientists

Hans G. Schlegel and Botho Bowien (Editors)
AUTOTROPHIC BACTERIA